TERRESTRIAL EARTHWORMS (*OLIGOCHAETA: OPISTHOPORA*) OF CHINA

TERRESTRIAL EARTHWORMS (OLIGOCHAETA: OPISTHOPORA) OF CHINA

NENGWEN XIAO

Chinese Research Academy of Environmental Sciences, Beijing, China

Academic Press is an imprint of Elsevier
125 London Wall, London EC2Y 5AS, United Kingdom
525 B Street, Suite 1650, San Diego, CA 92101, United States
50 Hampshire Street, 5th Floor, Cambridge, MA 02139, United States
The Boulevard, Langford Lane, Kidlington, Oxford OX5 1GB, United Kingdom

Notices
Knowledge and best practice in this field are constantly changing. As new research and experience broaden our understanding, changes in research methods, professional practices, or medical treatment may become necessary.

Practitioners and researchers must always rely on their own experience and knowledge in evaluating and using any information, methods, compounds, or experiments described herein. In using such information or methods they should be mindful of their own safety and the safety of others, including parties for whom they have a professional responsibility.

To the fullest extent of the law, neither the Publisher nor the authors, contributors, or editors, assume any liability for any injury and/or damage to persons or property as a matter of products liability, negligence or otherwise, or from any use or operation of any methods, products, instructions, or ideas contained in the material herein.

Library of Congress Cataloging-in-Publication Data
A catalog record for this book is available from the Library of Congress

British Library Cataloguing-in-Publication Data
A catalogue record for this book is available from the British Library

ISBN: 978-0-12-815587-5

For information on all Academic Press publications
visit our website at https://www.elsevier.com/books-and-journals

Working together
to grow libraries in
developing countries

www.elsevier.com • www.bookaid.org

Publisher: Charlotte Cockle
Acquisition Editor: Anna Valutkevich
Editorial Project Manager: Ruby Smith
Production Project Manager: Punithavathy Govindaradjane
Cover Designer: Christian Bilbow

Typeset by SPi Global, India

Contents

PART ONE

INTRODUCTION

PART TWO

SYSTEMATICS

Contributors

Xiaoqi Gao (Chapter 3 and Part II) Chinese Research Academy of Environmental Sciences, Beijing, China

Shengnan Ji (Chapters 1, 2, and 4) Chinese Research Academy of Environmental Sciences, Beijing, China

Junsheng Li (Part II) Chinese Research Academy of Environmental Sciences, Beijing, China

Gaohui Liu (Part II) Chinese Research Academy of Environmental Sciences, Beijing, China

Nana Shi (Chapter 2 and Part II) Chinese Research Academy of Environmental Sciences, Beijing, China

Qin Xu (Part II) Beijing Institute of Education, Beijing, China

Preface

Earthworms belong to the Oligochaeta of the Opisthopora, Annelida. There are about 4000 species globally, which are widely distributed in soil, freshwater, and marine environments. Earthworms plays a very important role in ecosystems, mainly accelerating the decomposition of organic matter, enhancing the C and N nutrition circulation, promoting the formation of soil, improving soil permeability, and improving the capacity of water storage and the efficiency of fertilizers. At the same time, earthworms can direct or indirectly control the composition of plant communities by accelerating the decomposition of carbon. Earthworms are the traditional indicator organisms for soil fertility and land use. In the food chain, earthworms are also a bridge between terrestrial organisms and soil organisms. Earthworms are a very important nontarget terrestrial soil organism for environmentally toxic substances. They can be used as an indicator to carry out ecotoxicological tests and in the monitoring of the ecological environment of soil.

Earthworm taxonomy has a long history. In the *Systema Naturae* (1758), Carl von Linné described the Lumbricus terrestrial, and then established the genus *Lumbricus*. Subsequently, the genus *Enterion* (1826), family Megascolecidae (1844), and family Lumbricidae (1876) were established. In 1867, Kinberg established the genera *Pheretima*, *Amynthas*, and *Perichaeta*. Sims and Easton (1972) and Easton (1979) revised the *Pheretima* twice, before Easton (1983) revised the *Pheretima* to 10 genera. In addition, Gates also revised some species of the family Lumbricidae. The history of earthworm taxonomy in China is relatively short, and the research is neither comprehensive nor systematic, with only a few scholars having done sporadic research. The earliest study was by Perrier (1872) who described the first earthworm species in China, *Perichaeta aspergillum*. After that, scholars such as Yi Chen, Yuanhui Zhong, Jiangping Qiu, and Xiaoyi Feng began research on earthworm taxonomy and resource investigation in China. Some scholars in Taiwan have made great breakthroughs in earthworm taxonomy. However, the classification data of earthworms in China are scattered, lacking, and the taxonomy is stagnant. The result has been great confusion in earthworm taxonomy in China. Therefore, this book has combed the study of earthworm taxonomy and collated all the literature and data in the country.

The author published *Terrestrial Earthworms of China* in 2012. Compared to *Terrestrial Earthworms of China*, the number of species is increased in this book. The habitat, etymology, distribution, deposition of types of each described species, and the difference between the similar species are supplemented. Citations of sources are provided before the description of each species, and moreover, each species is accompanied by a figure. The figure in this book is drawing by us from published authors, or the original introducer, and this is indicated in the figure caption. Figures that do not indicate the source are self-drawn according to the description and specimens of the species. The order of distribution of species in China is from Beijing to the surrounding area according to convention. The details are as follows: Beijing, Tianjin, Hebei, Liaoning, Jilin, Heilongjiang, Shanghai, Jiangsu, Zhejiang, Anhui, Fujian, Taiwan, Shandong, Jiangxi, Shaanxi (山西), Henan, Hubei, Hunan, Guangdong, Hong Kong, Aomen, Hainan, Guangxi, Chongqing, Sichuan, Guizhou, Yunnan, Xizang, Inner Mongolia, Shanxi (陕西), Gansu, Ningxia, Qinghai, Xinjiang. The county is usually the smallest unit and is described in the parentheses. Some special described the distributions by natural geographical unit, such as Mount Emei, the small Changshan Island, Changbai Mountain, etc.

This book is divided into two parts. The first part is the introduction and includes taxonomy, morphological structure, geographical distribution, and the sampling and conservation of earthworms. The second part is the classification and identification of terrestrial earthworms. The whole manuscript was completed by Nengwen Xiao and Qin Xu. Xiaoqi Gao wrote the sections on morphology and structure, and checked most of the species descriptions. Shengnan Ji wrote the sections on taxonomy, geographical distribution, and sampling and conservation of earthworms. Junsheng Li and Gaohui Liu checked the descriptions of a few species. Nana Shi produced the figures.

This book is financially supported by the project "Research and application on ecosystem evaluation technology based on multi-source data fusion"

(2016YFC0500200) and the project "the Program of Biodiversity Survey and Assessment" of the Ministry of Ecology and Environment.

This book can be used by ecologists researching soil organisms, master's students, doctoral students, and amateurs with an interest in earthworm taxonomy. Due to the limitations of data and time, deficiencies are unavoidable. We therefore welcome corrections and constructive criticism.

Nengwen Xiao
Beijing
February 2018

Abbreviations

ag/ac	accessory gland
amp	ampulla
cl	clitellum
cp	copulatory pouch
dl	dorsal lobe
dp	dorsal pore
dv/div	diverticulum
fp	female pore
gm	genital marking
gmg	genital marking gland
gp	genital papilla
gs	genital setae
gt	genital tumescence
ic	intestinal caecum
md	male disk
mp	male pore
np	nephropore
op	oval pad
p	porophore
pd	prostatic duct
pg	prostate gland
pp	prostatic pore
pr	prostomium
sc	seminal chamber
sep	septum
sg	seminal groove
sp/spp	spermathecal pore
sv	seminal vesicle
sw	skin wall
tp	tubercula pubertatis
ts	testis sac
vd	vasa deferens
ve	vas efferens
vp	ventral prostomium

Introduction

The study of earthworm taxonomy is as old as that of other biological groups. However, early research was limited to certain areas and was poorly communicated.

Scientific earthworm taxonomy dates back to the middle of eighteenth century. On January 1, 1758, Carl von Linné's *Systema Naturae* (10th edition) was published, becoming the prelude to the study of animal taxonomy around the world. On page 647, a common earthworm, *Lumbricus terrestris*, was named etymologically from the words terrestrial and lumbricus and since that time the scientific taxonomy of earthworms has been studied. All earthworms found by Linné' were classified in the genus *Lumbricus*. In 1826, a second genus, *Enterion*, was described by Savigny in Paris.

In 1844, the Megacoscolecidae were established by Templeton. In 1876, Lumbricidae was established. In 1867, the genera *Pheretima*, *Amynthas*, and *Pericaeta* were established by Kinberg. However, in later years, some taxonomists proposed that the genera *Amynthas*, *Pericaeta*, and *Rhodopis*, which had previously been used in the naming of other animals, were homonymous and couldn't be used again, with *Perichaeta* being invalid name for earthworms. After 1900, *Pheretima* (Kinberg, 1867) replaced the invalid *Pericaeta*, but *Pheretima* had the same Chinese name as *Pericaeta*. Early 20th century, *Pheretima* replaced the other genus with more structural similarities. After that point, the number of members of the genus *Pheretima* increased dramatically, and by 1972 the genus had reached more than 740 species. In 2010 *Pheretima* included 900 species.

Sims and Easton (1972) and Easton (1979) revised the *Pheretima* group twice by using a computer to numerically classify species. They restored some genera established by previous scholars, established some new genera of their own, and set some alternative genera to previously established general. Easton 1983 published the new genus *Begemius* based on new species from Australia which had the feature of the cecum beginning of the XXV. *Pheretima*, a previously mixed genus, was revised to 10 genera.

Around 1972, the famous earthworm taxonomist Gates published a series of papers to revising some species of Lumbricidae.

In 1976, *Nomenclatura Oligochaetologica, a catalogue of names, descriptions and type specimens of the Oligochaeta* was published by the University of New Brunswick in Canada by Reynolds and Cook, and then supplemented in 1981, 1989, and 1993, respectively. It went some way to perfecting the classification system, and the book more fully described global terrestrial earthworms. There are about 4000 species of terrestrial earthworms in 181 genera (Edwards and Lofty, 1977; Feng, 1985) belonging to 12 families.

It was only 120 years after Carl von Linne's published his *Systema Naturae* that China began to study the taxonomy of earthworms. The earliest record was of *Perichaeta aspergillum*, described by Perrier in 1872. By 2012, China had recorded 312 species and 14 subspecies of 31 genera in 9 families of terrestrial earthworms. In this book, 357 species (subspecies) of 31 genera in nine families are recorded.

INTRODUCTION

1

Taxonomy

Earthworms belong to the class Oligochaeta of the phylum Annelida. Many authors have produced classifications of the Oligochaeta, but it was not until 1900 that Michaelsen produced the system that is the basis of the modern taxonomy of this group. Michaelsen divided earthworms into 11 families, containing 152 genera, and 1200 species. In 1921, he revised his own classification into 21 families in two suborders. Stephenson (1930) simplified this into 14 families, an arrangement with little difference from Michaelsen's original grouping. A division of families into the Microdrili and Megadrili is still useful in terms of their definition, with the former consisting of small, mainly aquatic worms (including the terrestrial Enchytraeidae) while the latter are larger, mostly terrestrial worms. This book does not include the Microdrili because it is concerned mainly with the terrestrial worms.

Stephenson placed seven families in the Microdrili, namely the Aeolosomatidae, Naididae, Tubificidae, Pheodrilidae, Enchytraeidae, Lumbriculidae, and Branchiobdellidae. The remaining seven families were attributed to the Megadrili as follows:

1. Family Alluroididae.
2. Family Haplotaxidae.
3. Family Moniligastridae (Syngenodrilinae, Moniligastrinae).
4. Family Megascolecidae (Acanthodrilinae, Megascolecinae, Octochaetinae, Ocnerodrilinae).
5. Family Eudrilidae (Parendrilinae, Eudrilinae).
6. Family Glossoscolecidae (Glossoscolecinae, Sparganophilinae, Microchaetinae, Hormogastrinae, Criodrilinae).
7. Family Lumbricidae.

Since Stephenson (1930), some authors have tried to revise the classification of megadrile families, particularly the Glossoscolecidae, Megascolecidae, and the Moniligastridae (Jamieson, 1971a, b, c). For the Moniligastridae, Gates (1959a, b) revised its two subfamilies, of which the Syngenodrilinae was placed in the Alluroididae, thus the remaining subfamily was raised to family status.

In terms of the Glossoscolecidae, Gates raised all five subfamilies to family status because there was not a sufficiently close relationship for them to be included in a single family. However, Jamieson (1988) concluded that there was a total lack of affinity between the Criodrilinae, Sparganophilinae, and Glossoscolecinae.

The classification of the megascolecid earthworms has always been much more controversial than other oligochaete families. Omodeo (1958), Gates (1959a, b), Lee (1959), and Jamieson (1971a, b, c, 1985) proposed new systems of classification, all of which replaced Stephenson (1930). Omodeo raised one group to family status on the basis of the position and number of calciferous glands. Lee used the number and position of the male pores and the position of the nephridiopores as a key characteristic. Gates raised all the main groups to family status by considering important factors, such as structure of the prostatic glands and excretory system, and the position of the calciferous glands. Sims (1966) assessed the relative merits of the three earlier classifications of megascolecid genera using computer techniques and found that the pattern of megascolecid genera was in accordance to a large extent with the classification proposed by Gates.

The major divisions of the oligochaetes according to Gates (1959a, b) were as follows:

1. Family Moniligastridae.
2. Family Megascolecidae.
3. Family Ocnerodrilidae.
4. Family Acanthodrilidae.
5. Family Octochaetidae.
6. Family Eudrilidae.
   Subfamily Parendrilinae
   Subfamily Eudrilinae
7. Family Glossoscolecidae.
8. Family Sparganophilidae.
9. Family Microchaetidae.
10. Family Hormogastridae.
11. Family Criodrilidae.
12. Family Lumbricidae.

Jamieson (1971a, b) proposed an alternative classification to those of Omodeo, Gates, Lee, and Sims. He considered certain basic material used by the above workers together with morphology of the excretory system. Subsequently, Brinkhurst and Jamieson (1972) and Jamieson (1978) modified the classification twice. However, it is not within the scope of this book, which judges the individual merits of the various systems of classification that have been proposed. Jamieson (1988) later reviewed the overall phylogeny and higher classification of the Oligochaeta based on a cladistic analysis.

For convenience, the classification of terrestrial species in this book has the following divisions:

1. Family Haplotaxidae.
2. Family Moniligastridae.
3. Family Lumbricidae.
4. Family Ocnerodrilidae.
5. Family Acanthodrilidae.
6. Family Octochaetidae.
7. Family Megascolecidae.
8. Family Glossoscolecidae.
9. Family Microchaetidae.

The Megascolecidae and the Lumbricidae are the two most important families. The megascolecids and their close relatives comprise more than half of all known species. Worms in this group are distributed widely outside the Palearctic zone, in two genera, *Pheretima* and *Dichogaster*. Sims and Easton (1972) divided *Pheretima* into eight genera: *Archipheretima*, *Pitbemera*, *Ephemitra*, *Metapheretima*, *Planapheretima*, *Amynthas*, *Metaphira*, and *Pheretima*. However, the Lumbricidae was generally considered to be the most recently evolved family. Due to their ability to colonize new soils and become dominant, the Lumbricidae have followed the spread of human colonization around the world.

2

Geographical Distribution

The geographical distribution of terrestrial earthworms in China covers the Palaearctic and Oriental realms. The boundary between the two realms in China lies from 32° north latitudes, westward to east longitude 102°, southwest to 30° N and 101° E. Then, along the north latitude 30° line westward to east longitude 96°, passing through the north latitude 30° line southward. Turn southwest, go west along the southern slope of the Himalayas. North of the line is the Palaearctic realm and south of the line is the Oriental realm. At present, 46 species and 3 subspecies of 11 genera in 5 families have been recorded in the Palaearctic realm of China, and 285 species and 12 subspecies of 25 genera in 8 families have been recorded in the Oriental realm of China. In addition, 19 species of 6 genera in 3 families occur in both realms.

2.1 FAMILY MEGASCOLECIDAE

Family Megascolecidae is the largest family in China, with 333 species and 20 subspecies in 8 genera, accounting for more than 80% of China's total recorded earthworms. The Megascolecidae is distributed in both the Palaearctic and Oriental realms.

The typical earthworm of the Megascolecidae is a medium-sized species represented by *Metaphire asiatica* in the Palaearctic realm. It has spread from the center of the northern district in the Northeast China subregion to surrounding areas. To the west it has entered the Central Asia subregion of the Palaearctic realm, then the own unique native species have been formed. To the east it has extended far into the Korean peninsula and to the south it has merged with the *Metaphire* genera of the Oriental realm by crossing the southern boundary of the Palaearctic realm. There are 11 species of the *Metaphire* genus in the Palaearctic realm of China, of which 5 species are native species.

There are many species of the Megascolecidae in the Oriental realm. The dominant species are from the *Metaphire*, *Amynthas*, *Planapheretima*, *Pithemera*, and *Begemius* genera. In addition, the *Polypheretima*, *Laptio*, and *Perionyx* genera of the Megascolecidae also occur in the Oriental realm. The *Polypheretima* and *Perionyx* genera, each of them with a widespread species, only occur on Taiwan Island. A widespread species of the *Laptio* genus occurs in Hong Kong and Hainan. The genus *Begemius* has one native species on Taiwan Island and two native species in Guangdong.

2.2 FAMILY LUMBRICIDAE

Family Lumbricidae has 22 species and 3 subspecies of 9 genera in China, which mainly occur in the Palaearctic realm. In more detail, 20 species and 2 subspecies of 7 genera in the Lumbricidae belong to the Palaearctic realm of China and account for 92% of Chinese species in the Lumbricidae. Except for a few widespread species the Lumbricidae are all native to the Palaearctic realm of China. However, there are few native species in China. There are five species of the *Aporrectodea* genus and two species of the *Bimastos* genus from the Palaearctic realm of China. One species each from the *Dendrobaena*, *Dendrodrilus*, and *Lumbricus* genera occurs in Xinjiang and the northeast of China. There are two species of the *Octolasion* genus in Heilongjiang Province. Except for *Eisenia fetida*, which occurs in both realms, the other seven species and three subspecies of the *Eisenia* genus occur in the Palaearctic realm, of which four species and one subspecies are native to the Palaearctic realm of China. The *Eiseniella* genus has only one species, an exotic introduced through human activities to the island of Taiwan, in the Oriental realm.

2.3 FAMILY MONILIGASTRIDAE

Family Moniligastridae has 23 species and 4 subspecies of 2 genera in China. Among them, one species from the *Desmogaster* genus, occurring in Jiangsu Province, is a typical native species of the Oriental realm. In addition, 15 species and 4 subspecies of the *Drawida* genus are found in the Oriental realm.

2.4 FAMILY HAPLOTAXIDAE

There is only one species of the Haplotaxidae in China and it is a typical species of the Palaearctic realm. However, it is unable to expand to other areas of the Palaearctic and Oriental realms in China due to the obstacles provided by mountains and deserts.

2.5 OTHER FIVE FAMILIES

A total of 18 species from 11 genera belong to the families Ocnerodrilidae, Acanthodrilidae, Octochaetidae, Glossoscolecidae, and Microchaetidae, which are all distributed in the Oriental realm of China. In the Ocnerodrilidae, except for one species widely distributed outside the Oriental realm, the other three species of four genera only occur in parts of Hainan and Jiangsu Province. Family Acanthodrilidae has five species of three genera, which are sporadically distributed in Jiangsu, Sichuan, Hainan, Taiwan, and Yunnan provinces. There are five species of two genera of the Octochaetidae found in Taiwan, Hainan, and Fujian provinces. Family Glossoscolecidae has two

species in a single genus, with one being a native species from Guangdong Province, while the other, occurring in Guangxi, Hong Kong, and Taiwan provinces, may be an alien species.

The Palaearctic realm of China includes 19 provinces, municipalities, and autonomous regions in the Chinese administrative division, of which 13 provinces are entirely within the border. In contrast, the Oriental realm of China includes 21 provinces, municipalities and autonomous regions in the Chinese administrative division, of which 15 provinces are completely within the border. There are six provinces that span the two realms. The Palaearctic realm of China occupies more than two-thirds of China's territory, however, the number of species recorded is less than one-sixth of the total number of terrestrial earthworm species in China. In comparison, the Oriental realm of China accounts for less than one-third of China's territory, but the number of species recorded is about 90% of the total number of terrestrial earthworm species in China.

From the perspective of the administrative region (Fig. 1), in general, the abundance of terrestrial earthworms in the southern provinces of China is higher than

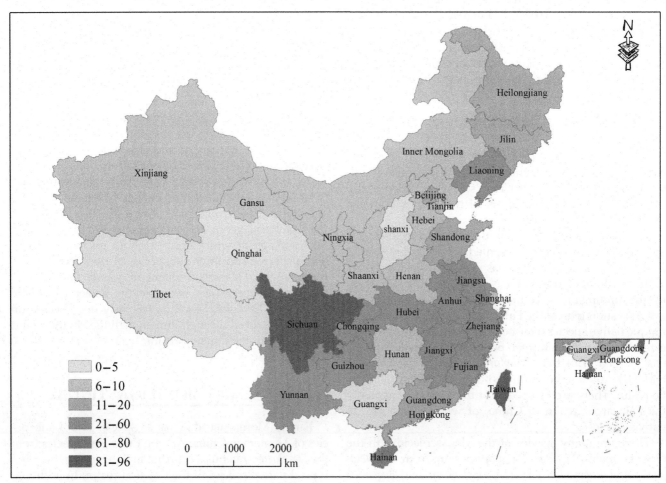

FIG. 1 Species richness of terrestrial earthworms in China.

that in the northern provinces. Similarly, the abundance of terrestrial earthworms in the eastern provinces is higher than that in the western provinces. The provinces with high levels of richness are Taiwan (96 species), Sichuan (82 species), Hainan (61 species), Chongqing (56 species), Hubei (37 species), Jiangsu (37 species), Yunnan (36 species), Guizhou (35 species), Jiangxi (34 species), Liaoning (34 species), Zhejiang (34 species), and Fujian (30 species). In terms of China's terrestrial earthworms, the provinces with high species richness per unit area are Hong Kong (20 species), Taiwan, Hainan, and Chongqing.

1. INTRODUCTION

3

Morphology

3.1 EXTERNAL STRUCTURE

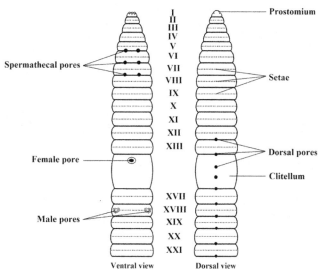

FIG. 2   External structure of an earthworm (*Metaphire asiatica*).

3.1.1 Body Shape

The body of the earthworm is divided externally into *segments* along the length of the body by *intersegmental furrows* that coincide with the positions of the *septa* dividing the body internally. External segments are often subdivided by secondary furrows into 2–9 *annuli* (secondary segmentation), particularly in the anterior part, but these are superficial divisions which are not reflected in the internal anatomy.

The length, width, and number of segments of earthworms vary, but are always within a certain range for the same species. Among the terrestrial earthworms of China are *Desmogaster sinensis* (Gates, 1930), a rather large species at 290–540 mm long, 8–12 mm wide, and with 360–588 segments, and *Amynthas infantilis* (Chen, 1938), a rather small species at 10–24 mm long, 1–1.8 mm wide, and with 58–82 segments.

Roman numerals, for example, I, II, III, IV, and V, indicate the segments of the body in an anteroposterior direction beginning with the peristomium in the morphological description of segments. A fraction is used to indicate intersegments, for example, 3/4 indicates the intersegment between the third segment and the fourth segment.

3.1.2 Prostomium

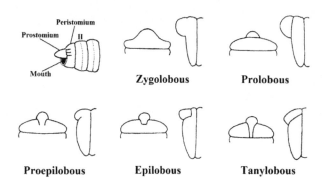

FIG. 3   Various forms of prostomium.

The first segment of an earthworm is called the *peristomium*. The *prostomium* is a protuberance anterior to and above the mouth from the peristomium. The way in which the peristomium and prostomium are joined differs between species and is a useful systemic character. It is divided into five types: zygolobous, prolobous, proepilobous, epilobous, and tanylobous, depending on the demarcation of the prostomium. If the prostomium is not in any way demarcated from the peristomium, it is termed *zygolobous*. The posterior prolongation into the region of the peristomium is called the tongue. If the prostomium is demarcated from the peristomium and is without a tongue, it is termed *prolobous*. If it encroaches the peristomium with a tongue, it is termed *proepilobous*. If this is more marked, it is termed *epilobous*. If the tongue goes back to the groove between the first segment and second segment, it is termed *tanylobous*.

3.1.3 Setae

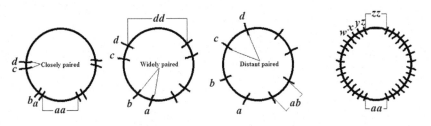

FIG. 4 Arrangement of setae.

The setae, which are bristle-like structures borne in follicles on the exterior of the body wall, provide traction for movement and help anchor and control the worm when it is moving through soil. They are present on each segment except for the first and last.

The setae are arranged in a single ring around the periphery of each segment, their number and distribution being typically termed either lumbricine or perichaetine. When there are eight setae per segment, it is termed a *lumbricine arrangement*. If the distance between the setae in each pair is very small, they are termed *closely paired*, and if they are wider apart they are termed *widely paired*. If they are very far apart so that the pairing is not obvious, they are termed *distantly paired*. When there are more than eight setae per segment in a more or less complete circle around the equator of a segment, it is termed a *perichaetine arrangement*. Here, usually with a larger or smaller break in the middorsal and midventral regions along the length of the body, the *middorsal line* and *midventral lines* are formed, respectively.

In the description of setae, *a* is the medianmost seta on each side of a segment, or the first seta from the midventral line, *b* is the seta next laterally to *a* or the second seta from the midventral line, *c* is the seta next laterally to *b*, *z* is the medianmost seta on the dorsum when the setae are perichaetin, *y* is the dorsal seta next laterally to *z*, *aa* indicates the ventral distance between the two *a* meridians, and *ab* indicates the distance between *a* and *b* meridians.

3.1.4 Surface Openings

The surface openings of earthworms include the dorsal pore, male pore, spermathecal pore, female pore, and nephridiopore.

Dorsal pores are small apertures situated in the intersegmental furrows on the middorsal line and communicate with the body coelom fluid. This is believed to permit moistening of the body surface and facilitate respiration. They are usually absent in the most anterior region of the body. The position of

the first dorsal pore is used as a systematic character at species level.

Earthworms are hermaphrodites and have both male and female genital openings. *Male pores* are openings to the exterior of the male ducts and may be associated with one or two pairs of prostatic pores, either superficial or invaginated into chambers confined to the parietes, sometimes reaching more or less extensively into the coelomic cavities serving as *copulatory pouches*, which are an important taxonomic character of *Metaphire*.

Female pores are external apertures of the female ducts, most commonly a single pair, situated either in an intersegmental furrow or on a segment, sometimes united into a single median pore. Their position often being diagnostic of a particular family.

Spermathecal pores are openings of a spermathecal diverticulum. The number, position, and shape are all important taxonomic characters. Earthworms usually have 2–7 pairs of spermathecal pores, sometimes just one pair is present, or they may be absent.

Nephridiopores are the external openings of the nephridia. They are very small and often difficult to see, usually situated immediately behind the intersegmental furrows on the lateral aspect of the body, and extend in a single series along the body on either side, but in some cases they are situated in more or less regular alternation some distance above and some distance below the lateral line of the body.

3.1.5 Clitellum and Genital Markings

A *clitellum* is a regional tumescence of the epidermis that is associated with cocoon production. It can only be seen when the worm is sexually mature. It is located near the anterior of the body and is saddle-shaped or annular. As the position and number of segments occupied by the clitellum is constant for each species, the position of the clitellum is used as a diagnostic character.

Earthworms possess various markings at sexual maturity, in the form of *tubercles*, *ridges*, and *papillae* on the anterior surface, and these differ greatly in number and form in different species of earthworms. A porophore is a protuberance or a special structure bearing a pore which usually is an opening of a spermatheca or an oviduct or a vasa deferens. These genital markings are of special interest to the systematist as they provide some of the most useful characters for the discrimination of species.

3.1.6 Color

The color of earthworms depends mostly on the presence or absence of pigment, which is either in the form of granules or in pigmented cells in the subcuticular muscle layer. The color of pigmented worms, when preserved in formalin, is fairly stable but the reds and pinks of unpigmented worms usually fade rapidly.

3.2 INTERNAL STRUCTURE

3.2.1 The Digestive System

The digestive system of an earthworm consists of the mouth, buccal cavity, pharynx, esophagus, crop, gizzard, intestine, and caecum (Fig. 5).

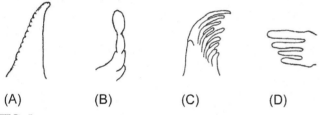

(A) (B) (C) (D)

FIG. 5   Four shapes of caecum.

The *pharynx*, after—and not always clearly demarcated from—the buccal cavity, extends back to about VI. The *esophageal* is the portion between the pharynx and the intestine, ending posteriorly in an esophageal valve. The *crop* is a widened portion that lacks the muscularity of a gizzard. In the lumbricidae the crop is in front of the gizzard. The *gizzard* is muscular and lined with a thicker cuticle than the crop. It is used as the "stomach" of earthworms. The number of gizzards is different in different species, Megascolecidae species commonly have 2–10 gizzards, while in *Metaphire californica* (Kinberg, 1867) the gizzard is absent. The *intestine* is a straight tube for most of its length and is the longest portion of the alimentary canal. It is slightly constricted at each septum. The caeca are blind diverticula branched anteriorly from the intestine. They are conical and are supposed to be important digestive glands which secrete digestive enzymes to help in the digestion of food. The position and shape of caeca are main diagnostic characters. Depending upon the shape, which is related to the

digestive enzyme that it produces, the caecum are divided into two types: the simple type (Fig. 5A and B) and the manicate type (Fig. 5C and D).

3.2.2 The Vascular System

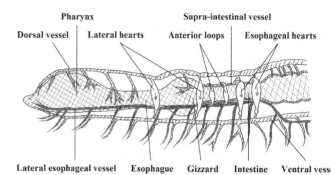

FIG. 6   Digestive and vascular systems of an earthworm (lateral view) (*Metaphire asiatica*).

An earthworm has a closed circulatory system that uses vessels to send blood through its body. There are three main vessels that supply the blood to organs within the earthworm. There is one *dorsal vessel*, which is the largest of the longitudinal vessels. It is associated closely with the gut for most of its length except in the most anterior portion. It takes blood from the back of the body to the front. There are two *ventral vessels* to take blood in the other direction, from front to back. Some families, for example, the Megascolecidae and Glossoscolecidae have a *supraintestinal vessel*. This lies along the dorsal wall of the gut in the anterior segments and is part of the complex of blood vessels serving the alimentary canal. Other longitudinal blood vessels are the *lateral esophageal vessels*, which lie along either side of the gut.

There are three main kinds of commissural vessels that pass around the body from the dorsal vessel to the ventral vessels. Some of the anterior commissural vessels that directly connect the dorsal vessel with ventral vessels are enlarged, contractile, and with valves termed *lateral heart* or *heart*. Some species have *esophageal hearts* that connect both dorsal and supraintestinal vessels with the ventral vessels. Some species have *anterior loops* that connect the supraintestinal vessels with the lateral esophageal vessels.

3.2.3 The Excretory System

The *nephridium* is the basic excretory organ of an earthworm. It is paired in almost every segment except for the anteriormost and the last ones. A typical nephridium has an internal funnel-like opening called the *nephrostome*, which is fully ciliated. The nephrostome is in a single segment and the rest of the tube is in the following segment. This tube has three distinct divisions. The first part, following the nephrostome, is termed the ciliated region. The next part is wider and is thrown into coils termed the glandular region as it has glands on its wall. The last part has neither cilia nor glands and is termed the muscular region. This region opens to the outside through an aperture called the *nephridiopore*.

Depending upon the size and number present in a segment, the nephridia are divided into three types: the *meganephridia* or *holonephridia*, which are large in size and occur in a pair at each segment; the *micronephridia* or *meronephridia*, which are small and numerous in each segment (it is believed that the micronephridia are nothing but broken or disintegrated meganephridia); and the *tufted nephridia*, which are derived from micronephridia or macronephridia, are incompletely branched, and are grouped together. These are usually found in one or several of the preclitellar segments of many species.

3.2.4 The Reproductive System

Earthworms are hermaphrodites, that is, male and female reproductive structures develop in the same individual.

The male reproductive system consists of testes, testis sacs, seminal vesicles, vasa deferens, prostate glands, and accessory glands. The basic male organs are *testes*, which are white digited structures and are made up of testicular lobules or follicles. These follicles contain spermatogonia. The *testis sac* is usually a closed-off coelomic space containing 1–2 testes and spermatic funnels of a large, thin-walled segment. *Seminal vesicles* are the largest and most conspicuous organs of the reproductive system. They are paired, with the anterior pair commonly included in testis sacs. Seminal vesicles connect to the testis sac, and from the testis sac the spermatogonia go to the seminal vesicle for maturation. After maturation they return to the sacs. Behind each testis and enclosed in the testis sac there is a ciliated *spermatic funnel*. The funnel leads into a slender duct, termed the *vasa deferens* or *spermatic duct*, which continues posteriorly and enters the prostate glands. *Prostate glands* are large, flat, and creamy white glands. Their function is to produce a fluid in which sperm can be transferred during copulation. *Accessory glands* function in association with prostate glands. They open to the exterior via a number of ducts through *genital papillae*. These glands produce a secretion which helps in copulation.

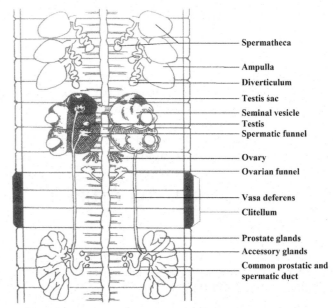

FIG. 7 Reproductive system of an earthworm (*Metaphire asiatica*).

The female reproductive system consists of ovaries, ovarian funnels, and spermathecae. *Ovaries*, which produce oocytes, are usually paired. The ovisacs are backward-facing evaginations of the anterior face of the septum immediately behind the ovaries and open into the dorsal wall of the *ovarian funnels* which lead into conical ducts behind. Sometimes two oviducts unite to form a common oviduct which opens to the outside via the female pore. Spermathecae, also known as seminal receptacles in former publications, are paired organs receiving sperms from a copulatory partner. The spermatheca consists of an *ampulla* and a *diverticulum* whose distal portion is usually enlarged and serves as a *seminal chamber*. The ampullar ducts open to the outside via spermathecal pores.

4

Sampling and Conservation

4.1 PREPARATION FOR SPECIMEN SAMPLING

4.1.1 The Sampling Plans

Before the investigation and sampling work are carried out, it is necessary to clarify the tasks and draw up practical plans. We should determine two points. These are (1) where to sample and (2) how to sample.

4.1.2 Literature Reviewing

(1) Specimen information. First of all, the relevant information should be known, such as if anyone has performed an investigation here in the past? How did they work? What is the time, route, and purpose of the sampling? What problems have been solved or remain unsolved? Moreover, the specimen information, which has already been obtained in or adjacent to the area, should be used as references for fieldwork if possible.

(2) Environment. The planning and arrangement of fieldwork will be facilitated by checking the environmental conditions and traffic situation in the area, such as geography, climate, and vegetation.

According to the above aspects, a specific sampling plan should be developed, including main purposes, missions, dates, routes, budgets, etc.

4.1.3 Sampling Tools

(1) The sampling bag (sampling bottles, finger tubes): It is usually carried by the shoulders and contains various small collection tools and annelid specimens.

(2) The sampling frame (28.5 cm × 8 cm in diameter): This is used for the quantitative collection of large-sized soil animals.

(3) The sampler (5 cm × 5 cm in diameter, 100 mL; 3.5 cm × 2.8 cm in diameter, 25 mL).

(4) The dry funnel: types of funnel are large, small, and portable ones. They are used to separate small and medium-sized annelids from the soil. A funnel contains a bulb (40–60 W) in the top and a metal sieve in the middle (mesh 1–2 mm in small funnels; mesh 3–5 mm in large funnels and portable funnels).

(5) The wet funnel: for gathering small-sized wetted annelids with a 60 mesh made of nylon yarn.

(6) GPS: for recording the height, longitude, and latitude of the collection site.

(7) Other tools: magnifying glasses, recording paper, pens, horned hoes, two-tooth rakes, large tweezers (20 cm), medium tweezers (12 cm), small spades, ditching hoes, plastic cloths, white porcelain plates, measuring cylinders, backpacks, gauze, geothermometers, aluminum boxes for soil sampling, specimen bottles.

4.1.4 Chemicals

Alcohol and formalin of various concentrations.

4.2 SAMPLING METHODS

4.2.1 Choose a Day to Sample

Because annelids live in the soil they are affected by the soil moisture but not the climate. In general, sampling could be done all year around. In addition, the earthworms will be more active if it has rained a few days before sampling, but it's better to choose a day with no rain predicted.

4.2.2 Choose a Site to Sample

The sampling site depends on the species of annelid and its geographical distribution. Therefore, it will facilitate sampling to know the distribution range of the various types of annelids in advance. Sampling plots are generally selected according to the following conditions: (1) slight slope, less stone; (2) basically no human activity; (3) not on the edge of habitats; (4) avoiding ant nests; (5) moist soil, flatter terrain.

13

4.2.3 Choose a Sampling Method

Earthworms put piles of feces around the openings to their burrow holes. These piles are called middens and are normally 2–5 cm in diameter and 1–2 cm high. In the center of the pile there will be a burrow hole. Normally only one deep burrowing earthworm lives in each burrow hole. Therefore, it is easier to get earthworms in such a place.

(1) Hand-sorting. The number of earthworms is underestimated because smaller and black individuals are often ignored when using this method. Only individuals with greater than 0.2 g of live weight can be found. Moreover, there are differences in efficiency between extractors.
① Use a trowel or shovel to dig a 50 cm × 50 cm, or 25 cm × 25 cm hole, and put the soil on a garbage bag. The hole should be 30 cm deep. Pick up a handful of soil from the pile on your garbage bag and look through it for worms. If you find a worm, put it in a container. Use different containers for the earthworms you collect from hand-sorting. Finally, discard your handful of soil after looking through it and do this until you have looked through all of the soil. This is the most common method for collecting large-scale annelids in soil.
② Put the circular sampling frame (28.5 cm × 8 cm in diameter) into the soil. The hole should be 5 cm deep. Then pick up the soil by hand or use sieves to find worms.
③ Use a soil sampler (8 cm × 15 cm in diameter), then pick up the soil by hand or use sieves to find worms.
(2) Dry-funnel methods. These are used to separate small and medium-sized annelids from the soil. The funnel contains a bulb (40–60 W) in the top and a metal sieve in the middle. Move the soil into a funnel, where a temperature gradient is created by a heating source, such as a light bulb. This causes living worms to move from the warmer side to the cooler side of the funnel, and finally into the collecting bottle of the device.
(3) Wet-funnel methods.
① Many of the smaller and more fragile soil-inhabiting earthworms, particularly those that inhabit the water film around soil particles, are not extracted from soil efficiently by using dry funnels because they become trapped before they can emerge from the soil. There are a number of wet-funnel techniques which work better for these hydrophilic invertebrates (Edwards, 1991).
② The apparatus consists of an 11-cm diameter funnel resting in a 9 cm hole in an asbestos board. A 9 cm wire-gauze sieve (30 meshes to the inch) rests in the funnel. The bottom of the funnel is closed by a screw clip on a piece of rubber tubing. Heat is supplied from a 60 W electric light bulb enclosed in a light metal cylinder with a diameter of 11 cm and a height of 18 cm. The bottom of the cylinder is 1.5 cm from the top of the funnel. There are 24 such units wired in parallel and the heating is controlled by means of a variable resistance. The intensity is increased gradually so that the surface reaches a temperature of approximately 45°C after 150 min, when extraction is complete. Then the worms are tipped out from the bottom of the funnel for counting and identification (O'Connor, 1955).
(4) Electrical extraction. 2 fork electrodes, 50 cm long, separated by 1 m, are inserted into the soil. The current can be adjusted from 2–4 A through a variable resistor under a voltage of 220 V. This method is used to collect earthworms in the soil, especially from deep layers. The worms are usually found between 20 cm and 1 m away from the electrodes. Since the worms are often killed by the current close to the electrodes, it is better to reduce the loss of worms by insulating all the electrodes. Moreover, there are two factors that affect the effectiveness of this method. The moisture content of the soil is 1 factor. If the soil is dry, the worms will be driven downwards. The second factor is the pH value of the soil. More worms are found in soil with a lower pH value. However, this method only limited to discover the worms in a fixed volume of soil (Edwards and Lofty, 1977).
(5) Chemical extraction. The common chemical reagents used are mercury oxide, potassium permanganate, formalin, and diluted mustard solution or soap water. The common concentration and dosage of potassium permanganate are 1.5 g/L and 6.8 L/m^2, respectively. Formalin solution is used at a concentration of 0.1%–0.5% (25 mL of 40% formalin mixed in 4.56 L of water for 0.36 m^2 of soil), or 0.4% of formalin solution for 40 L/m^2. Mercury oxide is used at a concentration and dosage of 0.82 mg/L, 1.7–2.3 L.
(6) Soil washing methods. The freshwater oligochaetes can be collected from the bottom sediments of water bodies by the sieves, remove the impurities and flush the dirt with water before obtaining the earthworms together in a group. After that, put the sieve with the 0.5 mm mesh into a magnesium sulfate solution. The earthworms will float on the surface and can be collected.

4.3 PRESERVING EARTHWORMS

4.3.1 Anesthetize the Earthworms

(1) Get as much dirt off the earthworms as possible.
(2) Put the earthworms into the alcohol singly (rather than a whole handful all at once), so they don't get tangled up into a big mess. After a few seconds, the

earthworm will become anesthetized and relax into a mostly extended position.

(3) Place the anesthetized earthworms into a vial. You can keep the earthworms in alcohol for up to 24 h (keeping them cool and out of direct sunlight). However, do not wait too long as after a day or so in alcohol they start to lose their form.

Within 24 h you need to:

4.3.2 Fix With Formalin

(1) Place the earthworms in a leakproof vial and cover them in Formalin, which is a cellular fixative that will prepare the specimens for long-term storage.

(2) Once the earthworms have been in formalin for at least 24 h they can be placed back into alcohol (70%–100% isopropyl, NOT ethyl as it bleaches any pigmentation) for long-term storage. This allows you to reuse the formalin over and over again.

P.S.: Formalin is a dangerous chemical and safety precautions must be taken. It should be used only in a well-ventilated area. It should not be breathed in, swallowed, or come into contact with bare skin or eyes. If external contact does occur, wash the area with copious amounts of water. If ingested contact a poison control center immediately. Formalin has been shown to cause cancer in laboratory animals.

SYSTEMATICS
Phylum: Annelida

Class: Oligochaeta

Head underdeveloped, no parapodia, seta direct in the body wall, hermaphroditism, reproductive ducts present, direct development.

Order: Opisthopora

Most species terrestrial, a few species aquatic; male pore usually 1 pair, opening behind testes, and funnel 1 segment or several segments.

Family: Acanthodrilidae Claus, 1880

Acanthodrilinae Claus, 1880. *Grundzüge der Zoologie.* 1:479.

Acanthodrilinae Stephenson, 1930. *The Oligochaeta.* 820.

Acanthodrilidae Gates, 1972. *Trans. Am. Phil. Soc.* 62(7):32–33.

Type genus: ?

External Characters

Clitellum, multilayered, including female pore segment. Prostatic pores 2 pairs, on segments XVII and XIX; spermathecal pore 2 pairs, in or near furrows 7/8/9; male pores on XVIII or by more or less complete disappearance of 1 pair of prostates and the corresponding pair of spermathecae, more or less purely microscolecine or balantine (*v.inf.*).

Internal Characters

Intestinal origin behind XIII. Hearts behind XI. Holandric nephridia. Calciferous glands mostly absent; if present, not, or not only, in IX or IX and X. Spermathecae, diverticulate. Ovaries fan-shaped and with several egg strings (?). (Ovisacs, small and lobed?) Ova, not yolky. Seminal vesicles, trabeculate. Prostates, tubular and of ectodermal origin. Meganephridial.

Habitat: Mostly terrestrial, rarely limnic or littoral.

Global distribution: Myanmar, China, Australia, Tasmania, New Caledonia, New Zealand (Auckland, Chatham, and subantarctic islands), United States, Mexico, Central America, South America, South Africa, Madagascar, Ceylon, India.

There are 2 genera, 3 species from China.

TABLE 2 Key to the Genera of the Family Acanthodrilidae From China

1. Clitellum, XIII–XVII, annular .... *Microscolex*	
Clitellum, XIII–XVII or XVIII, saddle shaped 2	
2. Clitellum, XIII–XVII or XVIII or 1/2–1/3XVIII .... *Pontodrilus*	
Clitellum, 1/2XIII–1/2 or 1/3XVIII .... *Plutellus*	

GENUS: MICROSCOLEX ROSA, 1887

Microscolex Rosa, 1887. *Boll. Mus. Zool. Univ. Torino.* 19(2):1.

Microscolex Stephenson, 1930. *The Oligochaeta.* 824.

Microscolex Gates, 1972. *Trans. Am. Phil. Soc.* 62(7):33–34.

Type species: *Microscoles modestus* Rosa, 1887 = *Lumbricus phosphorvus*, Dugès, 1837.

External Characters

Setae 8 per segment. (Dorsal pores, lacking?). Clitellum swollen, in XIII–XVII, annular. Male pores on XVII or XVIII; Prostatic pore 1 or 2 pairs, on XVII and XIX, or on XVII only. Spermathecae usually 1 or 2 pairs, the last pair opening in furrow 8/9. Nephropores (obvious?), in a single longitudinal rank on each side of the body.

Internal Characters

Septa all present from 5/6. Digestive system, (with an intestinal origin behind XV?), without a strong gizzard, calciferous and supraintestinal glands, intestinal caeca and typhlosoles. Vascular system, with unpaired dorsal, supraesophageal and ventral trunks but no subneural, with paired extraesophageals median to the hearts and united posteriorly at mV on ventral face of gut (associated with posterior lateroparietal trunks?), with lateroesophageal hearts in X–XII. Nephridia, holandric, present from II, vesiculate, posterior bladders ocarina-shaped with pointed ends mesially and with short ducts from ventral side laterally (provided with parietal sphincters?). Holandric. Spermathecae, diverticulate. Ovisacs, present. Ovaries, fan-shaped and with several egg strings. Two pairs free testes and funnels, in X and XI.

Global distribution: South Patagonia, Tierra del Fuego, West Argentina, Darien (?), Falkland Islands, South Georgia, Kerguelen, Marion Island, Cape Colony, Crozet, Campbell, Auckland, Antipodes, Macquarie Islands, and China.

There is 1 species from China.

1. *Microscolex wuxiensis* Xu, Zhong & Yang, 1990

Microscolex wuxiensis Xu, Zhang & Yang, 1990. *Acta Zootaxonomica Sinica*. 15(1)28–31.
Microscolex wuxiensis Xu & Xiao, 2011. *Terrestrial Earthworms of China*. 274.

External Characters

Length 16–24.5 mm, width 0.9–1.2 mm, segment number 40–61. Prostomium 1/2 prolobous. Dorsal pore absent. Clitellum swollen, in XIII–XVII, ring-shaped, smooth except there is a)(-shaped groove on the ventral side of XVII; setae present on the ventral side of clitellum. Setae 4 pairs per segment; each pair setae spaced widely; on preclitellar segments $aa = 1.5–1.8ab < bc = cd, ab \leqq cd, dd = 2–2.5cd, dd = 1/3–2/5$ body circumference apart; on posterior segments $ab < aa < cd, dd = 2.5–2.8cd, dd = 1/3$ body circumference. Copulatory setae on XVII which long, thin, and broken off usually. Male and prostatic pores open in common on XVII, between a, b copulatory setae. Spermathecal pores 1 pair, in 8/9 intersegment. Nephridiopores 1 pair at anterior edge near intersegmental furrow of each segment, between setae b and c, nearer to c. Female pores 1 pair open on XIV, before setae a.

Internal Characters

Septa 5/6–11/12 thickened. Pharyngeal glands developed to VII or IX. No typhlosole, caeca, gizzard. Esophageal pouches in VII, IX indistinct and weakly developed. Intestine beginning in XVI. Hearts 3 pairs, the last pair in XII. Front nephridia larger than post. Two pairs free testes and funnels, in X, XI. Seminal vesicles 2 pairs in XI, XII, developed and distinct. Prostate glands 1 pair, long mass-shaped, in XVII. Ovaries in usual situation. 1 pair spermathecae in IX with 1 or 2 diverticula usually.

Color: Preserved specimens translucent and reddish.
Distribution in China: Jiangsu (Wuxi).

Type locality: Jiangsu (Wuxi).
Etymology: The name of the species refers to the type locality.
Deposition of types: Department of Biology, Nanjing University.
Habit: Lives in dark, moist, and humus soil.
Remarks: This species is similar to *Microscolex phosphoreus* (Ant. Dug.) but color reddish, translucent, setae on the clitellar segments present, on the ventral side of XVII, there are a)(-shaped groove and 2 tuberculas where male and prostatic pores open in common and between the copulatory setae a, b. Prostate glands long mass-shaped in XVII.

GENUS: *PLUTELLUS* PERRIER, 1873

Plutellus Perrier, 1873. *Arch. Zool. Exp. Gen*. 2:250.
Plutellus Stephenson, 1930. *The Oligochaeta*. 833.
Plutellus Gates, 1972. *Trans. Am. Phil. Soc*. 62(7):37–39.
Plutellus Xu & Xiao, 2011. *Terrestrial Earthworms of China*. 274–275.

Type species: *Plutellus heteroporus* Perrier, 1873.
Setae 8 per segment. Dorsal pores, present behind the clitellum. Clitellum in XIII–XVIII, saddle-shaped, intersegmental furrows boliterated and dorsal pores occluded. Male pores paired or single, in XVIII, XIX, or XX. Female pores mostly paired; spermathecal pores end at furrow 8/9 or on segment IX, a single pair, or a series of 1–7 pairs, or 5 single pores. Genital apertures, all minute and superficial. Pigment, lacking.

Internal Characters

Septa, present from 5/6. Gizzard in the region of segments V–VII; without calciferous glands or with such glands (and without any consideration of internal structure) as follows: 1 pair intramural in XVI or extramural in XVII, 2 pairs in XIV–XV or in XV–XVI, 3 pairs in X–XII or in XI–XIII, 4 pairs in X–XIII, XII–XV or in XIII–XVI; 5 pairs in IX–XIII. Intestinal origin, in XIV, XV, XVI, XVII, or XVIII, etc. With or without a subneural trunk, with dorsal blood vessel double or single, with all hearts lateral some esophageal or lateroesophageal. Purely meganephridial. Prostates tubular, with simple unbranched canal. Spermathecae, 1–7 pairs, in some or all of VI–XII.

Global distribution: Ceylon, India, Burma, Australia (Tasmania), New Caledonia, New Zealand (Stewart and Auckland Islands), Queen Charlotte Island and the Pacific coastal strip of the United States, Central America and northern South America.

There is 1 species from China.

2. *Plutellus hanyuangensis* Zhong, 1992

Plutellus hanyuangensis Zhong, 1992. *Acta Zootax. Sinica.* 17(3):268–273.
Plutellus hanyuangensis Xu & Xiao, 2011. *Terrestrial Earthworms of China.* 275–276.

External Characters

Length 22–29 mm, width 0.8–1.0 mm, segment number 50–76. Prostomium prolobous. Dorsal pore absent. Clitellum in 1/2 XIII–1/2 or 1/3 XVIII, saddle-shaped, gland epidermis on ventral extension to *a* or *b*, or between *bc*. Setae 8 per segment, $ab < cd < bc \leqq aa, dd = 1/2$ circumference. Male pores 1 pair, on XVIII, at *b*, in a circular porophore, diameter about 0.25 mm, front and adjacent mating setae. Mating setae 2, located setae *b*. Spermathecal pore 2 pairs, superficial, small transverse slits, in line of *c*, on posterior portion of VII and VII, close to intersegmental furrow. Female pores 1 pair, in *b* line of XIV. Mating setae 2, located setae *b*.

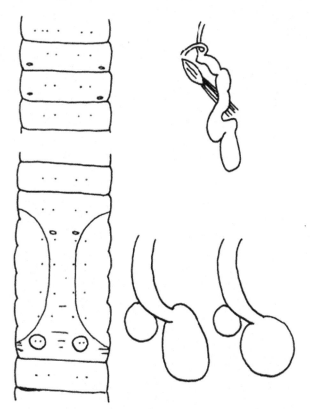

FIG. 8   *Plutellus hanyuangensis* (after Zhong, 1992).

Internal Characters

Septa 5/6 slightly thin, 6/7/8/9 thickened, behind 9/10, membranous. Gizzards absent. Intestine swelling in XIV. Hearts in XI and XII. Testis sacs 2 pairs in X, XI. Testis and sperm funnel separated from the body cavity. Seminal vesicles 1 pair, in XII. Prostate glands banded, in XVIII–XX or 1/5 XXI, M- or S-shaped; prostatic duct U-shaped. Penial setae 0.30–.0.32 mm long, 0.012 mm

thick, tip pointed, ornamented ectally by a row of small teeth. Spermathecae 2 pairs, in VII, IX, ampulla elliptic or spherical; ampulla duct stout and long, bending at the junction with ampulla; diverticulum spheroidal and sessile, just below ampulla.

> Color: Preserved specimens unpigmented. Live specimen yellowish pink. Clitellum light chestnut.
> Distribution in China: Sichuan (Hanyuan).
> Type locality: Sichuan (Hanyuan).
> Etymology: The name of the species refers to the type locality.
> Deposition of types: Department of Biology, Sichuan University.
> Remarks: This species differs from *Plutellus macer* Gates, 1955 by having 1 pair of seminal vesicles in XII and penial setae ornamented with small teeth.

GENUS: *PONTODRILUS* PERRIER, 1874

Pontodrilus Perrier, 1874. *Compt. Rend. Acad. Sci. Paris.* 78:1582.
Pontodrilus Stephenson, 1930. *The Oligochaeta.* 833–834.
Pontodrilus Gates, 1972. *Trans. Am. Phil. Soc.* 62(7):47.
Pontodrilus Xu & Xiao, 2011. *Terrestrial Earthworms of China.* 276.

Type species: *Pontodrilus marionis* Perrier, 1874 = *Lumbricus litoralis* Grube, 1855.

External Characters

Setae 8 per segment. Male pore paired (common openings of sperm and prostatic ducts), in XVIII, Female pores in XIV. Spermathecal pores, 2–4 pairs; at 7/8/9, the last in furrow 8/9.

Internal Characters

Septa all present from 4/5. Gizzard vestigial or absent. Intestinal origin behind XIII, but without calciferous and supraintestinal glands, intestinal caeca, and typhlosoles. With complete (single) and ventral trunks but without a subneural, with paired extraesophageal passing onto gut in XII–XIII and uniting mesially in XIV–XV, (a short supraesophageal?), hearts of V–IX slender, of X–XIII much thicker, all lateral (?) in X–XII. Excretory system, of holandric (avesiculate?) nephridia, lacking in preclitellar segments. 2 pairs of free testes and funnels. Prostates tubular, with simple unbranched canal. Biprostatic, metagynous, ovaries fan-shaped (? and with several egg strings? Ovisacs?).

Global distribution: Circumglobal, on seashores in the tropics and warmer parts of temperate zones in both hemispheres. Nothing is known as to the original home of the species (one possibility is the Australia-New Zealand region) or as to how the species have spread around the world.

There are 3 species from China.

TABLE 3   Key to the Species of the Genus *Pontodrilus* in China

1	Clitellum, XIII–XVII, XVIII, saddle-shaped; genital markings present .... 2 Clitellum, XIII–1/2 XVIII or XVIII, saddle-shaped; without genital marking .... *Pontodrilus sinensis*
2	Genital markings, unpaired, median, transversely elliptical, across 19/20 and less often 12/13, 13/14 .... *Pontodrilus bermudensis* Genital marking large, medioventral, transversely oval across 19/20, center depressed .... *Pontodrilus litoralis*

3. *Pontodrilus bermudensis* Beddard, 1891

Pontodrilus bermudensis Beddard, 1891. *Ann. Mag. Nat. Hist. Ser.* 6(7):96.
Pontodrilus bermudensis Chen, 1938. *Contr. Biol. Lab. Sci. Soc. China (Zool.).* 12(10): 379–381.
Pontodrilus bermudensis Gates, 1972. *Trans. Am. Phil. Soc.* 62(7):47–48.

External Characters

Length 32–120 mm, clitellum width 2–8 mm, segment number 78–120. Prostomium epilobous. Setae 8 per segment, ornamented ectally, *a*, *b*-XVII lacking, *ab* < *cd*, *aa* and *bc* may *ca* = *cd*, *dd* < 1/2 circumference. Nephropores, inconspicuous (?), at *c* (?). Dorsal pore none. Clitellum XIII–XVII, XVIII; saddle-shaped. Spermathecal pores 2 pairs in 7/8/9, at or lateral to *b*. Male pores paired in XVIII, minute, at *b*, on small papillae on lateral walls of longitudinal depressions median to longitudinal ridges (?). Genital markings, unpaired, median, transversely elliptical, across 19/20 and less often 12/13, 13/14. Female pores paired, medioventral in XIV, each anterior to seta *a*.

Internal Characters

Septa 5/6/7/8/9/11/12/13 muscular. Gizzard, none (or very rudimentary? And then in V?). Intestine from XVII. Esophageal hearts in VII–XIII. Nephridia, discoidal, avesiculate (?), lacking in I–XII and XIV, small in XIII, larger from XV, cells of peritoneal investment hypertrophied. Holandric. Seminal vesicles, acinous, in XI and XII. Sperm ducts, into prostatic ducts at parietes (? or more entally?) Prostate duct, to 2 mm long, curved into a crescent shape, with muscular sheen, narrowed at each end. Penial setae, none. Genital marking glands, none, epidermis thickened. Spermathecae 2 pairs in VII and IX, long enough to reach dorsal parietes, duct shorter than ampulla and emerging from parietes only at equators of VII, IX, diverticulum digitiform to club-shaped, from duct within parietes and emerging into coelom just behind 7/8/9.

Color: Preserved specimens none (unpigmented).
Type locality: Bermuda.
Distribution in China: Hainan.
Global distribution: Kadonkani (Pyapon), Burma, Andaman Islands, Vietnam, Borneo, Celebes, Aru, New Guinea, Australia, Palmyra Atoll, Fannin Island, Laysan, Texas, Mississippi, Louisiana, Florida, Virginia, Mexico, Bermuda, Bahamas, Dry Tortugas, Haiti, Jamaica, Virgin Islands, Mona Island, Colombia, Brazil, Congo, Angola, South Africa, Tanzania, Madagascar, Laccadive Islands, India (east and west coasts), Maldive Islands, Ceylon.

4. *Pontodrilus litoralis* (Grube, 1855)

Pontodrilus litoralis Easton 1984. *New Zealand Journal of Zoology.* 11:114.
Pontodrilus litoralis James, Shih & Chang 2005. *Jour. N. His.* 39(14):1022–1023.
Pontodrilus litoralis Chang, Shen & Chen, 2009. *Earthworm Fauna of Taiwan.* 148–149
Pontodrilus litoralis Xu & Xiao, 2011. *Terrestrial Earthworms of China.* 276–277

External Characters

Length 50–130 mm, clitellum width 1–2 mm, segment number 81–115. Prostomium epilobous. Setae lumbricine, *ab* absents in XVIII. Dorsal pore absent. Clitellum XIII–XVII, saddle-shaped, setae present. Spermathecal pores 2 pairs in 7/8/9, ventrolateral, in line with seta *b*. Female pores paired, medioventral in XIV, each anterior to seta *a*. Male pores minute, paired in XVIII, each on inner wall of a longitudinal depression, median to a longitudinal ridge extending the entire segment XVIII. Genital marking large, medioventral, transversely oval across 19/20, center depressed.

Internal Characters

Septa 5/6–12/13 thickened. Gizzard absent. Intestine from XVII. Esophageal hearts in VII–XIII. Spermathecae 2 pairs in VII and IX, tubular, diverticulum slender, narrower at the junction with ampulla. Accessory glands absent. Nephridia vesiculate, absent in I–XII and XIV, small in XIII, larger from XV. Testis sacs 2 pairs in X and XI. Seminal vesicles paired in XI and XII, thin, follicular. Prostate glands paired in XVIII, tubular, prostatic duct curved.

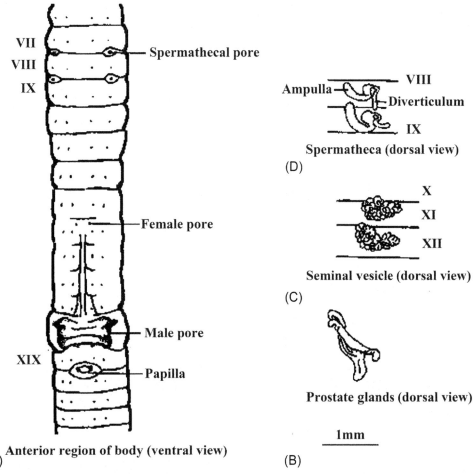

FIG. 9   *Pontodrilus litoralis* (after James et al., 2005).

Color: Preserved specimens pale, light brown around clitellum.
Type locality: South France.
Deposition of types: Humbolt Museum, Berlin, Germany.
Distribution in China: Taiwan (Penghu), Fujian (Jinmen Island).
Global distribution: Cosmopolitan, widely distributed on warm beaches throughout the world.
Remarks: *P. litoralis* dwells in sandy beaches, salty mud, or mangrove swamps of the intertidal zone.

5. *Pontodrilus sinensis* Chen & Xu, 1977

Pontodrilus sinensis Chen & Xu, 1977. *Acta Zool. Sinica.* 23(2): 178–179.
Pontodrilus sinensis Xu & Xiao, 2011. *Terrestrial Earthworms of China.* 277.

External Characters

Length 30.5–41 mm, width 1.2–1.8 mm, segment number 67–94. Prostomium 1/2 epilobous. Dorsal pore absent. Clitellum in XIII–XVIII or 1/2–2/3 XVIII, saddle-shaped, but gland in XIII and XVIII thin, seta *ab* clear. Setae 4 pairs per segment $aa = 2.5$–$3ab, bc = cd = 2ab, aa > bc, dd$ of dorsal distance less than 1/2 of body circumference apart. Male pores 1 pair, situated on ventral of XVIII, in front of setae *b*. Spermathecal pores 2 pairs, in 7/8/9, in line with *b*, pinprick-like. Female pores 1 pair, situated on ventral surface of XIV, in *a*.

Internal Characters

Septa 5/6–9/10 moderately thickened; 10/11/12/13 slightly thickened, and rest thin. No gizzard and calciferous glands. Intestine begins to swell in XIV or XV. Last heart in XII. Nephridis begin in XIII. Male funnels free in X, XI. Seminal vesicles 4 pairs, in IX–XII. Prostates with glandular portion long or short, in XIX–XXI or XX.

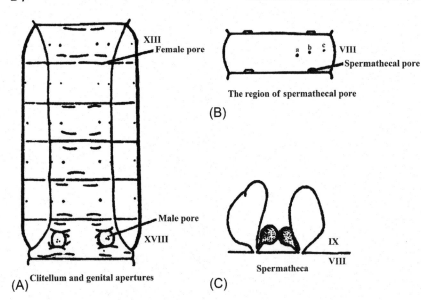

FIG. 10 *Pontodrilus sinensis* (after Chen & Xu, 1977).

Each vas deferens opens into the prostate duct at its termination and in a common pore. Spermathecae in VII–IX; ampulla thumb-like or irregular sac-like; duct very short, boundaries obvious; diverticulum shorter than main pouch, its seminal chamber enlarged, sphere-shaped.

Color: Preserved specimens anterior end gray-white, posterior end gray-brown. Clitellum yellowish pink.

Type locality: Yunnan (Kunming).
Distribution in China: Yunnan (Kunming).
Deposition of types: ?
Remarks: This species differs from the other species of *Pontodrilus* by the deferens opening into the prostate duct at its termination, and in a common pore, no gizzard, and spermathecal pores 2 pairs, in 7/8/9.

Family: Glossoscolecidae

Glossoscolecidae Michaelsen, 1900. *Das Tierreich.* 10:420.

Glossoscolecidae Stephenson, 1930. *The Oligochaeta.* 885–886.

Glossoscolecidae Gates, 1972. *Trans. Am. Phil. Soc.* 62(7):52–53.

Glossoscolecidae Brinkhurst & Jamieson, 1971. *Aquatic Oligochaeta of the World.* 723–725.

Type genus: *Glosssoscolex* Leuckart, 1835.

External Characters

Lateral lines present or absent. Dorsal pore absent. Clitellum multilayered, usually beginning behind segment XIV, and usually extensive, frequently occupying 10 or more segments. Setae usually sigmoid, simple pointed, 4 pairs per segment. Genital setae often present; sometimes longitudinally grooved. Male pore a pair, exceptionally 2 pairs, in the anterior of the clitellar region or anteclitellar, very rarely postclitellar. Spermathecal pore usually altogether behind testis segments. Female pores in XIV, or rarely, in XIII and XIV.

Internal Characters

Esophageal gizzards 1–3 in front of the testis segments, sometimes absent. Calciferous glands paired, in some of segments VII–XIV. Anterior end of the intestine with or without gizzard-like thickening of its musculature. Hearts variable in number (usually in VII–XI, XII); dorsal vessel single; ventral vessel well developed; median subneural and supraesophageal vessel present or absent. Nephridia holonephridia; exonephric with, rarely, some anteriorly enteronephric; rarely 2 pairs per segment (mononephry). Holandric; proandric or metandric; testis sacs present or absent. Ovaries in segment XIII or, rarely, in XII and XIII. Vasa deferentia (always?) concealed, in the body wall musculature; the ectal end usually simple but often with muscular copulatory sac, rarely associated with prostate glands. Prostate-like glands sometimes present in the vicinity of normal or genital setae. Spermathecae paired or transversely multiple; very rarely with diverticla.

Global distribution: Holarctic, Neotropical, Palaeotropical, and Malagasian.

There is 1 genus and 2 species from China.

GENUS: PONTOSCOLEX SCHMARDA, 1861

Pontoscolex Schmarda, 1861. *Neue Wirbellose Theire, Leipzig.* 1,2:11.

Pontoscolex Stephenson, 1930. *The Oligochaeta.* 895.

Pontoscolex Brinkhurst & Jamieson, 1971. *Aquatic Oligochaeta of the World.* 737.

Pontoscolex Gates, 1972. *Trans. Am. Phil. Soc.* 62(7):53–54.

Type species: *Lumbricus corethrurus* Müller, 1856.

External Characters

Setae in the hind region normally arranged in quincunx (*a b c* and *d* not forming 4 longitudinal rows, the 2 setae of a pair set somewhat widely apart and those of successive pairs alternating in position). Male pores intraclitellar.

Internal Characters

Calciferous glands 3 pairs, in VII–IX, panicled tubular sacs arising dorsally from the esophagus. Seminal vesicles very long, extensive. With spermathecae. Nephridia with termincter.

Global distribution: Bermuda, West Indies, South America, Venezuela, Suriname, and circummundane in the tropics.

There are 2 species from China.

TABLE 4 Key to the Species of the Genus *Pontoscolex* From China

1. Spermathecal pores at 5/6/7/8 .... *Pontoscolex guangdongensis*
Spermathecal pores at 6/7/8/9, clitellum XV, XVI–XXII, XXIII *Pontoscolex corethrurus*

6. *Pontoscolex Corethrurus* (Müller, 1856)

Lumbricus corethrurus Müller, 1856. *Abhandl. Naturgesch. Ges. Halle.* 4:26.

Pontoscolexcorethrurus Stephenson, 1916. *Rec. Indian Mus.* 12:349.

Pontoscolexcorethrurus Gates, 1972. *Trans. Am. Phil. Soc.* 62(7):54–55.

Pontoscolex corethrurus Chang, Shen & Chen, 2009. *Earthworm Fauna of Taiwan.* 14–15.

Pontoscolex corethrurus Xu & Xiao, 2011. *Terrestrial Earthworms of China.* 281-282.

External Characters

Length 60–120 mm, clitellum width 4–6 mm, segment number 167–220. Number of annuli (secondary segmentation) 2 in the middle portion of body, including a narrow anterior nonsetal annulus and a wide posterior setal annulus. Prostomium prolobous. Setae lumbricine. Dorsal pore absent. Clitellum XIV, XV–XXI, XXII (usually 7 segments), length 6.5–8.8 .... mm, segmented, saddle-shaped, setae absent. Tubercula pubertatis XV–XXI. Spermathecal pores invisible. Female pore invisible. Male pores paired in XVII, ventral, 0.06 circumference apart ventrally. Male pores invisible in some specimens. Nephridiopores distinct, paired, in intersegmental furrows, 2 longitudinal rows at the lateroventral sides of body in the anterior segments.

Internal Characters

Septa 6/7–9/10 thickened. Gizzard in IV–VI, large. Calciferous glands 3 pairs in VII, VII and IX, dorsolateral, oval-shaped. Intestine from 15. Intestinal caeca absent. Dorsal and lateral hearts greatly enlarged from X and XI. Spermathecae 3 pairs in VII–IX, small, ampulla small, duct slender, no diverticulum. Spermathecal ducts connected laterally to 6/7/8/9. Nephridial batteries paired in IV and V, nephridia holandric in remaining segments, with a straight nephridial duct and a slightly enlarged distal end. Testis sacs 1 pair in XII, large, extending anteriorly to XI. Seminal vesicles paired in XIII. Prostate glands absent.

Color: Preserved specimens white in color, clitellum light grayish brown. Live specimens pigment less, head portion pink to light purple, clitellum orange or grayish pink, body light bluish pink, tail white. 3 pairs of bright yellow spots (calciferous glands) observable externally in front of clitellum.

Distribution in China: Taiwan (Taibei, Pingdong, Gapxiong, Yilan, Xinzhu, Yaoyuan, Miaoli, Lanyu, Siaoliouciou, Turtle), Fujian (Kinmen Islands, Quanzhou), Hong Kong, Guangxi (Xining).

Global distribution: Cosmopolitan. This species was originally endemic to tropical America. At present, it has a pantropical distribution caused by human activity and is one of the most widespread earthworms in the world.

Etymology: The name of the species refers to the type locality.

Deposition of types: Itajahy, Brazil.

Remarks: *P. corethrurus* is a dominant earthworm in disturbed environments, such as parks and school campuses in cities, agricultural lands, and ditches along mountain roads in Taiwan. Its colonization probably occurred in the past 5 decades (Chen et al., 2004), and it is now causing serious problems to soil ecosystems. It can tolerate a variety of environmental changes, has high fertility, and is capable of both sexual and parthenogenetic reproduction. These factors contribute to its dominance in many localities and make this species a strong competitor against indigenous species.

7. Pontoscolex guangdongensis Zhang, Wu and Sun, 1998

Pontoscolox guangdongensis Zhang, Wu & Sun, 1998. *Sichuan Journal of Zoology.* 17(1):5-6.

Pontoscolex guangdongensis Xu & Xiao, 2011. *Terrestrial Earthworms of China.* 282.

External Characters

Length 90–157 mm, width 4–5 mm, segment number 103–210. Prostomium invisible. Dorsal pore absent. Clitellum XIV–1/2XXII or XXII, saddle-shaped, intersegmental furrows visible. Sex glandular ridge in 1/2 XVIII–1/2XXI; round swollen gland visible at the inside of *ab*. Setae long and stout; 8 per segment; front 4 pairs of seta on clitellum dense, $aa = 1.8–2.0ab$, arraying longitudinal; setae of posterior part of the body in quincunx arrangement. Male pores invisible in surface of body,

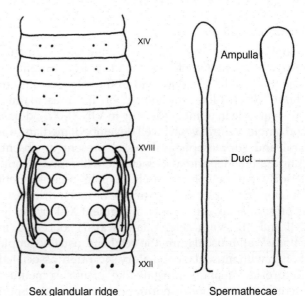

Sex glandular ridge   Spermathecae

FIG. 11   *Pontoscolex guangdongensis* (after Zhang et al., 1998).

on outside of XIX *b*. Spermathecal pores 3 pairs, in 5/6/7/8, on the rear edge of each segment, needle-like.

Internal Characters

Septa 8/9 absent, 5/6/7/8 thickened, 9/10 slightly thick, after 10/11 membranous. Pharynx developed. Gizzard bell-shaped to long bell-shaped, smooth on surface, in V. Calciferous glands 3 pairs in VI, VII, and VII; Last pair slightly developed, long bag-like, before 9/10. Intestine from XV. Intestinal caeca absent. Hearts 2 pairs, in X and XI. Testis sacs 1 pair in XI, thin and flat, enclosing part of the heart. Seminal vesicles 1 pair, indivisible lobe, short or very long, sometimes extending to XIII. Seminal vesicles very long extending to XVIII. No prostatic gland. Spermathecae 3 pairs in VII–IX, small, ampulla small, duct slender, no diverticulum. Spermathecal duct clavate, ampulla oval-shaped. No diverticulum.

Color: Surface without pigment, clitellum with orange or brown.
Type locality: Guangdong (Maoming).
Distribution in China: Guangdong (Maoming).
Etymology: The specific name indicates the type locality.
Deposition of types: Department of Biology, Hangzhou Normal College.
Remarks: This species is similar to *Pontoscolex corethrurus* (Müller, 1856); spermatheca 3 pairs, calciferous glands 3 pairs, setae of posterior part of the body in quincunx arrangement. However, significant differences of spermathecal pores in 5/6/7/8, male pores on outside of XIX, spermathecal duct clavate, and ampulla oval-shaped.

Family: Haplotaxidae (Claus, 1880)

Haplotaxidae Claus, 1880. *Grundzüge der Zool.* 1:482.
Haplotaxidae Brinkhurst & Jamieson, 1971. *Aquatic Oligochaeta of the World.* 286.
Haplotaxidae Gates, 1972. *Trans. Am. Phil. Soc.* 62(7):58–60.

Type genus: *Haplotaxis* Hoffmeister, 1843.

External Characters

Clitellum of a single layer of cells, mostly in the region of the genital pores. Setae single or closely paired, *S*-shaped or distally hooked, dorsals small or absent. Genital setae in some species. Male pore 2 pairs, ventrolateral or lateral, very small and frequently invisible externally, on segments XI and XII (or on X and XI) or both on XII. Spermathecae 1–4 pairs, anterior to the gonads, simple, without diverticula. Spermathecal pores frequently lateral or in the line of the dorsal setae. Female pores in 12/13, 12/13, and 13/14 (or 11/12 and 12/13). Male efferent ducts very short and slightly coiled, without atria or prostate glands, but distal end lined with cuticle.

Internal Characters

Some anterior septa may be thickened. Meganephirdia from a few pregonadal segments to the distal end except for the gonadal segments, pores near ventral setae. Gut simple with an eversible pharynx or muscles around the pharynx forming a gizzard. Septal glands in those species with eversible pharynx. Dorsal and ventral blood vessels connected by a single pair of long commissural vessels often in each segment. Testes in X and XI (or IX and X), ovaries in XII, or XII and XIII (or XI and XII).

Global distribution: Cosmopolitan.

There is 1 genera with 1 species from China.

GENUS: *HAPLOTAXIS* HOFFMEISTER, 1843

Haplotaxis gordioides Brinkhurst & Jamieson, 1971. *Aquatic Oligochaeta of the World.* 286–288.

Type species: *Lumbricus gordioides* Hartmann, 1821.

External Characters

Clitellum mostly in the region of the genital pores. Male pore 2 pairs on segments XI and XII (or on X and XI), or 1 pair both on XI. Female pores 1 pair in 12/13, or 2 pairs in 12/13 and 13/14.

Internal Characters

Gut simple with an eversible pharynx or muscles around the pharynx forming a gizzard. Septal glands in those species with eversible pharynx. Testes 2 pairs in X and XI, or 1 pair in X. Ovaries in XII, or XII and XIII (or XI and XII).

There is only 1 species from China.

Global distribution: United States, Denmark, Germany, Poland, Bohemia, Switzerland, China, France, Belgium, United Kingdom, Corsica, Italy, Yugoslavia, Hungary, Bulgaria, Russian Federation, Siberia, Japan, Poeloe Berhala, Australia, New Zealand, Auckland and Campbell Islands, Argentina, Paraguay, Peru, southern and equatorial Africa.

Distribution in China: Jiangxi, Guangdong, Hunan, Xinjiang, (Mt. A'ertai).

8. *Haplotaxis gordioides* (Hartmann, 1821)

Haplotaxis gordioides Brinkhurst & Jamieson, 1971. *Aquatic Oligochaeta of the World.* 289–291.
Haplotaxis gordioides Xu & Xiao, 2011. *Terrestrial Earthworms of China.* 43.

External Characters

Length 180–400 mm, width 0.3–2.0 mm, segment number more than 200. Prostomium long, with a transverse groove. Segment I short, anterior segments biannulate. Cuticle with unicellular glands. Clitellum XI–XVIII, annular. Setae mostly single, 4 per segment (with partially developed replacement setae ventrally). Dorsal setae small, varying in extent (absent behind XI or as far as XXX); ventral setae large, more or less sickle-shaped. Male pores 2 pairs in front of ventral setae of XI and XII, minute. Spermathecal pores 3 pairs, in 6/7/8/9, lateral, about 0.5 body circumference apart. Female pore 2 pairs, in 12/13 and 13/14 in line of ventral setae.

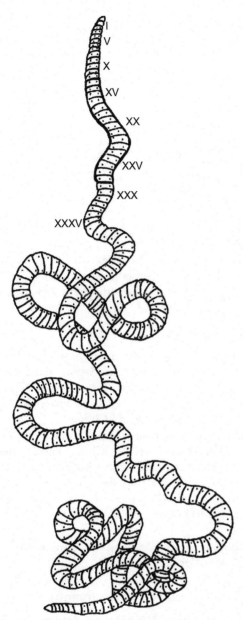

Internal Characters

   Fore-gut with a ring of circular muscle forming a simple gizzard in IV–V. No septal glands. Dorsal and ventral blood vessels linked by a pair of long commissural vessels in each segment. Nephridia IX–X, XV onwards, opening in front of ventral setae. Iimm's gland in XI–XVI or every segment, midventral beneath the nerve cord. Setal gland in XII–XIV, copulatory glands in XI–XIV. Testes and male funnels paired in X and XI. Male efferent ducts make several turns. Sperm sacs on 9/10 in IX, 10/11 in XI, 11/12 to XVIII. Spermathecae in VII to IX (? in VI), ducts short, not clearly separable from ampullae. Ovaries as long strings on posterior face of septum 13/14 and a pair of long egg sacs on dorsal side; ovaries and female funnels paired in XII, XIII. Egg sacs present, visible behind XVIII.

   Color: Light yellowish with orange at anterior end.
   Distribution in China: Jiangxi, Guangdong, Hunan, Xinjiang, (Mt. A'ertai).
   Global distribution: Holarctic.

FIG. 12   *Haplotaxis gordioides* (after Chen, 1959).

Family: Lumbricidae Claus, 1880

Lumbricidae Claus, 1880. *Ibid.* (ed. 4) 1:478.
Pontoscolex Stephenson, 1930. *The Oligochaeta.* 905.
Lumbricidae Gates, 1972. *Trans. Am. Phil. Soc.*
62(7):61–67.
Lumbricidae Reynolds, 1977. *The Earthworms of Ontario.* 34–35.

Type Genus: *Lumbricus* Linnaeus, 1758.

External Characters

Prostomium epilobous, prolobic, or tanylobic. Setae sigmoid, single-pointed, often finely ornamented, 8 per segment, in regular longitudinal rows. Dorsal pores present. Setae, sigmoid and single-pointed, 8 per segment, in regular longitudinal ranks, in genital tumescences elongated but slender and longitudinally grooved ectally. Dorsal pores, present. Clitellum usually saddle-shaped, beginning a greater or shorter distance behind the male pores. Male pore mostly on XV, rarely displaced on to 4 segments forward. Female pores as a rule on XIV. Frequently the setae on certain segments of the anterior part of the body are implanted on papillae and transformed into copulatory setae, with longitudinal ridge and intervening grooves in their distal portion.

Internal Characters

Esophagus with calciferous glands; a single well-developed gizzard at the beginning of the intestine. Meganephridial. Vascular system: with complete dorsal, ventral, and subneural (and lateroneural?) trunks, the latter adherent to nerve cord, extraesophageal trunk median to the hearts passing to dorsal trunk in region of X–XII, without supraesophageal and lateroparietal trunks. Hearts, lateral, the last pair anterior to segment XII. Two pairs of testes and funnels in X and XI; no prostates projecting freely into the body cavity; rarely prostate-like glandular cushions present. Ovaries in XIII. Spermathecae if present simple, without diverticula (often constricted by the septa in such a way that they appear to consist of 2 portions, ampulla and diverticulum), pore at intersegmental levels. Ovisacs, in XIV, small, lobed.

Global distribution: Mostly terrestrial, temperate and cold regions of the Northern Hemisphere; from Japan over Siberia, Central Asia and Europe to North America; in a southerly direction as far as Japan, Lake Baikal, Turkey, northern India, Palestine, Florida. However, many species have been transported into the temperate zone of the Southern Hemisphere, and a number also into tropical regions. Southeastern United States below the southern limit of glaciation and west from New Jersey (perhaps to the Mississippi River or somewhat beyond?), Europe below the southern limit of glaciation.

There are 9 genera, 21 species, and 4 subspecies from China.

TABLE 5 Key to the Genera of the Family Lumbricidae From China

1	Prostomium epilobous, prolobic, or tanylobic. Clitellum saddle-shaped..2
	Prostomium tanylobic, Clitellum, XXVI–XXXII, saddle-shaped ....*Lumbricus*
2	Male pore on XIII..*Eisenlella*
	Male pore on XV 3
3	Clitellum in XVIII, XXIX–XXXVII...................*Allolobophora*
	Clitellum before or behind XXXVII......................4
4	Clitellum in XXXI or XXXII or a part of XXXIII; or rarely XXXIV..5
	Clitellum before or behind XXXIII 7
5	Clitellum in XXXI or XXXII, setae widely paired...*Debdrodrilus*
	Clitellum in XXXI or XXXII or a part of XXXIII, setae closely paired 6
6	Tubercula pubertatis absent or indistinct, not specially sharply defined ....*Bimastus*
	Tubercula pubertatis in XVIII, XXIX, XX–XXI, XXXXII ....*Eisenia*
7	Setae closely paired..*Aporrectodea*
	Setae is or not closely paired..
8	Setae widely paired..*Dendrobaena*
	Setae on anterior end of body is closely paired, on posterior end is widely paired ....*Octolasium*

GENUS: ALLOLOBOPHORA EISEN, 1873

Enterion chloroticum Savigny, 1826.
Lumbricus riparius Hoffmeister, 1874.
Allolobophora Eisen, 1873. *Öfv. Vet. Akad. Förh. Stockholm.* 30(8):46.
Allolobophora Michaelsen, 1900. *Das Tierreich., Oligochaeta.* 10:480.
Allolobophora Michaelsen, 1910. *Ann. Mus. Zool. St. Petersburg.* 15:1.
Allolobophora Gates, 1972. *Trans. Am. Phil. Soc.* 62(7):68–69.
Allolobophora Eisen, 1874. *Ofvers. Vetensk.-Akad. Forhandl. Stockholm.* 30(8).
Allolobophora Gates, 1975. *Megadrilogica.* 2(3):3.
Allolobophora Reynolds, 1977. *The Earthworms of Ontario.* 35.
Allolobophora Feng, 1985. *Journal of Zoology.* 20:46–47.

Type species: *Lumbricus riparius* Hoffmeister, 1874 (=*Enterion chloroticum* Savigny, 1826).

External Characters

Prostomium usually epilobous, rarely tanylobous. Clitellum saddle-shaped, in XXVI–XXXIV. Setae more or less closely paired. Setae 4 pairs. Male pore on XV; spermathecal pore 2 pairs, in the line of setae *cd*. Nephropores, inconspicuous, behind the clitellum irregularly alternating between levels slightly above *b* and above *d*. Longitudinal musculature, pinnate.

Internal Characters

Gizzard taking up more than 1 segment. Testes and funnels free; seminal vesicles 4 pairs in IX–XII; those of X approximately as large as those of IX (? always). Calciferous gland, opening to gut through a pair of vertical sacs in X. Calciferous lamellae continued along lateral walls of sacs. Gizzard, mostly in XVII. Extraesophageal vessels, passing to dorsal trunk in XII. Hearts, in VI–XI. Nephridial bladders, J-shaped, closed end laterally, ducts passing into parietes near *b*.

Color: variable.
Discussion: *Allolobophora* was erected by Eisen (1873) without the designation of a type species and this situation was not corrected by Michaelsen (1900) in his revision of the Lumbricidae. Typification of the genus was by Omodeo (1956), who selected *Allolobophora chlorotica* to be the type. Additional species that Eisen included in his *Allolobophora* were: *arborea*, *fetida*, *muscosa*, *norvegica*, *subrubicunda*, and *turgida*, none of which is now referable to this genus.

There is 1 species from China.

9. *Allolobophora chlortica* (Savigny, 1826)

Enterion chlortica Savigny, 1826. *Mem. Acad. Sci. Inst. France.* 5:183.
Allolobophora chlorotica Gates, 1972. *Trans. Am. Phil. Soc.* 62(7):69–73.
Allolobophora chlorotica Reynolds, 1977. *The Earthworms of Ontario.* 36–39.
Allolobophora chlorotica Xu & Xiao, 2011. *Terrestrial Earthworms of China.* 285.

External Characters

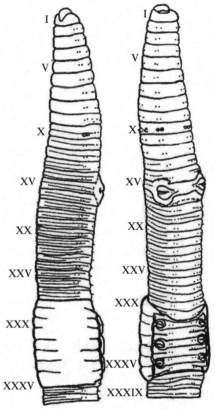

FIG. 13   *Allolobophora chlorotica* (after Reynolds, 1977).

Length 30–70mm, diameter 3–5mm, segment number 80–143. Prostomium epilobous, First dorsal pore 4/5 or 5/6. Body cylindrical. Clitellum in XVIII, XXIX–XXXVII, saddle-shaped. Tubercula pubertatis small, sucker-like discs on XXXI, XXXIII, and XXXV. Setae closely paired, *aa*>*bc*, *dd*=1/2 of circumference apart anteriorly; and *dd*<1/2 of circumference apart posteriorly. Setae *c* and *d* on X often on white genital tumescences. Spermathecal pores 3 pairs in 8/9/10/11, on level *cd*. Male pores 1 pair, in XV, with large elevated glandular papillae extending over XIV and XVI. Female pores paired in XIV.

Internal Characters

Septa 5/6/7/8/9/10 somewhat muscularized, 10/11/12/13/14 less so. Longitudinal musculature, fasciculate. Calciferous sacs, in X, digitiform to pyriform, anteriorly (and then often in contact with 9/10), antero-laterally or even dorsally directed, opening into gut ventrally about at level of insertion of anterior lamella of septum 10/11, calciferous lamellae continued to anterior ends of the sacs. Gut lumen, vertically slit-like in XI–XII, wider behind an internal constriction just in front of 12/13. Gizzard, mostly in XVII. Gut, narrowed and valvular in XIX or at insertion of 18/19. Intestine from XV. Typhlosole, beginning in region of XXI–XXIII, anteriorly compressed dorsoventrally so as to have an inverted T-shaped in cross section, with deep longitudinal grooves marking off 3–5 ridges on ventral face, ridges gradually disappearing posteriorly but T-shape of section retained much further back, ending abruptly in LXXXVI–XCV. Extraesophageal vessels, joining dorsal trunk in XII. Hearts in VII–XI. Nephridial bladders, J-shaped to almost U-shaped, closed end laterally, longer ectal limb posteriorly or ventrally. Holandric. Seminal vesicles 4 pairs, in IX–XII. Sperm ducts, without epididymis. Atrial glands, extending into XIV and to 16/17. Spermathecae 3 pairs in IX–XI, with shortly elliptical ampulla, coelomic portion of duct to as long as ampulla.

Color: Variable, frequently green but sometimes yellow, pink, or gray.
Distribution in China: Jiangsu, Anhui, Sichuan.
Global distribution: A native of the Palaearctic, *Allolobophora chlorotica* is known from Europe, Iran, North America, South America, North Africa, and New Zealand.
Type locality: France (Paris).
Deposition of types: Mus. Hist. Nat. Paris.
Etymology: The specific name derives from its body color.
Remarks: This species has been found in a variety of soil types, including gardens, field, pastures, forests, clay and peat soils, shores and stream banks, estuarine flats, and among all sorts of organic debris. The habitat preference is "wet and usually highly organic or polluted soil." (Eaton, 1942; Reynolds et al., 1974).

GENUS: APORRECTODEA ÖRLEY, 1885

Aporrectodea Örley, 1885. *Ertek. Term. Magyar Akad.* 15(18):22.
Allolobophora (part.) Michaelsen, 1900. *Das Tierreich., Oligochaeta.* 10:480.
Allolobophora (part.) Stephenson, 1930. *Oligochaeta.* 905–908.
Allolobophora Omodeo, 1956. *Arch. Zool. It.* 41:180.
Allolobophora Gates, 1972. *Trans. Amer. Philos. Soc.* 62(7):68–69.

Allolobophora Gates, 1972. *Bull. Tall Timbers Res. Stn.* 12:2.
Allolobophora Gates, 1975. *Megadrilogica.* 2(1):4.
Aporrectodea Reynolds, 1977. *The Earthworms of Ontario.* 40.
Aporrectodea Xu & Xiao, 2011. *Terrestrial Earthworms of China.* 285–286.

Type species: *Lumbricus trapezoides* Dugès, 1828.
Prostomium, epilobous. Calciferous gland, opening to gut through a pair of vertical sacs equatorially in X. Calciferous lamellae continued onto posterior walls of sacs. Gizzard, mostly in XVII. Extraesophageal vessels, passing to dorsal trunk in XII. Hearts, in VI–XI. Nephridial bladders, U-shaped, ducts passing into parietes near *b*. Nephropores, inconspicuous, irregularly alternating between levels slightly above *b* and above *d*. Setae paired. Longitudinal musculature, pinnate.

Color: Pigment, if present, not red.
Global distribution: Palaearctic, Western Europe, North America.
Discussion: This forgotten genus originally included *Enterion chloroticum* Savigny, 1826 and *Lumbricus trapezoides* Dugès, 1828. After Omodeo (1956) designated the former as the type species of *Allolobophora*, the latter automatically became the type for *Aporrectodea*. Bouche (1972) erected a new genus *Nicodrilus* with *Enterion caliginosum* Savigny, 1826 as the type and included *Lumbricus trapezoides* Dugès, 1828 in this new genus. Since *Aporrectodea* is a valid and available genus, *Nicodrilus* must be considered the junior synonym of *Aporrectodea*.

There is 1 species from China.

TABLE 6 Key to the Species of the Genus *Aporrectodea* From China

1	Clitellum in XXVII, XVIII, XXIX, XXX–XXXIV, XXXV; on 7–8 segments.............*Aporrectodea caliginosa*
	Clitellum in XXVII, XVIII, XXIX–XXXIV, XXV; on 8–9 segments..2
2	Clitellum in XXVII–XXXIV .... *Aporrectodea tuberculata*
	Clitellum in XXVII, XVIII, XXIX–XXXIII, XXXIV, XXV; on 9 segments 3
3	Clitellum in XXVII, XVIII–XXXIV, XXV; tubercula pubertatis in XXXI–XXXIII...................*Aporrectodea turgida*
	Clitellum in XXVII, XVIII, XXIX–XXXIII, XXXIV, XXV 4
4	Clitellum in XXVII, XVIII–XXXIV, XXV; tubercula pubertatis in XXXII–XXXIV ....*Aporrectodea longa*
	Clitellum in XXVII, XVIII–XXXIII, XXXIV; tubercula pubertatis in XXXI–XXXIII.............*Aporrectodea trapezoides*

10. *Aporrectodea caliginosa* (Savigny, 1826)

Aporrectodea caliginosa Chang, Shen & Chen, 2009. *Earthworm Fauna of Taiwan*. 16–17.

External Characters

Length 35–200mm, width 3.5–4.0mm. Segment number 117–246. Prostomium epilobous, Setae lumbricine. First dorsal pore 6/7–14/15. Clitellum in XXVII, XVIII, XXIX, XXX–XXXIV, XXXV, saddle-shaped. Spermathecal pores 2 pairs in 9/10/11, lateral in *cd* line. Female pores paired in XIV, lateral to seta *b*, small slits. Male pores 1 pair, in XV, large, slits, between setae *b* and *c*. Tubercula pubertatis in *bc* in XXXI and XXXIII, interrupted in XXXII, as 2 pairs of protuberances. Genital tumescences around *ab* in IX–XI, XXVII, XXX, XXXII–XXXIV, XXXV.

Internal Characters

Septa 5/6/7/8/9/10 slightly thickened. Gizzard in XVII. Calciferous glands in X. Intestine from XX. Typhlosole present from XXI–XXIV. Hearts in VI–XI. Spermathecae 2 pairs in X and XI, round, small. Nephridia holandric. Ovaries in XIII. Testes 2 pairs in X and XI. Seminal vesicles in IX–XII, small. Prostate glands absent.

> Color: Live specimens pink to light brownish red. Clitellum darker.
> Type locality: France (Paris).
> Deposition of types: Geneva, Switzerland.
> Global distribution: Cosmopolitan, including India, Pakistan, Middle East, Europe, North Africa, South Africa, Americas, Hawaii, Japan, China, New Zealand, Australia.
> Distribution in China: Taiwan.
> Etymology: The specific name *"caliginosa"* literally means "foggy or misty."

11. *Aporrectodea longa* (Ude, 1885)

Aporrectodea longa Gates, 1972. *Trans. Am. Phil. Soc.* 62(7):73–76.
Allolobophora longa Reynolds, 1977. *The Earthworms of Ontario*. 43–45.
Aporrectodea longa Xu & Xiao, 2011. *Terrestrial Earthworms of China*. 286.

External Characters

Length 90–150mm, width 6–9mm, segment number 150–222. Body cylindrical and dorsoventrally flattened posteriorly. Prostomium epilobous. First dorsal pore 12/13. Clitellum XXVII, XVIII–XXXIV, XXXV. Tubercula pubertatis XXXII–XXXIV. Setae lumbricine, closely paired, posteriorly *aa:ab:bc:cd* = 60:7:28:5; *a* and *b* in front of genital tumescences in IX, X, XI, XXXI, XXXIII, XXXV, and sometimes XII. Male pores 1 pair, in XV, with elevated

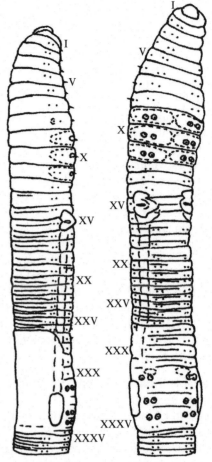

FIG. 14 *Aporrectodea longa* (after Reynolds, 1977).

glandular borders, sometimes extending to XIV and XVI. Spermathecal pores 2 pairs in 9/10/11, on level *c*. Female pore on XIV.

Internal Characters

Septa, none especially muscular, muscularity increasing 5/6/7/8/9, from 9/10 to 14/15 muscularity decreasing, 15/16 membranous. Pigment, brown in circle muscle layer. Longitudinal musculature, pinnate. Gizzard, mostly in XVII. Typhlosole, beginning in region of XXIII–XXIV, thicker ventrally, at first with regular transverse grooves and 1 median longitudinal ridge on the ventral face, posteriorly with a deep longitudinal groove on each side of the median ridge or with an appearance of 3 longitudinal ridges, ending abruptly in region of 115th–130th segments, leaving an atyphlosolate section extending through the last 56–88 metameres. Extraesophageal vessels, joining dorsal trunk in XI. Hearts, in VI–XI. Nephridial bladders, J-shaped, closed end laterally, reaching on parietes beyond *d*. Calciferous sacs, vertical, in X. Gut lumen, vertically slit-like in XI–XII, wider from XIII. Intestinal origin, in XV. Holandric. Seminal vesicles

4 pairs, in IX–XII, the anterior pair smaller. Sperm ducts, with epididymis, of 1 long hairpin loop, several short U-shaped loops, or in a ball of loops or coils. Spermathecae 2 pairs in IX and X, with short duct.

　　Color: Preserved specimens gray or brown with slight iridescence dorsally.
　　Type locality: Germany (garden, Göttingen).
　　Global distribution: Europe, North America, Central America, Africa, Australia, Asia, Iceland.
　　Distribution in China: Liaoning (Great Changshan Island, Little Changshan Island, Pikuo, Zhangzi Island), Heilongjiang, Shaanxi (Pingyao).
　　Remarks: In Europe Černosvitoc and Evans (1974) and Gerard (1964) reported this species from cultivated soil, gardens, pastures, and woodlands, and found it to be abundant in soils bordering rivers and lakes. According to Gates (1972) *Aporrectodea longa* is found in soils with a pH of 4.5–8.0, in greenhouses and botanical gardens, lawns, peat bogs, in compost and under manure, including chicken yards and cow yards, and in many other types of soils.

12. *Aporrectodea trapezoides* (Dugès, 1828)

Allolobophore caliginosa trapezoides 1931, Chen. *Contr. Biol. Lab. Sci. Soc. Chin. Zool.* 7(3):168–169.
Allolobophore caliginosa trapezoides Fang, 1933. *Sinensia* 3(7):179.
Allolobophore caliginosa trapezoides Chen, 1933. *Contr. Biol. Lab. Sci. Soc. China (Zool.).* 9:216-222.
Allolobophora caliginosa trapezoides Chen, 1936. *Contr. Biol. Lab. Sci. Soc. China (Zool.).* 11:270.
Allolobophora caliginosa trapezoides Chen, 1946. *J. West China. Border Res. Soc. (B).* 16:137.
Allolobophora longa Gates, 1972. *Trans. Am. Phil. Soc.* 62(7):76–79.
Aporrectodealonga Reynolds, 1977. *The Earthworms of Ontario.* 46–50.
Aporrectodea trapezoides Chang, Shen & Chen, 2009. *Earthworm Fauna of Taiwan.* 18–19.
Aporrectodea trapezoides Xu & Xiao, 2011. *Terrestrial Earthworms of China.* 286–287.

External Characters

　　Length 60–220mm, width 3–7mm, segment number 118–170. Prostomium 1/3 epilobous, narrow and cut off behind by a furrow. Body dorsoventrally flattened posteriorly so that a cross section nearly transversely rectangular with setal couples at the corners. Setae lumbricine, absent in the first and last segments; 4 pairs per segment, closely paired, moderate in size. Normal setae stout, nearly equal in size, sigmoid. First dorsal pore 8/9, very indistinct, that in 9/10 distinct, those on

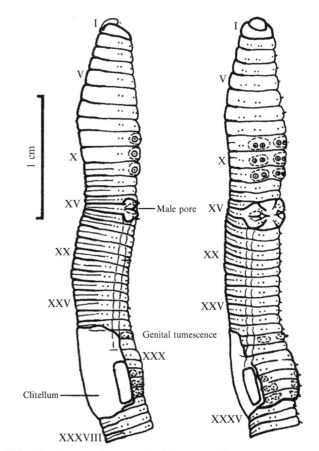

FIG. 15　*Aporrectodea trapezoides* (after Reynolds, 1977).

clitellum often invisible. Clitellum well-marked, saddle-shaped, glandular layer thicker on dorsal and lateral sides, less so on ventral side, commonly in XXVI–XXXIV, or in XXVII–XXXIV, or 1/2 XXXIV, or rarely 1/2 XXXV. Intersegmental furrows on ventral side of XXX–XXXIV not distinct, those in front clearly visible, its epidermis around setae *ab* of XXX, XXXII, and XXXIII often glandular in appearance. A "puberty wall" or glandular ridge present on each side of ventro-lateral side of XXX–XXXIII or 1/2 XXX–XXXIII or sometimes XXXI–XXXIII, just lateral to seta *b*, rather broad and slightly raised, each ridge continuous. Male pores 1 pair, in XV, between *bc*, in a wide and deep-cut transverse slit in front of anterior and posterior elevated lips extending from *c* to *b* or *a*, but its glandular character adjoined with that of opposite side ventrally and glandulated in adjacent parts of XIV and XVI; intersegmental furrows 14/15 and 15/16 on ventral side entirely invisible. Spermathecal pores 2 pairs in 9/10/11, between *cd*, slightly close to *c*, as small eye-like depressions, on each of which marks a small opening, these parts sometimes sinking down. Female pores 1 pair, as small transversely ovoid openings, on XIV, immediately lateral to setae *b*, its surrounding part not swollen.

Internal Characters

Septa 5/6 somewhat muscular, 6/7/8/9/10 increasingly muscular, maximum thickness in 9/10, muscularity decreasing posteriorly through 10/11–14/15, 15/16 membranous. Pigment, in circular muscular layer, reddish brown, or brown, seeming albinos with some slight color recognizable in free hand sections of the body wall. Longitudinal musculature, pinnate, muscular, 13/14/15 thinner. Gizzard in XV–XVIII, mostly in XVII. Gut, more or less valvular in region of insertion of 19/20 or in XX. Intestine begins in XV. Typhlosole small and thick in XXI. Hearts in VI–XI. Nephridial bladders, J-shaped, closed end of loop laterally and at about *cd* though other portions of the organ reach well into *dd*. Calciferous sacs, vertical, reaching slightly above and below esophageal level, wide and usually without marked constriction from the esophagus. Holandric. Seminal vesicles 4 pairs, small to almost rudimentary in IX, X, medium sized to smaller in XI, XII. Sperm ducts, with an epididymis comprising a single hairpin loop, shorter loops bound together in a ball, or spirally coiled as in a watch spring. Spermathecae 2 pairs in X and XI, ducts confined to body wall.

Color: Preserved specimens variable from light red to dark brown dorsally. In young forms, mostly reddish in life, in adult chocolate brown, deeper on anterior dorsum, darker along dorsomedian line; Live specimens dark gray or grayish brown. Clitellum cherry red or chocolate brown; pale or grayish buff ventrally.

Type locality: France (Montpellier).

Distribution in China: Beijing (Xicheng, Haidian, Dongcheng, Tongzhou, Changping, Sunyi, Huairou, Fangshan, Fengtai, Chaoyang, Miyun, Shijingshan), Hebei, Liaoning (Dalian, Shenyang, Dandong, Zhangzi Island, Haiyang Island), Shanghai, Jiangsu (Nanjing, Zhenjiang, Suzhou, Wuxi, Yixing, Yangzhou, Puzhen, Xuzhou,), Zhejiang (Cuanshan, Ningpo, Tonglu, Tiantai), Anhui (Anqing, Chuzhou), Taiwan, Shandong (Yantai, Weihai, Jinan), Jiangxi (Jiujiang, Dean), Henan (Xinxiang, Jiaozuo, Anyang, Boai, Yuanyang, Kaifeng, Yanling, Yucheng, Xuchang, Luoyang, Baofeng, Xishan, Zhechuan, Tanghe, Luoshan), Hubei (Yichang, Qiangjiang), Hunan (Yuezhou), Chongqing (Baixi, Beipei, Jiangbei, Luzhou, Nanchuan, Pingxian, Xufu), Sichuan

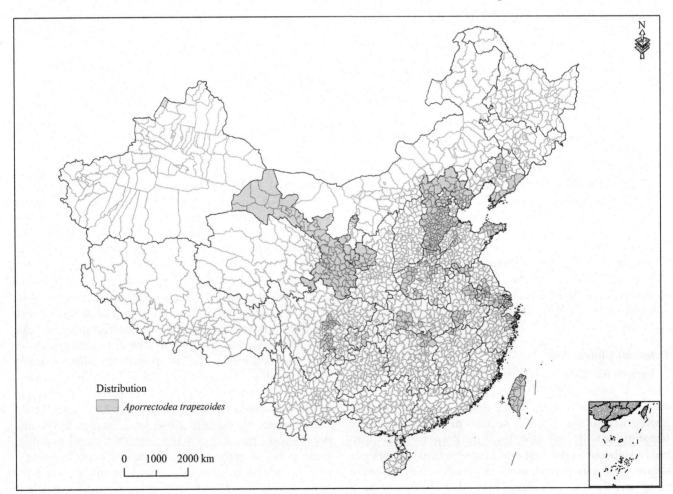

Distribution

Aporrectodea trapezoides

0   1000   2000 km

FIG. 16   Distribution in China of *Aporrectodea trapezoides*.

(Chengdou, Jiading, Pingshan), Gansu, Ningxia (Shizuishan, Zhongwei, Guyuan), Shaanxi (Xinjiang, Fushan, Qingxu), Shanxi, Xinjiang (Tacheng).

Etymology: The specific name indicates its trapezoid tail region.

Global distribution: Europe, North America, South America, Africa, Asia, Australia, Iceland. Korea (Gyungsangbuk-do, Jeollabuk-do, Jeollanam-do). Cosmopolitan. In Taiwan it is a very rare introduced species recorded at only a few localities.

Habitats: This species is found in a wide variety of habitats. It is found in the earth around the roots of potted plants, in gardens, cultivated fields, forest soils of various types, on the banks of streams, and sometimes in sandy soil. It has been recorded from caves in North America and Afghanistan, and in California and Arizona may occur at elevations of 1525 m or more (Gates, 1967, 1972a; Smith, 1917; Olson, 1928; Eaton, 1942; Reynolds, 1973a, b, c; Reynolds et al., 1974).

Remarks: For a long time *Aporrectodea trapezoides* was considered to be a variety of subspecies of *Allobophora caliginosa* but it is unlikely that all references to *Allolobophora* of the subspecies, variety, or form *trapezoides*, do in fact refer to *Aporrectodea trapezoides* (Gates, 1972).

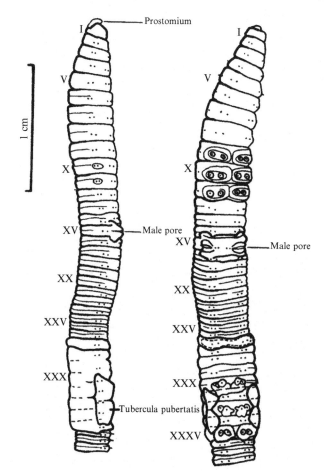

FIG. 17   *Aporrectodea tuberculata* (after Reynolds, 1977).

13. *Aporrectodea tuberculata* (Eisen, 1874)

Allolobophora longa Gates, 1972. *Trans. Am. Phil. Soc.* 62(7):79–84.

Aporrectodea tuberculata Reynolds, 1977. *The Earthworms of Ontario.* 50–55.

Aporrectodea tuberculata Hong, 2000. *Korean Jour. System. Zool.* 16(1):3–4.

Aporrectodea tuberculata Xu & Xiao, 2011. *Terrestrial Earthworms of China.* 287–288.

External Characters

Length 50–150mm, width 4–8mm, segment number 83–194. Body cylindrical, dorsoventrally flattened slightly. Prostomium epilobous. First dorsal pore 11/12 or 12/13. Clitellum XXVI, XXVII–XXXIV, XXV. Tubercula pubertatis, in XXX, XXXI–XXXIII, XXXIV between *b* and *c*. Setae lumbricine, closely paired; $ab \approx cd$, $aa > bc$, $dd = 1/2C$. Genital tumescences absent in XXXI and XXXIII, present in IX–XI, XXX, XXXII, and XXXIV and frequently in XXVI, surrounding *ab*, only 6 pairs. Male pore, at bottom near lateral end of an equatorial cleft almost reaching *b* but not so close to *c* in XV. Spermathecal pores 2 pairs, in 9/10/11, on level *c*. Female pore 1 pair on XIV, lateral to *b*.

Internal Characters

Septa 5/6–11/12 muscular, 13/14/15 thin. Pigment, none of usual sort present, sparse yellow flecks externally recognizable in older individuals. Longitudinal musculature, pinnate. Gizzard, mostly in XVII, circular muscle layer markedly narrowed anteriorly in XVII. Gut, more or less valvular in region of insertion of 19/20 or in XX. Typhlosole, beginning rather gradually in region XXI–XXIV. Hearts in VI–XI. Nephridial bladders, U-shaped, in *bc*, closed laterally. Calciferous sacs, vertical, reaching slightly above and below esophageal level, usually opening widely into the gut without any marked constriction. Calciferous gland, thickest in XI, with a slight constriction at 10/11, an internal and/or an external constriction near insertion of 11/12, behind which calciferous lamellae are narrower. Intestine begins in XV. Seminal vesicles 4 pairs, in IX–XII. Spermathecae 2 pairs in X and XI.

Color: Preserved specimens unpigmented, almost white or grayish, sometimes with light pigmentation on the dorsum.

Global distribution: Europe, North America, South America, Asia, Australia, Iceland.

Distribution in China: Liaoning (Haiyang Island, Zhangzi Island, Pikou), Ningxia (Guyuan), Taiwan. Habitats: The species lives in soils, turf, bogs, compost, ditches where there is a large concentration of organic matter (Gates, 1967, 1972; Reynolds et al., 1974). Remarks: *Aporrectodea tuberculata* lacks the pair of genital tumescences on segment XXXIII, which separates it from *Aporrectodea trapezoides*. It was first identified by Eisen as *Allolobophora cyanea* (Hoffmeister) (Gates, 1972).

14. *Aporrectodea turgida* (Eisen, 1873)

Allolobophoraturgida f. tuberculata Eisen, 1874. *Öfv. Vet. Akad. Förh. Stockholm.* 31(2):43.
Allolobophora similis Friend, 1911. *Zoologist, ser. 4,* 15:144.
Allolobophora tuberculata Gates 1972. *Bull. Tall Timbers Res. Stn.* 12:44–45.
Allolobophora longa Gates, 1972. *Trans. Am. Phil. Soc.* 62(7):84–86.
Aporrectodea trapezoides Reynolds, 1975. *Megadrilogica.* 2(3):3.
Aporrectodea turgida Reynolds, 1977. *The Earthworms of Ontario.* 56–60.
Aporrectodea turgida Xu & Xiao, 2011. *Terrestrial Earthworms of China.* 288.

External Characters

Length 60–85mm, width 1.5–5.0mm, segment number 130–168. Body cylindrical. Prostomium epilobous. First dorsal pore 12/13 or 13/14. Clitellum XXVII, XVIII, XXIX–XXXIV, XXXV, saddle-shaped. Tubercula pubertatis XXXI–XXXIII. Setae 4 pairs per segment, closely paired, $aa:ab:bc:cd:dd = 3:1:2:2/3:10$. Genital tumescences contain a and b only in IX–XI, XXX, XXXII–XXXIV, and frequently in XXVII. Male pores 1 pair, in XV, between b and c. Spermathecal pores 2 pairs in 9/10/11, between cd.

Internal Characters

Septa 5/6–9/10 muscular, though 6/7/8 not massively muscular and slightly thicker than the others, 10/11–13/14 strengthened. Pigment, of usual sorts lacking, sparse yellow flecks externally recognizable under the binocular microscope in older individuals, possibly representing, as also internally, remains of eleocytes. Longitudinal musculature, pinnate. Gizzard, mostly in XVII. Gut, more or less valvular in region of 19/20 or XX. Typhlosole, beginning rather gradually in region XXI–XXIV. Hearts in VI–XI. Nephridial bladders, J-shaped, closed end laterally, 1 limb about half the length of the other. Calciferous sacs, in X, vertical, reaching slightly above and below the esophageal level, usually opening widely into the gut without marked constriction. Esophagus, thickest in XI, with a slight constriction at 10/11, an

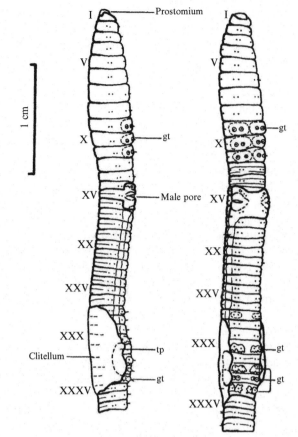

FIG. 18   *Aporrectodea turgida* (after Reynolds, 1977).

internal and/or an external constriction near insertion of 11/12, behind which calciferous lamellae gradually narrow. Seminal vesicles 4 pairs, in IX–XII. Spermathecae 2 pairs usually in X and XI, duct confined or almost so to the parietes.

Color: Preserved specimens unpigmented with the region anterior the crop flesh pink and remaining segments pale gray, or occasionally with light pigmentation on dorsal surface.
Global distribution: Europe, North America, South America, Asia, South Africa, Australia, Iceland.
Distribution in China: Heilongjiang, Shandong (Yantai).
Deposition of types: ?
Habitats: Gates (1972) records this species from a variety of habitats, including gardens, fields, turf, forest humus, compost, banks of springs and streams, wasteland, and from a cave in West Virginia (Eaton, 1942; Černosvitov and Evans, 1947; Murchie, 1956; Gerard, 1964; Reynolds et al., 1974).
Remarks: The species was first recorded in Ontario (Niagara County) by Eisen in 1874. There are 4 specimens labelled *Aporrectodea turgida*, but the other 3 appear to be *Aporrectodea tuberculata* Gates (1972).

GENUS: BIMASTOS MOORE, 1893

Bimastos Feng, 1985. *Journal of Zoology*. 20:46nal
Bimastos Reynolds, 1977. *The Earthworms of Ontario*. 61.

Type species: *Bimastos palustris* Moore, 1895.

External Characters

Setae, closely paired. Prostomium, epilobous. Dorsal pores present from region of 5/6. Clitellum saddle-shaped, in XXV–XXXII, or XXIV–XXXI. Tubercula pubertatis absent or indistinct, not particularly sharply defined. Male pores, equatorial in XV, in atrial chambers invaginated deeply into the coelom and bearing acinous glands. Spermathecal pore absent. Female pores, equatorial on XIV, shortly above *b*.

Internal Characters

Gizzard, mainly in XVII. Extraesophageal trunks, joining dorsal trunk in XII. Hearts, in VI–XI. Calciferous gland, without marked widening in XI–XII, opening to gut in X through paired vertical sacs. Calciferous lamellae continued onto posterior walls of sacs. Nephridial bladders, U-shaped, closed ends laterally, ducts passing into parietes near *b*. Nephropores, inconspicuous, irregularly alternating and with asymmetry, between levels somewhat above *b* and well above *d*. Holandric, testes and funnels free; seminal vesicles in XI–XII. Spermathecae, tubercula pubertatis gland, lacking. No spermathecae.

Color: Pigment red.
Global distribution: North America, Europe, Asia Minor, India, Pakistan, China, South America (except in tropical regions), South Africa, Hawaii (sometimes semiaquatic).

There are 4 species from China.

TABLE 7   Key to the Species of the Genus *Bimastos* From China

1	Clitellum in XXV or XXVI–XXI or 1/2 XXXII .... *Bimastos tenuis*
	Clitellum beginning before XXV..2
2	Clitellum in 1/2 XXIII, XXIII, XXIV–XXXI, XXXII, 1/3XXXII; small papillae absent in male pore region... *Bimastos beddardi*
	Clitellum in XXIII, XXIV–XXXI, XXXII, 1/3XXXII; with small papillae in male pore region..................... *Bimastos parvus*

15. *Bimastos beddardi* (Michaelsen, 1894)

Bimastus beddardi Hong, 2000. *Korean Jour. System. Zool.* 16(1):4–5.
Bimastus beddardi Xu & Xiao, 2011. *Terrestrial Earthworms of China*. 289.

External Characters

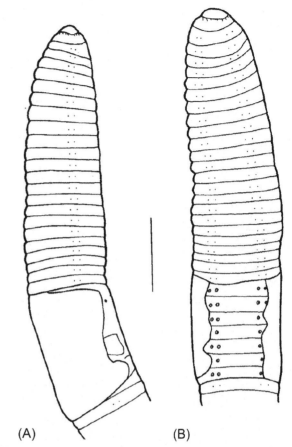

(A)   (B)

FIG. 19   *Bimastus beddardi* (after Hong, 2000).

Length 25–56 mm, width 1.5–2.1 mm, segment number 103–111. Prostomium epilobous. First dorsal pore 4/5. Clitellum XXIV–XXXI, saddle-shaped. Tubercula puertatis absent; genital tumescences lacking, if present XXV–XXVI, XVIII, XXX. Setae lumbricine, 4 pairs, closely paired. Male pores paired in XV, inconspicuous. Spermathecal pores absent. Female pores paired in XIV, lateral to *b*.

Internal Characters

Septa 5/6/7 thick, 7/8/9/10/11/12/13/14 thinner. Gizzard in XVI–XVIII. Calciferous glands in XI–XII. Intestine from XV. Hearts in IX–XI. Typhlosole small from XXIII. Spermathecae absent. Holandric. Ovaries in XIII, palmate. Testes 2 pairs in X and XI. Seminal vesicles in XI–XII. Prostate glands absent.

Color: Preserved specimens pink or red.
Distribution in China: Heilongjiang (Keshan, Harbin, Jiamusi, Mudanjiang), Jilin (Baicheng, Shanchengzhen), Xizang.
Global distribution: Korea (Mt. Keryong, Daejeon-shi).

16. *Bimastos parvus* (Eisen, 1874)

Allolobophora parua Eisen, 1874, 46.
Bimastus parvus Kobayashi, 1938. *Sci. Rep. Tohoku Imp. Univ.* 1938, 13(2):89–170.
Bimastus parvus Chen, 1931. *Contr. Biol. Lab. Sci. Soc. Chin. Zool.* 7(3):169–170.
Bimastus parvus Chen, 1946. *J. West China. Border Res. Soc. (B).* 16:137.
Bimastos parvus Gates, 1972. *Trans. Am. Phil. Soc.* 62(7):84–86.
Bimastos parvus Reynolds, 1977. *The Earthworms of Ontario.* 61–64.
Bimastos parvus Hong, 2000. *Korean Jour. System. Zool.* 16(1):6–7.
Bimastus parvus Chang, Shen & Chen, 2009. *Earthworm Fauna of Taiwan.* 20–21.
Bimastus parvus Xu & Xiao, 2011. *Terrestrial Earthworms of China.* 289–290.

External Characters

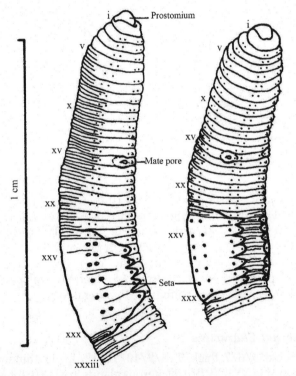

FIG. 20   *Bimastus parvus* (after Reynolds, 1977).

Length 17–85mm, width 1.4–3.0mm, segment number 85–124. Prostomium epilobous. First dorsal pore 4/5 or 5/6. Clitellum XXIII, XXIV, XXV, XXVI–XXX, XXXI, XXXII, XXXIII, saddle-shaped. Tubercula pubertatis absent; genital tumescences lacking, if present XXV–XVIII, or –XXIX, very indistinct. Setae lumbricine, 4 pairs per segment, $aa=cd$, aa a little more than bc, $aa=1/4$ ab or a trifle wider, dd about half circumference of body. Male pores as transverse slits, on XV, with distinctly elevated glandular whitish areas confined to the segment and each around on pore; the neighboring segments XIV and XVI inconspicuously thickened and glandular in character. Spermathecal pores absent. Female pores paired in XIV, small slits, from position of (missing) b. Female pores paired in XIV, distinct, lateral close to b.

Internal Characters

Septa 5/6/7 thick, 7/8/–13/14 thin, reduced after 16/17. Gizzard large in XVI–XVIII or XVII. Calciferous glands in XI–XII. Intestine from XV. Hearts in VII–XI, anterior 2 pairs weak. Holandric. Ovaries in XIII, palmate. Testes 2 pairs in X and XI. Seminal vesicles in X–XII. Prostate glands absent. Spermathecae absent.

Color: Live specimens dark to light red-brown on dorsum, pale on ventrum.
Distribution in China: Beijing (Xicheng, Haidian, Dongcheng, Miyun, Shijingshan), Tianjin, Hebei, Liaoning (Shenyang, Dashiqiao, Dandong, Huludao, Zhangzi Island, Haiyang Island), Jilin (Changchun, Jilin, Yanji, Tumen), Heilongjiang (Harbin, Jiamusi, Mudanjiang), Jiangsu (Nanjing), Fujian (Fuzhou), Taiwan, Shandong (Dezhou, Yantai, Weihai), Jiangxi (Dean), Saanxi (Qixian), Henan (Jiaozuo, Xinxiang, Boai, Wenxian, Kaifeng, Xishan, Xinyang), Hubei (Qianjiang), Chongqing (Fuling, Nanchuan, Beipei), Sichuan (Chengdu, Jiangbei, Luzhou, Xufu, Emeishan), Xizang, Shanxi, Ningxia (Shizuishan, Zhongwei, Guyuan, Jingyuan), Xinjiang. Global distribution: Europe, North America, South America, Asia, South Africa, Australia.
Cosmopolitan. Korea (Mt. Naejang, Jeongeup-shi, Jeollabuk-do).
Type locality: Mt. Lebanon, New York, USA.
Deposition of types: US National Museum, USA.
Etymology: The specific name *"parvus"* literally means "small or little" in Latin.
Remarks: This is a small worm, like *Bimastus beddardi*, and it is difficult to separate from the later by external characters. *Bimastus parvus* has the clitellum on segment XXIV, XXV–XXX, distinct from *Bimastus beddardi*, where it is on segment XXIV, XXV–XXXI.

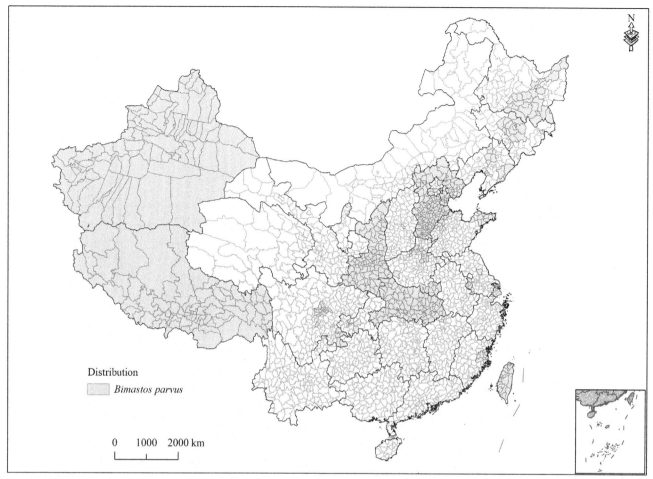

FIG. 21   Distribution in China of *Bimastos parvus*.

GENUS: DENDROBAENA EISEN, 1873

Dendrobaena Feng, 1985. *Journal of Zoology*. 20:47.
Dendrobaena Reynolds, 1977. *The Earthworms of Ontario*. 64.

Type species: *Dendrobaena boeckii* Eisen, 1873 (=*Enterion octoaedrum* Savigny, 1826).

External Characters

Integument mostly pigmented, red or purple. Prostomium mostly epilobous, seldom tanylobous. Clitellum usually saddle-shaped, in XXIV–XXXIII. Tubercula pubertatis usually on XXXI–XXXIII. Setae widely paired or separated, rarely closely paired. Male pore on XV; spermathecal pore in line with setae *c* or *d*, seldom absent, usually 2 pairs in furrows 9/10/11, rarely 1–2 additional pairs in the neighboring furrows. Nephropores, obvious, behind first few segments in single rank on each side, just above *b*. Setae, not closely paired. Longitudinal musculature, pinnate. Spermathecal pores 2 pairs, in 9/10/11.

Internal Characters

Gizzard mostly in XVII. Extraesophageal trunks, passing to dorsal trunk in vicinity of 9/10. Hearts, in VII–IX. Nephridial bladders, ocarina-shaped, with blunt end laterally and pointed end mesially, ventral side funnel-shaped and narrowing to pass into parietes at *b*. Calciferous gland, without sacs, opening into gut at vicinity of 10/11. Markedly moniliform in XI–XII. Calciferous sacs, lacking. Testes and funnel free; usually 3 pairs of seminal vesicles in IX, XI, XII; rarely (and only when the setae are widely paired) a fourth pair in X; these seminal vesicles in X are very small, much smaller than those of IX.

Color: Pigment red.
Global distribution: Europe, southern Siberia, northern India, Syria, Israel, the Caucasus, Iceland, Greenland, North America, southern South America, Egypt, China.
Discussion: The above description (Gates, 1972, 1975), is for a genus with *Dendrobaena octaedra* as the type species. However, another of Savigny's species,

Enterion rubidum, has been a congener of *Dendrobaena octoaedra* for decades, almost solely because of similarities in their genitalia. Based on more conservative somatic anatomy, Omodeo's subgenus *Dendrodrilus* (1956) has been elevated to full genus status with *Enterion rubidum* as the type species (Gates, 1975).

There is 1 species from China.

17. *Dendrobaena octaedra* (Savigny, 1826)

Enterion octaedrum Savigny, 1826. *Mém. Acad. Sci. Inst. Fr.* 5:183.
Lumbricus riparius (part.) Hoffmeister, 1845. *Regenwürmer.* 30.
Dendrobaena boeckii Eisen, 1873. *Öfv. Vet. Akad. Förh. Stockholm.* 30(8):53.
Helodrilus (Dendrobaena) octaedrus Michaelsen, 1900. *Das Tierreich., Oligochaeta.* 10:494.
Dendropaenaoctaedrus Chen, 1956, 56.
Dendrobaenaoctaedra Gates 1972. *Bull. Tall Timbers Res. Stn.* 15:16.
Dendrobaenaoctaedra Gates 1972. *Trans. Amer. Philos. Soc.* 62(7):89–92.
Dendrobaenaoctaedra Reynolds, 1977. *Life Sci. Misc. Publ. Roy. Ont. Mus.* 66–69.
Dendrobaena octaedra Reynolds, 1977. *The Earthworms of Ontario.* 65–68.
Dendrobaena octaedra Xu & Xiao, 2011. *Terrestrial Earthworms of China.* 290–291.

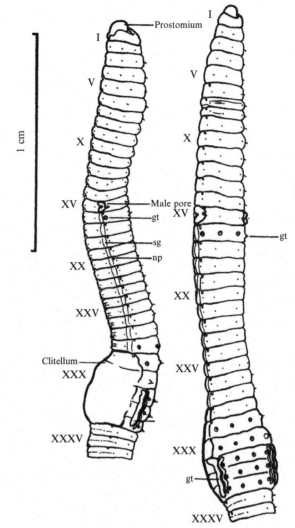

FIG. 22   *Dendrobaena octaedra* (after Reynolds, 1977).

External Characters

Length 17–60mm, width 3–5mm, segment number 60–100. Body cylindrical, with posterior portion octagonal. Prostomium epilobous. First dorsal pore in 4/5–6/7. Clitellum XXVII, XVIII, XXIX–XXXIII, XXXIV, saddle-shaped. Tubercula pubertatis usually on XXXI–XXXIII. Setae lumbricine, 4 pairs per segment, widely paired, $aa=ab=cd$ and dd is slightly greater. On XVI setae a or b are found on small genital tumescences. Spermathecal pores 3 pairs in 9/10/11/12, with long ducts on level with setae d. Male pores 1 pair, in XV, surrounded by small, often indistinct, glandular papillae.

Internal Characters

Septa present from 5/6, 6/7–9/10 gradually thickened. Gizzard in XVII. Calciferous glands in XI. Intestine from XX. Typhlosole present from XXI. Hearts in VI, VII–XI. Spermathecae 3 pairs in IX, X, and XI, with long duct.

Testes 2 pairs in X and XI. Seminal vesicles in IX, XI, and XII. Prostate glands absent.

Color: Preserved specimens red, dark red to purple.
Global distribution: Europe, North America, Colombia, Mexico, Asia, Northern Hemisphere, Iceland.
Distribution in China: Xinjiang (Tacheng).
Deposition of types: Itajahy, Brazil.
Habitats: *Dendrobaena octaedra* is mostly found in sites little affected by cultivation, including: in sods or under moss on stream banks, under logs and leafy debris, in cool, moist ravines and upland seepage (Murchie, 1956), under dung and in soil high in organic matter (Gerard, 1964), and on mountain tops and in caves.

GENUS: <u>DENDRODRILUS</u> OMODEO, 1956

Dendrobaena (Dendrodrilus) Omodeo, 1956. *Arch. Zool. It.* 41:175.
Dendrobaena Gates, 1972. *Trans. Amer. Philos. Soc.* 62(7):88.
Dendrodrilus Gates, 1975. *Megadrilogica.* 2(1):4.
Dendrodrilus Reynolds, 1977. *The Earthworms of Ontario.* 69.

Type species: *Enterion rubidum* Savigny, 1826.

External Characters

Prostomium, epilobous. Clitellum usually saddle-shaped, in XXIV–XXXIII. Tubercula pubertatis usually on XXXI–XXXIII. Setae usually to widely paired or separated, rarely closely paired. Male pore on XV; spermathecal pore in line with setae *c* or *d*, seldom absent, usually 2 pairs in furrows 9/10/11, rarely 1–2 additional pairs in the neighboring furrows. Nephropores, obvious, behind first few segments in single rank on each side, just above *b*. Setae, not closely paired. Longitudinal musculature, pinnate. Spermathecal pores 2 pairs, in 9/10/11, sometimes more than 1 or 2 pairs.

Internal Characters

Calciferous glands, opening into gut ventrally through a pair of sacs posteriorly just in front of insertion of 10/11. Calciferous lamellae continued along lateral walls of sacs. Gizzard, mainly in XVII. Extraesophageal trunks, passing to dorsal trunk in XII. Hearts, in VII–XI. Nephridial bladders, U-shaped loop. Nephropores, inconspicuous, alternating irregularly and with asymmetry on each side between a level above *b* and 1 above *d*. Setae, not closely paired. Longitudinal musculature, pinnate.

Color: Pigment, red.
Discussion: Species in *Dendrodrilus* formerly congeneric with species in *Dendrobaena* because of similarities in genital anatomy, are now separated based on differences in their more conservative somatic anatomy.

There is 1 species from China.

18. *Dendrodrilus rubidus* (Savigny, 1826)

Enterion rubidum Savigny, 1826. *Mavigny, 1826. m dusased.* 5:182.
Lumbricus rubidus Dugidus s 826*Ann. Sci. Nat. Ser.* 2, 8:17,23.
Helodrilus (Dendrobaena) rubidus Michaelsen, 1900. *Das Tierreich., Oligochaeta.* 10:490.

Dendrodrilus rubidus Reynolds, 1975. *Megadrilogica.* 2 (3):3.
Dendrodrilus rubidus Reynolds, 1977. *The Earthworms of Ontario.* 69–73.
Dendrodrilus rubidus Xu & Xiao, 2011. *Terrestrial Earthworms of China.* 291.

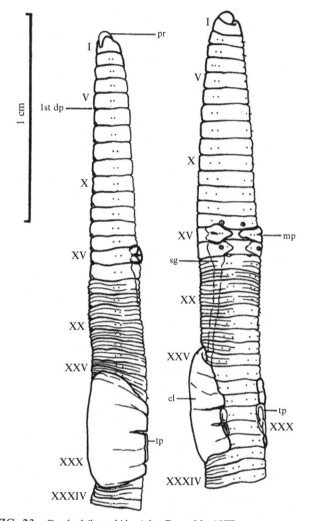

FIG. 23 *Dendrodrilus rubidus* (after Reynolds, 1977).

External Characters

Length 20–90mm, width 2–5mm, segment number 50–120. Body cylindrical. Prostomium epilobous. First dorsal pore 5/6. Clitellum in XXVI, XXVII–XXXI, XXXII, saddle-shaped. Tubercula pubertatis, if present, XVIII, XXIX–XXX. Setae 4 pairs per segment, widely paired, *ab<cd*, and *bc=2cd*. Male pores 1 pair, in XV, between *b and c*. Spermathecal pores 2 pairs in 9/10/11, on level with setae *c*. Female pores 1 pair, on XIV.

Internal Characters

Spermathecae 2 pairs in X and XI, with short duct. Testes 2 pairs in X and XI. Seminal vesicles in IX, X, and XII, small.

Color: Preserved specimens red, darker dorsally.
Global distribution: Europe, North America, South America, Asia, Africa, Australasia, Iceland.
Distribution in China: Xinjiang, Northeast China.
Habitats: *Dendrodrilus rubidus* has been found in a wide range of habitats including gardens, cultivated fields, stream banks, in moss in running water and wells and springs, peat, compost, and sometimes in manure, caves, greenhouses, botanical gardens, and the culture beds of earthworm farms (Gates, 1972) under the bark of old trees (Černosvitov and Evans, 1947; Gerard, 1964).
Dendrodrilus rubidus is facultatively parthenogenetic with male sterility and absence of spermathecae common (Gates, 1972; Reynolds, 1974).

GENUS: *EISENIA MALM, 1877*

Eisenia Malm, 1877. *Öalm, 1877. nogenetic with male st.* 1:45.
Eisenia (part.) Michaelsen, 1900. *Das Tierreich., Oligochaeta.* 10:474.
Eisenia Stephenson, 1930. *The Oligochaeta.* 820.
Eisenia Gates, 1969. *J. Nat. Hist. Lond.* 9:305.
Eisenia Gates, 1972. *Trans. Amer. Philos. Soc.* 62(7): 96.
Eisenia Reynolds, 1977. *The Earthworms of Ontario.* 74.
Eisenia Feng, 1985. *Journal of Zoology.* 20:47.
Eisenia Reynolds, 1977. *The Earthworms of Ontario.* 74.

Type species: *Enterion foetidum* Savigny, 1826.

External Characters

Setae, closely paired. Prostomium, epilobous. Longitudinal musculature, pinnate. Clitellum saddle-shaped. Setae closely to widely paired. Male pores on XV. Nephropores, inconspicuous, in 2 ranks on each side, alternating irregularly between a level just above *b* and 1 above *d*. Spermathecal pores 2–3 pairs, in furrows 8/9/10/11 or 9/10/11, above the line *d*, near or in the middle line.

Internal Characters

Gizzard, mostly in XVII. Hearts, in VII–XI. Nephridial bladders, sausage-shaped or digitiform, transversely placed. Calciferous gland, without sacs, opening into gut behind 10/11 through a circumferential of small pores. Calciferous sacs, lacking. Testes and funnels free.

Color: Pigment, red (Gates, 1972, 1975).
Global distribution: Europe, North America, Siberia, southern Russia, Israel.
Discussion: *Eisenia* was erected for 3 species, *Enterion foetidum* Savigny, 1826, and *Allolobophora norvegica* and *Allolobophora subrubicunda* Eisen, 1873 by Malm (1877) without the designation of type species. The 2 *Allolobophora* species are now synonyms of *Dendrodrilus rubidus*. Gates (1969) redefined *Eisenia* with *Eisenia foetida* as the type species, but another of Savigny's species, *Enterion roseum*, has been congeneric with *Eisenia foetida* since Michaelsen (1900) solely because of spermathecal pore location. If future revisions are based on the more conservative somatic anatomy, these 2 species will not remain congeneric. Most European workers have followed Pop (1941) and Omodeo (1956) and have transferred *Eisenia rosea* to *Allolobophora*. However, on the basis of somatic anatomy it is not reasonable to place *Eisenia rosea* in any genus of which *Enterion chloroticum* Savigny is the type species.
There are 9 species and 4 subspecies from China.

TABLE 8   Key to the Species of the Genus *Eisenia* From China

1	Clitellum in XXVI, XXVII–XXXIV ... *Eisenia nordenshioldi manshurica* Clitellum in or before XXXIII 2
2	Clitellum in XXXIII, beginning XXIII or XXII..3 Clitellum in XXXIII or XXXII, beginning XXIV–XXVII....................5
3	Clitellum in XXXIII, beginning XXIII..4 Clitellum in XXII–XXXII or XXXIII............................*Eisenia jeholensis*
4	Clitellum in XXIII–XXXIII ....*Eisenia dairenensis* Clitellum in XXIII or XXIV–XXXII or XXXIII.................*Eisenia hataii*
5	Clitellum in XXIV, XXV, XXVI–XXXII; tubercula pubertatis in XVIII–XXX .... *Eisenia fetida* Clitellum in XXXIII or XXXII; beginning XXV, XXVI or XXVII 6
6	Clitellum in XXV, XXVI–XXXII, XXXIII ....7 Clitellum in XXVI, XXVII–XXXIII...................................*Eisenia nordenshioldi nordenshioldi*
7	Clitellum in XXV, XXVI–XXXII, XXXIII; tubercula pubertatis in XXX–XXXI...................................*Eisenia veneta* Clitellum in XXV, XXVI–XXXII, XXXIII; tubercula pubertatis in XXIX–XXXI...8
8	Male pores on a horseshoe-shaped papilla............. *Eisenia harenensis* Male pores on a nonhorseshoe-shaped papilla 9
9	Male pores 1 pair, with elevated glandular papillae in XV, with male tumescences extending over XIV and XVI................................*Eisenia rosea rosea* Male pores 1 pair, without elevated glandular papillae in XV .... *Eisenia rosea macedonica*

19. *Eisenia dairenensis* (Kobayashi, 1940)

Allolobophora dairenensis Kabayashi, 1940. *Sci. Rep. Tohokuo Univ.* 15:291–293.
Eisenia dairenensis Xu & Xiao, 2011. *Terrestrial Earthworms of China*. 292–293.

External Characters

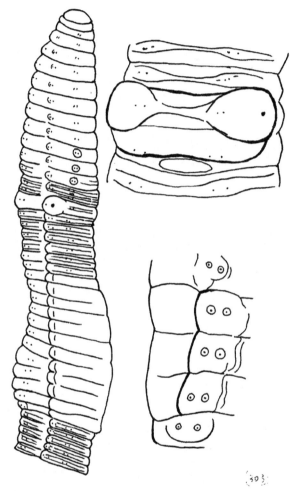

FIG. 24   *Eisenia dairenensis* (after Reynolds, 1977).

Length 80–111mm, width 3.5–5.5mm (clitellum), segment number 137–139. Except for the anterior end, body dorsoventrally flattened and somewhat 4-edged; behind the male pore region (except the clitellum) segments are triannular; both anterior and posterior ends are rather bluntly terminated. Prostomium proepilobous. First dorsal pore in 4/5. Clitellum saddle-shaped, in XXIII–XXXIII. Pubertatis tubercles in XXIX–XXXI distinct in outline being ridged by 3 prominent papillae. Setae 4 pairs per segment, closely paired; setal distance *aa* is nearly twice *bc*; *ab* is slightly greater than *cd*, *dd* is smaller than 1/2 of the circumference, *aa:ab:bc:cd:dd:* 1/2μ=83:8:38:6:132:160 in a segment immediately

posterior to the clitellum. Setae *cd* of X, XI, and XII (or most of them), and *ab* of IX (rarely on 1 side), XV, XVI, and of clitellar segments are situated on papillae. Male pores 1 pair, on XV, setae on prominent papillae, somewhat ovoidal, slightly encroaching into both of XIV and XVI; setae *ab* planted on medial border of the male pore papillae and *cd* are just dorsal to it. The ventral portion lying between the male pore papillae and the ventral surface of the anterior two-thirds of XVI are also elevated but to a slight degree. Thus, XV and the greater part of XVI appear to form an elevated rectangular plate. On XVI, the setae *ab* planted on the posterior margin of the elevation. Spermathecal pores are absent. Female pores 1 pair, on XIV, just lateral to setae *b*.

Internal Characters

Septa 6/7/8/9/10 moderately thickened and musculated, 5/6, 10/11, and 11/12 slightly thickened, the rest thin. Crop in XV–XVI. Gizzard in XVII–XVIII. Testes and funnels free, in X and XI. Seminal vesicles 4 pairs in IX–XII, whitish; the anterior 2 pairs are much smaller than the posterior ones; those in x are slightly smaller than those in IX. The setae *cd* of IX–XII, and *ab* of XV and XVI, and of the clitella segments are found through internal dissection to be contained in large sacs. Except for the proximal end which is slightly curved, they are almost straight, grooved, and moderately pointed. Those of XV are grooved at about distal 3/5–4/5, while the rest of the genital setae is grooved at about distal 1/2. Spermathecae are entirely absent.

Color: Preserved specimens uniformly pinkish.
Clitellum is fleshy red and resembles that of *Drawida gisti* Michaelsen.
Distribution in China: Liaoning (Dalian).
Remarks: Among the athecal species of the genus, this species resembles *Allolobophora prashadi* and *Eisenia hataii* but differs from these mainly in the color of the body, the body shape, the extension of the clitellum, the male pore aspect, and in the shape of the tubercle.

20. *Eisenia fetida* (Savigny, 1826)

Enterion fetidum (corr. foetidum) Savigny, 1826. *Mém. Acad. Sci. Inst. Fr.* 5:182.
Lumbricus foetidus Dugès, 1837. *Ann. Sci. Nat. Ser.* 2, 8:17,21.
Helodricus (Eisenia) foetidus Michaelsen, 1913. *Zool. Jb. Syst.* 34:551.
Eisenia foetida Kobayashi, 1940. *Sci. Rep. Tohokuo Univ.* 15:287.
Eisenia foetida Gates, 1972. *Trans. Amer. Philos. Soc.* 62(7):97–103.

Eisenia fetida fetida + *E. f. andrei* Bouche, 1972. *Inst. Natn. Rech. Agron.* 380–381.

Eisenia foetida Reynolds, 1977. *The Earthworms of Ontario.* 74–77.

Eisenia fetida Hong, 2000. *Korean Jour. System. Zool.* 16(1):7–8.

Eisenia fetida Xu & Xiao, 2011. *Terrestrial Earthworms of China.* 293.

External Characters

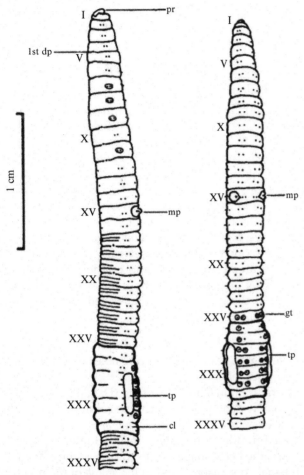

FIG. 25   *Eisenia fetida* (after Reynolds, 1977).

Length 35–130 mm (generally <70 mm), width 3–5 mm, segment number 80–110. Body cylindrical. Prostomium epilobous. First dorsal pore 4/5 (sometimes 5/6).

Clitellum XXIV, XXV, XXVI–XXXII, saddle-shaped, *cd* setae and dorsal invisible externally within clitellum. Tubercula pubertatis on XVIII–XXX. Setae lumbricine, 4 pairs per segment, *ab=cd, bc<aa*, anteriorly *dd=*1/2C but posteriorly *dd<*1/2C. Genital tumescence may be present around any of the setae on IX–XII, usually around setae *a* and *b* of XXIV–XXXII. Male pores 1 pair, near lateral margins of ventral in XV, with large glandular papillae. Spermathecal pores 2 pairs in 9/10/11.

Internal Characters

Septa usually thin. Testes and funnels in ventral paired sacs in X, XI. Seminal vesicles 4 pairs in IX–XII, sometimes in IX, X, XI–XII. Spermathecae 2 pairs in X and XI, each ampulla middle voluminous pouch, thick, duct short.

Color: Preserved specimens variable, purple, red, dark red, brownish red, sometimes alternating band of red-brown on dorsum with pigment less yellow in intersegmental areas.

Global distribution: Europe, North America, South America, Asia, Africa, Australasia, Iceland, Korea.

Distribution in China: Beijing (Dongcheng, Shijingshan), Tianjin, Hebei, Liaoning (Dalian, Pikou, Zhangzi Island, Haiyang Island), Jilin, Heilongjiang (Harbin), Shanghai, Jiangsu, Zhejiang, Anhui, Taiwan, Shandong (Yantai, Weihai), Shaanxi (Yuncheng, Jishan, Xinjiang, Fushan, Jiexiu, Pingyao, Qingxu), Henan (Xinxiang), Hubei (Qianjiang), Chongqing (Beibei), Sichuan (Chengdu), Shanxi (Wugong), Ningxia (Shizuishan, Helan), Xinjiang.

Etymology: The specific name indicates its trapezoid tail region.

Habitats: This species is found in manure and decaying vegetation (Olson, 1928), manure, compost heaps, and soil high in organic matter, as well as forests, gardens, under stones and leaves (Černosvitov and Evans, 1974; Gerard, 1964), under logs and debris, at roadside dumps (Reynolds et al., 1974), and in taiga, forests, and steppe habitats (Gates, 1972).

Eisenia fetida has been reared on earthworm farms and sold all over the world.

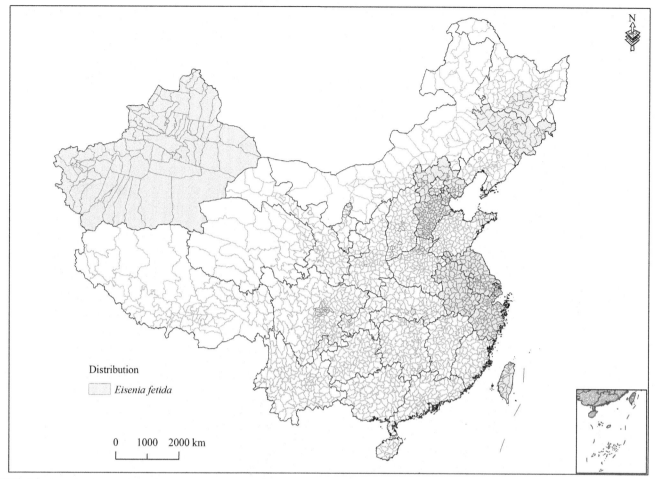

FIG. 26　Distribution in China of *Eisenia fetida*.

21. *Eisenia harbinensis* (Kobayashi, 1940)

Allolobophora harbinensis Kabayashi, 1940. *Sci. Rep. Tohokuo Univ.* 15:290–291.
Eisenia harbinensis Xu & Xiao, 2011. *Terrestrial Earthworms of China.* 293–294.

External Characters

Length 79–96mm, width 2.7–3.3mm (clitellum), segment number 134–144. Segments near the male pore region and those posterior to the clitellum are of 3 annuli, but the annulation is generally indistinct. Prostomium proepilobous. First dorsal pore in 4/5. Clitellum saddle-shaped, extending from XXV or 1/3–1/2XXV, or seldom XXVI, to XXXII or 1/3–1/2 XXXIII. Pubertatis tubercula, usually in XXIX–XXXI, or sometimes in XXIX–1/2 XXXI, has narrow and distinct groove with slightly elevated margin, which next to the side of papillae. Setae 4 pairs per segment, closely paired; setal distance *aa* is nearly twice *ab*; *ab* is slightly greater than *cd*, *dd* is less than 1/2 of the circumference, *aa*:*ab*:*bc*:*cd*:*dd*:1/2μ = 93:7.5:45:6:178:194 on a segment immediately posterior to the clitellum. Setae *ab* of XXVII–XXXII are situated on papillae, *ab* of IX, and XII and *cd* of IX, X and XII are also situated on papillae, but some of these are often absent. Male pores 1 pair, on XV, in the mid-area of the horseshoe. Spermathecal pores 2 pairs in 9/10/11, on the *cd* line. Female pores 1 pair, on XIV, just lateral to setae *b*.

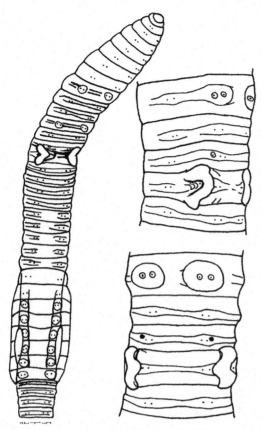

FIG. 27   *Eisenia harbinensis* (after Kabayashi, 1940).

Internal Characters

Septa 6/7/8/9 moderately, 9/10 slightly thickened, the rest thin. Gizzard in XVII–XVIII. Testes and funnels free, in X and XI. Seminal vesicles 4 pairs in IX–XII, darkened in color, the posterior 2 pairs are larger than the anterior ones, and are nodular in appearance or form a small number of small spherical lobules; those in X are nearly equal in size to those in IX. The ventral and lateral setae of IX, X, and XII are seen through internal dissection to be contained in large and somewhat triangular sacs. Spermathecae small, ampulla spherical, thin-walled, with a relatively long duct, through the thin body wall they are usually seen externally.

Color: Preserved specimens uniformly gray except the clitellum which is brownish yellow.
Distribution in China: Heilongjiang (Harbin).
Remarks: This species is closely allied to *Allolobophora prashadi and Eisenia hataii* but differs from these mainly in having spermathecae and in the aspect of the male pore.

22. *Eisenia hataii* (Kobayashi, 1940)

Allolobophora hatai Kabayashi, 1940. *Sci. Rep. Tohokuo Univ.* 15:288–289.
Eisenia hataii Xu & Xiao, 2011. *Terrestrial Earthworms of China.* 294–295.

External Characters

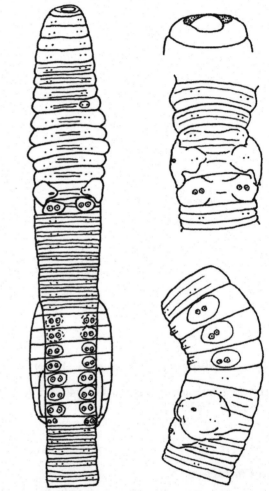

FIG. 28   *Eisenia hataii* (after Kabayashi, 1940).

Length 78–97mm, width 2–2.3mm (clitellum), segment number 134–142. Except the clitellum, segments behind male pore region with 3 annuli. Prostomium proepilobous. First dorsal pore in 4/5, sometimes in 5/6. Clitellum saddle-shaped, in XXIV–XXXII; it is not rare that it invades XXIII anteriorly and XXXIII posteriorly. Pubertatis tubercula, in XXIX–XXXI, ridged and distinct in outline, immediately lateral to the papillae in position. Setae lumbricine, 4 pairs per segment; setal distance *aa* is nearly twice *bc*; *ab* is greater than *cd*, *dd* is smaller than 1/2 of the circumference, *aa*:*ab*:*bc*:*cd*:*dd*:1/2μ=96:7:46:5:175:194 observed on a segment immediately posterior to the clitellum. Setae *cd* of X, XI, and XII, and *ab* of XV, XVI, and XXV–XXXII are situated on papillae; *ab* of IX may also rarely be situated on papillae. Of these papillae, those on XV and XVI are prominent and are constantly found. Male pores 1 pair, on XV, are seated on very prominent and somewhat irregularly circular papillae, encroaching anteriorly about half of XIV and posteriorly about one-third of XVI. On each side of XV, a papilla with setae *ab* is coalesced with the male pore papilla, and setae *cd* are situated on the dorsal part of the latter. Spermathecal pores absent. Female pores 1 pair, on XIV, were not definitely identified.

Internal Characters

Septa 6/7/8/9 moderately, 9/10 slightly thickened, the remaining ones thin. Crop in XV–XVI. Gizzard in XVII–XVIII. Intestine from XX. Testes and funnels free, in X and XI. Seminal vesicles 4 pairs in IX–XII, darkened in color, the posterior 2 pairs are larger than the anterior ones, and are either nodular in appearance or consist of a few small spherical lobules; those in X are nearly equal in size to those in IX. Setae *cd* of X–XII, and *ab* of XV, XVI, and XXV–XXXII, are found in the internal dissection to be contained in large sacs, each of somewhat triangular shape. These genital setae are almost similar in form to one another; almost straight but proximally slightly curved, grooved close to distal half, rather bluntly pointed at distal end; setae *cd* of X–XIII are slightly longer than others, about 0.7 mm long. Spermathecae are entirely absent.

Color: Preserved specimens uniformly gray except the clitellum, which is light yellow or light brownish yellow.
Distribution in China: Liaoning (Huludao, Yingkou).
Etymology: The specific name is dedicated to Dr. Shinkishi Hatai.
Remarks: This species closely resembles in many respects *Allolobophora prashadi* Stephenson, which was recorded from India. It differs from the latter in the constant occurrence of the genital papillae on XV and XVI, and in the aspect of the male pore (perhaps also in the characteristic features of the papillae of X–XII). No spermathecae were recognizable in the parietal wall.

23. *Eisenia jeholensis* (Kobayashi, 1940)

Allolobophora jeholensis Kabayashi, 1940. *Sci. Rep. Tohokuo Univ.* 15:293–295.
Eisenia jeholensis Xu & Xiao, 2011. *Terrestrial Earthworms of China.* 295.

External Characters

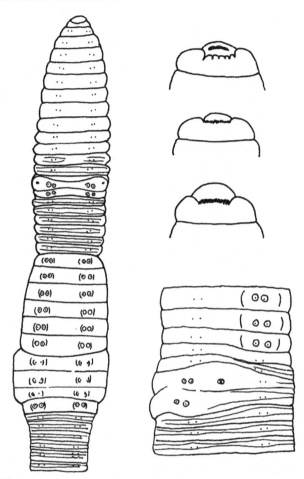

FIG. 29   *Eisenia jeholensis* (after Kabayashi, 1940).

Length 41–53mm, width 4.0–4.6mm (clitellum), segment number 132–140. Body cylindrical in form, each segment behind the male pore region with 3 annuli. Prostomium 1/4–1/3 epilobous or proepilobous. First dorsal pore in 4/5. Clitellum saddle-shaped, moderately swollen, in XXIII–XXXII or –XXXIII. Pubertatis tubercule distinctly ridged; the region just lateral to the genital papillae of XXIX–XXXI is thickened and distinctly stretched in a lateral direction. The ventral part of these segments is also a little thicker than the remaining part of the clitellum. Even in most of the clitellate specimens, the characteristic appearance of the clitellar region is only slightly developed. Setae lumbricine, 4 pairs per segment, closely paired; setal distance *ab* is larger than *cd*, *aa* is nearly twice *bc*; *dd* is a little smaller than 1/2 of the circumference; *aa:ab: bc:cd:dd*=40:3.8:22:2.6:78 in a segment immediately posterior to the clitellum. The setae *cd* of IX–XII (or more of them) and *ab* of XV, XVI, and clitellar segments, are planted on papillae; the papillae on the clitellar segments are less distinct than the rest. Male pores 1 pair, on XV, between *b* and *c*, slightly nearer to *b*; seated on prominent papillae of somewhat ovoidal shape, encroaching anteriorly into the greater part of XIV and posteriorly a very short way into XVI (setae *ab* of XVI are planted on the elevation); the midventral portion of XIV–XVI is also slightly elevated, thus the ventral part of these segments appears to be, as a whole, an elevated rectangular plate. Spermathecal pores absent. Female pores 1 pair, on XIV, just lateral to *b*.

Internal Characters

Septa 6/7/8/9/10 moderately thickened, the rest thin. Crop in XV–XVI. Gizzard in XVII–XVIII. Calciferous glands in 11. Intestine from 20. Typhlosole large and thick, in XX. Hearts in VII–XI. Testes and funnels free, in X and XI. Seminal vesicles 4 pairs in IX–XII, whitish, the anterior 2 pairs are smaller than the posterior ones; those in C are a little smaller than, or nearly equal to, those in IX. With internal dissection, the setae *cd* and *ab* on the genital papillae are found to be contained within the respective large sac; sacs in IX–XII are larger than those in others. They are similar in shape to one another; almost straight except for the proximal end which is slightly curved, and are grooved and moderately pointed distally. Spermathecae are entirely absent.

Color: Preserved specimens unpigmented, body appears generally pale and clitellum limey white.
Distribution in China: Shandong (Yantai), Neimonggu (Chifeng), Ningxia (Zhongning).
Remarks: This species differs from other athecal members of the genus mainly in the characteristic appearance of the clitellar region and the male pore region.

24. *Eisenia nordenskioldi nordenskioldi* (Eisen, 1878)

Helodrilus (Eisenia) nordenskioldi typica, Michaelsen, *Ann. Mus. Acad. Imp. Sci. St. Petersburg*, XV, 17–18.
Eisenia nordenskioldi nordenskioldi Kabayashi, 1940. *Sci. Rep. Tohokuo Univ.* 15:282–284.
Eisenia nordenskioldi nordenskioldi Xu & Xiao, 2011. *Terrestrial Earthworms of China.* 295–296.

External Characters

Length 50–110mm, width 3–6mm, segment number 124–165, generally in the range of 130–145. Prostomium 1/3–1/2 epilobous. First dorsal pore 4/5, distinct and functional. Clitellum saddle-shaped, usually in XXVI or XXVII–XXXIII; seldom 1/n XXVI–XXXIII or XXVI–1/n XXXIII, or very seldom XXVI–XXXII or XXVII–XXXII. Pubertatis tubercula, usually in XXIX–XXXI, seldom in 1/3 XVIII–XXXI or XXIX–1/3 XXXI or 1/3 XVIII–1/3 XXXI; usually in the form of a distinct groove immediately lateral to *b*-line, sometimes it may be of a slightly elevated ridge, or seldom, it may be rather indistinct. Setae lumbricine, 4 pairs per segment, closely paired, moderate in size, *ab* > *cd*, and *aa* > *bc*, or *aa* ≈ 7–8*ab* = 1.5–1.8*bc*, *dd* a little smaller than 1/2 of the circumference. Setae *ab* on XXV–XXXV or most of *cd* on X–XII or –XIII may be planted on whitish indistinct tumescences. These genital setae are not large, being about 0.5–0.6mm long, proximally curved slightly, distally nearly straight but with a keel-like curvature and longitudinal grooves in its distal 1/4–1/5. Male pores 1 pair, on XV, usually without marked glandular elevation, between *b* and *c*, nearer to *b*. Spermathecal pores 2 pairs in 9/10/11, close to middorsal line. Female pores 1 pair, on XIV.

Internal Characters

Septa not especially thickened, 6/7/8/9 slightly thickened. Crop in XV–XVI. Gizzard in XVII–XVIII. Hearts in VII–XI. Testes and funnels free, in X and XI. Seminal vesicles 4 pairs in IX–XII, the anterior 2 pairs on the faces of 9/10 and 10/11, and the posterior 2 pairs on the posterior face of 10/11 and 11/12. Vesicles in IX and X moderate in size and usually simple in feature; those in X are slightly (or sometimes clearly) smaller than those in XI. Those in XI and XII are larger than the anterior 2 pairs, and are in most cases slightly lobated. Spermathecae small and ball-like, each provided with an inconspicuous stalk.

Color: Preserved specimens somewhat resemble *Eisenia fetida*. In fully extended worms, intersegmental furrows may appear yellowish. Dorsally deep or slightly darkened violet-red, ventrally whitish to yellowish gray; dorsolateral surface of IX–XI faded. Clitellum fleshy or sometimes yellowish flesh.
Global distribution: Siberia, Eastern Russia, Northern Europe, North America, Hawaiian Islands.
Distribution in China: Liaoning (Shenyang, Huludao, Xifeng), Jilin (Changchun, Tumen, Jilin), Heilongjiang (Keshan, Qiqiha'er, Ha'erbin, Jiamusi, Boli, Mudanjiang), Neimonggu (Haila'er, Yiliekede, Chifeng), Xinjiang (Altai).
Remarks: The appearance of this species resembles that of *Eisenia fetida*. With regard to the shape of the pubertatis tubercle, the 2 types were found to be nearly identical to that given in Michaelsen's work, however, the usual type of specimens from the northeast of China agrees with that of northeast Siberian ones.

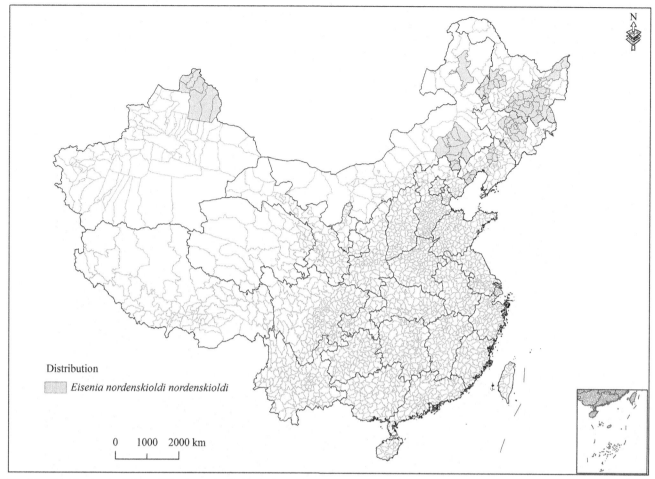

FIG. 30   Distribution in China of *Eisenia nordenskioldi nordenskioldi*.

25. *Eisenia nordenskioldi manshurica* Kobayashi, 1940

Eisenia nordenskioldi manshurica Kobayashi, 1940. *Sci. Rep. Tohokuo Univ.* 15:284–285.
Eisenia nordenskioldi manshurica Xu & Xiao, 2011. *Terrestrial Earthworms of China.* 296.

External Characters

Length 111–144mm, greatest diameter 6.5mm, segment number 54–175. Prostomium 1/3 epilobous. First dorsal pore in 4/5. Clitellum saddle-shaped, in XXVI– or XXVII–XXXIV. Pubertatis tubercula, in XXIX–XXXII; it is in the form of a distinct groove and is placed immediately lateral to *b*-line. Setae lumbricine, 4 pairs per segment, generally a little larger than those of *Eisenia nordenskioldi*; setal distance *ab* > *cd*, *aa* > *bc*, *dd* a little shorter than 1/2 of the circumference. Setae *ab* of XXII–XXXV (or most of them) slightly enlarged and planted on very small whitish tumescences, *cd* of XIV–XXXIV (or most of them) and *ab* of V–IX may be also slightly enlarged, but are not so markedly enlarged and are not planted on the tumescences as in the case of the former setae. Genital setae are about 0.8mm long, proximally only slightly curved; the distal 1/4 is slightly thinner than the proximal, possesses longitudinal grooves (examination of a ventral seta from the clitellar region). Male pores 1 pair, on XV, without marked glandular elevation, between *b* and *c*, nearer to *b*; its general appearance closely resembles that of *Eisenia nordenskioldi*. Spermathecal pores 2 pairs in 9/10/11. Female pores 1 pair, on XIV.

Internal Characters

Septa not especially thickened, 6/7/8/9 slightly thickened. Crop in XV–XVI. Gizzard in XVII–XVIII. Intestine from 20. Hearts in VII–XI. Testes and funnels free, in X and XI. Seminal vesicles 4 pairs in IX–XII, the anterior 2 are small and simple in shape; those in X are a little smaller than those in IX. The posterior 2 are much larger than the anterior ones, and each of the former is divided into 2–3 lobules in the anterior face. Spermathecae ball-like, each with an inconspicuous stalked portion which may be slightly longer than that of *Eisenia nordenskioldi*.

Color: Preserved specimens dorsally, from dusty red to blackish red with a faint purplish tinge, preclitellarly concentrated, and lateral and dorsolateral surfaces of IX–XI faded; ventrally dusty gray. Clitellum deep purplish or dusty brown. Its coloration is clearly different from that of *Eisenia nordenskioldi nordenskioldi*. Distribution in China: Liaoning (Huludao, Anshan), Heilongjiang (Mudanjiang).

Remarks: This form is related more closely to *Eisenia nordenskioldi lagodechiensis* than *Eisenia nordenskioldi nordenskioldi*, in the longer extension of the clitellum and in the features of the pubertatis tubercles. It differs from *Eisenia nordenskioldi lagodechiensis* in the absence of the glandular elevation around the male pore, in the coloration of the body, and in the relative size of the seminal vesicles. The male pore is similar to that of *Eisenia nordenskioldi nordenskioldi* in its general aspects.

26. *Eisenia rosea rosea* (Savigny, 1826)

Enterion roseum Savigny, 1826. *Mém. Acad. Sci. Inst. Fr.* 5:182.
Lumbricus roseus Dugès, 1837. *Ann. Sci. Nat. Ser.* 2, 8:17, 20.
Lumbricus communis anatomicus (part.) Hoffmeister, 1845. *Regenwürmer*. 28.
Allolobophora rosea Rosa, 1893. Mem. Acc. Torino, ser.2, 43:424, 427.
Eisenia rosea Michaelsen, 1900. *Das Tierreich., Oligochaeta.* 10:478.
Eisenia rosea typica Kobayashi, 1940. *Sci. Rep. Tohokuo Univ.* 15:285.
Allolobophora rosea Edwards and Lofty, 1972. *Biol. Earthworm*. 217.
Eisenia rosea Gates, 1972. *Trans. Amer. Philos. Soc.* 62(7):104–108.
Aporrectodea roaes Gates, 1976. *Megadrilogica.* 2(12):4.
Eisenia rosea Reynolds, 1977. *The Earthworms of Ontario.*78–83.
Eisenia rosea Hong, 2000. *Korean Jour. System. Zool.* 16(1):8–9.
Eisenia rosea rosea Xu & Xiao, 2011. *Terrestrial Earthworms of China.* 296–297.

External Characters

Length 25–85mm, width 2.6–5.0mm, segment number 119–150. Body cylindrical, except in clitellar region. Prostomium epilobous. First dorsal pore 4/5. Clitellum XXV, XXVI–XXXII, saddle-shaped, *cd* setae and dorsal invisible externally within clitellum. Tubercula pubertatis usually XXIX–XXXI. Setae lumbricine, 4 pairs per segment, closely paired, *ab*>*ab*>*cd*, *ab*>*bc*<*dd*, anteriorly

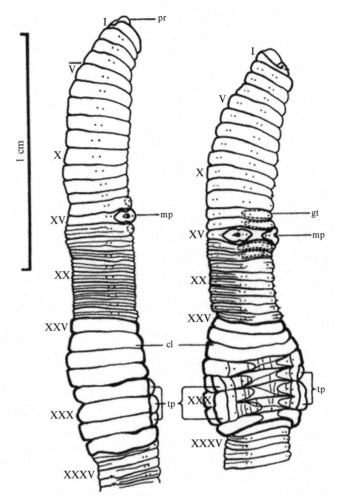

FIG. 31   *Eisenia rosea rosea* (after Reynolds, 1977).

dd=1/2 circumference, posteriorly *dd*=1/3 circumference. Male pores 1 pair, with elevated glandular papillae in XV, with male tumescences extending over XIV and XVI. Spermathecal pores 2 pairs in 9/10/11.

Internal Characters

Septa usually thin. Calciferous glands in 11. Intestine from 20. Typhlosole large and thick, in XX. Testes and funnels in ventral paired sacs in X, XI. Seminal vesicles 4 pairs in IX–XII, sometimes in IX, X, XI–XII. Spermathecae 2 pairs in X and XI, with duct short.

Color: Preserved specimens unpigmented but color appears rosy or grayish when alive and white when preserved.
Global distribution: Europe, North America, South America, Africa, Asia, Australia, Iceland, Korea. It has a cosmopolitan distribution although it is apparently absent from tropical lowlands.
Distribution in China: Beijing (Haidian, Chongwen, Dongcheng, Tongzhou, Changping, Huairou, Fangshan, Fengtai, Chaoyang, Miyun, Shijingshan,

Yanqing), Liaoning (Shenyang), Jilin, Heilongjiang (Harbin, Keshan), Xinjiang (Tacheng).

Habitats: *Eisenia rosea rosea* is one of the cosmopolitan species introduced by Europeans to all parts of the world. Its habitats include soil, fields, gardens, pastures, forests, and under leaves and stones. It is also found frequently on river banks and lake edges (Černosvitov and Evans, 1974; Geard, 1964), under logs, and on the virgin steppes of Russia (Gates, 1972c).

According to Thomson and Davies (1974), *Eisenia rosea rosea* produces surface casts despite some contrary statements in the literature. The species is parthenogenetic and biparental reproduction of anthropochorous morphs is unknown (Gates, 1974; Reynolds, 1974). Černosvitov (1930) reported atypical spermatogenesis. Evans and Guild (1948) reared isolated individuals to sexual maturity, which then produced fertile cocoons.

GENUS: *EISENIELLA* MICHAELSEN, 1900

Eiseniella Michaelsen, 1900. *Das Tierreich., Oligochaeta.* 10:471.
Eiseniella Gates, 1972. *Trans. Amer. Philos. Soc.* 62(7):108.
Eiseniella Reynolds, 1977. *The Earthworms of Ontario.* 83.

Type species: *Enterion tetraedrum* Savigny, 1826.

External Characters

Prostomium usually epilobous; rarely without backwardly projecting tongue. Setae closely paired, ventrolateral and dorsolateral. Clitellum begins at segment XXIII or farther forwards, and includes 4–8 segments; the tubercles at puberty fused to form regular ridges. Nephropores, inconspicuous, behind XV alternating irregularly and with asymmetry between a level just above *b* and a level above *d*. Longitudinal musculature, pinnate. Male pores on XV, or displaced 2–4 segments farther forwards. Spermathecal pore 2 pairs, between the line of setae d and the middorsal line.

Internal Characters

Calciferous sacs, in X, digitiform opening posteriorly into the gut ventrally in region of insertion of 10/11. Esophagus of nearly uniform width through XI–XIV, calciferous channels narrow, lamellae low and continued along the lateral walls of the sacs. Intestinal origin, in XV. Gizzard, in XVII, weak, 17/18 not fenestrated. Typhlosole, simply lamelliform. Extraesophageal trunks, joining dorsal vessel in XII. Hearts, in VII–XI. Nephridial bladders, short and sausage-shaped. Testes and funnels free; 4 pairs of seminal vesicles in IX–XII.

Discussion: Michaelsen (1900) proposed the new name *Eiseniella* for a genus erected by Eisen in 1873. Although designated only as a new name for *Allurus* Eisen, 1873, Michaelsen included a second genus of Eisen's (*Tetragonurus* Eisen, 1874) in his *Eiseniella*. Both of Eisen's generic names were preoccupied (i.e., already in use as generic names in other groups); *Allurus* Foerster, 1826 had been used as a genus of hymenopterous insects, while *Tetragonurus* Risso, 1810 had been employed as a genus of fish. The type species for both of Eisen's genera are synonyms of *Enterion tetraedrum* Savigny, 1826.

There is 1 species from China.

27. *Eiseniella tetraedra* (Savigny, 1826)

Enterion tetraedrum Savigny, 1826. *Mém. Acad. Sci. Inst. Fr.* 5:184.
Lumbricus tetraedrus Dugès, 1837. *Ann. Sci. Nat. Ser.* 2, 8:17,23.
Lumbricus agilis Hoffmeister, 1843. *Arch. Naturg.* 9 (1):191.
Eiseniella tetraedra Michaelsen, 1900. *Das Tierreich., Oligochaeta.* 10:471.
Eiseniella tetraedra f. *typica*, Černosvitov, 1937. *Rec. Indian Mus.* 39:107.
Eiseniella tetraedra Gates, 1972. *Trans. Amer. Philos. Soc.* 62(7):108–113.
Eiseniella tetraedra Reynolds, 1977. *The Earthworms of Ontario.* 88.
Eiseniella tetraedra Shen, Tsai, and Tsai, 2005. *Taiwania.* 50(1):16–17.
Eiseniella tetraedra Chang, Shen & Chen, 2009. *Earthworm Fauna of Taiwan.* 22–23.
Eiseniella tetraedra Xu & Xiao, 2011. *Terrestrial Earthworms of China.* 297–298.

External Characters

Length 30–63mm, width 2–4mm. Segment number 60–95. Prostomium epilobous or zygolobous. First dorsal pore 4/5 or 5/6. Clitellum XXII, XXIII–XXVI, XXVII, saddle-shaped, setae present. Tubercula pubertatis XXIII–XXV, XXVI. Setae closely paired, *aa:ab:bc:cd:dd*=3:1:3:1:6–8 posteriorly. Ventral setae on X, or IX and X modified into genital setae. Male pores paired in lateral clefts in XIII with slightly elevated glandular papillae in Černositov's "typical form." Spermathecal pores 2 pairs in 9/10 and 10/11, dorsolateral, distance between paired pores shorter than *dd*. Female pores paired in XIV, medioventral, each medial to seta *a*.

Internal Characters

Septa not especially thickened. Calciferous glands in X. Crops large in XV–XVII. Gizzard small in XVIII. Intestine

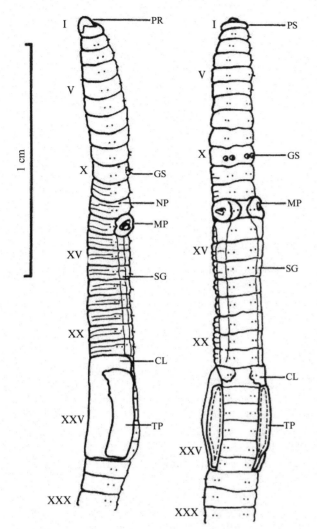

FIG. 32   *Eiseniella tetraedra* (after Reynolds, 1977).

from XVIII. Esophageal hearts in VII–XI. Spermathecae 2 pairs in X and XI, small, ampulla round, sessile, about 0.5mm wide, adiverticulate. Nephridia holandric. Ovaries in XIII. Testis sacs 2 pairs in X and XI. Seminal vesicles paired in XI and XII, small. Prostate glands absent.

Color: Live specimens reddish brown in color with yellowish orange clitellum, squarish toward posterior end. Preserved specimens reddish brown, yellowish orange around clitellum.
Distribution in China: Taiwan (Shalihsien Creek, Cijiawan Creek).
Type locality: Paris, France.
Deposition of types: Paris Museum, France.
Global distribution: Cosmopolitan, mainly in the temperate zone. Europe, North America, South America, Asia, Africa, Australasia (Gates, 1972). It also occurs in

Iceland (Backlund, 1949). It is another cosmopolitan species that has been carried around the world. Habitats: This species dwells in springs, lakes, mountain torrents, and swampy soil. It is the only recorded earthworm species living in freshwater habitats in Taiwan. When present, it is abundant.

GENUS: *LUMBRICUS* LINNAEUS, 1758

Lumbricus (part.) Linnaeus, 1758. *Syst. Nat.* (ed.10), 647.
Enterion (part.) Savigny, 1826. *Mém. Acad. Sci. Inst. Fr.* 5:179.
Lumbricus (part.) Hoffmeister, 1843. *Regenwürmer.* 4.
Lumbricus Michaelsen, 1900. *Das Tierreich., Oligochaeta.* 10:508.
Lumbricus Stephenson, 1930. *Oligochaeta.* 914.
Lumbricus Gates, 1972. *Trans. Amer. Philos. Soc.* 62(7):113–114.
Lumbricus Reynolds, 1977. *The Earthworms of Ontario.* 88.

Type species: *Lumbricus terrestris* Linnaeus, 1758.
Clitellum usually saddle-shaped, in XXVI–XXXII. Calciferous sacs, in X, digitiform to pyriform, opening into gut posteriorly and ventrally in region of insertion of 10/11, lamellae continued along lateral walls. Esophagus, widened and markedly moniliform in XI–XII, in those segments with a vertically slit-like lumen which widens as lamellae gradually narrow behind 12/13. Intestinal origin, in XV. Gizzard, mainly in XVII. Typhlosole, high, rather thick and nearly oblong vertically, grooves not continuous across ventral face. Extraesophageal trunks, joining dorsal trunk in region of IX–X. Hearts, in VII–XI. Nephridial bladders, J-shaped, closed end laterally, duct passing into parietes near *b*. Nephropores, obvious, behind XV irregularly alternating, with asymmetry, between levels just above *b* and well above *d*. Setae, closely paired. First dorsal pore, anterior to 10/11. Prostomium, tanylobic. Body compressed dorsoventrally behind clitellum and with a more or less transverse section. Longitudinal musculature, pinnate (Gates, 1972, 1975).

Color: Pigment, lacking immediately underneath intersegmental furrows, in circular muscle layer.
Discussion: The genus *Lumbricus* Linnaeus, 1758 originally contained only 2 species, *L. terrestris* and *L. marinus*. Since the latter was not a member of the Oligochaeta, the type species was declared to be *L. terrestris*. Discussions of what was really meant by *L. terrestris* Linnaeus, 1758 have raged ever since but are officially settled (Sims, 1973; Gates, 1973; Bouche, 1973).

The result of this action was the neotypification of *L. terrestris* Linnaeus by Sims (1973) with an expanded definition, and the deposition of type material from Sweden in the British Museum (Natural History).

There is 1 species from China.

28. *Lumbricus rubellus* Hoffmeister, 1843

Lumbricus rubellus Hoffmeister, 1843. *Arch. Naturg.* 9 (1): 187.
Digaster campestris (part.) Hutton, 1883. *N. Z. J. Sci.* 1:586.
Lumbricus rubellus Michaelsen, 1900. *Das Tierreich., Oligochaeta*. 10:509.
Lumbricus rubellus Stephenson, 1923. *Fauna British India Oligochaeta*. 508.
Lumbricus rubellus Gates, 1972. *Trans. Am. Philo. Soc.* 62(7):115–118.

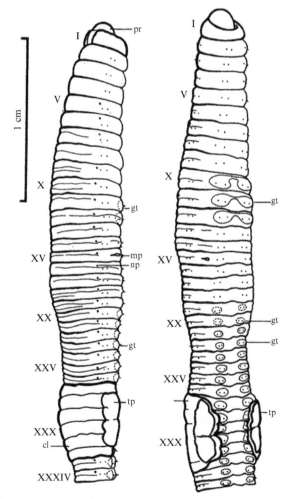

FIG. 33 *Lumbricus rubellus* (after Reynolds, 1977).

Lumbricus rubellus Reynolds, 1977. *The Earthworms of Ontario*. 94–98.
Lumbricus rubellus Xu & Xiao, 2011. *Terrestrial Earthworms of China*. 298.

External Characters

Length 50–150mm, width 4–6mm, segment number 70–120. Body cylindrical and sometimes dorsoventrally flattened posteriorly. Prostomium tanylobic. First dorsal pore 5/6–8/9. Clitellum saddle-shaped, in XXVI, XXVII–XXXI, XXXII. Tubercula pubertatis on XVIII–XXXI. Setae closely paired, $aa>bc$, $ab>cd$, $dd=1/2$ C posteriorly. Genital tumescences in VII–XII (less frequently on x), XX–XXIII, XXVI–XXXVI. Spermathecal pores 2 pairs in 9/10/11.Male pores 1 pair, in XV, inconspicuous, without glandular papillae on XV. Female pores 1 pair, as small transversely ovoid openings, on XIV.

Internal Characters

Seminal vesicles 3 pairs, in IX, XI, and XII+XIII. Spermathecae 2 pairs in X and XI, with short ducts.

Color: Preserved specimens ruddy brown or red-violet and iridescent dorsally, pale yellow ventrally.
Distribution in China: Northeast, Northwest.
Global distribution: Europe, Iceland, North America, Mexico, South Africa, New Zealand (Gates, 1972).
Habitat: *Lumbricus rubellus* has a wide range of habitats, including debris, stream banks, parks, gardens, pastures, under stones or old leaves, under logs, and in wood (Olson, 1928, 1936; Eaton, 1942; Černosvitov and Evans, 1947).

GENUS: OCTOLASION ÖERLEY, 1885

Octolasion (part.) Örley, 1885. *Ertek. Term. Magyar Akad.* 15(18):13.
Octolasium Michaelsen, 1900. *Das Tierreich., Oligochaeta*. 10:504.
Octolasium Stephenson, 1930. *Oligochaeta (Oxford)*. 914.
Octolasium Gates, 1972. *Trans. Amer. Philos. Soc.* 62(7):123.
Octolasium Bouché, 1972. *Inst. Natn. Rech. Agron.* 253.
Octolasion Gates, 1975. *Megadrilogica*. 2(1):4.
Octolasion Reynolds, 1977. *The Earthworms of Ontario*. 104.

Type species: *Octolasion lacteum* (Öerley, 1885) = (*Enterion tyrtaeum* Savigny, 1826).

External Characters

Prostomium mostly epilobous, rarely tanylobous. Longitudinal musculature, pinnate. Clitellum usually saddle-shaped. Setae mostly separated, rarely closely paired. Tubercula pubertatis fused to form longitudinal ridges. Spermathecal pores in the line of *c*, or between *c* and *d*, or somewhere below *c*.

Internal Characters

Calciferous sacs, in X, large, lateral, communicating vertically and widely with gut lumen though reaching beyond esophagus both dorsally and ventrally. Calciferous lamellae continued onto posterior walls of sacs. Intestinal origin, in XV. Gizzard, mostly in XVII. Extraesophageals, passing up to dorsal trunk posteriorly in XII. Hearts, VI–XI. Nephridial bladders, ocarina-shaped. Nephropores, obvious, behind XV in 1 regular rank on each side, just above *b* setae, behind the clitellum not closely paired. Testes and funnels usually enclosed in 2 pairs of testis sacs. If testis sacs absent, the septa of the testis segments connected together by horizontal bands, or fused by their margins so as to form narrow chamber. There are 4 pairs of seminal vesicles in IX–XII.

Discussion: There has been considerable confusion concerning the spelling of this genus name since the early 1900s. Michaelsen (1900) changed many of the Greek generic endings to Latin endings, that is, *Octolasion* to *Octolasium* and *Bimastos* to *Bimsatus*. According to the International Code of Zoological Nomenclature (ICZN), the original spelling is correct. Therefore, *Octolasion* Örley, 1885 and *Bimastos* Moore, 1893 are the correct and most current oligochaetologists are now employing them. The genus *Octolasion* Örley, 1885 is now considered to belong in the genus *Octodrilus* Omodeo, 1956 with *Lumbricus complanatus* Dugès, 1828 as the type (cf. Bouche, 1972; Gates, 1975a).

There are 2 species from China.

TABLE 9   Key to the Species of the Genus *Octolasion* From China

1	Tubercula pubertatis in 1/2 XXX–1/3 XXXV .... *Octolasion lacteum*
	Tubercula pubertatis in XXXIV–XXXVI .... *Octolasion tyrtaeum*

29. Octolasion lacteum (Öerley, 1885)

Octolasion lacteum Kabayashi, 1940. *Sci. Rep. Tohokuo Univ.* 15:302–303.
Octolasion lacteum Xu & Xiao, 2011. *Terrestrial Earthworms of China.* 299.

External Characters

Length 131 mm, width 4 mm, segment number 126. Prostomium 1/2 epilobous. Body cylindrical; secondary

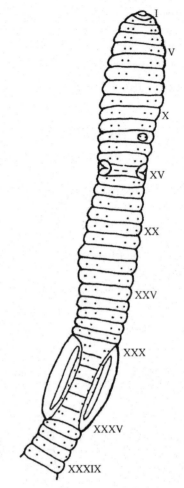

FIG. 34   *Octolasion lacteum* (after Reynolds, 1977).

annulation indistinct or may be said to be absent. First dorsal pore in 12/13, distinct; in 10/11 and 11/12 indistinct and nonfunctional pores are found. Clitellum saddle-shaped, in XXX–XXXV. Pubertatis tubercles in 1/2XXX–1/3XXXV, darkened, each with a long groove. Setae found on segments anterior to male pore are paired, $ab < bc > cd$ or $aa{:}ab{:}bc{:}cd{:}dd{:}1/2\mu = 29{:}6{:}16{:}4{:}108{:}95$ in IX; those posterior to male pore are widely paired or separated, $ab > bc > cd$ or $aa{:}ab{:}bc{:}cd{:}dd{:}1/2\mu = 27{:}11{:}10{:}7{:}63{:}73$ in a segment immediately posterior to the clitellum. Setae *a* and *b* of XII (left side only) are planted on a small but distinct genital papilla; on internal dissection, they are found to be contained within the usual setal sac; about 0.7 mm long, grooved on the distal half, nearly straight but proximally slightly curved, distally pointed rather bluntly. Male pores 1 pair, slit-like in XV, between *b* and *c*, nearer to *b*; each on a marked elevation which encroaches a little into both XIV and XVI. Spermathecal pores 2 pairs in 9/10/11, on *c*-line. Female pores 1 pair, on XIV, just lateral to setae *b*.

Internal Characters

Septa not especially thickened; 6/7/8/9 moderately thickened. Testis sacs present. Seminal vesicles 4 pairs, in IX–XII; those in IX and X are digitiform and clearly differ in shape from those in XI and XII. Prostate glands absent. Spermathecae spherical, each with a very inconspicuous stalk.

Color: Preserved specimens pinkish gray; unpigmented. Clitellum light flesh.
Distribution in China: Liaoning (Dandong), Heilongjiang (Haerbin).
Global distribution: Widely distributed, Europe, Austria, Hungary, Yugoslavia, Bulgaria, Switzerland, Spain, France, Germany, England, Czechoslovakia, Romania, Italy, western and southern Russia, Algeria, Azores, Canary Islands, Aland Islands, North America, Mexico, Uruguay, Australia, North India.

30. *Octolasion tyrtaeum* (Savigny, 1826)

Enterion tyrtaeum Savigny, 1826. *Mém. Acad. Sci. Inst. Fr.* 5:180.
Lumbricus tyrtaeum Dugès, 1837. *Ann. Sci. Nat. Ser.* 2, 8:17, 22.
Allolobophora profuga Rosa, 1884. *Lumbric. Piemonte*, 47.
Octolasium lacteum Edwards and Lofty, 1972. *Biol. Earthworm.* 216.
Octolasium lacteum Michaelsen, 1900. *Das Tierreich., Oligochaeta.* 10:506.
Octolasium tyrtaeum Gates, 1972. *Trans. Amer. Philos. Soc.* 62(7):125–128.
Octolasion tyrtaeum Reynolds, 1977. *The Earthworms of Ontario.* 108–111
Octolasion tyrtaeum Xu & Xiao, 2011. *Terrestrial Earthworms of China.* 300.

External Characters

Length 25–130mm, width 3–6mm. Segment number 75–150. Body cylindrical but slightly octagonal posteriorly. Prostomium epilobous. First dorsal pore 9/10–13/14, usually 11/12. Clitellum XXX–XXXV, saddle-shaped. Tubercula pubertatis XXXI–XXXIV. Setal pairings as in *Octolasion cyaneum*. Frequently setae *a* and/or *b* on XXII and occasionally on IX–XII, XIV, XVII, XIX–XXIII, XXVII, XXXVII, or XXVIII are on genital tumescences and modified into genital setae. Male pores on XV and on large glandular papillae extending over XIV and XVI, occasionally limited to XV. Spermathecal pores 2 pairs in 9/10 and 10/11, on level between *c* and *d*. Female pores paired in XIV.

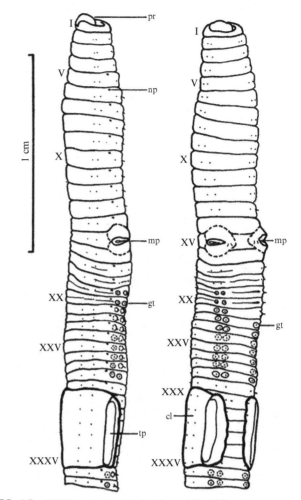

FIG. 35 *Octolasion tyrtaeum* (after Reynolds, 1977).

Internal Characters

Spermathecae 2 pairs in X and XI. Seminal vesicles 4 pairs in IX–XII, with pairs in XI and XII larger than pairs in IX and X. Prostate glands absent.

Color: Preserved specimens variable, milky white, gray, blue, or pink.
Distribution in China: Heilongjiang (Harbin).
Global distribution: Europe, North America, Asia, Africa, Australia.
Habitats: *Octolasion tyrtaeum* has been found under stones and logs, in peat, leaf mold, compost, forest litter, gardens, cultivated fields and pastures, bogs, stream banks, springs, around the roots of submerged vegetation, in debris and rocks, and by digging (Smith, 1917; Gates, 1972; Reynolds et al., 1974).

Octolasion tyrtaeum is an obligatory parthenogenetic species (Gates, 1973; Reynolds, 1974) and copulation occurs below the surface of the soil.

Family: Megascolecidae Templeton, 1844?

Megascolecidae Stephenson, 1930. *The Oligochaeta*. 818.
Megascolecidae Gates, 1972. *Trans. Amer. Philos. Soc.* 62(7):130–132.

Type genus: *Megascolex*

External Characters

Prostomium, prolobous. Clitellum unilayered (?), annular, intersegmental furrows not obliterated, beginning with or in front of segment XV. Male pores 1 pair, usually on either XVII or XVIII, seldom on XIX. Female pores paired, or 1 median pore, with exceptions on XIV. Setae, sigmoid and single pointed, 4 pairs per segment.

Internal Characters

Gizzards usually present. Testes 2 pairs in X and XI, or 1 pair only, in X or XI. One or 2 pairs of prostates; prostates rarely absent. Ovaries 1 pair in XIII.

Global distribution: Mostly terrestrial; a number limnic, a few littoral. Widely distributed over the globe, occurring throughout the whole of the Southern Hemisphere and the southern portion of the Northern; absent from North and West Asia, North Europe and northern North America, while only peregrine forms are found in Mid and South Europe and North Africa.

There are 8 genera, 21 species, and 4 subspecies from China.

TABLE 10 Key to the Genera of the Family Megascolecidae From China

1.	Clitellum in XIV–XVI, with depressed bodies	*Planapheretima*
	With cylindrical bodies	2
2.	Clitellum in XIII, XXIV–XVII	3
	Clitellum in XXIV–XVI	4
3.	Clitellum in XIII–XVII	*Perionyx*
	Clitellum in XIV–XVII	*Lampito*

Continued

TABLE 10 Key to the Genera of the Family Megascolecidae From China—cont'd

4.	Without caeca	*Polypheretima*
	With caeca	5
5..	Intestinal caeca present originating in XXII	*Pithemera*
	Intestinal caeca present originating behind XXII	6
6.	Intestinal caeca present originating in XXV	*Begemius*
	Intestinal caeca present originating in XXVII	7
7.	With copulatory pouches	*Metaphire*
	Without copulatory pouches	*Amynthas*

I. AMYNTHAS KINBERG, 1867

Amynthas Kinberg, 1867:97.
Amynthas Kinberg, 1867:101.
Nitoeris Kinberg, 1867:112.
Pheretima (*Pheretima*) (part) Michaelsen, 1928a:8.
Amynthas. Sims and Easton. *Biol. J. Soc.* 1972 4:211.

Type species. *Amynthas aeruginosus* Kinberg, 1867
Diagnosis: Megascolecidae with cylindrical bodies of varying sizes. Setae numerous, regularly arranged around each segment. Clitellum annular, XIV–XVI, rarely beginning on XIII. Male pore paired, discharging directly onto the surface of XVIII (rarely XIX). Female pore single, rarely paired, XIV. Spermathecal pore small or large, usually paired (bithecal) but occasionally numerous (polythecal) or single (monothecal) between 4/5 and 8/9.

Gizzard present septa between 7/8 and 9/10. Esophageal pouches absent. Intestinal caeca present originating in XXVII. Testes holandric or metandric. Prostatic glands racemose. Copulatory pouches absent. Ovaries paired in XIII. Spermathecae usually paired, rarely multiple or single. Meronephridial, nephridia rarely present on spermathecal ducts.

Global distribution: Oriental realm, ? Australasian realm, and ? introduced into the Oceanian realm. There are 225 species from China.

TABLE 11 Key to the Species Groups of the Genus *Amynthas* From China

1. Spermathecal pores absent .... *illotus* group
   Spermathecal pores present 2

2. Spermathecal pores 2 pairs, at VI and 5/6 .... *youngi* group
   Spermathecal pores at intersegment or at segment 3

3. Spermathecal pores at intersegment 4
   Spermathecal pores at segment 17

4. Spermathecal pores 2 pairs and 1, at 4/5/6/7 .... *swanus* group
   Spermathecal pores paired or single 5

5. Spermathecal pores 1, at 5/6 .... *zyosiae* group
   Spermathecal pores paired 6

6. Spermathecal pores 1 pair 7
   Spermathecal pores more than 1 pair 9

7. Spermathecal pores 1 pair, at 5/6 .... *minimus* group
   Spermathecal pores 1 pair, at 7/8 or 8/9 8

8. Spermathecal pores 1 pair, at 7/8 .... *zebrus* group
   Spermathecal pores 1 pair, at 8/9 .... *supuensis* group

9. Spermathecal pores 2 pairs 10
   Spermathecal pores more than two pairs 13

10. Spermathecal pores 2 pairs, at 4/5/6 .... *swanus* group
    Spermathecal pores 2 pairs, behind 4/5/6 11

11. Spermathecal pores 2 pairs, at 5/6/7 .... *morrisi* group
    Spermathecal pores 2 pairs, behind 5/6/7 12

12. Spermathecal pores 2 pairs, at 6/7/8 .... *tokioensis* group
    Spermathecal pores 2 pairs, at 7/8/9 .... *aeruginosus* group

13. Spermathecal pores 3 pairs 14
    Spermathecal pores more than 3 pairs 16

14. Spermathecal pores 3 pairs, at 4/5/6/7 .... *pauxillulus* group
    Spermathecal pores 3 pairs, behind 4/5/6/7 15

15. Spermathecal pores 3 pairs, at 5/6/7/8 .... *hawayanus* group
    Spermathecal pores 3 pairs, at 6/7/8/9 .... *sieboldi* group

16. Spermathecal pores 4 pairs, at 5/6/7/8/9 .... *diffringens* group
    Spermathecal pores 5 pairs, at 4/5/6/7/8/9 *megascolidioides* group

17. Spermathecal pores 2, at VI and VII .... *mamillaris* group
    Spermathecal pores paired 18

18. Spermathecal pores 1 pair, at VI .... *glabrus* group
    Spermathecal pores more than 1 pair 19

19. Spermathecal pores 2 pairs 20
    Spermathecal pores more than 2 pairs 21

20. Spermathecal pores 2 pairs, at VI and VII .... *canaliculatus* group
    Spermathecal pores 2 pairs, at VII and VII .... *pomellus* group

21. Spermathecal pores 3 pairs, at VI, VII, VII .... *bournei* group
    Spermathecal pores 4 pairs, at VI, VII, VII, IX .... *rimosus* group

There are 223 species and 13 subspecies from China.

aeruginosus Group

Diagnosis. Holandric, *Amynthas* with spermathecal pores 2 pairs, at ventral region of 7/8/9.
There are 22 species from China.

TABLE 12 Key to the Species of the *aeruginosus* Group of *Amynthas* From China

1. First dorsal pore 10/11 2
   First dorsal pore behind 10/11 3

2. Pimple-like papillae usually present, 1 in front of and
   another behind each spermathecal pore but more
   ventral .... *Amynthas triastriatus*
   Spermathecal pore on round flatted papilla behind setal
   zone .... *Amynthas kulingianus*

3. First dorsal pore 11/12 4
   First dorsal pore 12/13 11

4. Without papillae in front of clitellum, male pores on the tip of a
   smooth porophore moderately large, broadly rounded,
   ball-shaped .... *Amynthas kiangensis*
   With papillae front clitellum 5

5. Papillae between VII–X 6
   Papillae in VII and IX 8

6. There are 2–10 papillae on the ventral sides of VII–X *Amynthas cheni*
   Papillae between VII–IX 7

7. Several or many ampulla-like papillae scattered irregularly on
   ventral side of segments VII, VII and IX, infrequently in front of setae
   of X .... *Amynthas lautus*
   Genital papillae round and depressed in center, paired
   on entromedian side of VII–IX, postsetal, between c
   and e .... *Amynthas hexitus*

8. Genital papillae numbering 1–6, small, near spermathecal pores, in
   transverse rows .... *Amynthas aspergillum*
   With some small papillae in the spermathecal pores region 9

9. A papilla is found close to each pore, each papilla
   ovoid .... *Amynthas ultorius*
   The papillae round 10

10. Spermathecal diverticulum slender and very long, coiled or
    straight .... *Amynthas corrugatus*
    Spermathecal diverticulum slightly longer, its ental half distended as
    a cylindrical seminal chamber .... *Amynthas masatakae*

11. Without papillae front clitellum 12
    With papillae front or/and behind clitellum 14

12. Male pores situated on a large round and flat-topped papilla,
    which extend anteriorly to 1/2 XVII and posteriorly to
    1/2 XIX .... *Amynthas meioglandularis*
    Male pores on wart-like porophores 13

13. Wart-like porophores extending the entire length of the segment. At
    the base of the porophores, some ring-like furrows, on the median
    side of the tip some few (2–4?) vague slightly raised glandular
    papillae .... *Amynthas siemsseni*
    Wart-like porophores, with 1 or more slightly more or less clear
    circular furrows. On the tip of each male porophore there lies close to
    each male aperture and obliquely anteromedian to it, a tiny but
    slightly raised sex papilla .... *Amynthas fokiensis*

14. With a small number papillae front and behind clitellum 15
    With a large number papillae front and behind clitellum 17

15. With 5–8 pairs of papillae on the ventral side of XVIII–XXII or XXV;
    with 6–12 papillae on ventral side of V–VII *Amynthas longisiphous*
    With a small papilla behind clitellum 16

16. Male pores among which small papillae are interposing; Another
    similar papilla situated postsetally on XVII; with 2 rows papillae in

TABLE 12 Key to the Species of the *aeruginosus* Group of *Amynthas* From China—cont'd

VII and IX, about 22, or close to spermathecal pores .... *Amynthas kinfumontis*
With 1 or 2 similar papillae around male pore, 1 or 2 additional presetal papillae present medioventrally in XVIII; papillae paired in spermathecal pores and/or presetal in VII and IX or in VII and X .... *Amynthas robutus*

17. With more than 10 papillae in male pore region 18
    Male pores on top of papillae, each surrounded by 8 genital papillae in 4s on 2 sides of an isosceles triangle with the base turned externally; with 2 or 3 genital papillae on the posterior margin of the nest preceding segment .... *Amynthas takatorii*

18. With more than 20 papillae front and behind clitellum 19
    Male pores surrounded by 5–9 small papillae; with a cluster of 3–8 small papillae immediately adjacent to the margins between XVII and XVIII, and a cluster of 9–20 papillae arranged obliquely in presetal XIX; genital papillae tiny, round, 9–24 arranged in front of an arc stretching from postsetal VII to presetal VII and from postsetal VII to presetal IX .... *Amynthas kinmenensis*

19. Male pores surrounded by 2–9 small papillae, a large number of small, round papillae present between the paired male pores, numbering 19–31 and 15–22 presetal and postsetal, ventrally in XVIII; with genital papillae present in VII and IX, medioventral, numbering 10–23 .... *Amynthas polyglandularis*
    Male pores with 3–7 small papillae surrounding the depression; with papilla more or less rounded or wart-like, about 40–50 closely arranged in rows posteriorly in front of, or covering the setal zone of the ventral side of XI, or with 4–5 similar papillae situated at anterior part of ventral side of VII .... *Amynthas omeimontis*

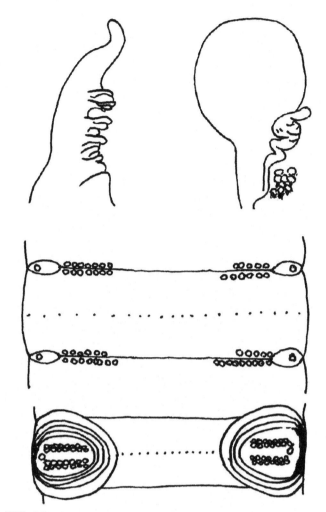

FIG. 36 *Amynthas aspergillum* (after Chen, 1938).

31. *Amynthas aspergillum* (Perrier, 1872)

Pheretima aspergillum Gates, 1935a. *Smithsonian Mis. Coll.* 93(3):7.
Pheretima (Pheretima) aspergillum Chen, 1938. *Contr. Biol. Lab. Sci. Soc. China (Zool.).* 12(10):382.
Pheretima aspergillum Gates, 1939. *Proc. U.S. Natr. Mus.* 85:420–425.
Amynthas aspergillum Chang, Shen & Chen, 2009. *Earthworm Fauna of Taiwan.* 24–25.
Amynthas aspergillum Xu & Xiao, 2011. *Terrestrial Earthworms of China.* 82.

External Characters

Length 117–416mm, clitellum width 8.5–11.3mm, segment number 109–153. Setae 51–67 (VII), 62–69 (XX), 12–19 between male pores. First dorsal pore 11/12. Clitellum XIV–XVI, annular, setae absent, dorsal pore absent. Male pores paired in XVIII, on top of a small, round papilla. Several papillae in 2–4 transverse rows medial to male pores, numbering 2–6 or more in each row, higher in the rows closer to setal lines. Male pores and the associated papillae surrounded by several circular folds. Spermathecal pore 2 pairs in 7/8/9. Genital papillae absent or present, if present, numbering 1–6, small, near spermathecal pores, in transverse rows. Female pore single in XIV, medioventral.

Internal Characters

Septa 5/6–7/8, 10/11–13/14 greatly thickened, 8/9/10 absent. Gizzard large in VII and IX. Intestine from XV. Intestinal caeca paired in XXVII, simple, extending anteriorly to XXIII or XXII. Last hearts in XIII. Spermathecae 2 pairs in VII and VII, ampulla peach-shaped, stalk short, stout; diverticulum stalk long, slender, coiled, seminal chamber oval, elongated. Accessory glands present. Ovaries paired in XIII. Testis sacs 2 pairs in x and XI, ventrally connected. Seminal vesicles paired in XI and XII, with dorsal lobes. Prostate glands large in XVIII, extending posteriorly to XXII, divided into several pieces. Accessory glands present, corresponding to external papillae in the male pore regions.

Color: Live specimens brownish red or dark gray on dorsum, light brown on venter. Preserved specimens purplish brown on dorsum, light gray on ventrum and clitellum.
Distribution in China: Fujian (Xiamen, Fuzhou), Taiwan (Taibei, Pingdong), Guangdong (Dongguan), Hong Kong (Jiulong), Hainan (Wanning), Guangxi.

Global distribution: China, Vietnam.

Deposition of types: Museum National d'Histoire Naturelle, Paris, France. (Paris Museum)

Etymology: The specific name indicates the arrangement of its male pore papilla.

Remarks: *Begemius paraglandularis* is so very similar to *Amynthas aspergillum* with regard to a number of structures of major systematic importance that there can be little if any doubt that the 2 are synonymous. However, in *Begemius paraglandularis*, according to Fang, the male pores are large slits; there are "moderate" copulatory chamber, each containing an elongate genital papilla; there are 2 pairs of testis sacs, the conjoined transverse pairs connected with each other anteroposteriorly. All these rather unusual characteristics are doubtless the result of errors in observation or interpretation. It is scarcely necessary to discuss all these errors. The gizzard is always in segment VII in the genus *Amynthas*. Fang has mistaken septa 5/6/7/8 for septa 6/7/8/9. The elongate genital papilla is doubtless the transverse ridge on the male area.

Fang's figure of a male genital area would do quite well for that of one of the Hamburg specimens if the transverse ridge were lobulated instead of smooth and with genital markings.

Chen examined Fang's specimens but failed to correct the errors in Fang's account. According to Chen the hearts of X are lacking; possibly the missing pair was overlooked owing to coverage by connective tissue.

32. *Amynthas cheni* Qiu, Wang & Wang, 1994

Amynthas cheni Qiu, Wang & Wang, 1994. *Sichauan Journal of Zoology* 13(4):143–145.
Amynthas cheni Xu & Xiao, 2011. *Terrestrial Earthworms of China*. 92–93.

External Characters

Body rather small, length 46–85 mm, width 2.0–3.0 mm, segment number 59–107. Prostomium 1/2 epilobous. First dorsal pore in 11/12. Clitellum on segment XIV–XVI, annular, setae absent. Setae enlarged conspicuously before VII, and fine after VII; setal number 24–32 (III), 27–34 (V), 34–36 (VII), 32–42 (XX), 31–44 (XXV); 8–12(VII), 11–18 (VII), between spermathecal pore, 6–9 between male pores. Male pores 1 pair on the ventrolateral side of XVIII, each situated on a tiny protuberance; 2 small papillae intersegmental on 17/18 and 18/19 and slightly to each male pore; about 1/3 of circumferences apart ventrally. Spermathecal pores 2 pairs in 7/8/9, or 3 pairs in 6/7/8/9, about 2/5 of circumferences apart ventrally; there are 2–10 papillae on the ventral sides of VII–X. Female pore single, medioventral in XIV.

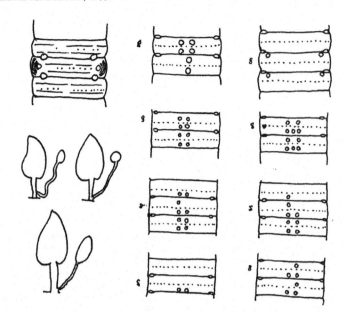

FIG. 37   *Amynthas cheni* (after Qiu et al., 1994).

Internal Characters

Septa 8/9/10 absent, 5/6/7/8 rather thick, other thin. Gizzard barrel-shaped. Intestine swelling in XVI. Caeca paired in XXVII, simple, extending anteriorly to XXIV. Last pair of hearts in XIII. Testis sacs 2 pairs in X and XI, communicated. Seminal vesicles 2 pairs in XI and XII. Prostate glands developed, massed, in XVI–XX. Accessory glands round chunk on 17/18 and 18/19, with short coiled ducts. Spermathecae 2 pairs in VII and IX or 3 pairs in VII, VII and IX; ampulla heart-shaped, about 2.0–2.5 mm long; diverticulum slight shorter than main pouch; seminal chambers round or globe-shaped or long sac-shaped; accessory glands round mass-like with short coiled ducts on the ventral side of VII–X.

Color: Preserved specimens with clitellum red brown.

Type locality: In the mosses of tree trunks or populated in the humus of a tree foot, tree crotch, and in rocks of Mt. Fanjing, Guizhou.

Deposition of types: Institute of Biology, Guizhou Academy of Sciences, Guiyang.

Distribution in China: Guizhou (Mt. Fanjing).

Remarks: This species is similar to *Amynthas jaoi* (Chen, 1946) in terms of appearance and habitat. Some individuals of this species have 3 pairs of spermathecal pores, in 6/7/8/9. However, it differs from *Amynthas jaoi* by the first dorsal pore in 11/12, the round elliptical or long sac-shaped seminal chamber, the rather small seminal vesicles, and the different arrangement of papillae on the body.

33. *Amynthas corrugatus* (Chen, 1931)

Pheretima (Pheretima) corrugata Chen, 1931. *Chin. Zool.* 7(3):131–137.
Amynthas corrugatus Xu & Xiao, 2011. *Terrestrial Earthworms of China.* 95–96.

External Characters

Length 100–140 mm, width 4–6 mm, segment number 88–110; postclitellar segment narrower, about 1/2 narrower than any of the segments just in front of clitellum; secondary annulation very indistinct, three annuli somewhat appearing behind segment VII or IX. Prostomium 1/2 or 1/3 epilobous. First dorsal pore 11/12. Clitellum XIV–XVI, annular, smooth, short, without setae but with indistinct intersegmental furrows, glandular part slightly raised. Its length about equal to the 4 following segments. Setae in continuous rings, longer in anterior few segments, Setae number 30–35 (III), 40–44 (VII), 38–44 (XI), 52–54 (XII), 58–65 (XXV); 18 between spermathecal pores, 12–16 between male pores. Male pores more than 1/3 of their ventral circumference apart, situated in a very shallow copulatory chamber, with single small papilla usually present medial to the pore surrounded by the 5 circular ridges, every adjacent 2 of the latter separated from each with a deep furrow. Spermathecal pores 2 pairs, in intersegmental furrows 7/8/9, about 5/18 of their ventral circumference apart, usually not visible externally, in a small eye-like pit, occasionally a small papilla posterior to each pore. Genital papillae generally absent in this region. Female pore single in XIV, medioventral.

Internal Characters

Septa 8/9/10 absent, but 8/9 traceable at ventral side, 4/5 thin, 5/6/7/8, 10/11/12/13/14 strongly musculated and brightly glittering on surface. Gizzard round, smooth and glittering. Intestine swelling in XVI. Intestine caeca simple, very long and slender in XXVI, extending anteriorly to XXIV or XXII. Hearts 4 pairs in X–XIII. Testis sacs comparatively large, in front of septa 10/11, 11/12, but situated in XI and XII. Anterior pair V-shaped, very narrowly connected medially; posterior pair large, more broadly connected; each sac communicated with its fellow of the opposite side. Sperm duct 2 each side meeting each other immediately behind the second pair of testis sacs. Testis small but seminal funnel very large, 2 pairs of vestigial ovoid bodies paired in 12/13, 13/14. Seminal vesicles small or moderately large; with a large triangular dorsal lobe, grayish in color; irregularly wrinkled on the surface of the main vesicle, or tubercular in appearance. Prostate glands large, in XVI–XX, fan-shaped, finely divided; duct long, curved at both ends, straight and stout at middle. Accessory glands, if present, with a bit of racemose gland part

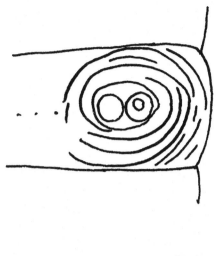

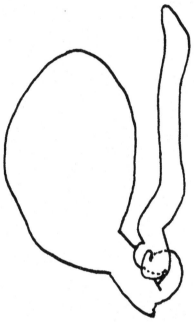

FIG. 38 *Amynthas corrugatus corrugates* (after Chen, 1931).

and long stalk. Spermathecae 2 pairs, in VII and IX; ampulla oval or round, thin-walled, its surface always wrinkled, its duct stout and rather short; diverticulum slender and very long, much longer than main part; coiled or straight; distended either in its middle region or at its free end. Accessory glands sometimes present near the base of ampulla duct.

Color: Preserved specimen light grayish pale, or rather pale on ventral side, becoming slightly greenish or pale yellowish towards the dorsolateral side; light chestnut brown on dorsomedial; a dark brownish streak along the dorsal pores, faint gray at anterior dorsal end. Clitellum fleshy or light cherry red. The whole animal colorless or pale if the dorsomedial streak is overlooked.

Distribution in China: Jiangsu (Yixing), Zhejiang (Xindeng), Jiangxi (Jiujiang), Hubei (Huangzhou), Sichuan (Jiading, Leshan).

Deposition of types: Mus. Biol. Lab. Sci. Soc. China.

Remarks: This species stands near *Amynthas fokiensis* Mich. Recently described, it differs from the former mainly by (1) smaller size, (2) different appearance of male pore region, (3) less constant of genital papillae less constantly present at spermathecal and male pore regions, (4) prostate glands and seminal vesicles more developed and testis sacs more closely placed, and (5) different shape of the ampulla and diverticulum.

In some respects, it is also similar to *Metaphire hesperidum* (Beddard, 1892), notably (1) the exact position and number of the spermathecae, the general shape of the ampulla and diverticulum, (2) the surface view of the gizzard, septa and prostatic duct, and (3) the shape and character of the seminal vesicles and prostatic glands.

34. *Amynthas fokiensis* (Michaelsen, 1931)

Pheretima fokiensis Michaelsen, 1931. *Peking Natural History Bulletin*. 5(2):19–20.
Amynthas fokiensis Xu & Xiao, 2011. *Terrestrial Earthworms of China*. 115–116.

External Characters

Length 125 mm, width 5.0–5.5 mm, segment number 136. First dorsal pore in 12/13. Prostomium epilobous. Clitellum XIV–XVI, annular, without setae. Setae on the anterior part of the body, except for segment II, or II and III, enlarged, especially ventrally. The setae rows are only vaguely and irregularly interrupted, far apart or close, according to the size of the setae. Setae number: 45 (V), 55 (IX), 65 (XIII), 58 (XVII); Male pores lateral to the setae zone of XVIII, on ventral side and not quite 1/2 the body circumference apart, on moderately large wart-like porophores, with 1 or more (more or less clear) circular furrows. On the tip of each male porophore there lies close to each male aperture and obliquely anteromedian to it, a tiny but slightly raised sex papilla. Spermathecal pores 2 pairs, lateral on intersegmental furrows 7/8/9, those of each pair almost 1/2 of the body circumference from the other. Spermathecal sex papillae cannot be recognized, but after examining the internal anatomy, there is directly at the end of the ampulla duct a stalked gland whose numerous small tubules probably open on a small papilla hidden in the crack of the spermathecal aperture. Female pore single in XIV, medioventral.

Internal Characters

Septa 5/6/7 somewhat thickened, 7/8 still a little thickened, 8/9/10 apparently abortive, 10/11/12 moderately strongly thickened. Gizzard large, behind septum

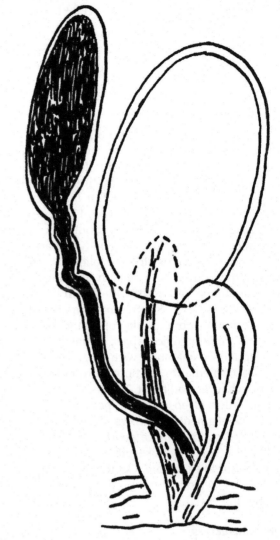

FIG. 39 *Amynthas fokiensis* (after Michaelsen, 1931).

7/8. Intestinal caeca slender club-shaped, extending straight in front, reaching about 4 segments, rather simple, with indentations, or slight outgrowths only the ventral edge. Testes sacs 2 pairs, small, ventral in X and XI. Seminal vesicles extending up one side of the alimentary canal increasing in breadth above. Prostate glands with the glandular part extending through almost 3 segments; the indentation lines arrange themselves almost radially; prostatic duct in the middle part thick, shiny, muscular, thinner at the end; on the whole buckle-shaped, in the middle part toward the posterior; the end part bent in a convex manner toward the anterior. Since the proximal bend lies entirely under the gland part, the duct appears to make a transverse S-shaped curve in situ. The duct opens immediately at the tip of the porophore. There are no copulation pouches formed. Spermathecae 2 pairs in VII and IX, ampulla egg-shaped, smooth, thin-skinned; Ampulla duct almost as long and a maximum of roughly one-third as thick as the ampulla, muscular, distally becoming somewhat thinner and with its end pressed far into the lumen of the ampulla. The distal end of the duct tapers somewhat less than the proximal end. At the distal end, a very short distance above the opening, a blind hose-like diverticulum opens into the ampulla duct. The diverticulum is somewhat longer than main part. The proxminal end of the diverticulum is widened out into a long oval thin-walled sac filled with the sperm ball, occupying about one-third of the entire length of the diverticulum. Distally this seminal chamber somewhat abruptly passes over into the hose-like diverticulum stalk, which describes various (irregular and even) short narrow coils.

Color: Preserved specimens in general gray (faded?), on the anterior part of the body yellowish.
Clitellum brown.
Type locality: Fujian (with no more precise location).
Distribution in China: Fujian.
Etymology: The name of the species refers to the type locality.
Deposition of types: Hamburg Museum, Germany.

35. *Amynthas hexitus* (Chen, 1946)

Pheretima hexita Chen, 1946. *J. West China. Border Res. Soc. (B).* 16:100–101, 139, 148.
Amynthas hexitus Sims & Easton, 1972. *Biol. J. Linn. Soc.* 4: 234.
Amynthas hexitus Xu & Xiao, 2011. *Terrestrial Earthworms of China.* 124.

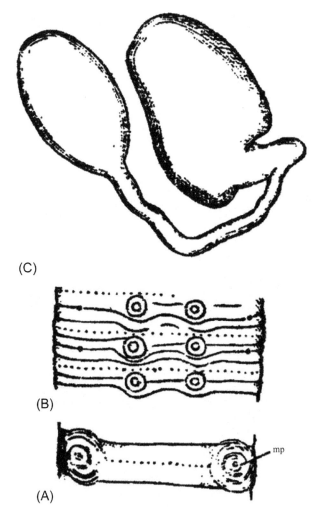

(C)

(B)

(A)

FIG. 40 *Amynthas hexitus* (after Chen, 1946).

External Characters

Medium-sized worm. Length 80 mm, width 4 mm, segment number ?. Segments moderately long, preclitellar portion about 14 mm in length. Prostomium 1/2 epilobous, its width narrower than length of segment II. First dorsal pore 11/12. Clitellum XIV–XVI, annular, smooth. Setae all fine and evenly distributed (ventral ones slightly closer), none enlarged. Setal chain rather closed. Setae number 34 (III), 56–58 (IX), 58–60 (XXV); 22 between spermathecal pores, 14 between male pores, $aa = 1.1$–$1.2ab$, $zz = 2zy$. Male pores paired in XVIII, about 1/3 of circumference ventrally apart, each on a projected porophore, surrounded by 3 or 4 circular ridges. No genital papillae. Spermathecal pores 2 pairs in 7/8/9, about 3/7 of their ventral circumference apart. No genital marking around each pore. Genital papillae round and depressed in center, paired on entromedian side of VII–IX, postsetal, between *c* and *e*. Female pore single in XIV, medioventral.

Internal Characters

Septa 8/9/10 absent, 4/5 thin, 5/6/7/8 slightly musculated, 10/11/12/13 thicker. Nephridial tufts anteriorly on 6/7/8 thick. Gizzard in X. Intestine swelling in XVI. Caeca simple and elongate, extending anteriorly to XX, with shallow ventral indentations. Vascular loops in IX and X stout. Testis sacs in X, XI both as transverse flat bands, widely separate linked with fibrous strands. Seminal vesicles small (also in the cotype), about 1 mm in dorsoventral length including a comparatively large dorsal lobe. Prostate glands well developed in XVI–XX, its duct rather short, V-curved. Spermathecae 2 pairs, in VII and IX, Ampulla large and distended, with very short duct. Diverticulum longer than main pouch, sometimes curved in a zigzag manner, with a date-shaped seminal chamber. Accessory glands usually in a single mass or more and connected with respective papillae externally; surface rough, with short cord-like ductule.

Color: Preserved specimens pale, dark gray along middorsal side. Clitellum cinnamon red.
Distribution in China: Sichuan (Ebian), Guizhou (Fanjingshan).

Etymology: The specific name indicates the 6 genital papillae of VII–IX.

Remarks: This species is distinct from *Amynthas robustus* (Perrier, 1872) by (1) the large egg-shaped diverticulum, (2) the more uniformly distributed setae, and (3) the more constant and regular papillae.

36. *Amynthas kiangensis* (Michaelsen, 1931)

Pheretima kiangensis Michaelsen, 1931. *Peking Natural History Bulletin* 5(2): 21–22.
Amynthas kiangensis Xu & Xiao, 2011. *Terrestrial Earthworms of China.* 132–133.

External Characters

Length 125 mm, width 2.5–3.0 mm, segment number 110. First dorsal pore in 11/12. Prostomium 1/2 epilobous. Dorsal tongue very short and broad, separated behind by a cross furrow. Clitellum XIV–XVI, annular, without setae. Setae in general very delicate, at the anterior from segment III–VII (especially on segments V and VI), ventrally noticeably enlarged, dorsally less noticeably. Setae rows in the region of the enlarged setae ventrally regularly but narrowly interrupted (*aa* about 1.5–2*ab*), otherwise ventrally and dorsally not clearly interrupted. Setae number 29 (V), 41 (VII), 54 (XII), 62 (XXVI). Male pores ventral on XVIII, about 4–5/12 of the body circumference apart ventrally, each on the tip of a smooth porophore that is moderately large, broadly rounded, and ball-shaped. Spermathecal pores 2 pairs, invisible, on intersegmental furrow 7/8/9, those of 1 pair about 1/3 of the body circumference. No other genital marking. Female pore single in XIV, medioventral.

FIG. 41 *Amynthas kiangensis* (after Michaelsen, 1931).

Internal Characters

Septa 8/9/10 absent, 5/6/7/8 and 10/11/12 moderately thickened. Gizzard rather large, between septum 7/8 and 10/11. Intestine swelling in XIV; Intestinal caeca paired stretching from XXVII forward to XXV in both specimens carefully examined; they are rather broad at the base with a pointed blind end; the caeca are simple and show at the most vague septal constrictions. Testis sacs 2 pairs, rather small an irregularly egg-shaped, ventral in X and XI, those of one segment connected to each other by a short narrow cross tube just in front of the septum which bounds them behind. Two pairs of broad sac-like seminal vesicles extend from septum 10/11 and 11/12 laterally from the alimentary canal into XI–XII. They are reticular on the surface and their edge is irregularly notched. Each seminal vesicle is connected with the testis sac lying in front of it by a narrow tube piercing the septum between them. Glandular part of the prostate flat, broader than long, occupying about 4 segments (from XVII–XX), with a few (at times) deep indentations on the dorsal edge, for the rest, with small notches, reticular on the surface. Duct with a middle part which is much swollen, spindle-shaped, slightly bent, shining, and muscular. The thin proximal end is entirely hidden under the gland part, so that the duct appears to leave the glandular part as a tube of considerable thickness. The thin distal end of the duct makes a few short twists partly hidden in the body wall, and opens as a very thin tube directly at the compact and visibly uncontractible porophore. There is no trace of copulatory pouch or penis with penis pouch. Spermathecae 2 pairs in VII and IX, ampullae egg-shaped, mostly with narrow proximal pole. Ampulla duct sharply distinct, cylindrical, about two-thirds as long and one-quarter as thick as the ampulla. Into the distal end of the ampulla duct there opens a blind hose-like diverticulum, which when stretched out is not quite as long as the main pouch (ampulla plus duct), but in position scarcely protrudes beyond the base of the ampulla because of the irregular twisting. The diverticulum consists of a thin hose-like distal stalk about one-quarter of the length of the entire diverticulum, and a very distinct proximal cylindrical seminal chamber, about 3 times as long and more than twice as thick. Its blind proximal end broadly rounded sometimes appearing slightly swollen.

Color: Preserved specimens in general yellowish gray dorsally at anterior end, slightly darker, toward olive-gray-brown in the middle, and less clearly toward the posterior, the darker dorsal coloring is cut by a brown-gray median longitudinal line.
Clitellum brown.
Type locality: Jiangsu (Suzhou).
Distribution in China: Jiangsu (Suzhou), Sichuan (Leshan).
Etymology: The name of the species refers to the type locality.
Deposition of types: Hamburg Museum, Germany.

37. *Amynthas kinfumontis* (Chen, 1946)

Pheretima kinfumontis Chen, 1946. *J. West China. Border Res. Soc. (B).* 16:119–120, 140, 156.

Amynthas omeimontis kinfumontis Sims & Easton, 1972. *Biol. J. Linn. Soc.* 4: 234.

Amynthas domosus Xu & Xiao, 2011. *Terrestrial Earthworms of China.* 133.

External Characters

Length 50–110 mm, width 3.5–5.0 mm, segment number 80–110. Annulation indistinct. Prostomium 2/3 epilobous, tongue wider the length of segment II. First dorsal pore 12/13. Clitellum XIV–XVI, annular, smooth, rather elongate. Setae all fine, none particularly so, slightly closer ventrally; setal breaks slight; setae number 32–34(III), 40–42 (VI), 42–44(IX), 44–48(XXV), 21–23 (VII), 24–25 (IX) between spermathecal pores, 10–12 between male pores; $aa = 1.5ab$, $zz = 1.5$–$2zy$. Male pores paired on ventrolateral corners of XVIII, about 1/2 of ventral circumference apart, no copulatory pouches, each represented by a pore-bearing papilla surrounded by several concentric ridges as in *Amynthas morrisi* (Beddard, 1892), among which 2 small papillae are interposing. Another similar papilla situated postsetally on XVII. Spermathecal pores 2 pairs in 7/8/9, on anterior border of segment VII and IX, posterior pair about 1/2 circumference apart, anterior pair situated more ventrally about 1 or 2 setae space. Genital papillae occurring on IX, ventrally to pore region, either 2 (with 1 presetal and the other postsetal) or as many as 22, or similar ones frequently on VII, either on ventral side or close to each pore. Female pore single in XIV, medioventral.

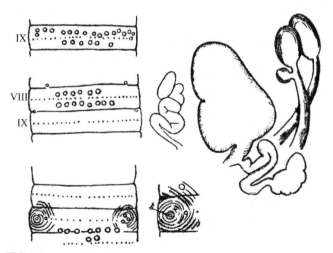

FIG. 42   *Amynthas kinfumontis* (after Chen, 1946).

Internal Characters

Septa thin and membranous, scarcely maculated, 9/10 absent, 8/9 present on ventral side, 10/11 comparatively thin, 11/12/13/14 very thin. Nephridial tufts present anteriorly on 5/6/7 close to gut. Gizzard barrel-shaped, in IX and X, not shining on surface. Intestine swelling in XVI. Caeca simple slender and very long, wrinkled on both dorsal and ventral sides, reaching anteriorly to XXII or a little further. Vascular loops in IX asymmetrically developed, those in X note found. Testis sacs united, anterior pair smaller, V-shaped, median connection narrow; posterior pair U-shaped, median connection wider. Vas deferens of each side large in caliber. Seminal vesicles very large, lying between IX and XIV, dorsal lobes small and indistinct. Prostate glands well developed, in XVI–XXII or XXIII, rather finely lobate; its duct long, ental half comparatively thick, ectal half thin, middle portion thicker. A loose whitish cushion of tissues around ectal end of duct. Accessory glands large and with long stalk, about 3–4 mm long, glandular portion round and smooth. Spermathecae 2 pairs, in VII and IX, posterior pair usually twice as large, ampulla 3.0–4.5 mm in diameter, its duct short. Diverticulum with long duct, ectal half coiled, seminal chamber indistinctly twisted or distinctly twisted in zigzag manner. Demarcation between its duct and chamber not well shown. Accessory glands similar in character to those in prostate region.

> Color: Greyish brown dorsally, pale ventrally. Clitellum chocolate red.
> Distribution in China: Chongqing (Daheba, Mt. Jinfo), Hubei (Lichuan).
> Etymology: The name of the species refers to the type locality.
> Remarks: This species, in its general body form and male pore region, appears very similar to the cosmopolitan species *Amynthas morrisi* (Beddard, 1892), and to a lesser degree like *Amynthas omeimontis* (Chen, 1931), especially in the characters of its papillae and stalked glands.

These specimens were all collected from the top of the mountain and presented several noteworthy features as follows:(1) Setae are less numerous, for example, 22–30 (III), 34–36 (VI), 34–43 (IX), 36–40 (XXV), 16–20 (VII), 17–20 (IX) between spermathecal pores, 8–10 between male pores. (2) The genital papillae are confined to the 9th segment, scarcely on the 8th. (3) Both the spermathecae and the accessory glands are much larger, duct of accessory gland about 3 mm long, 0.8 mm wide, ampulla about 5 mm in diameter.

38. *Amynthas kinmensis* Shen, Chang, Li, Chih & Chen, 2013

Amynthas kinmenensis Shen, Chang, Li, Chih & Chen, 2013. *Zootaxa.* 3599(5):473–475.

External Characters

Medium to large. Length 110–276 mm, width 0.67–5.23 mm, segment number 104–141. Prostomium epilobous. First dorsal pore in 12/13 or 13/14. Clitellum XIV–XVI, annular, setae and dorsal pores absent, length 4.4–8.0 mm and width 3.8–6.1 mm. Setae minute, number 47–69 (VII), 52–81 (XX); 11–12 between male pores. Male pores inconspicuous, paired in XVIII, 0.3–0.4 body circumference ventrally apart, each situated on a small round porophore surrounded by 5–9 small papillae, within a crescent-shaped or kidney-shaped area. Two groups of genital papillae in line with each male pore area: (1) a cluster of 3–8 small papillae immediately adjacent to the margins between XVII and XVIII, (2) a cluster of 9–20 papillae arranged obliquely in presetal XIX; often 3 or 4 skin folds stretching from postsetal XVII to postsetal XIX and enclosing the 2 groups of genital papillae together with each male pore area laterally. Spermathecal pores lateral, 2 pairs in intersegmental furrows of 7/8/9; distance between paired pores 0.35–0.41 body circumference ventrally apart. Genital papillae tiny, round, 9–24 arranged in front of an arc stretching from postsetal VII to presetal VII and from postsetal VII to presetal IX, medial to each spermathecal pore. Female pore single in XIV, medioventral.

Internal Characters

Septa 8/9/10 absent, 5/6/7/8 thick, 10/11–13/14 muscular. Nephridial tufts thick on anterior faces of 5/6/7. Gizzard large in VII–X. Intestine swelling in XV or XVI. Intestinal caeca paired in XXVII, extending anteriorly to XXIII–XXV, each simple, distal end straight or bent. Esophageal heart in X–XIII. Holandric, Testis sacs small, 2 pairs in ventrally joined X and XI. Seminal vesicles large, surface smooth, 2 in XI and XII, anterior pair rectangular-shaped, extending anteriorly to the posterior half or the entire compartment of segment X, with a round dorsal lobe at the middle of the dorsal end; posterior pair elongated oval, with a round, finely granulated dorsal lobe at the dorsoanterior end. Prostate glands large in XVI–XXI, wrinkled and lobed. Prostatic duct C-shaped or U-shaped in XVIII. Accessory glands short-stalked, mushroom-like or irregular-shaped, corresponding to external genital papillae. Spermathecae 2 pairs, in VII and XIX, or in VII and IX, each with a peach-shaped or elongated oval-shaped ampulla 1.8–3.3 mm long and 2.1–2.7 mm wide, and a short, stout spermathecal stalk 0.5–0.8 mm in length. Diverticulum 2.8–4.8 mm in total length, coiled with a short, slender and straight proximal part. Accessory glands long-stalked or short-stalked, mushroom-like, 0.6–1.3 mm in length, each corresponding to external genital papilla.

Color: Preserved specimens dark brown or dark reddish on dorsum, brown to light brown on ventrum, and brown to dark brown on clitellum. Live worm dark orange or dark brown.

Distribution in China: Fujian (Jinmen, Quanzhou).

Etymology: The name of the species refers to the type locality.

Deposition of types: The Invertebrate Zoology and Cell Biology Lab, Department of Life Sciences, National Taiwan, Taibei, Taiwan.

Remarks: *Amynthas kinmensis* is the most abundant earthworm and widely distributed on the main island of Jinmen. It is fairly similar to *Amynthas polyglandularis* (Tsai, 1964) from northern Taiwan in terms of both external and internal morphological characters. Both species are quadrithecal with spermathecal pores in 7/8/9 and have numerous small genital papillae. Molecular studies also suggest that the 2 species are closely related (unpublished data). However, *A. kinmensis* has 9–24 papillae arranged in front of an arc stretching across 2 annulets and medial to each spermathecal pore, whereas *A. polyglandularis* has 1–5 papillae in a transverse row on both the anterior and posterior margins of each spermathecal pore (Tsai1964). Two groups of genital papillae, a cluster of 3–8 immediately adjacent to the margins between XVII and XVIII, and a cluster of 9–20 arranged obliquely in presetal XIX, in line with each male pore area are present in *A. kinmensis*, while these

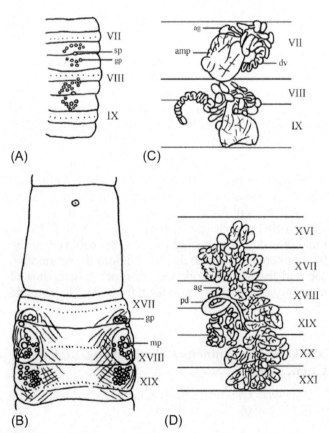

FIG. 43 *Amynthas kinmensis* (after Shen et al., 2013).

papillae arrangements are absent in *A. polyglandularis*. In addition, *A. polyglandularis* has groups of papillae on midventrum of segments VII, IX, and XVIII, while these papillae arrangements are absent in *A. kinmensis*. Papillae are more concentrated midventrally in *A. polyglandularis*, whereas the papillae of *A. kinmensis* are laterally distributed. It is possible that the 2 species derive from a common ancestor and have gone through allopatric speciation.

Amynthas omeimontis (Chen, 1931) from Sichuan also has spermathecal pores in 7/8/9 and numerous small genital papillae. However, it has manicate intestinal caeca, genital papillae absent around spermathecal region and presetal XIX, but about 40–50 small papillae closely arranged in rows in presetal XI (Chen, 1931). Also, it has a lower setal number with 38–46 in VI–VII and 40–44 in XX (Chen, 1931).

39. *Amynthas corrugata kulingianus* (Chen, 1933)

Pheretima corrugata kulingiana Chen, 1933. *Contr. Biol. Lab. Sci. Soc. China (Zool.).* 9:278–281.
Amynthas kulingiana Xu & Xiao, 2011. *Terrestrial Earthworms of China.* 96–97.

External Characters

Length 102–150 mm, width 4.0–4.5 mm, segment number 70–112. Prostomium 1/2 epilobous. First dorsal pore 10/11, its opening somewhat smaller but distinct. Clitellum in XIV–XVI, annular, smooth, without setae and other marking. Setae small and delicate, uniformly formed throughout whole body; ventral setae more distinct and slightly longer due to thinner epidermis. Both dorsal and ventral breaks slight, $aa = 1.2–1.5ab$; $zz = 1.5–3.0$ yz. Setae number 30–38 (III), 36–45 (VI), 40–54 (VII), 44–56 (XII), 48–64 (XXV); 21 between spermathecal pores, 12–14 between male pores. Male pores situated on ventrolateral sides of XVIII, a little more than 1/3 (=5/14) of body circumference apart, each pore on a very minute pore-papilla, surrounded by 5–9 concentric circular ridges, every adjacent 2 ridges separated by shallow groove, without papillae medial to pore-papilla. This portion specially rose in an egg-shape, its glandular skin extending anteriorly and posteriorly to segments XVII and XIX, medially to seta *e*, a part medial and lateral to concentric ridges smooth and pale in color, without any markings, about 3 setae present on median glandular part. Median bases of swellings extending almost to entire ventral side of the segment, leaving about 1/3 of ventral side nonglandular. Each swelling also extending to lateral side of body. Spermathecal pore 2 pairs, in intersegmental furrow 7/8/9, in line to male pore, marked by a small elliptical pad, not in eye-like depression, each pore usually not distinctly demonstrated. No genital

markings around each pore; 2 pairs of fairly papillae regularly paired on posterior side of segments VII and VII, a little nearer to furrow than to setal zone, behind seta *c*, about 1 mm apart. Each papilla round and flattened, but not particularly raised, dark grayish on top. Such papillae constantly present and even recognizable in 3 immature specimens. Female pore single, medioventral in XIV.

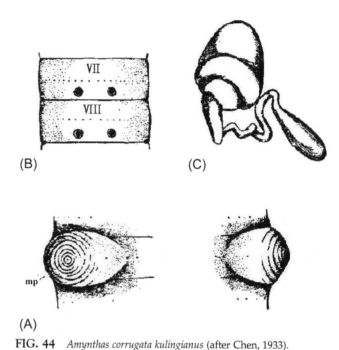

(B) (C)

mp

(A)

FIG. 44 *Amynthas corrugata kulingianus* (after Chen, 1933).

Internal Characters

Septa 9/10 absent, 8/9 present ventrally, 5/6/7, 10/11/12/13 or 13/14 moderately thickened but not so musculated nor tenacious as in *Amynthas corrugatus*, slightly shining (not glittering) on surface, 7/8, 13/14 less thickened, thin from 16/17. Gizzard barrel-shaped, its surface pale but not shining; anterior crop usually small. Intestine swelling in XVI. Intestinal caeca simple, in XXVI or XXVII, extending anteriorly to XXIII, with ventral edge tooth-shaped, same color with intestine. Hearts 4 pairs, dorsal vessel between gizzard and prostate enlarged. Testis sacs and seminal vesicle generally similar to *Amynthas corrugatus*. Testis sacs as large as main vesicle, well distended; anterior pair more widely placed with thin but broader membranous connection, posterior pair larger but more closely placed. Testis small and seminal funnel large. Sperm ducts of each side meeting behind septum 12/13. Seminal vesicles also with a large triangular dorsal lobe. Prostates very large, in XVI–XXII, rather grossly lobate, smooth on surface, rectangular in outline, each with a very long duct, deeply curved proximally, loosely coiled distally, its middle part slightly crooked, about 5 mm in total length, pale but not glittering on surface.

No accessory glands found in this region. Spermathecae 2 pairs, in VII and VII. Ampulla always rounded, about 2mm in diameter, feebly wrinkled or smooth, with a short duct, about 0.5mm in length. Diverticulum longer than main pouch, with a slender but very long duct, nearly as long as main pouch and an elongate seminal chamber, whitish and smooth, about 1mm in length; its enlargement of chamber gradually increasing from duct and pointed somewhat entally. Diverticular duct much more slender than ampullar duct, nearly as long as main pouch, but if coiled then shorter than main pouch.

Color: Preserved specimens brownish dorsally, darker anteriorly, light buff on posterior half of body, slaty posterior to prostate region, slightly deep along dorsomedian line, greenish pale ventrally. Clitellum cinnamon yellow.
Type locality: Jiangxi (Jiujiang).
Deposition of types: Mus. Zool. Cent. Univ. Nanjing.
Distribution in China: Jiangxi (Jiujiang).
Remarks: In its general external appearance the present form looks more like *Amynthas heterochaetus*, with which it was found, rather than the form *Amynthas corrugatus*. It is a slender, elongate and very active creature. In structure it is very similar to the form *Amynthas corrugatus* but differs in (1) possessing an egg-like swelling around the male pore region which is never retracted as a slit and has no papilla medial to the pore, (2) genital papillae regularly paired behind the setae on segments VII and VII. Such regular arrangement of papillae has never been so far found in *Amynthas corrugatus*. The minor differences are the character of gizzard and septa, among others.

40. *Amynthas lautus* (Ude, 1905)

Pheretima lauta Chen, 1933. *Contr. Biol. Lab. Sci. Soc. China (Zool.).*9:282–288.
Pheretima lauta Tsai 1964. *Quer. J. Taiwan Mus.* 17:25–26.
Amynthas lautus Sims & Easton, 1972. *Biol. J. Soc.* 4:234.
Amynthas lautus Xu & Xiao, 2011. *Terrestrial Earthworms of China.* 135–136.

External Characters

Length 120–196mm, width 4.0–8.5mm, segment number 100–130. Preclitellar segments moderately long, X longest, postclitellar segments suddenly shortened, about 10 or more segments posterior to XVIII particularly shortened, usually about 2.5–4 segments equal to longest preclitellar segment. Prostomium 1/2 or 1/3 epilobous. First dorsal pore 11/12. Clitellum in XIV–XVI, annular, in 3 full segments, broader than long, without setae and other markings. Setae rather conspicuous or sometimes small, longer, and more widely spaced ventrally, those on III–IX much longer and more prominent, setal zones usually elevated, ventral setae posterior to

clitellum also very distinct and on posterior part of body also prominent on account of thinner epidermis. Both dorsal and ventral breaks slight, $aa = 1.0$–$1.2ab$, $zz = 1.5$ or $2\, yz$, narrower on preclitellar segments; setae number 22–40 (III), 34–52 (VI), 38–55 (VII), 40–62 (XII), 50–70 (XXV); 12–23 between spermathecal pores, 12–22 between male pores. Male pores situated on ventrolateral side of XVIII, usually more than $1/3$ ($=7/18$) of circumference apart, each pore opening to exterior through a flat-topped and medium-sized papilla, its pore usually not marked, surrounded concentrically by several circular ridges extending to intersegmental furrows 17/18 and 18/19; the whole porophore roundish in outline, often raised as a large-based elevation and male papilla sometimes a little retracted as a shallow depression or protruded as a round teat or this portion not raised nor male papilla retracted, 1 small round papilla placed medially within circular ridges posterior to setal zone, or another in front, with a similar 1–4 sometimes found irregularly in front of setae on ventral side of XVIII, in some specimens 1 or 2 raised portions bearing 1–2 papillae surrounded by circular ridges, similar to male pore region placed medioventrally just in front of setae of XVIII but in this case no other scattered ones around them. Each papilla generally small, round and well distended, grayish in color, sometimes depressed. Spermathecal pores 2 pairs, in 7/8/9, about 4/9 of circumference apart, usually visible externally, single

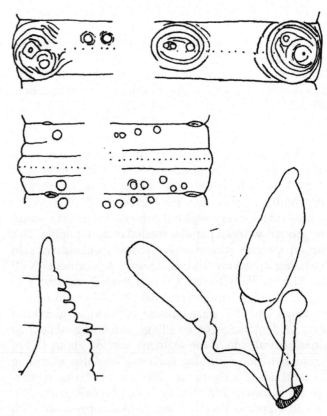

FIG. 45 *Amynthas lautus* (after Chen, 1933).

small flattened papilla found at margin of following segment and just immediately posterior to each pore which is sometimes marked as a whitish spot in the intersegmental furrow, this papilla absent in most cases; frequently several or many ampulla-like papillae scattered irregularly on ventral side of segments VII, VII, and IX, but not found before setae of VII and behind setae of IX, also infrequently in front of setae of X, about 2–9 found on VII and IX behind, and about 4–6 before setae respectively, fewer on VII (about 2); their shape similar to those present on XVIII, but no circular elevation bearing papillae found in this region. Female pore single, medioventral in XIV.

Internal Characters

Septa 9/10 absent, 8/9 traceable ventrally, 5/6/7 and 10/11/12/13/14 almost equally thickened, 7/8, 14/15 less so; these septa (except 7/8) with rich and well-developed musculature, convexly directed posteriorly, always with silky shine on surface specially on posterior faces of 10/11/12/13/14/15; those behind 15/16 very thin and nonmaculated. Gizzard always large and rounded, in IX, X, largely in X, anteriorly with an elongate large crop.

Intestine suddenly swelling in XVI or in a part of XV. Intestinal caeca simple, in XXVII, extending anteriorly to XXIII. Hearts 4 pairs, in X–XIII. Testis sacs usually small, widely placed but connected by a small bridge closely under ventral vessel; or moderately large, with a narrow and shorter connection, anterior pair V-shaped, partly in X, posterior pair in XI more broadly connected. Seminal vesicles usually small, rarely filling up each segment, not meeting dorsally, usually as narrow elongated and flat lobes lying upon posterior faces of septa 10/11, 11/12, smooth or irregularly grooved, each with a large dorsal lobe, round or elongate. Prostate glands not well developed, in XVI–XIX or XX, a little more than 3 segments, much divided yet smooth on individual pieces, its duct smooth and shiny on surface, slender and rather long, coiled proximally and loosely coiled distally. Accessory glands present corresponding to papillae outside, each gland part rather larger, with a long stalk, about 2 mm in length, free in body cavity. Spermathecae 2 pairs in VII and IX, or anterior pair in VII; ampulla fairly large, usually heart-shaped, rather elongate and pointed entally, smooth or weakly grooved but not wrinkled, usually well distended; ampullar duct stout and rather long but shorter than ampulla and sharply

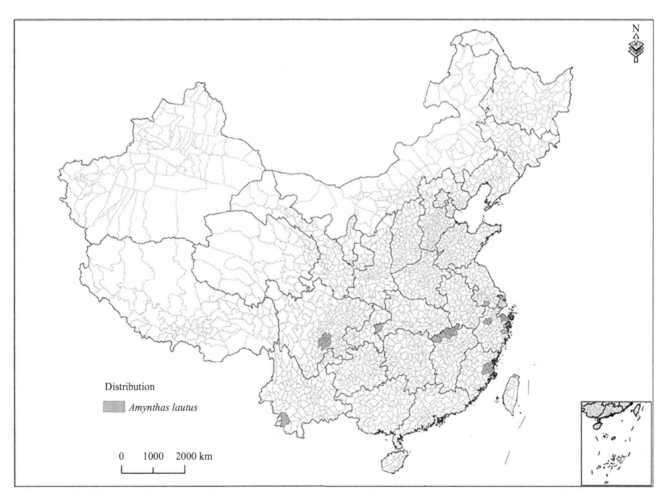

FIG. 46   Distribution in China of *Amynthas lautus*.

constricted from the latter. Diverticulum elongate with a rod-like or ovoid seminal chamber, slightly shorter the ampulla, its duct very slender and distinctly constricted from the chamber or sometimes the ental end of duct inserting on lower lateral side of seminal chamber. The whole usually longer than main pouch. Accessory glands similar to those in prostate region.

Color: Live specimens deep fleshy or chestnut red on anterior dorsum, greenish fleshy on posterior dorsum, with setal zones lighter. Preserved specimens from chocolate or chestnut to very faint grayish dorsally, pale ventrally. Clitellum liver brown.
Type locality: Fujian (Fuzhou).
Deposition of types: Mus. Zool. Cent. Univ. Nanking.
Distribution in China: Fujian (Fuzhou), Zhejiang (Ningbo, Zhoushan, Linhai, Xindeng, Tonglu, Fenshui), Jiangsu (Yixing), Jiangxi (Jiujiang), Taiwan (Taibei), Hubei (Lichuan), Sichuan (Leshan), Yunnan (Lancang).

Remarks: This specimen is similar to *Amynthas siemssensi* Michaelsen and *Amynthas forkiensis* Michaelsen. However, in both Michaelsen's species, the first dorsal is in 12/13 and their setae are slightly more numerous than mine. The points given here by Ude fit exactly into my description and therefore enable me to make a final determination.

41. *Amynthas longisiphonus* (Qiu, 1988)

Pheretima longisiphona Qiu, 1988b. *Sichauan Journal of Zoology* 7(1):2–3.
Amynthas longisiphonus Xu & Xiao, 2011. *Terrestrial Earthworms of China.* 140–141.

External Characters

Length 54–107 mm, width 2–3 mm, segment number 70–126. Prostomium 1/2 epilobous. First dorsal pore 12/13. Clitellum in XIV–XVI, annular, setae absent. Setae fine, setae number 28–37 (III), 34–35 (V), 53–57 (VII), 40–48 (XX), 43–51 (XXV); 26–29 (VII) between spermathecal pores, 9–11 (XVIII) between male pores. Male pores 1 pair, on ventrolateral side, about 1/3 of circumference ventrally apart; each on top of a raised cone, surrounded by concentric ridges, and with 5–8 pairs of papillae on the ventral side of XVIII–XXII or XXV. Spermathecal pores 2 pairs, in 7/8/9 intersegmental, about 1/2 circumference ventrally apart; having 6–12 papillae on ventral side of V–VII. Female pore single in XIV, medioventral.

Internal Characters

Septa 8/9/10 absent, other transparent membranous. Gizzard small, side flat. Intestine enlarged from XVI; caeca simple, digital, in XXVII, extending anteriorly to XXIV, smooth on both edges. Last pair of hearts in XIII.

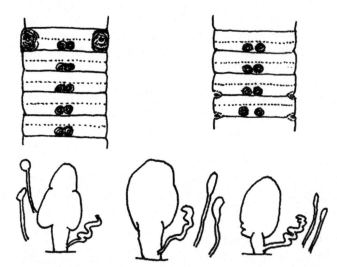

FIG. 47 *Amynthas longisiphonus* (after Qiu, 1988b).

Testis sacs communicated in X and XI, oblong, front pair V-shaped, back pair Λ-shaped, both sides communicated. Seminal vesicles very large, well developed, the second pair extending back to XIV, dorsal lobe distinct, smaller, triangular or oval-shaped, meeting dorsally. Prostate glands well developed, in XVI or XVII–XXI, or XXV. Prostatic duct slightly short, slightly S-shaped, there are 10–18 long-stalked accessory glands in this region. Spermathecae 2 pairs, ampulla long and round in shape; ampulla duct short and stout; diverticulum rather shorter than main part, slightly Z-shaped and twisted in the middle part with its seminal chamber slightly enlarged. Accessory glands developed, rod-like, about 6–12, located in the inner side of the seminal vesicle, with a long duct.

Color: Preserved specimens light in color, before XXV slightly pale pink, after XXV gray. Clitellum red brown.
Type locality: Guizhou (Mt. Taiyang, Shuiyang).
Distribution in China: Guizhou (Mt. Taiyang, Shuiyang).
Etymology: The specific name indicates the long-stalked accessory glands.
Deposition of types: Institute of Biology, Guizhou Academy of Science. Guiyang.
Remarks: This species is similar to *Amynthas robustus*. However, this species with developed prostate glands; with long tubular accessory gland, ental of spermathecal diverticulum is not bar-shaped enlargement and papillate present in XVIII–XXII or XXV.

42. *Amynthas masatakae* (Beddard, 1892)

Amynthas masatakae (Beddard), *Zool. Jb. Syst.* (1892) 6:755–766.
Pheretima masatakae Kobayashi, 1936c. *Sci. Rep. Tohoku Univ.* (4)11:337–340.

Amynthas masatakae Chuang & Chen, 2002. *Acta. Zool. Taiwan.* 13(2):73–79.
Amynthas masatakae Xu & Xiao, 2011. *Terrestrial Earthworms of China.* 145–146.

External Characters

Length 105–138 mm, width 4.0–7.5 mm, segment number 96–138. Prostomium epilobous. First dorsal pore in 11/12, distinct. Clitellum XIV–XVI, annular, setae absent. Setae beginning on II and rather large; both middorsal and midventral breaks slight, $aa = 1.2$–$1.9ab$, $zz = 1.4$–$2.2zy$; setae 34–41 in VII, 41–49 in XX, 13–15 between male pores. Male pores paired in XVIII, ventrolateral, about 1/3 of the circumference apart; each male pore with 2 papillae forming a flat triangle: male pore lateromedial, 1 pair of papillae ventroanterior to setal ring, the other pair of papillae ventroposterior to setal ring, and all of them surrounded by 2 or 3 circular folds. Spermathecal pores 2 pairs in 7/8/9, about 2/5 of the circumference ventrally apart. Just medially to each pore are 2 small genital papillae found facing each other on both borders of the intersegmental furrow. Each papilla much smaller than those of the male segment, conical in shape, with a minute opening on its tip; similar ones were also found on 9/10 in line with the former (right side only). Female pore single in XIV, medioventral.

Internal Characters

Septa 9/10 absent, 8/9 ventrally traceable, 5/6/7/8 and 10/11/12/13 much or very thickened and musculated, 13/14 slightly thickened. Gizzard in IX–X, peach-shaped; yellowish white. Intestine enlarged from XV. Caeca, simple, in XXVI, extending as far anteriorly as XXII, each with several serriformed outgrowths on ventral margin only. Hearts 4 pairs in X–XIII, moderate in caliber; dorsal vessel rather large. Testis sacs 2 pairs, in X and XI, rather small; anterior pair oval shaped, posterior pair larger than the anterior, fused with its fellow forming a transverse sac. Both sacs are separated by thick septum 10/11. Seminal vesicles, small, 2 pairs in XI and XII, somewhat darkened and weakly vesicular on surface. Testes and funnels usual in position and structure. Sperm ducts on each side meeting in XIII. Prostate glands absent or very small, only appear prostatic duct in XVIII, thin but musculated, shiny on surface, nearly straight or simply twisted, with nearly equal thickness throughout. Close and medially to the ectal end of the duct 2 large accessory glands with long stalk found corresponding to the external papillae. Spermathecae 2 pairs, in VII and IX; ampulla round or ovoid, about 1 mm in diameter; duct short but thick, distinctly marked off from the former; diverticulum slightly longer than the main portion, its ental half distended as a cylindrical seminal chamber, the ectal half slender. Close to the ectal end of each spermathecal duct present are 2 relatively large accessory glands with a moderately long stalk; saccular portions of the glands on each side appear to be arranged at almost equal distance from each other.

Color: Preserved specimens, dorsally dark brown, ventrally yellowish brown. Clitellum light russet.
Type locality: Myanmar (Teung Cong).
Distribution in China: Taiwan.
Deposition of types: Indian Museum.
Remarks: In most species of earthworm, the prostate gland is large, however, it is absent from *Amynthas masatakae*. This is a key character for identifying *Amynthas masatakae*. The male pore is another key character for recognizing this species: each male pore has along with 2 papillae which form a flat triangle. Exclusive of those 2 papillae, no other papilla exists around the male pore. This morphology has not been described for other earthworm species.

43. *Amynthas meioglandularis* Qiu, Wang & Wang, 1993

Amynthas meioglandularis Qiu, Wang & Wang, 1993. *Act Zootaxonomica Sinica.* 18(4):406–407.
Amynthas meioglandularis Xu & Xiao, 2011. *Terrestrial Earthworms of China.* 148–149.

External Characters

Length 28–51 mm, width 1.2–2.0 mm, segment number 66–97. Prostomium 1/3 epilobous. First dorsal pore 12/13. Clitellum XIV–XVI, annular, with 5–11 setae on

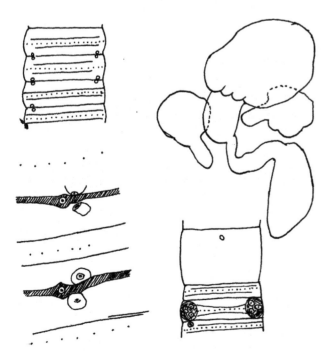

FIG. 48   *Amynthas masatakae* (after Kobayashi, 1936c).

ventral side of XVI. Setae fine; setae number 30–34 (III), 37–42 (V), 46–52 (VII), 45–52 (XX), 43–50 (XXV); 14–17 (VII) between spermathecal pores, 15–18 (XVII), 14–18 (XIX) between male pores, 5–8 (XVIII) between apertura of male pore. Male pores 1 pair, in ventrolateral side of XVIII, each situated on a large round and flat-topped papilla, which extends anteriorly to 1/2 XVII and posteriorly to 1/2 XIX; about 1/3 circumferences apart ventrally. Spermathecal pores 2 pairs in 7/8/9, about 1/3 circumferences apart ventrally; No other genital marking in this region. Female pore single, medioventral in XIV.

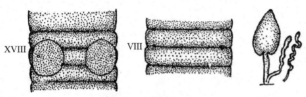

FIG. 49   *Amynthas meioglandularis* (after Qiu et al., 1993b).

Internal Characters

Septa 8/9/10 absent, others thin. Gizzard developed, barrel-shaped, in IX–X. Intestine from XVII. Caeca simple, smooth on surface, in XXVII, extending anteriorly to XXIII. Last heart in XIII. Testis sacs 2 pairs in X and XI, anterior pair oval-shaped, posterior pair round-shaped. Seminal vesicles 2 pairs in XI and XII, developed. Dorsal lobe small, long round-shaped. Prostate glands large, in XVI–XX, massive, divided into strip leaflets. Prostatic duct small, U-shaped twist. Accessory glands mass-like, located on the inside of the prostate tube. Spermathecae 2 pairs in VII–IX, ampulla heart-shaped about 2 mm long, ampulla duct stout and long; diverticulum shorter than the main part. No accessory gland in this region.

Color: Preserved specimens slightly red-brown on preclitellar, white-gray on postclitellar. Clitellum red-brown.
Distribution in China: Guizhou (Mt. Leigong), Sichuan (Leshan).
Type locality: Guizhou (Mt. Leigong).
Deposition of types: Institute of Biology, Guizhou Academy of Science. Guiyang.
Remarks: This species is similar to *Amynthas mammoporphoratus* Thai, 1982, *Amynthas dignus* (Chen, 1946), and *Amynthas homosetus* (Chen, 1938), but differs from them both by having only 2 spermathecal pores in 7/8 and 8/9, the males pores situated on the lateral part of the large round flat-topped papillae, and the diverticulum twisted in a wave-like fashion.

44. *Amynthas omeimontis* (Chen, 1931)

Pheretima (Pheretima) paraglandularis omeimontis Chen, 1931. *Chin. Zool.* 7(3):155–160.

Pheretima omeimontis Gates, 1935a. *Smithsonian Mis. Coll.* 93(3):12.
Pheretima omeimontis Gates, 1939. *Proc. U.S. Natr. Mus.* 85:455–456.
Pheretima omeimontis Chen, 1946. *J. West China. Border Res. Soc. (B).* 16:136.
Amynthas omeimontis Xu & Xiao, 2011. *Terrestrial Earthworms of China.* 159–160.

External Characters

Length 120–150 mm, width 3–5 mm, segment number ?–120. Prostomium 2/3 epilobous. First dorsal pore in 12/13. Clitellum elongate, in XIV–XVI, annular, longer than 3 longest preclitellar segments, without setae and intersegmental furrow; dorsal pore absent. Setae short and small, slightly longer in preclitellar segments. $aa = 1.5ab$; $zz = 2$–3 yz; setae number 32–34 (III), 38–42 (VI), 40–46 (VII), 46–50 (XII), 40–44 (XXV); 9–22 between spermathecal pores, 10–12 between male pores. Male pores paired situated very ventrolaterally in XVIII, a litter more than 1/3 of circumference ventrally apart, no copulatory chamber but a shallow depression often present, with 3–7 small papillae surrounding the depression, another group of 2 to 5 or 7 similar ones usually present on anteromedial side of the depression in front of setal circle. Spermathecal pores 2 pairs, in 7/8/9, about 5/12 of circumference ventrally apart, very small elliptical depression or eye-like. Genital papillae absent around spermathecal region, if present, ventral to the pores. Small papillae in a rectangular slightly raised area, each papilla more or less rounded or wart-like, about 40–50 closely arranged in rows posteriorly in front of, or covering the setal zone of the ventral side of XI and anteriorly to intersegmental furrow 10/11, about 4 mm in width. The rectangular patch of skin somewhat raised, whitish and glandular in character. Female pore single in XIV, medioventral.

Internal Characters

Septa 8/9/10 absent, but 8/9 traceable ventrally, 5/6/7, 10/11/12/13 thick but not very musculated, 7/8, 13/14 less thickened. Gizzard globular, narrower towards anterior end, its surface smooth, whitish and somewhat glittering, without a dilated crop. Intestine enlarged from middle part of XV; the part of the esophagus between gizzard and intestine brownish and glandular on surface, particularly glandular just behind gizzard. Intestine caeca lobulated, in XXVII, extending anteriorly to XXV, with 5 or more ventral diverticula, finger-like, the dorsal one longest, gradually decreasing in length toward the ventral side, rarely with ventral tooth-like indentations. Hearts 4 pairs, first pair close to septum 10/11, last pair in XIII. Testis sacs moderate or rather large in size, anterior pair in front of septum 10/11 partly extending to the posterior part of X. Seminal vesicles 2 pairs, fully occupied in XI and XII,

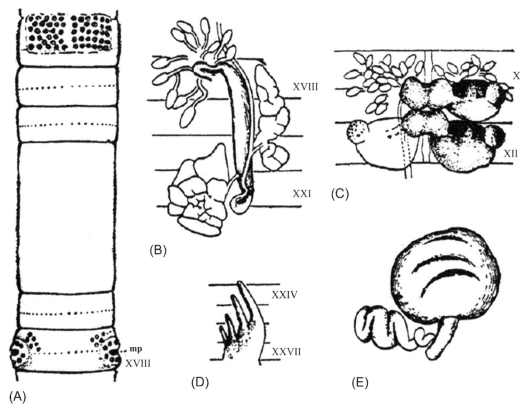

FIG. 50 *Amynthas omeimontis* (after Chen, 1931).

posterior pair extended to XIII, surface smooth and whitish, sometimes tuberculated and grayish in color, each with a large distinct lobe on its dorsal or anterodorsal side which is marked off from vesicle with a deep groove. Prostate glands with only a few large lobes, smooth on surface, usually directed posteriorly in XVIII–XXII, with a very long and stout duct which becomes slender and coils at both ends and whose middle part is about 5 mm long, straight, and with its surface smooth and whitish. Five or more glands present near or around distal end of duct corresponding to the papillae referred above. Each gland sac-like, whitish and smooth, round or oval in shape, about 0.7–1.0 mm in diameter with a rather long duct about 1.0 mm or longer. Spermathecae 2 pairs, in VII and IX, ampulla round, sac-like, about 2 or 3 mm in diameter, whitish smooth and often not well distended, alveolar appearance always lacking, the ampulla duct moderately stout, about half the length of ampulla, or much shorter. Diverticulum with ental one-third or two-thirds distended as seminal chamber, sometimes closely looped in a zigzag manner, its ectal end of the duct spirally coiled, half as long as the main part if it is closely coiled, as long as, or much longer than the latter, if loosely coiled, but in most cases appearing a little shorter.

Color: Preserved specimens chestnut brown on postclitellar and dark purplish brown on preclitellar dorsal side, light rufous towards the dorsolateral sides, whitish around each setal zone discontinuous at dorsomedian line where the dorsal break of the setal circle is marked; light brownish on preclitellar ventrum where the white rings are visible and not interrupted at the ventromedian line; postclitellar ventrum pale. Clitellum pinkish white or milky.

Distribution in China: Sichuan (Mt. Emei), Guizhou (Mt. Fanjing).

Etymology: The name of the species refers to the type locality.

Type locality: Sichuan (Mt. Emei).

Deposition of types: Museum of Science Society of China.

Remarks: This species is similar to *Begemius paraglandularis* Fang, but *Amynthas omeimontis* is small at about ?–150 mm in length, 3.5–5 mm in diameter when fully mature, while one specimen of B. paraglandularis is about equal in size to largest specimen (= L. 120 mm, D. 5 mm) of A. omeimontis is found perfectly immature sexual organs such as spermathecae, seminal vesicles and prostate glands are very rudimentary, being hardly visible. The position of the stalked glands and the additional pair of hearts is also different from that of *B. paraglandularis*.

45. *Amynthas polyglandularis* (Tsai, 1964)

Amynthas polyglandularis Tsai, 1964. *Quarterly Journal of the Taiwan Museum*. 17(1&2),30–34.
Amynthas polyglandularis Chang, Shen & Chen, 2009. *Earthworm Fauna of Taiwan*. 72–73.
Amynthas polyglandularis Xu & Xiao, 2011. *Terrestrial Earthworms of China*. 168–169.

External Characters

Length 134 mm, clitellum width 5.5 mm, segment number 88. First dorsal pore 12/13. Clitellum XIV–XVI, annular, setae absent, dorsal pore absent. Setae 64–65 (VII), 67–70 (XXV), 20–21 between male pores. Male pores 1 pair in 18, ventrolateral, round or transverse slit-like, situated on a flat-topped, heart-shaped tubercle surrounded by 2–9 small, round genital papillae with irregular arrangement, 2 oval, large, flat-topped genital pads situated medial to the pores and the round genital papillae, the above structures surrounded by 1–3 circular folds, a large number of small, round papillae present between the paired male pores, numbering 19–31 and 15–22 presetal and postsetal, respectively, forming a rectangular or irregular plate ventrally in XVIII. Spermathecal pores 2 pairs in 7/8/9, lateral, with 1–5 small, round genital papillae in a transverse row on both the anterior and posterior edges of the pores, a group of small, round genital papillae present in VII and IX, medioventral, numbering 10–23. Female pore single in XIV, medioventral.

Internal Characters

Septa 8/9/10 absent, 6/7/8 thickened, 10/11 thin. Gizzard in IX and X. Intestine from XV. Intestinal caeca simple in XXVII, extending anteriorly to XXIII. Lateral hearts in X–XIII. Testis sacs 2 pairs combined together within a single sac in XI. Seminal vesicles paired in XI and XII. Prostate glands paired in XVIII, large, extending anteriorly to XVI and posteriorly to XXI, subdivided into several lobes, prostatic ducts looped, with several accessory glands around its base, each corresponding to an external papilla in the male pore region. A group of accessory glands present ventrally between the paired prostate glands, with size and shape similar to those around spermathecae. Spermathecae 2 pairs in VII and IX, ampulla heart-shaped or peach-shaped, with short and stout stalk, diverticulum long and large, with horn-shaped seminal chamber and coiled slender stalk. Accessory glands present, 2–5 and numerous near the base of spermathecae and between the paired spermathecae, respectively, each corresponding to external papillae in preclitellar regions, each accessory gland round or heart-shaped, with a short, slender stalk. Ovaries paired in XIII.

Color: Preserved specimens purplish brown on dorsum and light gray on ventrum and clitellum.
Type locality: A small hill near the main campus of the National Taiwan University, Taibei City, Taiwan.
Deposition of types: Types are missing.
Distribution in China: Northern Taiwan.
Etymology: The specific name indicates its numerous accessory glands in the preclitellar and postclitellar regions.

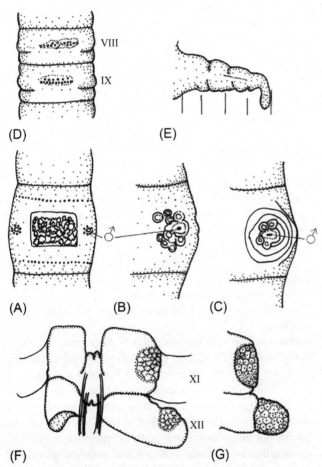

FIG. 51 *Amynthas polyglandularis* (after Tsai, 1964).

46. *Amynthas robustus* (Perrier, 1872)

Perichaeta robusta Perrier, 1872. *Nouvelles Archives Du Museum*. 112–118.
Pheretima robusta Gates, 1935b. *Lingnan J. Sci.* 14(3):453–454.
Pheretima robusta Chen, 1936. *Contr. Biol. Lab. Sci. Soc. China (Zool.).* 7(3):271.
Pheretima robusta Gates, 1939. *Proc. U.S. Natr. Mus.* 85:473–482.

Pheretima robusta Chen, 1946. *J. West China. Border Res. Soc. (B).* 16:136.

Pheretima robusta Tsai, 1964. *Quer. J. Taiwan Mus.* 17:26–29.

Pheretima robusta Gates, 1972. *Trans. Amer. Philos. Soc.* 62(7):216–218.

Pheretima robustus Sims & Easton, 1972. *Biol. J. Soc.* 4: 234.

Amynthas robusta Chang, Shen & Chen, 2009. *Earthworm Fauna of Taiwan.* 76–77.

Amynthas robusta Xu & Xiao, 2011. *Terrestrial Earthworms of China.* 174–175.

External Characters

Length 85–125 mm, clitellum width 4–7 mm, segment number 71–138. Prostomium epilobous. First dorsal pore 12/13. Clitellum XIV–XVI, annular, setae absent, dorsal pore absent. Setae 46–55 (VII), 15–21 between male pores. Male pores paired in XVIII, on a raised papillae with 1 or 2 similar papillae around it, surrounded by several circular folds, 1 or 2 additional presetal papillae present medioventrally in XVIII, if 2, slightly connected. Spermathecal pores 2 pairs in 7/8/9, ventrolateral, 0.5 circumference apart ventrally. Genital papillae paired in spermathecal pores and/or presetal in VII and IX, slightly medial to spermathecal pores, additional presetal papillae present in VII and X in some specimens. Female pore single in XIV, medioventral.

Internal Characters

Septa 8/9/10 absent, 6/7/8, 10/11/12/13/14 greatly thickened. Gizzard in VII and IX. Intestine from XIV or 1XV. Intestinal caeca paired in XXVI or XXVII, simple, extending anteriorly to XXIII or XXIV. Lateral hearts in X–XIII. Spermathecae 2 pairs in VII and IX, ampulla large, round or oval, stalk stout, straight, shorter than ampulla, diverticulum long, stalk slender, looped, seminal chamber rod-shaped. Ovaries paired in XIII. Testis sacs 2 pairs in X and XI. Seminal vesicles paired in XI and XII, with dorsal lobes. Prostate glands paired in XVIII, racemose, extending anteriorly to XVI and posteriorly to XX or XXI.

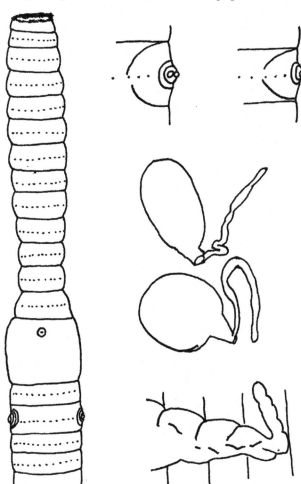

Color: Live specimens light red or reddish brown on dorsum, reddish white on ventrum, reddish white or dark red on clitellum.

Type locality: Mauritius and Philippines (Manila).

Deposition of types: Paris Museum, France.

Global distribution: Cosmopolitan, recorded in China, Korea, Japan, Taiwan, Philippines (Manila), Myanmar, India, Mauritius. In Taiwan, it is recorded in northeastern, northern, central, and southern Taiwan, and Lanyu Island.

Distribution in China: Jilin (Yongji), Jiangsu, Zhejiang, Fujian (Fuzhou), Taiwan (Taibei and Kaohsiung), Jiangxi, Hubei (Lichuan, Qianjiang), Hong Kong, Chongqing (Shapingba, Nanchuan, Pingshan, Dahebaba, Nanchuan, Beipei, Shapingba), Sichuan (Chengdu, Luzhou, Mt. Emei, Loshan, Ganxigou), Guizhou (Mt. Fanjing).

Etymology: The specific name indicates its robust body size and shape.

Remarks: *Amynthas robusta* is close to *A. aspergillum* from which it may be distinguished by the definitely smaller number of genital markings in the immediate vicinity of each male pore.

FIG. 52   *Amynthas robustus* (after Tsai, 1964).

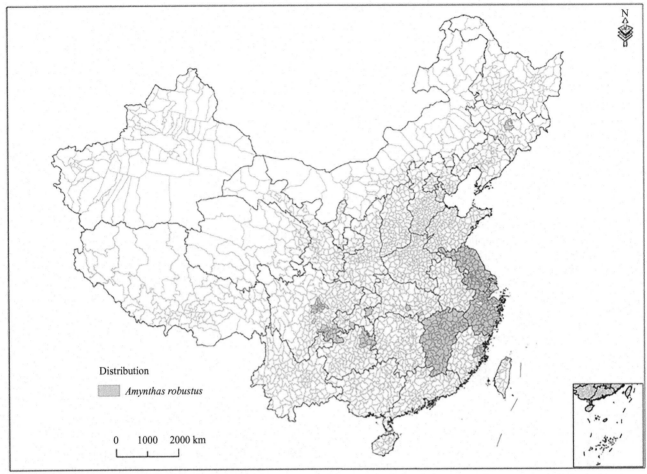

FIG. 53 Distribution in China of *Amynthas robusta*.

47. *Amynthas siemsseni* (Michaelsen, 1931)

Pheretima siemsseni Michaelsen, 1931. *Peking Natural History Bulletin* 5(2):17–19.
Amynthas siemsseni Xu & Xiao, 2011. *Terrestrial Earthworms of China.* 181.

External Characters

Length 120–200 mm, width 6–8 mm, segment number 136–146. First dorsal pore in 12/13. Prostomium epilobous. Clitellum XIV–XVI, annular, without setae. Setae rows dorsally continuous, ventrally regular at least in the anterior part of body, but occasionally interrupted ($aa=1.25ab$), Setae number 47 (V), 56 (VI), 63 (IX), 70 (XIII), 74 (XXVI). Male pores ventral on XVIII, about 1/3 of the body circumference apart ventrally; each on large wart-like porophores extending the entire length of the segment. At the base of the porophores, some ring-like furrows, on the median side of the tip some few (2–4?) vague slightly raised glandular papillae, perhaps marked by a circular ring. Spermathecal pores 2 pairs, ventrolateral in intersegmental furrow 7/8/9, those of 1 pair about 1/3 of the body circumference apart ventrally. Spermatheca papillae not present. Female pore single in XIV, medioventral.

Internal Characters

Septa 8/9/10 absent, 4/5 slightly thickened, 5/6/7/8 and 10/11/12/13 strongly thickened, 13/14 slightly thickened, those following delicate. Gizzard large, between septum 7/8 and 10/11. Intestine swelling in XIV; Intestinal caeca surprisingly long, reaching alongside the intestine from segment XXVII to XXII, simple, constricted somewhat by the septa, base broad, gradually narrowing toward the anterior, with pointed and somewhat bent anterior end. Testes sacs 2 pairs, rather large, long, not like sperm sacs, ventral in X and XI, the 2 on each side completely fused together forming a thick ball-like space. The testis sacs of the anterior pair in X are united through a short low broad cross bridge, so that therefore all the testis sacs communicate with each other. Two pairs of funnels occupy nearly all the

FIG. 54   *Amynthas siemsseni* (after Michaelsen, 1931).

space of the testis sacs, leaving only a little space for the testis and the sex products. Sperm vesicles 2 pairs in XI and XII on the anterior of the septa median to the alimentary canal communicating at the base with the testis sacs in the segments in front. The sperm vesicles consist of a ventral narrow main part wider above, clearly outlined at the dorsal edge, and a long narrow outgrowth extending forward from this dorsal portion. While the main part of the sperm vesicle is finely reticular on the surface, the outgrowth is smooth. Prostate glands, the glandular portion about 3 segments, from XVIII–XX, flat, much broader than long, with many deep indentations on the widest proximal edge, otherwise with many small indentations and coils, reticular on the surface. The duct forms a narrow straight band extending backward. There are no copulatory pouches present, apparently no penis formation. Spermathecae 2 pairs in VII and IX, ampullae ball-shaped, sac-like, or egg-shaped, the duct sharply differentiated, cylindrical, somewhat more than half as long and one-third as thick as the ampulla, becoming thinner at the distal end. Into this distal end of the ampulla duct there opens a blind hose-like diverticulum about as long as the main pouch (ampulla plus duct). The diverticulum with its narrow stalk forming almost the whole of the distal half, passes over proximally into the rounded sperm space, gradually when only slightly full, but with a sharp line when completely full (slender pear-shaped to cylindrical but distally rounded). The diverticulum is in general only slightly bent, but sometimes forms in its middle a moderately broad narrow loop.

Color: Preserved specimens yellow gray (faded?) to brown gray.
Type locality: Fujian (Fuzhou).
Distribution in China: Fujian (Fuzhou).
Deposition of types: Hamburg Museum, Germany.

48. *Amynthas takatorii* (Goto & Hatai, 1898)

Pheretima takatorii Goto & Hatai, 1898. *Annot. Zool. Jap.* 2:76–77.
Amynthas takatorii Xu & Xiao, 2011. *Terrestrial Earthworms of China.* 186–187.

External Characters

Length 314 mm, width 8 mm, segment number 120. Prostomium epilobous. First dorsal pore 11/12. Clitellum XIV–XVI, annular, setae absent. Number of seta fewer in the anterior segments; in the spermathecal segments 51, behind them about 65. Male pores 1 pair in XVIII, each pore on top of papillae, each surrounded by 8 genital papillae in groups of 4 on the 2 sides of an isosceles triangle with the base turned externally. Spermathecal pores 2 pairs in 7/8/9, with 2 or 3 genital papillae on the posterior margin of the next preceding segment. Sperm duct pores on top of papilla, each surrounded by 8 genital papilla arranged in groups of 4 on the 2 sides of an isosceles triangle with the base turned externally. Female pore single, medioventral in XIV.

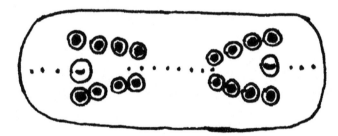

FIG. 55   *Amynthas takatorii* (after Goto & Hatai, 1898).

Internal Characters

Septa 8/9/10 absent, 5/6/7/8, 10/11/12/13 thickened. Gizzard in VII–IX. Intestine from XV. Caeca paired in XXVI, extending anteriorly to XXIII. Last hearts in XIII. Testis sacs 2 pairs, in X and XI; seminal vesicles in XI and XII. Prostate glands lobate, in XVII–XXI. Spermathecae 2 pairs in VII and IX, anterior pair with 2 finger-shaped, more or less winding diverticula of unequal lengths, and posterior pair of similar spermathecae and 3 pairs of accessory spermathecae without diverticula; the accessory spermathecae being situated internally to be well-developed ones.

Color: Preserved specimens very different ventrally and dorsally, the former being yellowish brown and the latter light gray. Clitellum uniform light yellow.
Type locality: Taiwan (Taibei).
Distribution in China: Taiwan (Taibei).

49. *Amynthas triastriatus* (Chen, 1946)

Pheretima triastriata Chen, 1946. *J. West China. Border Res. Soc. (B).* 16:97–98, 139, 146.
Amynthas triastriatus Sims & Easton, 1972. *Biol. J. Linn. Soc.* 4:234.
Amynthas domosus Xu & Xiao, 2011. *Terrestrial Earthworms of China.* 190–191.

External Characters

Large sized worm. Length 110 mm, width 7 mm, segment number 88. Segments stoutly built, segment VII being 3.5 mm long, equal to 2 postclitellar ones. Prostomium 1/2 epilobous, its tongue wider than length of II. First dorsal pore 10/11. Clitellum XIV–XVI, annular, not swollen, without setae, part on dorsal and lateral side of XVI not glandulate. Setae of anterior 8 segments appearing longer, none particularly enlarged, slightly shorter and closer on dorsal side, setae *a*, *b*, nearly as large as those on dorsal side. Setae number 34 (III), 36 (IX), 38 (XXV); 17 (IX) between spermathecal pores, 12 between male pores; $aa = 1.2$–$1.5ab$, $zz = 1.5$–$2.0zy$. Those on postclitellar dorsum scarcely visible. Male pores paired in XVIII, about 1/3 of circumference ventrally apart, each on a round papilla, 2 similar ones on its median side, pore papilla large and with lateral wrinkled skin. Spermathecal pores 2 pairs in 7/8/9, about 1/3 circumference ventrally apart, pimple-like papillae usually present, 1 in front and another behind each pore but more ventral. Female pore single in XIV, medioventral.

Internal Characters

Septa 8/9/10 absent, 4/5 thin, 5/6/7 slightly maculated, with thick nephridial tufts on anterior face, 7/8 also thin, 1011/12 /3/14 thicker and more muscular. Gizzard in 1/2 IX and 1/2 X, globular. Intestine swelling in XV. Caeca simple, smooth on both sides, extending anteriorly to XXIII. Vascular loops in IX and X asymmetrically developed, 1 limb very stout, those in X and XI equally developed, as stout as heart in XII and XIII. Testis sacs comparatively small, V-shaped anterior pair about 1 mm at widest diameter, median tunnel narrow; posterior pair twice as large, similarly shaped. Seminal vesicles filling about two-thirds of each segment, about 3 mm in dorsoventral length, dorsal lobe small, dorsally situated, hardly distinguishable from the vesicle. Small strands of vestigial vesicle found on posterior face of 12/13. Prostate

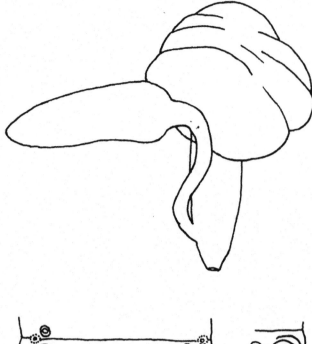

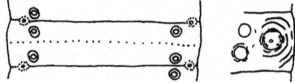

FIG. 56 *Amynthas triastriatus* (after Chen, 1946).

gland totally absent, its duct U-shaped, about equally thickened. Accessory gland round and rough on surface, each with cord-like ductule. Spermathecae 2 pairs, in VII and VII. Ampulla heart-shaped, large, ampulla duct wide, slightly shorter. Diverticulum longer than main pouch. Seminal chamber about 1.5 mm long, 0.6 mm wide, narrowed distally. Accessory glands similar in character.

Color: Grey dorsally, pale ventrally, setal circles whitish. Clitellum light chocolate red.
Type locality: Sichuan (Mt. Emei).
Distribution in China: Hubei (Lichuan), Sichuan (Mt. Emei), Guizhou (Mt. Fanjing).
Deposition of types: ?
Remarks: This species is close in appearance to *Amynthas leucocircus* (Chen, 1933) and *Amynthas pingi pingi* (Stephenson, 1925). However, it differs from the former in the number of spermathecae, the character of diverticulum, the setae, the absence of prostate, and from the latter by the number of spermathecae, the size relation, and the setal character. In both of the latter species, the setae on the anterior ventral side are longer and widely spaced. This species should be distinct from either of them.

50. *Amynthas ultorius* (Chen, 1935)

Pheretima ultoria Chen, 1935b. *Bull. Fan. Mem. Inst. Biol.* 6:42–47.
Amynthas ultorius Sims & Easton, 1972. *Biol. J. Soc.* 4:234.
Amynthas ultorius Xu & Xiao, 2011. *Terrestrial Earthworms of China*. 192–193.

External Characters

Length 80–98 mm, width 4.5–5.0 mm, segment number 79–119. Preclitellar segment relatively long while postclitellar short. Prostomium 1/2–2/3 epilobous, dorsal process narrow, its width equal to segmental length 1/2 II. First dorsal pore in 11/12. Clitellum XIV–XVI, short, annular, setal pits sometimes vaguely visible but setae totally absent, dorsal pore not visible, its length equal to a neighboring preclitellar or 4 postclitellar segments. Seta zones raised particularly in 10 anterior and 30 posterior segments; $aa = 1.2ab$, $zz = 1.5$–$2zy$. Setae number 25–28 (III), 27–32 (VI), 45–54 (VII), 50–54 (XII), 50–55 (XXV); 15–19 (VII), 18–24 (VII), 18–24 (IX) between spermathecal pores; 14–16 between male pores. Male pores 1 pair, on ventrolateral sides of XVIII, each pore situated on a round or ovoid papilla, about 1/3 of circumference ventrally apart, both of the pore papilla and 2 genital papillae surrounded by circular ridges appearing glandular on the part between them; one of the papillae situated medially on setal line or slightly behind, the other anteriorly between setal line and intersegmental furrow 17/18 but about midway between other genital papilla and pore papilla; each ovoid with glandular center, as large as pore papilla. Spermathecal pores 2 pairs, in 7/8/9, about 5/14 of circumference apart; each opening situated in center of an ovoid area, intersegmental in position; a papilla is found close to each pore, which is placed near anterior edge of succeeding segment, about 2 or 3 setae

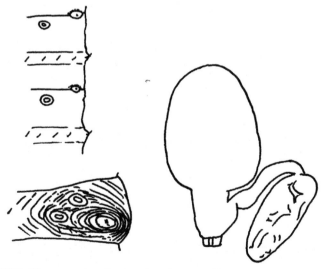

FIG. 57 *Amynthas ultorius* (after Chen, 1935b).

between it and spermathecal pore; each one ovoid, slightly smaller than those in male pore region. Female pore single in XIV, medioventral.

Internal Characters

Septa 9/10 absent; 8/9 membranous, present on ventral side only; 5/6/7/8, 10/11/12/13/14 thick but not strongly muscular. Gizzard barrel-shaped, in IX and X, its surface smooth and glittering. Intestine swelling in XVI. Intestinal caeca paired in XXVII, extending anteriorly to XXIV or XXIII, moderately broad and pointed, smooth both dorsally and ventrally. Hearts 4 pairs, first pair in X, stout. Testis sacs in X about 1.2 mm in diameter, connected medially in a thick V shape, those in XI broadly united; testis disc-like and compact, in usual position, seminal funnel as usual. Seminal vesicles in XI and XII, very large, about 2 mm in length (anteroposterior) 4 mm in width (dorsoventral) pale and weakly tubercular on surface; each with a large dorsal lobe, almost equal to one-third of main vesicle, situated anterodorsally, slightly granular on surface. Prostate glands large, in XVII–XIX or XVI–XX, compact, rather finely divided, prostatic duct short and stout, ental end about equally stout, loosely curved, about 1.5 mm in length. Accessory glands composite, finely tubercular, each with a stout and large stalk leading to respective papilla externally; strong ligaments in XVII–XIX extending from ventromedian side to lateral to partially cover small glands and nephridia in body cavity. Spermathecae 2 pairs, in VII and VII; ampulla round or ovoid, about 1 mm or more in diameter, smooth and distended; ampullar duct extremely short but stout, sharply marked off from ampulla; its diverticulum as long as, or longer than, main pouch; ental half distended as seminal chamber, cylindrical in outline, clearly distinguished from its slender duct which joins ectal portion of main duct; or seminal chamber about one-third of whole length, not clearly marked from its duct, or globular in shape and clearly marked from its duct.

Color: Preserved specimens brownish or light chestnut dorsally, more distinct at postclitellar region, with setal zones pale, narrow, and very vague; pale or grayish or light brownish ventrally. Clitellum light chocolate brown.
Type locality: Hong Kong.
Distribution in China: Hong Kong, Sichuan (Leshan).
Deposition of types: Mus. Fan. Inst. Biol. Beijing.
Remarks: It certainly differs from *Pheretima sluteri* (Horst, 1890) from Borneo. The latter is much larger (190 mm long, 135 segments) and does not possess genital papillae. It agrees with *Amynthas masatakae* (Beddard, 1892) from Japan in the characters of spermathecae and genital papillae around the spermathecal region but differs from the latter by its

smaller size (125 mm long, width 0.6 mm, 90 segments in *Amynthas masatakae*) and many important characters of the latter undescribed, for example, aspect of the male pore region, setae, etc. *Pheretima ornata* (Gates 1929) from Burma is also of large habit and the terminal end of prostatic duct has a small concavity. It differs also from *Amynthas siemsseni* (Michaelsen, 1931) and *Amynthas fokiensis* (Michaelsen, 1931) from Fujian in the general form of the body, the character of the genital papillae and testis sacs etc. It can be distinguished from *Amynthas corrugatus* (Chen 1931) by (1) genital papillae around male pore and spermathecal regions constantly present, (2) setae of anterior end distinctly enlarged, and (3) testis sacs more closely placed.

megascolidioides Group

Diagnosis. Holandric, male pores on XVIII, with spermathecal pores 5 pairs, at 4/5/6/7/8/9.

Global distribution: Oriental realm, ? Australasian realm and ? introduced into the Oceanian realm. There is 1 species from China.

51. *Amynthas dongfangensis* Sun and Qiu, 2010

Amynthas dongfangensis Sun, Zhao and Qiu, 2010, *Zootaxa*. 26–32.
Amynthas dongfangensis Xu & Xiao, 2011. *Terrestrial Earthworms of China*. 107.

External Characters

Length 44 mm, width 1.5 mm, segment number 70. Prostomium 1/2 epilobous. First dorsal pore in 11/12. Clitellum XIV–XVI, annular, swollen, setal and dorsal pore invisible. Setae numerous, 28 (III), 26 (IV), 34 (VII), 34 (XX), 36 (XXV); 10 between spermathecal pores, 5 between male pores, setal formula $aa = 1.2ab$, $zz = 2zy$. Male pores paired in VII, 1/3 body circumference ventrally apart, each on the center of a slightly raised conical porophore, with 3–5 circular folds. Spermathecal pores 5 pairs in 4/5/6/7/8/9, ventral, eye-like, 1/3 body circumference apart. Female pore single in XIV, midventral.

Internal Characters

Septa 8/9/10 absent, 6/7 slightly thickened. Gizzard in IX–X, long bucket-shaped. Intestine enlarged gradually from XV and enlarged suddenly from XXI. Intestinal caeca paired simple, smooth, in XXVII, extending anteriorly to XXV. Hearts in X–XIII. Holandric, testis sacs 2 pairs, developed, in X–XI, second pair enclosing first pair with seminal vesicle. Seminal vesicles paired in XI–XII, developed. Prostate glands in

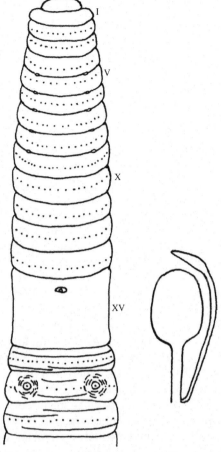

FIG. 58   *Amynthas dongfangensis* (after Sun and Qiu, 2010).

XVII–XXI, developed, lobule. Prostatic duct uniform, inverted U-shaped. Spermathecae 5 pairs in V–IX, ampulla oval-shaped. Ampulla oval and full, ampulla tube uniform and equal to the length of the ampulla. Ampulla duct longer than main pouch, ental serving as seminal chamber, pepper-shaped, seminal chamber about half the length of stalk.

Color: Preserved specimens grayish brown to purple dorsally, ventrally ficelle on preclitellar; deep brown dorsally, ventrally no pigment on postclitellar.
Clitellum red brown.
Distribution in China: Hainan (Mt. Jiefeng).
Deposition of types: Shanghai Natural Museum.

bournei Group

Diagnosis. *Amynthas* with spermathecal pores 3 pairs, at VI, VII, and VII.

Global distribution: Oriental realm, ? Australasian realm and ? introduced into the Oceanian realm. There are 3 species from China.

TABLE 13 Key to the Species of *bournei* Group of the *Amynthas* From China

1.	Spermathecal pores anteriorly segment VI–VIII, close to the intersegmental furrows; with 1 pair papillae on VII and VII presetal *Amynthas dolosus*
	Papillae behind of setal ring on VII & VIII 2
2.	Genital papillae unpaired ventromedially on VII, VII postsetal .... *Amynthas domosus*
	Genital papillae paired ventrally on VII and VII, postsetal .... *Amynthas mucrorimus*

52. *Amynthas dolosus* (Gates, 1932)

Pheretima dolosa Gates, 1932. *Rec. Ind. Mus.* 34(4):443–444.
Pheretima dolosua Gates, 1972. *Trans. Amer. Philos. Soc.* 62(7):181.
Amynthas dolosus Sims & Easton, 1972. *Biol. J. Soc.* 4: 234.
Amynthas dolosus Xu & Xiao, 2011. *Terrestrial Earthworms of China.* 105–106.

External Characters

Length 84 mm, width 3.5 mm, segment number ?. Prostomium epilobous ?. First dorsal pore 12/13, but there is a distinctly pore-like marking in 11/12 filled with a sort of whitish coagulum. Clitellum XIV–XVI, intersegmental furrows and dorsal pores lacking. There are 7 setae ventrally and ventrolaterally on ii; none dorsally. Setae number 42 (XX); 19 (VI) but with a midventral gap, 18 (VII) but with ventrolateral gap, between spermathecal pores; 15 (XVII), 6 (VII), 15 (XIX) between male pores. Male pores are minute, on small, widely separated, transversely oval discs, each disc about 1.5 intersetal distances wide transversely. There are 2 pairs of presetal preclitellar genital markings, one pair on VI, the other pair on VII. Each marking is transversely oval, flat, slightly raised and about 2 intersetal distances wide transversely; about 4–6 intersetal distances median to the spermathecal pore and about 5–6 intersetal distances apart from the other marking of the same segment. There are 2 pairs of postclitellar marking on XVIII. The circular presetal markings are closely paired, almost in contact at the midventral line, each marking protuberant, about 2 intersetal distances wide transversely and about 4 intersetal distances to the male pores. The transversely oval, protuberant, postsetal markings are slightly smaller than the presetal markings and are more widely separated, the lateral margin of each marking reaching to about the male pore line. Each marking preclitellar or postclitellar has a grayish central portion and a peripheral region of more whitish appearance. Spermathecal pores 3 pairs, anteriorly on VI–VII, close to the intersegmental furrows, pore minute. Female pore single in XIV, medioventral.

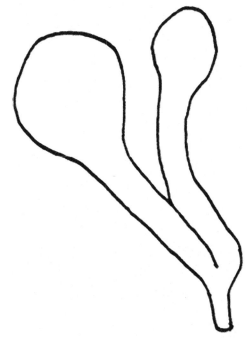

FIG. 59 *Amynthas dolosus* (after Gates, 1932).

Internal Characters

Septa 8/9/10 absent, 5/6/7/8 and 10/11/12 muscular; 12/13/14 strengthened and translucent. Gizzard ?. Intestine begins in XV. Caeca simple, in XXVII, extending anteriorly to XXIV. There is a pair of commissures belonging to segment IX. The last pair of hearts is in XIII. All hearts of IX–XIII pass into the ventral blood vessel. Testis sacs 2 pairs, each sac nearly spherical; the sacs are segmented from each other and apparently without transverse communication. Seminal vesicles cover the dorsal blood vessel in segment XI and XII. Prostates extend through segments XVII–XIX or XX; prostate duct is bent into a U- or C-shape, is about 2.5 mm in length, the ectal two-thirds slightly thicker than the entalmost portion. Spermathecae 3 pairs, in VI, VII, and VII; ampulla is round to ovoid, the duct longer than the ampulla and slender. Diverticulum may be longer or shorter than the combined lengths of the duct and ampulla; is slender, widened entally, and passes into the anterior face of the duct which is much narrowed in the parietes. There are glandular masses projecting into the coelom from the parietes over the genital markings.

Color: In preserved specimens the dorsum is light brownish. Clitellum reddish.
Type locality: Yunnan (Teung Cong).
Distribution in China: Yunnan (Mengmeng).
Deposition of types: Indian Museum.
Remarks: The genital markings of this species are somewhat like the smaller genital markings of

Amynthas exiguus. The preclitellar and the presetal marking of XVIII correspond roughly to similarly placed markings in *A. exiguus*. The latter species, however, never has postsetal genital markings while *Amynthas dolosus* has 1 pair on XVIII.

53. *Amynthas domosus* (Chen, 1946)

Pheretima domosa Chen, 1946. *J. West China. Border Res. Soc. (B).* 16:102–103.
Amynthas domosus Sims & Easton, 1972. *Biol. J. Linn. Soc.* 4:234.
Amynthas domosus Xu & Xiao, 2011. *Terrestrial Earthworms of China.* 106–107.

External Characters

Large-sized worm. Length 116mm, width 5.5mm, segment number 90. Segment of anterior body rather long, 3 annuli distinct after 7th segment. Prostomium 1/4 epilobous, narrow, not shown when retracted. First dorsal pore 12/13. Clitellum XIV–XVI, annular, smooth and swollen. Setae all fine hardly visible on postclitellar segments, barely visible on anterior ventral side; setae number 46 (IX) ?, 52 (XXV); 12 (V), 20 (VII) between spermathecal pores, 12 between male pores (between 2 papillae); setal chain close, $aa = 1.2ab$, $zz = 1.5$–$2zy$. Male pores paired on setal line of XVIII, about 1/4 of circumference ventrally apart, each situated laterally in a glandular area, an ill-defined large papilla on XVIII and part on XIX. Spermathecal pores 3 pairs, segmental on anterior edge of VI–VII (that or right VI lacking), about 1/4 of their ventral circumference, each close to furrow but definitely segmental represented by small pimple-like porophore. Genital papillae unpaired ventromedially on VII, VII postsetal. Both indistinctly outlined. Represented by a glandular ovoid patch of skin, reddish in appearance, transversely grooved in center. Female pore single in XIV, medioventral.

Internal Characters

Septa 8/9/10 absent, 4/5/6/6/7/8 thin and membranous (7/8 a little thicker), 10/11/12 thicker than preceding ones, 13/14 slightly muscular, thin after 15/16. Gizzard rounded, in 1/2 IX and X. Intestine swelling in XV. Caeca simple and pointed, in XXVII, extending anteriorly to XXII or XXIII. Vascular loops in IX asymmetrical, in X lacking or rudimentary. Testis sacs also large, those in X about 2.5mm in diameter, 2 sacs closely placed but not connected. Seminal vesicles large, in XI, XII, each with a large dorsal lobe, prostate pair similar in shape but closer; not communicated. Prostate glands large, coarsely lobate, in XVII–XIX, its duct thick, U-curved. Accessory glands sessile, as big patch around each prostatic duct. Spermathecae 3 pairs, in VI–VII. Ampulla 2mm long, round or with a pointed apex, its duct as long. Diverticulum as long as main pouch chamber spherical. Accessory glands of similar character.

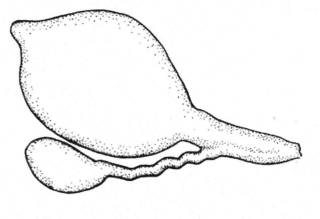

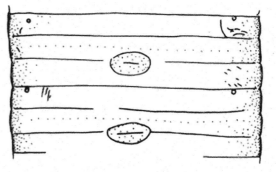

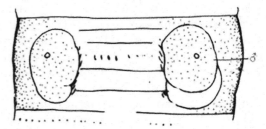

FIG. 60 *Amynthas domous* (after Chen, 1946).

Color: Preserved specimens light greyish dorsally, whitish ventrally. Clitellum chocolate brown.
Distribution in China: Sichuan (Mt. Emei).
Remarks: The spermathecal pores are not strictly segmental and distinguish it from *Amynthas mucrorimus* (Chen, 1946), *Amynthas pomellus* (Gates, 1935) etc.

54. *Amynthas mucrorimus* (Chen, 1946)

Pheretima mucrorima Chen, 1946. *J. West China. Border Res. Soc. (B).* 16:108–109, 139, 150.
Amynthas mucrorimus Sims & Easton, 1972. *Biol. J. Linn. Soc.* 4:234.
Amynthas mucrorimus Xu & Xiao, 2011. *Terrestrial Earthworms of China.* 153.

External Characters

Length 105–118mm, width 5.5–6mm, segment number 99–100. Prostomium 1/2 epilobous, not cut behind.

First dorsal pore 12/13. Clitellum XIV–XVI, annular, without setae. Setae all uniformly built and almost equally distributed, scarcely visible on dorsum of ii. Setal chain rather closed. Setae number 40–42 (III), 66–75 (VII), 55–60 (XXV); 28 (VI, VII), 32 (VII) between spermathecal pores, 14 between male pores; $aa = 1.1ab$, $zz = 1.2$–$2zy$. Male pores paired, superficial, each on a minute elevation at venterolateral corner of XVIII; single ovoid papilla constantly present in front and usually another behind each pore; both papillae and porophore surrounded by horseshoe glandular wall, open on median side. Spermathecal pores 3 pairs, segmental on VI–VII, about 4/9 of their ventral circumference, each one minute; its region slightly elevated, a transverse groove medial to pore, not glandular nor other genital marking, in front of 14th setae, closer to setae than intersegmental furrow; genital papillae paired ventrally on VII and VII, postsetal, somewhat depressed in middle. Female pore single in XIV, medioventral.

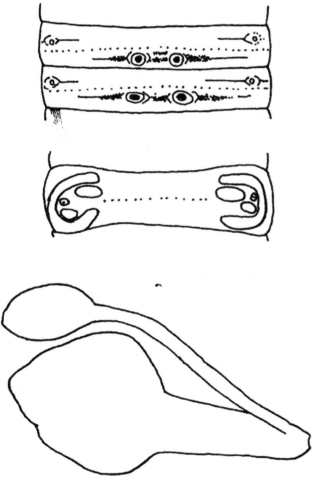

FIG. 61   *Amynthas mucrorimus* (after Chen, 1946).

Internal Characters

Septa 8/9/10 absent, 4/5/6/7/8 thick, 10/11/12/13 equally thickened. Nephridial tufts on 5/6/7 thick. Intestine swelling in XV. Caeca simple and small, in XXVII, extending anteriorly to XXIII. Vascular loops in X vestigial, those in IX stout. Testis sacs in X and XI large, anterior pair 2 mm in diameter, posterior a little smaller, both pairs widely separate (about 2 mm) not connected. Seminal vesicles small in both cases observed, first pair larger, about 4 mm in vertical length, 2 mm in width; dorsal lobe of each vesicle large, surface smooth; second pair about half the size. Prostate gland coarsely lobate, in XVII–XX; prostate duct stout and in a V shape. Accessory glands sessile and fair in size. Spermathecae 3 pairs, in VI, VII, and VII, ampulla round or heart-shaped, its duct short and stout, main pouch about 2 mm long. Diverticulum as long, with terminal ovoid seminal chamber. Accessory glands also sessile in round patch.

> Color: Preserved specimens brownish gray on dorsal side, darker anteriorly, pale ventrally. Clitellum brick brown.
> Distribution in China: Sichuan (Mt. Emei).
> Remarks: The very characteristic structure of this species is the position of the spermathecal pores, which are strictly segmental.

canaliculatus Group

Diagnosis. *Amynthas* with spermathecal pores 2 pairs, at VI and VII.
There is 1 species from China.

55. *Amynthas benigmus* (Chen, 1946)

Pheretima benigma Chen, 1946. *J. West China. Border Res. Soc. (B).* 16:100–101, 139, 148.
Amynthas benigmus Sims & Easton, 1972. *Biol. J. Linn. Soc.* 4: 234.
Amynthas benigmus Xu & Xiao, 2011. *Terrestrial Earthworms of China.* 84–85.

External Characters

Large-sized worm. Length 130 mm, width 5 mm, segment number 100. Prostomium 1/3 epilobous. First dorsal pore 12/13. Clitellum XIV–XVI, annular, no setae, nor swollen. Setae rather brittle, ventral ones slightly longer and closer; setae number 16 (III), 28 (VII), 29 (VII), 32 (XVII), 35 (XX); 18 (VI), 17 (VII), between spermathecal pores, 8 (VII), 14 (XIX) between male pores; $aa = 1.1$–$1.5ab$, $zz = 2zy$. Male pores paired in XVIII, on setal line, about 1/3 of circumference ventrally apart, each lateral to a large and flat papilla; outer glandular skin extending to small part of neighboring segments. No copulatory pouch. Spermathecal pores 2 pairs on VI, VII, as small pimple in middle of first annulus. No other genital marking. The distance between spermathecal pores of first pair slightly over half circumference of segment; second pair more ventral, less than 1/2 of circumference. Female pore single in XIV, medioventral.

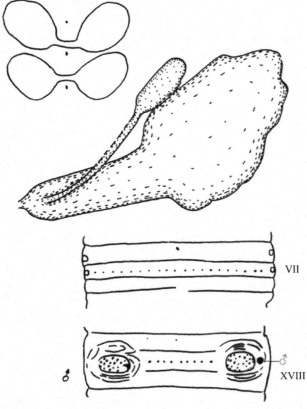

FIG. 62 *Amynthasbenigmus* (after Chen, 1946).

vesicles in XI and XII, about 4 mm in dorsoventral length, dorsal lobe about half as large. A rudimentary pair on XIII, 2.2 mm, in vertical length. Prostate glands well developed in XVII–XX, duct stout in middle part, slender ectally. Accessory glands each in a large patch, sessile. Spermathecae 2 pairs, in VI and VII, ampulla spatulate, with a duct equally long, indistinctly marked off. Diverticulum shorter than main pouch. Seminal chamber round elongate ovoid.

Color: Brownish dorsally, setal circle lighter, ventral side pale. Clitellum chocolate.
Distribution in China: Sichuan (Ebian Cizhuping), Guizhou (Mt. Fanjing).
Etymology: The name of the species refers to the type locality.
Remarks: This species resembles *Amynthas pingi pingi* (Stephenson, 1925) in general appearance. The setae on anterior venter are large but not irregularly spaced. The shape of the ampulla and diverticulum is nearly identical. It differs essentially in the position and number of spermathecal pores and the aspect of the male pore region. It also differs from *A. mucrorimus* (Chen, 1946) and *A. cupreae* (Chen, 1946) in the position and number of spermathecal pores, the character of setae, and the close testis sacs. The species' name signifies the locality, that is, "Tz-Tzo-Ping" literally meaning "benevolent bamboo flat" where it was first found. The particular kind of bamboo is symbolic of Beddard's Mercy, to which the common name was due.

Internal Characters

Septa 8/9/10 absent, 6/7/8 slightly musculated, 10/11/12/13/14 more muscular and thicker. Gizzard in IX. Intestine swelling in XV. Caeca simple, smooth on both sides, extending anteriorly to XXII. Vascular loops in IX small, in X and XI very stout. Testis sacs in X and XI, those in X egg-shaped, 2.5 mm in diameter, closely placed, but very narrowly connected, in XI round, also narrowly connected and communicated. Seminal

diffringens Group

Diagnosis. Holandric, with spermathecal pores 4 pairs, at 5/6/7/8/9.
There are 50 species from China.

TABLE 14 Key to the Species of *diffringens* Group of the Genus *Amynthas* From China

1. Spermathecal pores in the dorsum	2
Spermathecal pores not in the dorsum	7
2. Clitellum in 9/10 XIV–7/10 XVI, annular	*Amynthas dorsualis*
Clitellum in XIV–XVI, annular	3
3. Male pores situated on XVIII, on conical porophores	4
Male pores situated on XVIII, genital markings extending or on XVII	5
4. Male pores situated on large conical porophores, surrounded by several circular folds, with an oval pad within the copulatory pouch or near the opening apart ventrally	*Amynthas formosae*
Male pores situated on low conical porophores within a copulatory pouch, a longitudinally orientated genital marking extending anteriorly from the aperture, the male pore area surrounded by circular folds, laterally bordered by a skin lip, partially covering male porophores	*Amynthas hengchunensis*
5. One or 2 pairs of large concave genital papillae behind or also in front of male pore	*Amynthas dyeri*
Male pores situated on XVIII, genital markings also on XVII	6

TABLE 14 Key to the Species of *diffringens* Group of the Genus *Amynthas* From China—cont'd

6. Rounded angular genital pad longitudinally oriented from 17/18 to equator of XVIII, surrounded by epidermal folds, lateral folds closest to male pores enlarged to form flap adjacent to or partially covering genital pad; rarely with paired oval genital markings presetal XVII .... *Amynthas kaopingensis*
   Genital markings, paired, about 4–6 intersetal intervals wide, just median to male porophores, across (?) 17/18 and/or 18/19 (rarely 19/20) .... *Amynthas rodericensis*

7. Spermathecal pores in the ventrum, without papillae front clitellum 8
   With papillae front or/and behind clitellum 34

8. Without papillae in male pore region 9
   With papillae and/or genital markings in male pore region 19

9. Male pores on a round, slightly elevated porophore .... *Amynthas taiwumontis*
   Male pores on glandular area 10

10. Male pores in center of a roundish glandular area 11
    Male pores not in center of a round glandular area 12

11. Without folds around the roundish glandular area .... *Amynthas homochaetus*
    Male porophore round, surrounded by 2 to 4 slight circular or diamond-shaped folds .... *Amynthas penpuensis*

12. Male pores on a nipple-like porophore, its base with concentric ridges, body wall of XVI–XXV in line with porophore on each side raised as skin flap, always directed and partly folded toward ventral side .... *Amynthas lacinatus*
    Male pores not on a nipple-like porophore 13

13. Male pores on middle of a crescent or comma-shaped papilla, a large ovoid and centrally depressed papilla placed medial to each pore papilla, median 2 in close approximation .... *Amynthas lunatus*
    Male pores not on a comma-shaped porophore 14

14. Male pores on circular-shaped markings 15
    Male pores not on a circular-shaped marking 16

15. Male pores on widely separated, circular discs; there is a pair of genital markings on XVIII; each marking is transversely elongate, with rounded ends, a raised rim and a concave central portion .... *Amynthas manicatus*
    Male pores flanked by 2 circular genital markings closely median, lateral to male pores .... *Amynthas nanrenensis*

16. Male pores on the center of a raised elliptical glandular area.
    Male pores on oval area 18

17. Male pores on the center of a raised elliptical glandular flat top, conical, surrounded by a folded pad .... *Amynthas fuscus*
    Male pores on the center of a slightly raised, glandular and approximately elliptical porophore, bordered by two folds on lateral margin. Genital markings look like a peanut in a shell, that is, a rectangle with enlarged rounded ends, concave-topped, paired behind 17/18 and on 18/19 .... *Amynthas mirifius*

18. Male pores a trifle lateral to the center of a male pore marking; the latter is nearly but not quite circular in shape, slightly protuberant but with a rather flat surface and surrounded by a slightly but definite circumferential furrow; just median to each male pore marking is a single transversely oval genital marking with a conspicuously protuberant, whitish rim and a depressed, grayish, central ares.
    Male pores at the center of a small, transversely oval area in the setal circle of XVIII. The single genital marking is transversely oval, with a concave, nonprotuberant rim. The markings extend anteroposteriorly through the whole length of XXII and may extend slightly onto XXIII and (or) XXI, and laterally to the male pore lines; or has 2 markings, one on XXII on the left side, the other on XXIII on the right side. Each of these markings ends abruptly with the midventral line .... *Amynthas labosus*

19. Male pores on the center of a slightly elevated round porophore. A large rounded rectangular genital marking within XVIII glandular region .... *Amynthas pulvinus*
    With papillae in male pore region 20

20. With 1 pair papillae in male pore region 21
    With more than 1 pair papillae in male pore region 25

21. The papillae round 22
    The papillae not round 23

22. Male pores on a round porophore with concave center. A pair of round genital papillae in XVIII, closely adjacent to the male porophore at the medioposterior portion immediately posterior to the setal line; both porophore and genital papilla surrounded by 1–2 skin folds *Amynthas biorbis*
    Male pores on a crescent or semicircular shaped area with a large papilla immediately medial to it. The whole male area including the genital papilla protuberant. Each papilla round .... *Amynthas wujhouensis*

23. Male pores on middle of a crescent or comma-shaped papilla, a large ovoid and centrally depressed papilla placed medial to each pore papilla, median 2 in close approximation .... *Amynthas lunatus*
    The papillae neither crescent nor comma-shaped 24

Continued

TABLE 14   Key to the Species of *diffringens* Group of the Genus *Amynthas* From China—cont'd

24. Male pores within shallow circular invagination of body wall, in each of which one oblong longitudinally orientated genital marking is folded in half transversely .... *Amynthas ailiaoensis*
     Male pores round on setal line with elevated center surrounded by 2-3 circular folds. Genital papillae ovate, flat-topped, surrounded by epidermal folds, paired on 17/18, 18/19 median to male pores .... *Amynthas montanus*

25. With 2 pairs papillae in male pore region 26
     With more than 2 pairs papillae in male pore region 23

26. Male pores on a teat-like porophore on a flat, diamond-shaped disc, on which 4 genital papillae are arranged in the form of a cross around the porophore: 1 anterior, 1 posterior, 1 lateral, and 1 medial. This male disc is surrounded by 3–4 diamond-shaped skin folds ...... *Amynthas cruxus*
     The papillae between XVII–XX 27

27. The papillae in XVII and XVIII or XIX 28
     The papillae in XVII or XVIII, XIX or XX 30

28. Male pores on low papillae in XVIII, each on a tubercle, laterally covered by a skin fold appearing like a half-closed eyelid; a raised papilla placed in front on segment XVII and another behind on XIX in line with male porophore. Genital papillae 4 pairs of groups, in V–VII; minute, circular, diamond-shaped depressions lying in the posterior portion of the segments; and 2 pairs of groups, front setae-ring of VII and VII .... *Amynthas mirabilis*
     The papillae in XVII, XVIII, and XIX 29

29. Male pores on a small porophore on ventrolateral side of XVIII, large papilla median to each porophore on XVIII but also on XVII, somewhat posterior to setal line, surrounded by papilla-like glandular ridge either laterally or posteriorly, 2 comma-like ones on anterior of XVIII, similar ones on XVII .... *Amynthas szechuanensis vallatus*
     Male pores on the elliptic, glandular porophore with slightly raised center, without surrounding epidermal wrinkles. Genital papillae ovate, flat-topped, paired above setae annulet on XVIII and XIX; below setae annulet on XVII and XVIII, arranged in longitudinal row slightly medial to male pore .... *Amynthas genitalis*

30. The papillae in XVII and XIX 31
     The papillae in XVII, or XVIII and XIX or XX 32

31. Paired papillae posteriorly on XVII, XIX .... *Amynthas yunlongensis*
     Male pores on a tubercle, laterally covered by a skin fold appearing like a half-closed eyelid; a raised papilla placed in front on segment XVII and another behind on XIX in line with male porophore .... *Amynthas rhabdoidus*

32. The papillae in XVII and XIX or XX 33
     Genital papillae large, round, usually 2 pairs located slightly medially to the porophore in presetal XVIII and XIX, and an additional pair in XX, occasionally 1 pair only in presetal XVIII or XIX, or 2 pairs in XVIII with 1 pair anteromedial and the other posteromedial to male porophores, or 2 pairs in XVIII together with 1 pair in presetal XIX .... *Amynthas hongyehensis*

33. A pair of genital papillae slightly medial to male pore present in each of XVII and XIX, each papilla oval with a depressed center, in posterior annuli between setal line and posterior intersegmental furrow, occasionally an additional pair or 1 papilla present in XX, or 1 missing in XVII or XIX .... *Amynthas wulinensis*
     Male pores on a coniform glandular disc surrounded by a round pad. Genital papillae postsetal, single or paired in XVII, XIX, and XX; number and segments variable; or in XVII, paired, or unpaired to the right; or in XIX, paired, or unpaired to the right; or in XX, paired, or unpaired to the right. Each papilla small, round, slightly convex .... *Amynthas stricosus*

34. With papillae only front clitellum 35
     With papillae front and behind clitellum 36

35. Genital papilla at anterior edge of VI–IX .... *Amynthas fornicatus*
     Genital papillae on VII, VII, and IX, in pairs .... *Amynthas divergens yunnanensis*

36. Genital papilla between VI–X, and between XVII–XX 37
     Genital papilla between VII–X, and between XVII–XX 41

37. Male pores round or oval on setal line, surrounded by 5–7 circular folds. Genital papillae postsetal, paired in XVII and XIX, or only the first papilla in XIX, or having an additional papilla or an additional pair of papillae in XX, each papilla small, oval, with concave center. Two pairs of papillae present in each of VII and VII, 1 presetal and 1 postsetal, or lacking papillae to varying extent, or having no spermathecal papillae, or with an additional postsetal papilla in VI, each papilla small, round .... *Amynthas meishanensis*
     Genital papilla between VII–X, and between XVII–XX 38

38. Genital papillae postsetal, paired in XIX, an additional pair in XVII, an additional pair in XX, or 2 additional pairs in XVII and XX, each papilla oval, with a concave center surrounded by a few circular folds. Genital papillae present or absent, if present, papillae presetal, postsetal, or both, presetal papillae 1–4 pairs in VII–X, each papilla large, round, disc-like, sometimes only 1 papilla present on some segments, postsetal papillae 2 pairs in VII and VII, similar to presetal ones but smaller, sometimes only 1 papilla present on some segments .... *Amynthas lini*
     Genital papilla VII, VII–IX, and between XVII–XX 39

TABLE 14   Key to the Species of *diffringens* Group of the Genus *Amynthas* From China—cont'd

39. Male pores with 1–8 genital papillae at medioanterior and/or medioposterior portion outside the skin fold. Postclitellar papillae similar in shape and size to those in the preclitellar region, also in transverse patches between male pores; presetal papillae 20–61 in XVIII, 8–59 in XIX and 0–1 in XX; postsetal papillae 0–8 in XVII, 0–5 in XVIII and 0–3 in XX. Genital papillae tiny, round, tubercle-like, and highly variable in number and positions; presetal papillae in transverse patches between setal line and intersegmental furrow; numbering 5–37 in VII and 5–40 in IX .... *Amynthas pavimentus*
    Genital papillae on VII–IX and XVII–XX 40

40. Genital papillae present in, XVII–XX, in 2 longitudinal rows fairly similar to arrangement in the preclitellar region, number on each side presetal 0–1 and postsetal 0–5 in XVII, presetal 1–6 and postsetal 1–4 in XVIII, presetal 1–6 and postsetal 0–2 in XIX. Sometimes, an additional postsetal papilla present medioventrally in XVIII or an additional pair of presetal papillae present in XX. Each papilla small, round, center flat or slightly concave, about 1/2 distance between setal line and segmental furrow, surrounded by a circular fold. Genital papillae present in VII–IX, in longitudinal rows slightly medial to spermathecal pores, number on each side presetal 0–4 and postsetal 0–2 in VII, presetal 1–5 and postsetal 0–2 in VII, presetal 0–6 and postsetal 0–2 in IX. Or an additional presetal papilla present medioventrally in VII and IX or 1 papilla immediately behind each spermathecal pore in 5/6/7/8. Each papilla small, round, tubercle-like, surrounded by a circular fold, its size about 1/3 to 1/4 distance between setal line and segmental furrow .... *Amynthas exiguus aquilonius*
    Genital papillae in horizontal rows closely along the setal lines in XVII, XIX, and rarely XX, similar to those in the spermathecal region. In addition, horizontal rows of smaller genital papillae closely along intersegmental furrows, 1 anterior and 1 posterior to each of 17/18, 18/19 and rarely 19/20. The number of presetal papillae 0–11 in XVII, and 0–14 in XIX; number of postsetal 0–10 in XVII, 5–12 in XIX and 0–7 in XX. Total number 19–34 in 17/18, 23–39 in 18/19 and 0–22 in 19/20. Genital papillae tiny, round, tubercle-like, arranged in horizontal row, 1 presetal and/or postsetal, closely along the setal lines in VII–IX. The number of presetal papillae 0–5 in VII and 4–14 in IX; number of postsetal 0–4 in VII, 9–15 in VII and 5–15 in IX. No preclitellar papilla in vicinity of the spermathecal pore and along intersegmental furrow .... *Amynthas libratus*

41. The genital marking are 3 pairs of transversely oval, slightly raised areas, 1 pair on each of 18/19, 19/20, and 20/21. Each marking has grayish margin, and is slightly concave. In addition, on each of segments VII and VII there is a transversely elongated, smooth, glistening area with bluntly rounded ends; the areas in the region of the setal circles and marking a break therein .... *Amynthas longiculatus*
    Genital papilla between VII–X, and between XVII–XIX 42

42. Male pores on flat-topped papilla, and a similar 1 nearby, also on XVII and XIX postsetally in line with male pore, papillae center concave. Spermathecal pores on small eye-like region, with a pair of similar papillae ventral postsetal on VII, sometimes absent or only singular .... *Amynthas exilens*
    Genital papilla between VII–X, and XVIII and/or XIX 43

43. Genital papilla between VII–X, and XVIII and XIX 44
    Genital papilla between VI–X, and XVIII or XIX 45

44. Male pores with a lateral papilla closely adjacent, both male pore and papilla surrounded by 2–3 slight circular folds. Genital papillae present or absent, if present, widely paired, presetal in XVIII and XIX, slightly medial to male pores, paired, 1 on right, or 1 on left, each papilla round, similar in structure to those in the preclitellar region. Genital papillae present or absent in VII–IX, if present, in VII, presetal papillae 1, medial ventral; in VII, presetal papillae 1 median, paired, 1 median with 1 on right or left, 1 median with 1 on both sides, 2 medians with 1 on either side, or 6 papillae, postsetal papillae 1 on right or left or 2; in IX, presetal papillae 1 on left or 2; each papilla round, center concave .... *Amynthas nanshanensis*
    Genital papillae rarely absent, 1–3 pairs (occasionally up to 5 pairs) usually placed around male pore region; 2 pairs on segment XVIII in front and behind setae and 1 pair on segment XIX in front of setae. The postsetal pair on XVIII often closer to male papilla while the antesetal pair on both segments either situated laterally or medially. Genital papillae beside those near pores less constantly present, often paired on segments VII and IX, those on VII placed either antesetally or postsetally .... *Amynthas pingi pingi*

45. Paired papillae found ventrally on VII–VII or XIX .... *Amynthas mediocus*
    Paired papillae found ventrally on VI–VII and XVIII 46

46. Male pores on tubercles of XVIII, with 4–5 minute papillae on anterior, posterior, and lateral sides. Spermathecal pores on a small elliptical area with 1 minute papilla immediately in front, another papilla placed anteriorly but slightly medially, 1 or sometimes 2 placed (on succeeding 1) posteriorly but rather laterally, the area between pore and papillae anteriorly and posteriorly often glandular in character, a similar 1 found on ventromedian line of segment VII behind setae .... *Amynthas directus*
    Genital papilla between VI–X, and XVIII 47

47. Male pores lateral 2 pairs small circular genital markings presetal, postsetal in XVIII, Genital papillae presetal VII–IX, additionally in VII, or VII and VI; postsetal VII, VII, or VII only or none postsetal, either in line with spermathecal pores or just median to line of pores .... *Amynthas corticus*
    With less or more than papillae one pair or two pairs or absent in male pore region 48

48. Male pores on a large and rounded papilla, 1–2 genital papillae often present close to each male papilla, 1 often posterior and lateral or slightly medial to male papilla, while the other anterior and lateral to, or often in line with, the posterior papilla. Genital papillae in spermathecal pores region divided into 2 groups: 1 close to spermathecal pores and the other on ventral side. One small elevated and flat-topped papilla placed either anteriorly or posteriorly to each pore, more often to its anterior side, sometimes not all 4 on each side constantly present, 1 or more disappearing. Two or 3 pairs often paired on ventral side of segments VII, VII, IX, usually widely separated or rather in close approximation, about 2 mm apart, each slightly elevated with a thick circular margin, flattened or depressed in center. Those on VII and VII more constant, or 3 pairs totally absent in a few cases .... *Amynthas heterochaetuss*
    With fewer than 2 pairs papillae in male pore region 49

Continued

2. SYSTEMATICS

TABLE 14   Key to the Species of *diffringens* Group of the Genus *Amynthas* From China—cont'd

49. Male pores on a round porophore surrounded by 3–5 circular folds, occasionally with 1 small papilla anteromedial to it. Spermathecal pores eye-like, Genital papillae round, variable in number, position and arrangement. Presetal papillae 1 median, closely paired or widely paired in VII–IX, occasionally also in X, immediately adjacent to intersegmental furrows, with or without a concave center .... *Amynthas amis*
   With fewer than 2 pairs papillae in spermathecal pore region 50

50. Male pores on a top of papilla surrounded by several circular ridges, with a small papilla with depressed center at its medial side. Genital papillae in this region 2 pairs in VII and VII, occasionally 3 pairs in VII, VII, and IX, or only 1 pair in VII, or absent. Each papilla rounded, with slightly depressed center, situated in front of setal line in a shallow furrow between setal annulus and anterior annulus, medial to spermathecal pore at a distance of about 2 or 3 setal intervals .... *Amynthas diffringens*
   Male pores on a depressed porophore, with a small papilla adjacent laterally, both surrounded by 2–3 circular folds, 2 postsetal genital papillae present in XVIII, each adjacent medially to male pore and close to setal line, about 10–11 intersetal distances apart, or 2 closely arranged in middle between the paired male pores, each papilla round, similar in structure to those in the preclitellar region. Genital papillae present postsetal in VII and presetal in IX, widely paired or with an additional 1 on medioventrum, occasionally, additional presetal papillae present in VII, each papilla round, center flat or slightly concave .... *Amynthas uvaglandularis*

56. *Amynthas ailiaoensis* James, Shih & Chang, 2005

Amynthas ailiaoensis James, Shih & Chang, 2005. *J. Nat. Hist.* 39(14):1020–1021.
Amynthas ailiaoensis Xu & Xiao, 2011. *Terrestrial Earthworms of China.* 79.

External Characters

Length 215–310 mm, width 9.0–12.5 mm, segment number 110–140. Prostomium epilobous, with tongue open. First dorsal pore in 12/13. Clitellum XIV–XVI, annular, setae invisible externally. Setae regularly distributed around segmental equators, size uniform; *aa:ab yz: zz* = 1:1:1:2.5 at XXV, setae number 86–96 (VII), 90–94 (X), 126–140 (XX); 20 between male pores. Male pores within shallow circular invagination of body wall, in each of which 1 oblong longitudinally orientated genital marking is folded in half transversely; about 1/5 of body circumference apart. Spermathecal pores 4 pairs, in midlateral to slightly ventral of midlateral 5/6/7/8/9. Female pore 1 in XIV, medioventral.

Internal Characters

Septa 8/9/10 absent, 5/6/7/8 and 10/11/12/13/14 muscular. Gizzard in VII. Intestine swelling in 1/2 XV. Caeca simple, margins smooth, in XXVII, extending anteriorly to XXIII. Hearts in X–XIII. Testes, funnels in ventrally joined sacs in XI, with large dorsal lobes; Seminal vesicles large in XI, with large dorsal lobes; other contents of IX sometimes enclosed in thin sac. Prostate glands in XVIII, 7 main lobes, each lobe served by 2–3 small ductulets radiating fan-like from ental end of prostatic duct; ducts stout, muscular, narrowing towards body wall; vasa deferentia join duct at duct-glandular portion junction; vasa deferentia nonmuscular. Spermathecae 4 pairs, in VI–IX, ampulla pear-shaped, duct much shorter than ampulla, diverticulum with small ovate chamber, stalk slender, tightly convoluted in hairpin loops enclosed in membrane; no nephridia on spermathecal duct.

Color: Preserved specimens brown pigment on dorsal third, intensity variable.
Type locality: Taiwan (Wutai, Pingdong).
Distribution in China: Taiwan (Wutai, Pingdong).
Etymology: This species is named after the Ailiao River of the Wutai region.
Deposition of types: National Museum of Natural Science, Taichung, Taiwan.
Remarks: The fourth octothecal proandiric *Amynthas* is much more similar to Gates' (1959) material from Chao-Chow (=Chaojhou), Pingdong, and Green

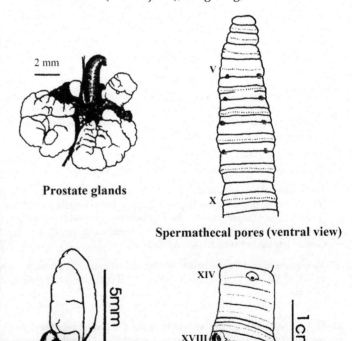

Prostate glands

Spermathecal pores (ventral view)

Spermatheca

Male pore (ventral view)

FIG. 63   *Amynthas ailiaoensis* (after James et al., 2005).

mountain (=Yangmingshan), Taibei than the others. Differences from Gates' material are few. It is possible that these are the same species. *Amynthas ailaoensis* has paired genital markings within a shallow invagination of the male field, lacks septum 8/9, has much shorter caeca, and the spermathecal pores are at or above midlateral rather than "well towards mL," which might mean below midlateral but close. Many small nematodes were found in the body cavity, mainly around the caeca, prostates and seminal vesicles.

57. *Amynthas amis* Shen, 2010

Amynthas amis Shen, 2012. *J. Nat. Hist.* 46(37–38): 2261–2267.

External Characters

Small to medium; Length 53–183 mm, width 3.5–4.4 mm (clitellum), segment number 77–115.

Prostomium epilobous. First dorsal pore in 11/12 or 10/11. Clitellum XIV–XVI, annular. Setae number 28–42 (VII), 40–48 (XX), 8–12 between male pores. Male pores paired in XVIII, a little less than 0.23–0.27 body circumference ventrally apart. Each pore on a round porophore surrounded by 3–5 circular folds, occasionally with 1 small papilla about 0.25 mm in diameter anteromedial to it. No other papilla in the postclitellar region. Spermathecal pores 4 pairs in intersegmental furrows of 5/6/7/8/9, distance between paired pores 0.25–0.32 body circumference apart. Genital papillae round, variable in number, position, and arrangement. Presetal papillae 1 median, closely paired or widely paired in VII–IX, occasionally also in X, immediately adjacent to intersegmental furrows. Each papilla 0.3–0.7 mm in diameter, with or without a concave center. No preclitellar papilla in the postsetal portion and in the vicinity of spermathecal pores. Female pore single in XIV, medioventral.

FIG. 64 *Amynthas amis. From Shen, H.P., 2012. Three new earthworms of the genus Amynthas (Megascolecidae: Oligochaeta) from eastern Taiwan with redescription of Amynthas hongyehensis Tsai and Shen. J. Nat. His. 46(37–38), 2259–2283.*

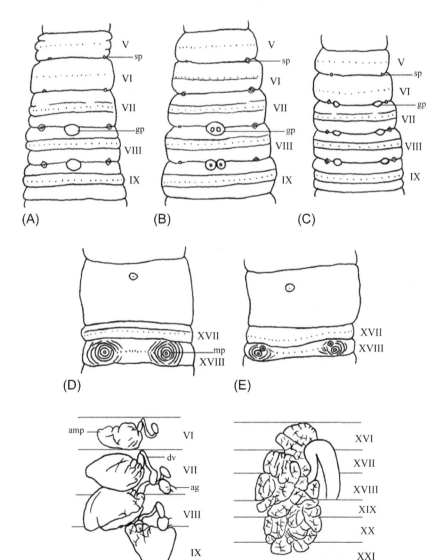

Internal Characters

Septa 8/9/10 absent, 5/6/7/8 and 10/11/12/13/14 thick. Nephridial tufts thick on anterior faces of 5/6/7 septa. Gizzard large, round in VII–X. Intestine enlarged from XVI. Intestinal caeca paired in XXVII, extending anteriorly to XXIII–XXIV, each simple, slender, straight or slightly bent at distal end. Hearts in X–XIII. Holandric, testes large, oval, 2 pairs in ventrally joined sacs X–XI. Seminal vesicles large, smooth, 2 pairs in XI–XII, occupying the full segmental compartment or 1.5 segments, with a round dorsal lobe. Prostate glands large, lobed with fillicular surface, occupying 3–5 segments in XVI–XX. Prostatic duct long, C-shaped in XVIII or U-shaped in XVII–XVIII, distal end enlarged. Accessory glands sessile, round, corresponding to external genital papilla anteromedial to the male pore if present. Spermathecae 4 pairs in VI–IX, varied in size and shapes. Each ampulla round or elongated oval-shaped, surface wrinkled 0.68–3.1 mm long and 0.45–1.82 mm wide, with a slender to stout spermathecal stalk chamber of 0.15–0.8 mm in length. Diverticulum with an iridescent, oval-shaped seminal chamber of 0.15–0.8 mm long and a slender stalk of 0.3–1.2 mm in length. Accessory glands short-stalked, round or slightly round, 0.45–0.85 mm in total length, corresponding to external genital papillae.

Color: Preserved specimens brown to dark brown on dorsum, light brown on ventrum, and light orange-brown to dark brown on clitellum.
Distribution in China: Taiwan (Changliang Forest Road, Jhuohsi, Hualien County; Yenping Forest Road near Hongyeh, Taidong County).
Etymology: This species' name "amis" is given because most of its distribution range overlaps with areas inhabited by the native Amis tribe living in the East Rift Valley and Pacific Coastal Plain east of the Central Mountain Range.
Deposition of types: Taiwan Endemic Species Research Institute, Jiji, Nantou, Taiwan.
Habitats: In elevations of 650–700 m, along the Changliang Forest Road; and at an elevation of 1470 m along the Yenping Forest Road, Taidong County.
Remarks: The genital papillar arrangement in A. amis (Shen, 2010) of widely paired papillae present in VII–IX for some individuals, and simple male pore structure with or without an anteromedial papilla are fairly similar to that of A. corticis (Kinberg, 1867), A. divergens (Michaelsen, 1892) from Japan, A. divergens yunnanensis (Stephenson, 1912) from Yunnan, China, and A. nipponicus (Beddard, 1938) from Korea. These 6 species are also members of the corticis species group. All of them have small or less developed seminal vesicles. Considerable variations in the structure and size of spermathecae and prostate glands were found in A. divergens (Ohfuchi 1937),

considerable variations in the size and shape of spermathecae were reported in A. nipponicus (Ohfuchi 1937), prostate gland absent or poorly developed was recorded for A. oyamai (Ohfuchi 1937; Kobayashi 1938a), and prostate gland (absent in most specimens) was found in A. corticis (Chen 1931, 1933), A. nipponicus (Ohfuchi 1937), and A. morii (Kobayashi 1938b). The single specimen of A. divergens yunnanensis has no prostate glands either (Stephenson, 1912). In contrast to the reproductive organ degeneration in the species, A. amis has large seminal vesicles and prostate, and iridescence on all 4 pairs of its spermathecae.

The simple male pore structure of A. amis is also similar to that of A. toriii (Ohfuchi, 1941) from Japan. A. toriii is much smaller (37–43 mm in length), has no genital papillae, and has a higher setal number than A. Amis with 43–45 in VII and 56–58 in XX, and slender, straight prostatic ducts (Ohfuchi, 1941). Therefore, the 2 species can be easily distinguished. A. toriii was also placed in synonymy of A. corticis in Easton (1981). It is certainly a valid species as discussed in Tsai et al. (2007).

58. Amynthas biorbis Tsai & Shen, 2010

Amynthas biorbis Tsai & Shen, 2010. J. Nat. Hist. 44(21–24):1260–1263.

External Characters

Small earthworm; Length 61–86 mm, width 2.63–3.32 mm (clitellum), segment number 72–97. Prostomium epilobous. First dorsal pore in 11/12. Clitellum XIV–XVI, annular, length 2.05–2.4 mm. Setal number 33–40 (VII), 41–48(XX); 8–12 between male pores. Male pores paired on setal line in XVIII, 0.24–0.3 body circumference ventrally apart. Each pore on a round porophore about 0.3 mm in diameter with concave center. A pair of genital papillae in XVIII, each papilla round, about 0.5 mm in diameter (larger than male porophore), closely adjacent to the male porophore at the medioposterior portion immediately posterior to the setal line; both porophore and genital papilla surrounded by 1–2 skin folds. Spermathecal pores 4 pairs, each small, not easily detectable, buried deeply in intersegmental furrows of 5/6/7/8/9; distance between paired pores 0.28–0.29 body circumference ventrally apart. No genital papillae in the preclitellar region. Female pore 1 in XIV, medioventral.

Internal Characters

Septa 8/9/10 absent, 5/6/7/8 and 10/11/12/13/14 thick. Nephridial tufts thick on anterior faces of 5/6/7. Gizzard large, round in VII–X. Intestine swollen from XV. Intestinal caeca paired in XXVII, extending anteriorly to XXIII–XXIV, each simple, slightly folded, the end straight or bent dorsally. Hearts in XI–XIII. Holandric, Testes small

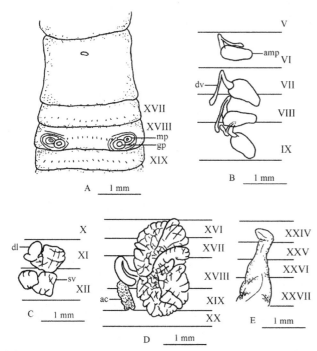

FIG. 65   *Amynthas biorbis* (after Tsai & Shen, 2010).

or large, 2 pairs in ventrally joined sacs in X and XI. Vas efferens connected in XII on each side to form a vas deferens. Seminal vesicles 2 pairs in XI and XII, each small, with a prominent and round or peach-shaped dorsal lobe. Prostate glands large, deeply divided into 5–6 lobes with follicular surface in XVI–XX. Prostatic duct U-shaped, in XVIII or XVII–XVIII, proximal portion enlarged. A large sessile accessory gland with folliculated patch extending longitudinally near the proximal end of the prostatic duct, corresponding to each external genital papillar. Spermathecae 4 pairs in VI–IX, each with a conical or peach-shaped ampulla of 0.9–1.11 mm long and 0.45–0.68 mm wide, and a slender to stout spermathecal stalk of 0.45–0.79 mm that is about one-half to two-thirds of ampullar length. Diverticulum with a small, oval-shaped seminal chamber and a slender stalk 0.35–0.75 mm in length, about the same length as spermathecal stalk.

Color: Preserved specimens pinkish brown on head and dorsum, light brown on ventrum, and dark pinkish brown on clitellum.
Distribution in China: Taiwan (Hsinwu and Hongyeh, Taidong County).
Etymology: The name *biorbis* is given to this species to indicate the large, paired genital papillae adjacent to the male porophores.
Deposition of types: Taiwan Endemic Species Research Institute, Jiji, Nantou, Taiwan.
Remarks: *A. biorbis* (Tsai & Shen, 2010) is endemic to the eastern slopes of the Central Mountain Range at elevations of 400–1000 m in southern Taiwan. *A. biorbis*

differs from *A. uvaglandularis* in having no genital papilla in the preclitellar region, and only 1 pair of large and round papillae in XVIII located posterior to setal line and closely adjacent to male pore porophores. *A. uvaglandularis* has genital papillae in VII, IX, and XVIII, whose number and positions are highly variable (Shen et al., 2003).

A. biorbis is also fairly similar to *A. corticis*. However, the former is smaller, with a pair of large papillae medioposterior to male pores and no genital papilla in the preclitellar region. These characters are consistent among *A. corticis* individuals and distinctly different from the variable papillae arrangements in the preclitellar and male pore region of *A. corticis*.

59. *Amynthas corticus* (Kinberg, 1867)

Perichaeta corticis Kinberg, 1867:102.
Amynthas corticus Sims & Easton, 1972. *Biol. J. Soc.* 4: 235.
Amynthas corticis Blakemore, 2003, *Organisms Diversity & Evolution.* 3(3): 241–244.
Amynthas cortices James, Shih & Chang, 2005. *J. Nat. His.* 39(14):1023–1024.
Amynthas corticus Chang, Shen & Chen, 2009. *Earthworm Fauna of Taiwan.* 38–39.
Amynthas cortices Xu & Xiao, 2011. *Terrestrial Earthworms of China.* 97–98.

External Characters

Length 96–184 mm, clitellum width 3.6–3.8 mm, segment number 93–118. Prostomium epilobous. First dorsal pore 11/12. Clitellum XIV–XVI, annular, setae absent, dorsal pore absent. Setae regularly distributed around segmental equators, setal number: 36–40 (VII), 40–46 (XXV); 10–14 between male pores. Male pores pairs in XVIII, on depressed corny pads, lateral 2 pairs small circular genital markings presetal and postsetal, 0.24 circumference apart ventrally. Spermathecal pores 4 pairs in 5/6/7/8/9, 0.28 circumference apart ventrally. Genital papillae presetal VII–IX, additionally in VII, or VII and VI; postsetal VII, VII, or VII only or none postsetal, either in line with spermathecal pores or just median to line of pores. Female pore 1 in XIV.

Internal Characters

Septa 8/9/10 absent, 5/6/7/8 and 10/11/12/13/14 thinly muscular. Gizzard in VII–X. Intestine from XVI. Intestinal caeca paired in XXVII, simple, long and slender, extending anteriorly to XXIII. Hearts paired in XI–XIII. Testis sacs 2 pairs in X and XI, ventrally joined. Seminal vesicles paired in XI and XII, large, with dorsal lobe. Prostate glands paired in XVIII, 3–4 deeply incised main lobes covering XVI–XX; prostatic duct thick, muscular in long hairpin loop forward to XV, vasa deferentia join at duct-glandular portion junction, vasa deferentia nonmuscular; genital marking gland lacking in XVIII. Spermathecae 4 pairs in VI–IX, ampulla ovoid, diverticulum blunt ovoid, with straight stalk; diverticulum axis shorter than ampulla axis; small sessile genital marking glands in spermathecal segment.

Color: Live specimens unpigmented or slight greenish brown shading on dorsalmost one-third. Preserved specimens unpigmented or slight greenish brown shading on dorsalmost one-third.
Type locality: Oahu. Types in Stockholm.
Deposition of types: Stockholm Museum, Sweden.
Global distribution: Cosmopolitan. This species is the most widely distributed species of the pheretimoid group. It has been recorded in many countries in east and southeast Asia, Australia, New Zealand, Europe, UK, USA, South America, and South Africa. The indigenous range of the species is believed to be in east and southeast Asia: Nepal, northern Pakistan, India, Myanmar, and southern China. It is also found in Taiwan, Korea, and Japan.
Distribution in China: Taiwan (Pingdong, Kinmen).
Remarks: Michaelsen (1900: 275) included the prior *Pheretima corticis* (Kinberg, 1867) in possible synonymy under *Pheretima indica* (Horst, 1883), and now it is uncertain whether all or any of the *indica* subspecies follow it into synonymy of *corticis* or whether they assume separate specific status. These subspecies are listed by Sims and Easton (1972: 235) as: *A. indicus cameroni* (Stephenson, 1932) from the Malay Peninsula; *A. indicus ceylonicus* (Michaelsen, 1897) from Sri Lanka; and *A. perkinsi* (Beddard, 1896) from Halemanua and Kauai, Hawaii, which was included by Michaelsen (1900: 276) as a "variety" of *indica*. *A. corticis* has been reported under the names of its junior synonyms as *Pheretima diffringens* or *P. peregrina* or, in earlier papers, as *P. heterochaeta*, and sometimes the name is misspelled as "corticus."

60. *Amynthas cruxus* Tsai & Shen, 2007

Amynthas cruxus Tsai & Shen, 2007. *J. Nat. Hist.* 41 (5–8):368–371.
Amynthas cruxus Chang, Shen & Chen, 2009. *Earthworm Fauna of Taiwan.* 40–41.
Amynthas cruxus Xu & Xiao, 2011. *Terrestrial Earthworms of China.* 98–99.

External Characters

Medium-sized earthworms. Length 100–170 mm, width 3.32–4.76 mm (clitellum), segment number 91–120. Prostomium epilobous. First dorsal pore in 11/12. Clitellum XIV–XVI, setae and dorsal pore absent. Setal number 27–37 (VII), 39–52 (XX); 11–13 between male pores. Male pores paired in XVIII, 0.24–0.26 body circumference

ventrally apart, each on a teat-like porophore on a flat, diamond-shaped disc (male disc), on which 4 genital papillae are arranged in the form of a cross around the porophore: 1 anterior, 1 posterior, 1 lateral, and 1 medial. This male disc is surrounded by 3–4 diamond-shaped skin folds. Each of the genital papillae about 0.2 mm in diameter, white in color, with a slightly depressed center. No other genital papillae in the postclitellar region. Spermathecal pores 4 pairs in intersegmental furrows of 5/6/7/8/9, each lip-like, about 0.3 body circumference ventrally apart. No genital papillae in the preclitellar region. Female pore 1 in XIV, medioventral.

Internal Characters

Septa 8/9/10 absent, 5/6/7/8 and 10/11/12/13/14 thick. Gizzard in IX–X, large, bell-shaped. Intestine swelling in XV or XVI. Intestinal caeca paired in XXVII, each simple, stocky, extending anteriorly to XXIV–XXV. Hearts enlarged from XI–XIII. Holandric; testis sacs paired in X and XI, round, shiny. Seminal vesicles paired in XI and XII, large, finely folliculated, each with a round or cone-shaped dorsal lobe. Prostate glands paired in XVIII, lobed, wrinkled, occupying 3–4 segments in XVII–XX. Prostatic duct unusually large, U-shaped, occupying 3 segments in XVI–XVIII, proximal half slender and enlarged. No accessory glands in the postclitellar region. Spermathecae large in VI–IX, ampulla oval- or pear-shaped, 1.9–2.57 mm long, 1.4–1.83 mm wide, with a slender to stout spermathecal stalk of 0.6–0.95 mm in length. Diverticulum with a round or oval, iridescent seminal chamber, 0.45–0.8 mm in length, and a slender stalk of 0.92–1.15 mm in length. Diverticulum stalk slightly longer than spermathecal stalk. No accessory glands in the preclitellar region.

Color: Preserved specimens light greyish brown on dorsum, light grey on ventrum and brown around clitellum.
Distribution in China: Taiwan (Taoyuan, Kaohsiung County).
Etymology: The specific name *cruxus* (=cross) is given to the species with reference to the cross formation of the genital papillae around the male pore.
Deposition of types: Taiwan Endemic Species Research Institute, Chichi, Nantou, Taiwan.
Type locality: The specimens were collected at an elevation of 1500 m from Tengchih, Taoyuan, Kaohsiung County.
Remarks: *A. cruxus* Tsai & Shen, 2007 is closely related to *A. pingi* (Stephenson, 1925) of China (Chen, 1933) and *A. hatomajimensis* (Ohfuchi, 1957) of the Ryukyus. These 3 species share a unique character in having a slightly elevated, flat, and smooth or granular male disc, on which there is a teat-like male porophore associated with genital papillae in a well-arranged pattern, and then surrounded by a groove or by one to a few circular skin folds (rings). This male disc in the 3

FIG. 66   *Amynthas cruxus* (after Tsai et al., 2007).

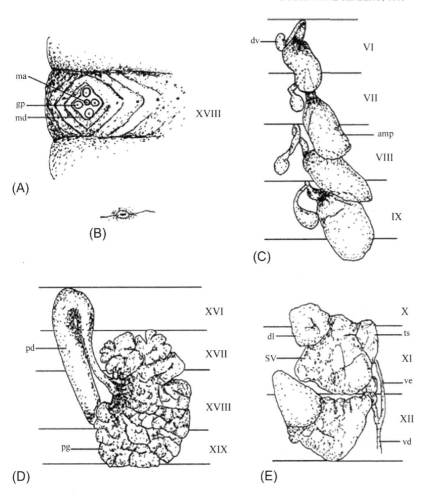

species is easily distinguishable from the male pore region of *A. corticis* (Kinberg, 1867), that has a teat-like or small disc-like elevated porophore, on which the male aperture is present, but its associated genital papillae if present (1 or 2) are located in an irregular pattern (not consistent arrangement) on first and second skin folds surrounding the porophore (not directly on the disc-like porophore). This flat male disc was noted for *hatomajimensis* by Ohfuchi (1957), and also for *pingi* by Chen (1933).

A. *cruxus* is easily distinguishable from *A. pingi* and *A. hatomajimensis* by the character of the genital papillae, length of spermathecal stalks, and shape and structure of the prostatic duct. *Amynthas cruxus* has 4 genital papillae on the male disc in the form of a cross (1 anterior, 1 posterior, 1 lateral, and 1 medial, the latter 2 on setal line), whereas *hatomajimensis* has 3 papillae (1 anterior, 1 posterior, and 1 medial on setal line), and *pingi* has 2 (antero-medial and posteromedial). Also, *cruxus* has an unusually large, thick, muscular, U-shaped prostatic duct, covering 3 segments from XVI to XVII, differing significantly from *hatomajimensis* and *pingi* that have a short duct covering the posterior half of XVII to the anterior half of XVIII (Chen, 1933; Ohfuchi, 1957). *Amynthas*

cruxus and *A. hatomajimensis* have a spermathecal stalk of which the length is about one-third of the ampullar length, whereas for *pingi* the stalk is along as the ampullar length. In addition, accessory glands associated with the genital papillae are absent in *cruxus*, present near sper-mathecae in *hatomajimensis*, and often present in VII, IX (occasionally in X), XVIII, and XIX in *pingi*. Also, *hatoma-jimensis* is shorter in length and has a lower segment num-ber than *cruxus* and *pingi*.

A. *cruxus* is also similar to *A. nanrenensis* (James et al., 2005) in having 4 pairs of spermathecae in VI–IX and small genital papillae around each male pore. However, *nanrenensis* has a much higher setal number than *cruxus* with 60–64 in VII and 62–72 in XX for the former, and 27–37 in VII and 39–52 in XX for the latter.

61. *Amynthas diffringens* (Baird, 1869)

Megascolex diffringens Baird, 1869. *Proc. Zool. Soc. Lond.* 1869:40–43.
Pheretima diffringens Gates, 1935a. *Amer. Mus. Novitates.* 141:6–9.
Pheretima diffringens Gates, 1939. *Proc. U.S. Natr. Mus.* 85:430–431.

Pheretima diffringens Chen, 1946. *J. West China. Border Res. Soc. (B).* 16:135.

Pheretima diffringens Tsai, 1964. *Quer. J. Taiwan Mus.* 17:2–4.

Amynthas diffringens Sims & Easton 1972. *Biol. J. Soc.* 4: 235.

Amynthas diffringens Xu & Xiao, 2011. *Terrestrial Earthworms of China.* 102–103.

External Characters

Length 114–184mm, width 4.5–5.0mm (clitellum), segment number 93–109. Prostomium prolobous or 1/3 epilobous. First dorsal pore in 12/13, occasionally in 11/12. Clitellum XIV–XVI, annular, smooth, without setae and dorsal pores. Setae in preclitellar segments, especially on the ventral surface, larger and more widely distributed, so that only about 4 setae in larger specimens and 6–8 in small specimens between genital papillae in spermathecal region. Setae number 24 (II), 28 (III), 29 (IV), 34 (VI), 37 (VII), 50 (XX); 18–32 between spermathecal pores, 11–14 between male pores. Male pores 1 pair, on lateroventral side of XVIII, each pore situated on a top of papilla surrounded by several circular ridges, with a small papilla with depressed center at its medial side. This portion slightly raised. Spermathecal pores 4 pairs in 5/6/7/8/9, situated rather ventrally, eye-like, 1/3 body circumference apart. Genital papillae 2 pairs in VII and VII, occasionally 3 pairs in VII, VII, and IX, or only 1 pair in VII, or absent. Each papilla rounded, with slightly depressed center, situated in front of setal line in a shallow furrow between setal annulus and anterior annulus, medial to spermathecal pore in a distance of 2 or 3 setal intervals. Female pore single in XIV, medioventral, in a small ovoid tubercle.

Internal Characters

Septa 8/9/10 absent, 5/6/7/8 and 10/11–14/15 thickened. Gizzard large, in IX–X, bell-shaped. Intestine swelling in XIV. Intestinal caeca simple, in XXVII, extending anteriorly to XXIV, smooth. Last hearts in XIII. Testis sacs 2 pairs, in X–XI, each pair connected on ventromedial sides into a V shape, Sperm ducts meeting in XII. Seminal vesicles paired in XI–XII, anterior pair small only occupying one-third of segment, each vesicle divided transversely and completely into 2 lobes, anterior lobe larger than posterior, with a large elongated dorsal lobe. Posterior pair large, occupying full segment, each vesicle divided incompletely into 2 parts by a transversal lateral wide groove, with a large elongated dorsal lobe, each vesicle subdivided into many small pieces on surface by irregular shallow grooves, white in color, and each dorsal lobe smooth on surface and yellow or white in color. Prostate gland, rudimental or absent. If present, its shape is variable, or with right 1 well developed only, occupying 3 segments and is divided into small pieces by narrow grooves on surface; or it is rudimental with variable

structure and absent from 1 side, their forms usually round in shape, only occupying segment XVIII, or separated distally in to 2 lobes connected by branches of prostatic duct, occasionally with transparent vestigial parts attaching on its border; or it is completely absent from both sides. Prostatic duct enlarged and hooked. Accessory gland is of compact round mass, usually concealed in the connective tissue of skin. Spermathecae 4 pairs in VI–IX, anterior pair with oval ampulla, distal borders serrated or smooth, and a slender and straight stalk being longer than the length of ampulla, its diverticulum with oval-shaped seminal chamber and a slender and straight stalk, as long as the stalk of spermatheca. The last pair, with heart-shaped ampulla with serrated or smooth distal border, and a short stalk being shorter than the length of ampulla, its diverticulum with oval shaped seminal chamber and a slender and straight stalk.

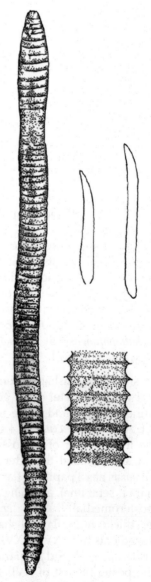

FIG. 67 *Amynthas diffringens* (after Gates, 1939).

Color: Preserved specimens brilliant purplish brown on dorsal surface except setal line which is light gray as well as the color on ventral surface, darker in preclitellar segment. Clitellum yellow or light gray in color. The dorsomedial line with a dark purplish-brown longitudinal line.

Type locality: Plas Machynlleth, North Wales.

Distribution in China: Jilin (Gongzhuling), Jiangsu, Zhejiang, Anhui, Fujian, Taiwan (Taibei), Jiangxi, Hubei (Lichuan), Hong Kong, Hainan, Chongqing (Nanchuan, Shapingba, Beipei, Banhe, Jiyunshan, Mt. Jingfo, Pingsha), Sichuan (Chengdu, Leshan, Mt. Emei , Pingshan), Guizhou (Mt. Fanjingshan).

Deposition of types: British Museum.

Remarks: Stephenson's specimen of *A. divergens* is quite clearly *A. diffringens*. One of the 3 paratypes of Chen's *A. corrugatus* rather obviously refers to *A. diffringens*.

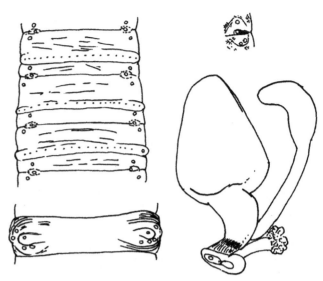

FIG. 68　*Amynthas directus* (after Chen, 1935b).

62. *Amynthas directus* (Chen, 1935)

Pheretima directa Chen, 1935b. *Bull. Fan. Mem. Inst. Biol.* 6:42–47.
Amynthas directus Xu & Xiao, 2011. *Terrestrial Earthworms of China.* 103–104.

External Characters

Length 96–102 mm, width 4.0 mm, segment number 96–138. Prostomium 1/3 epilobous. First dorsal pore in 12/13. Clitellum XIV–XVI, smooth and very short, its length equal to 2 immediate preclitellar and 3.5–5.0 post-clitellar segments with no intersegmental furrows and setal. No setae particularly enlarged, those on ventral side of V and VI slightly longer but not widely spaced. $aa = 1.2ab$, $zz = 2$–$2.5zy$; setae number 28–32 (III), 40–44 (VI), 42–44 (VII), 41–44 (XII), 40–48 (XXV); 13–16 (VI), 15–18 (VII), 15 (VII) between spermathecal pores; 10–12 between male pores. Male pores 1 pair, on a partly raised transverse tubercle of XVIII, about 1/3 of circumference ventrally apart, with 4 or 5 minute flat-topped papillae on anterior, posterior, and lateral sides, 2–3 on posterior, 1–2 on either lateral or anterior side. Spermathecal pores 4 pairs, slightly closer ventrally than male pores, about 7/24 of circumference apart, in 5/6/7/8/9 intersegmental, each on a small elliptical area with 1 minute papilla immediately in front, another papilla placed (on preceding segment) anteriorly but slightly medially, 1 or sometimes 2 placed (on succeeding 1) posteriorly but rather laterally, the area between pore and papillae anteriorly and posteriorly often glandular in character, a similar one found in ventromedian line of segment VII behind setae. Female pore 1 in XIV, medioventral.

Internal Characters

Septa 9/10 absent; 8/9 present but very thick, 5/6/7/8 rather thick, 10/11/12/13 slightly thickened, 13/14 less so, rest of septa membranous. Gizzard small, barrel-shaped, very slightly narrower anteriorly, smooth and glittering on surface, longitudinal capillaries small; portion of esophagus anterior to gizzard muscular and distended. Intestine swelling in XV. Intestinal caeca paired in XXVII, extending anteriorly to XXIV, smooth both sides. Hearts 4 pairs, first pair slender, about half the size of following pair. Testis sacs in X and XI, anterior pair large and round, connected medially, in thick V shape; posterior pair broadly connected. Testis round and compact, seminal funnel very large. Seminal vesicles in XI and XII, large, nearly meeting dorsally, each with a dorsal lobe which is elongate and large, about the size of the vesicle. Prostate glands fairly large, in XVI–XIX or sometimes XV–XXI, subdivided into lobules; prostatic duct slender at ental fifth, equally stout throughout, or slightly larger near ectal end. No copulatory chamber. Accessory glands very large in comparison with small papillae externally. Each papilla associated with a round patch of gland, glandular portion granular or finely tubercular on surface, with a common stalk composed of separate cords. Spermathecae 4 pairs, in VI–IX; ampulla usually heart-shaped, or elongate ovoid, smooth on surface, about 1.2 mm long and 1.0 mm wide, its duct wide but shorter than ampulla; diverticulum as long as, or longer than, main pouch; its ental third (or sometimes half) dilated, forming seminal chamber, which is usually straight and finger-shaped, distinguished from its duct. The latter generally longer than main duct, about one-third to one-fifth of the width of chamber, always straight.

Color: Preserved specimens chestnut brown on dorsal side, dark chocolate anteriorly; setal zone whitish, interrupted on dorsomedian line, brownish gray ventrally in middle body, paler at both ends. Clitellum chocolate brown.

Type locality: Hong Kong

Distribution in China: Hong Kong

Deposition of types: Mus. Fan. Inst. Biol. Beijing.

Remarks: This species is close to *Amynthas glandulosus* (Rosa 1896) and *Amynthas hippocrepis* (Rosa 1896), and to some extent to *Amynthas modiglianii* (Rosa 1889). All 3 species came from the islands close to Sumatra. Beddard in his revision of the genus (1900) united the former 2 and Michaelsen (1922) agreed. *Amynthas hippocrepis* differs from *Amynthas glandulosus* by absence of setae on clitellum (with trace of setal circle), greater setal number per segment, fewer papillae at male pore and absence of them in spermathecal pore region. *Amynthas modiglianii* was less adequately described and possessed no genital papillae. *Amynthas jacobsoni* is, however, distinguished by much closer approximation of both male and spermathecal pores. The essential difference between the present form and Rosa's species lies in the character of the diverticulum. The diverticulum in all of Rosa's 3 species is tubular and twisted in a zigzag manner. In *Amynthas glandulosus*, according to this figure, it is coiled about 5 times without a distinct seminal chamber. In the present form, the duct and seminal chamber are distinct and never coiled. Besides, the setal number seems to be smaller and the clitellum is always smooth and without setae. The male pore aspect is less than that in the Sumatran forms (up to 80 in *Amynthas hippocrepis*). The probably like that of *Amynthas hippocrepis* and the spermathecal aspect is exactly like that of *Amynthas glandulosus*. If the points of difference as described above can be clarified by further study of more material from the present locality, this species may prove to be synonymous with that distributed in Sumatra.

63. *Amynthas divergens* (Michaelsen, 1892)

Perichaeta divergens Michaelsen, 1892. *Arch. F. Naturg.* 243.

Amynthas divergens Beddard, 1900. *Proc. Zool. Soc. London.* 1900:625.

Amynthas divergens divergens Sims & Easton 1972. *Biol. J. Soc.* 4: 235.

Amynthas divergens divergens Xu & Xiao, 2011. *Terrestrial Earthworms of China.* 104–105.

External Characters

Length 120 mm, width 3.0 mm, segment number 120. Prostomium 1/2–1/3 epilobous. First dorsal pore in 12/13. Clitellum XIV–XVI. Setae of anterior segments stronger, medium size, setae number 11–13 (V), 11–14 (VI), 12–15 (VII), 13–15 (VIII) between spermathecal pores; 8–10 between male pores. Male pores 1 pair, about 1/3 of circumference ventrally apart. Spermathecal pores 4 pairs, in 5/6/7/8/9, about 1/3 of circumference apart, paired papilla on VII–IX, to which correspond internally stalked glands; also on XVII–XX.

Internal Characters

Septa 9/10 absent. Intestine swelling in XV. Testis sacs in X and XI, anterior pair large and in thick V shape; posterior pair with rectangular shape. Testis round and compact, seminal funnel very large. Seminal vesicles in XI and XII, oval, quite small, each with a relatively large dorsal lobe, dorsal lobe obvious contraction, smooth. Prostate glands of medium size, in XVI–XIX, subdivided into lobules and quite smooth; prostatic duct thin, medium length, C-shaped ring. Spermathecae 4 pairs, in VI–IX, small.

Color: ?

Distribution in China: Yunnan (Tengyuan).

64. *Amynthas divergens yunnanensis* (Stephenson, 1912)

Pheretima divergens yunnanensis Stephenson, 1912. *Records of the Indian Museum.* (VII):274–276.

Amynthas divergens yunnanensis Sims & Easton, 1972. *Biol. J. Soc.* 4:235.

Amynthas divergens yunnanensis Xu & Xiao, 2011. *Terrestrial Earthworms of China.* 105.

External Characters

Length 95 mm, width 3 mm, segment number 108. Prostumium 1/3 epilobous. No dorsal pores visible anterior clitellum. Clitellum XIV–XVI, annular, there are a few setae ventrally on XVI, otherwise the clitellum is without setae. Setae form closed ring; those on the anterior segments as far as VII or VII are enlarged somewhat, but not markedly; the intervals between the setae are approximately the same all round the ring; setae number 86 (VII), 47 (XIII), 46 (XVII), and 50–60 in the middle region of the body; 7–8 between spermathecal pores, 11–12 between male pores. Male pores in XVIII, about 1/3 of body circumference ventrally apart, in the line of the ring of setae; no setae immediately to the inner side of these latter. Spermathecal pores 4 pairs, in 5/6/7/8/9, intersegmental, small, sometimes on right side and only those in 5/6/7 visible. Genital papillae are present on VII, VII, and IX, in pairs, on the anterior part of the segment between the setal ring and the anterior boundary of the segment. Midventrally, in the line of the setal ring of VI–IX and XI–XIII, there are appearances which might possibly represent faintly marked copulatory areas, but more probably are due to postmortem changes, or to the specimen having been rubbed. Female pore 1 in XIV, medioventral.

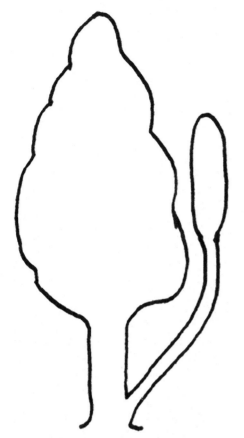

FIG. 69 *Amynthas divergens yunnanensis* (after Stephenson, 1912).

Internal Characters

Septa 8/9/10 absent, 5/6/7/8 moderately thickened, 10/11/12 considerably thickened, 12/13/14 slightly so. Gizzard in VII and IX. Intestine swelling in XVI, there is a well-marked typhlosole; caeca in XXVI, paired, large, conical. Last hearts in XIII. Testes and seminal funnels are closed in testicular sacs, of moderate size, paired, quite separate from each other, in segments X and XI. The 2 vasa deferentia of each side unite at the posterior boundary of XI. Seminal vesicles in XI and XII are paired, of comparatively small size, irregularly lobulated, with in every case a fairly distinct mesially projecting lobe. Prostate glands absent. The terminal portion of the male duct on each side is much thickened and looped. Spermathecae 4 pairs, in VI–IX; ampulla are of an inverted pear shape (the broader end below), the duct is thick and short, 1/3 of the ampulla in length; from the distal end of the duct arises the diverticulum, thin and tubular for most of its extent but swollen at its proximal end; the length of the diverticulum varies; it is mostly 1/2–3/4 as long as the ampulla and duct.

Color: Preserved specimens yellowish brown.
Type locality: Yunnan (Tengyuan).
Distribution in China: Yunnan (Tengyue).

Etymology: The subspecific name indicates the type locality.
Deposition of types: Indian and British Mus.
Remarks: This specimen has certain fairly well-marked differences from the typical form in size, in the presence of seta on the clitellum, and in the details of the spermathecal apparatus.

65. *Amynthas dyeri* (Beddard, 1892)

Amynthas dyeri Beddard, 1892. *P. Z. S.* 157.
Amynthas dyeri Beddard, *Proc. Zool. Soc. London.* 1900:623–624.
Amynthas dyeri Sims & Easton, 1972. *Biol. J. Soc.* 4: 235.
Amynthas dyeri Xu & Xiao, 2011. *Terrestrial Earthworms of China.* 108.

External Characters

Length 126 mm, segment number 104. Prostomium epilobous. First dorsal pore 11/12. Clitellum XIV–XVI, annular, setae absent. Setae larger anteriorly, these setae being ornamented; setae number: 35 (V), 50 (XXI). Male pores 1 pair on XVIII. One or two pairs of large concave genital papillae behind or also in front of male pores. Spermathecal pores 4 pairs in 5/6/7/8/9, spermathecal pore situated very dorsally. Female pore 1 in XIV.

Internal Characters

Septa 8/9/10 absent. Spermathecae 4 pairs in VI–IX, diverticulum often moniliform.

Color: ?
Distribution in China: Fujian (Fuzhou).
Remarks: This species is similar to *sinensis* and *monilicystis*, mainly differing in color. The moniliform diverticulum is not a specific character but an occasional condition. Some examples exist with 1 or 2 pairs of papillae, and some in which there was an asymmetry, with only one of the anterior pair being present. The position of the spermathecal orifices is unusual, but is paralleled in *A. trinitatis*.

66. *Amynthas dorsualis* Sun & Qiu, 2013

Amynthas dorsualis Sun & Qiu, 2013. *J. Nat. Hist.* 47(17–20):1147–1151.

External Characters

Length 121–? mm, width 2.5–2.7 mm (clitellum), segment number 116–?. Body cylindrical in cross section, gradually tapered towards head and tail. Prostomium 1/2 epilobous. First dorsal pore in 13/14. Clitellum annular in 9/10 XIV–7/10 XVI, smooth, setae and dorsal pore invisible, but traces of dorsal pore present within clitellum. Setae numbering, 30–32 (III), 20–24 (V), 28–36

(VII), 32–36 (XX), 32 (XXV); 9–15 (VI), 11–15 (VII), 11–20 (VII) between spermathecal pores on dorsum, 6–7 between male pores. Male pores paired in XVIII, 0.33 body circumference ventrally apart, each on the center of a slightly raised, tiny conical glandular porophore, surrounded by 5 or 6 elliptical or circular folds. Genital markings invisible externally. Spermathecal pores 4 pairs in 5/6/7/8/9 on dorsum, eye-like, 0.6 body circumference ventrally apart. Female pore single in XIV, midventral.

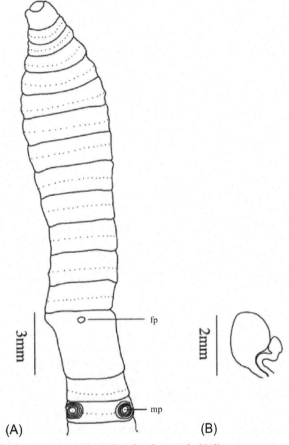

FIG. 70   *Amynthas dorsualis* (after Sun et al., 2013).

Internal Characters

Septa 8/9/10 absent, 10/11/12/13 slightly thickened, 6/7/8 thinner than 10/11/12/13. Dorsal blood vessel single, continuous onto pharynx; hearts enlarged from X–XIII. Gizzard in VII–X, barrel-shaped; intestine enlarged gradually from XV and full size from XX; intestinal caeca simple, in XXVII, extending anteriorly to 1/2 XXIV, surface smooth, finger-shaped sac with and an obvious incision on dorsal and ventral margins, where the septa constrict. Holandric, testis sacs 2 pairs, ventral in X and XI, anterior pair larger than posterior pair, ovate, separated ventrally, posterior pair small, silver, connected ventrally; seminal vesicles paired in XI–XII, anterior pair developed, fleshy. Prostates in XVI–XX, coarsely lobate, prostatic duct inverted U-shaped, distal end appreciably enlarged. Accessory glands absent. Spermathecae 4 pairs in VI–IX, about 2.2 mm long; ampulla heart-shaped, straight duct as long as 0.4 of ampulla; diverticulum shorter than main spermathecal axis by 0.2, slender, terminal 0.2 dilated into ovoid plump seminal chamber; no nephridia on spermathecal duct; or with a longer diverticulum stalk, which is longer than main spermathecal axis terminal 0.17 dilated into ovoid seminal chamber.

Color: Preserved specimens dark gray on dorsum before clitellum, dark brown on dorsum after clitellum, light gray on ventrum before VII, no pigment on ventrum after VII. Clitellum brown.
Type locality: Hainan (Mt. Diaoluo).
Distribution in China: Hainan (Mt. Diaoluo).
Etymology: This species is named after the location of its spermathecal pores, which are situated on the dorsum.
Deposition of types: Shanghai Natural Museum.
Type locality: at 930 m elevation, yellowish-brown soil under bamboo and camphor trees. Mt. Diaoluo, Hainan Province.
Remarks: *A. dorsualis* is similar to *Amynthas homosetus* (Chen, 1938) with respect to setae number between spermathecal pores, being octothecal in VI–IX, no genital papillae within male pore region, the intestinal caeca being simple, and the diverticulum shorter than main spermathecal axis with ovoid seminal chamber and no accessory glands near prostates. In contrast, *A. dorsualis* is smaller than *A. homosetus*, and the pigment is a little different although comparatively dark. The first dorsal pore in *A. dorsualis* is located at 13/14, versus 12/13 in *A. homosetus*; the clitellum occupies fewer than 3 segments in *A. dorsualis*; the anterior setae are uniformly sized and spaced in *A. dorsualis*, whereas in *A. homosetus* the setae of II–IV are prominent, shorter, and more closely spaced on the ventral side; spermathecal pores situated above midlateral and not on the ventrum in *A. dorsualis*; the distance between male pores is 0.33 body circumference and the porophore is surrounded by 5 or 6 elliptical circular folds in *A. dorsualis*, but 0.25 body circumference and no folds in *A. homosetus*.

Comparing *A. dorsualis* to *Amynthas corticis* (Kinberg, 1867), which is the typical species in the *corticis* group (Kinberg, 1867), they share several common characters: body size, pigment, clitellum extent, and setal number. However, most of the characters are different. The first dorsal pore in *A. dorsualis* is 13/14, but never in *A. corticis*. The spermathecal pores are situated on the dorsum in *A. dorsualis*, but on the ventrum and with normal spacing of 0.33 body circumference in *A. corticis*. Styles of male pore porophore of both species are totally different, being a slightly raised, glandular porophore surrounded by 5 or 6 elliptical circular folds in *A. dorsualis*, but a small, circular to transversely elliptical disc in *A. corticis*. The diverticulum of *A. dorsualis* is shorter than the main spermathecal axis compared with *A. corticis*, which has a long stalk. Moreover, *A. dorsualis* has genital papillae neither in the spermathecal pore region nor in the male pore region, whereas *A. corticis* always exhibits genital markings in the spermathecal pore region and occasionally in the male pore region.

67. *Amynthas exiguus aquilonius* Tsai, Shen & Tsai, 2001

Amynthas exiguus aquilonius Tsai, She & Tsai, 2001. *Zoological Studies.* 40(4), 277–279.
Amynthas exiguus aquilonius Chang, Shen & Chen. 2009. *Earthworm Fauna of Taiwan.* 42–43
Amynthas exiguus aquilonius Xu & Xiao, 2011. *Terrestrial Earthworms of China.* 110–111.

External Characters

Length 39–63 mm, clitellum width 1.9–2.6 mm, segment number 70–84. No secondary segmentation, setal line clearly elevated in preclitellar region. Prostomium prolobous or epilobous. First dorsal pore 6/7. Clitellum XIV–XVI, annular, dorsal pore absent, setae absent, length 2.0–3.2 mm. Setae 26–35 (VII), 28–38 (XX), 5–9 between male pores. Male pores paired in XVIII, lateroventral, 0.23–0.30 circumferences apart, porophore large, round, smooth, slightly elevated, with a male aperture inconspicuous on lateral concave area. Genital papillae present in XVII–XX, in 2 longitudinal rows fairly similar to arrangement in the preclitellar region, number on each side presetal 0–1 and postsetal 0–5 in XVII, presetal 1–6 and postsetal 1–4 in XVIII, presetal 1–6 and postsetal 0–2 in XIX. Sometimes, an additional postsetal papilla present medioventrally in XVIII or an additional pair of presetal papillae present in XX. Each papilla small, round, center flat or slightly concave, 0.2–0.3 mm in diameter, about 1/2 distance between setal line and segmental furrow, surrounded by a circular fold. Spermathecal pores 4 pairs in 5/6/7/8/9, ventrolateral, each in depressed furrow, about 0.45 circumferences apart ventrally. Genital papillae present in VII–IX, in longitudinal rows slightly medial to spermathecal pores, number on each side presetal 0–4 and postsetal 0–2 in VII, presetal 1–5 and postsetal 0–2 in VII, presetal 0–6 and postsetal 0–2 in IX. An additional presetal papilla present medioventrally in VII and IX or 1 papilla immediately behind each spermathecal pore in 5/6/7/8 for some specimens. Each papilla small, round, tubercle-like, surrounded by a circular fold, its size about 1/3 to 1/4 distance between setal line and segmental furrow. Female pore 1, medioventral in XIV.

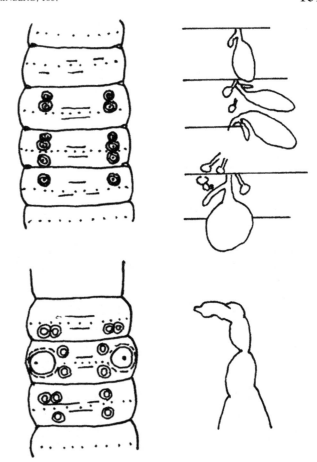

FIG. 71 *Amynthas exiguus aquilonius* (after Tsai, et al., 2001).

Internal Characters

Septa 8/9/10 absent, 10/11–12/13 thickened. Gizzard round in IX and X. Intestine from XV. Intestinal caeca paired in XXVII, simple, surface slightly wrinkled, extending anteriorly to XXIV–XXII. Hearts in XI–XIII. Testis sacs 2 pairs in X and XI, small, round. Seminal vesicles paired in XI and XII, large, surface wrinkled, follicular, with dorsal lobes. Prostate glands paired in XVIII, large, wrinkled, extending anteriorly to XVI and posteriorly to XX, prostatic duct C-shaped, distal end enlarged. Accessory glands present, similar in structure to those in the spermathecal region. Spermathecae 4 pairs in VI–IX, ampulla peach-shaped, stalk straight, much shorter than ampulla, diverticulum vestigial, short, seminal chamber rudimentary or absent, length shorter to slightly longer than spermathecal stalk, straight or slightly bent, originating 1/3 distance from spermathecal pore to ampulla. Accessory glands present, each round, stalked, corresponding to each external genital papilla. An accessory gland present at proximal end of spermathecal stalk near spermathecal pore, indicating presence of a genital papilla submerged within each spermathecal pore, but undetectable outside.

Color: Preserved specimens dark reddish brown on dorsum, light gray on ventrum, light grayish brown around clitellum.

Description: Endemic to Taiwan, recorded at medium and high elevations in central and southern areas.

Etymology: The name "*aquilonius*" was given to this subspecies to indicate that it has a distribution north of *A. exiguus exiguus* (Gates).

Type locality: Taiwan (Mt. Hehuan [elevation 3000 m], Hualien County, along Rd. 14A near the border with Nantou County).

Deposition of types: Taiwan Endemic Species Research Institute, Jiji, Nantou County, Taiwan.

Distribution in China: Taiwan (Mt. Hehuan).

Remarks: *A. exiguus* (Gates) is a small holandric, octothecal earthworm belonging to the *cortices* group. Gates (1932) divided this species into 2 subspecies, the northern form *P. exigua* and the southern form *P. exigua austrina*, based on the position of the spermathecal pores and the structure and arrangement of the genital papillae.

A. exiguus aquilonius has fewer setae and much shorter diverticula compared to those of *exiguus*. It has small usually widely paired presetal and/or postsetal papillae in both preclitellar and postclitellar regions. In contrast, *exiguous* has no postsetal papillae, but has closely paired or median presetal papillae in the preclitellar region, and closely or widely paired ones in some segments in the postclitellar region. Both subspecies have spermathecal pores in intersegmental furrows and have stalked accessory glands. *A. exiguus aquilonius* differs greatly from the above 2 subspecies by having intrasegmental spermathecal pores just behind the segmental furrows. It has large, closely paired postsetal papillae usually in the 18/19 furrow from the setal line in XVIII to near the setal line in XIX, occasionally occurring in 17/18 and 19/20. Also, its accessory glands are sessile.

68. *Amynthas exilens* (Zhong & Ma, 1979)

Pheretima exilens Zhong & Ma, 1979. *Acta Zootax. Sinica.* 4(3):228–229.
Amynthas exilens Xu & Xiao, 2011. *Terrestrial Earthworms of China.* 111.

External Characters

Length 47–98 mm, width 3–5 mm, segment number 59–110. Prostomium 1/2 epilobous. First dorsal pore 12/13. Clitellum entire, XIV–XVI, annular, with setae ventrally (or only XIV). Setae fine and close, evenly distributed, lightly thick in IV–VII, setal breaks of dorsomedial small; $aa = 1.2ab$, setae number 43–50 (III), 60–70 (VII), 64–71 (IX), 54–57 (XX); 23–26 (VI), 24–26 (VII) between spermathecal pores, 11–14 between male pores.

Male pores 1 pair, on ventrolateral side of XVIII, about 2/5 of circumference ventrally apart, each on flat-topped papilla, and a similar one nearby, also on XVII and XIX postsetally in line with male pore, papillae center concave. Spermathecal pores 4 pairs, intersegmental in 5/6/7/8/9, on small eye-like region, about 1/2 of circumference ventrally apart; with 1 pair of similar papillae ventral postsetal on VII, sometimes absent or singular.

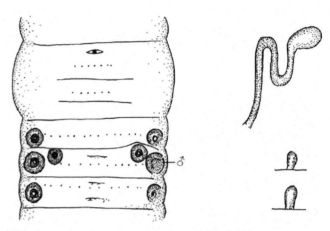

FIG. 72 *Amynthas exilens* (after Zhong & Ma, 1979).

Internal Characters

Septa 8/9/10 absent. Caeca simple, in XXVII, extending anteriorly to XXIV. Testis sacs all small, 0.8 mm in diameter and widely placed. Seminal vesicles elongate, in XI and XII. Prostates rarely developed, confined in XVIII, XIX, if present. Accessory glands racemose with short ductule. Spermathecae 4 pairs, all vestigial except in one case where the crooked seminal chamber is present.

Color: Preserved specimens grayish.

Distribution in China: Sichuan (Mt. Emei).

Etymology: The specific name comes from its small spermathecae.

Deposition of types: Department of Biology, Sichuan University.

Remarks: This species is similar to *Metaphire exilis* (Gates, 1935) in the shape of diverticulum and the arrangement of papillae, but may be distinguished from the latter by the presence of 4 pairs of spermathecal pores on 5/6/7/8/9 (*Metaphire exilis* has only 2 pairs on 5/6/7).

69. *Amynthas formosae* (Michaelsen, 1922)

Pheretima formosae Gates, 1959. *Amer. Mus. Novitates.* 1941:9–13.
Amynthas formosae Sims & Easton, 1972. *Biol. J. Soc.* 4:235.

Amynthas formosae James, Shih & Chang, 2005. *J. N. His.* 39(14):1017.
Metaphire formosae Chang, Shen & Chen, 2009. *Earthworm Fauna of Taiwan.* 110–111.
Amynthas domosus Xu & Xiao, 2011. *Terrestrial Earthworms of China.* 106–107.

External Characters

Length 159–393 mm, clitellum width 8.0–11.0 mm, segment number 103–176. Prostomium prolobous or epilobous. Setae 130–150 (VII), 126–138 (X), 104–120 (XXV), 19–32 between male pores. First dorsal pore 12/13. Clitellum XIV–XVI, annular, dorsal pore absent, setae absent. Male pores (opening of copulatory pouch) paired, situated on setal line close to lateral border of XVIII, on large conical porophores, surrounded by several circular folds, with an oval pad within the copulatory pouch or near the opening, 0.25–0.33 circumference apart ventrally. Spermathecal pores 4 pairs in 5/6/7/8/9, dorsal, intrasegmental, 0.875–0.94 body circumference apart ventrally. No genital papillae in the preclitellar region. Female pore single, medioventral in XIV.

Internal Characters

Septa 5/6–7/8 thickened, 8/9 thin, 9/10 absent, 10/1–13/14 greatly thickened. Gizzard in VII. Intestine from XV. Intestinal caeca paired in XXVII, simple, extending anteriorly to XXIII. Lateral hearts in XII and XIII. Spermathecae 4 pairs in VI–IX, ampulla large, oval, duct stout, half the length of ampulla, diverticulum stalk long, tightly coiled, enclosed within membrane, with a small oval seminal chamber on the tip. Nephridia tufted, attached to the postsegmental septa, surrounding the segmental chambers in V and VI. Ovaries paired in XIII, medioventral, close to the 12/13 septum. Testis sacs 1 pair in X, oval, smooth, medioventral in front of 10/11. Seminal vesicles paired in XI, large, each with a folliculate dorsal lobe. Prostate glands paired in XVIII, large, lobular, extending anteriorly to XVII and posteriorly to XIX.

Color: Preserved specimens purplish brown on dorsum, light brown on ventral.
Type locality: Taiwan (Chiahsien, Kaohsiung County).
Deposition of types: National Natuurhistorisch Museum, Leiden, Netherlands.
Distribution in China: Taiwan (Taibei, Pingdong, Gaoxiong).
Etymology: The specific name "*formosae*" refers to the type locality.
Remarks: *M. formosae* is a member of the *M. formosae*-species group. It lives in mountains where the vegetation is broadleaf forest and can be found in both virgin and secondary forests. This species is anecic with permanent vertical burrows 30 cm or more below the ground.

The spermathecal pores are intrasegmental and very close to the middorsal line, in contrast to representations made in Gates (1959) for other material. The prostatic duct structure of the type is clearly generally similar to *Amynthas henchunensis* and the other proandric species described below.

70. *Amynthas fornicatus* (Gates, 1935)

Pheretima fornicata Gates, 1935a. *Smithsonian Mis. Coll.* 93(3):9.
Pheretima fornicata Chen, 1936. *Contr. Biol. Lab. Sci. Soc. China (Zool.).* 11:296–298.
Pheretima fornicata Gates, 1939. *Proc. U.S. Natr. Mus.* 85:434–436.
Amynthas fornicatus Xu & Xiao, 2011. *Terrestrial Earthworms of China.* 116–117.

External Characters

Length 78–94 mm, width 4–6 mm, segment number 90–105. Tri-annuli on each segment. Prostomium 1/3 epilobous. First dorsal pore 11/12 or 12/13. Clitellum well marked, in XIV–XVI, annular, setae absent. Ventral setae of IV–VII a little longer but not widely spaced, a wide dorsal gap in the setal circle of II; setae number 31–36 (III), 43–50 (VI), 46–52 (VII), 42–52 (XII), 45–55 (XXV); 19–22 (VI), 19–20 (VIII), between spermathecal pores, 9–11 between male pores. Male pores superficial, on circular to transversely oval, disc-shaped porophores. No genital markings. Spermathecal pores minute and superficial, 4 pairs, in 5/6/78/9, about 1/3 of circumference ventrally apart, each on a papilla at anterior edge of VI–IX. Female pore single in XIV, medioventral.

Internal Characters

Septa 9/10 absent, 8/9 present but membranous, 5/6/7/8 thickly muscular. Intestine from XV. Intestinal caeca simple, in XXVII, extending anteriorly to XIII. First hearts not observed. Testis sacs connected ventrally and communicated (?), testis sacs of X and XI unpaired and horseshoe-shaped. Seminal vesicles paired in XI and XII, small and elongate. Prostate thick at ectal half, extending through some or all of segments XVI–XXI; prostatic duct is 3–5 mm long, bent in a U-shape, the ectal limb much thicker than the ental limb. Spermatheca 4 pairs, ampulla about 1 mm long, diverticulum with a long, slender stalk and spheroidal or asymmetrical seminal chamber.

Color: Preserved specimens Clitellum dark chestnut.
Type locality: Sichuan (Kangding)
Distribution in China: Sichuan (Kangding, Xikang), Xizang.
Deposition of types: Smithsonian Institution.

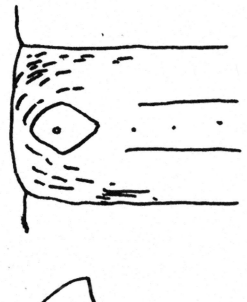

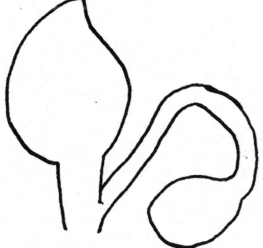

FIG. 73   *Amynthas fornicates* (after Gates, 1935a).

Remarks: Distinguished from *Amynthas hongkongensis* by the gap in the setal circle of II, the exclusion of seminal vesicles of XI from the posterior testis sac, and the absence of genital markings.

A.s fornicatus is distinguished from *A.s hongkongensis*, Michaelsen, 1910, by dorsal gap in the setal gap in the setal circle of II, the absence of genital markings, and the exclusion of the anterior seminal vesicles from the testis sac of XI. *A. fornicatus* is clearly distinguished from *Amynthas pingi* by the horseshoe-shaped testis sac of X and XI.

71. *Amynthas fuscus* Qiu & Sun, 2012

Amynthas fuscus Qiu & Sun, 2012. *Zootaxa*. 33458:152-153.

External Characters

Length 95–149 mm, width 4–5 mm (preclitellum), segment number 93–96. Body cylindrical in cross section, gradually tapered towards head and tail. Prostomium 1/2 epilobous. First dorsal pore 11/12 or 12/13. Clitellum 1/8 XIV–7/8 XVI, annular, swollen, and dorsal pore invisible. Setae number 30–34 (III), 28–36 (V), 32–40 (VII), 42–50 (XX), 36–50 (XXV); 12–16 (VI), 213–16 (VII), 13–16 (VII) between spermathecal pores, 8–10 between male pores; setal formula $aa = 1$–$1.3ab$, $zz = 1.3$–$2zy$. Male pores paired in XVIII, 0.33 body circumference ventrally apart; each on the center of a raised elliptical glandular flat top, conical, surrounded by a folded pad. Genital marking not visible externally. Spermathecal pores 4 pairs in 5/6/7/8/9, eye-like, 0.33 circumference ventrally apart. Female pore 1 in XIV, medioventral.

Internal Characters

Septa 8/9/10 absent, 5/6/7 thick and muscular, 10/11/12/13 slightly thickened. Gizzard in VII–X, ball-shaped. Intestine swelling in XVI and distinctly from XXI. Intestinal caeca simple, in XXVII, extending anteriorly to XXIV, surface smooth, finger-shaped sac with 2 slight incisions on terminal dorsal margin and 1 on ventral margin. Dorsal blood vessel single, continuous onto pharynx; Hearts enlarged from X–XIII. Holandric, Testis sacs 2 pairs, ventral in X–XI, connected on ventrum between left and right. Seminal vesicles paired in each of XI and XII, developed, connected on ventrum between left and right. Prostates developed, in XVI–XX, coarsely lobate with 3–4 major lobes. Prostatic duct inverted U-shaped, distal end stouter. Accessory glands absent. Ovaries in XIII. Spermathecae 4 pairs in VI–IX, about 3.2 mm long; ampulla irregularly heart-shaped, gradually slender duct as long as 0.5 ampulla. Diverticulum shorter than spermathecal axis by 0.33, slender and coiled, terminal 0.25 dilated into an elongated chamber; no nephridia on spermathecal ducts. Rarely has a degenerated spermatheca in the left of VI, without ampulla and its duct.

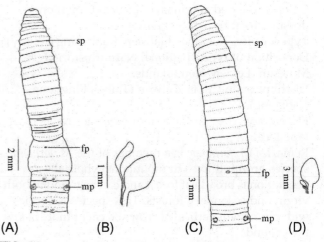

FIG. 74   *Amynthas fuscus* (after Sun et al., 2012).

Color: Preserved specimens dark brown on dorsum, light brown on ventrum, unclear middorsal.
Clitellum brown.
Distribution in China: Hainan (Mt. Jianfeng and Mt. Diaoluo).
Etymology: The species is named after its brown pigment.
Deposition of types: Shanghai Natural History Museum.
Remarks: *A. fuscus* is closely related to *Amynthas homosetus* (Chen, 1938) from China, *Amynthas baemsagolensis* Hong and James, 2001 from Korea, and *Amynthas sangumburi* Hong and Kim, 2002 from Korea. These 4 species share a unique character combination in having no genital papillae either around the male pore region or the spermathecal pore region, having 4 pairs of spermathecal pores intersegmental in 5/6/7/8/9, simple intestinal caeca and developed prostates.

A. fuscus is easily distinguishable from *A. homosetus* by the characters of the clitellum, male pore, distance between spermathecal pores transversely, and body pigment. As for *A. fuscus*, the clitellum occupies 1/8 XIV–7/8 XVI with invisible setae, and the spermathecal and male pores are about 0.33 body circumference ventrally apart. In *A. homosetus* the clitellum is longer, occupying XIV–XVI with a few barely visible setal pits ventrally on XVI, and the spermathecal and male pores are about 0.25 body circumference ventrally apart. Otherwise, the diameter of the male pore glandular region in *A. fuscus* smaller than that in *A. homosetus*.

Differs from *A. baemsagolensis* primarily in body size, and locations of clitellum and porophores of male pores. *A. fuscus* has a smaller body size, clitellum extending less than 3 segments, male pores on the center of a raised elliptical glandular flat top with a folded pad, but not on domed circular to oval porophores with 4–6 furrows.

Compared with *A. sangumburi*, *A. fuscus* has a bigger body size, more setae between male pores, smaller porophores of male pores, clitellum not regularly occupying XIV–XVI, and the stalk and seminal chamber of the diverticulum has a clear boundary, not widening gradually.

A. fuscus is much more similar to *A. fornicatus* (Gates, 1935) with respect to body size, the porophore of males, 4 pairs of spermathecal pores intersegmental in 5/6/7/8/9, same transverse distance between male pores and spermathecal pores, and the shape of spermathecae. However, *A. fuscus* has a rare clitellum situation occupying 1/8 XIV–7/8 XVI, and has a papilla at the anterior edge of VI–IX which has been mentioned in *A. fornicatus* (Chen, 1936). Another difference is that the diverticulum of *A. fuscus* is shorter than that of *A. fornicatus*.

72. *Amynthas genitalis* Qiu & Sun, 2012

Amynthas genitalis Qiu & Sun, 2012. *Zootaxa*. 33458:155-158

External Characters

Medium size. Length 83–97 mm, width 2.3–2.5 mm (preclitellum), segment number 99–109. Body cylindrical in cross section, gradually tapered towards head and tail. Prostomium 1/2 epilobous. First dorsal pore 12/13. Clitellum XIV–XVI, annular, setae can be seen ventrally in XVI; dorsal pore absent. Setae number 30–36 (III), 22–26 (V), 32–36 (VII), 38–46 (XX), 35–56 (XXV); 8 (VI, VII, VII) between spermathecal pores, 11–12 between male pores; setal formula $aa = 1.1ab$, $zz = 1.2–1.6zy$. Male pores paired in XVIII, 0.33 body circumference apart ventrally; each on the elliptic, glandular porophore with slightly raised center, without surrounding epidermal wrinkles. Genital papillae ovate, flat-topped, diameter 0.4–0.5 mm, paired above setae annulet on XVIII and XIX; below setae annulet on XVII and XVIII, arranged in longitudinal row slightly medial to male pore. Spermathecal pores 4 pairs in 5/6/7/8/9, about 0.33 circumference apart ventrally. Female pore 1 in XIV, medioventral.

Internal Characters

Septa 8/9/10 absent, 6/7/8 and 10/11/12/13 slightly thickened. Gizzard in VII–X, long bucket-shaped. Intestine swelling in XV. Intestinal caeca simple, in XXVII, extending anteriorly to XXIV, finger-shaped sac with 2 indentations on dorsal margin and 1 indentation on ventral margin. Dorsal blood vessel single, continuous onto pharynx; Hearts enlarged from X–XIII. Holandric, Testis sacs 2 pairs, ventral in X–XI, posterior pair bigger in size. Seminal vesicles paired in each of XI and XII, separated on ventrum between left and right. Prostates in XVII–XX, coarsely lobate with 3 major lobes, stout muscular duct U-shaped, distal end enlarged. Accessory glands absent. Ovaries in XIII. Spermathecae 4 pairs in VI–IX, about 1.4 mm long; ampulla ovoid, small, straight duct as long as 0.6 ampulla. Diverticulum longer than main spermathecal axis by 0.4, terminal 0.29 dilated into a rod-shaped chamber, bright and white; no nephridia on spermathecal ducts.

Color: Preserved specimens no pigment, unclear middorsal line. Clitellum brown.
Distribution in China: Hainan (Mt. Diaoluo).
Etymology: The species is named after its genital in the male pore region.
Deposition of types: Shanghai Natural History Museum.
Habitat: Mt. Diaoluo (elevation 930 m), yellow soil under herbaceous vegetation beside road, and yellow

FIG. 75 *Amynthas genitalis* (after Sun et al., 2012).

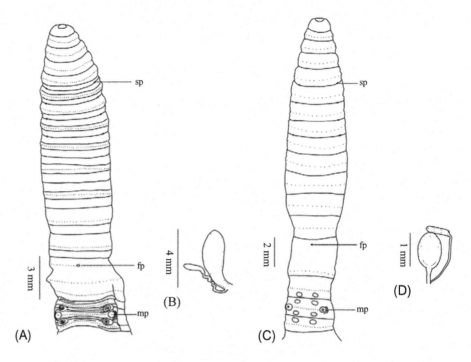

cinnamon soil in bamboo, camphor, and beech forest, Hainan.

Remarks: *Amynthas genitalis* is somewhat similar to *A. montanus* and *A. exiguus aquilonius* in the characters of their male pore region.

The body size of *A. genitalis* is between *A. montanus* and *A. exiguus aquilonius*. The clitellum setae in *A. genitalis* can be seen ventrally in XVI, but exist in XIV–XVI in *A. montanus* and are absent in *A. exiguus aquilonius*. The setae of *A. genitalis* are fewer than *A. montanus*. The porophore of *A. genitalis* is clearly smaller than *A. exiguus aquilonius*. The locality of genital papillae of the 3 species is not the same as each other, particularly in the male pore region. The diverticum of *A. genitalis* is longer than the main spermathecal axis compared with a shorter one in *A. montanus* and *A. exiguus aquilonius*. Accessory glands were not seen in *Amynthas genitalis*, but were observed in *A. exiguus aquilonius*.

73. *Amynthas hengchunensis* James, Shih & Chang, 2005

Amynthas hengchunensis James, Shih & Chang, 2005. *J. Nat. Hist.* 39(14):1015-1016.
Metaphire paiwanna hengchunensis Chang, Lin, Chen, Chung & Chen, 2009. *Earthworm Fauna of Taiwan*. 120–121.
Amynthas domosus Xu & Xiao, 2011. *Terrestrial Earthworms of China*. 122.

External Characters

Length 200–252 mm, clitellum width 8.0–9.5 mm, segment number 138–148. Number of annuli (secondary segmentation) per segment 3 in 6–9, 5 in 10–13, 3 in body segments behind 17. Prostomium epilobous. First dorsal pore 12/13. Clitellum 14–16, annular, dorsal pores absent, setae absent. Setae regularly distributed around segmental equators, size uniform; setal formula *aa:ab:yz zz* = 1.7:1:1:3 at XXV, numbers: 120–170 (VII), 140–164 (X), 170–208 (XXV), 26–28 between male pores. Male pores minute on low conical porophores at posterior extension of speculate longitudinally orientated genital marking surrounded by epidermal folds, lateral fold close to male pores enlarged to form flap or hood partially covering male porophore. Spermathecal pores 4 pairs in 5/6/7/8/9, dorsolateral, 0.60–0.63 circumference apart ventrally. No genital papillae in the preclitellar region. Female pore 1, medioventral in XIV.

Internal Characters

Septa 9/10 absent, 8/9 thin, 5/6/7/8 thickened, 10/11/12/13/14 thickly muscular. Gizzard in VII. Intestine from 1/2 XV. Intestinal caeca paired in XXVII, conical, simple, extending anteriorly to XXIII. Hearts in X–XIII. Testis sacs 1 pair in X, oval, smooth, medioventral in front of 10/11. Seminal vesicles paired in XI, large, with a small dorsal lobe. Prostate glands paired in XVIII, large, 4 main lobes, each lobe served by 2–5 small ductulets radiating fan-like from ental end of prostatic duct; duct stout; straight, muscular, narrowing towards body wall;

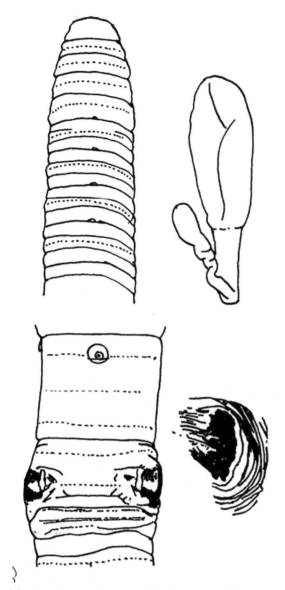

FIG. 76   *Amynthas hengchunensis* (after James et al., 2005).

vasa deferent join duct at duct-glandular portion junction; vasa deferentia nonmuscular. Spermathecae 4 pairs in VI–IX, ampulla oval, duct shorter than ampulla, diverticulum small, oval, stalk slender, composed of hairpin loops not enclosed in membrane.

Color: Preserved specimens light brown.
Type locality: Taiwan (Nanrenshan, Kending, Pingdong County).
Deposition of types: National Museum of Natural Science, Taichung, Taiwan. Cat. no. NMNS 4054-006 for the holotype.
Etymology: The species was named after the Hengchun Peninsula, where it was discovered.
Distribution in China: Taiwan.

Remarks: *A. hengchunensis* is similar to *A. formosae* (Michaelsen, 1922) with respect to proandry, the enclosure of the contents of XI in a sac, being octothecal in VI–IX, the general form of the spermathecae, the body size, the intestinal origin, and the very large number of setae in the anterior segments. In contrast, *A. hengchunensis* has many more setae, especially in the postclitellate segments, and has spermathecal pores more ventrally placed. Other differences from *A. hengchunensis* include the male field possessing a flap partially covering the male porophore, much shorter caeca, and lack of membranous covering of the diverticulum stalk.

74. *Amynthas heterochaetus* (Michaelsen, 1869)

Pheretima (Pheretima) heterochaeta Chen, 1931. *Chin. Zool.* 7(3):123–125.
Pheretima heterochaeta Chen, 1933. *Contr. Biol. Lab. Sci. Soc. China (Zool.).* 9:234–238.
Pheretima heterochaeta Chen, 1936. *Contr. Biol. Lab. Sci. Soc. China (Zool.).* 11:270–271.
Pheretima (Pheretima) heterochaeta Chen, 1938. *Contr. Biol. Lab. Sci. Soc. China (Zool.).* 12(10):382.
Pheretima heterochaeta Chen, 1946. *J. West China. Border Res. Soc. (B).* 16:136.
Amynthas heterochaetus Xu & Xiao, 2011. *Terrestrial Earthworms of China.* 122–123.

External Characters

Length 100–158mm, width 3–5mm, segment number 92–120. Prostomium 1/2 epilobous. First dorsal pore in 11/12, first distinct, those on clitellum usually obsolete. Clitellum in XIV–XVI, annular, smooth, without setae, setal pits often recognizable on ventral side. Setae often small and short, both dorsal and ventral breaks slight, $aa = 1.2–1.5ab$, $zz = 1.2–2.0$ yz. Ventral setae of preclitellar segments (particularly ventromedian ones) characteristically modified, enlarged and widely spaced, a, b, c, d of III or IV–VI or VII particularly enlarged and widely placed, a the longest and stoutest, decreasing in length and intervals laterad, d or e laterad resuming normal shape; on VII and IX only a, b enlarged and on X–XIII only a enlarged but not so marked as in preceding segments; no setae posterior to clitellum so modified. Intervals of these modified setae on III–VII often markedly wide, proportional to their length, that is, $aa = 1.2–1.5ab = 2.5$ $bc = 2.8–3.0$ $cd = 3$ de; $aa = 2.5$ zz, intervals gradually shortened toward lateroventral side. Male pores paired in XVIII, each pore on a large and rounded papilla, which is raised or depressed in setal line, its diameter about 1/3 (or 1/4) of a segmental length, sometimes transversely elongate coalesced by 2 papillae, perforated on its top, the pore is indistinctly marked with a whitish spot, male papilla generally surrounded by 3 or more concentric ridges separated from each other by shallow grooves, occasionally indistinct; 1–2 genital papillae often present close to each male papilla, quite variable in their arrangement, one often posterior and lateral or slightly medial to male papilla, while the other anterior and lateral to, or often in line to the posterior papilla. Spermathecal pores 4 pairs, in intersegmental furrows 5/6/7/8/9, about 4/7 of body circumference ventrally apart, usually inconspicuous or marked by a small transversely ovoid tubercle on which opens the duct; the tubercle often depressed or sometimes surrounded by glandular skin. Genital papillae in this region divided into 2 groups: one close to spermathecal pores and the other on ventral side. One small elevated and flat-topped papilla placed either anteriorly or posteriorly to each pore, more often to its anterior side, sometimes not all 4 on each side constantly present, one or more disappearing. Two or 3 pairs often on ventral side of segments VII, VII, and IX, usually widely separated or rather in close approximation, about 2 mm apart, each slightly elevated with a thick circular margin, flattened or depressed in center. Those on VII and VII more constant, or 3 pairs totally absent in a few cases. Female pore single in XIV, medioventral.

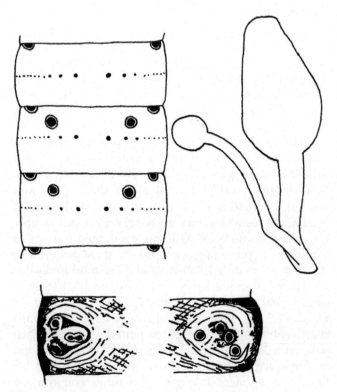

FIG. 77   *Amynthas heterochaetus* (after Chen, 1931).

Internal Characters

Septa 8/9/10 absent but 8/9 traceable ventrally, 5/6/7 much thickened and musculated, 7/8 less thickened, 10/11/12 also thickened, 12/13 less so. Gizzard moderate in size, usually with a crop-like dilation anteriorly. Intestine swelling in XVI. Caeca simple, conical in shape, in XXVII, extending anteriorly to XXIV or XIII, ventral edges rather smooth, buff or dark in color. Hearts only 3 pairs, absent in X, all small in caliber; commissural loops in front of septum 10/11 present, asymmetrically developed. Testis sacs large, anterior pair on posterior part of X and XI, V-shaped, usually bridged between the pair by a narrow membranous tube, often widely placed; posterior pair slightly on posterior side of XI and XII larger than anterior pair, rounded, more broadly connected or united into a transverse band. Testis large, about 0.8–1.0 mm in diameter, occupying nearly half of the sac; its funnel in irregular folding. Seminal vesicles in XI and XII, usually small and flattened, irregularly shaped with or without grooves on surface, smooth or finely granular, wedge-shaped, each with its dorsal portion projecting up on both sides of dorsal lobe which is elongate or triangular, smooth and pale on surface. Prostate glands always totally absent, rarely present, sometimes only found as a vestigial part; if present, as a round thick patch, generally on one side; if totally absent, each with its lobular ducts attached to body wall by a few fibers of connective tissue. Accessory glands not connected with male papilla but with other papillae. Each gland portion roundish, soft but compact, rather smooth and whitish on surface, close to body wall, with a very short duct. Ovaries racemose, very large, extending almost to the whole segment. Spermathecae 4 pairs in VI–IX, often with diverticulum in preceding segment; ampulla generally ovoid, or narrower but round at free end, smooth and pale on surface, its duct as long as, or shorter than, ampulla; moderately thickened, distinctly marked off from the latter. Diverticulum usually shorter than main pouch (duct + ampulla), generally with a round or elongate oval seminal chamber, well-developed, smooth and brilliant white, with a long slender duct joining to the main duct near body wall.

Color: Preserved specimens chestnut brown with purplish iridescence on dorsal side, whitish around setal zones interrupted along dorsomedian line; pale ventrally. Clitellum chocolate brown. In life brownish orange dorsally and brownish yellow on clitellum. Distribution in China: Liaoning (Dandong), Jilin, Jiangsu (Nanjing, Suzhou, Yixing), Zhejiang (Lanxi, Tonglu, Linhai, Anhui, Xindeng), Ahui (Anqing), Fujian (Xiamen), Taiwan, Jiangxi (Nanchang, Jiujiang, Dean), Hong Kong, Hainan, (Wanning), Chongqing (Beibei), Sichuan (Emei, Chengdu, Luzhou), Guizhou Deposition of types: Mus. Biol. Lab. Sci. Soc. China.

75. *Amynthas homochaetus* (Chen, 1938)

Pheretima (Pheretima) homochaeta Chen, 1938. *Contr. Biol. Lab. Sci. Soc. China (Zool.).* 12(10):409, 414–416.
Pheretima (Ph.) homochaeta Chen, 1938. *Contr. Biol. Lab. Sci. Soc. China (Zool.).* 12(10):414–416.
Amynthas homochaetus. Sims & Easton, 1972. *Biol. J. Soc.* 4:235.
Amynthas homochaetus Xu & Xiao, 2011. *Terrestrial Earthworms of China.* 125.

External Characters

Length 116mm, width 5.2mm, segment number 85. Prostomium 1/2 epilobous. First dorsal pore 12/13. Clitellum entire, in XIV–XVI, annular, with a few barely visible setal pits ventrally on XVI. Setae prominent, shorter on II–IV, closer on ventral side; those on ventral side of V–IX neither enlarged nor widely spaced; ventral break indistinct, dorsal break $zz = 1.2$–1.5 yz behind clitellum; setae number 40 (VI), 44 (VII), 56 (XVII), 48 (XXV); 11 (VI), 12 (VII), 13 (VII), 11 (IX) between spermathecal pore, 9 (VII), 18 (XVII) between or in line with male pores. Male pores 1 pair in XVIII, about 1/4 of circumference ventrally apart, each in center of a roundish glandular area which is about 1.5mm in diameter. No other genital papillae. Spermathecal pores 4 pairs in 5/6/7/8/9, about 1/4 of circumference ventrally apart, each as a tiny simple intersegmental. No genital papillae either around each pore or ventral side in this region. Female pore 1 in XIV, medioventral.

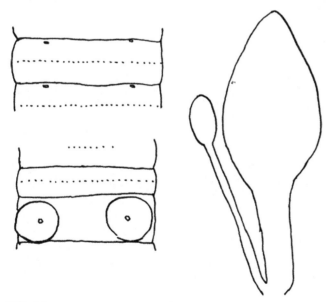

FIG. 78   *Amynthas homochaetus* (after Chen, 1938).

Internal Characters

Septa 9/10 absent, 8/9 very thin, in front gizzard, 5/6 thin, 6/7 thicker, 7/8 muscular and well thickened, 10/11/12/13 equally thickened. Gizzard in 1/2 IX–X.

Intestine enlarged from XV. Caeca simple, slender and long, in XXVII, extending anteriorly to XXIV. Hearts 3 pairs, in XI–XIII. Testis sacs in X and XII, both pairs broadly united and communicated ventrally. Seminal vesicles in XI and XII, each pair very large, filling whole segmental cavity, dorsal lobe small, thumb-shaped. Prostate gland well developed, coarsely lobate, in XVI–XXI, its duct U-shaped, ectal 2/3 thicker. No visible accessory glands around base of each duct. Spermathecae 4 pairs, in VI–IX; posterior 2 pairs large; ampulla heart-shaped, indistinctly marked off from its duct; main pouch about 3mm long. Diverticulum about 2/5 as long, seminal chamber ovoid and whitish.

Color: Preserved specimens dark chocolate on anterior dorsum, gray on other parts of body. Clitellum dark brown.
Distribution in China: Hainan (Xingangcun), Sichuan (Leshang), Yunnan (Lancang).
Remarks: This species is distinguished from *Amynthas heterochaetus* (Mich.) by many essential characters. The ventral setae on anterior end in *heterochaetus* are always enlarged and widely spaced. However, in *homochaetus* the ventral setae instead of being widely spaced are noticeably closer. In *heterochaetus* the spermathecal pores are widely situated (about 4/7 circumference), while in *homochaetus* they are more ventral in position. Furthermore, there are no genital papillae found in any part of the body. In *heterochaetus*, however, large papillae always occur ventrally on VII–IX and smaller ones in or around each of the spermathecal and male pore.

76. *Amynthas hongkongensis* (Michaelsen, 1910)

Pheretima hongkongensis Michaelsen, 1910. *Mitt. Naturhist. Mus. Hamburg.* 27:107.
Pheretima hongkongensis Gates, 1935a. *Smithsonian Mis. Coll.* 93(3):10.
Pheretima hongkongensis Gates, 1939. *Proc. U.S. Natr. Mus.* 85:446–448.
Amynthas hongkongensis Xu & Xiao, 2011. *Terrestrial Earthworms of China.* 126.

External Characters

Medium-sized earthworm; length 100mm, width ? mm, segment number 150. Prostomium epilobous. First dorsal pore in 11/12. Clitellum XIV–XVI, annular, dorsal pore and intersegmental furrows lacking or not clearly indicated; setae present ventrally. The clitellum is dull and roughish, not smooth and glistening, apparently not fully developed. Setae small, closely and regularly spaced; they begin on II, on which segment there is a complete circle; setal number 21 (VI), 20 (VII), 8+ (VIII), 7 (XVIII), 58+ (XX). Male pores paired in XVIII, minute, each pore very slightly lateral to the center of a male pore marking; the latter is nearly but not quite circular in shape, 1.5–2 intersetal intervals wide transversely, slightly protuberant but with a rather flat surface and surrounded by a slight but definite circumferential furrow; just median to each male pore marking is 1 transversely oval genital marking with a conspicuously protuberant whitish rim and a depressed, grayish, central area; 2–3 intersetal intervals wide transversely. Genital marking is not in actual contact with the male pore area though close to it. The spermathecal pores 4 pairs, in 5/6/7/8/9, minute, transverse slits, each pore at the center of a very small, smooth, transversely oval area. Female pore single in XIV, medioventral.

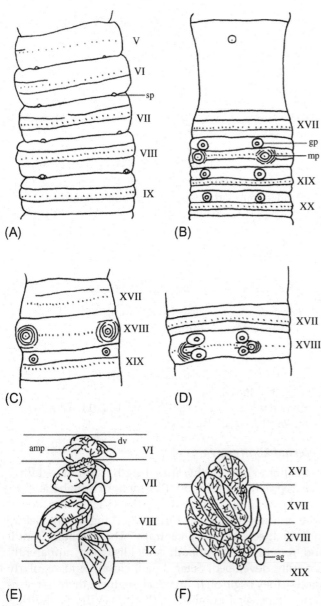

(A)   (B)

(C)   (D)

(E)   (F)

FIG. 79   *Amynthas hongyehensis* (after Tsai & Shen, 2010).

Internal Characters

Septa 8/9 present at least as a ventral rudiment. Intestinal caeca simple, without marginal incisions or setal constrictions. The last heart in XIII. Testis sac of XI is U-shaped, the limbs of the U reaching to the dorsal blood vessel. The seminal vesicles of XI are within the testis sac of XI, surrounded by a thin layer of testicular coagulum. There is only a small quantity of testicular coagulum in the testis sac of X. The seminal vesicles are medium-sized vertical bodies, each with a deep dorsoventral groove on the posterior face. Each vesicle is provided with an elongate, more or less finger-like, primary ampulla, the base of which is sunk deeply into a cleft in the dorsal margin. Prostate ducts 6–8 mm in length, muscular, but uniformly slender throughout, that is, without any particular thickening of an ectal portion. Spermathecae 4 pairs in VI–IX, are flattened out on the ventral parietes; the duct is much shorter than the ampulla, almost triangular in outline; The diverticulum, which passes into the anterior face of the duct just at or within the parietes is a slender, elongate-tubular structure. In an ectal stalk portion of the diverticulum (about one-half or more of the length) the lumen is narrow; the wall of the lumen smooth or ridged transversely. In the remaining ental portion of the diverticulum the lumen gradually widens until the wall becomes very thin. This ental portion is doubtless the seminal chamber, but it is not noticeably wider than the stalk nor marked off from the stalk. Within the seminal chamber is an elongate, opaque, firm mass with no spermathecal iridescence, the mass composed of corpuscular bodies and smaller, homogeneous, spheroidal to ovoidal particles. Dorsal to each genital marking a glandular mass projects through the parietes and very slightly into the coelomic cavity, the glandular material just median to the ectal end of the prostatic duct.

Color: ?
Distribution in China: Hong Kong.
Etymology: The name of the species refers to the type locality.
Type locality: Hong Kong
Deposition of types: Hamburg Museum. Germany.
Remarks: The clitellar granularity is almost certainly not fully developed and this together with the small quantities of testicular coagulum in the testis sacs indicates that the worm is either not quite normal (also note spermathecae) or not completely sexual (presexual or postsexual).

77. *Amynthas hongyehensis* Tsai & Shen, 2010

Amynthas hongyehensis Tsai & Shen, 2010. *J. Nat. Hist.* 44(21–24):1263–1266.
Amynthas hongyehensis Shen, 2012. *J. Nat. Hist.* 46(37–38):2277–2280.

External Characters

Medium-sized earthworm; length 129–197 mm, width 4.48–7.17 mm (clitellum), segment number 85–138. Three annulets per segment in VII–XIII. Prostomium epilobous. First dorsal pore in 11/12. Clitellum XIV–XVI, annular, setae and dorsal pore absent, length 4.32–6.77 mm. Setae number 46–73 (VII), 59–82 (XX), 10–18 between male pores. Male pores paired in XVIII, 0.22–0.29 body circumference ventrally apart. Each pore on a round, white porophore 0.55–0.8 mm in diameter, surrounded by 1–3 shallow skin folds. Genital papillae large, round, usually 2 pairs located slightly medially to the porophore in presetal XVIII and XIX, and an additional pair in XX, occasionally 1 pair only in presetal XVIII or XIX, or 2 pairs in XVIII with 1 pair anteromedial and the other posteromedial to male porophores, or 2 pairs in XVIII together with 1 pair in presetal XIX. Each papilla 0.45–0.98 mm in diameter with depressed center. Spermathecal pores 3–4 pairs, small, buried deeply in the intersegmental furrows of 6/7/8/9, distance between paired pore 0.26–0.29 body circumference apart ventrally. Genital papillae absent in the preclitellar region. Female pore single in XIV, medioventral.

Internal Characters

Septa 5/6/7/8 and 10/11 thick. 11/12/13/14 muscular. Nephridial tufts thick on anterior faces of 5/6/7 septa. Gizzard large, in VII–X, yellowish in color. Intestine enlarged from XVI. Intestinal caeca paired in XXVII, extending anteriorly to XXII–XXIV, each simple, slender, slightly wrinkled. Hearts in X–XIII. Holandric, testes large, 2 pairs in ventrally joined sacs X–XI. Vas efferens connected in XII on each side to form a vas deferens. Seminal vesicles 2 pairs in XI and XII, large, occupying the full compartment, usually anterior pair larger, each vesicle with a round, finely folliculated lobe, darker in color. Prostate glands large, lobed, folliculated, in XVI–XIX, XVI–XX, or XVII–XX. Prostatic duct long, U-shaped in XVII–XVIII. Accessory glands sessile, each oval-shaped, 0.55–0.75 mm long, corresponding to external genital papilla. Spermathecae 3 pairs in VII–IX, or 4 pairs in VI–IX, ampulla large, oval-shaped or peach-shaped, surface wrinkled, each 1.38–3.91 mm long and 1.4–2.54 mm wide, with a short, stout spermathecal stalk 0.35–0.57 mm in length. Diverticulum with an oval-shaped seminal chamber 0.7–1.1 mm long and a slender stalk 0.81–1.65 mm in length.

Color: Preserved specimens purple-brown on head and dorsum, light brown on ventrum, and darkish purple-brown around clitellum. Setal ring white, distinctive.
Distribution in China: Taiwan (Hongyeh, Taidong County).
Etymology: This species is named after its location.

Deposition of types: Taiwan Endemic Species Research Institute, Jiji, Nantou, Taiwan.?
Remarks: *Amynthas hongyehensis* morphologically fairly similar to peregrine *Amynthas hupeiensis* (Michaelsen, 1895). *A. hongyehensis* is distinguishable from *A. hupeiensis* in that it has paired presetal genital papillae between the setal and intersegmental furrows in XVIII and XIX (or an additional pair in XX), whereas *A. hupeiensis* has paired genital papillae in the intersegmental furrows of 17/18 and 18/19 (Chen 1933; Tsai 1964). Furthermore, *A. hupeiensis* has a much higher setal number than *A. hongyehensis* with 100–121 in VII and 79–88 in XX for the former (Tsai 1964) and 52–57 in VII and 65–78 in XX for the latter. *A. hupeiensis* has a very long diverticulum (Chen 1933), while that of *A. Hongyehensis* is much shorter than its ampulla.

78. *Amynthas kaopingensis* James, Shih & Chang, 2005

Amynthas kaopingensis James, James, Shih & Chang, 2005. *J. Nat. His.* 39(14):1017–1020.
Amynthas kaopingensis Xu & Xiao, 2011. *Terrestrial Earthworms of China*. 92.

External Characters

Length 170–300 mm, width 10–14 mm, segment number 160–177. Prostomium epilobous, with tongue open. First dorsal pore in 12/13, or 13/14. Clitellum XIV–XVI, annular, setae invisible externally. Setae regularly distributed around segmental equators, size uniform; setae number 130–170 (VII), 126–170 (X), 104–126 (XXV); 32–40 between male pores. Male pores minute at posterior end of seminal grooves extending to center of ovate to rounded angular genital pad longitudinally oriented from 17/18 to equator of XVIII, surrounded by epidermal folds, lateral folds closest to male pores enlarged to form flap adjacent to or partially covering genital pad; rarely with paired oval genital markings presetal XVII slightly median to male pore line; about 1/3 of circumference apart ventrally. Spermathecal pores 4 pairs, on 5/6/7/8/9, about 0.29–0.32 of circumference apart dorsally. Female pore 1 in XIV, medioventral.

Internal Characters

Septa 9/10 absent, 8/9 membranous, 5/6/7/8 and 10/11–13/14 thickly muscular. Gizzard in VII. Intestine swelling in 1/2 XV. Caeca simple, margins smooth, in XXVII, extending anteriorly to XXIII. Hearts in X–XIII. Testes, funnels in ventrally joined sacs in X; seminal vesicles large in XI, with small fine-textured dorsal lobe; seminal vesicles contents of XI enclosed in thin sac. Prostate glands large in XVIII, 3–5 main lobes, each lobe served by 2–5 small ductulets radiating fan-like from ental end

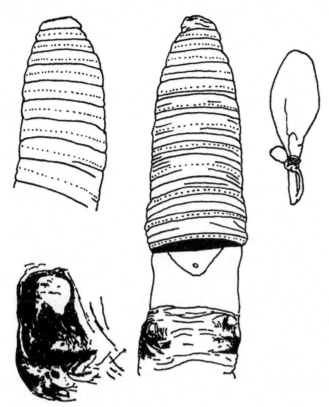

FIG. 80   *Amynthas kaopingensis* (after James et al., 2005).

of prostatic duct; ducts stout, straight, muscular, narrowing towards body wall; vasa deferentia join duct at duct-glandular portion junction; vasa deferentia nonmuscular. Spermathecae 4 pairs, in VI–IX, ampulla pear-shaped, duct shorter than ampulla; diverticulum small ovate chamber, stalk slender, either straight or kinked, about same length as duct; no nephridia on spermathecal duct.

Color: Preserved specimens unpigmented to dorsal brown pigment of variable darkness.
Type locality: Taiwan (Pingdong County).
Distribution in China: Taiwan (Pingdong).
Etymology: This species is named after the combination of prefixes, "Kaoping," derived from its localities of Kaohsiung County and Pingdong County.
Deposition of types: National Museum of Natural Science, Taichung, Taiwan.
Remarks: *A. kaopingensis* is closest to *Amynthas hengunensis* in all respects, including details of internal anatomy, such as the structlets of the prostates, the membrane enclosing segment XI, and the presence of septum 8/9. It differs from *A. hengchunensis* in features of the male field, a more dorsal placement of spermathecal pores, and the structure of the spermathecal diverticulum. There seems to be considerable morphological unity among *A. kaopingensis*, *A. hengchunensis*, and *A. formosa*e.

79. *Amynthas labosus* (Gates, 1932)

Pheretima dolosa Gates, 1932. *Rec. Ind. Mus.* 34(4):543–544.
Amynthas domosus Xu & Xiao, 2011. *Terrestrial Earthworms of China*. 134.

External Characters

Length 70 mm, width 4 mm, segment number?. Prostomium ?. First dorsal pore 12/13. Clitellum XIV–XVI. The setae begin on II, on which segment there is a complete setal circle. The setae are small and closely crowded; the setal circles without middorsal or midventral breaks, except on XXII where the setal circle is interrupted ventrally by the genital marking; setae number 22–26 (VI), 23–26 (VII), 22–28 (VII), 79–89 (XXV); 11–19 between male pores. Male pores are minute, on XVIII, each at the center of a small, transversely oval area in the setal circle of XVIII. The single genital marking is transversely oval, 9–12 intersetal distances wide transversely, with a concave, grayish central portion and whitish, nonprotuberant rim. The markings extend anteroposteriorly through the whole length of XXII and may extend slightly onto XXIII and (or) XXI, and laterally to the male pore lines. Rarely has 2 markings, one on XXII on the left side, the other on XXIII on the right side. Each of these markings ends abruptly with the midventral line. Spermathecal pores 4 pairs in 5/6/7/8/9, pore minute. No other genital marking. Female pore 1 in XIV, medioventral.

Internal Characters

Septa 9/10 absent, 8/9 is usually present as a ventral rudiment but rarely complete, 10/11 thin, 11/12/13 slightly thickened but membranous. Intestine swelling in XV. Caeca simple. There is a pair of commissures belonging to IX, rarely the single commissure of IX is on the left side, or the right side. The last pair of hearts is in XIII. Testis sacs unpaired. Seminal vesicles are always well developed and cover the dorsal blood vessel in XI and XII, the posterior vesicles displacing 12/13 and 13/14 posteriorly. Prostates are confined to XVIII; its duct is slender and straight, or bent into a hairpin loop. Spermathecae 4 pairs, in VI–IX, they are rudimentary but projecting through the parietes into the coelom; ampulla is represented by a slight swelling of the end of the duct; diverticulum is a straight, slender tube, longer than the combined lengths of duct and ampulla and passing into the anterior face of the duct.

Color: Preserved specimens unpigmented.
Type locality: Myanmar (Nam Hpen Noi and Pang Wo).
Distribution in China: Yunnan (Mengmeng, Lancang).
Deposition of types: Indian Museum.

Remarks: There is a slight, whitish, and somewhat lobed ridge on the esophagus just behind the gizzard that looks as if it were the rudiment of a postgizzard glandular collar.

80. *Amynthas lacinatus* (Chen, 1946)

Pheretima lacinata Chen, 1946. *J. West China. Border Res. Soc. (B).* 16:98–99, 139, 148.
Amynthas lacinatus Sims & Easton, 1972. *Biol. J. Linn. Soc.* 4:236.
Amynthas lacinatus Xu & Xiao, 2011. *Terrestrial Earthworms of China.* 134–135.

External Characters

Length 92–95 mm, width 3.5–5.0 mm, segment number 106. Annulation triannuli appearing behind VII. Prostomium epilobous. First dorsal pore 11/12. Clitellum XIV–XVI, annular, setae absent. Setae all fine, none enlarged, those behind clitellum shedding. Both dorsal and ventral breaks slight; $aa=1.1–1.5ab$, $zz=1.1–1.2zy$; Setae number 57–76 (III), 54–100 (VI), 38–66 (XXV); 22–40 (VI), 22–44 (VII) between spermathecal pores, 8–14 (XVII), 14–18 (XIX) between or in line with male pores. Male pores paired in XVII, about 1/3 of circumference apart ventrally, each on a nipple-like porophore, its base with concentric ridges, body wall of XVI–XXV in line with porophore on each side raised as skin flap, always directed and partly folded toward ventral side. No genital papillae. Spermathecal pores 2 pairs in 5/6/7, or 3 pairs in 5/6/7/8, or 4 pairs in 5/6/7/8/9; intersegmental, hardly visible externally, about 4/9 of circumference ventrally apart. No genital marking. Female pore 1 in XIV, medioventral.

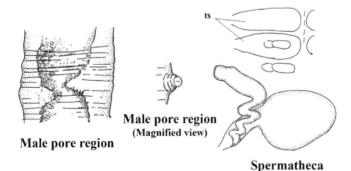

Male pore region

Male pore region
(Magnified view)

Spermatheca

FIG. 81 *Amynthas lacinatus* (after Chen, 1946).

Internal Characters

Septa 3/4/5/6/7/8 very thin, 8/9 slightly thick, 9/10 thick, all shoved back behind the gizzard, 10/11/12 thickest, 12/13 nearly as thick as 9/10, 13/14/15/16 as thick as 8/9, others thin. Gizzard large, globular, in IX and X. Intestine swelling in XV. Caeca small and simple, extending anteriorly to XXIV. Testis sacs very large, in X and XI, meeting or overlapping dorsally but disconnected ventrally. Seminal vesicles small, about 1.5 mm in ventral length, dorsal lobe about one-third as large, first pair small, enclosed in testis sacs, or large. Spermathecae 2 pairs in VI and VII, or 3 pairs in VI–VII, or 4 pairs in VI–IX, large. Ampulla sac-like, distended, 1.2 mm long, with short and stout duct. Diverticulum as long or longer, duct slender and twisted, or straight; seminal chamber elongate spherical, or club-shaped, shining with seminal contents.

Color: Pale both dorsally and ventrally. Clitellum chocolate brown.
Distribution in China: Chongqing (Mt. Jinfo), Sichuan (Ebian).
Remarks: This species is characterized by (1) the peculiar shape of the prostomium, (2) the projected flap on which the porophore is situated, (3) the presence of the gizzard septa, and (4) the enlarged testis sacs.

81. *Amynthas libratus* Tsai & Shen, 2010

Amynthas libratus Tsai & Shen, 2010. *J. Nat. Hist.* 44(21–24):1256–1260.

External Characters

Small earthworm; Length 55–72 mm, width 2.31–3.03 mm (clitellum), segment number 62–71. Prostomium epilobous. First dorsal pore in 5/6. Clitellum XIV–XVI, annular, length 2.6–2.64 mm. Setal number 32–38 (VII), 34–41 (XX); 9–11 between male pores. Male pores paired in XVIII, 0.24–0.27 body circumference apart ventrally. Each pore on a round porophore surrounded by 1–3 shallow skin folds. No genital papillae in the vicinity of the male pores. Genital papillae in horizontal rows closely along the setal lines in XVII, XIX, and occasionally XX, similar to those in the spermathecal region. In addition, horizontal rows of smaller genital papillae closely along intersegmental furrows, 1 anterior and 1 posterior to each of 17/18, 18/19, and occasionally 19/20. The number of presetal papillae 0–11 in XVII, and 0–14 in XIX; number of postsetal 0–10 in XVII, 5–12 in XIX and 0–7 in XX. Total number (prefurrow and postfurrow papillae combined) 19–34 in 17/18, 23–39 in 18/19 and 0–22 in 19/20. No presetal or postsetal papillae in XVIII. Spermathecal pores 4 pairs, each small, oval in intersegmental furrows of 5/6/7/8/9; distance between paired pores 0.27–0.28 body circumference apart ventrally. Genital papillae tiny, round, tubercle-like, arranged in horizontal rows, 1 presetal and/or postsetal, closely along the setal lines in VII–IX. The number of presetal papillae 0–5 in VII and 4–14 in IX; number of postsetal 0–4 in VII, 9–15 in VII and 5–15 in IX. No preclitellar papilla in vicinity of the spermathecal pore and along intersegmental furrow. Female pore 1 in XIV, medioventral.

Internal Characters

Septa 8/9/10 absent, 6/7/8 and 11/12/13 thick. Nephridial tufts thick on anterior faces of 5/6/7. Gizzard large in VII–X, rectangular shaped, white in color. Intestine swelling from XV. Intestinal caeca paired in XXVII, extending anteriorly to XXIII–XXIV, each simple, slender, distal end either straight or bent dorsally. Hearts in XI–XIII. Holandric, testes 2 pairs in ventrally joined sac in X and XI. Seminal vesicles large, 2 pairs in XI and XII, folliculated with a round dorsal lobe, occupying 1.5 segments. Prostate glands large, lobed with follicular surface in XVII–XX. Prostatic duct short, C-shaped or U-shaped. Accessory glands corresponding to external genital papillar and post-clitellar region. Each small with a round head 0.13–031 mm long and a slender stalk 0.35–1.2 mm in length. Spermathecae 4 pairs in VI–IX, each with a round or oval-shaped ampulla 1.17–1.58 mm long and 0.59–1.12 mm wide, and a slender spermathecal stalk 0.19–0.49 mm in length. Diverticulum with a small, oval-shaped seminal chamber and a slender stalk 0.59–0.65 mm in length, longer than the spermathecal stalk.

Color: Preserved specimens pinkish brown on head and dorsum, light brown on ventrum, and dark pinkish brown on clitellum.

Distribution in China: Taiwan (Taidong).

Etymology: The name *libratus* is given to this species with reference to the horizontal line in the arrangement of its genital papillae.

Deposition of types: Taiwan Endemic Species Research Institute, Jiji, Nantou, Taiwan.

Remarks: *Amynthas libratus* Tsai & Shen, 2010 is distinguishable from *A. pavimentus* in the arrangement of genital papillae: the papillae of *A. libratus* are arranged in a single horizontal row at one side or both sides of setal lines or intersegmental furrows and no papilla around male pores, whereas the genital papillae of *A. pavimentus* are patched between setal line and intersegmental furrows, and often a few papillae are near the male pores. The arrangement of genital papillae along setals and intersegmental furrows of *Amynthas libratus* is fairly similar to that of the quadrithecate *Amynthas kinfumontis* (Chen, 1946) of China.

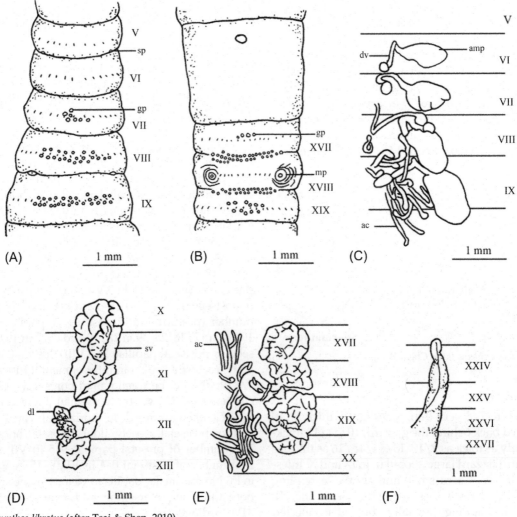

FIG. 82 *Amynthas libratus* (after Tsai & Shen, 2010).

82. *Amynthas lini* Chang, Lin, Chen, Chung & Chen, 2007

Amynthas lini Chang, Lin, Chen, Chung & Chen, 2007. *Organisms Diversity & Evolution*. 234–236.

Amynthas lini Chang, Lin, Chen, Chung & Chen, 2009. *Earthworm Fauna of Taiwan*. 54–55.

External Characters

Length 212–254 mm, width 7–9 mm (clitellum), segment number 117–129. Number of annuli per segment 3 in VI and beyond. Prostomium epilobous. First dorsal pore in 12/13. Clitellum XIV–XVI, annular, setae absent. Setae numerous, 20–26 (V), 33–45 (VII), 49–57 (X); 8–17 between male pores. Male pores paired in XVIII, 0.33 body circumference apart ventrally, porophores round or oval on setal line, with a concave center surrounded by 2–3 circular folds. Genital papillae postsetal, paired in XIX, an additional pair in XVII, an additional pair in XX, or 2 additional pairs in XVII and XX present in some specimens, each papilla oval, with a concave center surrounded by a few circular folds. Spermathecal pores 4 pairs in 5/6/7/8/9, ventrolateral, 0.4 circumference apart ventrally. Genital papillae present or absent, if present, papillae presetal, postsetal, or both, presetal papillae 1–4 pairs in VII–X, each papilla large, round, disc-like, with a concave center surrounded by a few circular folds, distance between paired genital papillae about 0.10–0.30 circumference apart ventrally, sometimes only 1 papilla present on some segments, postsetal papillae 2 pairs in VII and VII, similar to presetal ones but smaller, about 0.40 circumference apart ventrally, sometimes only 1 papilla present on some segments. Female pore single in XIV, medioventral.

Internal Characters

Septa 8/9/10 absent, 5/6/7/8 thickened, 10/11/12/13/14 greatly thickened. Gizzard round in X. Intestine from XV. Intestinal caeca paired simple, in XXVII, surface slightly folded with the septa, extending anteriorly to XXIII or XXII. Hearts in XI–XIV. Testis sacs 2 pairs, in X and XI, small, irregular. Seminal vesicles paired in XI and XII, large. Prostate glands paired in XVIII, large, extending anteriorly to XVI, with a thick straight duct.

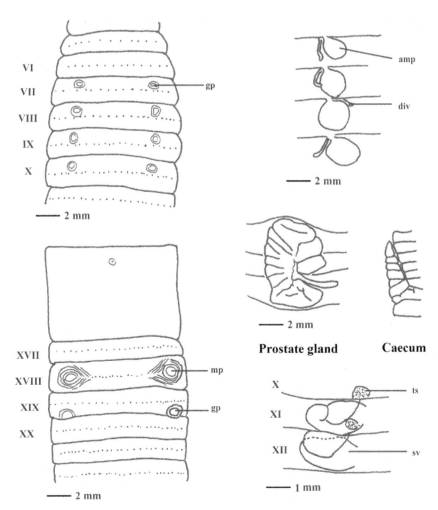

FIG. 83   *Amynthas lini* (after Chang et al, 2007).

Prostate gland   **Caecum**

Accessory glands paired in XIX, with positions corresponding to external papillae. Spermathecae 4 pairs in VI–IX, with a short thick stalk about 0.45 mm long, ampulla round, about 3.0 mm in diameter. Diverticulum with a small oval seminal chamber of 0.21 mm and a slender, straight stalk of 0.17 mm. Nephridia tufted, attached to postsegmental septa, surrounding segmental chambers anterior to septum 6/7. No nephridia on spermathecal duct. Ovaries paired in XIII, medioventral, close to septum 12/13.

> Color: Preserved specimens deep brown on dorsum and around clitellum, light yellow on ventrum and setal lines, forming a striped or banded appearance composed of a deep brown circular band and a light yellow band in sequence.
> Type locality: Wulai, Taibei County, Taiwan.
> Distribution in China: Taiwan (Taibei).
> Etymology: This species epithet was given in honor of the Taiwanese Zoologist Dr. Yao-Sung Lin, who promoted earthworm studies in Taiwan in the 1900s and 2000s.
> Deposition of types: Institute of Zoology, National Taiwan University, Taibei.
> Habitats: The specimens were collected in brown forest soil under roadside forests. Elevation from 400 to 3000 m.

83. *Amynthas longiculiculatus* (Gates, 1931)

Pheretima longiculiculata Gates, 1931. *Rec. Ind. Mus.* 33:395–400.
Pheretima longiculiculata Gates, 1932. *Rec. Ind. Mus.* 34(4):525–527.
Pheretima longiculiculata Gates, 1972. *Trans. Amer. Philos. Soc.* 62(7):163–164.
Amynthas longiculiculatus Xu & Xiao, 2011. *Terrestrial Earthworms of China.* 139–1140.

External Characters

Length 140–244 mm, width 7–10 mm, segment number 137–140. Prostomium 1/2 epilobous. First dorsal pore is in 12/13. Clitellum XIV–XVI, annular; no intersegmental furrows or dorsal pores. Setae begin on segment II, on which there are only 4, all ventral, *aa* = 2*ab*. There are 3 whitish lines present on the clitellum that indicate either presence of the setae or the former positions of setae now lost. The setae are closely crowded both dorsally and ventrally, a dorsal and ventral break lacking, except on segments VII and VII ventrally. Setae number 96–106 (XX); 36–39 between spermathecal pore line, 24–31 between male pores. Male pores are minute, on XVIII, each at center of a small, circular area, nearly 0.5 mm in diameter in the setal circle, the area surrounded by a narrow but deep and completely circumferential furrow. This circular

marking is situated on a slight, blunt protuberance from the parietes. The genital markings are 3 pairs of transversely oval, slightly raised areas, 1 pair on each of 18/19, 19/20, and 20/21. Each marking has grayish margin, a slightly concave, whitish center, and occupies a space represented by about 9–10 intersetal distances. The markings are 11 mm apart midventrally. Spermathecal pores 4 pairs in 5/6/7/8/9, each pore a minute opening on a tiny, transeversely oval, whitish area in the intersegmental furrow. The ventral distance between a pair of spermathecal pores decreases posteriorly. In addition on each of segments VII and VII there is a transversely elongated, smooth, glistening area with bluntly rounded ends; the areas in the region of the setal circles and marking a break therein. These areas are not in reality markedly different from the rest of the parietes and are without obvious demarcating grooves or lines. Female pore single in XIV, medioventral.

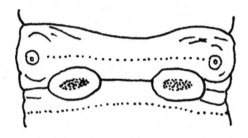

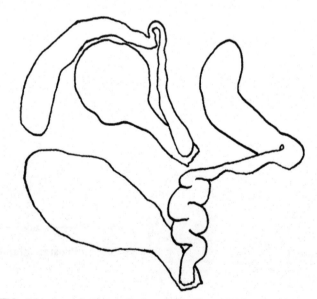

FIG. 84   *Amynthas longiculiculatus* (after Gates, 1931).

Internal Characters

Septa 8/9/10 absent, 4/5 membranous, 5/6 slightly muscular, 6/7/8 thicker still, 10/11/12 muscular, 12/13 and succeeding septa membranous. Gizzard elongate, narrowed anteriorly, posterior end enlarged and flange-shaped. Intestine swelling in XV. Caeca simple, flattened and strap-shaped, in XXVII, anteriorly into XXIV, where they are bent under the intestine but long enough to extend into XX. The last pair of hearts is in XIII. There is 1 (?) bilobed testis sac on the anterior face of 10/11. There is a 1 (?) elongate testicular chamber in XI, extending from 10/11 to 11/12. The seminal vesicles are paired in XI and XII, each vesicle in XII about twice the size of the corresponding vesicle in XI. The prostates are small and confined to segment XVIII, each gland composed of 3 major lobes of unequal size; The duct thick, muscular, and bent into a hairpin shape with the limbs in apposition. The duct relative to the size of the prostate is large and long. The vas deferentia of a side come into contact in XII and unite, the united ducts large and readily traced. Spermathecae 4 pairs, in VI–IX; ampulla rather small and roughly spherical, its duct short and stoutish, diverticulum much longer than the combined lengths of duct and ampulla, the ectal portion rather attenuated, the ental portion widened and elongately saccular. There are paired, whitish, glandular masses protruding into the coelom in segments XIX, XX, and XXI, the anterior pair of glands also projecting slightly into the posterior portion of XVIII.

> Color: Preserved specimens dorsally dark bluish to greyish blue, ventrally greyish. Clitellum dark grey-blue.
> Type locality: Myanmar (Kengtung state).
> Distribution in China: Yunnan (Mengmeng, Lancang).
> Deposition of types: Indian Museum.
> Remarks: In this species the anlage of the genital markings becomes visible externally long before the worm is fully grown, the development of the clitellar glandularity may be completed before some internal reproductive organs have attained full development.

84. *Amynthas lunatus* (Chen, 1938)

> *Pheretima (Pheretima) lunata* Chen, 1938. *Contr. Biol. Lab. Sci. Soc. China (Zool.).* 12(10):409, 411–412.
> *Amynthas lunatus* Xu & Xiao, 2011. *Terrestrial Earthworms of China.* 142–143.

External Characters

Length 145–270 mm, width 5–7 mm, segment number 170–179. Prostomium 1/2 epilobous. First dorsal pore 12/13. Clitellum in XIV–XVI, annular, well-marked, with setae on ventral side. Setae uniformly built, those on dorsal closer, none particularly enlarged; both dorsal and ventral breaks indistinct; Those on II and III much shorter, hardly visible; setae number 58–86 (IV), 66–82 (VI), 70–74 (IX), 56–66 (XIX), 64–65 (XX); 15–19 (VI), 19–20 (IX) between spermathecal pores, 4 (VII) between 2 papillae to male

pore. Male pores 1 pair in XVIII, about 1/3 of circumference apart ventrally, each on middle of a crescent or comma-shaped papilla, a large ovoid and centrally depressed papilla placed medial to each pore papilla, median 2 in close approximation (less than 2 mm at interval). Spermathecal pores 4 pairs, intersegmental, in 5/6/7/8/9, about or a little less than 1/4 circumference ventrally apart; each in minute slit (0.4 mm wide) on anterior boarder of segments VI–IX. No genital papillae in this region. Female pore 1 in XIV, medioventral.

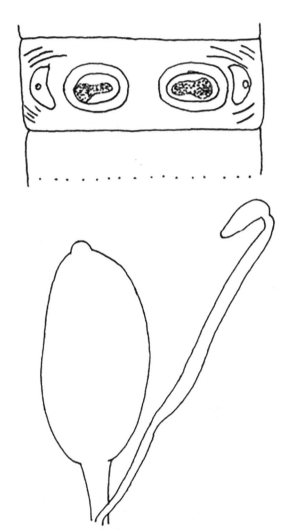

FIG. 85   *Amynthas lunatus* (after Chen, 1938).

Internal Characters

Septa 8/9/10 absent, 5/6/7/8 and 10/11/12 equally thickened, 12/13 less so. Gizzard small, in IX–X. Intestine enlarged from XV. Caeca simple, horn-shaped, in XXVII, extending anteriorly to XXV, smooth on both edges. Hearts in X–XIII, first pair stout. Testis sacs in X broadly connected ventrally, those in XI wholly united. Seminal vesicles in XI larger, each with a large dorsal lobe. Prostate gland finely lobate, in XVII–XX or XXI, its duct 4 mm long, in a deep curve, slender throughout (0.3 mm thick). Accessory glands in a round patch, sessile and velvety. Spermathecae 4 pairs, in VI–IX, ampulla 3 mm long, sac-like, with a short and distinct duct. Diverticulum tubular, 1.6 mm long, its ectal 1/4 slender as duct, no particularly enlarged seminal chamber.

Color: Preserved specimens chocolate dorsally, pale ventrally. Clitellum dark brown.
Distribution in China: Hainan (Sanya).
Etymology: The species was named after its crescent- or comma-shaped papilla.

85. *Amynthas manicatus decorosus* (Gates, 1932)

Pheretima manicata decorosa Gates, 1932. *Rec. Ind. Mus.* 34(4):528–529.
Amynthas manicatus decorosus Sims & Easton, 1972. *Biol. J. Soc.* 4:235.
Amynthas manicatus decorosus Zhong & Qiu, 1992. *Guizhou Science.* 10(4):40.
Amynthas manicatus decorosus Xu & Xiao, 2011. *Terrestrial Earthworms of China.* 144–145

External Characters

Length 40 mm, width 2.5 mm, segment number 60. Prostomium ?. There are pore-like marking from 7/8 posteriorly. Clitellum XIV–XVI, annular; with intersegmental furrows and dorsal pores; setae are present ventrally only; there are 17 setae on XVI. The setae begin on II, on which segment there is a complete setal circle. There are no middorsal or midventral breaks in the setal circles; setae number 20 (XVII), 19 (XIX), 47 (XX); 28 between spermathecal pores; 1 (XVIII) between male pores; Male pores are minute, on XVIII, on widely separated, circular discs, each disc about 1 intersetal distance wide. There is 1 pair of genital markings on XVIII, on the posterior half of the segment as indicated by their position behind the single male setae of the segment, but taking up a considerable portion of the anteroposterior length of the segment. Each marking is transversely elongate, with rounded ends, a raised rim and a concave central portion. The markings are about 8 intersetal distances wide transversely, almost reaching to the midventral line

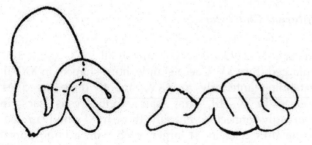

FIG. 86   *Amynthas manicatus decorosus* (after Gates, 1932).

and laterally coming into contact with the male pore discs. Spermathecal pores 4 pairs in 5/6/7/8/9. Female pore 1 in XIV, medioventral.

Internal Characters

Septa 8/9/10 absent, 10/11 very thin. Gizzard ?. Intestine swelling in XV, but is narrower through segments XV–XIX than in XX and succeeding segments. Caeca compound, in XXVII, each caecum consisting of 6–8 long, finger-like secondary caeca directed anteriorly into XXII or XIII. The single commissure of IX is on the left side. The last pair of hearts is in XIII. Testis sacs 2 pairs, but there appear to be transverse communications between the sacs of a segment. Seminal vesicles cover the dorsal vessel in XI and XII. Prostates extend through segments XVII–XIX; its duct is bent into a hairpin loop, the ectal limb thicker than ental limb; the length of the duct about 1.5 mm. Spermathecae 4 pairs, in VI, VII, VII, and IX; ampulla duct is slender and of about the same length as the ampulla. The diverticula of the first 4 spermathecae are looped into regular zigzags. The diverticula of the spermathecae of VII and IX are not looped in a regular zigzag fashion. The diverticulum passes into the anterior face of the duct which is constricted within the parietes. There are glandular masses projecting into the coelom from the parietes over the genital marking on XVIII.

Color: In preserved specimens the dorsum is bluish red anterior to the clitellum, reddish to very light brownish posterior to the clitellum. Clitellum deep red.
Type locality: Myanmar (Teung Cong).
Distribution in China: Yunnan (Lancang).
Deposition of types: Indian Museum.
Remarks: No spermathecal pores were actually seen, either before or after dissection. There 4 pairs of spermathecae in VI–IX with ducts passing into the parietes anteriorly so that the pores should be in or near intersegmental furrows 5/6/7/8/9. The 2 anteriormost spermathecae were pulled carefully out of the parietes. The holes thus produced are on intersegmental furrow 5/6.

86. *Amynthas mediocus* (Chen and Hsü, 1975)

Pheretima medioca Chen, Hsü, Yang & Fong, 1975. *Acta Zool. Sinica.* 21(1):92.
Amynthas mediocus Xu & Xiao, 2011. *Terrestrial Earthworms of China.* 146.

External Characters

Length 104–143 mm, width 4–5 mm, segment number 125–156. Prostomium 1/2–1/3 epilobous. 5–7 annulets per segment in front of the clitellum. First dorsal pore in 12/13. Clitellum XIV–XVI, annular, glandular, and thick. Setae no special change, slightly wider on ventral area of II–VII, dorsal breaks distinct, $aa = 1.2ab$, $ab > cd$, $zz = 1.5$–$2.5zy$; setal number, 36–40 (III), 53–58 (VI), 50–55 (VII), 51–62 (XII), 45–50 (XXV). Male pores on round porophore surrounded by wrinkled glandular skin, a little elevated, in ventral of XVIII, about 1/3 of body circumference apart ventrally. Spermathecal pores 4 pairs in 5/6/7/8/9; intersegmental, almost 1/2 of body circumference apart ventrally. Paired papillae found ventrally on VII–VII or XIX. Female pore 1 in XIV, medioventral.

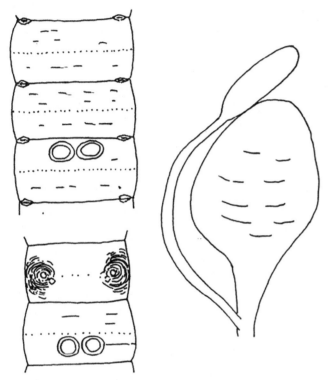

FIG. 87 *Amynthas mediocus* (after Chen et al., 1975).

Internal Characters

Septa 9/10 absent, 8/9 thin. 5/6/7/8 thick, and 10/11–13/14 slightly thick, behind 14/15 membranous. Gizzard large, barrel-shaped, in IX–X. Intestine swelling in XVI; caeca simple, in XXVII, extending anteriorly to XXII. Last hearts in XIII. Testis sacs 2 pairs, in X and XI, enclosing first pair of seminal vesicles. Seminal vesicles 2 pairs, anterior pair larger, dorsally jointed; posterior pair small, symmetrically or asymmetrically developed, dorsal lobe obviously. Prostate glands well developed, lobulated, in XVI–XXI or XVI–XX; Prostatic duct S-shaped, proximal end thin and distal end stout. Accessory glands on base with short cord duct. Spermathecae paired, in VI–VII, behind septum; ampulla oval or heart-shaped, clear boundaries with the duct; diverticulum longer than main pouch, ental end with date-shaped straight seminal chamber.

Color: Preserved specimens dark dorsally, brown-blue in preclitellar segment, gray-brown in postclitellar segment. Clitellum brown.
Type locality: Guangdong (Guangzhou).
Distribution in China: Guangdong (Guangzhou).
Remarks: This species differs from *Amynthas hongkongensis* and *Amynthas homochaetus* by the distribution of papilla and formation of male pores.

87. *Amynthas meishanensis* Chang, Lin, Chen, Chung & Chen, 2007

Amynthas meishanensis Chang, Lin, Chen, Chung & Chen, 2007. *Organisms Diversity & Evolution*. 236–239.
Amynthas meishanensis Chang, Lin, Chen, Chung & Chen, 2009. *Earthworm Fauna of Taiwan*. 56–57.

External Characters

Length 38–65 mm, width 2.7–3.5 mm, segment number 51–113. One annulus per segment on all segments. Prostomium epilobous. First dorsal pore 10/11. Clitellum XIV–XVI, annular, seta absent, Dorsal pore absent. Setae number 27–31 (V), 35–42 (VII), 41–48 (X); 5–8 between male pores. Male pores paired in XVIII, lateroventrally, about 0.35 circumference apart ventrally, porophores round or oval on setal line, surrounded by 5–7 circular folds. Genital papillae postsetal, paired in XVII and XIX, some specimens having only the left papilla in XIX, some specimens having an additional papilla or an additional pair of papillae in XX, each papilla small, oval, with concave center. Spermathecal pores 4 pairs in 5/6/7/8/9, lateral, about 0.5 circumference apart ventrally, 2 pairs of papillae present in each of VII and VII, 1 presetal and 1 postsetal, some specimens lacking papillae to varying extent, some specimens having no spermathecal papillae, some specimens have an additional postsetal papilla in VI, each papilla small, round, distance between paired genital papillae about 0.20 circumference apart ventrally. Female pore single, medioventral in XIV.

Internal Characters

Septa 8/9/10 absent, 10/11–13/14 thickened. Gizzard round in VII–X. Intestine swelling in XV. Caeca simple, in XXVII, extending anteriorly to XXIV. Lateral hearts in X–XIII. Testis sacs 2 pairs in X and XI, small, irregular. Seminal vesicles paired in XI and XII, large. Prostate glands paired in XVIII, large, extending anteriorly to XVII posteriorly to XX, with a thick duct. Spermathecae 4 pairs in VI–IX, with a short stalk about 0.2 mm long and a peach-shaped or oval ampulla about 0.6–1.0 mm long, diverticulum with a peach-shaped seminal chamber and a straight stalk about as long as seminal chamber. No nephridia on spermathecal duct. Ovaries paired in XIII, medioventral, close to septum 12/13.

FIG. 88 *Amynthas meishanensis* (after Chang et al., 2007).

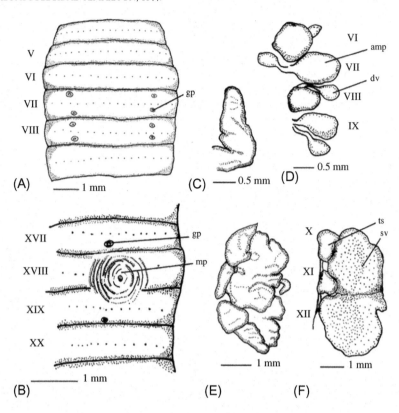

Color: Preserved specimens reddish brown on dorsum, light yellowish brown on venter.

Distribution in China: Taiwan.

Etymology: The specific epithet was given with reference to the type locality of Meishan, Chiayi County, Taiwan.

Type locality: Meishan, Chiayi County, south central Taiwan.

Deposition of types: Institute of zoology, National Taiwan University, Taibei.

88. *Amynthas mirifius* Sun and Qiu, 2013

Amynthas mirifius Sun & Qiu, 2013. *J. Nat. Hist.* 47(17–20):1151–1156.

External Characters

Length 174 mm, width 4.1 mm (clitellum), segment number 132. Body cylindrical in cross section, gradually toward head and tail; no secondary annulations. Prostomium 1/2 epilobous. First dorsal pore in 12/13. Clitellum XIV–XVI, annular, swollen, smooth, 2 and 12 setae in the venter of XIV and XVI respectively, dorsal pore invisible within clitellum. Setae numerous, 30 (III), 26 (V), 34 (VII), 42 (XX), 46 (XXV); 6 (VI), 10 (VII), 12 (VII) between spermathecal pores on venter; 19–23 between spermathecal pores, 13 between male pores. Male pores paired in XVIII, 0.33 body circumference apart ventrally, each on the center of a slightly raised, glandular, and approximately elliptical porophore, bordered by 2 folds on lateral margin. Genital markings look like a peanut in a shell, that is, a rectangle with enlarged rounded ends, concave-topped, paired behind 17/18 and on 18/19, 4 intersetal intervals between left and right genital markings. Spermathecal pores 4 pairs in 5/6/7/8/9, ventral, tiny, 0.33 body circumference apart. Female pore single in XIV, medioventral.

Internal Characters

Septa 8/9/10 absent, 5/6/7 and 10/11/12 slightly thickened, other septa thicker than usual case. Dorsal blood vessel single, continuous onto pharynx; esophageal hearts enlarged from X–XIII. Gizzard in VII–X, barrel-shaped. Intestine enlarged gradually from XV. Intestinal caeca simple, in XXVII, extending anteriorly to 1/2 XXIII, long finger-shaped sac, smooth on dorsal and ventral margins. Holandric, testis sacs 2 pairs, in X–XI, anterior pair ovate, separated from each other with posterior pair enclosing first pair of seminal vesicles; seminal vesicles paired in XI–XII, developed. Prostates in XVII–XIX, moderately developed, finger-shaped on the outside margin and coarsely lobate on other part, prostatic duct well developed, inverted U-shaped, distal end stout. Accessory glands absent. Spermathecae 4 pairs in VI–IX, about 1.7–1.8 mm long; ampulla heart-shaped or ovate, slender terminal duct as long as 0.4 ampulla; diverticulum longer than main spermathecal axis by fraction of 0.2, slender, terminal 0.33 dilated into narrow

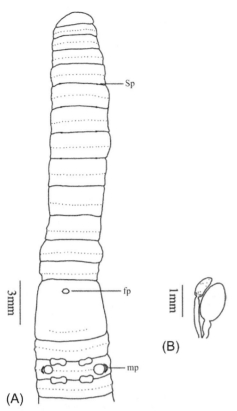

FIG. 89 *Amynthas mirifius* (after Sun et al., 2012).

ovoid seminal chamber; no nephridia on spermathecal ducts.

Color: Preserved specimens dark gray on dorsum before IX, brown on dorsum after IX, no pigment on venter, clear middorsal line after clitellum. Clitellum grey and brown.
Distribution in China: Hainan (Mt. Jianfeng).
Etymology: This species is named after its extraordinary male pore region papillae using the Latin *mirificus*, meaning marvelous or extraordinary.
Deposition of types: Shanghai Natural Museum.
Deposition of the cytochrome oxidase type 1 sequence: NCBI GenBank under no. JQ905265.
Type locality: Mt. Jianfeng (elevation 841 m), brown soil with rich organic matter, near the entrance to core area at the roadside, Hainan Province.
Remarks: *Amynthas mirifius* is closely related to *Amynthas genitals* discovered on Hainan Island. *A. mirifius* and *A. genital* share the spermathecal pores in 5/6/7/8/9, the distance between spermathecal pores being 0.33 body circumference, and even the appearance of the male porophore. Both of them have no genital papillae in the spermathecal pore region and have a similar setal number. The differences between these 2 species are as follows: *A. mirifius* has a bigger body size than *A. genital*; pigment in *A. mirifius* is dark

gray and brown on dorsum, whereas there is no pigment in *A. genital*; the genital papillae within the male pore region are shaped like a peanut in a shell, concave-topped and paired behind 17/18 and on 18/19 in *A. mirifius* versus ovate, flat-topped and paired presetal on XVIII and XIX, postsetal on XVII and XVIII in *A. genitals*. *A. mirifius* and *A. genitalis* have similar diverticulum characters, but the diverticulum of *A. mirifius* looks a little shorter than that of *A. genitalis*.

89. *Amynthas mirabilis* (Bourne, 1886)

Perichaeta mirabilis Bourne, 1886. *Proc. Zool. Soc. London.* 1886:668–669.
Pheretima mirabilis Gates, 1935a. *Smithsonian Mis. Coll.* 93(3):12.
Amynthas mirabilis Xu & Xiao, 2011. *Terrestrial Earthworms of China.* 149–150.

External Characters

Length 130 mm, width 8 mm, segment number 114. Prostomium 1/3 epilobous. First dorsal pore 12/13. Clitellum entire, in XIV–XVI, annular. Setae fine, none particularly enlarged, closer on dorsal and lateral sides, very fine and close on dorsal of anterior segments; both dorsal and ventral breaks indistinct; $aa = 1ab$, $zz = 1.2$–2.0 yz behind clitellum; setae number 74–84(III), 100–115(VI), 94–116(VII), 92–106(IX), 70–80(XX), 80–82(XXV); 34–48 (VI), 36–60(VII), 36–40(IX) between spermathecal pore, 20–22(VII) between male pores, 12 on XIX between 2 papillae. Male pores 1 pair, widely separated, situated on low papillae in XVIII, on ventrolateral corner of XVIII, about 1/3 of circumference apart ventrally, each on a tubercle, laterally covered by a skin fold appearing like half-closed eyelid; a raised papilla placed in front on segment XVII and another (usually 2 coalesced) behind on XIX in line with male porophore. Spermathecal pores 4 pairs in 5/6/7/8/9, each intersegmental and superficial appearing as a pinhole. Genital papillae 4 pairs of groups, in V, VI, VII, and VII; minute, circular, diamond-shaped depressions lying in the posterior portion of the segments; and 2 pairs of groups, front setae-ring of VII and VII. Female pore single in XIV, medioventral.

Internal Characters

Gizzard in X. Caeca simple, in XXVI, extending anteriorly to XXV. Spermathecae 4 pairs, in VI–IX, they present a single appendage. They open, as is usually the case, exactly between the segment in which they lie and the preceding segment. A single appendage, both ampulla and diverticulum small, main pouch about 2 mm long; ampulla sac-like about 1.8 mm wide, transversely wrinkled, clearly marked off from its duct which is short but wide. Diverticulum shorter than main pouch, with a very short duct (1/3 of ampullar duct) and an elongate club-shaped seminal chamber (1.5 mm long), whitish in appearance.

Color: ?.
Distribution in China: Sichuan.
Remarks: Gates thought *Amynthas mirabilis* (Bourne, 1886) was synonyms with *A. heterochaeta* and *A. divergens yunnanensis* (Gates, 1935). Michaelsen thought *Amynthas mirabilis* was the same species as *A. divergens yunnanensis* (Michaelsen, 1931).

90. *Amynthas montanus* Qiu & Sun, 2012

Amynthas montanus Qiu & Sun, 2012. *Zootaxa*. 33458:154-155.

External Characters

Big earthworm, Length ?–210 mm, width 4.5–7.5 mm (preclitellum), segment number ?–193, incomplete posterior amputee in paratype. Body cylindrical in cross section, gradually tapered towards head and tail. Prostomium 1/2 epilobous. First dorsal pore 12/13. Clitellum XIV–XVI, annular, smooth, setae can be seen externally in XIV–XVI; dorsal pore present; rarely has furrows. Setae number 52–74 (III), 72–120 (V), 80–142 (VII), 52–102 (XX), 60–96 (XXV); 26–36 (VI), 21–42 (VII), 24–40 (VII) between spermathecal pores, 12–18 between male pores; setal formula

$aa = 1.2–1.8ab$, $zz = 1.8–2zy$. Male pores paired in XVIII, 0.33 body circumference apart ventrally; each found on setal line with elevated center surrounded by 2–3 circular folds. Genital papillae ovate, flat-topped, diameter 0.7–0.8 mm, surrounded by epidermal folds, paired on 17/18, 18/19 median to male pores. Spermathecal pores 4 pairs in 5/6/7/8/9, about 0.33 circumference apart ventrally. Female pore single in XIV, midventral.

Internal Characters

Septa 8/9/10 absent, 5/6/7/8 and 10/11/12/13 muscular, 13/14 slightly thickened. Gizzard in VII–X, ball-shaped. Intestine swelling in XVI. Intestinal caeca simple, in XXVII, extending anteriorly to XXV, finger-shaped sac with smooth margin. Dorsal blood vessel single, continuous onto pharynx; Hearts enlarged from X–XIII. Holandric, Testis sacs 2 pairs, ventral in X–XI, anterior pair connected with membrane on venter between left and right side. Seminal vesicles paired in each of XI and XII, anterior pair bigger in size, the latter pair vestigial. Prostates in XVI–XX, coarsely lobate. Prostatic duct U-shaped, distal end appreciably enlarged. Accessory glands absent. Ovaries in XIII. Spermathecae 4 pairs in VI–IX, about 4.2 mm long; ampulla heart-shaped, stout duct as long as 0.75 ampulla. Diverticulum shorter than main spermathecal axis, terminal 0.4 dilated into a rod-shaped chamber. Stalk composed of entally widening hairpin loops; no nephridia on spermathecal ducts.

Color: Preserved specimens no pigment, clear middorsal line. Clitellum light red or light brown.
Distribution in China: Hainan (Mt. Diaoluo).
Etymology: The species is named after its mountain habitat.
Deposition of types: Shanghai Natural History Museum.
Type locality: Mt. Diaoluo (elevation 394 m), cinnamon sandy soil under banana and arbor forests; at an elevation of 930, yellow cinnamon soil in bamboo and camphor forests at the roadside, Hainan.
Remarks: *Amynthas montanus* is a big octothecal earthworm. The arrangement of male pore region genital papillae of *Amynthas montanus* is somewhat similar to that of *A. tetrapapillatus* from Hainan Island, *Amynthas diaoluomontis* from Hainan Island, and *A. wulinensis* from Taiwan.

Amynthas montanus has no pigment, first dorsal pore in 12/13, setae externally in XIV–XVI, 0.33 circumference ventrally apart between spermathecal pores and 4 pairs of spermathecae, which are 4.2 mm long. In contrast, *A. tetrapapillatus* has light maroon pigment, first dorsal pore in 11/12, no setae externally in XIV–XVI, spermathecal pores 0.17 circumference apart dorsally and only 1 pair of spermathecae, which is 1.8–2.0 mm long.

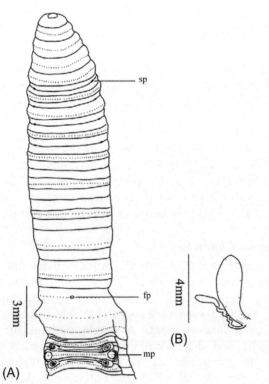

FIG. 90   *Amynthas montanus* (after Sun et al., 2012).

Amynthas montanus is easily distinguished from *A. diaoluomontis* by the number of spermathecae, transverse distance between spermathecal pores, prostomium, setal formula, characters of male pores, indentations of intestinal caeca, the size of spermathecae, and characters of the diverticulum.

Amynthas montanus and *A. wulinensis* can be separated on the basis of the variable position of genital papillae from XVII–XX and the character of the diverticulum. The genital papillae of *A. wulinensis* are usually located in posterior annulet between the setal line and the posterior intersegmental furrow, while there may occasionally be an additional pair or 1 papilla in XX, or 1 missing from XVII or XIX. There are other differences in the position of first dorsal pore, setae, dorsal area of clitellum, transverse distance of spermathecal pores, pigment, and characters of gizzard and accessory glands near prostates.

91. *Amynthas nanrenensis* James, Shih & Chang, 2005

Amynthas nanrenensis James, Shih & Chang 2005. *J. N. His.* 39(14):1008–1012.
Amynthas nanshanensis Chang, Shen & Chen, 2009. *Earthworm Fauna of Taiwan.* 62–63.
Amynthas nanshanensis Xu & Xiao, 2011. *Terrestrial Earthworms of China.* 155–156.

External Characters

Length 97 mm, clitellum width 4.2 mm, segment number 98. No secondary segmentation. Prostomium epilobous. First dorsal pore 11/12 or 12/13. Clitellum XIV–XVI, annular, setae absent, dorsal pores absent. Setae formula *aa:ab:yz:zz* = 2.5:1:1:3 at XXV, 60–64 (VII), 62–72 (XX), 10–12 between male pores. Male pores paired in XVIII, distance between the paired pores 0.21 circumference apart ventrally, each male pore flanked by 2 circular 0.2 mm genital marking closely median and lateral to male pores. Spermathecal pores 4 pairs in 5/6/7/8/9, 0.22 circumference apart ventrally. Female pore single, medioventral in XIV.

Internal Characters

Septa 8/9/10 absent, 5/6/7/8 greatly thickened, 10/11/12/13/14 thickened. Gizzard in VII–X. Intestine from XVI. Lymph glands present from XXVII. Intestinal caeca paired in XXVII, simple, extending anteriorly to XXIII, no incisions. Esophageal hearts paired in XII and XIII. Male sexual system holandric. Testis sacs 2 pairs in X and XI, ventrally jointed. Seminal vesicles paired in XI and XII, large, with dorsal lobe. Prostate glands paired in XVIII, 2 or 3 main lobes, duct thick, muscular, joins vasa deferentia distal to glandular portion. Spermathecae 4 pairs in VI–IX, ampulla ovoid, diverticulum

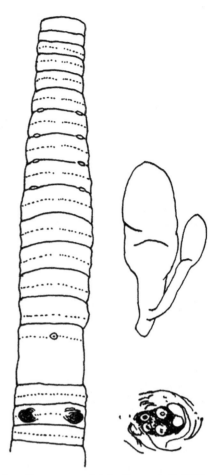

FIG. 91 *Amynthas nanrenensis* (after James et al., 2005).

short, reaching the middle of ampulla, with elongated ovoid- to almond-shaped seminal chamber and slender straight stalk. Ovaries in XIII.

Color: Unpigmented.
Type locality: Nanrenshan, Kending, Pingdong County, Taiwan.
Deposition of types: National Museum of Natural Science, Taichung, Taiwan.
Distribution in China: Taiwan (Pingdong).
Etymology: This species was named after its type locality, Nanrenshan or Mt. Nanren.
Remarks: *Amynthas nanrenensis* differs from *A. corticis* (Kinberg, 1867) as described in detail in Blakemore (2003) by the narrower spacing of the spermathecal pores. It differs from *A. corticis* in having spermathecal pores nearly midlateral, a different arrangement of genital markings in the male field, presence of genital markings in the spermathecal segments, and hearts in XI. Male function is clearly present in *A. nanrenensis*, with iridescent male funnels and iridescent spermathecal diverticulum chambers. Spermathecal segment genital markings are lacking in *A. nanrenensis*, unlike *A. corticis* which generally has them. Genital marking glands are

absent, another separation from *A. corticis*, and cannot, as has been done in other cases, be ascribed to parthenogenetic degradation of sexual characters.

Amynthas nanrenensis differs from all other known *Amynthas* by spermathecae in VI–IX and ventrally placed spermathecal pores with respect to the lack of hearts in X and XI. Plus the arrangement of genital markings in the male field.

92. *Amynthas nanshanensis* Shen, Tsai & Tsai, 2003

Amynthas nanshanensis. Shen, Tsai & Tsai 2003. *Zool. Stud.* 42:482–484.
Amynthas nanshanensis Chang, Shen & Chen, 2009. *Earthworm Fauna of Taiwan*. 64–65.
Amynthas nanshanensis Xu & Xiao, 2011. *Terrestrial Earthworms of China*. 156–157.

External Characters

Length 41–89 mm, clitellum width 2.2–3.0 mm, segment number 50–104. Number of incomplete annuli (secondary segmentation) per segment 2–3 in IX–XIII. Prostomium epilobous. First dorsal pore 5/6. Clitellum XIV–XVI, annular, 1.9–3.2 mm long, setae absent, dorsal pores absent. Setae 28–36 (VII), 34–44 (XX), 8–12 between male pores. Male pores paired in XVIII, 0.24–0.29 circumference apart ventrally, each with a lateral papilla closely adjacent, both male pore and papilla surrounded by 2–3 slight circular folds. Genital papillae present or absent, if present, widely paired, presetal in XVIII and XIX, slightly medial to male pores, paired, 1 on right, or 1 on left, each papilla round, center concave, about 0.2 mm in diameter, similar in structure to those in the preclitellar region. Spermathecal pores 4 pairs in 5/6/7/8/9, 0.33–0.34 circumference apart ventrally. Genital papillae present or absent in VII–IX, if present, in VII, presetal papillae single, medial ventral; in VII, presetal papillae 1 median, paired, 1 median with one on right or left, one median with one on both sides, 2 median with one on either side, or 6 papillae, postsetal papillae 1 on right or left or 2; in IX, presetal papillae one on left or 2; each papilla round, center concave, 0.2–0.4 mm in diameter. Female pore single, medioventral in XIV.

Internal Characters

Septa 5/6/7/8, 10/11–13/14 thickened, 8/9/10 absent. Gizzard round in IX and X. Intestine from XIV or XV. Intestinal caeca paired in XXVII, simple, surface slightly wrinkled, short, extending anteriorly to XXV. Esophageal hearts in XI–XIII. Testis sacs 2 pairs in XI, round. Seminal vesicles paired in XI and XII, large, each occupying about 1.5 segments, follicular. Prostate glands paired in XVIII, extending anteriorly to XVII and

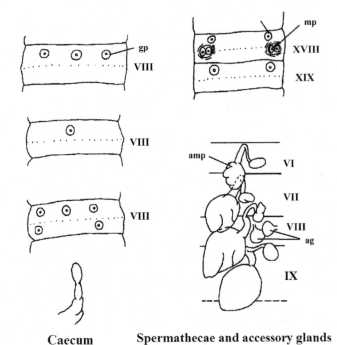

Caecum   Spermathecae and accessory glands

FIG. 92   *Amynthas nanshanensis* (after Shen et al., 2003).

posteriorly to XX, prostatic duct C-shaped. Accessory glands smaller than those in preclitellar region, slightly lobed, stalk 0.1–0.3 mm long, head 0.1–0.3 mm long. No gland associated with papilla lateral to male pore. Spermathecae 4 pairs in VI–IX, ampulla peach-shaped, 1.0–1.5 mm long, 0.6–1.5 mm wide, duct stout, 0.5 mm long, diverticulum stalk slender, about 0.5 mm long, seminal chamber oval, about 0.5 mm long. Accessory glands large, round, about 0.5 mm in diameter, stalk short, about 0.2 mm long, corresponding to external genital papillae.

Color: Preserved specimens purplish brown on dorsum, light gray on venter, yellowish gray around clitellum.
Distribution in China: Taiwan (Nantou).
Type locality: Mountain slope along Nanshan Creek (elevation 800–900 m), Jenai Township, Nantou County, Taiwan.
Deposition of types: Taiwan Endemic Species Research Institute, Jiji, Nantou County, Taiwan.
Holotype: coll. no. 1999-20-Shen.
Etymology: The name "*nanshanensis*" was given with reference to the type locality of Nanshan Creek.
Remarks: *Amynthas nanshanensis* has a small round porophore immediately adjacent laterally to a small genital papilla without an accessory gland, and also has widely paired genital papillae with stalked accessory glands in both spermathecal and male pore regions. These characters are easily distinguishable from *A. penpuensis*, *A. Uvaglandularis*, and *A. wulinensis*. *A. penpuensis* has no genital papillae, and *A. wulinensis* has no papillae in the spermathecal region, but has widely paired papillae with sessile

accessory glands in the male region (Tsai et al., 2001). *A. penpuensis* has large sessile folliculate accessory glands associated with papillae adjacent to each of the male porophores and also paired genital papillae in both spermathecal and male pore regions. Whereas *A. nanshanensis* has stalk accessory glands. Based on the position of the genital papillae, *A. nanshanensisis* more closely related to *A. uvaglandularis* than to *A. wulinensis* or *A. penpuensis*.

Amynthas nanshanensis differs from *A. youngtai* in being smaller and having lower segment and setal numbers. Also, the former has stalked accessory glands while the latter has sessile accessory glands.

93. *Amynthas pavimentus* Tsai & Shen, 2010

Amynthas pavimentus Tsai & Shen, 2010. *J. Nat. Hist.* 44(21–24):1252–1256.

External Characters

Small to medium. Length 50–121 mm, width 1.98–3.96 mm (clitellum), segment number 61–103. Prostomium epilobous. First dorsal pore in 5/6 or 6/7. Clitellum XIV–XVI, annular. Setal number 30–41 (VII), 40–51 (XX); 8–13 between male pores. Male pores paired in XVIII, 0.27–0.3 body circumference apart ventrally. Each pore on a round porophore about 0.35 mm in diameter, surrounded by 1–3 shallow skin folds, with 1–8 genital papillae at medioanterior and/or medioposterior portion outside the skin fold. Postclitellar papillae similar in shape and size to those in the preclitellar region, also in transverse patches between male pores; presetal papillar 20–61 in XVIII, 8–59 in XIX and 0–1 in XX; postsetal papillae 0–8 in XVII, 0–5 in XVIII and 0–3 in XX. Spermathecal pores 4 pairs, each small, lip-like elevation, in intersegmental furrows of 5/6/7/8/9; distance between paired pores 0.28–0.32 body circumference apart ventrally. Genital papillae tiny, round, tubercle-like, and highly variable in number and positions. Presetal papillae in transverse patches between setal line and intersegmental furrow; the number 5–37 in VII and 5–40 in IX. No preclitellar papilla in the postsetal portion and in the vicinity of spermathecal pores. Female pore single in XIV, medioventral.

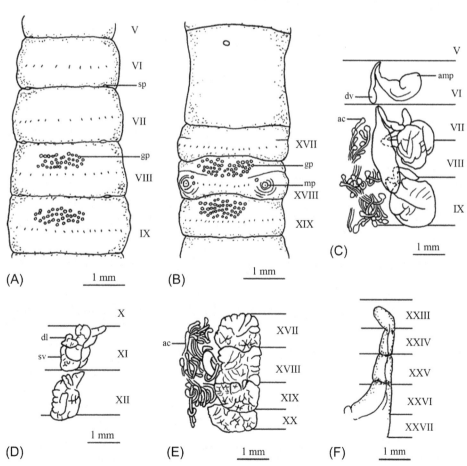

FIG. 93 *Amynthas pavimenta* (after Tsai & Shen, 2010).

Internal Characters

Septa 8/9/10 absent, 5/6/7/8 and 10/11–13/14 thick. Nephridial tufts thick on anterior faces of 5/6/7. Gizzard large, round in VII–X. Intestine swollen from XV. Intestinal caeca paired in XXVII, extending anteriorly to XXIII–XXIV, each simple, slender, slightly folded with white end that is bent toward the dorsum. Hearts in XI–XIII. Holandric, testes 2 pairs in ventrally joined sacs in X and XI; first pair large. Vas efferens connected in XI on each side to form a vas deferens. Seminal vesicles 2 pairs in XI and XII, normal, medium-sized or large, occupying the full compartment, follicular surface with a round dorsal lobe. Prostate glands large, mostly rectangular-shaped with follicular surface in XVI–XIX or XVII–XX. Prostatic duct short, C-shaped or U-shaped. Accessory glands stalked, transparently white in color, corresponding to external genital papillae; each gland small with a round head 0.15–0.47 mm long and a slender stalk 0.15–1.5 mm in length in both preclitellar and postclitellar regions. Spermathecae 4 pairs in VI–IX, varying in size and shape. Each with a round or oval-shaped ampulla 0.7–3.27 mm long and 0.4–2.18 mm wide, and a slender to stout spermathecal stalk 0.37–0.87 mm in length. Diverticulum with a small, oval-shaped seminal chamber and a slender stalk 0.6–1.0 mm in length that is as long as, or slightly longer than, the spermathecal stalk.

Color: Preserved specimens pinkish brown on head and dorsum, light brown on venter, and dark pinkish brown on clitellum.
Distribution in China: Taiwan (Hsiangyang, Taidong County; Meishan, Kaohsiung County).
Deposition of types: Taiwan Endemic Species Research Institute, Jiji, Nantou, Taiwan.
Type locality: Elevation 2100–2300 m near Hsiangyang, Taidong County.
Remarks: *Amynthas pavimentus* is easily distinguishable from the other 3 earthworms with multiple genital papillae reported from Taiwan, *A. papulosus sauteri* (Michaelsen, 1922), *A. tessellatus* Shen et al., 2002, and *A. polyglandularis* (Tsai, 1964). The former 2 species are sexthecate with 3 pairs of spermathecae in segments VI–VII, whereas *A. polyglandularis* is quadeithecate with 2 pairs of spermathecae in segments VII and IX.

94. *Amynthas penpuensis* Shen, Tsai & Tsai, 2003

Amynthas penpuensis, Shen, Tsai & Tsai, 2003. *Zoological Studies*. 42(4), 479–490.
Amynthas penpuensis, Chang, Shen & Chen, 2009. *Earthworm Fauna of Taiwan*. 70–71.
Amynthas penpuensis, Xu & Xiao, 2011. *Terrestrial Earthworms of China*. 106–107.

External Characters

Length 55–104 mm, clitellum width 2.2–3.2 mm, segment number 62–104. Number of incomplete annuli (secondary segmentation) per segment 2–3 in VII–XIII. Prostomium epilobous. First dorsal pore 5/6 or 6/7. Clitellum XIV–XVI, annular, 2.1–2.8 mm long, setae absent, dorsal pores absent or marked with shallow depression. Setae 27–37 (VII), 36–46 (XX), 8–11 between male pores. Male pores paired in XVIII, 0.25–0.28 circumference apart ventrally, porophore round, around 0.3 mm in diameter, surrounded by 2–4 slight circular or diamond-shaped folds. Spermathecal pores 4 pairs in 5/6/7/8/9, not visible externally. Female pore single, medioventral in XIV.

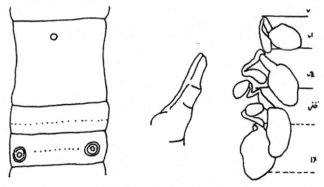

FIG. 94   *Amynthas penpuensis* (after Shen, et al., 2003).

Internal Characters

Septa 8/9/10 absent, 6/7/8, 11/12/13/14 thickened. Gizzard round in IX and X. Intestine from XVI. Intestinal caeca paired in XXVII, simple, surface slightly wrinkled, extending anteriorly to XXV–XXIV. Esophageal hearts in XI–XIII. Testis sacs 2 pairs in XI, round. Seminal vesicles paired in XI and XII, the posterior pair larger than the anterior pair, occupying 1.5 to nearly 2 segments, fairly smooth. Prostate glands paired in XVIII, extending from XVII to XX, prostatic duct slender, U-shaped. Spermathecae 4 pairs in VI–IX, ampulla peach-shaped, 1.0–1.3 mm long, 0.7–0.9 mm wide, duct stout, about 0.4 mm long, diverticulum stalk slender, straight or curved, 0.4–0.6 mm long, seminal chamber oval, 0.4–0.5 mm long.

Color: Preserved specimens gray on dorsum, whitish gray on venter, light brown around clitellum.
Type locality: A mountain slope along Penpu Creek, Jenai Township, Nantou County, Taiwan.
Deposition of types: Taiwan Endemic Species Research Institute, Jiji, Nantou County, Taiwan.
Holotype: coll. no. 1999-21-Shen. Paratypes: same collection data as for holotype.
Distribution in China: Taiwan (Nanshang, Pingpu).
Etymology: The name *"penpuensis"* was given with reference to the type locality of Penpu Creek.

Remarks: *Amynthas penpuensis* is fairly similar in structure to *A. formicatus* (Gates) of Sichuan, China, in having a simple male pore structure without genital papillae. However, *A. formicatus* has larger, disc-shaped or square male porophores and higher setal numbers (Gates 1935, Chen 1936), while *A. penpuensis* has small round porophores and lower numbers of setae. Also spermathecal pores are not visible externally on the papillae (porophores) on *A. formicatus*. In addition, the first dorsal pore is in 5/6 or 6/7 for *Amynthas penpuensis* whereas it is in 11/12 or 12/13 for *A. formicatus*.

95. *Amynthas pingi pingi* (Stephenson, 1925)

Pheretima pingi Stephenson, 1925. *Proc. Zool. Soc. London.* 1925:891–893.
Pheretima pingi Chen, 1933. *Contr. Biol. Lab. Sci. Soc. China (Zool.).* 9:228–234.
Pheretima ping Gates, 1935a. *Smithsonian Mis. Coll.* 93(3):14.
Pheretima pingi Gates, 1939. *Proc. U.S. Natr. Mus.* 85:465–469.
Pheretima pingi Chen, 1946. *J. West China. Border Res. Soc. (B).* 16:136.
Amynthas pingi pingi Xu & Xiao, 2011. *Terrestrial Earthworms of China.* 166–167.

External Characters

Length 160–340 mm, width 6–10 mm, segment number 110–179. Prostomium 1/2–2/3 epilobous, open behind. First dorsal pore in 12/13, those on clitellum usually obliterated. Clitellum entire, in XIV–XVI, annular, without setae and other markings, its length equal to about 3.5 immediate postclitellar segments. Setae generally stout, ventral setae of III–VII or IX noticeably long and widely spaced, $aa = 1.2–2.0ab$, $zz = 1.2–1.5\ yz$. Setae number 24–29 (III), 30–36 (VI), 34–57 (VII), 55–64 (XII), 47–72 (XXV); 13–18 (VI), 18–24 (VII) between spermathecal pores, 12–20 between or in line with male pores. Male pores paired in XVIII, about 1/3 of circumference apart ventrally, each pore as a whitish spot in center of a flattened, round or somewhat wedge-shaped papilla which is slightly raised or a little sunken with 2 or more shallow concentric grooves, its surface glandular and smooth, with less distinct margin. Genital papillae rarely absent, 1–3 pairs (occasionally 5 pairs) usually placed around male pore region; 2 pairs on segment XVIII in front and behind setae and 1 pair on segment XIX in front of setae the postsetal pair on XVIII often closer to male papilla while the antesetal pair on both segments either situated laterally or medially. Spermathecal pores 4 pairs, in intersegmental furrows 5/6/7/8/9, about 1/3 of body circumference apart ventrally, each pore minute, situated

on posterior side of a transversely oval papilla which is flat-topped, distinctly raised as large as other genital papillae or inconspicuous as a small tubercle. Genital papillae often paired on segments VII and IX, those on VII placed either antesetally or postsetally if only 1 pair present, on both sides of setae if 2 pairs present; 1 or occasionally 2 pairs on IX generally placed antesetally (very seldom present postsetally on IX and antesetally on X). These papillae paired either lateroventrally or medioventrally. Female pore single in XIV, medioventral.

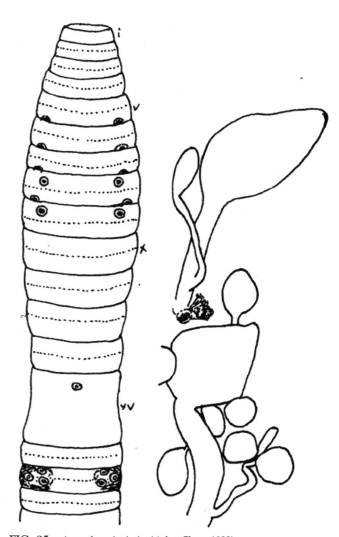

FIG. 95   *Amynthas pingi pingi* (after Chen, 1933).

Internal Characters

Septa 9/10 absent, 5/6/7/8 greatly thickened, 8/9 membranous, 10/11/12 or 12/13 also thickened, those behind 14/15 thin and less musculated. Gizzard in IX and X, globular, fairly large, with a slight crop-like dilation on its anterior side. Intestine swelling in XV. Caeca simple, conical in shape, nearly smooth on both dorsal and ventral edges, in XXVII, extending anteriorly to XXIV or XIII. Heart 3 pairs, in XI–XIII. Testis sacs in X and XI, large and conspicuous; anterior pair rounded, projecting into segment X, connected medially either in a narrow tube or broadly with its opposite fellow, usually in a thick U shape; posterior pair broadly connected or completely fused into a transverse sac, or slightly constricted medially. Seminal vesicles large, in XI and XII, usually tubercular on surface subdivided into 2 or more divisions, each with a small dorsal lobe, granular on surface. Prostate glands well developed, subdivided into transverse lobes, in XVI–XX or XXI, glandular parts of both sides nearly overlapping dorsally; each with a rather short but stout duct in a deep U-shaped curve, thicker at distal third. Accessory glands found in this region corresponding to papillae externally. Each papilla associated with 1 gland internally, which is roundish and granular on surface, probably as an aggregate of small glands, with a short stout duct apparently being composed of numerous cords. Spermathecae 4 pairs in V–IX, last 2 pairs placed between septa 7/8 and 8/9; ampulla heart-shaped, elongate pear-shaped or spatulate, smooth or sometimes transversely wrinkled on surface, its duct rather stout and long, ectally slightly narrower, usually marked off from the latter. Diverticulum about half length of both ampulla and ampullar duct, consisting of a long slender duct and an ental dilated portion which is ovoid or elongate spherical or date-shaped, often bright white in color.

Color: Preserved specimens dark chestnut or purplish chocolate dorsally, darker at anterior dorsum, uniformly pigmented except slightly whitish around setal circles, pale or grayish pale ventrally. Clitellum brownish.

Distribution in China: Beijing (Haidian, Huairou), Liaoning (Dandong), Shanghai, Jiangsu (Nanjing, Wuxi, Suzhou), Zhejiang (Zhoushan, Chuanshan, Ningbo, Shengxian, Lanxi, Tonglu, Xindeng), Anhui (Anqing, Chuzhou), Fujian, Shandong (Yantai, Weihai), Jiangxi (Nanchang, Jiujiang, Dean), Henan (Huixian, Xinyang, Shangcheng, Luoshan, Xixia, Xuchang, Luoyang), Hubei (Qianjiang), Hong Kong, Aomen, Chongqing (Shapingba), Sichuan (Chengdu, Leshan), Guizhou (Mt. Fanjing).

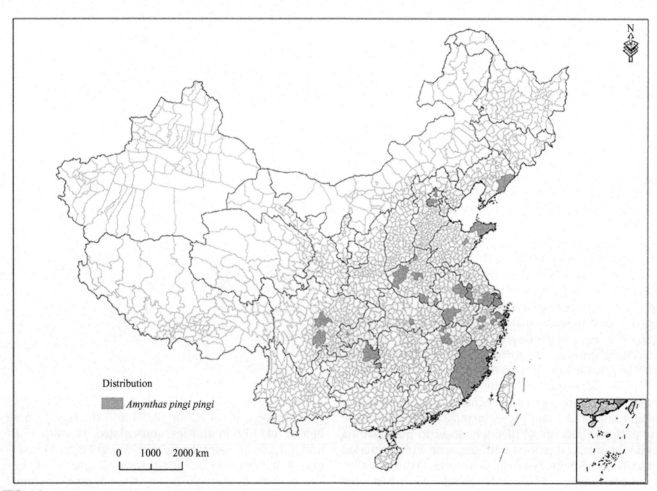

FIG. 96 Distribution in China of *Amynthas pingi pingi*.

Type locality: Jiangsu (Nanjing).

Deposition of types: British Museum.

Remarks: This endemic species is very common in the Yangtze valley, particularly abundant in South Jiangsu and Zhejiang. The abnormalities with an additional spermatheca in VI and absence of both spermathecae in V are occasionally noted. In Stephenson's original description, several points are not very clear.

96. *Amynthas pulvinus* Sun & Jiang, 2013

Amynthas pulvinus Sun & Jiang, 2013. *J. Nat. Hist.* 47(17–20):1156–1158.

External Characters

Length 93.5 mm, width 3.4 mm (clitellum), segment number 108. Body cylindrical in cross section, gradually tapered towards head and tail; no secondary annulations. Prostomium 1/2 epilobous. First dorsal pore in 12/13. Clitellum XIV–XVI, annular, swollen, smooth, several setae visible in the venter of each of XIV–XVI, dorsal pores invisible within clitellum. Setae number, 30 (III), 36 (V), 30 (VII), 54 (XX), 56 (XXV); 12 (VI), 14 (VII), 17 (VII) between spermathecal pores on venter; 0 between male pores. Male pores paired in XVIII, 0.33 body circumference apart ventrally, each on the center of a slightly elevated round porophore. A large rounded rectangular genital marking within 17/18–18/19, glandular. Spermathecal pores 4 pairs, in 5/6/7/8/9, ventral, tiny, difficult to find, 0.33 circumference apart ventrally. Female pore single in XIV, medioventral.

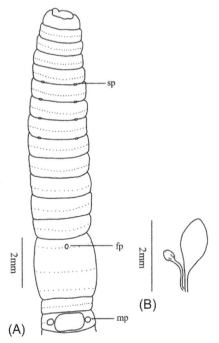

FIG. 97 *Amynthas pulvinus* (after Sun et al., 2013).

Internal Characters

Septa 8/9/10 absent, 5/6/7 thick, 7/8 and 10/11–14/15 slightly thickened. Dorsal blood vessel single, continuous onto pharynx; esophageal hearts enlarged from X–XIII. Gizzard in VII–X, barrel-shaped. Intestine enlarged gradually from XV. Intestinal caeca simple, in XXVII, extending anteriorly to XXIII, finger-shaped sac, an incision present on venter and dorsum respectively where the septa constrict. Holandric, testis sacs 2 pairs, in X–XI, anterior pair more developed than the posterior pair, separated from each other; seminal vesicles paired in XI–XII, developed and separated. Prostate glands developed, in XVII–XX, coarsely lobate, prostatic duct developed, inverted U-shaped, distal end stout. A pair of unstalked accessory glands clings to body wall, irregular in shape, and extended from XVII–XIX. Spermathecae 4 pairs in VI–IX, about 2 mm long; ampulla slim heart-shaped, duct as long as 0.54 ampulla; diverticulum shorter than main spermathecal axis by 0.35, slender, terminal 0.2 dilated into a small ovoid seminal chamber, silvery; no nephridia on spermathecal ducts.

Color: Preserved specimens buff on dorsum, no pigment on venter, clear middorsal line. Clitellum pale.

Distribution in China: Hainan (Mt. Limu).

Deposition of types: Shanghai Natural Museum.

Deposition of the cytochrome oxidase type 1 sequence: NCBI GenBank under no. JQ905266.

Type locality: In black soil under herb plant in a farm on Mt. Limu (elevation 641 m), Hainan Province.

Remarks: *Amynthas pulvinus* appears very unusual according to its male pore region. The most similar species the *diffringens* group is *A. homsetus* (Chen, 1938). *A. pulvinus* and *A. homosetus* are similar in the location of the first dorsal pore, clitellum locality, no genital papillae within the spermathecal pore region, and the form of the diverticulum. *Amynthas pulvinus* differs from *A. homosetus* in having smaller body size, having lighter pigment, and setae visible in the ventrum of XIV–XVI separately. *A. pulvinus* has a slightly elevated round porophore, but *A. homosetus* has a bigger roundish glandular porophore compared with *A. pulvinus*. The most special character in *A. pulvinus* is the large rounded rectangular genital marking within 17/18–18/19, whereas there are no genital papillae within the male pore region in *A. homosetus*. A large mass of accessory glands clings to the body wall near the prostates, corresponding to the male pore region papilla externally in *A. pulvinus*, whereas there are no accessory glands in *A. homosetus* in the same place.

97. *Amynthas rhabdoidus* (Chen, 1938)

Pheretima (Pheretima) rhabdoida Chen, 1938. *Contr. Biol. Lab. Sci. Soc. China (Zool.).* 12(10): 410–411.
Pheretima rhabdoidus Sims & Easton, 1972. *Biol. J. Soc.* 4:235.
Amynthas rhabdoidus Xu & Xiao, 2011. *Terrestrial Earthworms of China.* 173.

External Characters

Length 133–152 mm, width 5.5–6.5 mm, segment number 114–118. Prostomium 1/3 epilobous. First dorsal pore 12/13. Clitellum entire, in XIV–XVI, annular, with setae on ventral side. Setae fine, none particularly enlarged, closer on dorsal and lateral sides, very fine and close on dorsal of anterior segments; both dorsal and ventral breaks indistinct; $aa = 1.2ab$, $zz = 1.2$–2.0 yz behind clitellum; setae number 74–84 (III), 100–115 (VI), 94–116 (VII), 92–106 (IX), 70–80 (XX), 80–82 (XXV); 34–48 (VI), 36–60 (VII), 36–40 (IX) between spermathecal pores, 20–22 (VII) between male pores, 12 on XIX between 2 papillae. Male pores 1 pair, on ventrolateral corner of XVIII, about 1/3 of circumference apart ventrally, each on a tubercle, laterally covered by a skin fold appearing like a half-closed eyelid; a raised papilla placed in front on segment XVII and another (usually 2 coalesced) behind on XIX in line with male porophore. Spermathecal pores 4 pairs in 5/6/7/8/9, about 2/5 of body circumference apart ventrally; each intersegmental and superficial appearing as a pinhole. No genital papillae in this region. Female pore single in XIV, medioventral.

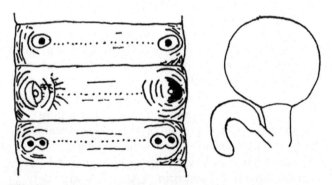

FIG. 98   *Amynthas rhabdoidus* (after Chen, 1938).

Internal Characters

Septa 9/10 absent, 8/9 thicker than 5/6 and 10/11 wrapping whole gizzard, 6/7/8 much thickened; 11/12/13 thin. Gizzard in IX–1/2 X. Intestine enlarged from XV. Caeca simple, long and pointed, in XXVII, extending anteriorly to XXV. Hearts in XI–XIII, first pair absent. Testis sacs in X elongate reaching upper half of septum, communicate ventromedially; those in XI entirely united ventromedially and enclosing first pair

of seminal vesicles, greatly distended. Seminal vesicles in XI small, those in XII larger, tubercular on surface. No ovoid bodies on next septa. Prostate gland rather compact, divided into main lobes, in XVII–XIX; its duct short (2.5 mm long) slightly enlarged at ectal end. Accessory glands present in corresponding to genital papillae outside, each small and sessile. Spermathecae 4 pairs, in VI–IX, both ampulla and diverticulum small, main pouch about 2 mm long; ampulla sac-like about 1.8 mm wide, transversely wrinkled, clearly marked off from its duct which is short but wide. Diverticulum shorter than main pouch, with a very short duct (1/3 of ampullar duct) and an elongate club-shaped seminal chamber (1.5 mm long), whitish in appearance.

Color: Preserved specimens darker on dorsal side, pale on lateral and verntral sides.
Distribution in China: Hainan (Wenchang).
Etymology: The specific name indicates the rod-like seminal chamber of the diverticulum.

98. *Amynthas rodericensis* (Grube, 1879)

Pheretima rodericensis Gates, 1972. *Trans. Am. Philos. Soc.* 62(7):218–219.
Amynthas rodericensis Xu & Xiao, 2011. *Terrestrial Earthworms of China.* 176–177.

External Characters

Length 55–150 mm, width 3–10 mm, segment number 80–100. Prostomium epilobous, tongue open. First dorsal pore 11/12 or 12/13. Clitellum entire, in XIV–XVI, annular, without externally recognizable setae. Setae number 30–40 (III), 23–38 (VII), 40–49 (XII), 42–48 (XX), 45–51 (XXX). Male pores, in XVIII, superficial, each in a small porophore. Genital markings, paired, about 4–6 intersetal intervals wide, just median to male porophores, across (?) 17/18 and/or 18/19 (rarely 19/20). Spermathecal pores 4 pairs in 5/6/7/8/9, pores minute, superficial, well within dorsum and about 1/5–1/6 circumference apart ventrally. Female pore single in XIV, medioventral.

Internal Characters

Septa 8/9/10 absent, 5/6/7/8 slightly muscular, 10/11–13/14 muscular. Intestine enlarged from XV. Caeca simple, in XXVII, extending anteriorly to XXIV. Holandric, testis sacs, unpaired and ventral; seminal vesicles in XI–XII, medium-sized to large, filling the coelomic cavities. Prostate gland in XVI–XXII; its duct 4–6 mm long, muscular, thicker ectally, coiled or looped. Spermathecae 4 pairs, in VI–IX, rather small, duct shorter than ampulla and narrowed in parietes, diverticulum from median face of duct at parietes, often longer than main axis, with slender stalk, wider and longer seminal chamber often markedly moniliform and with 4–7 constrictions. Accessory glands shortly stalked, coelomic, closely crowded, a number associated with each marking.

> Color: Preserved specimens in dorsum, reddish, reddish brown, rich brown, grayish, slate.
> Distribution in China: Jiangsu, Fujian (Fuzhou).
> Deposition of types: Breslau Mus. ? British Mus.

99. *Amynthas stricosus* Qiu & Sun, 2012

Amynthas stricosus Qiu & Sun, 2012. *Zootaxa*. 33458:150-152.

External Characters

Length 72–97 mm, width 2–2.8 mm (preclitellum), segment number 116–142. Body cylindrical in cross section, gradually tapered towards head and tail. Prostomium 1/2epilobous. First dorsal pore 11/12 or 12/13. Clitellum XIV–XVI, annular, setae can be seen in XIV–XVI; dorsal pore present or absent; some have furrows in clitellum. Setae number 30–54 (III), 50–76 (V), 62–72 (VII), 40–70 (XX), 38–52 (XXV); 22–28 (VI), 22–30 (VII), 23–29 (VII) between spermathecal pores, 10–12 between male pores; setal formula $aa = 1.1$–$1.2ab$, $zz = 1.2$–$2zy$. Male pores paired in XVIII, 0.33 body circumference apart ventrally; each on a coniform glandular disc surrounded by a round pad. Genital papillae postsetal, single or paired in XVII, XIX, and XX; Each papilla small, round, slightly convex. Spermathecal pores 4 pairs in 5/6/7/8/9, 0.4 circumference apart ventrally. Female pore single in XIV, medioventral.

Internal Characters

Septa 8/9/10 absent, 6/7/8 thick and muscular, 10/11–13/14 slightly thickened. Gizzard in VII–X, ball-shaped. Intestine swelling in XVI and distinctly from XXI. Intestinal caeca simple, in XXVII, extending anteriorly to XXIV, horn-shaped with smooth margins. Dorsal blood vessel single, continuous onto pharynx; Hearts in X–XIII. Holandric, Testis sacs 2 pairs, ventral in X–XI, anterior pair bigger in size, which is in close proximity to two sides on dorsum. Seminal vesicles paired in each of XI and XII, anterior pair bigger in size and connected with membrane on dorsum, the latter pair separated but closer. Prostates in XVI–XX, coarsely lobate. Prostatic duct S-shaped, distal end appreciably enlarged. Accessory glands absent. Ovaries in XIII. Spermathecae 4 pairs in VI–IX, about 1.6 mm long; ampulla heart-shaped, gradually slender duct as long as ampulla. Diverticulum as long as main spermathecal axis, slender, terminal 0.4 dilated into a band-shaped chamber; no nephridia on spermathecal ducts.

> Color: Preserved specimens no pigment; unclear middorsal line. Clitellum greyish white or light brown.
> Distribution in China: Hainan (Mt. Diaoluo).
> Etymology: The specific name referes to its lack of pigmention.
> Deposition of types: Shanghai Natural History Museum.
> Type locality: Dark cinnamon soil under meadow vegetation (Elevation 930 m); yellow cinnamon soil in bamboo and forest at roadside; brown soil under a dead and fallen tree (elevation 1008 m); yellow cinnamon soil under meadow vegetation (elevation 925 m) on Mt. Diaoluo, Hainan.
> Remarks: *A. stricosus* is similar to *A. homosetus* (Chen, 1938) in the locality and number of spermathecal pores, heart-shaped ampulla and developed prostate glands. However, *A. stricosus* can be distinguished from *A. homosetus* in that it lacks pigment, has a smaller body size, setae visible externally in XIV–XVI, 2 male pores more widely spaced, variable numbers of genital papillae near male pores, and a longer band-shaped diverticulum chamber. *A. homosetus* has dark chocolate pigment on anterior dorsum and gray on other parts of body. Body size rather larger, setae visible ventrally just in XVI, 0.25 of circumference ventrally separated male pores, no genital papillae near male pores, and a shorter diverticulum with an ovoid seminal chamber.

> Additionally, *A. stricosus* resembles *A. saccatus* Qiu and Wang, 1993 in the genital papillae of male pore region, but can be distinguished from the latter by many

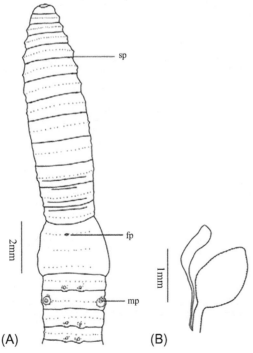

FIG. 99 *Amynthas stricosus* (after Sun et al., 2012).

essential characters. (1) There are 3 pairs of intersegmental spermathecal pores in 6/7/7/9 in *A. saccatus*, but 4 pairs of spermathecal pores in 5/6/7/8/9 in *A. stricosus*, (2) each male pore of *A. saccatus* is situated on a long round porophore with 2 small round pads without papillae nearby, (3) there are several genital papillae and accessory glands in the spermathecal pore region internally or externally in *A. saccatus*. However, none of them occur in *A. stricosus*, and (4) the diverticulum of *A. saccatus* is shorter than the main spermathecal axis and is twisted heavily on its middle part, but the diverticulum is equal to the main spermathecal axis and slenderer in *A. stricosus*.

100. *Amynthas szechuanensis vallatus* (Chen, 1946)

> *Pheretima szechuanensis vallata* Chen, 1946. *J. West China. Border Res. Soc. (B).* 16:103–105, 139, 150.
> *Amynthas szechuanensis vallatus* Sims & Easton, 1972. *Biol. J. Linn. Soc.* 4:236.
> *Amynthas szechuanensis vallatus* Xu & Xiao, 2011. *Terrestrial Earthworms of China.* 184.

External Characters

Large-sized worm. Length 120 mm, width 5 mm, segment number 107. Segments strongly built and long. Prostomium 1/2 epilobous, tongue wide. First dorsal pore 12/13. Clitellum XIV–XVI, annular, entire and smooth. Setae stout, closer on ventrolateral sides, widely spaced and enlarged from about 8 setae on anterior, ventral, ventromedian ones longest, more evenly distributed on postclitellar segments; setae number 26 (III), 38 (VI), 45 (IX), 40 (XXV); 17 (V), 22 (VI), 34 (VII) between spermathecal pores, 17 (VII), 20 (IX) between male pores; $aa = 1.2–1.5ab$, $zz = 1.5–2zy$. Male pores paired in XVIII, about 1/3 of circumference apart ventrally, each on a small porophore on ventrolateral side of XVIII, large papillar to the type species found median to each porophore on XVIII but also on XVII, somewhat posterior to setal line, surrounded by papilla-like glandular ridge either laterally or posteriorly, 2 comma-like ones on anterior of XVIII, similar ones on XVII. Spermathecal pores 4 pairs intersegmental in 5/6/7/8/9, about 3/7 of their ventral circumference, no genital markings Female pore single in XIV, medioventral.

Internal Characters

Septa 8/9/10 absent, rest slightly muscular and thin. Nephridial tufts thin. Pharyngeal glands in front of 5/6 thick and compact, acinose glands also found around ligaments in VI. Gizzard and caeca like *Amynthas szechuanensis*. Intestine swelling in XVI.

Caeca compound, in XXVII. Vascular loops in IX asymmetrical, 1 limb missing, those in X well developed. Testis sacs large, anterior pair V-shaped, posterior pair in transverse band, both pairs communicated. Seminal vesicles enormously developed, general appearance like the type species. Accessory glands in segments XVII and XVIII, very well developed, in very thick and compact masses, highly elevated but sessile, median to prostatic duct. Spermathecae 4 pairs, in VI–IX, rather small. Ampulla sac-like, its duct short but very thick, clearly marked from ampulla; main pouch about 3 mm long. Diverticulum also shorter but its seminal chamber subspherical, distinctly marked off from its duct.

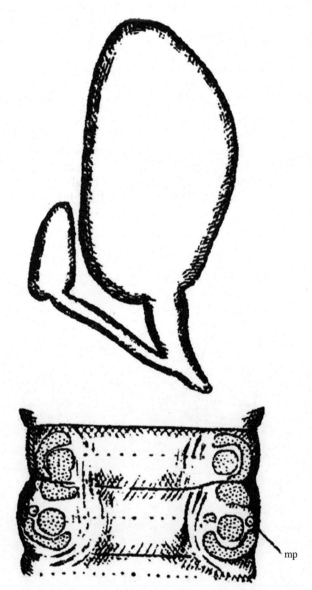

FIG. 100 *Amynthas szechuanensis vallatus* (after Chen, 1946).

Color: Preserved specimens dark purplish on dorsal side.

Distribution in China: Sichuan (Mt. Emei).

Etymology: The name of the species refers to the type locality.

Remarks: *Amynthas szechuanensis vallatus* is distinct from its type form in several important characters. It possesses 4 pairs of spermathecae instead of 3 pairs as always found in the latter. The accessory glands are very thick and massive, while in the latter they are thin and velvety in appearance in all cases observed. The genital papillae are also found on XVII but in the latter they are confined to the 18th segment except Gates recorded an example with papillae on XVII. The anterior septa comparatively thin, while in the latter they are strongly musculated.

101. *Amynthas taiwumontis* Shen, Chang, Li, Chih & Chen, 2013

Amynthas taiwumontis Shen, Chang, Li, Chih & Chen, 2013. *Zootaxa*. 3599(5):479–481.

External Characters

Small to medium; length 133–151 mm, width 3.32–4.77 mm (clitellum), segment number 131–142. Prostomium epilobous. First dorsal pore in 11/12. Clitellum XIV–XVI, annular, length 2.79–4.78 mm. setae and dorsal pore absent. Setae 74–97 (VII), 73–89 (XX); 16–22 between male pores. Male pores paired in XVIII, 0.22–0.29 body circumference apart ventrally, each inconspicuous on a round, slightly elevated porophore. Genital papillae absent. Spermathecal pores small, 4 pairs in intersegmental furrows of 5/6/7/8/9; distance between paired pore 0.25–0.28 body circumference apart ventrally. Genital papillae absent. Female pore single, medioventral in XIV.

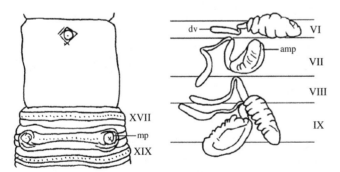

FIG. 101 *Amynthas taiwumontis* (after Shen, et al., 2013).

Internal Characters

Septa 8/9/10 absent, 5/6/7/8 thick, 10/11–13/14 thick and muscular. Nephridial tufts on anterior faces of 5/6/7. Gizzard large in VII–X. Intestine swelling in XV or XVI. Intestinal caeca paired in XXVII, simple, long, extending anteriorly to XXI–XXIII, or bent XXIV. Hearts in XI–XIII. Holandric, testes large, oval, 2 pairs in X and XII. Seminal vesicles small, 2 pairs in XI and XII, anterior pair occupying half to full segmental compartment and posterior pair vestigial, occupying half a segment, each vesicle with a round or elongated oval dorsal lobe. Prostate glands large, occupying 4–9 segments in XVI–XXIV, racemose, divided into lobules by groove. Prostatic duct stout, C- or S-shaped in XVIII or XVIII–XIX. Accessory glands absent. Spermathecae 4 pairs, in VI–IX, ampulla round or elongated oval–shaped, surface wrinkled, 0.8–2.3 mm long and 0.5–1.3 mm wide, spermathecal stalk stout, 0.3–0.9 mm in length. Diverticulum long, proximal part slender and slightly enlarged toward distal end, but no specially enlarged seminal chamber, 1.0–3.8 mm in length. Accessory glands absent.

Color: Preserved specimens grayish brown. Clitellum dark to light brown.

Distribution in China: Fujian (Jinmen, Quanzhou).

Etymology: The specific name refers to the type locality.

Deposition of types: The Invertebrate Zoology and Cell Biology Lab, Department of Life Sciences, National Taiwan, Tsipei, Taiwan.

Remarks: *Amynthas taiwumontis* was only found in areas around Mt. Taiwu on Jinmen Island. It is octothecal, and has a simple male pore structure and no genital papillae or genital markings. These characters are similar to *Amynthas fornicatus* (Gates, 1935) from the west border of Sichuan. However, the 2 species can be easily distinguished by their external characters. *A. fornicatus* is smaller (length 80–94 mm and segments 90–105) and has much lower setal number than *A. taiwumontis*, with 43–52 setae in VI–VII and 45–55 setae in XXV for the former (Chen 1936) and 74–97 setae in VII and 73–89 setae in XX for the latter.

Amynthas marenzelleri (Cognetti, 1906) from Japan is octothecal and also has a simple male pore structure and no genital markings. It differs from *A. taiwumontis* by the larger body size (length 160–190 mm and diameter 6–7 mm) and lower setal number (36–43 setae in VII and 50–54 setae in XXIV) (Cognetti, 1906; Kobayashi, 1938). In addition, *Amynthas marenzelleri* has small prostate glands confined in XVIII and small spermathecae without diverticulum, whereas *A. taiwumontis* has large prostate glands occupying 4–9 segments in XVI–XXIV and spermathecae with long diverticulum.

Amynthas sinabunganus (Michaelsen, 1923) from Sumatra and *Amynthas tertiadamae* (Michaelsen, 1934) from Gulf of Siam are also octothecal with simple male pore structure and without genital markings. However, the 2 species are distinguishable from *A. taiwumontis* simply by their much smaller body size and lower setal number. Both are 52 mm long and 2–3 mm wide, with 104 segments for *A. sinabunganus* (Michaelsen, 1923) and 120 segments for *A. tertiadamae* (Michaelsen, 1934), Setal numbers range from 23 to 40 per segment for *A. tertiadamae* (Michaelsen, 1934) and about 50 per segment for *A. sinabunganus* (Michaelsen, 1923).

102. *Amynthas uvaglandularis* Shen, Tsai & Tsai, 2003

Amynthas uvaglandularis Shen, Tsai & Tsai, 2003.
Zoological Studies. 42(4), 479–490.
Amynthas uvaglandularis Chang, Shen & Chen, 2009.
Earthworm Fauna of Taiwan. 98–99.
Amynthas uvaglandularis Xu & Xiao, 2011. *Terrestrial Earthworms of China.* 193–194.

External Characters

Length 62–113 mm, clitellum width 3.0–3.3 mm, segment number 75–115. Number of incomplete annuli per segment 2–3 in VII–XIII. Prostomium epilobous. First dorsal pore 10/11. Clitellum XIV–VII, annular, 2.5–3.6 mm long, setae absent, dorsal pore absent. Setae 33–44 (VII), 38–48 (XX), 11–13 between male pores. Male pores paired in XVIII, lateroventral, minute, on a depressed porophore, with a small papilla adjacent laterally, both surrounded by 2–3 circular folds, 0.25–0.30 circumference apart ventrally, 2 postsetal genital papillae present in XVIII, each adjacent medially to male pore and close to setal line, about X or XI intersetal distances apart, or 2 closely arranged in middle between the paired male pores, each papilla round, similar in structure to those in the preclitellar region. Spermathecal pores 4 pairs in 5/6/7/8/9, ventrolateral, each on a papilla-like porophore in segmental furrow, 0.31–0.35 circumference apart ventrally. Genital papillae present postsetal in VII and presetal in IX, widely paired or with an additional papilla on medioventrum, occasionally, additional presetal papillae present in VII, each papilla round, center flat or slightly concave, 0.3–0.4 mm in diameter, surrounded by a white rim. Female pore single, medioventral in XIV.

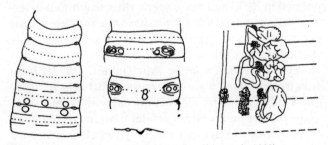

FIG. 102   *Amynthas uvaglandularis* (after Shen et al., 2003).

Internal Characters

Septa 8/9/10 absent, 11/12–13/14 thickened. Gizzard round in IX and X. Intestine from XV. Intestinal caeca paired in XXVII, simple, surface slightly wrinkled, extending anteriorly to XXIV or XXIII. Esophageal hearts in XI–XIII. Testis sacs 2 pairs in X and XI, large, round. Seminal vesicles paired in XI and XII, large, each occupying 1.5–2 segments, follicular, with a large granulated dorsal lobe. Prostate glands paired in XVIII, large, lobed, extending anteriorly to XVI and posteriorly to XX or XXI, prostatic duct C-shaped, distal end enlarged. Accessory glands present, similar in structure to those in the spermathecal region, large glands (about 0.8 mm long) corresponding to genital papilla medial to male pores, small ones (about 0.6 mm in length) at the base of prostatic duct corresponding to papilla lateral to male pore, both about 0.4 mm wide. Spermathecae 4 pairs in VI–IX, ampulla oval, 1.3–1.5 mm long, 0.7–1.1 mm wide, duct stout, around 0.4 mm long, diverticulum long and slender, 1.2–1.6 mm long, with or without distal oval seminal chamber. Accessory glands sessile, each in the form of a grape-like follicular mass, about 0.7 mm long and 0.5 mm wide, corresponding to each of the external genital papillae, additional accessory glands present at the base of each spermathecal duct, in a mass of smaller follicles.

Color: Preserved specimens dark brown on dorsum, light brown on venter, dark brown around clitellum.
Type locality: Rueyen Creek Nature Reserve, Nantou County, Taiwan.
Deposition of types: Taiwan Endemic Species Research Institute, Jiji, Nantou County, Taiwan.
Holotype: coll. no. 1999-24-Shen. Paratypes: same collection data as for holotype.
Etymology: The name "*uvaglandularis*" was given with reference to the grape-like accessory glands of this species.
Distribution in China: Taiwan (Nantou).
Remarks: *Amynthas uvaglandularis* has 4 pairs of spermathecae in VI–IX, and belong to the *diffringens* species group (Sims and Easton 1972). Its porophore structure resembles that of *Amynthas silvertrii* (Cognetii, 1909) from Hawaii. However, the median papilla is larger and lateral papilla smaller in *Amynthas uvaglandularis*. Also, *Amynthas uvaglandularis* has long diverticula and large prostate gland, whereas *Amynthas silvertrii* has short diverticulum stalks and small prostate glands.

103. *Amynthas wujhouensis* Shen, Chang, Li, Chih & Chen, 2013

Amynthas wujhouensis Shen, Chang, Li, Chih & Chen, 2013. *Zootaxa.* 3599(5):477–479.

External Characters

Large, easily broken; length 241–345 mm, width 5.2–6.5 mm, segment number 128–212. Prostomium epilobous. First dorsal pore in 11/12. Clitellum XIV–XVI, annular, length 6.1–9.6 mm; with setae on the ventral side, 3–6 (XIV), 0–8 (XV) and 7–11 (XVI); dorsal pore absent. Setae 68–78 (VII), 59–71 (XX); 8–10 between male pores. Male pores paired in XVIII, 0.26–0.31 body circumference apart ventrally, each on a crescent or semicircular-shaped area with a large papilla immediately medial to it. The whole male area including the genital papilla protuberant, 2.0–2.2 mm in width. Each papilla round, center depressed, 1.1–1.4 mm in diameter. Spermathecal pores 4 pairs, buried deeply in intersegmental furrows of 5/6/7/8/9; 0.24–0.3 body circumference apart ventrally. Genital papillae absent in the preclitellar region. Female pore single in XIV, medioventral.

Internal Characters

Septa 8/9/10 absent, 5/6/7/8 thick and muscular, 10/11–13/14 muscular. Nephridial tufts thick on anterior faces of 5/6/7. Gizzard large in VII–X. Intestine swelling in XV. Intestinal caeca paired in XXVII, extending anteriorly to XXIV–XXV, each simple, distal end either straight or bent. Hearts in X–XIII. Holandric, testis oval, 2 pairs in ventrally joined sac in X and XI. Seminal vesicles small, surface finely folliculated, 2 pairs in XI and XII, each vesicle with a round or elongated oval dorsal lobe. Prostate glands large, occupying more than 5 segments in XVI–XXI, divided into several lobules by grooves. Prostatic duct slender, U-shaped. Accessory glands large, sessile, amorphous, 0.5–1.1 mm wide, corresponding to external genital papillae. Spermathecae 4 pairs in VI–IX, each with a peach-shaped or elongated oval-shaped ampulla 2.0–4.2 mm long and 1.5–2.4 mm wide, and a slender or stout spermathecal stalk 0.4–1.45 mm in length. Diverticulum long, slender, tubular, 2.26–4.8 mm in length. Accessory glands absent in the preclitellar region.

Color: Preserved specimens pale, pink when alive.
Distribution in China: Fujian (Jinmen, Quanzhou).
Etymology: "Wujhou" is an ancient name for Jinmen and is used to name this species.
Deposition of types: The Invertebrate Zoology and Cell Biology Lab, Department of Life Sciences, National Taiwan, Taibei, Taiwan.
Remarks: *Amynthas wujhouensis* is the longest earthworm found in Jinmen. It is easily broken during the collecting process. The shape of the male porophore and the arrangement of papillae in the male pore area are fairly similar to *A. lunatus* (Chen, 1938) from Hainan Island. Both species are octothecal with spermathecal pores in 5/6/7/8/9, and both have similar numbers of setae and no genital papillae in the preclitellar region. However, the large, median papilla in XVIII of *A. lunatus* is not immediately adjacent to each male pore. Furthermore, the 2 papillae between male pores are in close approximation (less than 2 mm at interval) with only 4 setae between them in *A. lunatus* (Chen 1938). In *A. wujhouensis*, the median papillae are immediately adjacent to male pores and widely separated by 8–10 setae. In addition, *A. lunatus* has a small gizzard, while that of *A. wujhouensis* is large.

A. manicatus manicatus (Gates, 1931) from Burma also has a pair of genital markings internal to the male pore. However, the genital markings of *A. m. manicatus* are elongately oval, extending anteroposteriorly to 17/18 and 18/19, or slightly onto XVII and XIX (Gates 1931). In addition, there is a space of 3–4 setae between the marking and the male pore (Gates 1931). Also,

FIG. 103 *Amynthas wujhouensis* (after Shen et al., 2013).

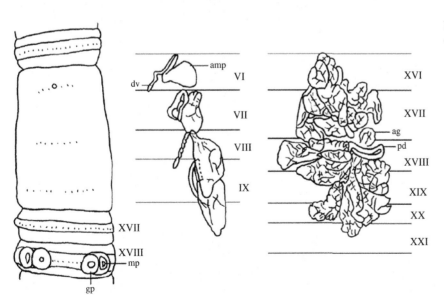

A. m. manicatus is smaller (56–111 mm long) than *A. wujhouensis*, has large seminal vesicles and diverticulum coiled into a spherical mass of loops (Gates 1931).

104. *Amynthas wulinensis* Tsai, Shen & Tsai, 2001

Amynthas wulinensis Tsai, Shen & Tsai, 2001. *Zoological Studies.* 40(4), 285–286.
Amynthas wulinensis Chang, Shen & Chen, 2009. *Earthworm Fauna of Taiwan.* 102–103.
Amynthas wulinensis Xu & Xiao, 2011. *Terrestrial Earthworms of China.* 197–198.

External Characters

Length 128–174 mm, clitellum width 5.6–6.1 mm, segment number 93–123. Number of annuli (secondary segmentation) per segment 3 after VI. Prostomium epilobous. First dorsal pore 11/12. Clitellum XIV–XVI, annular, length 5.5–5.8 mm, slightly shorter than width, dorsal pore absent, setae absent. Setae 42–45 (VII), 55–69 (XX), 13 between male pores. Male pores paired in XVII, lateroventral, 0.24–0.28 body circumferences apart ventrally, each round or oval, on setal line with depressed center (male aperture not visible), surrounded by 2 or 3 circular folds, usually a pair of genital papillae slightly medial to male pore present in each of XVII and XIX, each papilla oval with a depressed center, in posterior annuli between setal line and posterior intersegmental furrow, occasionally an additional pair or 1 papilla present in XX, or 1 missing in XVII or XIX. Spermathecal pores 4 pairs in 5/6/7/8/9, ventrolateral, about 0.29 circumference apart ventrally. No papillae in the preclitellar region. Female pore single, medioventral in XIV.

Internal Characters

Septa 8/9/10 absent, 5/6/7/8 thickened, 10/11–13/14 greatly thickened. Gizzard small, round in X. Intestine from XV. Intestinal caeca paired in XXVII, simple, extending anteriorly to XXII or XX. Esophageal hearts in XI–XIII. Testis sacs 2 pairs in X and XI, small, round. Seminal vesicles paired in XI and XII, each large, follicular, with a large, follicular dorsal lobe. Prostate glands paired in XVIII, large, racemose, follicular, extending anteriorly to XV and posteriorly to XX, prostatic duct coiled, hook-shaped. Accessory glands paired in XVII and XIX, each corresponding to external genital papillae, sessile (no stalk), flowery. Spermathecae 4 pairs in VI–IX, each with a very short, stout stalk, diverticulum with an oval seminal chamber and a slender, straight stalk originating from spermathecal stalk.

Color: Preserved specimens light brown on dorsum, light yellow on venter, grayish tan around clitellum.

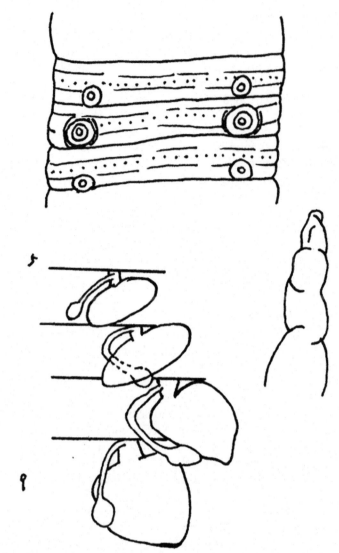

FIG. 104   *Amynthas wulinensis* (after Tsai et al., 2001).

Type locality: Wulin Natural Scenery Observatory at an elevation of 3200 m, Nantou County, Taiwan.
Deposition of types: Taiwan Endemic Species Research Institute, Jiji, Nantou County, Taiwan.
Holotype: coll. no. 1999-16A-Shen. Paratypes: same collection data as for holotype.
Etymology: This species was named after its type locality of Wulin.
Distribution in China: Taiwan (Nantou).
Remarks: *Amynthas wulinensis* is a holandric, octothecal earthworm belonging to the *diffringens* species group (Sims and Easton 1972). It has simple, superficial porophores in the male pore region, and paired genital papillae in XVII and XIX, whose number, structure, location, and arrangement are fairly similar to those of *Metaphire posthuma* (Vaillant) of Burma (Gates 1932) and Taiwan (Tsai 1964) and *M. quadripapillata* (Michaelsen) of Sumatra (Beddard, 1900) which have copulatory pouches. However, the

genital papillae of *A. wulinensis* are located in the postsetal annulets close to the segmental furrows 17/18 and 19/20, whereas the genital papillae of *M. posthuma* and *M. quadripapillata* are in the setal annulet (the middle annulet) in XVII and XIX.

105. *Amynthas yunlongensis* (Chen & Xu, 1977)

Pheretima yunlongensis Chen & Xu, 1977. *Acta Zool. Sinica.* 23(2):176.
Amynthas yunlongensis Xu & Xiao, 2011. *Terrestrial Earthworms of China.* 201.

Length 58–102mm, width 4.5mm, segment number 68–108. Prostomium 1/3–1/2 epilobous. First dorsal pore 12/13. Clitellum XIV–XVI, annular, with setae ventrally, 6–8 (XIV), 8–9 (XV), 12–15 (XVI). Setae fine, dense and uniform, ventral break not obvious, dorsal break wide. Setal number 32 (?)–48 (III), 59–87 (VI), 50–60 (VII), ?–52 (XXV). Male pores single pair, in ventral area of XVIII, nearly 1/3 of circumference apart ventrally, each on a plain papilla, paired papillae posteriorly on XVII, XIX. Spermathecal pores 4 pairs, in 5/6/7/8/9, about 2/5 of circumference apart ventrally; each on eye-shaped area, needle-shaped; no other genital marking in this region. Female pore single in XIV, medioventral.

Internal Characters

Septa 9/10 absent, 8/9 traceable ventrally, 5/6/7/8 thick, 10/11–14/15 slightly thick, 15/16 thin. Gizzard round ball-like. Intestine enlarged from XVI. Caeca simple, in XXVII, extending anteriorly to XXIII. Testis sacs with narrow conjunction ventrally and connected dorsally. Second testis sacs developed, wrapped to seminal vesicle and heart. Seminal vesicles anterior pair is large, posterior pair is very small, dorsal lobe on dorsal side, long leaf-like or triangular. Prostate without gland part; only with slender duct, slightly U-shaped. Accessory glands absent or undeveloped, or linked in XVII–XIX, no duct. Spermathecae undeveloped, only 4 pairs granulation-like genitals.

Color: Preserved specimens flesh brown, clitellum red-brown.
Type locality: Yunnan (Yunlong).
Distribution in China: Hubei (Lichuan), Guizhou (Mt. Fanjing), Yunnan (Yunlong).
Etymology: This species is named after its type locality.
Remarks: This species is different from *Amynthas hongkongensis* (Michaelsen, 1910) and *Amynthas mediocus* (Chen & Hsü, 1975) in its small shape, distribution characteristics of male pores, greater setal number, septa 8/9 traceable ventrally, testis sacs with long arm, and prostate and seminal vesicle underdeveloped.

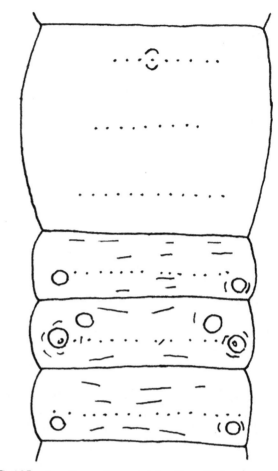

FIG. 105   *Amynthas yunlongensis* (after Chen, 1977).

glabrus Group

Diagnosis. *Amynthas* with spermathecal pores 1 pair, at VI, or absent.

Global distribution: Oriental realm, ? Australasian realm and ? introduced into the Oceanian realm. There are 2 species from China.

TABLE 15   Key to the Species of the *glabrus* Group of the Genus *Amynthas* From China

1. Spermathecal pores absent .... *Amynthas glabus*	
Spermathecal pores present .... *A. plantoporophoratus*	

106. *Amynthas glabrus* (Gates, 1932)

Pheretima glabra Gates, 1932. *Rec. Ind. Mus.* 34(4):395–396.
Pheretima glabra Gates, 1972. *Trans. Amer. Philos. Soc.* 62(7):187–188.
Amynthas glabrus Sims & Easton, 1972. *Biol. J. Soc.* 4:237.
Pheretima glabra Zhong & Qiu, 1987. *Sichuan Journal of Zoology.* 6(2):24–25.

Amynthas glabrus Zhong & Qiu, 1992. *Guizhou Science.*
10(4):40.
Amynthas glabus Xu & Xiao, 2011. *Terrestrial Earthworms of China.* 118.

External Characters

Length 58–110 mm, width 2–5 mm, segment number 114–119. First dorsal pore 12/13, rarely has a pore-like marking in 11/12. Clitellum in XIV–XVI, annular, swollen, intersegmental furrows lacking, no setae recognized, and dorsal pore-like marking visible. Setae begin on II, on which segment there is an unbroken circle; a midventral break is usually lacking in the setal circles but a slight middorsal break of variable width may be present. Setae are lacking ventrally on XVII and XVIII, and laterally on XIX in a region just behind each genital marking; setae number 6–10 (XIX), 48–54 (XX). Male pores paired in XVIII, each pore with tiny transverse slits at the posterior ends of grooves on the genital marking. Genital markings 2, each marking elongate, with rounded ends, flattened, but protuberant from the surface of the body (especially anteriorly), placed diagonally with the posterior end slightly nearer the midventral line than anterior end, and extending anteroposteriorly from mid-XVII to the region of 18/19; and the markings may be diagonal in position or parallel to the midventral line; each marking provided with central, anteroposterior groove, the posterior end of which turns laterally to pass into the male pore; the genital markings are about 4–6 intersetal distances wide transversely, there is a smooth glistening region midventrally extending from 16/17 to the setae of XIX on which intersegmental furrows and setae are lacking. Spermathecal pores absent. Female pore single in XIV, medioventral.

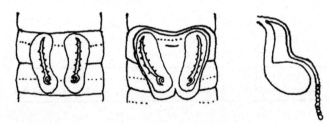

FIG. 106 *Amynthas glabus* (after Zhong & Qiu, 1987).

Internal Characters

Septa 9/10/11 absent, 4/5/6/7/8 are present, 5/6/7/8 slightly thickened and muscular, 8/9 is ventral rudiment only, 11/12 is thin, 12/13/14 and several succeeding septa are slightly strengthened. Gizzard in VII–X, ball-shaped. Intestine swelling in XX (XIX–XX); caeca simple, in XXVII, extending anteriorly to XVIII–XX. The last pair of hearts is in XIII. Testis sacs 2 pairs, the sacs of a side are in contact and project anteriorly in a diagonal fashion from the anterior face of 11/12 so that the anterior sac is further from the nerve cord than the posterior sac; the sacs of a segment are widely separated and without transverse communication or connection. Seminal vesicles 2 pairs, in XI and XII, large, covering the dorsal blood vessel in those segments, the anterior vesicles extending forward into contact with the gizzard, the posterior vesicles pushing 12/13 posteriorly into contact with 13/14. Prostates extend through segments XVII–XIX (XVII–XX); the prostatic duct is about 4.5 mm long, an ental portion about 1.5 mm in length is narrow but pinkish and firm, while the ectal two-thirds or three-quarters is much thickened. Spermathecae absent.

Color: Preserved specimens unpigmented. Clitellum reddish.
Distribution in China: Guizhou (Fanjingshan, Chishui, Jiangkou), Sichuan (Yibin, Xingwen Gaoxian), Yunnan (Lancang).
Type locality: Burma (Nam Hpen Noi).
Deposition of types: Indian Museum.
Remarks: This species bears some resemblances to *A. doliaria* but the genital markings and the testis sacs are quite different, while *A. doliaria* has paired hearts belonging to X as well as ventral setae on XVII and XVIII.

107. *Amynthas plantoporophoratus* (Thai, 1984)

Pheretima plantoporophoratus Zhong & Qiu, 1987.
Sichuan Journal of Zoology. 6(2):24.
Pheretima plantoporophoratus Zhong & Qiu, 1992.
Guizhou Science. 10(4):40.
Amynthas plantoporophoratus Xu & Xiao, 2011.
Terrestrial Earthworms of China. 167–168.

External Characters

Length 40–69 mm, width 2–2.5 mm, segment number 80–110. First dorsal pore 12/13. Clitellum in XIV–XVI, annular. Setae thin and dense; setae number 50–60 (III), 60–62 (V), 55–61 (XIII), 47–56 (XX); 5–6 between spermathecal pores. Male pores paired in XVIII, lateroventral, each on the anterior ends of large plantar shape papillae. Its anterior end slightly pointed or rounded, reaching 17/18 furrows, posterior end slightly narrower or same width as anterior, extending to 19/20 or XX; with a transverse groove through the anterior and posterior; the anterior end of the groove curved in a P shape. Spermathecal pores 1 pair, large, on VI postsetal, about 1/6 of circumference apart ventrally, transverse slit-shaped, front and

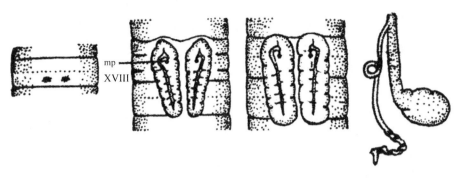

FIG. 107 *Amynthas plantoporophoratus* (after Zhong & Qiu, 1987).

rear wall lip-shaped. Female pore single in XIV, medioventral.

Internal Characters

Septa 8/9/10 absent. Intestine swelling in XVII; caeca simple, in XXVII–XXV. Testis sacs 2 pairs, in X and XI, anterior pair elongated, meeting but disconnected ventrally; wrapped in a thin film together with the heart of X, posterior pair elongated, no dorsal lobe, and packaged by the film together with the heart of XII. Prostates in XVI–XIX, developed; ampulla oval shaped. Spermathecae 1 pair in VI, duct long; diverticulum slender, longer than main pouch. The boundary of the seminal vesicle is not obvious.

Color: ?.
Distribution in China: Sichuan (Chengdu, Yibin, Ya'an, Hanyuan), Hubei (Badong), Guizhou (Fanjingshan, Chishui, Jiangkou).
Type locality: Hanoi, Vietnam.
Deposition of types: Hanoi, Vietnam.

hawayanus Group

Diagnosis. *Amynthas* with spermathecal pores 3 pairs, at 5/6/7/8.

Global distribution: Oriental realm, ? Australasian realm and ? introduced into the Oceanian realm. There are 29 species from China.

TABLE 16 Key to the Species of the *hawayanus* Group of the Genus *Amynthas* From China

1. Male pores in a shallow pouch (a skin fold in reality), with 2 very large and flat-topped papillae extending to segments XVII and XIX respectively, placed medially to each pore, 1 in front of and another behind setal line. Spermathecal pores in an ellipsoid area of dorsum .... *Amynthas fluxus*
Spermathecal pores in the ventrum 2
2. Genital papillae paired, presetal, on XI–XIV .... *Amynthas papilliferus*
With or without papillae front and/or behind clitellum 3
3. With papillae front and behind clitellum 4
Without papillae front and/or behind clitellum 14
4. Genital papillae present, patched, medioventral, presetal papillae 11–21 in XVII, 14–25 in XVIII, 10–25 in XIX, 0–16 in XX, 0–3 in XXI, postsetal papillae 1–2 in XIX, 1 in XX, with shape and arrangement similar to those in the preclitellar region; round genital papillae present ventromedially, presetal papillae 0–3 in VI 0–13 in VII, 16–24 in VII, 2–19 in IX, postsetal papillae absent or present, if present, 1–3 in VII, 1 in IX .... *Amynthas tessellates*
Genital papillae between XVII–XX and between V–X 5
5. Genital papillae between XVII–XIX and between V–X 6
Genital papillae between XVIII–XX and between V–X 10
6. Genital papillae between XVII–XIX and between V–X 7
Genital papillae XVII and between V–X 8
7. Genital papillae arranged in a longitudinal series, medioventral, 3–9 papillae from XVII to XXI, similar to those in the preclitellar region in size, shape, position and arrangement. Genital papillae medioventral in V–IX, in a longitudinal series like a chain, number and position highly variable: 3 presetal papillae in VI–VII and 2 postsetal papillae in VII and VII or 0–9 papillae in V–IX. Generally, 1 presetal papilla and/or 1 postsetal papilla in a segment, each occupying the entire width of an annulus adjacent to intersegmental furrows. Occasionally, 2 papillae joined closely in an annulus. Each papilla round with a concave center, associated with a round, stalked accessory gland internally .... *Amynthas catenus*
Numerous small presetal and postsetal discs present in transverse rows in XVII–XIX. Genital papillae present in VII–IX, presetal and postsetal small discs, in transverse rows .... *Amynthas papulosus*

Continued

TABLE 16   Key to the Species of the *hawayanus* Group of the Genus *Amynthas* From China—cont'd

8. Postsetal genital papillae present in a horizontal row in right and left lateroventral regions of XVII, each row with 4 papillae adjacent to setal line, each papilla round, center concave. Spermathecal pores surrounded by a genital papilla anteriorly and 1 or 2 papillae posteriorly, each papilla round, center concave .... *Amynthas wangi*
Genital papillae XVII and VII or IX and X 9

9. Male pores on a tiny glandular area, usually guarded anteriorly and posteriorly by a small papilla; 2 large elongate generally placed on ventral side of XVII, postsetal and medial to male porophores. Round papillae paired on ventral side of IX, close and postsetal, sometimes in similar position of X .... *Amynthas brevicingulus*
One pair of genital papillae in XVII, postsetal, large, round, flat-topped, in line with male pore; or, an additional pair just medial to the first, the 2 pairs sometimes partly fused. Papillae present or absent. If present, 1 pair present presetal and 1 postsetal in VII, presetal pair slightly medial to spermathecal pores, postsetal pair more medial than presetal pair .... *Amynthas rockefelleri*

10. Three pairs of papillae placed ventrally on XVIII–XX, posterior 2 pairs slightly medial to male porophores, first pair more ventromedially situated. Genital papillae paired ventrally on VII–VII, in front of 7th–9th setae .... *Amynthas limpidus*
Genital papillae between XVIII, and/or XIX and between VII–VII 11

11. Male pores close to the lateral margin, opening on low, large papillae projecting outwards when seen from the ventral side. Segments XVIII with 2 pairs of genital papillae, an inner and outer; the inner lying close to the median line and front of the chaetal line, the outer close to the male pores, internal to them, and behind the setal line; segment XIX with 1 pair of genital papillae exactly similar in position to the inner pair papillae of the segment XVIII. Segments VII, with a pair of genital papillae near the median line, in front of the aetel lines .... *Amynthas carnosus*
Genital papillae between VII, and VII and/or VII 12

12. Two pairs of genital papilla situated above the male pores, beneath XVII setae ring, near 17/18. A pair of genital papillae at VII ventrally .... *Amynthas hsinpuensis*
Genital papillae between XVIII, and VII and VII 13

13. Male pores opening on a small papilla somewhat raised surrounded by 5 or more circular ridges or whorls often with 1–5 small pit-like papillae medially and a little behind the setal circle of segment XVIII, arranged in 1 or 2 short transversely or oblique series, 2 or 3 being more frequent; these papillae often present on ventromedial side of the same segment, but always behind the setae; 2 cases out of 40 specimens without these papillae. Each papilla with a pigmented and depressed center, as large as a needle point, with a circular margin, pale and slightly raised. Sometimes this margin indistinct or absent, then the papilla only represented by a depressed pit or a round ampulla due to the swelling of the underlying tissues with the cuticle pushed up. Genital papillae usually paired on ventral side of segment VII and VII, medial to spermathecal pores or single on ventromedian line, but always behind setal circle. These papillae in many cases missing, those in VII segment being more constantly present. The character of these papillae same as those around male apertures but each with its circular margin not distinct probably due to the thick epidermis .... *Amynthas hawayanus*
Male pores superficial, on small round porophore, at centers of small, oval areas, surrounded by 4–5 small papillae. Genital papillae very small, paired on ventral side VII, in front of setae, near setae b and c, or occasionally on VII either paired or unpaired .... *Amynthas tuberculatus*

14. Without papillae front and behind clitellum 15
Without papillae front or behind clitellum 18

15. Male pore and papillae absent, or rarely have single male pore and a large papilla in XVIII ventrally, left or right side. Caeca manicata .... *Amynthas dandongensis*
Without papillae front or behind clitellum, Caeca simple 16

16. Male pores on a nipple-like porophore, its base with concentric ridges, body wall of XVI–XXV in line with porophore on each side raised as skin flap, always directed and partly folded toward ventral side. Caeca simple .... *Amynthas lacinatus*
Male pores no on a nipple-like porophore 17

17. Male disc somewhat quadranglar or bean-shaped with longitudinal slit found along its medium line .... Amynthas phaselus
Male pores round or oval, surface smooth, slightly convex, with or without a shallow horizontal slit (depression) in middle, surrounded by 2–3 circular folds, male aperture not visible .... *Amynthas proasacceus*

18. Without papillae behind clitellum 19
Without papillae front clitellum 20

19. Male pores on a small penis in glandular area. Spermathecal pores are shaped like the eye of a needle, or short transverse split-like, close to posterior edge of V–VII .... *Amynthas areniphilus*
Male pores invisible externally. Single large bean-shaped papilla placed medial to each male pore, which is on its lateral edge. Spermathecal pores superficial, skin around them glandular, with row of wrinkled skin along anterior and posterior side of the pore .... *Amynthas muticus*

20. Male pores towards the lateral margins of the genital markings. The genital marking are transversely oval areas of slight tumesence, not sharply demarcated, extending anteroposteriorly to about 17/18 and 18/19 and about 12 intersetal distances wide transversely; although there are 12 male setae between the genital markings on XVIII, the midventral region between the 2 markings is narrow .... *Amynthas balteolatus*
Genital marking are not transversely oval areas 21

TABLE 16 Key to the Species of the *hawayanus* Group of the Genus *Amynthas* From China—cont'd

21. Male pores are minute, on XVIII. There is a single, transversely elongate, genital marking with slightly rounded ends on XVIII, extending anteroposteriorly slightly onto XVII and XIX (approximately, intersegmental furrows 17/18 and 18/19 not visible ventrally and laterally). The marking is protuberant and is circumscribed by a definite circumferential furrow. On this marking are 2 lateral regions of particular prominence, not sharply delineated, separated by a less protuberant midventral region. The male pores are towards the lateral margins of these 2 prominences .... *Amynthas bellatulus*
Genital marking without slightly rounded ends 22

22. Male pores on top a conical porophore, surrounded by 3–5 skin folds. Two large oval genital papillae on the inner side of the male pores anterosetal and posterosetal. Spermathecal pores each eye-like .... *Amynthas bouchei*
Genital papillae not oval 23

23. Genital papillae flat-topped 24
Genital papillae are not flat-topped 25

24. Male pores on top and surrounded by 1 or 2 indistinct wrinkles of skin. Single pair of flat-topped papillae to the inner side of the male pore anterosetal in XIX .... *Amynthas edwardsi*
Male pores as a shallow crescent-shaped pouch covering the male pore and a large-sized papilla immediately medial to the pouch, which is low and flat-topped, its anteroposterior border extending partially to segments XVII and XIX respectively .... *Amynthas magnificus*

25. Male pores on small circular porophores, lateral to 1 or more pairs of small postsetal genital markings (small discs) .... *Amynthas gracilis*
Male pores not on small circular porophores 26

26. Male pores on papilla-like or disc-like porophore 27
Male pores on round-shaped or cone-shaped porophore 28

27. Male pores small, papilla-like or indistinct, male apertures invisible, but each with 2 genital papillae, 1 anterior and 1 posterior, or with 3 papillae, 1 anterior, 1 posterior and 1 lateral, and then surrounded by 4–6 circular folds, each papilla round, center flat or slightly convex, surrounded by a circular fold. A horizontal row of genital papillae, 1–3 (usually 2) in number, just medial to each of the male pore regions, immediately posterior to setal line, each papilla similar in size and structure to those associated with the male pores .... *Amynthas hohuanntis*
Male pores on an oval or round papilla-like or disc-like porophore, surrounded by 3–5 skin folds, often with a horizontal groove anteriorly, so that it looks like an eye. Spermathecal pores varied in number from 3 pairs (sexthecate) to absent (athecate), invisible externally .... *Amynthas shinanmontis*

28. Male pores on top of a round-shaped porophore, surrounded by 1 or 2 indistinct wrinkles of skin. Single pair of large oval papillae to the inner side of the male pore in 18/19 .... *Amynthas omodeoi*
Male pores on a raised cone-shaped porophore, surrounded by circular ridges; basal width of each cone about two-thirds the length of segment XVIII; both porophores everted. No trace of penis-like structure .... *Amynthas pongchii*

108. *Amynthas areniphilus* (Chen & Hsü, 1975)

Pheretima areniphila Chen, Hsü, Yang & Fong, 1975.
Acta Zool. Sinica. 21(1):91–92.
Amynthas areniphilus Easton, 1979. *Bull. Br. Mus. Nat. Hist. (Zool).* 35:124.
Amynthas areniphilus Xu & Xiao, 2011. *Terrestrial Earthworms of China.* 80–81.

External Characters

Length 111–158 mm, width 5–7 mm, segment number 145–158. Prostomium prolobous. First dorsal pore in 12/13. Clitellum XIV–XVI, annular, setae absent. Setae all fine and numerous; $aa = 1.0$–$1.2ab$, $zz = 1.0$–$1.5zy$; setal number, 80–104 (III), 106–130 (V), 108–126 (VII), 50–63 (VII), 58–69 (XXV). Male pores in ventral area of XVIII, about 1/4 of body circumference apart ventrally, each on a small penis in glandular area. Spermathecal pores 3 pairs in 5/6/7/8, each pore is shaped like the eye of a needle or a short transverse split, close to posterior edge of V–VII, about 1/4 of body circumference apart ventrally. No other genital markings. Female pore single in XIV, medioventral.

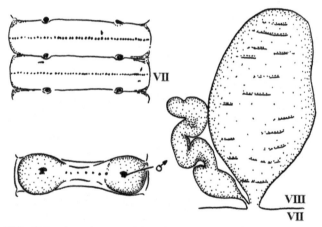

FIG. 108 *Amynthas areniphilus* (after Chen et al., 1975).

Internal Characters

Septa 9/10 absent, 8/9 thick, 4/5/6/7/8 very thick, and 10/11 membranous. Gizzard barrel or long ball-shaped, in IX–X. Intestine swelling in XV; caeca thin and long, simply, distal end sharp. Last hearts in XIII. Testis sacs 2 pairs, anterior pair with bilateral symmetry or asymmetrically developed, connected with a narrow commissure, posterior pair not easy to divide, does not include anterior pair seminal vesicles. Posterior pair seminal vesicles developed, pushing into XIII or XIV, most of the dorsal lobe not obvious. Prostate glands divided into stout lobes, in XVI–XIX; prostatic duct slender, distal end swollen and U-shaped. Accessory glands absent. Spermathecae 1 pair, in V–VII, ampulla long sac-shaped, its duct short, boundaries with ampulla not obvious; diverticulum shorter than main pouch, with 3–4 zigzags, or straight, as a seminal chamber, its duct short. No accessory glands.

Color: Preserved specimens opaque, gray-brown on dorsal, grayish on ventral. Clitellum brownish.
Type location: Riverside sand of Losuo River, Mengla, Yunnan.
Distribution in China: Yunnan (Mengla).
Etymology: The name of the species refers to its habit of swallowing sandy soil.
Remarks: This species is similar to Metaphire abidita having a septum and 3 pairs of spermathecal pores, but it is easily distinguished by the habitat, setal number, shape of seminal chamber, and male pores region.

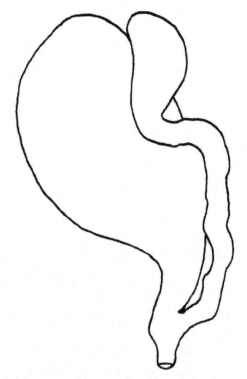

FIG. 109   *Amynthas balteolatus* (after Gates, 1932).

109. *Amynthas balteolatus* (Gates, 1932)

Pheretima balteolata Gates, 1932. *Rec. Ind. Mus.* 34(4):426–427.
Amynthas balteolatus Sims & Easton, 1972. *Biol. J. Soc.* 4:235.
Amynthas balteolatus Zhong & Qiu 1992. *Guizhou Science.* 10(4):40.
Amynthas balteolatus Xu & Xiao, 2011. *Terrestrial Earthworms of China.* 83–84.

External Characters

Length 89 mm, width 3 mm, segment number 100. Prostomium ?. First dorsal pore 12/13. Clitellum XIV–XVI, annular, intersegmental furrow 16/17 absent, or setae, dorsal pore and intersegmental furrows absent. There are a few scattered setae on segment II (2–8); no midventral breaks in the setal circles but may be a slight middorsal break on some of the segments behind the clitellum; posteriorly no middorsal break; setae number 96 (XX); 22–25 (VI), 23–28 (VII) between spermathecal pores, 22–24 (XVII), 10–14 (VII), 24–26 (XIX) between male pores. Male pores paired in XVIII, minute, towards the lateral margins of the genital markings. The genital markings are transversely oval areas of slight tumescence, not sharply demarcated, extending anteroposteriorly to about 17/18 and 18/19, and about 12 intersetal distances wide transversely; although there are 12 male setae between the genital markings on XVIII, the midventral region between the 2 markings is narrow. Spermathecal pores 3 pairs, in 5/6/7/8; pores minute. Female pore single in XIV, medioventral, transversely slit-like.

Internal Characters

Septa 8/9/10 absent, 5/6/7/8 and 10/11 muscular, 11/12/13 slightly strengthened and translucent. Gizzard ?. Intestine swelling in XV. Caeca simple, in XXVII, extending anteriorly to XXII. Last pair hearts in XIII. Testis sacs 2, unpaired. Seminal vesicles large, covering the dorsal blood vessel in XI and XII; the posterior pair of vesicles pushing 12/13 posteriorly into contact with 13/14. Prostates extend through segments XVII–XX; its duct slender, looped, about 2–2.5 mm in length. Spermathecae 3 pairs, in VI, VII, and VII; ampulla duct slender, shorter than the ampulla which is elongate saccular; diverticulum fastened by very delicate connective tissue to the lateral or median margin of the duct and ampulla is a slender tube, slightly widened at the ental end. The diverticulum passes into the anterior face of the spermathecal duct; the duct constricted within the parietes.

Color: Preserved specimens unpigmented. Clitellum yellowish.
Type locality: Myanmar (Teung Cong, Pang Wo).
Distribution in China: Yunnan (Lancang).
Deposition of types: Indian Museum.

110. *Amynthas bellatulus* (Gates, 1932)

Pheretima bellatula Gates, 1932. *Rec. Ind. Mus.* 34(4):427–428.

Pheretim bellatua Gates, 1972. *Trans. Amer. Philos. Soc.* 62(7):170.

Amynthas bellatulus Sims & Easton, 1972. *Biol. J. Soc.* 4: 235.

Amynthas bellatulus Zhong & Qiu, 1992. *Guizhou Science.* 10(4):40.

Amynthas ballatulus Xu & Xiao, 2011. *Terrestrial Earthworms of China.* 84.

External Characters

Length 72 mm, width 3.5 mm, segment number?. Prostomium ?. First dorsal pore 11/12. Clitellum XIV–XVI, annular; setae, intersegmental furrows and dorsal pores lacking. The setae on II on which segment there is a nearly complete setal circle, a slight middorsal gap present. A midventral break in the setal circles is lacking throughout but middorsal break is present constantly; setae number 38 (XX); 9 (VI), 11 (VII), between spermathecal pores, 12 (XVII), 0 (VII), 10 (XIX) between male pores; Male pores are minute, on XVIII. There is a single, transversely elongate, genital marking with slightly rounded ends on XVIII, extending anteroposteriorly slightly onto XVII and XIX. The marking is protuberant and is circumscribed by a definite circumferential furrow. On this marking there are 2 lateral regions of particular prominence, not sharply delineated, separated by a less protuberant midventral region. The male pores are towards the lateral margins of these 2 prominences. Spermathecal pores 3 pairs in 5/6/7/8, pore minute. No other genital marking. Female pore single in XIV, medioventral.

Internal Characters

Septa 8/9/10 absent. Gizzard ?. Intestine swelling in XV. Caeca simple, short, and blunt; in XXVII, extending anteriorly to XXV or XXIV. The single commissure of IX is on the left side. The last pair of hearts is in XIII. Testis sacs 2 pairs, relatively large; the anterior sac with a bilobed anterior margin; the posterior sac with a bilobed posterior margin, the 2 posterior lobes pushing 12/13 into 2 pocket-like invaginations into XIII. Seminal vesicles are large, covering the dorsal blood vessel in segment XI and XII; the posterior vesicles push 12/13 into contact with 13/14. Prostates extend through segments XVII–XX; its duct slender, looped, the ectal portion slightly thicker than the ental portion; length about 2 mm. Spermathecae 3 pairs, in VI, VII, and VII; ampulla duct is shorter than the ampulla. Diverticulum passes into the anterior face of the duct and has a slender ectal portion in which the lumen is narrow and a wider ental portion in which the lumen is wider; the widest part of the diverticular lumen is slightly more than half the length from the duct.

Color: Preserved specimens with dorsum very light brownish. Clitellum yellowish brown.
Type locality: Myanmar (Teung Cong).
Distribution in China: Yunnan (Lancang).
Deposition of types: Indian Museum.

111. *Amynthas bouchei* Zhao & Qiu, 2009

Amynthas bouchei Zhao & Qiu, 2009. *J. Nat. Hist.* 43(17–20):1029–1031.

Amynthas bouchei Xu & Xiao, 2011. *Terrestrial Earthworms of China.* 87–88.

External Characters

Length 225–286 mm, width 7–7.8 mm (clitellum), segment number 212–286. Prostomium 1/5 epilobous. First dorsal pore in 12/13. Clitellum 3/4 XIV–XVI, annular, markedly glandular; with 12 setae ventral in XVI, dorsal pore and annulus visible. Setae intensive, regularly distributed around segmental equators; Setae number 54–64 (III), 80–86 (V), 84–90 (VII), 90 (XX), 79–88 (XXV); 34–40 (VI), 35–39 (VII) between spermathecal pores, 17–18 between male pores, setal formula $aa = 1$–$1.6ab$, $zz = 1.3$–$2zy$. Male pores paired in XVIII, a little less than 0.33 body circumference apart, ventrally placed, each on

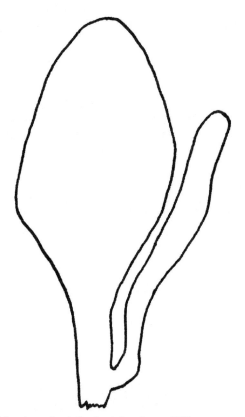

FIG. 110 *Amynthas bellatulus* (after Gates, 1932).

top a conical porophore (diameter 1.8–2.2mm), surrounded by 3–5 skin folds. Two large oval genital papillae on the inner side of the male pores anterosetally and posterosetally. Spermathecal pores 3 pairs, ventral, in intersegmental furrows of 5/6/7/8, each eye-like and obvious, distance between paired pores about 0.4 body circumference apart. Genital markings absent. Female pore single in the center of XIV ventrally, ovoid, brown.

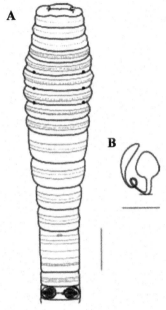

FIG. 111   *Amynthas bouchei* (after Zhao et al., 2009).

Internal Characters

Septa 8/9/10 absent, 5/6/7/8, 10/11 and 11/12 comparatively thick and muscular; 12/13/14 a little thicker than those following. Gizzard ball-shaped, in IX–X. Intestine swelling in XV. Intestinal caeca simple, in XXVII, extending anteriorly to XXIV, dark brown, with distinct dorsal tooth-shaped diverticula, smooth ventrally. Hearts in X–XIII. Holandric, Testis sacs 2 pairs in X and XI, those in X elongate reaching to dorsomedian, those in XI enclosing the first pair of seminal vesicles. Seminal vesicles paired in XI and XII, stout. Prostate glands in XVII–XIX, small, coarsely lobed and divided into 3 main lobes, with a short S-shaped duct, its ectal limb stouter. Spermathecae 3 pairs in VI–VII, ampulla elongate ovoid, duct stout, about two-thirds of ampulla, diverticula a little longer than main pouch and coiled, the ental half dilated, serving as seminal chamber, or diverticula a little shorter than main pouch and nearly straight, ental third enlarged, serving as a seminal chamber.

> Color: Preserved specimens lacking pigment dorsally before clitellum and ventrally; light olive-brown dorsally after clitellum. Lutescens on the clitellum. Distribution in China: Hainan (Mt. Jianfengling).

Deposition of types: Shanghai Museum of Natural History.

Type locality: Jianfengling National Nature Reserve (elevation 900m), Hainan.

Remarks: *Amynthas bouchei* is similar to *A. brevicingulus* (Chen, 1938) in having 3 pairs of spermathecal pores in 5/6/7/8 and simple intestinal caeca, but it has 2 large oval glandular papillae on the inner side of the male pores anterosetally and posterosetally, while *A. brevicingulus* has 2 large elongate papillae generally placed on ventral side of XVII, postsetal and medial to male porophores. What is more, the range interval between spermathecal pores of *A. bouchei* is about 2/5 body circumference apart compared with 1/8 body circumference of *A. brevicingulus*. *A. bouchei* with body length 225–286mm is much larger than *A. brevicingulus*. The 2 species also differ in the position of the first dorsal pore: 12/13 in *A. bouchei* versus 11/12 in *A. brevicingulus*.

112. *Amynthas brevicingulus* (Chen, 1938)

Pheretima (Pheretima) brevicingulus Chen, 1938. *Contr. Biol. Lab. Sci. Soc. China (Zool.).*12(10):401–403.
Amynthas brevicingulus Sims & Easton, 1972. *Biol. J. Soc.* 4:235.
Amynthas brevicingulus Xu & Xiao, 2011. *Terrestrial Earthworms of China.* 88–89.

External Characters

Length 80–100mm, width 3.4–4.3mm, segment number 94–115. Prostomium 1/3 epilobous. First dorsal pore 11/12. Clitellum entire, in XIV–XVI, annular, without setae; its length about 2mm, very short. Setae fine and uniformly distributed, none particularly enlarged, more numerous on preclitellar segments; both dorsal and ventral breaks nearly indistinct; $aa = 1.2ab$, $zz = 1.2–1.5$ yz; setae number 46–64 (III), 62–70 (VI), 57–74 (VII), 40–44 (XXV); 6–8 (VI), 6–7 (VII) between spermathecal pore, 6–8 between male pores. Male pores 1 pair in XVIII, about 1/4 of circumference apart ventrally, each on a tiny glandular area, usually guarded anteriorly and posteriorly by a small papilla; 2 large elongate papillae generally placed on ventral side of XVII, postsetal and medial to male porophores. Spermathecal pores 3 pairs in 5/6/7/8, about 1/8 of circumference apart ventrally, close to ventromedian line; each pore tiny appearing on anterior border of segments VI–VII, in intersegmental furrows. Round papillae paired on ventral side of IX, close and postsetal, sometimes in similar position in X. Female pore single in XIV, medioventral.

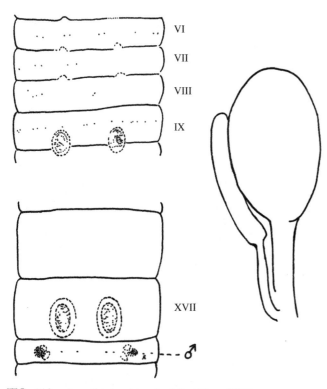

FIG. 112 *Amynthas brevicingulus* (after Chen, 1938).

Internal Characters

Septa 8/9/10 absent, 6/7/8 moderately thickened, 10/11 membranous in front of testis sacs, 11/12 also thin, 12/13 equally thin. Gizzard barrel-shaped, in X. Intestine enlarged from XV. Caeca simple, short and conical, confined to 1 segment, in XXVII, extending anteriorly to XXVI. Hearts 3 pairs, first pair in X absent. Testis sacs 2 pairs, anterior pair of testis sacs very large, about 1.5 mm in anteroposterior width, very narrowly constricted and communicated on ventromedian side; second pair empty. Seminal vesicles in XI small, each with a round dorsal lobe about 0.7 mm in diameter, vesicle as a band; second pair more rudimentary in appearance. Prostate gland coarsely lobate, compact, in XVII–XIX; with a long U-shaped duct, its ectal limb stouter. Accessory glands each in a round patch, sessile on body wall. Spermathecae 3 pairs, in VI–VII, ampulla ovoid, distended, about 1.2 mm long, 1.0 mm wide, its duct thick, about 1.0 mm long. Diverticulum shorter than main pouch, its duct slender and short, with a long club-shaped seminal chamber.

> Color: Preserved specimens fleshy pale both dorsally and ventrally. Clitellum cinnamon brown.
> Distribution in China: Hainan.
> Etymology: The specific name refers to the short clitellum.

113. *Amynthas carnosus carnosus* (Goto & Hatai, 1899)

Perichaeta carnosa Goto & Hatai, 1899. *Annot. Zool. Jap.* 2:15.
Amynthas carnosus Chang, Shen & Chen, 2009. *Earthworm Fauna of Taiwan.* 32–33.
Amynthas carnosus. Sims & Easton, 1972. *Biol. J. Soc.* 4:235.
Amynthas carnosus Xu & Xiao, 2011. *Terrestrial Earthworms of China.* 89–90.

External Characters

Length 143–153 mm, clitellum width 5–8 mm, segment number 106–126. Setae 55 in the spermathecal region. 14 between male pores. First dorsal pore 13/14. Clitellum XIV–XVI, annular. Male pores paired in XVIII, close to the lateral margin, opening on low, large papillae projecting outwards when seen from the ventral side. Segments XVIII with 2 pairs of genital papillae, an inner and outer; the inner lying close to the median line and front of the setal line, the outer close to the male pores, internal to them, and behind the setal line; segment XIX with 1 pair of genital papillae exactly similar in position to the inner pair of the preceding segment. Spermathecal pores 3 pairs in 5/6/7/8. Segments VII and VII, with a pair of genital papillae near the median line, in front of the setal lines. Female pore single in XIV, medioventral.

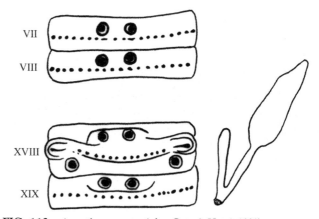

FIG. 113 *Amynthas carnosus* (after Goto & Hatai, 1899).

Internal Characters

Septa 8/9/10 absent, 4/5/6/7/8 and 10/11–14/15 thickened. Gizzard in VII–IX. Intestine from XV. Intestinal caeca paired in XXVI, extending anteriorly to XXIII. Ovaries present. Seminal vesicles paired in XI and XII, with dorsal surface divided into 3 lobes. Prostate glands large, in XVI–XX. Internally there is a large glandular patch of a circular form in the median line, lying equally in segments XVIII and XIX, and therefore corresponding

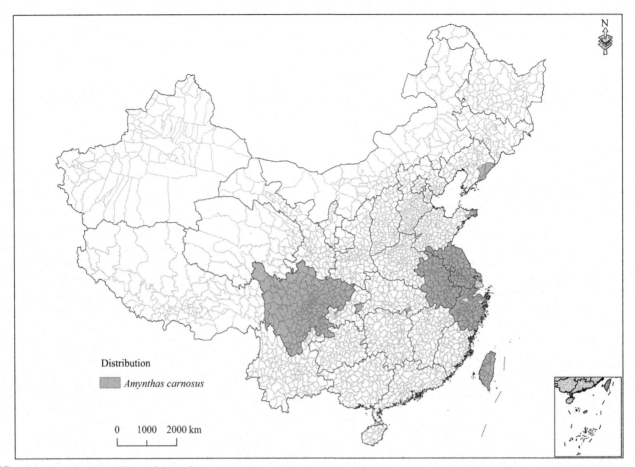

FIG. 114   Distribution in China of *Amynthas carnosus*.

to the region surrounded by genital papillae already described. Spermathecae 3 pairs in VII–IX, with straight appendages half as long as the main sac and duct.

Color: Coloration dark brown on dorsum, light gray on ventrum, with metallic luster.
Type locality: Japan (Tokyo).
Distribution in China: Liaoning (Dandong), Jiangsu, Zhejiang, Anhui, Taiwan, Shandong (Weihai), Hubei (Lichuan), Hong Kong, Chongqing (Beibei), Sichuan.

Remarks: *Amynthas carnosus* was first described by Goto and Hatai (1899) as a species with 3 pairs of spermathecae, and thus Sims and Easton (1972) assigned it to the *hawayanus* species group. However, Ohfuchi (1937) made a detailed examination of the species and indicated that 4 pairs of spermathecae are the typical form of *A. carnosus* should be placed within the *diffringens* species group as *A. nanshanensisis*. When *Amynthas carnosus* of Japan, described by Ohfuchi (1937), is compared with *Amynthas youngtai* of Korea described by Hong and James (2001), it is clear that they share similar characters. Accordingly, *Amynthas*

carnosus (Hong and James, 2001) is synonymous with *Amynthas carnosus* (Goto and Hatai, 1899).

114. *Amynthas catenus* Tsai, Shen & Tsai, 2001

Amynthas catenus Tsai, Shen & Tsai 2001. *Zool. Stud.* 40(4):279–282.
Amynthas catenus Chang, Shen & Chen, 2009. *Earthworm Fauna of Taiwan.* 34–35.
Amynthas catenus Xu & Xiao, 2011. *Terrestrial Earthworms of China.* 91–92.

External Characters

Length 61–106 mm, clitellum width 2.7–4.2 mm, segment number 85–103. Number of annuli (secondary segmentation) per segment 3 in IX–XIII, occasionally in X–XIII or VII–XIII. Prostomium epilobous. First dorsal pore in 11/12 or 12/13. Clitellum XIV–XVI, annular, setae absent, dorsal pore absent. Setae minute, numbering 29–38 (VII), 41–47 (XX), 7–10 between male pores. Male pores paired, ventral in XVIII, each porophore round, circular tubercle on setal annulus, 0.23 circumferences apart ventrally. Genital papillae arranged in a longitudinal series, medioventral, 3–9 papillae from XVII–XXI, similar to those in the preclitellar region in size, shape, position and arrangement. Spermathecal pores ventrolateral, number highly variable: no pore, or 3 pairs in 5/6/7/8, 3 pairs but lacking a right pore in 7/8, or lacking a right pore in 6/7 and a left pore in 7/8, or only a single left pore in 6/7; 0.22 circumferences apart ventrally. Genital papillae medioventral in V–IX, in a longitudinal series like a chain, number and position highly variable: or 3 presetal papillae in VI–VII and 2 postsetal papilla in VII and VII or 0–9 papillae in V–IX. Generally, 1 presetal papilla and/or 1 postsetal papilla in a segment, each occupying the entire width of an annulus adjacent to intersegmental furrows. Occasionally, 2 papillae joined closely in an annulus. Each papilla round with a concave center. Female pore single, medioventral in XIV.

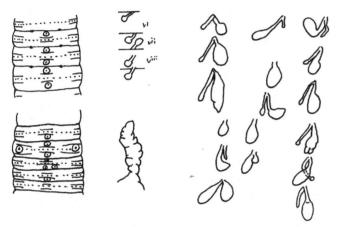

FIG. 115   *Amynthas catenus* (after Tsai et al., 2001).

Internal Characters

Septa 8/9/10 absent, 5/6/7/8 thickened, 10/11–13/14 greatly thickened. Gizzard in VII–X, large, cylindrical. Intestine from XVI or XV. Intestinal caeca paired in XXVII, simple, extending anteriorly to XX or XXII. Esophageal hearts enlarged from XI–XIII. Testis sacs paired in XI, large. Seminal vesicles paired in XI and XII, small, irregular, follicular. Occasionally, small vestigial seminal vesicles (pseudovesicles) in XIII. Prostate glands paired in XVIII, large, racemose, extending anteriorly to XVI and posteriorly to XXI, asymmetric in position and size between left and right; prostatic duct hook-shaped, if vestigial, only ducts present. Each genital papilla in both spermathecal and male pore regions associated with a white, round, stalked accessory gland. Spermathecae vestigial or absent, number highly variable in 6–8: 6 (3 pairs), 5, 4, 1, or no spermathecae, if present, size and structure highly variable: ampulla large to small, round to irregular, with or without stalk; diverticulum present, vestigial, or absent, if present, stalk long, straight, with normal seminal chamber, if vestigial, seminal chamber absent with vestigial stalk.

Color: Preserved specimens whitish purple on dorsum, whitish gray on ventrum, light grayish tan on clitellum. Type locality: Mt. Hohuan near the border between Hualien and Nantou counties in central Taiwan. Etymology: The name *"catenus"* refers to the "chainlike" arrangement of genital papillae in the medioventral portion. Deposition of types: Taiwan Endemic Species Research Institute, Jiji, Nantou County, Taiwan. Holotype: coll. no. 1999-17-Shen. Paratypes: same collection data as for holotype. Distribution in China: Taiwan (Mt. Hohuan). Remarks: This species may be parthenogenetic. The arrangement of genital papillae of *Amynthas catenus* is fairly similar to that of *A. monoserialis* (Chen 1938) of Hainan. However, both species are easily distinguishable by the characters of the spermathecae, setal number, and accessory glands. *A. catenus* has spermathecae which are highly variable in number, size, and structure. They show different vestigial stages in parthenogenetic degeneration, but with 3 pairs in VI–VIII as the original form. *A. monoserialis* has 2 pairs of normal spermathecae in VI and VII. Parthenogenetic degeneration of the spermathecae in *A. catenus* is fairly similar to those of *A irregularis* (Goto and Hatai, 1899; Ohfuchi 1938, 1939) of Japan and *A. varians* (Chen, 1938) of Hainan. Also, 2 of the specimens dissected lacked prostate glands, while other specimens had large but asymmetrical prostate glands, indicating the presence of parthenogenetic degeneration.

115. *Amynthas edwardsi* Zhao & Qiu, 2009

Amynthas edwardsi Zhao & Qiu, 2009. *J. Nat. Hist.* 43(17–20):1038–1040.
Amynthas edwardsi Xu & Xiao, 2011. *Terrestrial Earthworms of China.* 109.

External Characters

Length 53–56 mm, width 1.8–2 mm, segment number 94–97. Prostomium 1/2 epilobous. First dorsal pore in 12/13. Clitellum XIV–XVI, annular, markedly glandular, smooth, setae and dorsal pore invisible. Setae uniformly distributed; Setae number 25–26 (III), 32–34 (V), 46 (VII), 32–38 (XX), 38–40 (XXV); 12 (VI), between spermathecal pores, 8–9 between male pores, setal formula $aa = 1.2ab$, $zz = 1.2$–$1.3zy$. Male pores paired in XVIII, 0.33 body circumference apart ventrally, each on top of a pulvinate porophore, smooth on top and surrounded by 1 or 2 indistinct wrinkles of skin. One pair of flat-topped papillae to the inner side of the male pore anterosetal in XIX (diameter 0.7–0.8 mm). The distance between them is 1/3 of circumference apart. Spermathecal pores 3 pairs in 5/6/7/8, about 0.33 body circumference apart ventrally. Female pore single in XIV, medioventral, ovoid.

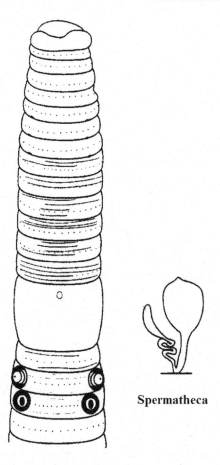

Anterior of body (ventral view)

FIG. 116   *Amynthas edwardsi* (after Zhao et al., 2009).

Spermatheca

Internal Characters

Septa 8/9/10 absent, 7/8 comparatively thick; 10/11/12 a litter thicker than those following. Gizzard barrel-shaped, in IX–X. Intestine swelling in XVI. Intestinal caeca simple, in XXVII, extending anteriorly to 1/3 XXV, smooth on both sides. Hearts 3 pairs, in XI–XIII, conjoint with supraesophageal vessel. A pair of circular vessels in X, slender, conjoint with ventral vessel. Holandric, testis sacs 2 pairs, in X and XI, the first pair stouter, the second pair small, silvery white, both separate ventrally. Seminal vesicles 2 pairs, in XI and XII, the first pair is stouter. Prostate glands very well developed, in 1/2XVI–XXIV, finger-shaped lobate, with a U-shaped duct, its ectal limb stouter, the ental limb rather long. Spermathecae 3 pairs in VI–VII, ampulla heart-shaped (0.5–0.7 mm long), duct about four-fifths of ampulla, uniform in caliber. Diverticulum is about three-quarters of main pouch, with middle in zigzag fashion, the ental one-third club-shaped and enlarged, forming seminal chamber.

Color: Preserved specimens brown both dorsally and ventrally. Brown on clitellum.

Distribution in China: Hainan (Mt. Diaoluo).

Deposition of types: Shanghai Museum of Natural History.

Type locality: the chocolate-brown soil near streams in the tropical rainforest on Mt. Diaoluo (elevation 940 m), Hainan.

Remarks: This species is similar *Amynthas limpidus* (Chen, 1938) in having 3 pairs of spermathecal pore in 5/6/7/8 and by the shape of diverticulum. Nevertheless, it only has 1 pair of flat-topped papillae to the inner side of the male pores anterosetal in XIX while 3 pairs of papillae are positioned ventrally on XVIII–XX in *A. limpidus*. Compared with the simple and small prostate gland in *A. limpidus*, *A. edwardsi* has well-developed prostate gland. Also the second pair of testis sacs does not enclose the first pair of seminal vesicles in *A. edwardsi*. Moreover, *A. edwardsi* is much smaller than *A. limpidus*, with a length of 53–56 mm.

116. Amynthas fluxus (Chen, 1946)

Pheretima fluxa Chen, 1946. *J. West China. Border Res. Soc. (B).* 16:133–134, 141, 160.

Amynthas fluxus Sims & Easton, 1972. *Biol. J. Linn. Soc.* 4:235.

Amynthas luxus Xu & Xiao, 2011. *Terrestrial Earthworms of China.* 115.

External Characters

Medium-sized worm. Length 100 mm, width 5 mm, segment number 116. Segment behind clitellum shortened, 4 or 5 annuli distinct. Prostomium epilobous. First dorsal pore 12/13. Clitellum entire in XIV–XVI, annular, setal pits visible ventrally but no setae ever found. Setae very fine and evenly distributed. Both dorsal and ventral breaks indistinct. Setae number 60 (III), 74 (VI), 76 (VII), 64 (IX), 70 (XXV); 43 (VI), 42 (VII) between spermathecal pores, 20 (XVII), 16 (VII), 18 (XIX) between or in line with male pores. Male pores paired in XVIII, about 1/3 of circumference apart ventrally, each in a shallow pouch (a skin fold in reality), with 2 very large and flat-topped papillae extending to neighboring segments XVII and XIX respectively, about 1.5 mm in diameter, placed medially to each pore, 1 in front of and another behind setal line. Spermathecal pores 3 pairs in 5/6/7/8, about 4/7 of circumference apart ventrally, each in an ellipsoid area. No genital marking in this region. Female pore single in XIV, medioventral.

Internal Characters

Septa 5/6–9/10 very muscular, about equally thickened, 10/11 and those following thin and membranous; thick nephridial tufts found anteriorly on 5/6/7. Gizzard round and moderate in size, in front of septum 8/9, lying in 1/2 IX and X. Intestine swelling in XVI. Caeca simple and small, finger-shaped, extending anteriorly to XXIII. Vascular

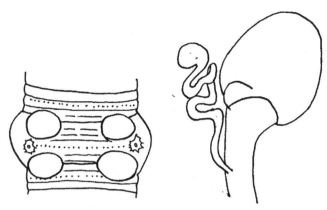

FIG. 117 *Amynthas fluxus* (after Chen, 1946).

loops in X small, partly enclosed in testis sacs. Testis sacs 2 pairs, well developed, elongate, and communicated dorsally; ventral bridge of posterior pair very narrow, that of anterior pair broader and communicated. Seminal vesicles 2 pairs, first pair smaller and enclosed in posterior pair of testis sacs; second pair larger, occupying segments XII and XIII, each with a large dorsal lobe. Prostate glands developed, lobate, in XVI–XIX, its duct slender and uniformly thickened. Accessory glands as cotton-like cushions, sessile, about 2 mm in diameter in each mass. Gland lobules absent. Spermathecae 3 pairs, small, main pouch about 2 mm long. Ampulla sac-like, about 1 mm wide. Diverticulum longer than main pouch, its duct slender and loosely coiled, ended with an ovoid seminal chamber.

Color: Greenish, deep green along dorsomedian line, pale ventrally. Skin rather slippery, lubricated. Clitellum chocolate brown.
Distribution in China: Chongqing (Beipei).
Etymology: The specific name refers to the type habitat of sandy ground on the wet bed of a creek at Hsia-Ch'i-K'ou.

117. *Amynthas gracilis* (Kinberg, 1867)

Amynthas gracilis Sims & Easton, 1972. *Biol. J. Soc.* 4:235.
Amynthas gracilis James, Shih & Chang, 2005. *J. N. His.* 39(14):1024–1025.
Amynthas gracilis Shen & Yeo, 2005. *Raf. Bul. Zool.* 53(1):21.
Amynthas gracilis Chang, Shen & Chen, 2009. *Earthworm Fauna of Taiwan.* 46–47.
Amynthas gracilis Xu & Xiao, 2011. *Terrestrial Earthworms of China.* 118–119.

External Characters

Length 60–158 mm, clitellum width 3.8–5 mm, segment number 71–98. Prostomium epilobous. Setae

37–46 (VII), 56–59 (XX), 14–19 between male pores. First dorsal pore 10/11 or 11/12. Clitellum XIV–XVI, annular, setae absent, dorsal pore absent. Male pores paired in XVIII, 0.33 circumference apart on tenth setal line, on small circular porophores, lateral to 1 or more pairs of small postsetal genital markings (small discs). Spermathecal pores 3 pairs in 5/6/7/8, 0.25–0.3 circumference apart ventrally. Female pore single, medioventral in XIV.

Internal Characters

Septa 8/9/10 absent, 6/7/8 thickened, 10/11–13/14 slightly thickened. Gizzard in VII–X. Intestine from XV. Intestinal caeca simple in XXVII, extending anteriorly to XXIV or XXV, 6–7 small pockets on ventral margin. Esophageal hearts paired in X–XIII or XI–XIII. Testis sacs 2 pairs in X and XI, ventrally joined. Seminal vesicles paired in XI and XII, with small dorsal lobe. Prostate glands paired in XVIII, occupying several segments, prostatic duct straight. Ovaries paired in XIII. Spermathecae 3 pairs in VI–VII, ampulla pear-shaped, duct shorter than ampulla, diverticulum with small, ovate seminal chamber and slender stalk.

Color: Live specimens dark red with white clitellum.
Type locality: Rio de Janeiro, Brazil.
Etymology: The specific name *"gracilis"* literally means gracile in Latin.
Deposition of types: Leiden Museum, Netherlands.
Global distribution: Cosmopolitan. Taiwan (Pingdong), Fujian (Jinmen).
Remarks: According to Gates (1972) *Amynthas gracilis* is native to China, and was introduced to Hawaii and California before 1852. It is a cosmopolitan species and has often been confused with *Amynthas morrisi* (Gates, 1968).

Differences between *A. gracilis* and *A. wangi* are primarily in the male reproductive organs and the locations of genital markings. Genital markings are present in the spermathecal segments in *A. wangi*, but not in this material of *A. gracilis*, and in *A. wangi* on XVII in line with the male porophores, rather than in XVIII medial to the male pores. The testes sacs of *A. wangi* are paired, while on the other hand the testes sacs in X of *A. gracilis* are joined ventrally.

118. *Amynthas hawayanus* (Rosa, 1891)

Perichaeta hawayana Beddard, 1896. *Pro. Zool. Soc. London.* 1895:201–203.
Pheretima hawayana Stephenson, 1912. *Rec. Indian Mus.* 7:276–278.
Pheretima hawayana Chen, 1931. *Chin. Zool.* 7 (3):142–148.

Pheretima hawayana Chen, 1933. *Contr. Biol. Lab. Sci. Soc. China (Zool.).* 9: 238.

Pheretima hawayana Chen, 1936. *Contr. Biol. Lab. Sci. Soc. China (Zool.).* 11:270.

Pheretima hawayana Gates, 1939. *Proc. U.S. Natr. Mus.* 85:445–446.

Pheretima hawayana Chen, 1946. *J. West China. Border Res. Soc. (B).* 16:135.

Pheretima hawayana Tsai, 1964. *Quer. J. Taiwan Mus.* 17:9–11.

Pheretima hawayana Gates 1972. *Trans. Amer. Philos. Soc.* 62(7):189–190.

Amynthas hawayanus Xu & Xiao, 2011. *Terrestrial Earthworms of China.* 121.

External Characters

Length 100–150 mm, width 3.5–6.0 mm, segment number 66–92. Annuli not visible until VII or IX, represented by secondary groves on each side of setal circle. Prostomium 1/2 epilobous. First dorsal pore 10/11. Clitellum always not entire, fewer than 3 segments (3/4 XIV–2/3 XVI); without intersegmental grooves; setae only present on ventral side of XVI, about 5–7 in most cases, those in the preceding 2 segments not visible, or merely marked as pits; dorsal pore absent. Setae small and delicate, longer and closer on ventral side, those in anterior segments II–VI, slightly enlarged and widely spaced, but in VII resuming the normal aspect. Ventral break small but evident, dorsal break greater; $aa = 12$–$1.5ab$, $zz = 2$–$3zy$; setae number 16–22 (III), 18–22 (VI), 32–42 (VII), 42–54 (XII), 48–60 (XXV); 10–18 between male pores. Male pores each exposed, represented by a slight ridge on ventrolateral side of segment XVII, about 1/3 of ventral circumference apart; without a copulatory chamber, its opening on a small papilla somewhat raised surrounded by 5 or more circular ridges or whorls often with 1–5 small pit-like papillae medially and a little behind the setal circle of segment XVIII, arranged in 1 or 2 short transversely or oblique series, 2 or 3 being more frequent; these papillae often present on ventromedial side of the same segment, but always behind the setae; seldom without these papillae. Each papilla with a pigmented and depressed center, as large as a needle point, with a circular margin, pale and little raised, about 0.2 mm in diameter. Sometimes this margin indistinct or absent, then the papilla only represented by a depressed pit or a round ampulla which is due to the swelling of the underneath tissues with the cuticle pushed up. Spermathecal pores 3 pairs, in intersegmental grooves 5/6/7/8, about 1/3 of ventral circumference apart, always not quite visible externally, or sometimes recognized by the slight eye-like markings. Genital papillae usually paired on ventral side of segment VII and VII, medial to spermathecal pores or single on ventromedian line, but always behind setal circle. These papillae in many cases missing, those in VII segment being more constantly present. Female pore single in a large depression on ventromedial side of XIV.

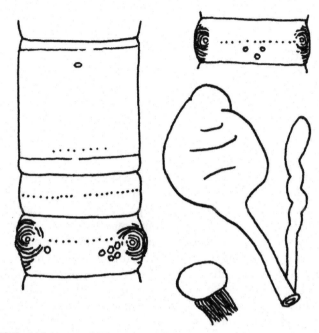

FIG. 118 *Amynthas hawayanus* (after Chen).

Internal Characters

Septa 8/9/10 absent, 5/6/7/8, 10/11–13/14 thickened, 5/6/7 and 11/12/13 particularly thickened. Gizzard globular, narrower anteriorly, usually in IX, X, surface smooth, whitish and glittering. Intestine swelling in XV. Intestine caeca simple, in XXVII, extending anteriorly to XXV or XXVI, with small ventral indentations, smooth or broadly wrinkled on their dorsal edges, lying lateral to intestine, directed upward or downward anteriorly. Hearts 4 pairs in X–XIII. Testis sacs in X and XI; anterior pair large, rounded but sometimes flattened, in front of septum 10/11, connected medially with each other through a narrow passage. Posterior pair in front of septum 11/12 and in contact with posterior side of the anterior pair, closely under anterior vesicle, communicating with each other. Sperm duct meeting ventral to septum 12/13. Vestigial ovoid bodies and elongate strand on posterior faces of septa 12/13, 13/14. Seminal vesicles very large in XI and XII, often with irregular cuts on surface or in some cases tuberculated, posterior pair extended into XIII, with distinct nipple-like dorsal lobes, sometimes large about a quarter of the whole vesicle, granular on surface and grayish in color sharply constricted off from the main vesicle. Prostate glands usually in XVII–XXI or XVI–XXII, seldom in XVI–XX; with rather large lobules, smooth on surface; each prostatic duct not very long, straight and stouter at its middle, usually with a deep loop at its proximal end and loose loop at distal, bending several time near the body wall before opening to the exterior, its distal looped part stouter and shorter than other end. Its surface also similar to that of gizzard. Several large masses of glands present medial to the duct, each with a long large stalk or cord leading out to its papilla. The gland proper being soft and whitish, roundish or very irregular in shape, granular on surface. The largest gland about 1.5 mm in diameter and with a very long stalk about 2 mm in length. Such large-sized glands quite inproportional to the small papillae as referred above. Spermathecae 3 pairs in VI–VII; ampulla large about 2.5 mm in length, 2 mm in width, elongate sac-like or round, usually with an apical knob, surface whitish and smooth or wrinkled, ampulla duct long and slender as long as, or longer than, the ampulla, about 0.3 mm in width, but stout and enlarged at its ental third, with a definite line marked off itself from the ampulla; diverticulum usually shorter than the main part, half as long as the latter, or sometimes longer, ental two-thirds enlarged with a terminal oval seminal chamber; the whole slightly twisted or straight, its duct slender joining to the main duct at the body wall. Stalked glands usually occurring in VII and VII to correspond with the papillae outside, their character being same as those around prostate region. Generally 1 cap-like gland connected with 1 papilla outside, but in few cases, 2 or 3 glands occurring in same place leading to 1 small papilla, in this case the glands naturally smaller.

Color: Preserved specimen rich brown or chestnut on anterodorsal side, buff brown or grayish fleshy at the rest of dorsal side, ventral side pale. Clitellum reddish brown or dark cherry red, lighter ventrally.

Distribution in China: Jiangsu, Zhejiang (Zhoushan, Linhai, Tiantai, Chuanshan, Shengxian, Fenshui), Fujian (Fuzhou, Xiamen), Hubei (Qianjiang), Hong Kong, Chongqing (Banhe, Shapingba, Nanchuan, Jiangbei, Luzhou), Sichuan (Chengdu, Jiading, Leshan), Yunnan (Tengyuan).

Etymology: The specific name refers to the type locality.

Type locality: Hawaii.

Deposition of types: Vienna Museum. Mus. Biol. Lab. Sci. Soc. China.

Remarks: Michaelsen and others maintain this form as *forma typica* with the other related form, *A. barbadensis*, as a subspecies. Stephenson later found that there are intermediate forms discovered in some Yunnan and Indian specimens and combined these 2 into a single species named *A. hawayana* (Rosa).

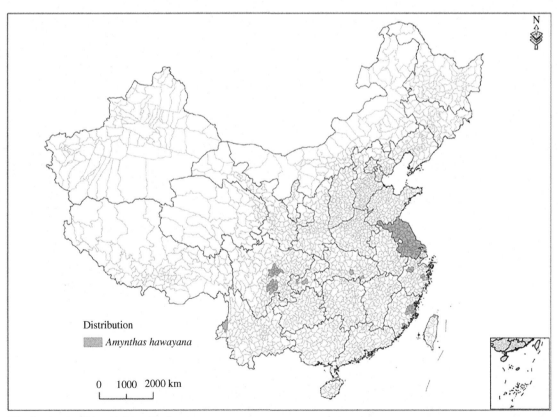

FIG. 119 Distribution in China of *Amynthas hawayanus*.

119. *Amynthas hohuanmontis* Tsai, Shen & Tsai, 2002

Amynthas hohuanmontis Tsai, Shen & Tsai, 2002. *J. Nat. Hist.* 36, 757–765.
Amynthas hohuanmontis Xu & Xiao, 2011. *Terrestrial Earthworms of China.* 124–125

External Characters

Length 73–113 mm, clitellum width 3.4–4.4 mm, segment number 85–103. Annuli (secondary segmentation) 3 per segment in VI–XIII. Prostomium tanylobous. Setae 32–41 (VII), 42–46 (XX), 9–11 between male pores. First dorsal pore 12/13. Clitellum XIV–XVI, annular, 2.5–5.3 mm long, 0.72–1.21 times longer than width, dorsal pores absent, setae absent. Male pores paired in XVIII, ventrolateral, 0.26–0.29 circumference apart ventrally. Porophores small, papilla-like or indistinct, male apertures invisible, but each with 2 genital papillae, 1 anterior and 1 posterior, or with 3 papillae, 1 anterior, 1 posterior and 1 lateral, and then surrounded by 4–6 circular folds, each papilla round, center flat or slightly convex, 0.2–0.3 mm in diameter, surrounded by a circular fold. A horizontal row of genital papillae, 1 to 3 (usually 2) in number, just medial to each of the male pore regions, immediately posterior to setal line, each papilla similar in size and structure to those associated with the male pores. Spermathecal pores 3 pairs in 5/6/7/8 or absent. Genital papillae absent in preclitellar region. Female pore single, medioventral in XIV.

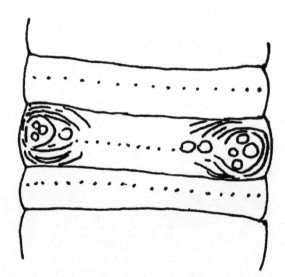

FIG. 120   *Amynthas hohuanmontis* (after Tsai et al., 2002).

Internal Characters

Septa 8/9/10 absent, 5/6–7/8, 10/11–12/13 thickened. Gizzard round in IX and X. Intestine from XV. Intestinal caeca paired in XXVII, simple, surface slightly wrinkled, extending anteriorly to XXIV–XXII. Esophageal hearts in XI–XIII. Testis sacs 2 pairs in XI, or anterior pair in 10/11 (partly in X and XI) and posterior in XI, each small, round or oval. Seminal vesicles paired in XI and XII, surface smooth or folliculate, with a granulate dorsal lobe. Prostate glands vestigial, small, nodule-like or absent, but prostatic ducts normal, large, C- or S-shaped. Accessory glands round with stalk, corresponding to genital papillae in XVIII. Spermathecae variable 3 pairs in VI–VII or absent. Accessory glands absent in preclitellar region.

Color: Preserved specimens light grayish brown on dorsum, light grey on ventrum, dark brown around clitellum.
Etymology: The name *hohuanmontis* was given to this species with reference to its type locality of Mt. Hohuan in central Taiwan.
Type locality: Mt. Hohuan (elevation 3000 m), along Rd. 14A in Hualien County, near the border with Nantou County, Taiwan.
Deposition of types: Taiwan Endemic Species Research Institute, Jiji, Nantou County, Taiwan.
Holotype: coll. no. 1999-17-Shen. Paratypes: same collection as holotype.
Distribution in China: Taiwan (Nantou, Hualien).
Remarks: This species may be parthenogenetic. *Amynthas hohuanmontis* is easily distinguishable from other nominal species of the *A. illotus* species group by body size, porophore structure, number and arrangement of genital papillae, and size of prostate glands. Based on size, genital papillae, and male pore structure, *A. hohuanmontis* is closely related to *A. sheni* of Hong Kong (Chen, 1935). However, *A. hohuanmontis* has 2 or 3 postclitellar papillae at each of the male pore regions and a horizontal row of 1–3 postsetal papillae medial to the pore, whereas *A. sheni* has a single postclitellar papilla medial to each male pore (Chen, 1935). Also, the former has no preclitellar genital papillae, whereas the latter has paired preclitellar papillae in VIII. The prostate glands of *A. hohuanmontis* are small and vestigial, or absent, while *A. sheni* has normal, large prostate glands (Chen, 1935). The other 3 species, *A illotus*, *A. assacceus*, and *A. oyuensis*, have no genital papillae.

120. *Amynthas limpidus* (Chen, 1938)

Pheretima (Pheretima) limpida Chen, 1938. *Contr. Biol. Lab. Sci. Soc. China (Zool.).* 12(10):405–407.
Amynthas limpidus Sims & Easton, 1972. *Biol. J. Soc.* 4:236.
Amynthas limpidus Xu & Xiao, 2011. *Terrestrial Earthworms of China.* 138–139.

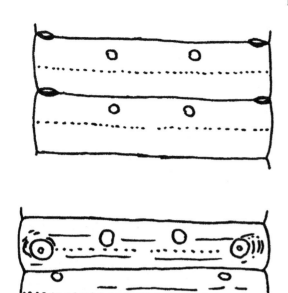

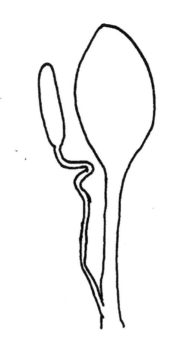

FIG. 121   *Amynthas limpidus* (after Chen, 1938).

External Characters

Length 150 mm, width 6 mm, segment number 218. Preclitellar segments rather long (IX–XI longer), shorter after XIV (especially short after about XXX). Prostomium proepilobous. First dorsal pore 12/13. Clitellum just beginning to appear in XIV–XVI, not distinctly glandular, annular; setae present but shorter in all segments. Setae extremely fine and numerous, none particularly enlarged; both dorsal and ventral breaks indistinct; $aa = 1.2ab$, $zz = 1.5$ yz behind clitellum; setae number 82 (III), 128 (VI), 142 (VII), 96 (XXV); 46 (VI), 50 (VII) between spermathecal pores, 22 between male pores. Male pores 1 pair in XVIII, superficial, about 1/3 of body circumference apart ventrally, each on a small tubercle. Three pairs of papillae placed ventrally on XVIII–XX, posterior 2 pairs slightly medial to male porophores, first pair more ventromedially situated (about 4 setae ventral). Spermathecal pores 3 pairs in 5/6/7/8, about (or more than) 3/8 of circumference apart ventrally; each appearing as an elliptical area on anterior border of VI–VII, intersegmental and superficial. Genital papillae paired ventrally on VII–VII, in front of 7th–9th setae. Female pore single in XIV, medioventral.

Internal Characters

Septa 8/9/10 absent, 5/6/7/8 well thickened, 10/11/12/13 also thick, 13/14 slightly so. Gizzard small and round, in X. Intestine enlarged from XVI. Caeca simple and small, in XXVII, extending anteriorly to XXV. Hearts in X–XIII, first pair absent. Testis sacs small, partially filled; those in X close to nerve cord, connected ventromedially, probably not communicated both dorsally and ventrally. Seminal vesicles small, whole (vesicle + dorsal lobe) 3 mm long, 1 mm wide; dorsal lobe smooth, elongate, about 2/5 size of vesicle; narrow whitish bands present on septa 12/13 and 13/14 respectively. Prostate gland portion very small, 3.5 mm long, with a long duct about 4.2 mm long, uniform in thickness (except thinner at ental end). Accessory glands small and sessile. Spermathecae 3 pairs, in VI–VII, main pouch 3 mm long, ampulla spatulate, 1 mm wide, with a thin but long duct. Diverticulum slightly longer than main pouch, with very thin and long duct and an elongate ovoid seminal chamber which is whitish and iridescent in appearance.

Color: Preserved specimens fleshy pale throughout, Clitellum brownish
Distribution in China: Hainan (Wanning).

121. *Amynthas magnificus* (Chen, 1936)

Pheretima magnifica Chen, 1936. *Contr. Biol. Lab. Sci. Soc. China (Zool.).* 11:283–286.
Amynthas magnificus Sims & Easton, 1972. *Biol. J. Soc.* 4:236.
Amynthas magnificus Xu & Xiao, 2011. *Terrestrial Earthworms of China.* 143–144.

External Characters

Length 240 mm, width 7.0 mm, segment number 124. Prostomium with a small round dorsal lobe, about 1/3 epilobous, completely cut off behind. First dorsal pore 11/12, minute. Clitellum in XIV–XVI, annular; setae much shorter and scarce, will probably disappear in the course of development. Setae more prominent on preclitellar segments, about equal in length, but slightly wider in interval on medioventral side; both dorsal and ventral breaks slight, $aa = 1.5ab$, $zz = 1.2$ yz; setae number 61 (III), 67 (VI), 65 (VII), 48 (XII), 32 (XXV); 36 (VI), 33 (VII) between spermathecal pores; 9 between male pores. Male pores 1 pair, in XVIII ventrally, about 1/4 of circumference apart ventrally; each as a shallow crescent-shaped pouch covering the male pore and a large-sized papilla immediately medial to the pouch, which is low and flat-topped, about 2 mm in diameter, its anteroposterior border extending partially to segments XVII and XIX respectively; distance between 2 papillae about 2 mm. Spermathecal pores 3 pairs, in 5/6/7/8, each on a minute tubercle, intersegmental in position, posterior pair about 1/2 of circumference apart ventrally, anterior 2 pairs moved ventrally one after another, second pair about 3

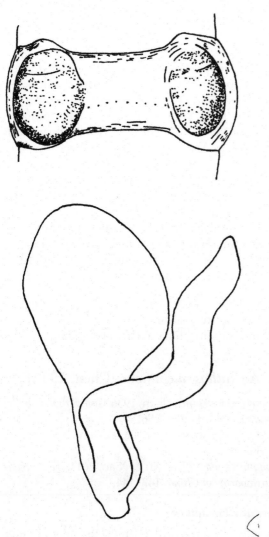

FIG. 122 *Amynthas magnificus* (after Chen, 1936).

setae in distance ventral to third pair; first pair about 2 setae to second pair. No genital papillae thereabouts. Female pore single in XIV, medioventral.

Internal Characters

Septa 5/6–10/11 very thick, especially 6/7–9/10; 11/12/13 comparatively thick, rest of septa very thin. Gizzard in IX, very small, thinly muscular, soft, in front of septum 8/9. Intestine enlarged from XVI. Intestinal caeca simple, slender and round, with circular wrinkles, in XXVII ?, extending anteriorly to XXIII. Hearts in X–XIII, first pair moderately stout. Testis sacs 2 pairs in X and XI: anterior pair very narrowly connected and also communicated, posterior pair united. Seminal vesicles in XI and XII, small but extending to nearly dorsal side of intestine, about 1.5 mm long, 4 mm high, each with a distinct large dorsal lobe which is slightly grayish and granular on surface. Prostate glands small, in XVII and XVIII, its duct short, thick at middle part. Accessory glands in a large low patch, sessile and smooth on surface. Spermathecae 3 pairs, in VI, VII, VII; not well developed; ampulla roundish about 1 mm in length, with a short but distinct duct; diverticulum club-shaped, a little narrower entally, with a distinct duct, whole length slightly shorter than main pouch.

> Color: Preserved specimens greenish on dorsal side generally, grayish on ventral side.
> Type locality: Chongqing.
> Distribution in China: Chongqing, Sichuan.
> Deposition of types: Mus. Biol. Lab. Sci. Soc. China.
> Remarks: This species is distinct from any hexathecal species distributed in this region. The type specimen is not fully mature, however, it is recognizable as a distinct species.

122. *Amynthas muticus* (Chen, 1938)

Pheretima (Pheretima) mutica Chen, 1938. *Contr. Biol. Lab. Sci. Soc. China (Zool.).*12(10):403–405
Amynthas muticus Sims & Easton, 1972. *Biol. J. Soc.* 4:236.
Amynthas muticus Xu & Xiao, 2011. *Terrestrial Earthworms of China.* 154.

External Characters

Large-sized. Length ? mm (mutilated), width 8 mm, segment number ?. Prostomium ?. First dorsal pore 12/13. Clitellum glandular, but not swollen, entire, in XIV–XVI, annular, whitish along setal circle; setae nearly invisible. Setae all fine numerous, none particularly enlarged, similarly spaced on both dorsal and ventral sides, those on anterior venter slightly wider at intervals; setal breaks; $aa = 1.2ab$, $zz = 1.5$ yz before clitellum; setae number 73 (III), 92 (VI), 100 (VII), 142 (XIX), 146 (XXV); 21 (VI), 22 (VII) between spermathecal pores, 14 between male pores. Male pores 1 pair in XVIII, invisible externally, about 1/3 of body circumference. Single large bean-shaped papilla placed medial to each male pore which is on its lateral edge, usually indistinct externally. Spermathecal pores 3 pair in 5/6/7/8, intersegmental, about 1/4 of circumference apart ventrally. Each pore superficial, skin around it glandular, with 1 row of wrinkled skin along anterior and posterior side of the pore. Glandular skin rather inconspicuous. No other genital papillae in this region. Female pore single in XIV, medioventral.

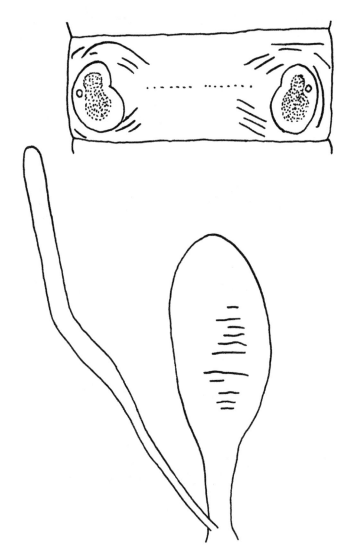

FIG. 123 *Amynthas muticus* (after Chen, 1938).

Internal Characters

Septa 9/10 absent, 8/9 membranous on ventral side, 5/6/7 well thickened, 10/11–13/14 as thick, 7/8 less muscular. Gizzard in IX–1/2 X, rounded, shining on surface. Intestine enlarged from XV. Caeca simple and slender, in XXVII, extending anteriorly to XXIV. Hearts in X–XIII.

Testis sacs in X and XI, containing nothing expect whitish testis and funnel; anterior pair more widely situated (about 2 mm) but connected by membrane. Seminal vesicles small, vesicle tubercular on surface, about 1.5 mm broad, 2 mm high, its dorsal lobe as large. A pair of elongate bands (1 mm wide) attached to posterior surface of 12/13, those on 13/14 considerably larger (about 2 mm wide, 4.2 mm high). Prostate small; gland portion finely granular, in XVII–XIX, with a long duct (about 5 mm long), uniform in caliber. Accessory glands not visible. Spermathecae 3 pairs, in VI–VII, small. Ampulla ovoid and dull pale, with a short duct indistinctly marked off, whole length 3 mm, width 1.5 mm. Diverticulun much longer than main pouch (about 4.5 mm long), ental 4/5 wider but seemingly empty, its duct slender, indistinctly marked off from seminal chamber.

Color: Preserved specimens chestnut on dorsal side, pale ventrally, with setal circles paler.
Distribution in China: Hainan (Wanning).

123. Amynthas omodeoi Zhao & Qiu, 2009

Amynthas omodeoi Zhao & Qiu, 2009. *J. Nat. Hist.* 43(17–20):1031–1038.
Amynthas omodeoi Xu & Xiao, 2011. *Terrestrial Earthworms of China.* 160.

External Characters

Length 78 mm, width 2 mm, segment number 123. Prostomium 1/2 epilobous. First dorsal pore in 11/12. Clitellum XIV–XVI, annular, markedly glandular, smooth, setae and dorsal pore invisible. Setae sparse, uniformly distributed; Setae number 24–26 (III), 26–27 (V), 30–36 (XX), 34–36 (XXV); 7–9 (VI), 14–15 (VII) between spermathecal pores, 5–6 between male pores, setal formula $aa = 1.8$–$2ab$, $zz = 1$–$1.3zy$. Male pores paired in XVIII, 0.33 body circumference apart ventrally, each on top of a round-shaped porophore, surrounded by 1 or 2 indistinct wrinkles of skin. One pair of large oval papillae (diameter 0.6–0.8 mm) to the inner side of the male pore in 18/19, about 0.25 circumference apart. Spermathecal pores 3 pairs in 5/6/7/8, about 0.4 body circumference apart. Female pore single in XIV, medioventral, ovoid.

Internal Characters

Septa 8/9/10 absent, 6/7/8 comparatively thick; 10/11/12/13 a litter thicker than those following. Gizzard long barrel-shaped, in IX–X. Intestine swelling in XVI. Intestinal caeca simple, in XXVII, extending anteriorly to XXIV, with a distinct ventral lobation, smooth dorsally. Hearts 3 pairs, in XI–XIII, conjoint with supraesophageal vessel; the last the biggest. A pair of circular vessels in X, slender, conjoint with ventral vessel.

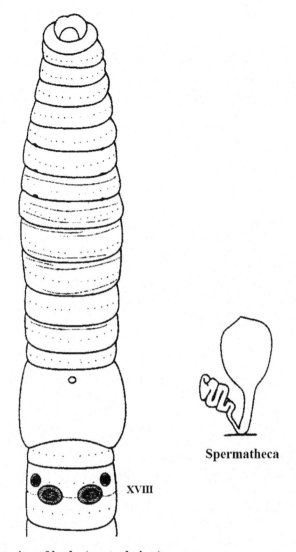

Anterior of body (ventral view)

Spermatheca

XVIII

FIG. 124　*Amynthas omodeoi* (after Zhao & al., 2009).

Holandric, testis sacs 2 pairs, in X–XI, the first is stouter. Seminal vesicles 2 pairs, in XI and XII, the first pair more developed. Prostate glands well developed, in XVII–XXIII, finger-shaped diverticulum, with U-shaped duct, its ectal limb stouter. Accessory glands in XVIII–XIX, well developed, about 0.1–0.3 mm long lobed, with a long restiform duct. Spermathecae 3 pairs in VI–VII, ampulla ovoid, a little peaked at the top, duct thin, about two-thirds of ampulla. Diverticulum about three-fifths of main pouch, ectal one-third straight, ental two-thirds twisted in zigzag fashion, forming seminal chamber.

Color: Preserved specimens nonpigmented ventrally. Olive brown dorsal. Brown on clitellum.
Distribution in China: Hainan (Mt. Diaoluo).
Deposition of types: Shanghai Museum of Natural History.

Type locality: Elevation 1008 m, the brown soil under a dead and fallen tree on Mt. Diaoluo, Hainan.
Remarks: *Amynthas omodeoi* closely resembles *Amynthas oculatus* (Chen, 1938) in the characteristic of the male pores, genital papillae, and septa. Both also have simple intestinal caeca and well-developed prostate gland and accessory glands. Furthermore, the diverticulum is in the same zigzag fashion. However, they are easily separated on the basis of spermathecal pores. *A. omodeoi* has 3 pairs in 5/6/7/8, but *A. Oculatus* has 2 pairs in 5/6/7. Moreover, the first pair of seminal vesicles is not enclosed in the second pair of testis sac in *A. omodeoi*, whose body length is 78 mm, more than twice the size of *A. oculatus*. Finally, the position of the first dorsal pore is different. In *A. omodeoi*, it is in 11/12, while in *A. oculatus* it is in 12/13.

A. omodeoi differs greatly from *A. brevicingulus*, *A. limpidus*, and *A. muticus* by the characteristics in the male pore region. It differs from *Amynthas sinuosus* in having 3 spermathecal pores.

124. *Amynthas papilliferus* (Gates, 1935)

Pheretima papillifera Gates, 1935a. *Smithsonian Mus. Coll.* 93(3):13.
Pheretima papillifera Chen, 1936. *Contr. Biol. Lab. Sci. Soc. China (Zool.).* 11:300–301.
Pheretima papillifera Gates, 1939. *Proc. U.S. Natr. Mus.* 85:459–460.
Pheretima papillifera Chen, 1946. *J. West China. Border Res. Soc. (B).* 16:137.
Amynthas papilliferus Sims & Easton, 1972. *Biol. J. Soc.* 4: 236.
Amynthas papilliferus Xu & Xiao, 2011. *Terrestrial Earthworms of China.* 162.

External Characters

Length 100–105 mm, width 4.0–4.5 mm, segment number 103. Prostomium 2/3 epilobous. First dorsal pore 11/12. Clitellum in XIV–XVI, annular, not sharply distinguished, without setae. Setae very fine, disappearing in some parts, setal pits observable on II–III, a few on VI, those on other parts of body missing locally, more evident on dorsal side; $aa = 1.5ab$, $zz = 2.0\ yz$; setae number 36 (VI), 40 (VII), 39 (XII), 40 (XXV); 19 (VII) between spermathecal pores, 12 between male pores. Male pores 1 pair in XVIII, on a rectangular papilla, slightly raised and flat-topped. Spermathecal pores 3 pairs in 5/6/7/8, minute and superficial, slightly on anterior edge of segments, about 2/5 of circumference apart ventrally. Genital papillae paired ventrally on XI–XIV. Female pore single in XIV, medioventral.

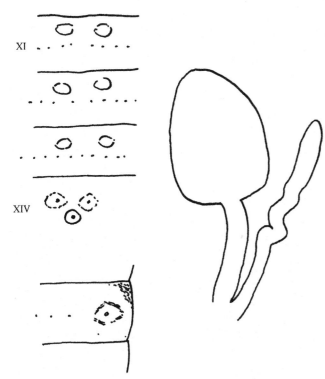

FIG. 125   *Amynthas papilliferus* (after Chen, 1936).

Internal Characters

Septa 8/9/10 absent, 5/6/7 very thick. Gizzard ? Intestine swelling in XV. Intestinal caeca simple, narrow and long, in XXVII, extending anteriorly to XXIII. Last hearts in XIII. Testis sacs paired in X and XI, separate ventrally. Seminal vesicles are either attached to the dorsal surface of these testis sacs or are contained within the sacs. Prostate glands compact, in XVII–XIX or XX, its duct is about 5.5 mm long and coiled. Accessory glands racemose, sessile, 1 connected with each papilla externally. Spermathecae 3 pairs, in VI, VII, and VII; ampulla heart-shaped, distinct from its long duct; diverticulum with a short muscular stalk and a longer, more irregular seminal chamber. Genital marking gland sessile on the parietes.

Color: Clitellum chocolate reddish.
Type locality: Sichuan (Ya'an).
Distribution in China: Sichuan (Ya'an, Mt. Emei).
Deposition of types: Smithsonian Institution.
Remarks: Distinguished from sexthecal Chinese species of *Pheretima* with spermathecal pores in 5/6/7/8 by the absence of setae on II–III and dorsally on IV and by the location of the genital markings. Distinguished from *Amynthas omeimontis* by the copulatory chambers, and from *Metaphire schmardae* by the large size, the invagination of the spermathecal pores, and the presence of genital markings. It is distinguished from *Metaphire abdita* by the superficial

male pores, from *A. tuberculatus* (Gates, 1935a) by the simple intestinal caeca, and from *A. hawayanus* by the genital markings on XI–XIV.

125. *Amynthas papulosus* (Rosa, 1896)

Pheretima papulosa Gates, 1935a. *Amer. Mus. Novitates.* 141:18–19.
Pheretima papulosa Gates, 1972. *Trans. Amer. Philos. Soc.* 62(7):206–207.
Amynthas papulosus Sims & Easton, 1972. *Biol. J. Soc.* 4: 236.
Amynthas papulosus Zhong & Qiu, 1992. *Guizhou Science.* 10(4):40.
Amynthas papulosus Chang, Shen & Chen, 2009. *Earthworm Fauna of Taiwan.* 68–69.
Amynthas papulosus Xu & Xiao, 2011. *Terrestrial Earthworms of China.* 163

External Characters

Length 45–78 mm, clitellum width 3–5 mm, segment number 90–119. Prostomium epilobous. First dorsal pore 12/13. Clitellum in XIV–XVI, annular, setae usually present ventrally in XIV–XV, XVI. Setae number 54–60 (V), 61 (XI), 60(XII), 62–66(XIII), 4–6 (XVI), 56–62 (XIX). Male pores paired on XVIII, 0.25 circumference apart ventrally. Numerous small presetal and postsetal small discs present in transverse rows in XVII–XIX. Spermathecal pores in 5/6/7/8, 0.25 circumference apart ventrally. Genital papillae present in VII–IX, presetal and postsetal small discs, in transverse rows. Female pore single on XIV, medioventral.

Internal Characters

Septa 8/9/10 absent, 5/6/7/8 thickened. Gizzard in VII–IX. Intestine from XVI. Intestinal caeca paired in XXVII, extending anteriorly to XXII. Lateral hearts in X–XIII. Testis sacs paired in X and XI. Seminal vesicles paired in XI and XII. Prostate glands paired in XVIII, racemose, extending anteriorly to XVI and posteriorly to XXI. Spermathecae paired in VI–VII, duct slender, longer than ampulla. Diverticulum slender, reaching the base of ampulla. Nephridia meroic. Ovaries paired in XIII.

Color: Live specimens pink or reddish pink. Preserved specimens white.
Type locality: Balighe, Sumatra.
Deposition of types: Genoa, Italy.
Distribution in China: Zhejiang, Taiwan (Gaoxinpo Village), Hong Kong, Yunnan (Shuangjiang).
Etymology: The species was named after its papulous papillae.

126. *Amynthas phaselus* (Hatai, 1930)

Pheretima phaselus Hatai, 1930. *Sci. Rep. Tohoku Imp. Univ.*, Biol. 5(4):659–661.
Pheretima phaselus Kobayashi, 1938. *Ann. Zool. Jop.* 3 (2):410–411.
Pheretima phaselus Song & Paik, 1969. *Korean J. Zool.* 12(1):16.

External Characters

Length 100 mm, width 5.5–6.0 mm, segment number 109. Prostomium 1/2 epilobous. First dorsal pore in 12/13. Clitellum entire, in XIV–XVI, annular, without setae and other markings. Setae moderate in size, those on about III–IX sometimes may be slightly enlarged, ventral ones slightly more closely set than the dorsal, and nearly equal in size to the latter; setae number 32–34 (III), 41–46 (V), 48–52 (VI), 50–56 (VII), 58–64 (XVII), 60–68 (XX); 13–16 (V), 14–16 (VI), 14–17 (VII), 15–18 (VII) between spermathecal pores, 13–14 between male pores. Male pores 1 pair on ventrolateral side of XVIII, about 1/4 of the circumference apart ventrally; each male

disc somewhat quadrangular or bean-shaped with longitudinal slit found along its medium line. The slit is outwardly curved in the middle. Single genital papilla constantly present, close to its posterior side on segment XVIII. No other genital papillae. Spermathecal pores 3 pairs, in intersegmental furrows 6/7/8/9, about 1/4 of body circumference apart ventrally, each with a large genital papilla closely behind. None found near spermathecal region, nor on ventral side of those segments. Female pore single in XIV, medioventral.

Internal Characters

Septa well-developed, 8/9/10 absent, 5/6/7/8 and 10/11/12/13 slightly thickened but 8/9 traceable ventrally. First pair of hearts not found, probably absent. Gizzard globular, fairly large, in segments IX and X, with a slight crop-like dilation on its anterior side. Intestine swelling in XV or posterior part of XV. Caeca simple, conical in shape, nearly smooth on both dorsal and ventral edges, in XXVII, extending anteriorly to XXIV or XIII. Heart 3 pairs, in XI–XIII, first pair in X entirely absent but commissural present, asymmetrically developed, connecting dorsal and ventral vessels. Testis sacs large and conspicuous anterior pair rounded, in X and XI, projecting into segment X, connected medially either by a narrow tube or broadly with its opposite fellow, usually in a thick U shape; posterior pair broadly connected or completely fused into a transverse sac, or slightly constricted medially. Seminal vesicles very large, in XI and XII. Usually tubercular on surface subdivided into 2 or more divisions, each with a small dorsal lobe, granular on surface, partly concealed or exposed, situated on anterior dorsal side of each vesicle. Prostate glands large, well developed, subdivided into transverse lobes, in XVI–XX, or XXI, glandular parts of both sides nearly overlapping dorsally; each with a rather short but stout duct in a deep U-curve, thicker at distal third. Accessory glands found in this region corresponding to papillae externally. Each papilla associated with 1 gland internally which is roundish and granular on surface, probably as an aggregate of small glands, with a short stout duct apparently being composed of numerous cords. Spermathecae 3 pairs in VII–IX, rather small; ental third of each diverticulum club-shaped and whitish. last 2 pairs placed between septa 7/8 and 8/9; ampulla heart-shaped, elongate pear-shaped or spatulate, smooth or sometimes transversely wrinkled on surface, its duct rather stout and long, ectal slightly narrower, usually marked off from the latter. Diverticulum about half length of both ampulla and ampullar duct, comprising a long slender duct and an ental dilated portion which is ovoid or elongate spherical or date-shaped, often bright white in color. These saccules may also be found on the ampulla and diverticulum.

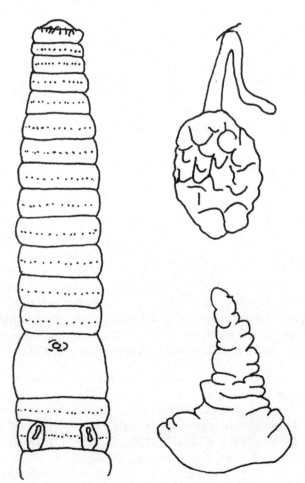

FIG. 126   *Amynthas phaselus* (after Kobayashi, 1938).

Color: Preserved specimens dark chestnut or purplish chocolate dorsally, darker at anterior dorsum, uniformly pigmented except slightly whitish around setal circles, pale or grayish pale ventrally. Clitellum brownish.

Global distribution: Korean, China.

Distribution in China: Jiangsu (Nanjing, Wuxi, Yixing, Suzhou), Zhejiang (Zhoushan, Chuanshan, Ningbo, Chenxian, Lanyu, Tonglu, Xindeng), Anhui (Anqing, Chuzhou), Jiangxi (Nanchang, Jiujiang, Dean), Chongqing (Shapingba, Baixi).

Remarks: *A. phaselus* is very close to *A. kamitai*, and differs only in aspects of the male pore. *A. kamitai* may be a special form of the present species if the testis sacs are similarly annular in shape.

127. *Amynthas pongchii* (Chen, 1936)

Pheretima pongchii Chen, 1936. *Contr. Biol. Lab. Sci. Soc. China (Zool.).* 11:279–281.

Pheretima pongchii Sims & Easton, 1972. *Biol. J. Soc.* 4:236.

Amynthas pongchii Xu & Xiao, 2011. *Terrestrial Earthworms of China.* 170.

External Characters

Length 100 mm, width 5 mm, segment number 105. Prostomium proepilobous. First dorsal pore 12/13. Clitellum highly glandular, XIV–XVI, annular, smooth, its length equal to 5 preclitellar or 5 postclitellar segments. Setae conspicuous and numerous, none particularly enlarged or widely spaced; setal zone slightly raised, more noticeable on preclitellar region; Those on middle part of body less conspicuous; both dorsal and ventral breaks very slight; setae number 78 (III), 76 (VI), 78 (VII), 70 (XII), 60 (XX); 32 (VI), 34 (VII) between spermathecal pores; 12 between male pores. Male pores 1 pair, in XVIII ventrally, about 1/3 of circumference apart ventrally; each on a raised cone-shaped porophore, surrounded by circular ridges; basal width of each cone about two-thirds the length of segment XVIII; both porophores everted. No trace of penis-like structure. No genital papillae. Spermathecal pores 3 pairs, in 5/6/7/8, intersegmental, about 1/2 circumference apart ventrally; no genital markings thereabout. Female pore single in XIV, medioventral.

Internal Characters

Septa 5/6–9/10 very thick, 10/11/12 less thickened; those behind 11/12 very thin and membranous. Gizzard very small, thinly muscular, in front of septum 8/9, in VII. Intestine enlarged from XVI, chestnut brown in color. Intestinal caeca simple, elongate and smooth on both sides, in XXVII, extending anteriorly to XXIII, darker chestnut. Hearts 4 pairs, each with a

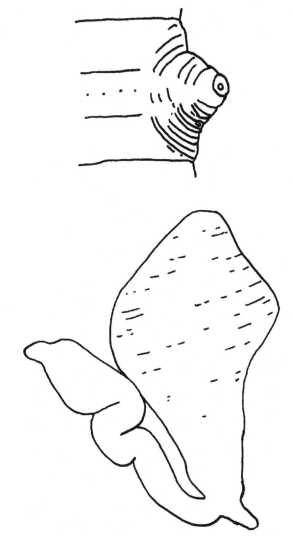

FIG. 127   *Amynthas pongchii* (after Chen, 1936).

section whitish near its origin, such whitish sections also appearing in the region of each septum along dorsal vessel. Testis sacs in X very large, filling up the whole segment, meeting on dorsal side, and also coalesced ventrally with a slight constriction; those in XI closely applied to inner surface of first pair of seminal vesicles, not well distended, with seminal mass on ventral side of each sac; both sacs united medioventrally. Sperm duct of each side meeting in front of segment XII. Seminal vesicles paired very large, anterior pair in XII but in front of septum 11/12, hardly separable from the testis sac, septa 11/12 and 12/13 very thin and closely applied to the vesicles, posterior pair larger, in XIV–XV, with septum 13/14 fastening at middle of the vesicle, about 3 mm in anteroposterior length, dorsal lobe large but not distinct from vesicle in appearance, pale and smooth on surface. Prostate glands only in XVI–XIX, with coarse lobules, its duct short, about 2 mm long, very slender, with a U-shaped curve. Spermathecal 3 pairs, in VI, VII, VII; ampulla heart-shaped, with a knob-like apex, usually transversely wrinkled, about 1.4 mm in width, it ducts very broad at ental portion, slightly distinct from ampulla; diverticulum about length of main pouch, its duct short, about 1 mm long, 0.25 mm wide, ental two-thirds slightly distended and filled with seminal products, pale in color.

Color: Preserved specimens greenish on dorsal side, grayish ventrally, a deep purplish green along dorsomedian line.
Type locality: Chongqing
Distribution in China: Chongqing (Beipei), Sichuan (Leshan).
Deposition of types: Mus. Biol. Lab. Sci. Soc. China.
Remarks: This species is very close to *Metaphire abdita* (Gates, 1935a) in body size, setal character, and male organs. However, it lacks the genital papillae in the male pore region and the penis.

128. *Amynthas proasacceus* Tsai, Shen & Tsai, 2001

Amynthas proasacceus Tsai, Shen & Tsai, 2001, *Zoological Studies.* 40(4), 282–285.
Amynthas proasacceus Chang, Shen & Chen, 2009. *Earthworm Fauna of Taiwan.* 74–75.
Amynthas proasacceus Xu & Xiao, 2011. *Terrestrial Earthworms of China.* 170–171.

External Characters

Length 39–76 mm, width 2.9–4.0 mm, segment number 57–106. Annuli (secondary segmentation) 3 per segment after III in preclitellar region. Prostomium with a large mouth opening surrounded by a small, soft, thick semicircular dorsal lip and a large, thick, white ventral lip. First dorsal pore 11/12. Clitellum XIV–XVI, annular, length 1.4–2.5 mm. Setae 33–40 (VII), 43–51 (XX), 6–9 between male pores. Male pores paired in XVIII, 0.22–0.28 circumference apart ventrally, each round or oval, surface smooth, slightly convex, with or without a shallow horizontal slit (depression) in middle, surrounded by 2–3 circular folds, male aperture not visible. No genital papillae in postclitellar region. Spermathecal pores not visible. No genital papillae in preclitellar region. Female pore single, medioventral in XIV.

Internal Characters

Septa 8/9/10 absent, 5/6–7/8, 10/11–13/14 thickened. Gizzard in VII–X, cylindrical. Intestine from XV. Intestinal caeca paired in XXVII, simple, extending anteriorly to XX, XXI or XXIII. Esophageal hearts 3 pairs in XI–XIII. Testis sacs 2 pairs in XI and partly in X or/and XII, each round or triangular. Seminal vesicles paired in XI and XII, each small, irregularly-shaped with a small, oval dorsal lobe. Prostate glands paired in XVIII, racemose, occupying 3–6 segments in XVI–XXI, prostatic duct

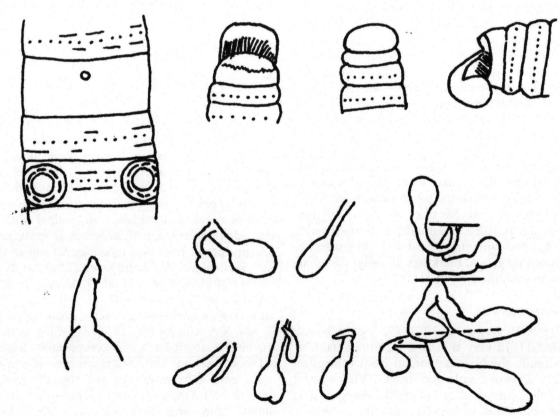

FIG. 128   *Amynthas proasacceus* (after Tsai et al., 2001).

hook-shaped, seldom lacking prostate glands but with prostatic ducts. Spermathecae 6 (3 pairs), 5, 4, or 2 in VI–VII, structure highly variable: for normal spermathecae, ampulla oval-shaped, 0.8–1.4 mm long, with slender spermathecal stalk 0.9–1.3 mm long, diverticulum small with oval seminal chamber, about 0.3 mm long with a slender stalk about 0.6 mm long; for vestigial spermathecae, diverticula often lacking seminal chambers, stalk vestigial or absent, some spermathecae lacking diverticula, and some diverticula lacking seminal chambers. Ovaries paired in XIII, each large with follicular surface.

Color: Preserved specimens whitish pink on dorsum, whitish gray on ventrum, light grayish tan on clitellum.
Type locality: Mt. Hohuan (elevation 3000 m), near the border between Hualien and Nantou counties in central Taiwan.
Deposition of types: Taiwan Endemic Species Research Institute, Jiji, Nantou County, Taiwan.
Holotype: coll. no. 1999-17-Shen. Paratypes: same collection data as for holotype.
Distribution in China: Taiwan (Hualien, Nantou).
Etymology: The name "*proasacceus*" indicates that this is a species closely related to the ancestral form of *A. asacceus*.
Remarks: *Amynthas proasacceus* shares similar external characters of body size, segment number, setal number, coloration, and genital papillae with *A. asacceus* (Chen 1938) of Hainan and *A. pusillus* (Ohfuchi 1956) (synonymous to *A. asacceus*) (Tsai et al., 2001) of the Ryukyu Islands. *A. proasacceus* and *A. asacceus* also share the occurrence of parthenogenetic degeneration but at different levels in the reproductive organs, such as spermathecae, seminal vesicles, and prostate glands. *A. proasacceus* usually has 3 pairs of spermathecae, but there is a trend of reduction in number and size, and also occurrence of structural deformation. These suggest that *A. proasacceus* was originally a sexthecal earthworm. On the other hand, *A. asacceus* has no spermatheca, the most advanced (final) stage of spermathecal degeneration, and is a member of the athecal *illotus* species group (Sims and Easton, 1972; Easton, 1981). Furthermore, *A. proasacceus* has a pair of large, circular porophores, but their edges do not reach the intersegmental furrows 17/18 and 18/19, while those of *A. asacceus* (Chen, 1938; Ohfuchi, 1956) reach the furrows. The former is considered a more primitive character than the latter.

129. *Amynthas rockefelleri* (Chen, 1933)

Pheretima rockefelleri Chen, 1933. *Contr. Biol. Lab. Sci. Soc. China (Zool.).* 9:238–244.

Pheretima rockefelleri Gates, 1935b. *Lingnan J. Sci.* 14(3):454–455.
Pheretima rockefelleri Tsai, 1964. *Quer. J. Taiwan Mus.* 17:8–9.
Pheretima rockefelleri Sims & Easton, 1972. *Biol. J. Soc.* 4:236.
Amynthas rockefelleri Chang, Shen & Chen, 2009. *Earthworm Fauna of Taiwan.* 78–79.
Amynthas rockefelleri Xu & Xiao, 2011. *Terrestrial Earthworms of China.* 175–176.

External Characters

Length 82–130 mm, clitellum width 3.0–4.2 mm, segment number 108–142. Number of annuli per segment 3 after V. Prostomium epilobous. First dorsal pore 11/12. Clitellum XIV–XVI, annular, glandular layer thick, swollen, smooth, and evenly formed on dorsal and lateral side, but less glandulated or usually roughened on ventral side, often with irregular grooves, extending to lateroventral side; setae present ventrally, slightly enlarged and wider in space on ventral side of 3 clitellar segments, more numerous posteriorly, 6–8 on XIV, 10–14 on XV, 12–16 on XVI, each seta more straight and blunt at tip, slightly longer than those in other segments; dorsal pore present or absent. Setae number 38–54 (III), 58–70 (VI), 58–75 (VII), 46–68 (XII), 49–59 (XX), 52–56 (XXV); 20–26 between spermathecal pores, 10–16 between male pores. Male pores paired in XVIII, ventrolateral, on top of a small, round papilla surrounded by several circular folds; about 1/3 body circumference apart. One pair of genital

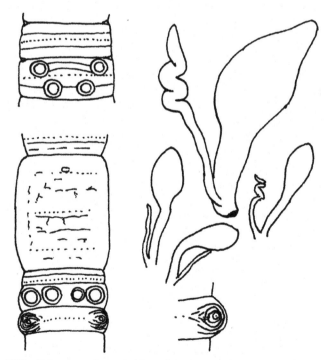

FIG. 129 *Amynthas rockefelleri* (after Chen, 1933).

papillae in XVII, postsetal, large, round, flat-topped, in line with male pore; in some specimens, an additional pair just medial to the first, the 2 pairs sometimes fused partially. Spermathecal pores 3 paired in 5/6/7/8, ventrolateral, about 3/11 of body circumference apart, usually invisible, sometimes with a minute dark porophore in intersegmental furrow; Papillae present or absent. If present, 1 pair present presetal and 1 postsetal in VII, presetal pair slightly medial to spermathecal pores, postsetal pair more medial than presetal pair. Female pore single, medioventral in XIV.

Internal Characters

Septa 8/9/10 absent, 5/6/7/8 and 10/11/12/13 thickened. Gizzard in IX and X, round. Intestine from XV, XVI or XVII. Intestinal caeca in XXVI or XXVII, simple, horn-shaped, extending anteriorly to XXIII or XXIV. Hearts 4 pairs, in X–XIII, first pair very slender or vestigial, or even absent. Ovaries paired in XIII. Testis sacs 2 pairs in X and XI, large, occupying full segments. Seminal vesicles paired in XI and XII, small, anterior pair included in posterior testis sacs. Prostate glands absent with only prostatic ducts. Accessory glands present, large, corresponding to external papillae. Spermathecae 3 pairs in VI–VII, ampulla round to rod-shaped, stalk slender. Diverticulum slender, rod-shaped, seminal chambers indistinguishable.

Color: Live specimens pale in general appearance, pinkish at anterior 5 or 6 segments, whitish before clitellum, grayish pale behind, pale or grayish pale posteriorly, greenish gray along dorsal pores, clitellum whitish or milky, slightly paler on ventral side. Preserved specimens white with light brownish yellow clitellum; or pale over whole body except if clitellum light chocolate.
Type locality: Zhejiang (Linhai, Tiantai, Shengxian).
Deposition of types: National Museum of Natural History, USA. Mus. Zool. Cent. Univ. Nanking.
Distribution in China: Zhejiang (Linhai, Tientai, Shengxian), Fujian (Quanzhou, Jinmen), Taiwan (Taibei), Hong Kong.
Etymology: This species was named after Rockefeller, after whom the Rockefeller Foundation was named.
Remarks: *Amynthas rockefelleri* has long been considered as a parthenogenetic morph of *A. papulosus* and thus a junior synonym by many authors. However, they are very different in coloration and body shape when alive.

130. *Amynthas shinanmontis* Tsai & Shen, 2007

Amynthas shinanmontis Tsai & Shen, 2007. *J. Nat. Hist.* 41(5–8):358–362.

Amynthas shinanmontis Chang, Shen & Chen, 2009. *Earthworm Fauna of Taiwan.* 82–83.
Amynthas nanshanensis Xu & Xiao, 2011. *Terrestrial Earthworms of China.* 156–157.

External Characters

Medium-sized earthworms. Length 86–187 mm, width 2.5–5.2 mm (clitellum), segment number 75–114. Prostomium epilobous. First dorsal pore in 11/12–13/14. Clitellum XIV–XVI, annular. Setae minute, 31–43 (VII), 38–55 (XX); 6–11 between male pores. Male pores paired in XVIII, ventrolateral, 0.20–0.28 body circumference apart ventrally, each on an oval or round papilla-like or disc-like porophore, surrounded by 3–5 skin folds, often with a horizontal groove anteriorly, like an eye. Genital papilla absent in both preclitellar and postclitellar regions. Spermathecal pores varied in number from 3 pairs (sexthecate) to absent (athecate), invisible externally. Female pore single in XIV, medioventral.

Internal Characters

Septa 8/9/10 absent, 5/6/7/8 and 10/11/12/13/14 thickened. Nephridial tufts thick on anterior faces of 5/6/7 septa. Gizzard large in IX–X. Intestine swelling in XV or XVI. Intestinal caeca paired in XXVII, simple, slender, extending anteriorly to XXII–XXV. Hearts enlarged from XI–XIII. Holandric, testis sacs small, 2 pairs in X and XI or both in XI. Vas efferens connected in XII or XIII in each side to form a large, straight vas deferens and then to prostatic duct in XVIII. Seminal vesicles 2 pairs in XI and XII, variable in size: large (normal) seminal vesicle with pink, follicular surface, and a small, reddish-brown dorsal lobe, occupying an entire segmental compartment; medium vesicle with highly folliculated surface and a large dorsal lobe, occupying about half of the segmental compartment; vestigial vesicle and its dorsal lobe highly folliculated, occupying less than half of the compartment; and nodule-like vesicle, a small, irregular nodule without dorsal lobe. Prostate glands variable in size from normal to absent (aprostatic): normal in XVI–XVIII; vestigial only in XVIII or XVII–XVIII; nodule-like or absent. Prostatic duct thick, swollen, U-shaped or coiled. Rarely has no left prostatic duct and left male porophore. Accessory glands absent in both preclitellar and postclitellar regions. Spermathecae varied in number from 3 pairs (sexthecate) to absent (athecate); 3 pairs in VI–VII. Sizes and structure of spermathecae variable from large (normal), vestigial, nodule to absent: large spermatheca with a peach-shaped ampulla with a very short, stout stalk or almost no stalk, and a diverticulum with a small, oval-shaped seminal chamber and slender stalk; vestigial spermatheca with a small degenerated ampulla and stalk, and no diverticulum; nodule-like spermatheca with a small, degenerated ampulla without a stalk or diverticulum.

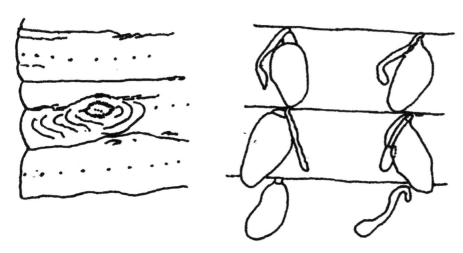

FIG. 130 *Amynthas shinanmontis* (after Tsai & Shen, 2007).

Color: Preserved specimens greyish brown on head and dorsum, dark grey on middorsum, and light brown on ventrum. Clitellum light to dark grey.
Distribution in China: Taiwan (Nantou).
Deposition of types: Taiwan Endemic Species Research Institute, Chichi, Nantou, Taiwan.
Remarks: *Amynthas shinanmontis* has a continuous spectrum of degeneration in spermathecae and male reproductive organs (prostate gland, prostatic ducts, and seminal vesicles). Most of the specimens examined have certain levels of degeneration in at least 1 of the organs. For spermathecae, the degeneration is faster for those in VII than VII and VI, and those in VI are the final ones to be lost.

A. shinanmontis somewhat resembles *A. gracilis* (Kinberg, 1867), but is easily distinguishable by genital papillae at the male pores absent in *Amynthas shinanmontis*, and 1 to several postsetal papillae arranged in 1 or 2 oblique rows medial to each of the male pores in *A. gracilis*. Also, the former is endemic while the latter is peregrine.

131. *Amynthas tessellates* Shen, Tsai & Tsai, 2002

Amynthas tessellates tessellates Shen, Tsai & Tsai, 2002. *The Raffles Bulletin of Zoology*. 50(1):2–7.
Amynthas tessellates tessellates Chang, Shen & Chen, 2009. *Earthworm Fauna of Taiwan*. 92–93.
Amynthas tessellates Xu & Xiao, 2011. *Terrestrial Earthworms of China*. 188–189.

External Characters

Length 52–103 mm, clitellum width 2.5–3.4 mm, segment number 66–114. Prostomium epilobous. First dorsal pore 11/12. Clitellum XIV–XVI, annular, 1.7–2.9 mm long, setae absent. Setae 30–40 (VII), 35–48 (XX), 5–11 between male pores. Male pores paired in XVIII,

porophore simple, round, papilla-like, surrounded by 2–3 incomplete circular folds, about 0.28 circumference apart ventrally. Genital papillae present, patched, medioventral, presetal papillae 11–21 in XVII, 14–25 in XVIII, 10–25 in XIX, 0–16 in XX, 0–3 in XXI, postsetal papillae 1–2 in XIX, 1 in XX, with shape and arrangement similar to those in the preclitellar region. Spermathecal pores invisible. Patches of small, round genital papillae present ventromedially, presetal papillae 0–3 in VI 0–13 in VII, 16–24 in VII, 2–19 in IX, postsetal papillae absent or present, if present, 1–3 in VII, 1 in IX. Female pore single, medioventral in XIV.

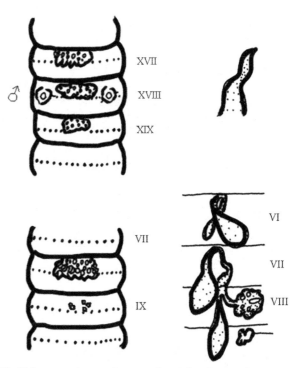

FIG. 131 *Amynthas tessellates tessellates* (after Shen et al., 2002).

Internal Characters

Septa 8/9/10 absent, 5/6/7/8 thickened, 10/11–13/14 greatly thickened. Gizzard in IX–X, round. Intestine from XVI. Dorsal typhlosole from XXVII, with height about one-third of that of intestine. Intestinal caeca paired in XXVII, simple, extending anteriorly to XXIII. Esophageal hearts 3 pairs in XI–XIII. Nephridia tufted in intersegmental spaces, anteriorly to 6/7. Testis sacs 2 pairs in XI. Seminal vesicles paired in XI and XII, with a round dorsal lobe. Prostate glands paired in XVIII, large, follicular, extending anteriorly to XVI and posteriorly to XX, prostatic duct U-shaped with the slender proximal end connecting to prostate gland and the enlarged distal end connecting to male pore. Accessory glands present in presetal and postsetal areas in the spermathecal and male pore regions, round, mostly sessile on body wall, few with very short stalks of about 0.2 mm long, number much less than that of external genital papillae in the corresponding segments. Spermathecae 3 pairs in VI–VII, ampulla oval, about 1.3 mm long, 0.9 mm wide, with a stout stalk of about 0.9 mm long, diverticulum with an oval seminal chamber about 0.6 mm long and a slender stalk of about 1.1 mm long.

> Color: Preserved specimens whitish olive on dorsum, olive-gray on ventrum, dark brown on clitellum.
> Type locality: elevations between 1000–3200 m at Meichi, North Dongyan, Rueyen, Hohuan, and Wulin in the Central Mountain Range of Taiwan.
> Deposition of types: Taiwan Endemic Species Research Institute, Jiji, Nantou County, Taiwan.
> Holotype: coll. no. 1999-24-Shen. Paratypes: same collection data as for holotype.
> Distribution in China: Taiwan (Nantou).
> Etymology: The name "*tessellatus*" refers to the character of the "mosaic formation" of its genital papillae in both preclitellar and postclitellar regions.
> Remarks: *Amynthas tessellates* has patches of genital papillae in VII and usually in VII and IX in the preclitellar region and in XVII, XVIII, and XIX in the postclitellar region. Accessory glands are sessile or mostly sessile including a few with very short stalks.

132. *Amynthas tuberculatus* (Gates, 1935)

Pheretima tuberculata Gates, 1935a. *Smithsonian Mus. Coll.* 93(3):18.
Pheretima tuberculata Chen, 1936. *Contr. Biol. Lab. Sci. Soc. China (Zool.).* 11:302–303.
Pheretima tuberculata Gates, 1939. *Proc. U.S. Natr. Mus.* 85:494–497.
Pheretima tuberculata Chen, 1946. *J. West China. Border Res. Soc. (B).* 16:137.
Amynthas tuberculatus Xu & Xiao, 2011. *Terrestrial Earthworms of China.* 191–192.

External Characters

Length 80–110 mm, width 3–5 mm, segment number 80–110. Prostomium 1/2 epilobous. First dorsal pore 10/11 or 12/13. Clitellum smooth all around, in XIV–XVI, annular, but not fully in 3 segments. Setae rough and more widely spaced on III–IX, ventral break indistinct; $aa = 1.1ab$, dorsal break wider $zz = 2.0yz$; setal number 22–27 (III), 26–33 (VI), 32–38 (VII), 38–49 (XXV); 10–13 (VI), 13–14 (VII) between spermathecal pores, 10–12 between male pores. Male pores 1 pair in XVIII, superficial, on small round porophore, surrounded by 4–5 small papillae. Spermathecal pores minute 3 pairs, in 5/6/7/8, superficial and intersegmental, about 1/3 of circumference apart ventrally. Genital papillae very small, paired on ventral side VII, in front of setae, near setae b and c, or occasionally on VII either paired or unpaired. Female pore single in XIV, medioventral.

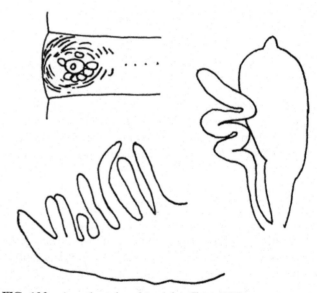

FIG. 132   *Amynthas tuberculatus* (after Chen, 1936).

Internal Characters

Septa 8/9/10 absent, none of the septa are thickly muscular, though 5/6/7/8 and some or all of 10/11//2/13 are strengthened and with muscular fibres. Intestine swelling in XV. Intestinal caeca in XXVII, extending anteriorly to XXIV, manicate, with rather long ventral tooth-like diverticula, about 8 in number, nearly equal in length. First pair hearts small. Testis sacs of X and XI unpaired, not united ventrally. Seminal vesicles are fairly large, filling segments XI and XII, those of a segment in contact transversely over the dorsal blood vessel. Prostate glands with small lobules, posterior branches extending much further back, in XVII–XXIII. Accessory glands stalked. Spermathecal ampulla about 2 mm long, sharply distinguished from its short duct; diverticulum club-shaped, with a muscular stalk and an elongately tubular seminal chamber, the latter loops in a regularly zigzag fashion. Genital marking gland stalked and coelomic, stalks erect in the coelom.

Color: ?
Type locality: Sichuan (Yibin).
Distribution in China: Sichuan (Yibin, Ebian, Mt. Emei).
Deposition of types: Smithsonian Institution.
Remarks: Distinguished from sexthecal species of *Amynthas* by having spermathecal pores in 5/6/7/8 by the manicate intestinal caeca.

133. *Amynthas wangi* Shen, Tsai & Tsai, 2003

Amynthas wulinensis Shen, Tsai & Tsai, 2003. *Zoological Studies.* 42(4), 489–490.
Amynthas wulinensis Chang, Shen & Chen, 2009. *Earthworm Fauna of Taiwan.* 100–101.
Amynthas wulinensis Xu & Xiao, 2011. *Terrestrial Earthworms of China.* 195–196.

External Characters

Length 62 mm (amputated), clitellum width 3.4 mm, segment number 70 (amputated). Number of incomplete annuli (secondary segmentation) per segment 2–3 in VII–XIII and XVII. Prostomium epilobous. First dorsal pore 11/12. Clitellum XIV–XVI, annular, 2.7 mm long, setae absent, dorsal pore absent. Setae 34 (VII), 36 (XX), 10 between male pores in VII. Male pores paired in XVIII, about 0.28 circumference apart ventrally, male aperture minute, in a transversely oval porophore of about 0.5 mm in diameter, surrounded by a few slight oval folds; postsetal genital papillae present in a horizontal row in right and left lateroventral regions of XVII, each row with 4 papillae adjacent to setal line, each papilla round, center concave, 0.2–0.3 mm in diameter, breadth of the 4 papillae about 0.9 mm. Spermathecal pores 4 pairs in 5/6/7/8, about 0.3 circumference apart ventrally, each surrounded by a genital papilla anteriorly and 1–2 papillae posteriorly, each papilla round, center concave, about 0.2 mm in diameter. Female pore single, medioventral in XIV.

Internal Characters

Septa 8/9/10 absent, 5/6–7/8 and 10/11–12/13 thickened. Gizzard round in IX and X. Intestine from XV. Intestinal caeca paired in XXVII, simple, surface slightly wrinkled, short, extending anteriorly to entire segment of XXIV. Esophageal hearts in XI–XIII. Testis sacs 2 pairs in X and XI, small, round. Seminal vesicles paired in XI and XII, large. Prostate glands paired in XVIII, extending anteriorly to XVI and posteriorly to XXI, lobed, prostatic duct C-shaped, distal end enlarged. Accessory glands stalked, a large gland or paired small glands associated with each of the genital papillae, stalk about 0.2 mm and head

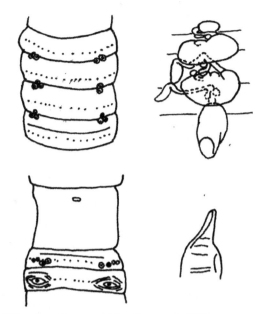

FIG. 133   *Amynthas wangi* (after Shen et al., 2003).

0.1–0.2 mm long. Spermathecae 3 pairs in VI–VII, ampulla large, elongated oval, 2.2–2.9 mm long, 1.2–1.5 mm wide, duct 0.9–1.4 mm long, diverticulum stalk slender, 0.8–1.1 mm long, seminal chamber oval, 0.8–1.1 mm long. Accessory glands stalked, 0.3–0.7 mm long, corresponding to each of the papillae around spermathecal pores.

Color: Preserved specimens whitish olive on dorsum, whitish gray on ventrum, light brown around clitellum.
Type locality: Rueyen Creek Nature Reserve (elevation 2300 m), Nantou County, Taiwan.
Deposition of types: Taiwan Endemic Species Research Institute, Jiji, Nantou County, Taiwan.
Holotype: coll. no. 1999-24-Shen.
Etymology: The name *"wangi"* was given to this species in honor of Dr. Yuhsi Wang, late professor and head of the Department of Zoology, National Taiwan University between 1955 and 1973, who made great contributions to the early taxonomy of the millipedes and centipedes of Taiwan.
Distribution in China: Taiwan (Nantou).
Remarks: The genital papillae arrangements around the spermathecal pores and in XVII, and the paired accessory glands in XVII are unique characters that are easily distinguishable from other species in the *gracilis* species group.

illotus Group

Diagnosis: *Amynthas* without spermathecal pores.
Global distribution: Oriental realm, ? Australasian realm and ? introduced into the Oceanian realm. There are 13 species from China.

TABLE 17   Key to Species of the *illotus* Group of the Genus *Amynthas* From China

1. Without papillae front and behind clitellum 1
   With papillae front and/or behind clitellum 3

2. Male pores are minute, transverse slits or depression on the posterior ends of the genital marking. Genital markings are elongate, longitudinal, raised, whitish areas, each area sharply delimited by a definite circumferential furrow, the ends of the markings rounded, the median margins straight, the lateral margins concave, the posterior ends slightly wider than the anterior ends, both ends wider than middle portion of the marking on XVII. Each marking extends anteroposteriorly from 18/19 to about the region of 15/16 and is about 8 intersetal distances wide transversely on XVIII. Along the middle of each marking there is an anteroposterior groove which on the posterior portion of the marking turns laterally to pass into the male pore. Intersegmental furrows 17/18 and 18/19 are lacking ventrally but 17/18 seem to be continued across the genital marking .... *Amynthas tenellulus*
   Genital markings are not elongate, longitudinal, raised areas 5

3. Male pores at the centers of the genital markings. The genital markings are paired on XVIII, each marking a slightly protuberant, circular area extending anteroposteriorly just across 17/18 and 18/19. A central portion of the marking, transversely oval in shape, is delimited from the rest of the marking by a slight furrow .... *Amynthas illotus*
   Genital markings are not a protuberant, circular area 4

4. Male pores in a glandular patch, roundish, extending to neighboring segments; male pore appearing in center .... *Amynthas asacceus*
   Male pores on an oval or round papilla-like or disc-like porophore, surrounded by 3–5 skin folds, often with a horizontal groove anteriorly, resembling an eye .... *Amynthas shinanmontis*

5. Without papillae in front of clitellum, with papillae behind clitellum 6
   With papillae front and behind clitellum 7

6. Male pores small, oval, disc-like porophore, with a small round or slightly oval-shaped genital papilla with a depressed center, that is equal in size to or slightly smaller than that of the male porophore, on setal line immediately medial to porophore, and then surrounded by 3–4 skin folds .... *Amynthas chilanensis*
   Male pores, Porophores small, papilla-like or indistinct, but each with 2 genital papillae, 1 anterior and 1 posterior, or with 3 papillae, 1 anterior, 1 posterior and 1 lateral, and then surrounded by 4–6 circular folds, each papilla round, center flat or slightly convex, surrounded by a circular fold. A horizontal row of genital papillae, 1–3 (usually 2) in number, just medial to each of the male pore regions, immediately posterior to setal line, each papilla similar in size and structure to those associated with the male pores .... *Amynthas hohuanmontis*

7. The papillae between V–IX and between XV–XXI 8
   The papillae between VI–IX and between XVII–XIX 9

8. Male pores as a cone-shaped porophore, with a penis-like structure on its tip, probably acting as copulatory organ. Genital papillae in this region usually on XV–XVIII or XIX, more often on XVII, rarely totally absent, small and numerous, up to 14 in a transverse row, frequently about 6–10, arranged presetally and postsetally; in small forms only 2 large ones present presetally on XVII in line with male porophores; in the type none on XVIII, few on XVII, some on XV and XVI; similar small and numerous papillae on VII and VII .... *Amynthas varians*

TABLE 17   Key to Species of the *illotus* Group of the Genus *Amynthas* From China—cont'd

Male pores porophore round, circular tubercle on setal annulus. Genital papillae arranged in a longitudinal series, medioventral, 3–9 papillae from XVII–XXI, similar to those in the preclitellar region in size, shape, position and arrangement. Genital papillae medioventral in V–IX, in a longitudinal series like a chain, number and position highly variable: 3 presetal papillae in VI–VII and 2 postsetal papillae in VII and VII or 0–9 papillae in V–IX. Generally, 1 presetal papilla and/or 1 postsetal papilla in a segment, each occupying the entire width of an annulus adjacent to intersegmental furrows. Occasionally, 2 papillae joined closely in an annulus. Each papilla round with a concave center, associated with a round, stalked accessory gland internally .... *Amynthas catenus*

9. Male pores on an elongate transverse ridge, the center part of the ridge slightly extending anteroposteriorly. Genital papillae large, widely paired in postsetal XVII and often in presetal XIX, and occasionally an additional pair in postsetal IX. Number and arrangement of papillae variable in different individuals; rarely widely paired papillae in presetal VI and VII .... *Amynthas amplipapillatus*
   The papillae between VI–IX and XVIII or XIX 10

10. Genital papillae in XIX, presetal, each slightly medial to male pore: sometimes examined, or 1 pair, or 1 on left, or 1 on right, or absent. The structure of the papillae similar to those the preclitellar region. Genital papillae presetal, closely in 2 parallel and longitudinal row on midventrum of VI–IX with numbers variable among specimens and segments. Rarely examined, in VI 1 on right, or 1 on left, or paired; in VII 1 on right, or paired; in IX paired; in X paired. Each papilla round, center concave, adjacent to setal line .... *Amynthas bilineatus*
    The papillae between VII–IX and XVIII 11

11. Male pores on a rectangular glandular patch, surrounded by several concentric ridges, minute papillae found among the ridges, a small papilla found presetally on VII .... *Amynthas vividus*
    The papillae in VII, or VII and IX and XVIII 12

12. Genital papillae present beside each male pore, most commonly 1 medial and 1 anterior; number variable. 3 postsetal genital papillae, arranged in transverse row in VII, small. Tubercle-like, number variable, or 0, or 1, or 2, or 3 .... *Amynthas obsoletus*
    Genital papillae paired (or frequently with an unpaired papilla placed ventromedially) posteriorly on VII, not infrequently only 1 on either side before setal line, and occasionally on posterior part of IX .... *Amynthas sheni*

134. *Amynthas amplipapillatus* Shen, 2010

Amynthas amplipapillatus Shen, 2012. *J. Nat. Hist.* 46(37–38):2274–2277.

External Characters

Small to medium; Length 67–101 mm, width 2.97–3.9 mm (clitellum), segment number 90–129. Prostomium epilobous. First dorsal pore in 11/12. Clitellum XIV–XVI, dorsal pores absent, 0–12 setae on ventral XVI, 1.84–3.45 mm in length. Setae number 44–61 (VII), 50–63 (XX), 9–11 between male pores. Male pores paired in XVIII, 0.24–0.3 body circumference apart ventrally,

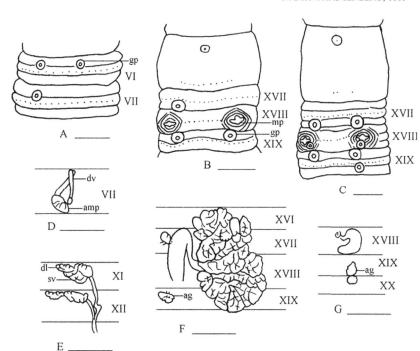

FIG. 134 *Amynthas amplipapillatus* (after Shen, 2012).

each pore on an elongate transverse ridge, the center part of the ridge slightly extending anteroposteriorly. Genital papillae large, widely paired in postsetal XVII and often in presetal XIX, and occasionally an additional pair in postsetal XIX. Number and arrangement of papillae variable in different individuals, 1–3 pairs in postsetal XVII, presetal XIX or postsetal XIX. Each papilla round, 0.4–0.8 mm in diameter with concave center, slightly medial to the male pore. Rarely an additional pair of smaller genital papillae closely medial to the male porophore and posterior to the setal line in XVIII, each papilla round 0.35–0.45 mm in diameter with concave center. Spermathecal pores absent. Genital papillae absent in the preclitellar region, but rarely widely paired papillae in presetal VI and VII. Female pore single in XIV, midventral.

Internal Characters

Septa 8/9/10 absent, 5/6/7/8 and 10/11–13/14 thick. Nephridial tufts thick on anterior faces of 5/6/7 septa. Gizzard large, round in VII–X. Intestine enlarged from XVI. Intestinal caeca paired in XXVII, extending anteriorly to XXII–1/2XXV, each simple, slightly folded, the end straight or bent. Hearts in XI–XIII. Holandric, testes small or large, 2 pairs in ventrally joined sacs X and XI. Vas efferens connected in XII on each side to form a vas deferens. Seminal vesicles 2 pairs in XI and XII, each small, about one-half to two-thirds of the segmental compartment, with a prominent and round or elongated oval dorsal lobe. Prostate glands variable in size from normal in XVI–XIX to absent, lobed with follicular surface.

Prostatic duct U-shaped, distal end enlarged from XVII–XVIII, or short, C-shaped, in XVIII. Accessory gland pad-like or mushroom-like, sessile or short-stalked 0.32–0.45 mm wide, corresponding to external genital papilla. Spermathecae absent, but a single spermatheca in left VII. Spermatheca small, with an oval ampulla of 0.68 mm long and 0.4 mm wide, and a long spermathecal stalk 0.7 mm in length. Diverticulum with a small, round seminal chamber and a slender stalk of 0.7 mm in length, the same length to spermathecal stalk.

Color: Preserved specimens dark to grayish brown on dorsum and clitellum, light brown on ventrum.
Distribution in China: Taiwan (Hualien).
Deposition of types: Taiwan Endemic Species Research Institute, Jiji, Nantou, Taiwan.
Habitat: The specimens were collected at an elevation of 931 m, along the Shakadang Forest Road; and at an elevation of 675 m along the Wanrung Forest Road, Hualien County.
Remarks: *Amynthas amplipapillatus* is endemic to eastern Taiwan. It has large genital papillae widely paired in postsetal XVII and often in presetal XIX, and occasionally an additional pair in postsetal XIX. Similar papillar arrangements can also be found for *A. mutus* (Chen, 1938) and *A. tetrapapillatus* Quan and Zhong, 1989 from Hainan Island, *A. fluxus* (Chen, 1946) from Sichuan, China, *A. micronarius* (Goto and Hatai, 1898) from Japan, and *A. obtusus* (Ohfuchi, 1957) from the Ryukyu Islands. Both *A. Micronarius* and *A. obtusus* are *octothecate* with setal numbers 26–39 in VII and 33–51 in XX for the

former (Ohfuchi 1937), and 33–37 in VII and 43–50 in XX for the latter (Ohfuchi 1957). *A. fluxus* is sexthecate with 3 pairs of spermathecal pores in 5/6/7/8 and 73 and 70 setae in VII and XXV, respectively (Chen, 1946). *A. tetrapapillatus* is bithecate with a pair of spermathecal pores near the dorsomedian line in 5/6; it is much larger with body length 152–165 mm long and has much higher setal number (93–107 in VII and 86–102 in XX) (Quan and Zhong 1989). *A. amplipapillatus* is easily distinguishable from the above 4 species. As for *A. mutus*, its body size (50–80 mm in length and 85–135 segments) and setal number (41–68 in VII and 32–44 in XXV) are fairly similar to *A. amplipapillatus*. However, *A. mutus* has 1 small papilla in XIX behind each male pore, occasionally another pair in XVII in line with those in XIX, 1 pair of spermathecal pores in dorsolateral 5/6 and hearts in X–XIII (Chen 1938).

135. *Amynthas asacceus* (Chen, 1938)

Pheretima (Pheretima) asaccea Chen, 1938. *Contr. Biol. Lab. Sci. Soc. China (Zool.).* 12(10):382–383.
Amynthas asacceus Sims & Easton 1972. *Biol. J. Soc.* 4:236.
Amynthas asacceus Xu & Xiao, 2011. *Terrestrial Earthworms of China.* 81.

External Characters

Length 35–60 mm, clitellum width 2.0–2.5 mm, segment number 69–90. Prostomium 1/2 epilobous. First dorsal pore 12/13. Clitellum XIV–XVI, annular, setae absent. Setae similarly built, none specially enlarged, both dorsal and ventral breaks alight, $aa = 1.2ab$, $zz = zy$; setae number 38 (III), 37 (VI), 42 (VII), 32 (XVII), 32 (XX), only 1 seta between 2 glandular patches of male pore region, 9 (XVII), 9 (XIX) in line with male pores respectively. 12–19 between male pores. Male pores on XVIII, about 1/3 of circumference apart ventrally, each in a 0.7 mm glandular patch, roundish and dark in color, extending to neighboring segments. More pores appearing in center. No other genital markings. Spermathecal pores absent. No genital markings in this region.

Internal Characters

Septa 8/9/10 absent. 6/7/8 thicker, 5/6 thinner but as thick as 10/11 and 11/12. Gizzard pushed backward in X and 1/2 XI. Intestinal swelling in XVI. Intestinal caeca paired in XXVII, conical shaped, simple, conical-shaped, extending anteriorly to XXV. First pair of hearts in X present. Testis sacs small, in X and XI, both pairs widely separate. Seminal vesicles small, as elongate bands attached to posterior faces of septa 1011 and 11/12. Each

with a rather large dorsal lobe. Prostate well developed, with large lobules, in XVI–XXII; its duct long and coiled. No obvious accessory glands, velvety structures slightly thicker than nephridial tubules found around base of prostatic duct. Ovaries and oviduct-funnels usual in character. Spermatheca absent.

Color: Preserved specimens pale generally. Clitellum reddish.
Distribution in China: Taiwan, Hainan, Sichuan (Leshan).
Remarks: Although they are fully sexually mature, there is no trace of spermatheca. The absence of the structure is probably a character of this species.

136. *Amynthas bilineatus* Tsai & Shen, 2007

Amynthas bilineatus Tsai & Shen, 2007. *J. Nat. Hist.* 41(5–8):366–368.
Amynthas bilineatus Chang, Shen & Chen, 2009. *Earthworm Fauna of Taiwan.* 26–27.
Amynthas bilineatus Xu & Xiao, 2011. *Terrestrial Earthworms of China.* 86–87.

External Characters

Medium earthworms. Length 96–153 mm, width 3.49–5.03 mm (clitellum), segment number 89–104. Number of incomplete annulets (secondary segmentation) 2–3 per segment in VI–XIII, XVII, and XVIII. Prostomium epilobous. First dorsal pore in 11/12. Clitellum XIV–XVI, annular, smooth, setae and dorsal pores absent. Setae number 44–52 (VII), 52–56 (XX); 8 between male pores. Male pores paired in XVIII, lateroventral, 0.23–0.26

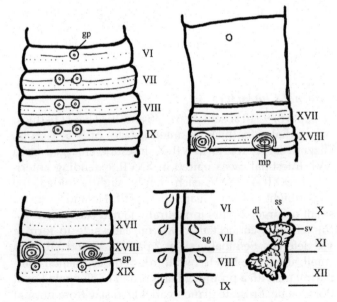

FIG. 135   *Amynthas bilineatus* (after Tsai & Shen, 2007).

circumference apart ventrally, and each pore minute, superficial, on a round or transversely oval porophore surrounded by 3–4 circular folds. Genital papillae in XIX, presetal, each slightly medial to male pore: sometimes examined, or 1 pair, or 1 on left, or 1 on right, or absent. The structure of the papillar similar to those the preclitellar region. Spermathecal pores absent. Genital papillae presetal, closely paired in 2 parallel and longitudinal rows on midventrum of VI–IX with numbers variable among specimens and segments. Rarely examined, in VI or VII 1 on right, or 1 on left, or paired; in IX or x paired. Each papilla round, center concave, 0.25–0.5 mm in diameter, adjacent to setal line. Female pore single in XIV, medioventral.

Internal Characters

Septa 8/9/10 absent, 5/6/7/8 and 10/11/12/13/14 thickened. Gizzard round in IX and X. Intestine enlarged from XVI. Intestinal caeca paired in XXVII, each simple, short, surface slightly wrinkled, extending anteriorly to full XXV or to two-thirds of XXIV. Hearts enlarged from XI–XIII, the second and third hearts enlarged. Holandric; testis sacs two pairs in X and XI, each small and round. Seminal vesicles 2 pairs in XI and XII, small, follicular, surface slightly wrinkled, yellow in color, each with a small oval dorsal lobe. Prostate glands paired in XVIII, wrinkled, extending to XVII and XIX. Prostatic duct U-curved, occupying 2 segments XVII and XVIII, distal end enlarged. Accessory glands in XIX corresponding to external genital papillae, each slightly divided into 2–3 round lobes with stalk 0.24–0.31 mm length.

Spermathecae absent. Accessory glands corresponding to external genital papillae in the preclitellar region; each round, nearly sessile or with stalk to 0.26 mm in length.

Color: Preserved specimens tinted pink in preclitellar region, light brown in the postclitellar region, and dark brown around clitellum.
Distribution in China: Taiwan (Nantou).
Deposition of types: Taiwan Endemic Species Research Institute, Chichi, Nantou, Taiwan.
Habitat: The specimens were collected at an elevation of 1000 m from a hill slope along Dongyan Creek.
Remarks: *Amynthas bilineatus* is closely related to *A. hohuanmontis* of central Taiwan and *A. sheni* (Chen, 1935a, b) of Hong Kong and *A. chilanensis* of northeastern Taiwan. *A. bilineatus* has paired, longitudinal series of ventral genital papillae in VI–IX that are easily distinguishable from a single pair in VII for *sheni*, and none for *chilanensis* and *hohuanmontis*. Also, *bilineatus* has a pair of ventral genital papillae in XIX and none in XVIII, differing from *sheni*, *chilanensis*, and *hohuanmontis* that have no genital papilla in XIX, but do have them in XVIII.

114. *Amynthas catenus* Tsai, Shen & Tsai, 2001 (*v.ant.* 114)

137. *Amynthas chilanensis* Tsai & Tsai, 2007

Amynthas chilanensis Tsai & Tsai, 2007. *J. Nat. Hist.* 41(5–8):362–366.
Amynthas chilanensis Chang, Shen & Chen, 2009. *Earthworm Fauna of Taiwan*. 346–347.
Amynthas chilanensis Xu & Xiao, 2011. *Terrestrial Earthworms of China*. 93–94.

External Characters

Medium earthworms. Length 133–168 mm, width 4.71–4.84 mm (clitellum), segment number 88–116. Prostomium epilobous. First dorsal pore in 11/12. Clitellum XIV–XVI, annular, constricted. Setae minute, 32–35 (VII), 47–50 (XX); 11–12 between male pores. Male pores paired in XVIII, ventrolateral, 0.26–0.32 body circumference apart. Each pore on a small, oval, disc-like porophore, with a small round or slightly oval-shaped genital papilla with a depressed center, that is equal in size to or slightly smaller than that of the male porophore, on setal line immediately medial to porophore, and then surrounded by 3–4 skin folds. Spermathecal pores absent. No genital papillae in the preclitellar region. Female pore single in XIV, medioventral.

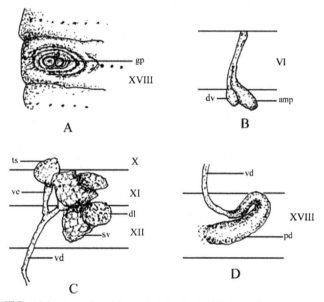

FIG. 136 *Amynthas chilanensis* (after Tsai & Tsai, 2007).

Internal Characters

Septa 8/9/10 absent, 5/6/7/8 and 10/11–13/14 thickened. Nephridial tufts thick on anterior faces of 5/6/7 septa. Gizzard large in IX–X for clitellates but small in X for aclitellate. Intestine enlarged from XV. Intestinal caeca paired in XXVII, each simple with a wide base, extending anteriorly to XXIII–XXIV. Hearts enlarged from XI–XIII. Holandric, testis sacs large, 2 pairs in X and XI. Vas efferens short, thick (swollen), joining in XII or XIII on each side to form a large, straight vas efferens, connecting to prostatic duct in XVIII. Seminal vesicles 2 pairs in XI and XII, small (degenerated), follicular surface, pink in color, each with a large dorsal lobe either reddish brown or whitish in color. Prostate glands absent, but prostatic duct U-curved, distal end enlarged, silver colored. Accessory glands absent. Spermathecae absent, but a single vestigial spermatheca in right VI. The vestigial spermatheca small, with an oval ampulla and a long, straight stalk, and its diverticulum small, with a small oval seminal chamber and short, straight stalk attached in the middle of the spermathecal stalk. Ovaries in XIII, paired, large with follicular surface.

> Color: Preserved specimens purplish brown on dorsum and light brown on ventrum. Clitellum dark chocolate.
> Distribution in China: Taiwan (Chilan, Ilan County).
> Deposition of types: Taiwan Endemic Species Research Institute, Chichi, Nantou, Taiwan.
> Habitat: Collected along the Forest Road at an elevation of 1325 m in Chilan, Ilan County.
> Remarks: *Amynthas chilanensis* is closely related to *A. sheni* from Hong Kong (Chen 1935). However, *A. sheni* are typical athecate A morph with well-developed male terminalia, prostate glands, preclitellar genital papillae, and large accessory glands.

138. *Amynthas illotus* (Gates, 1932)

Pheretima illotus Gates, 1932. *Rec. Ind. Mus.* 34(4):397.
Pheretima illota Gates, 1972. *Trans. Amer. Philos. Soc.* 62(7):196.
Amynthas illotus Sims & Easton, 1972. *Biol. J. Soc.* 4:236.
Amynthas illotus Zhong & Qiu, 1992. *Guizhou Science.* 10(4):40.
Amynthas illotus Xu & Xiao, 2011. *Terrestrial Earthworms of China.* 128–129.

External Characters

Length 149–160 mm, width 5–6 mm, segment number 110–120. Prostomium ?. First dorsal pore 12/13. Clitellum XIV–XVI, annular; setae, intersegmental furrows and dorsal pores absent. Setae begin on segment II but are present only dorsally and dorsolaterally on that segment. Setal circles are without a midventral break but there may be a slight middorsal break of variable width. Setae are small and crowded; setae number 16–17 (II), 18–20 (XVII), 18–19 (XIX), 94–103 (XX); Male pores are minute, on XVIII, at the centers of the genital markings and about 2 mm apart. Genital markings are paired on XVIII, each marking a slightly protuberant, circular area extending anteroposteriorly just across 17/18 and 18/19. A central portion of the marking, transversely oval in shape, delimited from the rest of the marking by a slight furrow. The markings are 11–13 intersetal distances wide transversely and are more closely approximated midventrally than the markings of other species from the same locality. Spermathecal pore absent. Female pore single in XIV, medioventral.

Internal Characters

Septa 9/10 absent, 5/6/7/8 muscular, 8/9 present as a ventral rudiment only, 10/11/12 muscular, 12/13/14 and several succeeding septa slightly strengthened and translucent. Gizzard ?. Intestine enlarged from XV. Caeca simple, in XXVII, extending anteriorly to XXII. Single commissure of IX on the right side. Last pair of hearts in XIII. A single testis sac on the anterior face of 10/11 and a similar sac in XI. Seminal vesicles small, vertical bodies deep down in segments XI and XII, each vesicle with a conspicuous indentation or cleft in the dorsal margin within which is seated a large spherical primary ampulla. Prostates glands extend through segments XVII–XXI (and cleft into 6–10 major lobes). Prostatic duct on the parietes, bent into a shape like a question mark, 3–4 mm in length, of about the same thickness throughout. Spermathecae absent (nor traces thereof buried in the parietes).

> Color: Preserved specimens dorsum very light brownish, Clitellum reddish.
> Type locality: Myanmar (To Noi).
> Distribution in China: Yunnan (Lancang).
> Deposition of types: Indian Museum.

139. *Amynthas obsoletus* Qiu & Sun, 2013

Amynthas obsoletus Qiu & Sun, 2013. *J. Nat. Hist.* 47(17–20):1144–1147.

External Characters

Length 118–143 mm, width 3.8–5.2 mm, segment number 80–116. Body cylindrical in cross section, gradually tapered towards head and tail. Prostomium 1/2 epilobous. First dorsal pore 11/12. Clitellum XIV–XVI, annular, swollen, smooth; setae and dorsal pore invisible, but traces of dorsal pore present within clitellum. Setae thick and long, especially before; setae number: 18–24 (III), 22–24 (V), 36 (VII), 42–56 (XX), 46–60 (XXV); 12–14 between male pores. Male pores paired in XVIII, 0.33 body circumference apart ventrally; each on the center of a raised elliptical pad, conical, with 2–3 circular folds. Genital papillae present beside each male pore, most commonly 1 medial and 1 anterior; number variable.

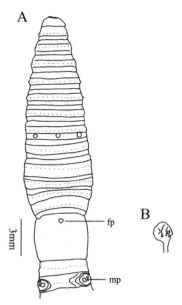

FIG. 137   *Amynthas obsoletus* (after Sun et al., 2013).

Spermathecal pores absent. 3 postsetal genital papillae, arranged in transverse row in VII, small. Tubercle-like, number variable, or 0, or 1, or 2, or 3. Female pore single in XIV, medioventral.

Internal Characters

Septa 8/9/10 absent, 5/6/7 thick and muscular, 10/11–13/14 slightly thickened. Dorsal blood vessel single, continuous into pharynx; esophageal hearts enlarged from III. Gizzard in VII–X, ball-shaped. Intestine enlarged from XVI. Caeca simple, in XXVII, extending forward to 1/3 XXIII, 3 incisions on both dorsal and ventral margins. Holandric, testis sacs 2 pairs, ventral in X, XI, separated from each other; seminal vesicles paired in each of XI and XII, undeveloped. Prostates in XVII–XX, large, composed of 3 parts, finger-shaped lobe in middle and block-like lobe in upper and lower, prostatic duct U-shaped, distal end appreciably enlarged. No accessory glands were seen near the prostatic ducts. Spermathecae absent; a stalked heart-shaped accessory gland with coarse surface on the right of VII.

Color: Preserved specimens grayish on dorsum before clitellum, changing from grayish to brownish on dorsum after clitellum, no pigment or light tawny on ventrum, middorsal line distinctly colored compared with the rest of the dorsal surface. Clitellum dark red.
Type locality: Hainan (Mt Diaolio).
Distribution in China: Hainan (Mt Diaolio).
Deposition of types: Shanghai Natural Museum.
Habitat: The specimens were collected at an elevation of 915 m from cinnamon sandy soil under a meadow near a pond on Mt Diaoluo, Hainan Province.
Remarks: *Amynthas obsoletus* seems most closely related to *Amynthas hohuanmontis* Tsai, 2002 (Tsai et al., 2002). Both species have no spermathecae and are similar in their pigment, setae number, and male pore area. In contrast, *A. obsoletus* is bigger than *A. hohuanmontis*, with dorsal pores beginning at 11/12 not 12/13, and distance between male pores is greater than in *A. hohuanmontis*. Spermathecal pore region papillae are absent in *A. hohuanmontis*. Prostate glands of *A. obsoletus* are normal and large, whereas those of *A. hohuanmontis* are vestigial or absent. *A. obsoletus* has no accessory glands in the prostate gland region, whereas *A. hohuanmontis* has round, stalked ones. Another related athecate earthworm in Hainan is *A. asacceus* (Chen, 1938). They share some similar characters: distance between 2 male pores; normal and developed prostates (Chen 1938), and no accessory glands near prostates. However, the differences are obvious. For example, the body size of *A. obsoletus* is larger than *A. asacceus*; the pigment of *A. obsoletus* is greyish or brownish on dorsum and light tawny on ventrum, but pale in *A. asacceus*; some genital papillae are present in *A. obsoletus*, but none in *A. asacceus*; and the male porophore in *A. obsoletus* is smaller than that of *A. asacceus*.

140. *Amynthas sheni* (Chen, 1935)

Pheretima sheni Chen, 1935b. *Bull. Fan. Mem. Inst. Biol.* 6:38–42.
Amynthas sheni Xu & Xiao, 2011. *Terrestrial Earthworms of China.* 179.

External Characters

Length 120–160 mm, width 5–7 mm, segment number 108–118. Prostomium 1/2 epilobous. First dorsal pore in 11/12. Clitellum XIV–XVI, annular, setae absent, dorsal pore not clearly visible, no intersegmental grooves. Setae stout on anterior end II–VII, notably enlarged and widely spaced on ventral side of V and VI; setae number 20–23 (III), 22–24 (VI), 36–42 (VII), 44–51 (XII), 45–56 (XXV); 11–15 between male pores. Male pores 1 pair, on ventrolateral sides of XVIII, about 1/3 of circumference apart ventrally, each on large ovoid papilla, usually associated with a single papilla medial to it and slightly behind setal line, surrounded by a few circular ridges. Spermathecal pores absent. Genital papillae paired (or frequently with an unpaired papilla placed ventromedially) posteriorly on VII, not infrequently only 1 on either side before setal line, and occasionally on posterior part of IX. Female pore single in XIV, medioventral.

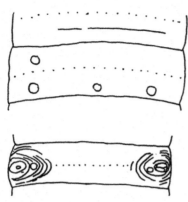

FIG. 138   *Amynthas sheni* (after Chen, 1935b).

Internal Characters

Septa 9/10 absent, 8/9 exceedingly thin, 5/6/7/8, 10/11 equally thickened; 7/8, 11/12/13 also thickened. Gizzard large, in IX–X. Intestine enlarged from XVI. Intestinal caeca simple, paired in XXVII, extending anteriorly to XXII, elongate and cylindrical, smooth on both sides. Hearts 4 pairs, first pair in X. Testis sacs 2 pairs in X and XI, first pair roundish, narrowly connected on median side in a thick V shape; second pair similar in size but more broadly connected. Seminal vesicles in XI and XII, usually very small, each with a distinct dorsal lobe. Small pseudovesicles on 12/13, not 13/14. Prostate glands well developed, very large, usually meeting on dorsal side, glandular subdivided into transverse lobes and glandular on surface, in XVI–XX; each with a rather long duct, U-curved, proximal limb more slender, distal limb fairly stout. Accessory glands composite, with a short but thick stalk. Spermathecae absent.

Color: Preserved specimens brownish or dark chestnut on dorsal side, purplish chocolate on anterior dorsum, setal circles pale or whitish; pale ventrally. Clitellum brownish or dark chocolate.
Type locality: Hong Kong.
Distribution in China: Hubei, Hong Kong, Sichuan (Mt. Emei).
Deposition of types: Mus. Fan. Inst. Biol. Beijing.
Remarks: This species is very similar to *Amynthas pingi*. However, there are several points which may distinguish it specifically from the latter:

(1) First dorsal pore in 12/13 in *A. pingi* but invariably in 11/12 in the present form.
(2) Dorsal process of prostomium much wider in *A. pingi*, wider than segmental length of II, while in the present form it is equal.
(3) Aspect of male pore region different in the 2 species; in *A. pingi* pore papilla similar in shape and size to other papillae, in the present form pore tubercle simply a swelling of skin, much larger than a genital

papilla which is never large in size, nor widely placed, nor more than 1 on each side.
(4) Genital papillae on ventral side of spermathecal region more frequently in front of setae in *A. pingi* but behind setae in the present form. Character of papilla also different: in *A. pingi* larger, cup-shaped and highly glandular in center, but in the present form usually ampulla-like and small in size.
(5) First pair of hearts absent in *A. pingi* but normally present in the present form.
(6) Lymph glands on side of dorsal vessel saccular, large, and beginning around XVI in *A. pingi*, but branched and beginning around caecal region in the present form.

130. *Amynthas shinanmontis* Tsai & Shen, 2007 (*v.ant.* 130)

141. *Amynthas tenellulus* (Gates, 1932)

Pheretima tenellula Gates, 1932. *Rec. Ind. Mus.* 34(4):398–401.
Amynthas tenellulus Sims & Easton, 1972. *Biol. J. Soc.* 4:237.
Amynthas tenellulus Xu & Xiao, 2011. *Terrestrial Earthworms of China.* 187–188.

External Characters

Length 77 mm, width 1.5–3.0 mm, segment number 114–121. Prostomium ?. First dorsal pore 13/14 but there is a definitely pore-like marking in 12/13. Clitellum in XIV–XVI, annular, intersegmental furrows, dorsal pores and setae absent; a midventral region of the body between the genital markings on XVII and XVIII and anteriorly on XIX has exactly the same color and appearance as the clitellum, as if the clitellar glandularity were continued posteriorly as a narrow strip between the genital markings onto XIX where the region of special glandularity is slightly widened to extend laterally on each side just behind the genital markings. The posterior and lateral boundaries on XIX of this modified epidermal region are not sharply delineated. Setae begin on II, without middorsal or midventral breaks. Setae lacking ventrally on XVII and there is a break in the setal circle of XIX just behind each of the genital marking. There are 3 male setae on XVIII, 8 male setae on XIX. Male pores minute, transverse slits or depression on the posterior ends of the genital marking. Genital markings are elongate, longitudinal, raised, whitish areas, each area sharply delimited by a definite circumferential furrow, the ends of the markings rounded, the median margins straight, the lateral margins concave, the posterior ends

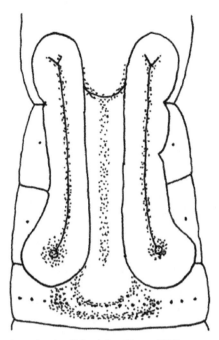

FIG. 139 *Amynthas tenellulus* (after Gates, 1932).

slightly wider than the anterior ends, both ends wider than middle portion of the marking on XVII. Each marking extends anteroposteriorly from 18/19 to about the region of 15/16 and is about 8 intersetal distances wide transversely on XVIII. Along the middle of each marking there is an anteroposterior groove, which on the posterior portion of the marking turns laterally to pass into the male pore. Intersegmental furrows 17/18 and 18/19 are lacking ventrally but 17/18 continued across the genital marking. Spermathecal pores absent. Female pore single or 1 pair in XIV, medioventral.

Internal Characters

Septa 9/10/11 absent, 5/6/7/8 muscular, 11/12 thin, 12/13–15/16 slightly strengthened but membranous. Gizzard in VII–X, ball-shaped. Intestine swelling in XXI; caeca simple, in XXVII, extending anteriorly to XVIII. Last pair of hearts is in XIII. Seminal vesicles of XII much larger than the vesicles of XI and covering the dorsal blood vessel in XII and pushing 12/13 back into contact with 13/14 and at the same time displacing 11/12 anteriorly; anterior of seminal vesicle is on the anterior face of 11/12. Prostates large, extend through segments XVII–XXIII; prostatic duct long and extending though XVIII–XX or XXII; each duct bent into a hairpin loop, ental limb of the loop slender, ectal limb much thicker. Spermathecae absent.

Color: Preserved specimens unpigmented. Clitellum bright yellowish, or reddish yellow, or orange. Global distribution: Burma (Kwa Yeh).

Distribution in China: Yunnan (Lancang).
Type locality: Burma (Kwa Yeh).
Deposition of types: Indian Museum.

142. *Amynthas varians* (Chen, 1938)

Pheretima (Pheretima) varians Chen, 1938. *Contr. Biol. Lab. Sci. Soc. China (Zool.).* 12(10): 385–389.
Amynthas varians Sims & Easton, 1972. *Biol. J. Soc.* 4:236.
Amynthas varians Xu & Xiao, 2011. *Terrestrial Earthworms of China.* 194–195.

External Characters

Length 15–130 mm, width 1–4 mm, segment number 78–148. Prostomium 1/3 epilobous, cut off behind, tongue open behind. First dorsal pore 12/13. Clitellum XIV–XVI, annular, either smooth or irregularly glandular on ventral side, ventral setae visible in most cases, especially when the papillae are present. Setae all fine and dense, evenly distributed; those on preclitellar segments more numerous; both dorsal and ventral breaks slight; $aa = 1.1–2ab$, $zz = 1.5yz$ behind clitellum; setae number 45–76 (III), 52–90 (VI), 54–96 (VII), 34–42 (IX), 40–54 (XXV); 21–40 (VI) between spermathecal pores, 8–16 between male pores. Male pores 1 pair in XVIII, about 1/2 of circumference apart ventrally, each as a cone-shaped porophore, with a penis-like structure on its tip, probably acting as copulatory organ; penis muscular, about 0.5 mm long. Genital papillae usually on XV–XVIII or XIX, more often on XVII, rarely totally absent, small and numerous, up to 14 in a transverse row, frequently about 6–10, arranged presetally and postsetally; in small forms only 2 large ones present presetally on XVII in line with male porophores; in the type none on XVIII, few on XVII, some on XV and XVI. Spermathecal pores 1 pair in 5/6, or sometimes 2 pairs in 5/6/7, either irregularly present or totally missing; a little less than half of circumference apart ventrally; each as a slit, intersegmental, with a semilunar flap on its posterior side, which is sometimes very conspicuous, often inconspicuous; similar small and numerous papillae on VII and VII.

Internal Characters

Septa 8/9/10 absent, 5/6/7/8 muscular and very thick, 10/11/12 less thickened. Gizzard in IX and X, round, not shining on surface. Intestine swelling in XVI. Caeca simple, in XXVII, extending anteriorly to XXV. First pair of hearts in X much smaller, concealed in testis sacs. Testis sacs in X and XI, first pair narrowly (not communicated) connected ventrally, but well distended and united dorsally, second pair similarly shaped. Seminal vesicles very small, as large as vesicle, first pair enclosed in second pair of testis sacs, each with a dorsal lobe. Sperm duct of each side join its prostatic duct when

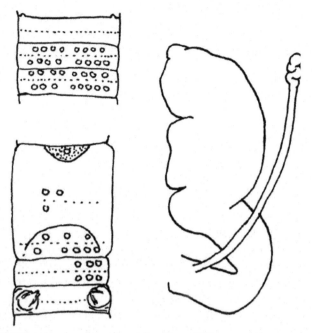

FIG. 140   *Amynthas varians* (after Chen, 1938).

male pore distinct. However, if the latter is absent, sperm duct becoming swollen in segment XV. Prostate gland very large, coarsely lobate, in XVI–XX, its duct long, U-curved. Prostate and its duct totally absent in many cases even if there is a male porophore. These naturally disappearing if there is no male pore. Accessory glands in small or large patches, several associated with each papilla, gland portion composed of small lobules enclosed within thin membrane, ducts cord-like, fairly long. Spermathecae vestigial in most cases, rather small. Main pouch about 1 mm long; ampulla heart-shaped, with an equally long duct. Diverticulum tubular, or with ectal 2/3 dilated. More developed in some of the smallest specimens, main pouch about 2 mm long. In a larger specimen with external apertures perfectly developed, spermathecae vestigial without distinction between diverticulum and ampulla. Diverticulum frequently vestigial. Accessory glands similar to prostate region.

Color: Preserved specimens grayish generally. Clitellum brownish.
Distribution in China: Hainan (Baopeng, Shamojue, Tengqiao), Yunnan (Lancang).
Remarks: This is an extremely variable species. Many of the characters are found to be inconsistent among the species observed. The penis-like structure on the terminal portion of each male genital tract appears to be a good diagnostic feature. Porophore variable in number and position: totally absent, or present only on 1 side. The number of spermathecal pores variable: 1 pair in 5/6, or 2 pairs in 5/6/7, or absent. The nature of

the genital papillae extremely variable with numerous small ones, or only 2 large ones, or no such papillae at all. The size of the body differs to extremes.

The small and large forms referred to may be considered as 2 distinct varieties. The 2 forms have several diagnostic features in common. These are (1) the presence of setae on the first segment, (2) the character of the male porophore and penis-like structure, (3) the position of the spermathecal pores, and (4) the enlarged testis sacs which enclose both the seminal vesicle and heart.

143. *Amynthas vividus* (Chen, 1946)

Pheretima vivida Chen, 1946. *J. West China. Border Res. Soc. (B).* 16:123–124, 140, 158.
Amynthas vividus Sims & Easton, 1972. *Biol. J. Linn. Soc.* 4:237.
Amynthas vividus Xu & Xiao, 2011. *Terrestrial Earthworms of China.* 195.

External Characters

Medium-sized worm. Length 80 mm, width 4–8 mm, segment number 78. Annulation not distinctly shown. Prostomium 1/2 epilobous. First dorsal pore 11/12. Clitellum XIV–XVI, annular, smooth, no other markings. Setae evenly distributed, none enlarged or modified, slightly closer on ventral side; Setae number 26–28 (III), 28–35 (VI), 34–42 (IX), 30–40 (XXV); 10 between male pores; $aa = 1.1ab$, $zz = 1.2 zy$. Male pores paired in XVIII, about 1/2 of circumference apart ventrally, each on a

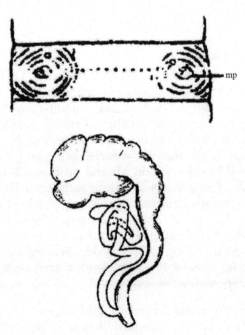

FIG. 141   *Amynthas vividus* (after Chen, 1946).

rectangular glandular patch, surrounded by several concentric ridges, minute papillae found among the ridges. Spermathecal pores usually absent, or 2 pairs in 6/7/8, about 2/7 of dorsal circumference apart, a small papilla found presetally on VII. No other marking. Female pore single in XIV, medioventral.

Internal Characters

Septa 9/10 absent but traceable ventrally, 8/9 membranous, all rest thin, scarcely musculated, 10/11–13/14 as thin as anterior ones. Gizzard elongate, barrel-shaped, in IX and X. Intestine enlarged from XV. Caeca simple, elongate, tooth-edged on both dorsal and ventral sides, in XXVII, extending anteriorly to XXIII. Vascular loops in IX asymmetrically developed, those in X not observed, in XI well developed. Testis sacs also developed, anterior pair widely separated, posterior pair less so, communicated (?). Seminal vesicles well developed, second pair extending to XIII, each with a distinct dorsal lobe. Prostate gland large, in XVI–XXI, its duct S-curved, enlarged at middle, soft whitish connective tissue found around its base. Accessory glands sessile. Spermathecae absent or 2 pairs in VII and VII, or lacking posterior right one; ampulla as irregular sac with duct twice as long; diverticulum present only in the anterior pair, ental end coiled, longer than main pouch if extended, seminal chamber not marked. Accessory glands in 3 lobules connected with a single papillae, sessile, cord-like ductules present.

> Color: Preserved specimens reddish purple dorsally, whitish around the setal circles, most pale ventrally. Clitellum chocolate brown.
> Distribution in China: Hubei (Lichuan), Chongqing (Nanchuan).
> Remarks: This species is very like *Metaphires chmardae* (Horst, 1883). It is easily caught in hands when alive and easily autotomized upon being caught in hands. It can also be distinguished from *A. diffringens* (Baird, 1869) by the characters of setae, genital papillae, diverticulum, etc.

mamillaris Group

Diagnosis: *Amynthas* with spermathecal pores 2, at VI and VII.

Global distribution: Oriental realm, ? Australasian realm and ? introduced into the Oceanian realm. There is 1 species from China.

144. *Amynthas mamillaris* (Chen, 1938)

Pheretima (Pheretima) mamillaris Chen, 1938. *Contr. Biol. Lab. Sci. Soc. China (Zool.).* 12(10):413–414.

Amynthas mamillaris Sims & Easton, 1972. *Biol. J. Soc.* 4:244.
Amynthas mamillaris Xu & Xiao, 2011. *Terrestrial Earthworms of China.* 144.

External Characters

Length 70–95 mm, width 2.2–2.5 mm, segment number 156–165. Prostomium epilobous. First dorsal pore 11/12. Clitellum entire, in XIV–XVI, annular, without setae, portion in front of female pore less glandulated. Setae all fine, evenly distributed both dorsally and ventrally. Preclitellar ones more numerous. Both dorsal and ventral breaks slight, almost indistinguishable in front of clitellum. Setae number 50–64 (III), 70–96 (VI), 70–84 (VII), 40–48 (XXV). Male pores 1 pair on ventral side of XVIII, about 1 mm apart, each on median side of an ovoid skin swelling which is smooth on surface and highly elevated, sharply marked off from its surrounding skin; each swelling starting from lateral side of XVII and XVIII and converging ventrad to bear male aperture on its ventromedian corner, these 2 swellings also prominent in young forms, no gland associated internally. Spermathecal pores 2, unpaired, on ventromedian line of VI and VII, in front of setae, distance between it and setal circle, an intersegmental furrow equal to that between it and setal circle, a transverse pit on a raised papilla-like ridge. No other genital markings. Female pore single in XIV, medioventral.

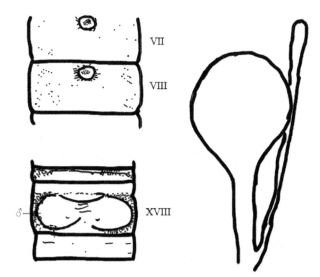

FIG. 142   *Amynthas mamillaris* (after Chen, 1938).

Internal Characters

Septa 8/9 absent, 9/10 very thin, 5/6/7/8 muscular, well thickened, 10/11/12 membranous and nonmuscular. Gizzard barrel-shaped, large, in X–1/2 XI or X–XI. Intestine enlarged from XVI. Caeca simple, small, in XXVII, extending anteriorly to XXV. Hearts 3 pairs, in

XI–XIII. Testis sacs well developed, in X and XI, first pair placed behind membranous septum 9/10, closely situated, with a ventromedian constriction and communicated; second pair more broadly connected, in front of septum 11/12. Seminal vesicles large, in XI and XII, posterior pair extending nearly to XIV; dorsal lobe one-quarter as large. Prostate coarsely lobate, in XVII–XIX, its duct rather short, weakly curved at proximal end, it distal portion not much thickened. Spermathecae 2, unpaired, in VI and VII, close to ventral nerve cord; ampulla sac-like, roundish, not distinctly marked off from its duct which is about 1 mm long. Diverticulum slender, about 2.4 mm long, sometimes twice as long as main pouch, with no distinct seminal chamber, slightly enlarged at ental end.

Color: Preserved specimens pale generally. Clitellum brownish.
Distribution in China: Hainan (Baopeng).

minimus Group

Diagnosis: *Amynthas* with spermathecal pores 1 pair, at 5/6.
Global distribution: Oriental realm, ? Australasian realm and ? introduced into the Oceanian realm. There are 14 species from China.

TABLE 18 Key to Species of the *minimus* Group of the Genus *Amynthas* From China

1. Spermathecal pores on dorsum 2
   Spermathecal pores on ventrum 3

2. About 1/6 of dorsally body circumference apart .... *Amynthas tetrapapillatus*
   About 2/5 of dorsally body circumference apart ..... *Amynthas limellus*

3. Male pores in XIX, each on a round papilla, with indistinct concentric ridge; sometimes 3 papillae placed ventromedially on XIX–XXI, but in most cases placed ventrolaterally, with 1 in front and another behind each male pore on XVIII and XX; rarely with no papillae in this region. Genital papillae paired on VI, in front of setae, or rarely on v; or with median or no papillae .... *Amynthas infantilis*
   Male pores in ventral of segment XVIII 4

4. Without papillae front and behind clitellum 5
   With papillae front and/or behind clitellum 8

5. Male pores on ear-like papilla .... *Amynthas conchipapillatus*
   Male pores not on ear-like papilla 6

6. Male pores a pear-shaped gland-pad ventrally, posterior part of glandular pad broader .... *Amynthas bisemicircularis*
   Male pores on pear-shaped glandular pad 7

7. Male pores on long round pad, glandular pad large; spermathecal ampulla round .... *Amynthas jishanensis*
   Male pores in a round, conical porophore, spermathecal ampulla elongated oval .... *Amynhas minimus*

TABLE 18 Key to Species of the *minimus* Group of the Genus *Amynthas* From China—cont'd

8. With papillae front and behind clitellum 9
   With papillae behind clitellum, without papillae front clitellum 11

9. Male pores closely paired, the pore regions with elliptical skin pads like butterfly wings extending from anterior border of XVII to the setal line of XIX, porophores located at the posterior end of longitudinal slits of the pads. Genital papillae 1 pair in V, postsetal, anterior to the spermathecal pores, each circular in shape with a slightly convex center .... *Amynhas papilio*
   Male pores regions, elliptical skin pads unlike a butterfly 10

10. Genital papillae on XV–XVIII or XIX, more often on XVII, rarely totally absent, small and numerous, up to 14 in a transverse row, frequently about 6–10, arranged presetally and postsetally; in small forms only 2 large ones present presetally on XVII in line with male porophores; in the type none on XVIII, few on XVII, some on XV and XVI; small and numerous papillae on VII and VII .... *Amynhas varians*
    Male pores on a cone-shaped porophore, 1 large papilla or 2 medial to but in front of male porophore, placed between XVII and XVIII; paired genital papillae present ventrally and postsetally on VI, or sometimes on V .... *Amynthas funginus*

11. Papillae in XVII 12
    Papillae between XVII–XIX 13

12. Male pores in the center of a raised papilla. Another similarly large and round papilla on XVII, postsetal raised but depressed in the center, in front of each pore papilla .... *Amynthas limellulus*
    Male pores on a slightly small papilla; pore papilla surrounded by 2–3 sharply defined circular ridges; genital papillae found only on ventral side of XVII: 1 pair closely behind setal circle and slightly medial to male pore region; another unpaired in front of setae and in close contact with papilla of posterior pair .... *Amynhas wui*

13. Male pores on a small flat-topped tubercle surrounded by a few concentric ridges; 1 small ampulla-like papilla on XIX behind each male pore, occasionally another on XVII in line with the former .... *Amynthas mutus*
    Male pores as 2 large projected nipple-like porophores, on tip of a small vertical teat-like protuberance, situated on top of a large-based porophore, its base extending to or over intersegmental furrows 17/18 and 18/19, slightly narrowing toward the free and more or less flat-topped end where bears a small teat-like protuberance .... *Amynthas zyosiae*

145. *Amynthas bisemicircularis* Ding, 1985

Amynthas bisemicircularis Ding, 1985. *Acta Zootaxa Sinica*. 10(4):354–356.
Amynthas bisemicircularis Xu & Xiao, 2011. *Terrestrial Earthworms of China*. 87.

External Characters

Body rather small, length 61–94 mm, width 2–2.5 mm (clitellum), segment number 91–125. Prostomium 1/2 epilobous. First dorsal pore in 12/13. Clitellum XIV–XVI, annular, without setae. Setae fine; Setae number 45–56 (II), 62–70 (VII), 51–73 (XII), 51–67 (XXI); 6–7 (V) between spermathecal pores. Male pores paired in ventral of XVIII, XVIII segment with a pear-shaped

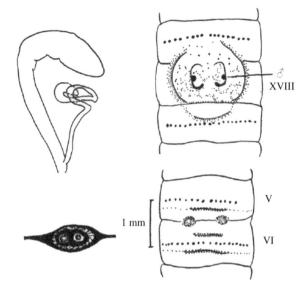

FIG. 143   *Amynthas bisemicircularis* (after Ding, 1985).

glandular pad ventrally, posterior part of glandular pad broader and more elevated than anterior part; male pore opening on ellipsoid furrow. Spermathecal pores 1 pair in 5/6, about 1/7 of circumference apart ventrally, each bedded in glandular lips anteriorly and posteriorly. Female pore single in the center of XIV ventrally.

Internal Characters

Septa 8/9/10 absent, 5/6/7/8 thick, 10/11/12/13 slightly thin. Gizzard ball-shaped. Intestine swelling in XVIII; caeca simple, in XXVII, extending anteriorly to XXV, smooth both dorsally and ventrally. Hearts 4 pairs, last pair in XIII. Testis sacs 2 pairs in X and XI, slightly small, separated from each other. Seminal vesicles paired developed, in XI and XII. Prostate glands fan-shaped, divided into many irregular leaflets, in XVI–XIX; prostate duct slender, about 2 mm long, U-shaped. Spermathecae 1 pair in VI, ampulla as elongate eggplant-shaped, with a few wrinkles on surface; diverticulum longer than main part, with ovoid seminal chamber.

Color: Preserved specimens white-gray, slightly transparent, anterior brownish. Clitellum red-brown.
Distribution in China: Sichuan (Chengdu, Yibin, Leshan), Guizhou (Chishui, Jiangkou).
Type locality: Chongqing (Fuling), Sichuan (Chengdu).
Deposition of types: Sichuan Institute of Natural Resources.
Remarks: This species is similar to *A. hainanicus* (Chen, 1938) in having a larger male genital area and shape of spermatheca, but spermathecal pores 1 pair, glandular pad area pear-shaped, elevated not cavate.

146. *Amynthas conchipapillatus* Qiu & Sun, 2010

Amynthas conchipapillatus Sun, Zhao & Qiu, 2010, *Zootaxa*. 2680: 26–32.

External Characters

Length 42–49 mm, clitellum width 1.5–2.0 mm, segment number 106–115. Prostomium 1/2 epilobous. First dorsal pore 12/13. Clitellum XIV–XVI, annular, raised, setal annulet particularly obvious on ventrum while setae invisible. Setae number 40 (III), 54–64 (V), 50–52 (VIII), 38–48 (XX), 40–50 (XXV); 8 between spermathecal pores, 5 between male pores; $aa = 1.2$–$2ab$, $zz = 1.3zy$. Male pores paired in XVIII, each situated on ventral side of a shellfish-like, convex and smooth surface. Spermathecal pores 1 pair in 5/6, eye-like, 1/5 circumference apart ventrally. Genital markings absent. Female pore single, medioventral in XIV.

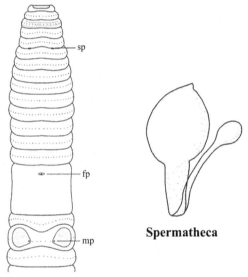

FIG. 144   *Amynthas conchipapillatus* (after Sun, Zhao & Qiu, 2010).

Internal Characters

Septa 8/9/10 absent, 5/6/7/8 thick, 10/11/12/13 slightly thickened. Gizzard in VIII–X, ball-shaped. Intestine enlarged distinctly from XX. Intestinal caeca simple, in XXVII, extending anteriorly to 1/3 XXIV, a bigger incision on dorsal margins. Esophageal hearts in X–XIII. Holandric, testis sacs 2 pairs in X–XI, anterior pair developed, second pair silvery; separated on ventrum. Seminal vesicles paired in XI–XII, anterior pair more developed than posterior pair. Prostate glands undeveloped, extending from XVII–XIX, thin, closely adherent to body wall. Prostatic duct n-shaped. Accessory glands absent. Spermathecae paired in VI with a gradually tapering slender duct, ampulla heart-shaped or oval-shaped, about 1.2 mm, spermathecal duct as long as ampulla; diverticulum shorter than main pouch by 1/5, no kinks, terminal 1/4 dilated into a pear-shaped chamber; no nephridia on spermathecal ducts.

Color: Preserved specimens lacking pigment on dorsum and ventrum, middorsal line clear because of the distinct vessel after XXII. Clitellum orange.
Distribution in China: Hainan (Mt. Diaoluo).
Etymology: This species is named after its papilla, which looks like a conch.
Locality and habitat: Collected from cinnamon sandy soil under banana and arbor forests on Mt. Diaoluo, Hainan province. Type locality: Hainan (Mt. Diaoluo).
Deposition of types: Shanghai Natural History Museum.
Remarks: This species is similar to *A. mutus* (Chen, 1938) and *A. infantilis* (Chen, 1938) which both have 1 pair of spermathecal pores in 5/6. *A. conchipapillatus* has distinct characteristics such as a special pair of papillae in male pore region, closely arranged spermathecal pores, small prostate glands and so on, whereas *A. mutus* has only 1 small papilla on XIX behind each male pore and occasionally another on XVII in line with the former, and the distance between spermathecal pores is slightly longer than half of the body circumference. The papillae in the male pore region of *A. infantilis* have the following characteristics: 3 papillae placed ventromedially on XIX–XXI, but in most cases placed ventrolaterally, with 1 in front of and another behind each male pore on XVIII and XX. Prostate glands well developed.

147. *Amynthas funginus* (Chen, 1938)

Pheretima (Pheretima) fungina Chen, 1938. *Contr. Biol. Lab. Sci. Soc. China (Zool.).* 12(10):389–391.
Amynthas funginus Sims & Easton, 1972. *Biol. J. Soc.* 4:236
Amynthas funginus Xu & Xiao, 2011. *Terrestrial Earthworms of China.* 117–118

External Characters

Length 55–65 mm, width 2.5–3.0 mm, segment number 121–126. Prostomium 1/3 epilobous. First dorsal pore 12/13. Clitellum entire, in XIV–XVI, annular, dorsal pore visible, with setae on ventral side. Setae closely spaced but distinct on each segment, more numerous on preclitellar segments, slightly wider at intervals ventrally (more noticeable on preclitellar segments). Both dorsal and ventral breaks indistinct; $aa = 1.2ab$, $zz = yz$; setae number 44–67(III), 86–90 (VI), 90–96 (VII), 51–65 (XXV); 28–36 (VI) between spermathecal pore, 8–13 (VII) between male pores. Male pores 1 pair in XVIII, about 1/3 of circumference apart ventrally, each on a

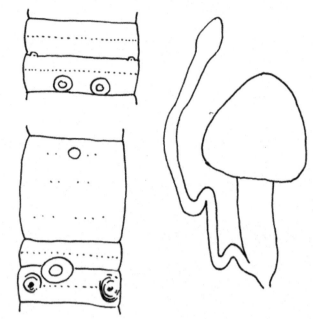

FIG. 145 *Amynthas funginus* (after Chen, 1938).

cone-shaped porophore, usually raised, 1 large papilla or 2 medial to but in front of male porophore, placed between XVII and XVIII. Spermathecal pores 1 pair in 5/6, intersegmental, about 2/5 of circumference apart ventrally; each guarded closely behind by an ampulla-like tubercle; paired genital papillae present ventrally and postsetally on VI, or sometimes on V.

Internal Characters

Septa 8/9/10 absent, 5/6/7/8 muscular and thicker, 10/11/12/13 thin but slightly thicker than those following. Gizzard large, globular, in X and 1/2 XI. Intestine enlarged from XVI. Caeca simple, in XXVII, extending anteriorly to XXIV, finger-shaped. First pair of hearts small, closely applied to anterior face of septum 10/11. Testis sacs in X large, sac-like, with a narrow connection medioventrally, those in XI more broadly connected. Seminal vesicles in XI small, as elongate band enclosed in second pair of testis sacs, those in XII also small. Prostate gland large, in XVII–XIX, with long coiled duct, uniform in caliber. Accessory glands in round patches, raised, each with whitish gland portion and short cord-like duct. Spermathecae 1 pair, in VI, ampulla either round or heart-shaped, thick-walled, with a duct equally long or as twice as long as ampulla, whole pouch about 2 mm long. Diverticulum slender, nearly twice as long as main pouch, ental 2/3 probably serving as seminal chamber.

Color: Preserved specimens pale on both dorsal and ventral sides.
Distribution in China: Hainan (Wanning).

148. *Amynthas infantilis* (Chen, 1938)

Pheretima (Pheretima) infantilis Chen, 1938. *Contr. Biol. Lab. Sci. Soc. China (Zool.).* 12(10):392–394.
Amynthas infantilis Xu & Xiao, 2011. *Terrestrial Earthworms of China.* 130.

External Characters

Length 10–14 mm, width 1.0–1.8 mm, segment number 58–82. Prostomium 1/3 epilobous. First dorsal pore 12/13. Clitellum entire, in XIV–XVI, annular, without setae. Setae uniformly built, closer on preclitellar segments, their breaks evident; $aa = 2ab$, $zz = 1.5yz$ before and $=2yz$ behind clitellum; setae number 37–45 (III), 48–55 (V), 50–70 (VII), 35–44 (XXV); 21 (VI) between spermathecal pore, 8–10 (VII) between male pores. Male pores 1 pair in XIX, about 1/3 of circumference apart ventrally, each on a round papilla, with indistinct concentric ridge; sometimes 3 papillae placed ventromedially on XIX–XXI, but in most cases placed ventrolaterally, with 1 in front and another behind each male pore on XVIII and XX; rarely with no papillae in this region. Spermathecal pores 1 pair in 5/6, intersegmental, nearly about 1/2 of body circumference apart ventrally, hardly visible externally, often entirely absent. Genital papillae paired on VI, in front of setae, about 7 setae at interval, or occasionally on V. These papillae present in specimens with median or no papillae, or absent in nearly all specimens with lateral papillae in male pore region.

Internal Characters

Septa 8/9/10 absent, 5/6/7/8 well thickened, 10/11/12 relatively thick. Gizzard large, in VII and 1/2 X. Intestine enlarged from XV, its portion in XVI–XX narrow, suddenly enlarged from about XXI. Caeca simple, in XXVII, extending anteriorly to XXV. First pair of hearts invisible. Testis sacs moderate in size, both pairs separate on ventromedian side, sometimes well developed, meeting and communicated dorsally. Seminal vesicles in XI and XII, rudimentary in all cases observed, each as a narrow strand fastening on posterior side of respective septa. Prostate gland well developed, in XVI–XX; its duct slender, deeply curved, ental end thinner. Accessory glands in minute lobules, almost indistinct from nephridial tubules. Spermathecae 1 pair, in VI, usually totally absent; if present fairly developed; main pouch about 1 mm length, ampulla elongate ovoid, ampullar duct slender, as long as ampulla. Diverticulum shorter than main pouch, with a spherical seminal chamber, clearly marked off from a slender duct.

> Color: Preserved specimens light gray generally. Clitellum brownish.
> Distribution in China: Hainan (Baopeng).
> Remarks: This species is smaller than others in this genus. Most of the specimens are about 10 mm or slightly longer and are fully sexually mature. It is distinguished by its male pore constantly on segment XIX.

149. *Amynthas limellulus* (Chen, 1946)

Pheretima limellula Chen, 1946. *J. West China. Border Res. Soc. (B).* 16:127–129, 141, 158.
Amynthas limellulus Sims & Easton, 1972. *Biol. J. Linn. Soc.* 4:236.
Amynthas limellulus Xu & Xiao, 2011. *Terrestrial Earthworms of China.* 137.

External Characters

Length 50–55 mm, width 1.5–1.8 mm, segment number 90–117. Prostomium proepilobous, small. First dorsal pore 11/12. Clitellum XIV–XVI, annular, swollen and smooth, with shorter setae on ventral side of XVI, markedly elongate, its length equal to 5 preclitellar or 9 postclitellar segments. Setae fine and numerous at anterior end, closer ventrally; Setae number ?–44 (III), 58–73 (VI), 56 (IX), 35–38 (XXV); 24–30 (V), 29–33 (VI) between spermathecal pores, 10 between male pores; $aa = 1.2\,ab$, $zz = 2.0zy$. Male pores paired in XVIII, about 1/3 of circumference apart ventrally, each in the center of a raised papilla. Another similarly large and round papilla on XVII, postsetal raised but depressed in the center, in front of each pore papilla. Spermathecal pores 1 pair in 5/6, about 1/2 of their ventral circumference apart, or a little

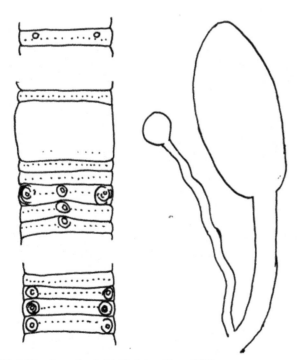

FIG. 146   *Amynthas infantilis* (after Chen, 1938).

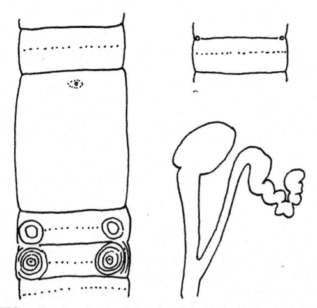

FIG. 147   *Amynthas limellulus* (after Chen, 1946).

over, each pore in an eye-like depression, intersegmental but on posterior margin of V. No other genital markings. Female pore single in XIV, medioventral, in a square depression.

Internal Characters

Septa 9/10 absent, 5/6/7/8 thick and muscular, 6/7 thickest, 8/9 traceable ventrally, 10/11/12 thin and membranous but thicker than those following. Nephridial tufts anteriorly on 5/6/7 fairly thick. Pharyngeal glands well developed in front of 5/6. Gizzard globular, fairly large, lying in 1/2 IX and X. Intestine swelling in XVI. Caeca simple, short and auricular in shape, in XXVII, extending anteriorly to only a small part of XXV (or to part of XXIV). Vascular loops in IX asymmetrically developed, those in X well developed. Testis sacs very developed, about twice as large as the vesicle, each sac measuring 1.1 mm in dorsoventral length, separate ventrally but communicated, nearly meeting dorsally. Seminal vesicles small, those in XII about 0.6 mm in dorsoventral length including the dorsal lobe; those in XI smaller and enclosed in the posterior pair of testis sacs. Prostate gland large, coarsely lobate, in XVI–XXII, its duct about 1.2 mm long, coiled entally and curved ectally, straight at middle portion. Accessory glands in minute lobules or racemose, sessile. Spermathecae 1 pair, in VI, main pouch about 2.0 mm long. Ampulla ovoid, not well differentiated from its duct, which is very long, about 1.3 mm. Diverticulum twice as long as main pouch, ental half coiled, no particular enlargement, or with an ovoid chamber, ental two-thirds serving as seminal receptacle.

Color: Pale generally, greyish along dorsomedian line. Clitellum chestnut-red.

Distribution in China: Hubei (Qianjiang), Chongqing (Shapingba, Beipei).

Remarks: This species is similar to *Amynthas limellus* (Gates, 1935), once recorded from this region, in the prostomium, setae, position of the spermathecal pores, genital papillae in the male pore region, characters of testis sacs, etc. However, it is distinguished as a distinct species mainly by absence of the gizzard septa and a different form of diverticulum. Minor differences are also found, such as (1) the male pore not situated lateral to its respective papilla and with no shallow pouch, (2) size much smaller, (3) setae less numerous, (4) setae present on ventral side of XVI, and (5) no papilla found on clitellum.

150. *Amynthas limellus* (Gates, 1935)

Pheretima limella Gates, 1935a. *Smithsonian Mus. Coll.* 93(3):11–12.

Pheretima limella Chen, 1936. *Contr. Biol. Lab. Sci. Soc. China (Zool.)*. 11:299–300.

Pheretima limella Gates, 1939. *Proc. U.S. Natr. Mus.* 85:451–453.

Pheretima limella Chen, 1946. *J. West China. Border Res. Soc. (B)*. 16:136.

Amynthas limellus Sims & Easton, 1972. *Biol. J. Soc.* 4:236.

Amynthas limellus Xu & Xiao, 2011. *Terrestrial Earthworms of China*. 137–138

External Characters

Length 60–85 mm, width 2.5–5.0 mm, segment number 106–110. Prostomium proepilobous. First dorsal pore in 12/13. Clitellum glandular and entire, in XIV–XVI, annular, setae absent. Setae fine and numerous on each segment, evenly distributed all over body, none particularly enlarged. Setal ring closed both dorsally and ventrally; $aa = 1.1ab$, $zz = 1.2yz$ behind clitellum. Those on II as numerous as on III. Setae number 42–56 (III), 62–72 (VI), 64–75 (XXV); 44–48 (V), 44–49 (VI) between spermathecal pores (ventrally), 14–18 between male pores. Male pores at the centers of a small, oval, smooth surfaced tubercles; about 1/3 circumference apart ventrally, a thin fold of tissue at the lateral margin of each male porophore can be drawn mesially over the tubercle like an eyelid. Genital markings paired and presetal on ventral side of XVII and XVI, in line with the male porophore; those on XVI less constant, frequently unpaired but seldom absent. Spermathecal pores minute and superficial, 1 pair, in 5/6, dorsal in position, about 3/5 circumference apart ventrally; each pore on a small tubercle, situated on posterior edge of V, but in the furrow. No other genital papillae in this region. Female pore single in XIV, medioventral.

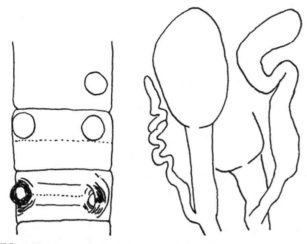

FIG. 148 *Amynthas limellulus* (after Chen, 1936).

Internal Characters

Septa 6/7/–9/10 very muscular and well thickened, 5/6 less thickened, 4/5 thin, 10/11/12 slightly muscular. Gizzard small in IX. Intestine enlarged from XV. Intestinal caeca simple, in XXVII, extending anteriorly to XIII, slightly wrinkled on both sides. Hearts 4 pairs, in X–XIII, first pair fairly large. Testis sacs of X and XI unpaired ?, anterior pair united and 2 funnels nearly meeting; posterior pair united ventrally. Seminal vesicles paired moderate or small in size, in XI and XII, upper half as dorsal lobes; those in XI enclosed in a membranous sac in common with testis sac. Prostate glands moderate in size, in XVII–XIX; prostatic duct short and slender, opening directly to the papilla referred above. Spermathecae 1 pair in V, ampulla fairly large, heart-shaped, clearly marked off from the duct which is either longer than ampulla or considerably shorter. Spermathecal diverticulum varying in length and shape; with a muscular stalk and an elongate seminal chamber, the ectal portion of the latter looped, the ental portion ovoidal. Accessory glands sessile, each in a round patch corresponding to respective papilla.

Color: Preserved specimens greenish gray on dorsal side, usually grass green along dorsomedian line; pale ventrally. Clitellum brownish.
Type locality: Chongqing (Xufu).
Distribution in China: Chongqing (Beipei, Shapingba, Fuling), Sichuan (Chengdu, Yibin, Xufu, Leshan).
Deposition of types: Smithsonian Institution
Remarks: Distinguished from *A. zoysiae* by the presence and muscularity of septa 8/9/10 and by the presence of genital marking.

151. Amynthas minimus (Horst, 1893)

Perichaeta minima Horst, 1893. *Zoologische Ergebnisse einer Reise in Niederländisch Ost-Indien*: 66.

Pheretima minima Gates, 1972. *Trans. Amer. Philos. Soc.* 62(7):201–202.
Amynthas minimus Sims & Easton, 1972. *Biol. J. Soc.* 4:236.
Amynthas minimus Shen & Yeo, 2005. *Raf. Bul. Zool.* 53(1):21–24.
Amynthas minimus Chang, Lin, Chen, Chung & Chen, 2009. *Earthworm Fauna of Taiwan*. 58–59.
Amynthas minimus Xu & Xiao, 2011. *Terrestrial Earthworms of China*. 149.

External Characters

Length 22–32 mm, width 1.5–2.5 mm, segment number 74–100. Prostomium epilobous. First dorsal pore 12/13. Clitellum XIV–XVI, annular, dorsal pores absent, setae absent or 1–3 on ventral XVI. Setae number: 40–55 (II), 40–47 (III), 54–58 (VII), 46–52 (XII) 38–44 (XX); 6–9 between male pores. Male pores paired in XVIII, each on a round, conical porophore about 0.3 mm in diameter, about 0.26 circumference apart ventrally,. Genital papillae absent in both preclitellar and postclitellar regions. Spermathecal pores 1 pair in 5/6, about 0.48 circumference apart ventrally. Female pore single, medioventral in XIV.

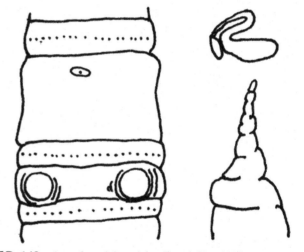

FIG. 149 *Amynthas minimus* (after Shen & Yeo, 2005).

Internal Characters

Septa 8/9/10 absent, 5/6/7/8 thickened. Gizzard in IX–X. Intestine swelling in XIV or XV. Caeca simple, long, in XXVII, extending anteriorly to XXIII. Hearts in X–XIII. Testis sacs 2 pairs in X and XI. Seminal vesicles paired in XI and XII, vestigial, first pair included in testis sacs. Prostate glands paired in XVIII, large, extending anteriorly to XVI or XVII and posteriorly to XIX or XXI; prostatic duct large, U-shaped. Accessory glands absent. Spermathecae present 1 pair in VI, ampulla elongated oval, about 0.5 mm long, 0.27 mm wide; with a slender stalk about 0.45 mm length; diverticulum with a slender stalk about 0.4 mm long with an elongated oval seminal chamber about 0.28 mm in length. Ovaries paired in XIII.

Color: Live specimens red to reddish white. Preserved specimens whitish gray with yellowish tan clitellum.
Global distribution: Cosmopolitan, recorded in East and Southeast Asia, Burma, Singapore, USA, Hawaii, South Africa, Madagascar, Samoa, and Australia.
Distribution in China: Taiwan, Fujian (Quanzhou, Jinmen).
Etymology: The name of this species refers to its small body size.
Type locality: Tjibodas, Java.
Deposition of types: Leiden Museum, Netherlands.
Remarks: *Amynthas minimus* is a small, holandric and bithecal earthworm belonging to the *minimus* species group of the genus *Amynthas* (Sims & Easton, 1972). Gates (1942) described *Amynthas humilis* as a small earthworm with 1 pair of spermathecae in VI. It has 6–12 setae on ventral XVI (clitellum) but no genital markings. Gates (1961) later found that its spermathecal diverticula were similar to that of *A. minimus* figured by Horst (1893). After examination of the Hawaiian worms, Gates (1961) considered that the indistinctness of some markings and the condition of associated tissues warrant a suspicion that glands as well as the markings may be disappearing in some lineages, and individuals without genital markings can be expected. Therefore, Gates (1961, 1972) considered *A. humilis* as a synonym of *A. minimus*.

Easton (1981) placed *A. zoysiae* (Chen, 1933) and *A. ishikawai* (Ohfuchi, 1941) in the brace of species synonyms of *A. minimus*. *A. minimus* is fairly similar to *A. zoysiae* of China in external characters, such as body size, segment number, and male porophore structure. However, *A. zoysiae* has higher setal number and larger prostate glands occupying 6–7 segments in XV–XXII. The prostate glands of *A. minimus* were described as well developed in the original description (Horst, 1893). However, Gates (1961) stated that prostates appear confined within XVII–XIX after examining the type specimen.

A. ishikawai was reported from a cave in Japan (Ohfuchi, 1941). It has 1 pair of spermathecae in VI, and its male porophore structure is similar to *A. minimus*. Gates (1961, 1972) and Easton (1981) considered it synonymous to *A. minimus*. However, *A. ishikawai* is of a large size, and has a lower setal number that shows a slightly increasing trend from the anterior region toward the posterior region, whereas the setal number of *A. minimus* is higher in the preclitellar region and lower in the postclitellar region. Furthermore, *A. ishikawai* has well-developed seminal vesicles and large prostate glands (occupying 7 segments) with straight, muscular ducts, whereas *A. minimus* has small, vestigial seminal vesicles and normal prostate glands (occupying 3–4 segments) with U-shaped ducts. Therefore, *A. ishikawai* is considered as a valid species different from *A. minimus* (Shen and Yeo, 2005).

152. *Amynthas mutus* (Chen, 1938)

Pheretima (Pheretima) muta Chen, 1938. *Contr. Biol. Lab. Sci. Soc. China (Zool.).* 12(10):391–392.
Amynthas mutus Sims & Easton, 1972. *Biol. J. Soc.* 4:236.
Amynthas mutus Xu & Xiao, 2011. *Terrestrial Earthworms of China.* 154–155.

External Characters

Length 50–80 mm, width 2.5–3.0 mm, segment number 85–135. Prostomium 1/3 epilobous, its dorsal process with a longitudinal median fissure. First dorsal pore 12/13. Clitellum entire, in XIV–XVI, annular, without setae. Setae uniformly built, closer on dorsal side; both dorsal and ventral breaks indistinct, none particularly enlarged; setae number 35–54 (III), 38–70 (VI), 41–68 (VII), 32–44 (XXV); 30–42 (VI) between spermathecal pores, 9–10 (VII) between male pores, 12 on XIX between 2 papillae. Male pores 1 pair in XVIII, about 1/3 of circumference apart ventrally, each on a small flat-topped tubercle surrounded by a few concentric ridges; 1 small ampulla-like papilla on XIX behind each male pore, occasionally another on XVII in line with the former. Spermathecal pores 1 pair in 5/6, intersegmental and superficial, a little more than half of body circumference apart ventrally. No other genital markings in this region.

Internal Characters

Septa 8/9/10 absent, 5/6/7/8 thickened, 10/11/12/13 less so. Gizzard in 1/2 IX and X. Intestine enlarged from XVI. Caeca simple, in XXVII, extending anteriorly to XXV. Hearts in X–XIII. Testis sacs in X large, with a ventromedian constriction, not communicated (?), often meeting dorsally; those in XI communicated ventrally, and often communicated dorsally. Seminal vesicles in XI small, enclosed in testis sacs if latter overdeveloped; those in XII also small, attached to septum 11/12. Prostate gland well developed, with coarse lobules, in XVII–XX; its duct long and coiled, nearly uniform in caliber. Accessory glands 3–4 minute balls of gland cells connected with a cord-like duct leading to each papilla externally; scarcely visible in most cases. Spermathecae 1 pair, in VI, ampulla very large, usually filled with whitish mass or minute globules, roundish or heart-shaped, about 1.2 mm long. Main duct wide, about 1 mm long. Diverticulum slightly shorter than main pouch; seminal chamber date-shaped or ovoid, full and whitish, its duct slender and long.

Color: Preserved specimens gray on dorsal side, pale on lateral and ventral sides.

Distribution in China: Hainan (Wenchang).

Remarks: In considering the character of testis sacs and position of the spermathecal pore the present species is similar to *A. varians*. However, it is clearly distinguished by (1) having no setae on the first segment, and (2) a flat-topped papilla as male porophore.

153. *Amynthas papilio* (Gates, 1930)

Pheretima papilio Gates, 1930. *Rec. Ind. Mus.* 32:316–318.
Amynthas papilio papilio Sims and Easton, 1972. *Biol. J. Linn. Soc.* 4:236.
Amynthas papilio Chen and Chang, 2003. *Emdemic Species Research* 5:89–94.
Amynthas papilio Chang, Shen & Chen, 2009. *Earthworm Fauna of Taiwan.* 66–67.
Amynthas papilio Xu & Xiao, 2011. *Terrestrial Earthworms of China.* 161–162.

External Characters

Length 40–70 mm, width 2–6 mm, segment number 87–119. Prostomium epilobous. Setae 37–41 (VII), 11 (XVII), 80 (XX), 6–8 between male pores. First dorsal pore 12/13. Clitellum XIV–XVI, annular, 2 mm long, setae absent, dorsal pore absent. Male pores closely paired, ventral in XVIII, about 0.18 circumference apart ventrally, the pore regions with elliptical skin pads like butterfly wings extending from anterior border of XVII to the setal line of XIX, porophores located at the posterior end of longitudinal slits of the pads. Spermathecal pores 1 pair in 5/6, ventral, not easily visible, 0.21 circumference apart ventrally. Genital papillae 1 pair in V, postsetal, anterior to the spermathecal pores, each circular in shape with a slightly convex center. Female pore single, medio-ventral in XIV.

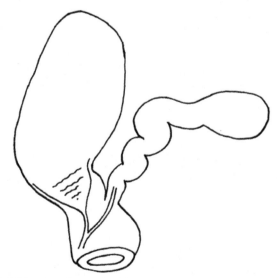

FIG. 150   *Amynthas papilio* (after Gates 1930).

Internal Characters

Septa 8/9/10 absent, 5/6/7 thickened. Gizzard in IX–X, peach-shaped. Intestine from XV. Intestinal caeca paired in XXVI, simple, surface wrinkled, extending anteriorly to XXIV. Testis sacs 1 pair in X, ventral, small, not easily visible. Seminal vesicles paired in XI and XII, large, with dorsal lobe. Prostate glands paired in XVIII, large, extending anteriorly to XV and posteriorly to XXI, ducts 2 mm long, L-shaped. Spermathecae 1 pair in VI, ampulla oval, about 1.5 mm long, 0.7 mm wide, stalk slender, 1.5 mm in length, equivalent to that of ampulla, diverticulum with a long striped seminal chamber 0.7 mm long and a stalk about 1.2 mm long.

Color: Preserved specimens white in color, clitellum yellow.
Type locality: San Hlan, Burma.
Deposition of types: Zoological Survey of India, India.
Global distribution: Burma, Iriomote Island in the Ryukyus, Japan, and Taiwan.
Distribution in China: Taiwan (Xinzhu).
Etymology: The name of this species refers to its papilionaceous male pore region.
Remarks: The characters of *A. papilio* collected from Taiwan are compared with those described by Gates (1930, 1972) from Burma and those reported by Ohfuchi (1956) from the Ryukyu Islands. They share similarity in most of the characters. In the original description, Gates (1956) mentions that the first dorsal pore of *A. papilio* is in 5/6, but in the review paper of Gates (1972), he corrects it to 12/13, similar to the case found in specimens from Taiwan, Also, *A. papilio* from Burma has a pair of spermathecal pores in 5/6 (Gates 1930, 1972), but that from the Ryukyu Islands is in 6/7 (Ohfuchi 1956; Kaminhira 1973). Due to the above difference in location of spermathecal pores, Gates (1972) speculates that the specimens from Ryukyu (Ohfuchi 1956) may be a different species from *A. papilio* of Burma (Gates 1930). The specimens from Taiwan have the spermathecal pores in 5/6, similar to that from Burma (Gates 1930). As the spermathecal pores are not easily visible externally, Ohfuchi (1956) might be in error regarding this character. A further verification on this character for the Ryukyu specimens is required to clarify their taxonomic status and discover whether they are different species from *A. papilio* of Burma as Gates (1972) speculates, or even a new species as we suspect.

Gates (1932, 1961, 1972) divides *A. papilio* into 3 subspecies: *A. papiliopapilio*, *A. papilioinsignis*, and *A. papilio hiulcus*. According to the characters of male pores, prostate glands, caeca, and spermathecal pores, the specimens from Taiwan were found to belong to the subspecies

A. Papilio papilio, whose only known locality was Thaton, Burma (Gates 1930).

154. *Amynthas tetrapapillatus* Quan & Zhong, 1989

Amynthas tetrapapillatus Quan & Zhong, 1989. *Act Zootaxonomica Sinica.* 14(3):274–275.
Amynthas tetrapapillatus Xu & Xiao, 2011. *Terrestrial Earthworms of China.* 189–190.

External Characters

Length 152–165 mm, width 4.5–5.0 mm, segment number 128–136. Prostomium 1/2 epilobous. First dorsal pore 11/12. Clitellum in XIV–XVI, annular, without setae. Setae fine; setae number 82–88 (III), 93–107 (VII), 88–100 (IX), 89–93(XI), 86–102 (XX); 12–13 (VI) between spermathecal pores, 8–10 between male pores. Male pores 1 pair, on ventrolateral side of XVIII, about 1/4 of circumference apart ventrally, each on a large projected porophore, surrounded by 1–3 circular ridges. Near the porophore, 2 large, slightly ovoid papillae situated between 17/18 and 18/19 respectively. Spermathecal pores 1 pair, in 5/6, near the dorsomedian line, horizontal slit-like, on a small circular papilla, about 1/6 of body circumference apart dorsally. Female pore single in XIV, medioventral.

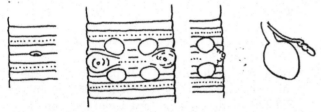

FIG. 151   *Amynthas tetrapapillatus* (after Quan & Zhong, 1989).

Internal Characters

Septa 8/9/10 absent, 5/6/7/8 thickened, muscular, 10/11/12 slightly thickened. Intestine enlarged from XV. Caeca simple, in XXVII, extending anteriorly to XXV. Last hearts in XIII. Testis sacs paired, connected ventrally. Seminal vesicles as transverse band, with dorsal lobe, the first pair of seminal vesicles are contained in the second pair of testis sacs. Prostates, in XVII–XX, duct curved. Spermathecae 1 pair, ampulla heart-shaped, about 1.8–2.0 mm, diverticulum as long as or longer than the main part, zigzag looped near seminal chamber. Seminal chamber ovoid.

Color: Preserved specimens light brown, dorsal and ventral consistent.
Distribution in China: Hainan (Ledon).

Type locality: Hainan (Jianfen Ridge, Ledon County).
Deposition of types: Institute of Tropical Forestry, Chinese Academy of Forestry Science, Guangzhou.
Remarks: In considering the character of the first pair of spermathecal pores in 5/6, ampulla heart-shaped, seminal chamber ovoid, and the first pair of seminal vesicles contained in the second pair of testis sacs, the present species is similar to *A. mutus* (Chen, 1938). However, it can be clearly distinguished from the latter by the spermathecal pores near the dorsomedian line and near the porophore, and 2 large slightly ovoid papillae situated between 17/18 and 18/19 respectively.

144. *Amynthas varians* (Chen, 1938) (v.ant. 144)

155. *Amynthas wui* (Chen, 1935)

Pheretima wui Chen, 1935a. *Contr. Biol. Lab. Sci. Soc. China (Zool.)* 11:109-113.
Amynthas wui Sims & Easton, 1972. *Biol. J. Soc.* 4:236.
Amynthas wui Xu & Xiao, 2011. *Terrestrial Earthworms of China.* 196–197.

External Characters

Length 50 mm, width 2.0 mm, segment number 98. Prostomium 1/3 epilobous. First dorsal pore in 12/13. Clitellum distinguishable in XIV–XVI, feebly glandular, without setae on dorsal and lateral sides, but with 5–8 on each segment of ventral side. Setae very uniformly distributed and about equally numerous in all segments; those on second segment also fine and numerous; no difference in size and interval between dorsal and ventral ones; dorsal break very slight; ventral break almost unnoticeable, before clitellum *aa* = 1/5 *ab*, behind clitellum *aa* = *ab*; setae number 50 (III), 62 (VI), 58 (VII), 40 (XXV); 22 (V), 24 (VI) between spermathecal pores; 8 between male pores. Male pores 1 pair, on ventrolateral sides of XVIII, about 1/3 of circumference apart ventrally; each on a slightly raised small papilla, whitish around pore; pore papilla surrounded by 2–3 sharply defined circle ridges; genital papillae found only on ventral side of XVII: 1 pair closely behind setal circle and slightly medial to male pore region; another unpaired papilla in front of setae and in close contact with a papilla of posterior pair. Spermathecal pores 1 pair, in 5/6, intersegmental placed, slightly less than half of body circumference apart ventrally; each on a small and distinctly raised nipple-like tubercle; no genital marking around this region. Female pore single in XIV, medioventral.

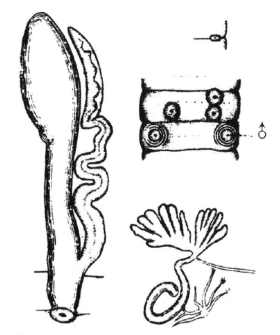

FIG. 152 *Amynthas wui* (after Chen, 1935a).

Internal Characters

Septa 8/9/10 absent; 4/5 very thin, 5/6 thicker, 6/7/8 much thickened, 10/11/12 also thick but less so than 6/7/8, rest septa membranous. Gizzard large, smooth and globular, in 1/3 IX and X. Intestine swelling in XVI. Intestinal caeca short and conical, in XXVII, extending anteriorly to 1/2 XXV. Hearts 4 pairs, first pair in X equally stout. Testis sacs closely paired, connected by short strand and communicated. Seminal vesicles small and elongate, about 1.2 mm length, 5 mm width, dorsal half larger. Prostate glands in XVII–XIX, about 2 segments, with elongate lobes; its duct very long, nearly uniform in caliber except its ental end which is thinner, smooth on surface, and without a dilation at its tail end. Copulatory chamber absent. Accessory glands with 4–7 long ducts in each group arising from the coelomic side of each papilla; ducts tapering toward free ends and closely applied to inner surface of body wall, smooth and shining on surface. Glandular portion not present, but usually with fibrous tissues fastening on the body wall. Spermathecae 1 pair, in VI; ampulla elongate and tapering toward free end, ectally narrowing gradually toward its duct which is rather broad and long, ampulla 0.63 mm long, 0.3 mm broad; demarcation between the 2 indistinct; diverticulum 1.25 mm long, 0.09 mm broad, nearly as long as main pouch (ampulla + duct); most of its ental end slightly twisted and about equally broad throughout the length; without an apparent seminal chamber, its ental third probably serving its purpose.

Color: Preserved specimens grayish on both dorsal and ventral sides, slightly darker dorsally.
Type locality: Fujian (Xiamen).
Distribution in China: Fujian (Xiamen).
Deposition of types: Mus. Biol. Lab. Sci. Soc. China.
Remarks: In several points, it resembles *A. nugalis* (Gates) which is found in Burma in terms of (1) body size, (2) the position and character of the spermathecal pores, and (3) the characters of septa, caeca etc. However, it differs from the latter in many important characters, such as the form of the spermathecae, position of the first dorsal pore, presence of the clitella setae, character of the setae, and the presence between 2 spermathecal pores in *A. nugalis*is which shorter. There is probably some difference in the aspect of the male pore region.

156. *Amynthas zyosiae* (Chen, 1933)

Pheretima zyosiae Chen, 1933. *Contr. Biol. Lab. Sci. Soc. China (Zool.).* 9:288–294.
Pheretima zyosiae Gates, 1935b. *Lingnan Journ, Sci.* 14(3): 456–457.
Amynthas zyosiae Sims & Easton, 1972. *Biol. J. Soc.* 4:236.
Amynthas zyosiae Xu & Xiao, 2011. *Terrestrial Earthworms of China.* 202–203.

External Characters

Length 20–45 mm, clitellum width 1.5–2.5 mm, segment number 70–108. Segments IX and X longest, XVIII as long as XIII, those behind XVIII slightly shorter. Secondary annulation weak. Prostomium 1/3 epilobous. First dorsal pore 11/12. Clitellum distinct, in XIV–XVI, annular, smooth, and without setae, dorsal pore absent. Setae minute and more numerous at anterior end; Setae number 34–42 (III), 42–52 (VI), 40–54 (VII), 34–43 (XII), 34–44 (XXV); 19 between spermathecal pores, 5–10 between male pores. Male pores 1 pair on ventrolateral sides of XVIII, very minute, as 2 large projected nipple-like porophores, about 1/3 of body circumference apart, on tip of a small vertical teat-like protuberance, smooth and conical in shape, situated on top of a large-based porophore which is a raised portion of body wall surrounded by circular wrinkled grooves 4–5 in number, its base extending to or over intersegmental furrows 17/18 and 18/19, slightly narrowing toward the free and more or less flat-topped end where it bears a small teat-like protuberance; ventromedian distance of XVIII between bases of 2 porophores about 1 mm, intervening 6 shorter and indistinct setae, these setae usually appearing shorter or absent in some part, no setae on the raised porophore. Spermathecal pores 1 pair, in intersegmental furrow 5/6, about 4/11 or a little more of body circumference apart, through a transversely ovoid tubercle attached to external end of duct, covered anteriorly and posteriorly by lip-like structures, posterior lip more distinct, resembling an eye shape. No genital papillae present in both regions. Female pore single, medioventral in XIV, the pore indistinct.

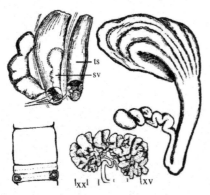

FIG. 153 *Amynthas zyosiae* (after Chen, 1933).

Internal Characters

Septa 8/9/10 absent, 5/6/7/8 and 10/11/12 nearly equal in thickness, 12/13 also thick, 13/14 less so, 14/15 comparatively thick, thin after 15/16 posteriorly. Gizzard very large, in VII–X, globular or vase-shaped, no distinct crop. Intestine enlarged from XVI. Intestinal caeca simple, short and conical in shape, in XXVI, extending anteriorly to XXIV (or anterior in XXVII) pale, smooth on both dorsal and ventral sides. Hearts 4 pairs, in X–XIII, first pair large in caliber and close to the septum. Testis moderate in size, on anterior lower side of each sac; seminal funnel small and simple. Sperm ducts of each side united in XII. Seminal vesicles in XI and XII, posterior pair larger, closely applied to posterior face of septum 11/12, about 0.5–0.8 mm in length, about 0.25 mm in width, its dorsal portion about 1/3 or 2/3 of whole part smooth like dorsal lobes but not clearly defined from its lower portion; lower half often constricted at edges and slightly roughened on surface. First pair 1/2 or 1/3 smaller than second pair, or reduced into 2 small strands. Testis sacs 2 pairs, in X and XI, inconstantly formed, both sacs of each pair are often disproportionally developed as in the case of spermathecae. The portion resting on dorsal side of esophagus transversely elongate and well distended, surface smooth and brownish yellow in color. Prostate glands always developed with many large lobules, the whole rather broad and much elongate, in XVI or XV–XXI or XXII, occupying; with finely lobules. Prostatic duct long and stout, ectal two-thirds stouter and straight, ental end slender and looped, with about 3 small tributaries from gland part. No accessory glands. Spermathecae 1 pair in VI, or occasionally ampulla lying under septa 6/7 and 7/8 if ventrally placed. In most specimens, no trace of spermathecal elements being found. some having spermathecae but both spermathecae not symmetrically developed; those with only 1 spermatheca often abnormally developed; ampulla enormously large in comparison with those of smaller worms, 1.2 mm long, 0.88 mm wide,

thick-walled and irregularly corrugated or wrinkled on surface, sac-like in outline, broader on both ectal and ental ends or slightly narrower entally and pear-shaped, at least about 1 mm in length; its ectal end not clearly marked off from its duct which is also very stout and thick-walled, as long as, or shorter than the ampulla. Diverticulum half the length of main pouch, with its ental two-thirds closely looped more or less on 1 plane and ended by a small round bulb which is often slightly enlarged. In a few specimens ental end constricted into a moniliform shape; its ectal third straight and slender, joining the distal part of main duct. Seminal chamber +m diverticular duct 1 mm long, duct about 0.09 mm in width.

Color: Live specimens light buff on dorsal side, pinkish around anterior 7 or 8 segments, yellowish at posterior end, grayish at middle region of body. Clitellum whitish or fleshy. Seminal vesicle, gizzard and prostate regions paler due to internal organs; ventral side dark pale. Preserved specimens generally grayish, grayish pale before and behind clitellum, grayish red at anterior end. Clitellum light chocolate.

Type locality: Zhejiang (Linhai).

Deposition of types: Mus. Zool. Cent. Univ. Nanking.

Distribution in China: Zhejiang (Linhai).

Etymology: The name of this species refers to its occurrence with the common low creeping grass, *Zoysia pungens* Willd var. *japonica* Hack.

morrisi Group

Diagnosis: *Amynthas* with spermathecal pores 2 pairs, at 5/6/7.

Global distribution: Oriental realm, ? Australasian realm and ? introduced into the Oceanian realm. There are 28 species from China.

TABLE 19 Key to Species of the *morrisi* Groups of the Genus *Amynthas* From China

1. Spermathecal pores on dorsum 2
Spermathecal pores on ventrum 3
2. Spermathecal pores about 1/4 of circumference dorsally apart; with a pair of large disc-like papillae placed in XVIII/XIX behind each male porophore .... *Amynthas oculatus*
Spermathecal pores closer to dorsomedian line; a raised papilla placed in front on XVII and another behind on XIX; ventrolateral region of XVII–XIX elevated .... *Amynthas sinuosus*
3. With papillae front and/or behind clitellum 4
Without papillae front clitellum, with papillae behind clitellum 14
4. Male pores in a round papilla, around 3–4 circular ridges. Paired or unpaired plain papillae found ventrally on IX, in front setal circle of the segment .... *Amynthas mediparvus*
With papillae front and behind clitellum 5
5. Genital papillae placed ventromedially on VII and VII, both in front of and behind setal circle; papillae placed ventromedially on XVI–XX, generally posterior, rarely absent ... *Amynthas monoserialis*
With papillae between V–VII, and between XVII–XIX 6

TABLE 19 Key to Species of the *morrisi* Groups of the Genus *Amynthas* From China—cont'd

6. 2 small papillae situated in front of each spermathecal pore tubercle (V, VI), occasionally 1 behind and 1 in front; Male pores represented by a small papilla on a rather prominent and raised porophore and surrounded by 3–6 small papillae either placed all around, or more often on median side if few .... *Amynthas incongruus*
   Without papillae at V 7

7. With papillae at VI, VII, and/or VII 8
   With papillae at VII and/or VII 7

8. With papillae at VI, VII, and VII 9
   With papillae at VI–VII or VII–VII 10

9. Unpaired papillae ventromedially on VI–VII, presetal, medium-sized and round; male pores porophore accompanied by 2 small, round-toped papillae on its median side .... *Amynthas tripunctus*
   1 small papilla sometimes present at anterior edge of VII; 1 similar papilla constantly present on ventromedial side of VII, before the setal circle; or 3 papillae in similar place of segments VI, VII, and VII each in 1 segment placed ventromedially; or, either in VI, VII or VII, VII; very rarely all papillae around male pore region absent; 2 small round flat-topped immediately medial to it, 1 in front and 1 behind the setal zone, only in 1–2 cases is 1 of them missing .... *Amynthas morrisi*

10. With papillae at VI and VII 11
    With papillae at VII and/or VII 12

11. Genital papillae on ventral side of VI and VII, 3 on VI, postsetal, 1 median and 1 ventral to pore on each side, 4 or 5 on VII. 3 presetal and 1–2 postsetal, similarly placed; Several genital papillae constantly present on ventral side of XVII–XIX; 2–3 on XVII, postsetal, 1–2 on XVIII, medially placed, 3–5 found on XIX placed both presetally and postsetally .... *Amynthas gravis*
    Genital papillae round, usually pairs in presetal VII or VI–VII, rarely in presetal VII and VII, and rarely in postsetal VI, present or absent. 2 types of arrangement of genital papillae in the male pore area for this species: (1) male pore situated on a papilla-like porophore with a genital papilla anterior to it, surrounded by 3–4 skin folds; (2) male pore on a flat-topped papilla-like porophore, with a genital papilla on or slightly behind the setal line immediately medial to it and anterior or anteromedial to it, surrounded by 3–4 skin folds. In addition, 1 papilla in line with or slightly medial to the right, left or each male pore in presetal XIX for about half of the specimens. Rarely, 4 papillae arranged transversely in presetal XVIII and XIX. Each papilla round .... *Amynthas mutabilitas*

12. Small and numerous papillae on VII and VII; Genital papillae in this region usually on XV–XVIII or XIX, more often on XVII, rarely totally absent, small and numerous, up to 14 in a transverse row, frequently about 6–10, arranged presetally and postsetally; in small forms only 2 large ones present presetally on XVII in line with male porophores; in the type none on XVIII, few on XVII, some on XV and XVI .... *Amynthas varians*
    With papillae at VII 13

13. A pair of round markings anterior to setae circle of VII; rarely absent. Irregular papillae present anterior to setae circle of XVII, XVIII, and XIX, which constitute 5–7 large gland protuberances .... *Amynthas dinganensis*
    There are a pair of largish papillae near the anterior margin of the VII; there are 8 largish papillae on the XVIII segment, each surrounded by a series of circular ridges upon the skin; 2 of these papillae form on each side with the male pore of their side a triangle; the remaining 4 form a line across the segment above the line of the setae; on XIX, on the left side of the body, is a single similar papilla .... *Amynthas insulae*

TABLE 19 Key to Species of the *morrisi* Groups of the Genus *Amynthas* From China—cont'd

14. Without papillae front and behind clitellum 15
    Without papillae front clitellum, with papillae behind clitellum 19

15. A marked I-shaped depression on ventral side 1/2 XVII–1/2 XIX; each pore on inner wall of the depression; 2 rather deep transverse grooves posteriorly on XVII and anteriorly on XIX; intersegmental furrows 17/18 and 18/19 entirely obliterated on ventral side .... *Amynthas hainanicus*
    Not I-shaped 16

16. Male pores situated in center of 2 coalesced flat-topped papillae, surrounded by several concentric ridges; the papillae with surrounding glandular area slightly raised .... *Amynthas dignus*
    A marked not coalesced flat-topped papilla 17

17. Male pores on a nipple-like porophore, its base with concentric ridges, body wall of XVI–XXV in line with porophore on each side raised as skin flap, always directed and partly folded toward ventral side .... *Amynthas lacinatus*
    Not nipple-like porophore 18

18. Male pores on a raised cone, slightly pointed at tip, with 2 glandular areas on median side near the tip and circular at base .... *Amynthas lubricatus*
    Male pores on a horseshoe-shaped porophore. It is usually invaginated, surrounded by 3–4 skin folds .... *Amynthas nanulus*

19. Male pores paired in XVIII, each situated in center of 2 coalesced flat-topped papillae, surrounded by several concentric ridges; the papillae with surrounding glandular area slightly raised .... *Amynthas choeinus*
    Genital papilla between XVII–XXI 20

20. Genital papilla at segment furrow 21
    Genital papilla on segments 23

21. Male pores on a conical papilla, with a pair of large disc-like papillae placed in 18/19 behind each male porophore .... *Amynthas oculatus*
    Papilla more than 1 pair 22

22. Male field genital papillae ovate, flat-topped, paired on 17/18, 18/19 median to male pore .... *Amynthas diaoluomontis*
    Male pores on the center of a slightly raised, conical, glandular porophore, with 1–2 circular folds. Paired oval papillae on 17/18/19/20/21 median to level of male pores .... *Amynthas octopapillatus*

23. Male pores on an ovoid porophore; rarely 2 large ovoid genital papilla exist to anterior and posterior of the ectal part of male pore .... *Amynthas tenuis*
    Genital papilla between XVII–XIX 23

24. Genital papilla on XVIII 24
    Genital papilla between XVII–XIX 23

25. Male pores on a rather pointed porophore, with a small papilla in front on XVII, rarely or some absent .... *Amynthas puerilis*
    Male pores on a porophore laterally surrounded by ridges, a large and flat-topped papilla placed ventral to each porophore .... *Amynthas sapinianus*

26. Male pores on the center of a slightly raised, conical porophore, with 2–3 circular folds. Male field genital papillae ovate, flat-topped; paired above setae annulet on XVIII and XIX .... *Amynthas lingshuiensis*
    Male pores on the top center of a slightly raised, conical porophore, with 2 circular folds not very clear. In XVII and XIX, in line with male pores, there are paired large round papillae .... *Amynthas zhangi*

Continued

2. SYSTEMATICS

157. *Amynthas choeinus* (Michaelsen, 1927)

Pheretima choeina Michaelsen, 1927. *Boll. Lab. Zool. Portici.* 21:85.
Pheretima choeina Gates, 1939. *Proc. U.S. Natr. Mus.* 85:427.
Amynthas choeinus Sims & Easton, 1972. *Biol. J. Soc.* 4:236.
Amynthas choeinus Xu & Xiao, 2011. *Terrestrial Earthworms of China.* 94–95.

External Characters

Length ? mm, width ? mm, segment number ?. Clitellum XIV–XVI, annular, smooth, setae absent. Setae fine, evenly distributed, both dorsal and ventral breaks indistinct. Setae number 24 (V), 32 (IX), 35 (XVII), 36 (XXV); *aa* slightly larger than *ab*, *zz* = 1–2*yz*; setae absent or small on segment X. Male pores paired in XVIII, about 1/3 of circumference apart ventrally, each situated in center of 2 coalesced flat-topped papillae, surrounded by several concentric ridges; the papillae with surrounding glandular area slightly raised. No other genital papillae. Spermathecal pores 2 pairs in 5/6/7, or 3 pairs in 4/5/6/7, intersegmental, about 1/3 of circumference apart ventrally. No genital marking. Female pore single in XIV, medioventral.

Internal Characters

Septa 8/9/10 absent, 3/4/5 very thin, 5/6/7/8 very slightly muscular, 10/11/12/13/14 thinner and membranous. Nephridial tufts in front of 5/6/7 close to gut moderately thick. Gizzard round and smooth on surface, in 1/2 IX, X. Intestine swelling in XVI. Caeca simple, cone-shaped, in XXVII, extending anteriorly to XXII. Vascular loops in IX asymmetrical, 1 limb stout, those in X both very stout. Anterior pair of testis sacs as large as seminal vesicles communicated both ventrally and dorsally. Seminal vesicles in XI, XII, fairly developed, dorsal lobes distinct and clearly marked off, about 0.8 mm in diameter. Second pair enclosed in testis sacs. Prostate glands large, in XVII–XXI, with finely divided lobules; its duct straight, uniform in caliber throughout the length, its ectal end slightly curved. Accessory glands not found, slightly velvety around the base of prostatic duct. Spermathecae 2 pairs in VI and VII, or 3 pairs in V–VII, main pouch about 1.5 mm long. Ampulla heart-shaped, about 0.7 mm long. Diverticulum longer than or as long as main pouch, ental end with an ovoid seminal chamber, its duct slender, ental four-fifths greatly coiled.

Color: Greyish generally, slightly darker on dorsal side. Clitellum reddish.
Type locality: Yunnan (Laoqiaozhen).
Distribution in China: Yunnan (Laoqiaozhen).
Etymology: The specific name refers to the type locality.

Deposition of types: Hamburg Museum.
Remarks: The seminal vesicles of XI and XII are small, little if at all larger than the pseudovesicles of XIII. The left anterior vesicle is smaller than the others. The appearance of the seminal vesicles and in particular that of the left side of XI together with the absence of the setae on X.

158. *Amynthas diaoluomontis* Qiu & Sun, 2009

Amynthas diaoluomontis Qiu & Sun, 2009. *Revue Suisse De Zoologie.* 116 (2): 290–293.

External Characters

Length 135–189 mm, clitellum width 3.9–4.8 mm, segment number 213–237. Secondary annulations conspicuous in V–XXXVI. Prostomium combined prolobous and 1/3 epilobous. First dorsal pore 12/13. Clitellum XIV–XVI, annular, setae visible externally, particularly evident on ventrum. Setae number 56–78 (III), 72–94 (V), 72–106 (VIII), 44–46 (XX), 48–66 (XXV); 18–32 between spermathecal pores, 9–13 between male pores; *aa* = 1.1–1.2*ab*, *zz* = 1.2–2*zy*. Male pores paired in XVIII, 0.33 circumference apart ventrally, each on the center of a slightly raised, conical, glandular porophore, without circular folds. Genital papillae ovate, flat-topped, diameter 0.8–1.2 mm, paired on 17/18, 18/19 median to male pores. Spermathecal pores 2 pairs in 5/6/7, eye-like, 1/4 circumference apart ventrally. Genital markings absent. Female pore single, medioventral in XIV, in a small ovoid tubercle.

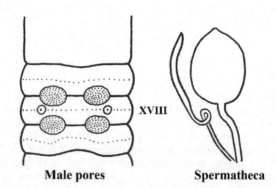

Male pores **Spermatheca**

FIG. 154 *Amynthas diaoluomontis* (after Qiu & Sun, 2009).

Internal Characters

Septa 8/9/10 absent, 5/6/7/8 thick and muscular, 10/11–14/15 slightly thickened. Gizzard in VIII–IX according to septum 7/8, ball-shaped. Intestine enlarged distinctly from XVI. Intestinal caeca simple and smooth, paired in XXVII, extending anteriorly to XXVI, with 2 incisions on dorsal margin. Esophageal hearts in X–XIII. Holandric, testis sacs 2 pairs, well developed, ventral in X–XI, in close proximity, but separated. Seminal vesicles paired in XI–XII, anterior pair bigger, none enclosed in testis sacs. Prostate glands developed, extending from XVI–XX, coarsely lobate. Prostatic duct U-shaped, slightly thicker at the distal part. No accessory glands present. Spermathecae 2 pairs in VI and VII, ampulla ovoid, about 2 mm long with a slender duct about equally as long. Diverticulum a little shorter than main pouch, slender, a distorted circle, terminal 3/5 dilated into a band-shaped chamber, partially filled, milky white.

Color: Preserved specimens lacking pigment on dorsum and ventrum, female pore lighter color than surroundings. Clitellum reddish brown.
Distribution in China: Hainan (Mt. Diaoluo).
Etymology: This species is named for its locality.
Type locality: Hainan (Mt. Diaoluo).
Deposition of types: Shanghai Natural History Museum and the Natural History Museum of Geneva.
Remarks: This species is similar to *A. tetrapapillatus* (Quan & Zhong, 1989). However, they differ with respect to the number of spermathecae, shape of spermathecal diverticulum, location of first dorsal pores, pigmentation, and enclosure of testes in sacs. *A. diaoluomontis* has 2 pairs of spermathecae in VI and VII, a straight diverticulum stalk, the first dorsal pore in 12/13, no pigmentation and the first pair of seminal vesicles is not enclosed in the second pair of testes sacs. In contrast, *A. tetrapapillatus* has only 1 pair of spermathecae in VI, a zigzag-looped diverticulum stalk near the seminal chamber, the first dorsal pore in 11/12, light maroon pigment on ventrum and dorsum, and the first pair of seminal vesicles contained in the second pair of testis sacs.

159. *Amynthas dignus* (Chen, 1946)

Pheretima digna Chen, 1946. *J. West China. Border Res. Soc. (B)*. 16:132–133, 160.
Amynthas dignus Sims & Easton, 1972. *Biol. J. Linn. Soc.*4:236.
Amynthas dignus Xu & Xiao, 2011. *Terrestrial Earthworms of China*. 103.

External Characters

Length 60–70 mm, width 2–3 mm, segment number 90–91. Prostomium 1/3 epilobous, tongue narrow, open behind. First dorsal pore 12/13. Clitellum XIV–XVI, annular, smooth, slightly extending over the intersegmental furrows 13/14 and 16/17, swollen, no setal pits on ventral

side. Setae all fine, evenly spaced, no difference in interval and length between dorsal and ventral ones, slightly longer on anterior ventral side. Setae number 38–42 (III), 50–56 (VI), 52–58 (IX), 35–44 (XXV); 16–17 (VI) between spermathecal pores, 8–12 between male pores; setal circles nearly closed, $aa = 1.2ab$, $zz = 1.5zy$. Male pores paired in XVIII, about 1/3 of circumference apart ventrally, each situated in center of 2 coalesced flat-topped papillae, surrounded by several concentric ridges; the papillae with surrounding glandular area slightly raised. No other genital papillae. Spermathecal pores 2 pairs in 5/6/7, or 3 pairs in 4/5/6/7, intersegmental, about 1/3 of circumference apart ventrally. No genital marking. Female pore single in XIV, medioventral.

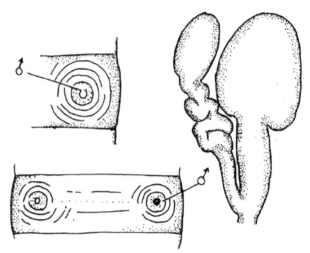

FIG. 155   *Amynthas dingus* (after Chen, 1946).

Internal Characters

Septa 8/9/10 absent, 3/4/5 very thin, 5/6/7/8 very slightly muscular, 10/11–13/14 thinner and membranous. Nephridial tufts in front of 5/6/7 close to gut moderately thick. Gizzard round and smooth on surface, in 1/2 IX, X. Intestine swelling in XVI. Caeca simple, cone-shaped, in XXVII, extending anteriorly to XXII. Vascular loops in IX asymmetrical, 1 limb stout, those in X both very stout. Testis sacs 2 pairs, anterior pair of testis sacs as large as seminal vesicles communicated both ventrally and dorsally. Seminal vesicles in XI, XII, fairly developed, dorsal lobes distinct and clearly marked off, about 0.8 mm in diameter. Second pair enclosed in testis sacs. Prostate glands large, in XVII–XXI, with finely divided lobules; its duct straight, uniform in caliber throughout the length, its ectal end slightly curved. Accessory glands not found, slightly velvety around the base of prostatic duct. Spermathecae 2 pairs in VI and VII, or 3 pairs in V–VII, main pouch about 1.5 mm long. Ampulla heart-shaped, about 0.7 mm long. Diverticulum longer than or as long as main pouch, ental end with an ovoid seminal chamber, its duct slender, ental four-fifths greatly coiled.

> Color: Greyish generally, slightly darker on dorsal side. Clitellum reddish.
> Distribution in China: Chongqing (Shapingba, Beipei).
> Remarks: This species differs from *A. sapinianus* (Chen, 1946) in the clitellum, the total absence of clitellar setae, more numerous setae, and thinner septa.

160. *Amynthas dinganensis* Qiu & Zhao, 2013

Amynthas dinganensis Qiu & Zhao, 2013. *Zootaxa*. 3619 (3):385–386.

External Characters

Length 77–91.5 mm, width 3.5–4.4 (clitellum), segment number 107–138; body cylindrical in cross section, gradually tapered towards head and tail. Prostomium epilobous. First dorsal pore in 11/12. Clitellum XIV–XVI, annular, setal absent externally. Setae number 40–56 (III), 46–56 (V), 48–60 (VII), 40–48 (XX), 42–44 XXV); 30–32 between spermathecal pores, 10–13 between male pores, setal formula $aa = 1.0$–$1.2ab$, $zz = 1.0$–$1.5zy$. Male pores paired in XVIII, a little less than 0.5 body circumference apart ventrally, surrounded by 4–6 circular folds. Irregular papillae present anterior to setae circle of XVII, XVIII and XIX, which constitute 5–7 large gland protuberances. Spermathecal pores 2 pairs in 5/6/7, ventral, eye-like, less than 0.5 body circumference apart. A pair of round markings anterior to setae circle of VII, about 0.14 circumferences apart; rarely absent. Female pore single in XIV, medioventral.

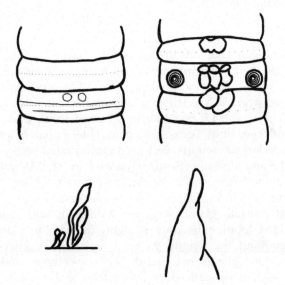

FIG. 156   *Amynthas dinganensis* (after Zhao et al., 2013).

Internal Characters

Septa 8/9/10 absent, 5/6/7/8 comparatively thick and muscular, 10/11 thick (10/11 rarely thin), 11/12/13 thin. Gizzard in IX–X, ball-shaped. Intestine swelling in XVI. Intestinal caeca simple, in XXVII or XVIII, extending anteriorly to XXIV or XX, 1–2 big incisions on dorsal margins, smooth on ventrum. Dorsal blood vessel single, continuous onto pharynx; hearts in X–XIII, first pair small. Holandric, Testis sacs 2 pairs in X–XI. Seminal vesicles paired in XI–XII, developed, separate ventrally, rarely not developed. Prostates in XVIII or XVII–XVIII, degenerate, closely adherent to body wall. Prostatic duct stout and long, U-shaped. 4 ovoid accessory glands present in segment XVII–XVIII, attach to the body wall, without stalk. Spermathecae 2 pairs in VI–VII with a gradually tapering slender duct; ampulla elongated, heart-shaped, spermathecal duct as long as 0.67–0.75 main pouch. Diverticulum as long as 0.66 to 0.8 of, or a little longer than, or equal to, main pouch, terminal 0.33–0.5 enlarged as a clavate seminal chamber; stipitate accessory glands present.

> Color: Preserved specimens dorsally light brown before clitellum and behind clitellum, grey ventrally. Few dorsally without pigmentation before clitellum and light red-brown behind clitellum, pale ventrally.
> Distribution in China: Hainan (Ding'an).
> Etymology: The name of this species refers to its type location.
> Deposition of types: Station Biologiqué, Université de Rennes 1, France.
> Type locality: The specimens were collected under a rubber plantation of Longmeng Town, Ding'an County, Hainan Province, China.
> Remarks: This species is similar to *A. varians* (Chen, 1938) and *A. hainanicus* (Chen, 1938). They share the following characters: 2 pairs of spermathecal pores in 5/6/7, septa also absent in 8/9/10, spermathecae heart-shaped, intestinal caeca simple. They differ from *A. dinganensis* in body size, genital marking, and genital papillae. The irregular papillae anterior to the setae circle of XVII, XVIII, and XIX make *A. dinganensis* different from the others. Furthermore, *A. dinganensis* is nearly 2 times larger than *A. varians* and *A. hainanicus*, and it has 1 pair of round genital markings anterior to the setae circle in VII. In addition, it differs from *Amynthas hainanicus* because of its degenerative prostate gland, presence of accessory glands, short spermathecal diverticulum, and body pigment. The sequence of a 16S gene fragment of the holotype of *A. dinganensis* (F-HN201116-04A) has been deposited in GenBank under the Accession No. JQ904530.

161. *Amynthas endophilus* Zhao & Qiu, 2013

Amynthas endophilus Zhao and Qiu, 2013 *J. Nat. Hist.* 47(33–36):2176–2183.

External Characters

Length 96 mm, width 3 mm (clitellum), segment number 159. Prostomium 1/2 epilobous. First dorsal pore in 12/13. Clitellum XIV–XVI, ring-shaped, markedly glandular; setae, dorsal pore, and annulus invisible. Setae dense intensive, numbers: 56 (III), 72 (V), 84 (VII); 25 (VI) between spermathecal pores, 2 (XVIII) between male pores; setal formula $aa = 1.4ab$, $zz = 2zy$. Male pores paired in XVIII, a little less than 0.33 body circumference apart ventrally, each on top of a round glandular porophore, surrounded by 4 skin folds. Genital papillae absent. Spermathecal pores 2 pairs, ventral, in 5/6–6/7, each eye-shaped and obvious, distance between paired pores about 0.4 body circumference apart. Genital markings absent. Spermathecal pores 2 pairs, ventral, in 5/6–6/7, about 0.4 body circumference apart. No genital markings in this region. Female pore single in the center of XIV ventrally, ovoid, brown.

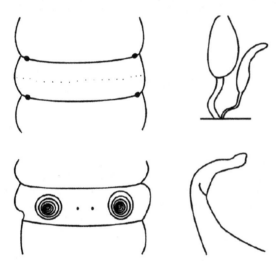

FIG. 157 *Amynthas endophilus* (after Zhao et al., 2013).

Internal Characters

Septa 8/9/10 absent, 5/6/7/8 thick and muscular, 10/11/12/13 a little thicker than those following. Gizzard ball-shaped, in IX–X. Intestine enlarged in XV. Intestinal caeca simple, originating in XXVII, extending anterior to XXIV, dark brown. Esophageal hearts 4 pairs, in X–XIII, the last 3 more developed. Testis sacs paired in X and XI, both separate ventrally. Seminal vesicles paired in XI and XII, stout and developed, both separate ventrally. Prostate gland in XVII–XXI, developed, racemose, and

divided into 5–6 main lobes, with a 6- or U-shaped duct, uniform in caliber. Spermathecae 2 pairs, in VI–VII, 2.3 mm long; ampulla elongated ovoid, duct a little shorter than ampulla, diverticulum 1.8 mm long, ental 0.6 dilated, serving as zonal seminal chamber, silvery white.

Color: Preserved specimens no pigmentation. Clitellum light greyish.
Distribution in China: Hainan (Diaoluoshan National Forest Park).
Deposition of types: Shanghai Natural Museum.
Type locality: The specimens were collected from black loam near the river behind the hotel in Diaoluoshan National Forest Park, southeast Hainan.
Remarks: *Amynthas endophilus* is most similar to *Amynthas morrisi* (Beddard, 1892). Both of them are moderately sized species. Their spermathecae are similar: both have an ovoid ampulla and a slender duct which is a little shorter than the ampulla; the diverticulum is shorter than the main pouch with a zonal seminal chamber. Furthermore, both of them have 2 pairs of spermathecal pores in 5/6/7, no septa in 8/9/10, racemose prostate gland and simple intestinal caeca. However, they differ in the pigmentation, genital markings, genital papillae, first dorsal pore, and gizzard. *A. endophilus* has no pigmentation, while *A. morrisi* is brownish dorsally and grey ventrally. *A. endophilus* has neither genital markings nor genital papillae. In contrast, genital markings are sometimes present in *A. morrisi* and *A. morrisi* always has genital papillae anterior and posterior male pores. The first dorsal pore of *A. endophilus* is in 12/13, while that of *A. morrisi* is in 11/12. The gizzard of *A. endophilus* is in IX–X, while that of *A. morrisi* is in IX or VII–IX.

162. *Amynthas fluviatilis* Zhao & Sun, 2013

Amynthas fluviatilis Zhao and Sun, 2013. *J. Nat. Hist.* 47(33–36):2183–2185.

External Characters

Length 121 mm, width 3.2 mm (clitellum), segment number 119. Prostomium 1/2 epilobous. First dorsal pore in 11/12. Clitellum in XIV–XVI, ring-shaped, purplish, markedly glandular; setae invisible. Setae sparsely paired, regularly distributed around segmental equators, numbers: 52 (III), 56 (V), 40 (VII), 46 (XX), 55 (XXV); 22 (VI) between spermathecal pores, 5 (XVIII) between male pores; setal formula $aa = ab$, $zz = 1.3zy$. Male pores paired in XVIII about 0.33 body circumference apart ventrally, each on the center of a slightly raised, tiny conical porophore, surrounded by 5 skin folds. A small papillae posterior to the left male pore. Another small papillae

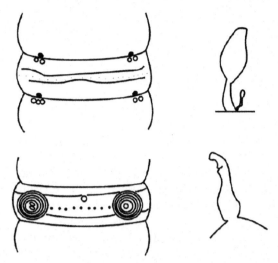

FIG. 158 *Amynthas fluviatilis* (after Zhao et al., 2013).

present in middle of XVIII ventrally above the setae line. Spermathecal pores 2 pairs in 5/6/7, ventral, eye-shaped, distance between them about 0.33 body circumferences apart. 2 tiny genital markings posterior to each pore, but 3 tiny markings posterior to the left pore in 6/7. Female pore single in the ventral center of XIV, ovoid, brown.

Internal Characters

Septa 8/9/10 absent, 6/7/8 comparatively thick and muscular; 10/11/12/13 thin but slightly thicker than those following. Gizzard barrel-shaped, in IX–X. Intestine enlarged from XV. Intestinal caeca simple, small, originating in XXVI, extending anterior to 1/2 XXIV, dark brownish, with a slight incision on terminal dorsal margin, smooth ventrally. Hearts 4 pairs, in X–XIII. Testis sacs paired in X and XI. Seminal vesicles paired in XI and XII, the first pair slender, the second more developed. Prostate gland in 1/2 XVII–1/3 XX, coarsely lobated, composed of a grey-brown vesicular capsule, its duct well-developed, U-shaped, and proximally appreciably enlarged. An accessory present in VII. Spermathecae 2 pairs, in VI–VII; ampulla ovoid, duct slender, a little shorter than ampulla, diverticulum short, as long as 0.25 of main pouch, ental 0.25 swollen, serving as seminal chamber, whitish in appearance.

Color: Preserved specimens light brownish dorsally; no pigmentation before clitellum, but light brownish after clitellum ventrally. Clitellum purplish.
Distribution in China: Hainan (Jianfengling National Nature Reserve).
Deposition of types: Shanghai Natural Museum.
Remarks: *A. fluviatilis* (Zhao and Sun, 2013) is somewhat similar to *A. incongruus* (Chen, 1933). Both of them are moderately-sized species with

pigmentation. They have 2 pairs of spermathecal pores in 5/6/7, simple intestinal caeca, and no septa in 8/9/10. Furthermore, their spermathecae are similar. They have an ovoid ampulla with a duct a little shorter than the ampulla. Their diverticulum is small; there is an accessory gland around the prostate. However, *A. fluviatilis* has no accessory gland around the spermathecae, and its seminal vesicles are developed. A pair of small papillae is present in the middle of XVIII ventrally above the setae line in *A. fluviatilis*, which is similar to some individuals of *A. incongruus*. However, there is only a small papilla posterior to the left male pore in *A. fluviatilis*, while 3–7 small genital papillae occur around the male pores in *A. incongruus*. In the spermathecal regions there are only 2–3 small genital markings posterior to each spermathecal pore in *A. fluviatilis*, while there are usually small paired genital markings presetal and/or postsetal in VI and/or VII in *A. incongruus*. Occasionally, there are no genital markings in *A. incongruus*.

163. *Amynthas fucatus* Zhao & Sun, 2013

Amynthas fluviatilis Zhao, Sun, Jiang and Qiu, 2013.
J. Nat. Hist. 47(33–36):2185–2187.

External Characters

Length 137 mm, width 4 mm (clitellum). Prostomium epilobous. First dorsal pore in 11/12. Clitellum XIV–XVI, ring-shaped, markedly glandular; ventral setae invisible, but dorsal pore visible. Setae intensive, their numbers: 44 (III), 52 (V), 60 (VII), 40 (XX), 48 (XXV); 26 (V), 24 (VI), 26 (VII), 30 (VII) between spermathecal pores, 8 (XVIII) between male pores; setal formula $aa = 1$–$2.5ab$, $zz = zy$. Male pores paired in XVIII about 0.4 body circumference apart ventrally, each on top of a conical porophore, 2 ovoid papillae in the inner side of male pore, both male pore and papillae are surrounded by 3–5 skin folds. Another small pair of round papillae positioned in the middle of XVIII ventrally above setae line. Spermathecal pores 2 pairs in 5/6/7, ventral, each eye-shaped and obvious, distance between paired pores about 0.4 body circumference apart. Genital markings absent. Female pore single in the ventral center of XIV, ovoid, brown.

Internal Characters

Septa 8/9, 9/10 absent; 5/6/7/8 comparatively thick and muscular; 10/11/12 thin. Gizzard ball-shaped, in IX–X. Intestine enlarged from XIV. Intestinal caeca simple, originating in XXIII, extending anterior to XXVI, dark brown, smooth ventrally and dorsally. Hearts 4 pairs, in X–XIII. Holandric, Testis sacs paired in X and XI. Seminal vesicles

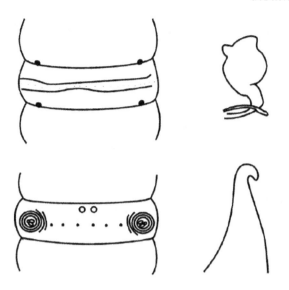

FIG. 159 *Amynthas fucatus* (after Zhao et al., 2013).

paired in XI and XII. Prostate gland in XV–XXI, developed, coarsely lobated, with a stout U-shaped duct. Spermathecae 2 pairs, in VI–VII; ampulla elongated ovoid, duct about 0.5 of ampulla, distinctly separated; diverticulum about 0.33 of main pouch and straight, ental 0.75 enlarged, serving as virgulate seminal chamber which is whitish in appearance.

Color: Preserved specimens dark red-brown dorsally and light red-brown ventrally before clitellum, pale after clitellum both dorsally and ventrally. Clitellum red-brown.
Distribution in China: Hainan (Limushan Nature Reserve).
Deposition of types: Shanghai Natural Museum.
Habitat: The specimens were collected in tropical rainforest on Quling Mountain in Limushan Nature Reserve, at an elevation of 739 m.
Remarks: *A. fucatus* Zhao and Jiang, 2013 is somewhat similar to *A. incongruus* (Chen, 1933). They are moderate-sized species with pigmentation, the first dorsal pore in 11/12, 2 pairs of spermathecal pores in 5/6/7, a developed prostate and simple intestinal caeca. Genital papillae occur around the male pores and a small pair of papillae is placed medioventrally in XVIII in both species. However, only *A. fucatus* has 2 ovoid papillae on the inner side of each male pore and no genital markings. Although the diverticulum is small in both species, it has no distinct seminal chamber in *A. fucatus*. Furthermore, there is no accessory gland in *A. fucatus*. In addition, *A. fucatus* is also similar to *A. lubricatus* (Chen, 1936) to some extent. They are moderate-sized species, their first dorsal pore is in 11/12, and they have 2 pairs of spermathecal pores in 5/6/7, simple intestinal caeca, an indistinct seminal chamber and neither accessory glands nor genital markings.

However, they are clearly different in the male pore region, pigmentation, prostate, and diverticulum. *A. fucatus* has pigmentation and genital papillae. Its prostate is developed but diverticulum is small.

164. *Amynthas gravis* (Chen, 1946)

Pheretima gravis Chen, 1946. *J. West China. Border Res. Soc. (B)*. 16:129–130, 141, 160.
Amynthas gravis Sims & Easton, 1972. *Biol. J. Linn. Soc.*4:236.
Amynthas gravis Xu & Xiao, 2011. *Terrestrial Earthworms of China*. 119–120.

External Characters

Length 80–88 mm, width 2.5–3.0 mm, segment number 110–112. Prostomium nearly tanylobous. First dorsal pore 12/13. Clitellum XIV–XVI, annular, less glandular on ventral side of XIV, with setae ventrally on XIV–XVI, dorsal and lateral sides glandular and smooth. Setae very fine and numerous, especially close on anterior end, no difference in size and interval both dorsally and ventrally. Setae number 80 (III), 124 (VI), 96 (X), 62 (XXV); 40–42 between spermathecal pores, 12–14 between male pores; $aa = 2.5$–$3\ ab$, $zz = 1.2$–$1.5zy$. Male pores paired in XVIII, about 1/3 of circumference apart ventrally, each on a small papilla. Several genital papillae constantly present on ventral side of XVII–XIX; 2 or 3 on XVII, postsetal, 1 or 2 on XVIII, medially placed, 3–5 found on XIX placed both presetally and postsetally. Spermathecal pores 2 pairs in 5/6/7, about 1/3 of their ventral circumference apart. Genital papillae on ventral side of VI and VII, 3 on VI, postsetal, 1 median and 1 ventral to pore on each side, 4 or 5 on VII, 3 presetal and 1–2 postsetal, similarly placed. Female pore single in XIV, medioventral.

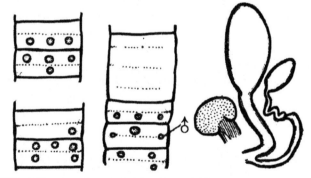

FIG. 160 *Amynthas gravis* (after Chen, 1946).

Internal Characters

Septa 8/9/10 absent, 5/6/7/8 thicker and muscular, 10/11/12 rather membranous. Nephridial tufts in front of 5/6/7 moderately thick. Gizzard round, moderate in size, in 1/2 IX and X. Intestine swelling in XV. Caeca simple, conical and smooth on both sides, extending anteriorly to XXIV. Vascular loops in IX very stout, enclosed partly in the testis sacs. Testis sacs in X moderately developed, narrowly connected ventrally; those in XI similar in size, connected ventrally. Both pairs connected and probably communicated dorsally (as there are strands of tissue connected). Seminal vesicles small usually as narrow strands, each with a distinct globular lobe dorsally, first pair enclosed in posterior pair of testis sacs. Prostate gland lobate in XVII–XIX, its duct long, ectal half slender. Accessory glands as small patch of gland tissue, with large but short cord-like ductule, sessile. Spermathecae 2 pairs, in VI and VII, main pouch about 2.0 mm long. Ampulla sac-like and small, diverticulum shorter than main pouch; seminal chamber date-shaped, ental third of duct generally coiled. Accessory glands similar in character to those in the prostate region.

Color: Light greyish dorsally, pale ventrally.
Distribution in China: Chongqing (Shapingba, Beipei, Fuling).

165. Amynthas hainanicus (Chen, 1938)

Pheretima (Pheretima) haiananica Chen, 1938. *Contr. Biol. Lab. Sci. Soc. China (Zool.)*. 12(10):396–398.
Amynthas hainanicus Sims & Easton, 1972. *Biol. J. Soc.* 4:236.
Amynthas haiananicus Xu & Xiao, 2011. *Terrestrial Earthworms of China*. 120.

External Characters

Length 50 mm, width 1.8 mm, segment number 110. Prostomium 1/3 epilobous. First dorsal pore 12/13. Clitellum entire, in XIV–XVI, annular, markedly glandular, without setae. Setae uniformly built, ventral setae of anterior segments slightly wider at interval (more noticeable on V–VII) but not enlarged; setal breaks $aa = 2ab$, $zz = yz$ at preclitellar, $=2yz$ at postclitellar region; setae number 34 (II), 56 (V), 46 (VII), 44 (VII), 44 (XXV); 7 (VI), 8 (VII) between spermathecal pores, 9 (XVII), 10 (XIX) between male pores. Male pores 1 pair, on ventrolateral side of XVIII, about 1/3 of circumference apart ventrally. A marked I-shaped depression on ventral side 1/2 XVII–1/2 XIX, glandular in appearance; each pore on inner wall of the depression; 2 rather deep transverse grooves posteriorly on XVII and anteriorly on XIX; intersegmental furrows 17/18 and 18/19 entirely obliterated on ventral side. Spermathecal pores 2 pair in 5/6/7, about 1/5 of circumference apart ventrally, each on anterior side of an ovoid tubercle projecting anteriorly from segments VI and VII into intersegmental furrows. Female pore 1 pair, in a small transverse ovoid tubercle, medioventral in XIV, 2 pores about 0.2 mm at interval.

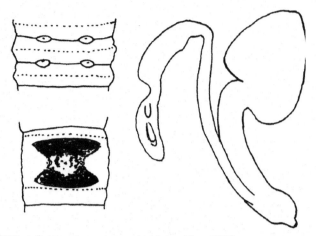

FIG. 161　*Amynthas haiananicus* (after Chen, 1938).

Internal Characters

Septa 8/9/10 absent, 4/5/6/7/8 thickened and well muscular, 10/11/12/13 slightly thickened. Gizzard moderate in size, in 1/2 VII–1/2 X. Intestine enlarged from XVI (?). Caeca simple, short and blunt, in XXVII, extending anteriorly to XXV or 1/2 XXIV, whitish in appearance. First pair of hearts in X present, attached to septum 10/11. Testis sacs enormously developed, communicated both dorsally and ventrally, second pair enclosing first pair of seminal vesicle. Seminal vesicles small, anterior pair about 0.7 mm in dorsoventral length, dorsal lobe two-thirds in size, second pair similar in size, vesicle larger. Prostate gland well developed, in XVI–XXI, coarsely lobate, its duct long, thinner at ectal end, without gland connected with ectal portion. Unicellular glands numerous embedded among muscle fibers of thickened ventral body wall. Setigenous sacs normal in shape but much enlarged, about 0.6 mm long, 0.035 mm thick. Spermathecae 2 pairs, in VI and VII; ampulla large, heart-shaped, about 0.5 mm long, with a long stout duct, about 0.75 mm long. Diverticulum longer than main pouch, seminal chamber thin walled, partially filled, 0.7 mm long, its duct thick and long, about 1 mm long, 0.08 mm thick.

Color: Preserved specimens pale generally.
Distribution in China: Hainan
Etymology: The name of the species refers to the type locality of Hainan province. There is no information regarding the specific locality on the island. The present specific name signifies its indefinite occurrence.

166. Amynthas incongruus (Chen, 1933)

Pheretima incongrua Chen, 1933. *Contr. Biol. Lab. Sci. Soc. China (Zool.)*. 9:270–274.

Pheretima incongrua Gates, 1935b. *Lingnan Journ. Sci.*
14(3):452–453.
Pheretima incongrua Gates, 1959. *Amer. Mus. Novitates.*
141:13–15.
Pheretima incongrua Tsai, 1964. *Quer. Jou. Taiwan Mus.*
17:19–20.
Amynthas incongruus Sims & Easton, 1972. *Biol. J. Soc.*
4:236.
Amynthas incongruus James, Shih & Chang, 2005. *Jour.
N. His.* 39(14):1025.
Amynthas incongruus Xu & Xiao, 2011. *Terrestrial
Earthworms of China.* 129–130.

External Characters

Length 82–207 mm, width 4.2–5.5 mm, segment number 157–167. First dorsal pore 11/12. Clitellum in XIV–
XVI, annular, distinct in all mature specimens, glandular
portion well marked, whole length equal to 3 immediate
preclitellar segments or to 5 or more postclitellar segments, smooth on surface except on ventral side where
there are some indistinct grooves, especially on intersegmental and setal zones, setae absent, dorsal pore absent.
Setae rather weak, no setae specially modified; setae
number 43–59 (III), 46–68 (VII), 50–55 (XII), 46–53
(XXV); 22–24 between spermathecal pores, 9–12 between
male pores. Male pores paired in XVIII, about 1/3 of circumference apart ventrally, each represented by a small
papilla on a rather prominent and raised porophore
and surrounded by 3–6 small papillae either placed all
around or, more often, on median side if few. Shape of
genital papillae very similar to male papilla, which is distinguished by its more pointed tip and rather indistinct
concentric ridges. All genital papillae and male papilla
situated on a raised porophore surrounded by concentric
ridges gradually lowering down to body surface. Each
raised porophore with its surrounding ridges confined
to segment XVIII, with its median profile steeper. No
setae found on porophore and its profile. Spermathecal
pores 2 pairs in 5/6/7, about 1/3 of circumference apart,
each pore on a very minute tubercle raised in furrow, 2
small papillae situated in front of each pore tubercle,
occasionally 1 behind and 1 in front. Genital papillae so

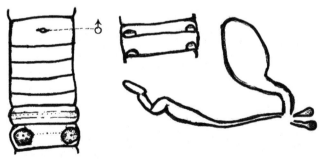

FIG. 162   *Amynthas incongruous* (after Chen, 1933).

far as observed in my specimens not found in other
regions just mentioned. Female pore single, in usual
position and in a rather larger depression.

Internal Characters

Septa 8/9/10 absent, 5/6 thick, 6/7/8 musculated and
notably thickened, 10/11/12/13 much less thickened,
thin behind 13/14. Gizzard in IX and X, globular and
slightly elongate, with smooth shining surface. Intestine
enlarged from XVI. Intestinal caeca paired in XXVII, simple, extending anteriorly to XXIV, slender and pointed
anteriorly, ventral edge slightly wrinkled. Hearts 4 pairs,
first pair in X very rudimentary, the rest also small in caliber. Testis sacs very conspicuous, both pairs nearly equal
in size, each pair almost occupying the whole segment,
anterior pair in X, posterior in XI, fully filled with sex
products. Seminal vesicles in XI and XII small and flattened, being exceedingly small while testis sacs enormously enlarged; concealed under and close to
posterior surface of septa 10/11 and 11/12 respectively,
each weakly wrinkled on surface, often narrow on lower
end but clearly marked off from its dorsal lobe, which is
globular or elongate, glandular on surface, somewhat
paler, generally larger than vesicle proper, and sometimes
3–4 times larger. Prostate glands very inconsistently
developed, gland portion entirely absent in most cases
with connective tissue attached to body wall, few well
developed, in XVII–XXI or XXII. Prostatic duct short,
with smooth and glistening surface, or relatively long
its ectal two-thirds stouter. Accessory glands found in
corresponding to papillae outside, each with a cluster of
gland and a stalk for each. Spermathecae 2 pairs in VI
and VII, ampulla ovoid or large sac-shaped, generally
smooth on surface; ampullar duct either long and slender,
or short and stout, not clearly marked off from its sac.
Diverticulum very inconsistently developed, generally
longer than main pouch, its duct slender; seminal chamber as an elongate sac, sometimes enormously enlarged
with some part slightly constricted. Accessory glands
similar to those found in prostate region but more
exposed to coelomic surface.

Color: Preserved specimens purplish brown dorsally
and pale gray ventrally.
Type locality: Zhejiang (Linhai).
Deposition of types: Mus. Zool. Cent. Univ. Nanjing.
Distribution in China: Zhejiang (Linhai), Taiwan
(Taibei, Pingdong), Chongqing (Fuling), Sichuan
(Mt. Emei).
Remarks: This species is close to *A. morrisi* in some
respects but is easily distinguished by: (1) clitellum in 3
full segments without setae, (2) medially and laterally
paired genital papillae absent, (3) male pore region
with more and smaller papillae, usually very close to
the pore papillae, (4) first dorsal pore in 11/12, not in

10/11, (5) the characteristic shape of testis sacs and the exceedingly small size of seminal vesicles, (6) the prostates inconsistently present, while the spermathecae and diverticula often disproportionally developed, and (7) it is comparatively slender and elongate in form.

167. *Amynthas insulae* (Beddard, 1896)

Perichaeta insulae Beddard, 1896. *Proc. Zool. Soc. London.* 1896:204–205.

External Characters

Length 103 mm, segment number 95. Clitellum annular, in XIV–XVI, with setae on XVI. Male pores are separated by a moderate distance, on XVIII, ventral. There are 8 largish papillae on the XVIII segment, each surrounded by a series of circular ridges upon the skin; 2 of these papillae form on each side with the male pore of their side a triangle; the remaining 4 form a line across the segment above the line of the setae; a single similar papilla on the left side of XIX. In addition to these papillae developed in the neighborhood of the male pores, there is a pair near the anterior margin of segment VII like those of *Amynthas indicus*. Spermathecal pores 2 pairs in 5/6/7. Female pore single in XIV, medioventral.

Internal Characters

Septa 4/5/6/7/8 are not very much thickened, 10/11/12/13 stouter. Gizzard is rather bell-shaped, diminishing in transverse diameter anteriorly, but

FIG. 163   *Amynthas insulae* (after Beddard, 1896).

truncated posteriorly, where it has a thickened rim. Intestine swelling in XV, at about the middle of that segment. Caeca simple, in XXVII, extending anteriorly to XXV. Last heart in XIII. Testis sacs in XI and XII, compact in form. Prostate gland also rather compact, commencing in XVII, and extending as far as back as XX; prostatic duct is stout and S-shaped. Spermathecae 2 pairs, in VI and VII. Diverticulum about half its own length ampulla, of an elongated oval form.

Distribution in China: Hong Kong.

80. *Amynthas lacinatus* (Chen, 1946) (*v.ant. 80*)

168. *Amynthas lingshuiensis* Qiu & Sun, 2009

Amynthas lingshuiensis Qiu & Sun, 2009. *Revue Suisse De Zoologie.* 116(2):296–298.

External Characters

Length 76–113 mm, clitellum width 2.7–3.1 mm, segment number 123–153. Prostomium 1/2 epilobous. First dorsal pore 12/13. Clitellum XIV–XVI, annular, swollen, setae and dorsal pores absent. Setae numerous, 44–60 (III), 44–54 (V), 48–52 (VIII), 38–46 (XX), 44–50 (XXV); 19–23 between spermathecal pores, 5–7 between male pores; $aa = 1$–$1.3ab$, $zz = 1.3$–$2.2zy$. Male pores paired in XVIII, 1/3 circumference apart ventrally, each on the center of a slightly raised, conical porophore, with 2–3 circular folds. Genital papillae ovate, flat-topped, diameter 0.3 mm; paired above setae annulet on XVIII and XIX. The first pair 0.17 circumference apart and the second pair 0.25. Spermathecal pores 2 pairs in 5/6/7, eye-like, 0.33 circumference apart ventrally. Genital markings absent. Female pore single, medioventral in XIV, in a small ovoid tubercle.

Internal Characters

Septa 8/9/10 absent, 6/7/8 thick and muscular, 10/11/12/13 slightly thickened. Gizzard in VIII–IX according to septum 7/8, long barrel-shaped. Intestine enlarged gradually from XVI to XX and enlarged suddenly from XXI. Intestinal caeca paired in XXVII, simple, smooth, extending anteriorly to XXV. Esophageal hearts in X–XIII. Holandric, testis sacs 2 pairs, developed, in X–XI, separated on ventrum, second pair enclosing first pair of seminal vesicle. Seminal vesicles paired in XI–XII, small. Prostate glands developed, extending from 1/2 XVI–2/3 XX, composed of 3 parts. The first 2 parts are bigger than the last, which is finger-shaped. Prostatic duct inverted U-shaped. Accessory glands absent. Spermathecae 2 pairs in VI–VII, ampulla heart-shaped, about 1.9 mm with about equal duct. Diverticulum longer than main pouch by 1/5, 2/5 base curved, terminal 3/5 dilated into a band-shaped distal seminal chamber.

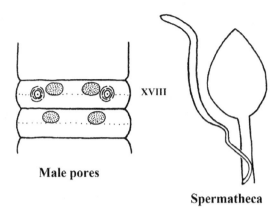

Male pores

Spermatheca

FIG. 164   *Amynthas lingshuiensis* (after Qiu & Sun, 2009).

Color: Preserved specimens lacking pigment on dorsum, light brown on ventrum, middorsal line purple. Clitellum light reddish.

Distribution in China: Hainan (Lingshui).

Etymology: The name of the species refers to the type locality.

Type locality: Hainan (Lingshui).

Deposition of types: Shanghai Natural History Museum and the Natural History Museum of Geneva.

Remarks: In comparison to the other species of the *morrisi* group reported from China and Southeast Asia (Sims & Easton, 1972), *A. lingshuiensis* is similar to *A. hainanicus* (Chen, 1938) in having spermathecal pores in 5/6/7, diverticulum longer than main pouch, and a thin-walled seminal chamber. However, it is easy to distinguish *A. lingshuiensis* from *A. hainanicus* by the larger body size, 2 pairs of papillae in 19/20/21, the larger interval spacing between spermathecal pores of a segment, and the lack of I-shaped depression in the male field. *A. hainanicus* has an I-shaped depression on ventral side of 1/2 XVII–1/2 XIX which is glandular in appearance and the male pore is on the inner wall of the depression.

169. *Amynthas lubricatus* (Chen, 1936)

Pheretima lubricata Chen, 1936. *Contr. Biol. Lab. Sci. Soc. China (Zool.)*. 11:281–283.

Pheretima lubricata Chen, 1946. *J. West China. Border Res. Soc. (B)*. 16:85–86, 142.

Amynthas lubricatus Sims & Easton, 1972. *Biol. J. Soc.* 4:236.

Amynthas lubricatus Xu & Xiao, 2011. *Terrestrial Earthworms of China*. 142.

External Characters

Length 85–110 mm, width 5.5–7.0 mm, segment number 105–112. Annulation of segments very distinct, 5 annuli on VII–XIII, 3 annuli on middlemost region of body, about last 20 segments very short and devoid of annulus. Prostomium with a very inconspicuous dorsal lobe, 1/2 epilobous, cut off completely behind. First dorsal pore 11/12. Clitellum in XIV–XVI, annular, smooth. Setae mostly deformed and very indistinct, generally absent on dorsal side, number of each segment unable to be counted, about 23–28 (VII) between spermathecal and 12–16 between male pores. Male pores 1 pair in XVIII ventrally, about 1/3 of circumference apart; each on a raised cone, slightly pointed at tip, with 2 glandular areas on median side near the tip and circular at base. No other genital marking; genital papillae absent. Spermathecal pores 2 pairs, in 5/6/7, intersegmental, about 2/5 circumference apart ventrally; each on a small tubercle, situated close to the preceding segment in the furrow; genital papillae absent. Female pore single in XIV, medioventral.

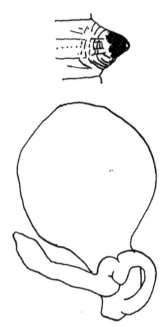

FIG. 165   *Amynthas lubricatus* (after Chen, 1936).

Internal Characters

Septa 5/6/7/8/9 very thick, 9/10/11 less thickened. Gizzard in IX and X, large and round. Intestine enlarged from XV. Intestinal caeca simple, in XXVII, extending anteriorly to XXV, smooth on both edges. Hearts 4 pairs, in X–XIII, all bifurcated from their origin on dorsal side: 1 branch connected with subintestinal vessel and the other with lateroesophageal vessel. First pair very slender, last 2 pairs stouter. Testis sacs in X and XI, completely united. Seminal vesicles small, each with a rather large dorsal lobe. Prostate glands compact and small, in XVII–XIX, its duct long, coiled and equally stout, with digitate branches connecting the glandular portion. Spermathecae 2 pairs in VI–VII, ampulla usually round, or subspherical, about 1 mm in diameter, ampullar duct about half length of ampulla, stout. Diverticulum elongate, club-shaped, with a short and slender duct, ental two-thirds straight and slightly enlarged; whole length about length of main pouch or slightly shorter.

Color: Preserved specimens pale both dorsally and ventrally; skin smooth and very slippery on surface.
Type locality: Chongqing
Distribution in China: Chongqing (Shapingba, Baixi, Beipei), Sichuan (Chengdu, Yibin).
Deposition of types: Mus. Biol. Lab. Sci. Soc. China.
Remarks: This species differs from all Chinese species known to date by its lubricated skin and degenerated setae. The skin is very much like that of *Desmogaster sinensis* Gates, 1930 or *Drawida sinica* (Chen, 1933). It is probably supplied with numerous glands in the hypodermis. This species is generally found in sandy soil.

170. *Amynthas mediparvus* (Chen & Xu, 1977)

Pheretima parva Chen & Xu, 1977. *Acta Zool. Sinica.* 23(2):177–178.
Pheretima mediparva Nakamura, 1999. *Edaphologia.* 64:3, 26.
Amynthas mediparvus Xu & Xiao, 2011. *Terrestrial Earthworms of China.* 146–147.

External Characters

Length 26–43 mm, width 2 mm, segment number 67–95. Prostomium 1/2 epilobous. First dorsal pore 13/14. Clitellum XIV–XVI, annular, setae absent. Setae fine and close; setal number 40–49 (III), 60–65 (VI), 63–64 (VII), 46–54 (XXV); $aa = 1.0–1.5ab$, $zz = 1.0–1.5yz$ at preclitellar, $=1.5–2.0yz$ at postclitellar region. Male pores 1 pair in XVIII, nearly 1/3 of circumference apart ventrally, each with a small penis in a round papilla, around 3–4 circular ridges. Spermathecal pores 2 pairs,

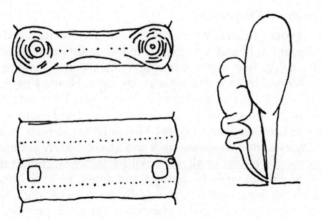

FIG. 166   *Amynthas mediocus* (after Chen & Xu, 1977).

in 5/6/7, close to anterior edge of VI, VII, nearly 1/3 of circumference apart ventrally. Paired or unpaired plain papillae found ventrally on IX, in front setal circle of the segment. Female pore single in XIV, medioventral.

Internal Characters

Septa 8/9/10 absent, 5/6/7/8 and 10/11/12/13 thick, 13/14/15 slightly thick, 15/16 membranous. Gizzard ball-shaped. Intestine enlarged from XVI. Caeca simple, cone-shaped, in XXVII, extending anteriorly to XXIV. Hearts in X–XIII, hearts in X, XI not included in the testis sacs. Testis sacs well developed, meeting dorsally but not communicated. First pair seminal vesicles contained in second pair of testis sacs, dorsal lobe obviously triangular. Prostate gland well developed, lobulated, in XVI–XX, its duct not curved, stout, about 0.5 mm long. Spermathecae 2 pairs in VI–VII, behind septum, ampulla long oval-shaped or long bag-shaped, 0.3–0.4mm in diameter, as long as duct; diverticulum slightly shorter than main part, as long as half the main duct, ental two-thirds bending, diverticulum with an enlarged and loosely seminal chamber. Accessory glands cauliflower-like.

Color: Preserved specimens gray-brown, a little light ventrally. Clitellum yellow-brown.
Type locality: Yunnan (Menglun).
Distribution in China: Yunnan (Menglun).
Remarks: This specimen differs from *A. puerilis* (Chen, 1938) and *A. oculatus* (Chen, 1938) by its small penis and spermathecae.

171. *Amynthas monoserialis* (Chen, 1938)

Pheretima (Pheretima) monoserialis Chen, 1938. *Contr. Biol. Lab. Sci. Soc. China (Zool.).* 12(10):399–401.
Amynthas monoserialis Sims & Easton, 1972. *Biol. J. Soc.* 4:236.
Amynthas monoserialis Xu & Xiao, 2011. *Terrestrial Earthworms of China.* 151–152.

External Characters

Length 52–150 mm, width 3–4 mm, segment number 136–146. Prostomium 1/3 epilobous. First dorsal pore 12/13. Clitellum in XIV–XVI, annular, less glandular, usually flattened on ventral side in fixed specimens; with distinct setae on ventral, but less on dorsal and lateral sides. Setae slightly larger and distinctly wider in intervals on ventral side, much denser on dorsal side of II–IX; preclitellar ventral setae widely spaced; setal breaks; $aa = 1.2–1.5ab$, $zz = 1.2yz$; setae number 70–74 (III), 107–122 (VI), 92–104 (VII), 50–64 (X), 36–44 (XXV); 28–44 (VI), 28–44 (VII), 20–36 (VII) between spermathecal pores, 7–9 between male pores. Male pores 1 pair in XVIII, about 1/3 of circumference apart ventrally, each on a small glandular area, papillae placed ventromedially on

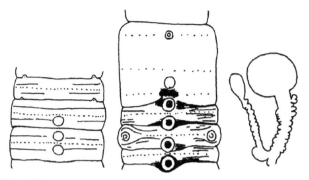

FIG. 167 *Amynthas monoserialis* (after Chen, 1938).

XVI–XX, generally posterior, rarely absent. Spermathecal pores 2 pairs in 5/6/7, about 1/3 of circumference apart ventrally, each on a tiny ampulla projecting out from anterior edge of VI and VII into intersegmental furrows. Genital papillae placed ventromedially on VII and VII, both in front of and behind setal circle; always smaller than those found in male pore region.

Internal Characters

Septa 8/9/10 absent, 6/7/8 well thickened, 10/11/12/13 slightly thicker and a little musculated. Gizzard in X, rounded. Intestine enlarged from XVI. Caeca simple, in XXVII, extending anteriorly to XXIV. Hearts in X–XIII. Testis sacs in X and XI, anterior pair well developed, narrowly connected on ventral side, posterior pair small, also narrowly connected. Seminal vesicles in XI enclosed in testis sacs, invisible on surface, those in XII larger; dorsal lobe larger than vesicle. Prostate gland large, coarsely lobate, in XV–XXI, with a short and straight duct, uniform in caliber. Accessory glands about 20 or more associated with a single papilla, each with a cord-like duct. Spermathecae 2 pairs, in VI and VII, small; main pouch about 2 mm long, ampulla globular with a long duct, usually with saccular structures either on ampulla or its duct. Diverticulum as long, or shorter; seminal chamber ovoid, light gray, also wrinkled on duct.

Color: Preserved specimens pale generally. Clitellum brownish.
Distribution in China: Hainan (Baopeng).
Remarks: This species appears to be very common in the Baopeng area as a large number of specimens were included in the collection.

172. Amynthas morrisi (Beddard, 1892)

Perichaeta morrisi Beddard, 1892. *P. Z. S.* 166.
Pheretima morrisi Chen, 1931. *Contr. Biol. Lab. Sci. Soc. China (Zool).* 7:148–155.
Pheretima morrisi Chen, 1933. *Contr. Biol. Lab. Sci. Soc. China (Zool).* 9:267–270.

Pheretima morrisi Chen, 1936. *Contr. Biol. Lab. Sci. Soc. China (Zool.).* 11:270.
Pheretima (Pheretima) morrisi Chen, 1938. *Contr. Biol. Lab. Sci. Soc. China (Zool.).* 12(10):382.
Pheretima morrisi Gates, 1939. *Proc. U.S. Natr. Mus.* 85:453–454.
Pheretima morrisi Chen, 1946. *J. West China. Border Res. Soc. (B).* 16:135.
Pheretima morrisi Tsai, 1964. *Quer. Jou. Taiwan Mus.* 17:17–18.
Pheretima morrisi Gates, 1972. *Trans. Amer. Philos. Soc.* 62(7):202–203.
Amynthas morrisi Sims & Easton, 1972. *Biol. J. Soc.* 4:236.
Amynthas morrisi Chang, Lin, Chen, Chung & Chen, 2009. *Earthworm Fauna of Taiwan.* 60–61.
Amynthas morrisi Xu & Xiao, 2011. *Terrestrial Earthworms of China.* 152–153.

External Characters

Length 75–144 mm, width 3.0–5.5 mm, segment number 64–156. Secondary annulation inconspicuous from VII posteriorly. Prostomium 1/2 epilobous. First dorsal pore in 10/11. Clitellum distinctly swollen, in XIV–XVI, annular, rather short, little longer than 2 immediately anterior segments; some specimens bearing a few very obscure setae on ventral side of XVI, also still small smaller setae in XIV. In some cases, only setal pits visible and others perfectly smooth. Setae small and delicate, shorter around the middle region of the body, setae at anterior segments quite uniformly arranged; dorsal setae inconspicuous due to the pigmentation of skin; their intervals on both dorsal and ventral sides about equal; *aa* equal to, or a little wider than *ab*; $zz = 1.5$–$2.5yz$. Setae number 26–34 (III), 38–42 (VI), 44–50 (VII), 42–56 (XII), 44–56 (XXV); 20–24 between spermathecal pores, 12–16 between male pores. Male pores 1 pair in ventrolateral side of XVIII segment, about 1/3 circumference apart ventrally, without a copulatory chamber, each on an elongate flat-topped papilla, the latter sometimes representing the fusion of 2 into 1 through which the prostatic duct opens. 2 small round flat-topped papillae immediately medial to it, 1 in front and 1 behind the setal zone, only in 1–2 cases is 1 of them missing. Circular ridges around the papillae and slightly elevated as a porophore; setae absent in the region. No papillae present. Spermathecal pores 2 pairs in 5/6/7, intersegmental, about 1/2 of circumference apart, usually a small glandular patch of skin around each pore; the glandular part often little depressed, pore sometimes as an eye-like depression with a whitish spot in the center. 1 small papilla similar to those around the male pore sometimes present at anterior edge of VII, about 1 mm ventromedial to the pore; 1 similar papilla constantly present on ventromedial side of VII, before the setal circle; in 1 individual examined 3

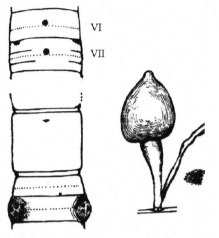

FIG. 168   *Amynthas monoserialis* (after Chen, 1938).

papillae in the similar place of segments VI, VII, VII each in 1 segment placed ventromedially. In some cases, either in VI, VII or VII, or VII. Very rarely all papillae around this region absent. Female pore single in XIV, medioventral.

Internal Characters

Septa 8/9/10 absent, 5/6/7/8, 10/11–13/14 thickened and musculated, sometimes 10/11 and 13/14 less thickened. Gizzard round bulb-like, smooth and glittering on surface, and somewhat transparent in appearance. Intestine enlarged from XV. Intestinal caeca paired simple, short, and small, in XXVII, extending anteriorly to XXV or rarely to XXIV. Hearts 3 pairs, in XI–XIII enlarged. Testis sacs in X and XI, large, seldom small, each pair entirely separated, sometimes widely placed; anterior pair in front of septum 10/11 and occupying segment X entirely, more widely separated than posterior pair; posterior pair also large, in contact with anterior pair, not communicated. Seminal vesicles in X and XI, a round middle lobe situated at anterior middle side of the vesicle, grayish and granular, marked off from the main vesicle, a ridge extending from the lobe downward to testis sac. Prostate glands small, usually in XVII–XX or XXI, racemose and finely divided into pieces, granular and compact in appearance; prostatic duct large at its middle, straight or S-shaped, coiled at both ends, directly opening into the larger papilla. Spermathecae 2 pairs in VI and VII, rarely the posterior pair in 2 segments, their ampulla lying in VII, while in the anterior pair 1 ampulla in VI, and 1 on opposite side in VII. Ampulla distended oval or heart-shaped, often with an apical knob, smooth and pale on surface, about

2.5 mm in length and 2 mm in width; its duct as long as itself, very stout entally, about 1.2 mm wide gradually narrowing towards the distal end about 0.3 mm in width, diverticulum slender, tube-like, a little shorter than main part, its duct occupying one-third the entire length, about 0.3 mm in diameter, slightly larger and roundish at ental end. In many cases, the diverticulum slightly twisted, with oval seminal chamber, much longer than the main part. Whitish stalked glands in a place close to the papillae externally; the form and character of the glands being same as those present in the region of prostatic glands.

Color: Preserved specimens dark gray or slaty gray with a violet tinge on anterodorsal side, brownish gray or dark buff-gray on posterodorsal side, with a chestnut streak along dorsal pores, grayish ventrally. Clitellum reddish brown or dull cherry.

Distribution in China: Jiangsu, Zhejiang (Lanxi, Tonglu, Fenshui, Linhai), Fujian (Xiamen), Taiwan (Taibei), Hong Kong, Hainan (Wanning), Chongqing (Nanchuan, Shapingba, Fuling, Jiangbei), Sichuan (Chengdu, Emei, and Leshan), Guizhou (Mt. Fanjing).

Type locality: Penang

Etymology: The name of the species refers to Mr. Morris.

Deposition of types: British Museum, England. Mus. Bio. Lab. Sci. Soc. China.

Remarks: This species is very similar to *Amynthas hawayanus*, but they differ chiefly as follows:

(1) The testis sacs in this species are entirely separate; (2) The seminal vesicle has a middle lobe, while in *A. hawayanus* it is absent; (3) The first pair of hearts is definitely absent, while in *A. hawayanus* they are present although small; (4) The prostate glands are finely granular, while in *A. hawayanus* they are coarsely lobate and smooth on the surface; (5) Accessory glands are small and irregularly shaped, while in *A. hawayanus* large and regularly ball-shaped; (6) Only 2 pairs of spermathecae present, while in *A. hawayanus* there are 3 pairs; (7) The papillae are confined to the male pore region and some (in VII or VI and VII) in front of setal circle, while in *A. hawayanus* they are not confined to that region and are all behind the setal circle; (8) The setae in II–VI are usually not widely spaced and spermathecal pores often marked externally, while in *A. hawayanus* this is not the case.

In Beddard's work, he placed the species *A. morrisi* synonymous with *A. hawayanus* but Chen considered it necessary to make them specifically different.

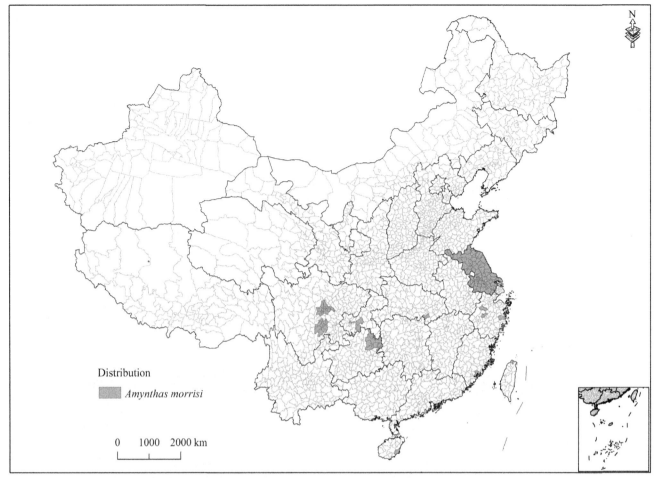

Distribution

Amynthas morrisi

0   1000   2000 km

FIG. 169   Distribution in China of *Amynthas morrisi*.

173. *Amynthas mutabilitas* Shen, 2012

Amynthas mutabilitas Shen, 2012. *J. Nat. Hist.* 46(37–38):2259–2283.

External Characters

Small to medium earthworms. Length 64–127 mm, width 2.8–4.24 mm (clitellum), segment number 72–110. Prostomium epilobous. First dorsal pore in 11/12 or 13/14. Clitellum XIV–XVI, annular, length 1.63–5.59 mm, setae and dorsal pores absent. Setal number, 35–48 (VII), 44–53 (XX); 5–8 between male pores. Male pores paired in XVIII, 0.19–0.3 body circumference apart ventrally. 2 types of arrangement of genital papillae in the male pore area for this species: (1) male pore situated on a papilla-like porophore with a genital papilla 0.25–0.45 mm in diameter anterior to it, surrounded by 3–4 skin folds; (2) male pore on a flat-topped papilla-like porophore, with a genital papilla on or slightly behind the setal line immediately medial to it and anterior or

anteromedial to it, surrounded by 3–4 skin folds. The 2 types of papillar arrangements can be observed in the same individual. In addition, 1 papilla in line with or slightly medial to the right, left or each male pore in presetal XIX for about half of all the specimens. Rarely, 4 papillae arranged transversely in presetal XVIII and XIX. Each papilla round, 0.3–0.45 mm in diameter with concave center. Spermathecal pores invisible to 2 pairs in intersegmental furrows of 5/6/7; distance between paired pores 0.29–0.37 body circumference apart ventrally, preclitellar genital papillae usually paired in presetal VII or VI–VII, rarely in presetal VII and VII, or in postsetal VI. Each papilla round, 0.3–0.5 mm in diameter, present or absent. Female pore single in XIV, medioventral.

Internal Characters

Septa 8/9/10 absent, 5/6/7/8 and 10/11–13/14 thick. Nephridial tufts thick on anterior faces of 5/6/7 septa. Gizzard large in VII–X. Intestine enlarged from XVI. Intestinal caeca paired in XXVII, extending anteriorly to

FIG. 170   *Amynthas mutabilitas* (after Shen, 2012).

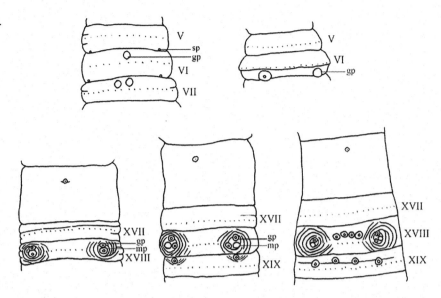

XXIII–XXV, each simple, distal end either straight or bent. Hearts enlarged from XI–XIII. Holandric, testes small or large, 2 pairs in ventrally joined sacs in X and XI. Vas efferens connected in XII on each side to form a vas deferens. Seminal vesicle small to large, 2 pairs in XI and XII, follicular with a round or elongate oval dorsal lobe, large vesicle occupying full segmental compartment while small vesicle about half of the segmental compartment. Prostate glands from large in XVI–XXI, normal in XVII–XIX, small in XVII–XVIII, to vestigial in XVIII only, wrinkled and lobed. Prostatic duct long, U-shaped in XVII–XVIII, distal end enlarged. Accessory glands round, sessile, or stalked and 0.35–1.4 mm in length, corresponding to external genital papillae. Spermathecae varied in number, from absent to 2 pairs in VI and VII (quadrithecate), each with a round or oval-shaped ampulla 0.82–2.15 mm long and 0.45–1.55 mm wide, and a long, slender to stout spermathecal stalk 0.35–1.7 mm in length. Diverticulum with an iridescent, oval-shaped seminal chamber of 0.38–0.85 mm long and a slender stalk of 0.9–1.5 mm long. Accessory glands long-stalked or short-stalked, 0.45–1.35 mm in length, each corresponding to external genital papilla.

Color: Preserved specimens dark brown on dorsum, light brown on ventrum. Clitellum brown to dark brown.
Distribution in China: Taiwan (Guanshan, Taidong County; Jhuohsi, Hualien County).
Deposition of types: Taiwan Endemic Species Research Institute, Jiji, Nantou, Taiwan.
Type locality: The specimens were collected at an elevation of 750 m, along the Hongshih Forest Road, and at an elevation of 750–840 m in Hualien County.
Remarks: *A. mutabilitas* is morphologically fairly similar to *A. morrisi* (Beddard, 1892). *A. morrisi* is a

quadrithecate earthworm belonging to the *morrisi* species group of the genus *Amynthas* (Sims and Easton, 1972). It has setae on ventral side of XVI, 12–16 setae between male pores, prostate glands in XVIII–XX or XVIII–XXI, and prostatic duct slender and coiled at distal end (Chen 1931, 1933). However, *A. mutabilitas* has variable numbers of spermathecae from 0 to 4 pairs in VI–VII, no setae on clitellum for fully mature individuals, 5–8 setae between male pores, prostate glands from in XVI–XXI to vestigial in XVIII, and distal end of prostatic duct straight and enlarged. The arrangement of genital papillae in the male pore area of *A. mutabilitas* is also similar to that of quadrithecate *A. ultorius* (Chen, 1935b) from Hong Kong, but the latter has 2 pairs of spermathecal pores in 7/8/9 (Chen, 1935b) and so belongs to the *aeruginosus* species group (Sims and Easton, 1972).

Like *A. mutabilitas*, *A. varians* (Chen, 1938) from Hainan Island also has variable numbers of spermathecae from totally absent to 2 pairs in VI–VII. The 2 species can be easily distinguished by their male pore structure and genital papillar arrangements. The characteristics of long penis on a small tubercle in the male pore area, and 1 presetal and 1 postsetal row of genital papillae in XVI–XVIII and VII–VII of *A. varians* (Chen, 1938) are distinctly different from those of *A. mutabilitas*.

174. *Amynthas nanulus* (Chen & Yang, 1975)

Pheretima nanula Chen, Hsü, Yang & Fong, 1975. *Acta Zool. Sinica.* 21(1):89–90.
Amynthas nanulus Xu & Xiao, 2011. *Terrestrial Earthworms of China.* 155.

External Characters

Length 45–50 mm, width 2.3–3.0 mm (clitellum), segment number 74–88. Prostomium 1/2 epilobous. First dorsal pore in 12/13. Clitellum XIV–XVI, annular, short, setae absent. Setae fine and close, VII–IX slightly close, dorsal and ventral breaks more obvious in postclitellar; $aa = 1.3–1.5ab$, $zz = 1.5–2.0zy$; Setal number, 34–40 (III), 46–57 (VII), 48–54 (XII), 41–48 (XXV); 13–16 (V), 14–16 (VI, VII) between spermathecal pores, 6–8 between male pores. Male pores 1 pair in ventral of XVIII, about 1/5 of body circumference apart ventrally, each on a horseshoe porophore, usually invaginated, surrounded by 3–4 skin folds. Spermathecal pores 2 pairs in 5/6/7, intersegmental; about 1/3 of body circumference apart ventrally, each on porophore front of VI–VII. No other genital markings. Female pore single in XIV, medioventral.

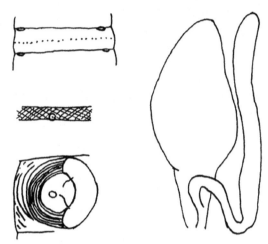

FIG. 171 *Amynthas nanulus* (after Chen et al., 1975)

Internal Characters

Septa 8/9/10 absent, 5/6/7/8 and 10/11/12/13 thick. Gizzard ball-like, in IX–X. Intestine swelling in XV; caeca stout and short, smooth on ventral. Last hearts in XIII. Testis sacs 2 pairs in X–XI, well developed, anterior large, separate ventrally but connected dorsally, posterior pair connected ventrally and dorsally, in large sac including seminal vesicles. Seminal vesicles 2 pairs in XI and XII, anterior pair small, posterior pair developed. Prostate glands large, in XVII–XX, widely fan-shaped, prostatic duct S- or C-shaped, and distal end stout. Accessory glands absent. Spermathecae 2 pairs, in VI–VII, behind septum; ampulla pear-shaped, its duct stout; diverticulum longer than main pouch with straight seminal chamber.

Color: Body without pigment, live specimens slightly light gray. Clitellum pale flesh red.
Type locality: Zhejiang (Xingdeng).
Distribution in China: Zhejiang (Xingdeng).

Remarks: This species is similar to *A. ralla* (Gates, 1936) but latter has clitellum with setae on ventral side; seminal chamber ovoid or round ball-shaped; male pore small, and on a transversal oval-shaped slightly swollen white area.

175. *Amynthas octopapillatus* Qiu & Sun, 2009

Amynthas octopapillatus Qiu & Sun, 2009. *Revue Suisse De Zoologie*. 116 (2):293–295.

External Characters

Length 123–138 mm, clitellum width 3.0–3.5 mm, segment number 139–205. Secondary annulations conspicuous in both anterior segments and others. Prostomium ? epilobous. First dorsal pore 12/13. Clitellum XIV–XVI, annular, swollen, setae and intersegmental furrows visible externally. 30 of them can be clearly seen on ventrum. Setae numerous, 56–70 (III), 74–92 (V), 68–90 (VIII), 64–66 (XX), 56–68 (XXV); 13–19 between spermathecal pores, 14–20 between male pores, $aa = 1.1–1.4ab$, $zz = 1.3–2zy$. Male pores paired in XVIII, 1/3 circumference apart ventrally, each on the center of a slightly raised, conical, glandular porophore, with 1–2 circular folds. Paired 0.6–0.8 mm oval papillae on 17/18–20/21 median to level of male pores. Spermathecal pores 2 pairs in 5/6/7, eyelike, 0.25 circumference apart ventrally. Genital markings absent. Female pore single, medioventral in XIV, in a small ovoid tubercle.

Internal Characters

Septa 8/9/10 absent, 6/7/8 thick and muscular, 10/11–13/14 slightly thickened. Gizzard in VIII–IX, barrel-shaped. Intestine enlarged distinctly from XVI. Intestinal caeca simple, smooth, in XXVII, extending anteriorly to XXIV. Esophageal hearts in X–XIII. Holandric, testis sacs 2 pairs, in X–XI, anterior pair larger, sacs of a segment very close ventrally, second pair enclosing first

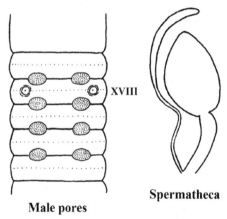

Male pores   Spermatheca

FIG. 172 *Amynthas octopapillatus* (after Qiu & Sun, 2009).

pair of seminal vesicles. Seminal vesicles paired in XI–XII, well developed. Prostate glands developed, extending from 1/3 XVI–1/3 XX, coarsely lobate. Prostatic duct U-shaped, slightly thicker at the distal part. Accessory glands absent. Spermathecae 2 pairs in VI–VII, ampulla heart-shaped, about 1.8 mm long with a long duct. Diverticulum longer than main pouch by 1/5, no kinks, terminal 1/2 dilated into a tube-shaped chamber, partially filled, silvery white.

> Color: Preserved specimens lacking pigment on dorsum and ventrum, middorsal line clear because of dorsal vessel. Clitellum brownish.
> Distribution in China: Hainan (Mt. Diaoluo).
> Etymology: The name of this species refers to its 8 papillae.
> Type locality: Hainan (Mt. Diaoluo).
> Deposition of types: Shanghai Natural History Museum and the Natural History Museum of Geneva.
> Remarks: In appearance, *A. octopapillatus* is somewhat similar to *A. diaoluomontis* (Qiu & Sun, 2009). Both species have 2 pairs of spermathecal pores in 5/6/7, no pigment, and the shape of the papillae on the ventrum. However, *A. octopapillatus* is distinguished from *A. diaoluomontis* by its smaller body size, its 2 additional pairs of papillae on 19/20/21, and by having the first pair of seminal vesicles enclosed in the second pair of testis sacs.

176. *Amynthas oculatus* (Chen, 1938)

Pheretima (Pheretima) oculata Chen, 1938. *Contr. Biol. Lab. Sci. Soc. China (Zool.).* 12(10):398–399.
Amynthas oculatus Sims & Easton, 1972. *Biol. J. Soc.* 4:236.
Amynthas oculatus Xu & Xiao, 2011. *Terrestrial Earthworms of China.* 158–159.

External Characters

Length 27–40 mm, width 1.2–1.8 mm, segment number 74–86. Prostomium 1/3 epilobous. First dorsal pore 12/13. Clitellum entire, in XIV–XVI, annular, with setal pits on ventral side of XVI (some on XV). Setae uniformly distributed, or very slightly closer ventrally, none particularly enlarged; both dorsal and ventral breaks noticeable, $aa = 1.5$–$2.5ab$, $zz = 1.5$–$2.0yz$; setae number 22–30 (III), 28–30 (VI), 28–30 (VII), 32–34 (XXV); 12–14 (VII) between spermathecal pores, 9–11 between male pores. Male pores 1 pair in XVIII, about 1/3 of circumference apart ventrally, each on a conical papilla, with a pair of large disc-like papillae placed in XVIII/XIX behind each male porophore; interval between them less than diameter of each. Spermathecal pores 2 pair in 5/6/7, intersegmental, about 3/4 of circumference apart ventrally. No

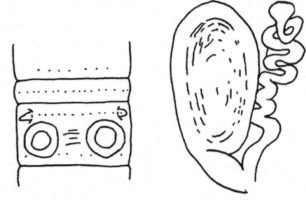

FIG. 173　*Amynthas oculatus* (after Chen, 1938).

genital markings thereabouts. Female pore single in XIV, medioventral, ovoid.

Internal Characters

Septa 8/9/10 absent, 5/6 thin, 6/7/8 comparatively thick, 10/11/12/13 a little thicker than those following. Gizzard in IX–X. Intestine enlarged from XVI. Caeca simple, in XXVII, extending anteriorly to XXV. First pair of hearts present. Testis sacs well developed, separate ventrally, meeting but not united dorsally. Seminal vesicles about 1/3 size of testis sac, those in XI enclosed in testis sac of that segment; dorsal lobe larger than vesicle proper. Prostate gland well developed, in XVI–XXII, with coarse lobules, its duct slender, uniform in caliber, deeply curved. Accessory glands in a whitish mass; each with a long cord associated with a bit of glandular tissue. Spermathecae 2 pairs, in VI and VII, ampulla as an elongate sac, distended, with a short and stout duct, about 1/5 of whole pouch. Diverticulum coiled, in zigzag fashion, its ental 4/5 whitish as seminal chamber, ectal 1/4 or 1/5 serving as duct; whole longer than main pouch in coiled condition.

> Color: Preserved specimens pale generally. Clitellum reddish.
> Distribution in China: Hainan (Baopeng).

177. *Amynthas puerilis* (Chen, 1938)

Pheretima (Pheretima) puerilis Chen, 1938. *Contr. Biol. Lab. Sci. Soc. China (Zool.).* 12(10):394–396.
Amynthas puerilis Sims & Easton, 1972. *Biol. J. Soc.* 4:236, 245.
Amynthas puerilis Xu & Xiao, 2011. *Terrestrial Earthworms of China.* 171–172.

External Characters

Length 20–37 mm, width 1–2 mm, segment number 47–72. Prostomium 1/3 epilobous. First dorsal pore 11/12. Clitellum entire, in XIV–XVI, annular, with setal pits on ventral side, often noticeable. Setae uniformly built, $aa = 1.2ab$, $zz = 1.2yz$; setae number: 30–32 (III), 32–42 (VI), 35–48 (VII), 30–36 (XXV); 20–22 (VI), 25–24 (VII) between spermathecal pores, 6 between male pores. Male pores 1 pair in XVIII, about 1/3 of circumference apart ventrally, each on a rather pointed porophore, with a small papilla in front on XVII, some absent. Spermathecal pores 2 pairs in 5/6/7, intersegmental; posterior pair about half of body circumference, anterior pair slightly ventral in position, about 2 setae at interval between this and posterior pair. No other genital markings in this region. Female pore single in XIV, medioventral.

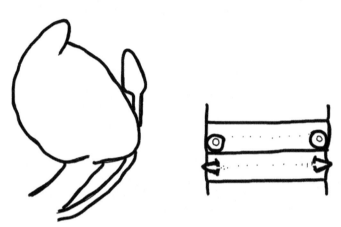

FIG. 174   *Amynthas puerilis* (after Chen, 1938).

Internal Characters

Septa 8/9/10 absent, 5/6/7/8 thickened, 10/11–13/14 about equally thickened, but thinner than pre gizzard ones, 14/15 a litter thicker than those following. Gizzard in 1/2 VII–IX. Intestine enlarged from XV. Caeca conical, in XXVII, extending anteriorly to XXV. First pair of hearts large. Testis sacs in X and XI, widely separate on ventromedian side, both pairs communicated dorsally. Seminal vesicles usually small, smaller than testis sacs if the latter developed; first pair not enclosed in testis sacs. Prostate gland well developed, coarsely lobate, in XVI–XXI; its duct about equal in caliber, in a deep U-shaped curve. Spermathecae 2 pairs, in VI and VII; ampulla large, heart-shaped, with a stout duct about equally long. Diverticulum shorter than main pouch, with short conical chamber, sharply marked off from its slender duct. Length of ampulla and its duct about 1 mm (longest about 2 mm).

Color: Preserved specimens pale generally. Clitellum reddish.
Distribution in China: Hainan (Baopeng).
Remarks: This species is similar to the small form of *A. vivians* in terms of body size and general appearance. However, it is distinct from the latter in

several important characters such as the first segment with no setae, the anterior pair of spermathecal pores more ventral in position, the male porophore as a flat topped papilla et al., etc.

178. *Amynthas sapinianus* (Chen, 1946)

Pheretima sapiniana Chen, 1946. *J. West China. Border Res. Soc. (B).* 16:130–132, 141, 160.
Amynthas sapiniana Sims & Easton, 1972. *Biol. J. Linn. Soc.*4:236.
Amynthas sapinianus Xu & Xiao, 2011. *Terrestrial Earthworms of China.* 178.

External Characters

Length 25–60 mm, width 1.5–2.5 mm, segment number 92–102. Prostomium 1/3 epilobous, tongue narrow, open behind. First dorsal pore 12/13. Clitellum XIV–XVI, annular, swollen, setal lines or short setae barely visible ventrally (more distinct on XV or XVI), its length equal to 8 immediately postclitellar segments. Setae not numerous and evenly distributed, slightly closer in ventrolateral sides. Setae number 26–28 (III), 30–34 (VI), 32–35 (IX), 30–32 (XXV); 8–12 (VI) between spermathecal pores, 4–6 between 2 papillae; 12 (XVII), 11 (XIX) between male pore line. Male pores paired in XVIII, about 1/3 of circumference apart ventrally, each on a porophore laterally surrounded by ridges, a large and flat-topped papilla placed ventral to each porophore. Spermathecal pores 2 pairs in 5/6/7, about 1/3 of ventral circumference apart. No genital papillae in this region. Female pore single in XIV, medioventral.

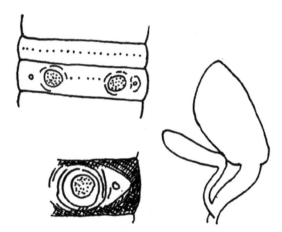

FIG. 175   *Amynthas apinianus* (after Chen, 1946)

Internal Characters

Septa 8/9/10 absent, 6/7/8 slightly muscular, 10/11/12/13 thicker. Nephridial tufts on anterior face of 5/6/7 thick. Pharyngeal glands lobular and developed in front of septum 5/6. Gizzard large, globular, in X. Intestine swelling in XV. Caeca simple and smooth on both sides, extending anteriorly to XXIII. Vascular loops in IX asymmetrical, those in X either asymmetrical developed or equally developed. Testis sacs well developed. Anterior pair in X meeting dorsally and connected by a membranous tube ventrally; posterior pair also connected on dorsal side and more broadly connected ventrally. Seminal vesicles confined in XI and XII, small, second pair larger, about 1 mm in dorsoventral length, each with a distinct dorsal lobe; first pair smaller enclosed in testis sacs. Prostate glands large, in XVI–XXII as coarse large lobules; its duct long but rather thin, in a loose coil. Almost uniform in caliber. Accessory glands with large lobules, racemose, sessile. Spermathecae 2 pairs, in VI and VII, rather large. Ampulla large, sac-like, its duct nearly as long, main pouch about 2 mm long. Diverticulum slightly shorter (or as long), ental half enlarged, finger-shaped serving as seminal chamber; ectal half as slender as duct. No accessory glands in this region.

Color: Pale generally, greyish along dorsomedian line. Clitellum bright chestnut-red.
Distribution in China: Chongqing (Beipei, Shapingba).
Etymology: The name of the species refers to the type locality.
Type locality: This species was collected by Chen from the wet ground under the moss coverage or amongst the grass roots.

179. *Amynthas sinuosus* (Chen, 1938)

Pheretima (Pheretima) sinuosa Chen, 1938. *Contr. Biol. Lab. Sci. Soc. China (Zool.).* 12(10):407–409.
Amynthas sinuosus Sims & Easton, 1972. *Biol. J. Soc.* 4:236.
Amynthas sinuosus Xu & Xiao, 2011. *Terrestrial Earthworms of China.* 181–182.

External Characters

Length 170–240 mm, width 4–5 mm, segment number 168–205. Prostomium 2/3 epilobous, with a longitudinal furrow at middle. First dorsal pore 12/13. Clitellum entire, in XIV–XVI, annular, with setae on ventral side. Setae all fine, *a–c* more widely spaced, closer on lateral and dorsal sides; both dorsal and ventral breaks slightly; $aa = 1.2ab$, $zz = 1.2$–$2.0yz$; setae number 52–70 (III), 64–76 (VI), 54–75 (VII), 46–54 (XX), 52–62 (XXV); 36–38 (VII) between spermathecal pores, 4–6 (VII) between male pores, 8 between papillae on either XVII or XIX. Male pores 1 pair on ventrolateral corner of XVIII, about 1/3 of circumference apart ventrally, each on a small conical porophore partially covered by a lateral skin fold, a raised papilla placed in front on XVII and another behind on XIX, in line with male porophore; ventrolateral region

of XVII–XIX elevated in preserved specimens. Spermathecal pores 2 pairs in 5/6/7, intersegmental, closer to dorsomedian than ventromedian line; each superficial, appearing like an elliptical depression on anterior border of segments VI–VII and a small part on preceding segments. Female pore single in XIV, medioventral.

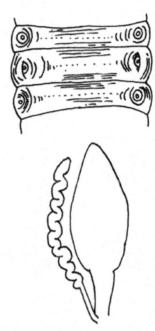

FIG. 176   *Amynthas sinuosus* (after Chen, 1938).

Internal Characters

Septa 8/9/10 absent, 5/6/7/8 thick, well muscular, 10/11/12 thick but more muscular than preceding ones. Gizzard in 1/2 IX–1/2 X. Intestine enlarged from XV. Caeca conical, in XXVII, extending anteriorly to XXV, smooth on both sides. Testis sacs very large, those in X about 3 mm in transverse width, narrowly communicated; those in XI united ventrally, with a common membranous sac to enclose testes, funnels and seminal vesicles, each side separable into dorsal portion (vesicle) and ventral portion (testis sac). Seminal vesicles small, close to respective septum, each with a dorsal lobe. No ovoid bodies found in following septa. Prostate coarsely lobate, in XVII–XX; with a very long (10 mm) duct, uniform in caliber. Accessory glands inconspicuous. Spermathecae 2 pairs, in VI and VII; ampulla elongate and hardened, about 3 mm in length, with a duct about 1 mm long. Diverticulum as long, its duct slender, longer than main duct, seminal chamber twice or more as long at its duct, twisted in a zigzag manner on 1 plane, about 10 or more turns, whitish in appearance.

Color: Preserved specimens purplish brown on dorsal side, pale ventrally.
Distribution in China: Hainan (Lingshui).

Etymology: The specific name indicates the twisting condition of the diverticulum.

Remarks: The present species is related to *A. rhabdoida* with respect to its general appearance, setal character, male pore region, and character of the clitellum and testis sacs. However, it differs essentially in (1) having only 2 pairs of spermathecae, (2) the diverticulum twisted in a zigzag fashion, (3) septum 8/9 absent, (4) the setae between male pores less numerous, and (5) the spermathecal pore more dorsally situated.

180. *Amynthas tenuis* Zhao & Qiu, 2013

Amynthas tenuis Zhao & Qiu, 2013. *Zootaxa.* 3619 (3):386–387.

External Characters

Length 48–56 mm, width 2 mm, segment number 72–89; body cylindrical in cross section, gradually tapered towards head and tail. Prostomium 1/2 epilobous. First dorsal pore in 12/13. Clitellum XIV–XVI, annular; 4 setae present in XVI or no setae ventrally. Setae number 12–28 (III), 30–36 (VII), 24–32 (XX), 28–38 (XXV); 10 between spermathecal pores, 8 between male pores, setal formula *aa* = 1.3–2*ab*, *zz* = 1.3–1.5*zy*. Male pores 1 pair in XVIII, about 0.33 body circumference apart ventrally, placed on an ovoid porophore. Genital papilla absent; some exist two large ovoid genital papillae before or after the ectal part of male pore. Spermathecal pores 2 pairs in 5/6/7, a little less than 0.33 body circumference apart ventrally. Genital markings absent. Female pore single in XIV, medioventral.

Internal Characters

Septa 8/9/10 absent, 5/6/7 comparatively thick, 7/8, 10/11–13/14 slightly thickened. Gizzard in XI–X,

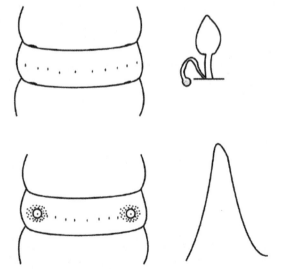

FIG. 177   *Amynthas tenuis* (after Zhao et al., 2013).

elongated ball-shaped. Intestine swelling in XV. Intestinal caeca simple, originating in XVII and extending anteriorly to XXIV, smooth, both on ventrum and dorsum. Dorsal blood vessel single, continuous onto pharynx; hearts in X–XIII, first pair small. Holandric, Testis sacs 2 pairs in X–XI, well developed, separate ventrally. Seminal vesicles paired in XI–XII, developed, separate ventrally. Prostate glands in XVI–XX, developed, prostatic duct in XVIII, n-shaped. Accessory glands present in XVIII, with a stalk. Ovaries in XIII. Spermathecae 2 pairs in VI–VII, about 1.5–1.9 mm; ampulla heart-shaped, about 1.1 mm, spermathecal duct long and straight, distinctly separate from ampulla; length of diverticulum a little less than or 2/3 of the main pouch, terminal 1/10 enlarged as an irregular chamber or terminal 0.67 enlarged as virgulate chamber; no nephridia on spermathecal ducts.

Color: Preserved specimens from light purple to light yellow-brown or pale on dorsum, no color on ventrum. Clitellum no color or pale.

Distribution in China: Hainan (Mt. Wuzhishan).

Etymology: This species is named after its slim body shape.

Deposition of types: Shanghai Natural History Museum.

Type locality: The specimens were collected from a core area in Wuzhishan National Nature Reserve, tropical rainforest, Hainan Province, China.

Remarks: *A. tenuis* is most similar to *A. infantilis* (Chen, 1938). Both species lack septa in 8/9/10 and first dorsal pore are is 12/13. Both of them have simple intestinal caeca and an accessory gland around the well-developed prostate glands. Both of them have an elongate ovoid spermathecal ampulla, and the diverticulum is shorter than the main pouch and with a spherical seminal chamber. However, they have distinct differences. *A. tenuis* is more than 3 times larger than *A. infantilis*. It has 2 pairs of spermathecal pores in 5/6/7, while *A. infantilis* has only 1 pair in 5/6. *A. tenuis* has no genital markings, while *A. infantilis* has both genital papillae and markings. The male pore is in XVIII in *A. tenuis* while it is in XIX in *A. infantilis*. The sequence of a 16S gene fragment of the holotype of *A. tenuis* (C-HN201109-08) has been deposited in GenBank under the Accession No. JQ904531.

181. *Amynthas tripunctus* (Chen, 1946)

Pheretima tripuncta Chen, 1946. *J. West China. Border Res. Soc. (B).* 16:85–86, 142.

Amynthas tripunctus Sims & Easton, 1972. *Biol. J. Linn. Soc.* 4:236.

Amynthas tripunctus Xu & Xiao, 2011. *Terrestrial Earthworms of China.* 191.

External Characters

Medium-sized worm. Length 83 mm, width 4 mm, segment number 91. Annulation in anterior segments indistinct. Prostomium 1/3 epilobous, width of tongue as length of segment II. First dorsal pore 12/13. Clitellum XIV–XVI, annular, smooth, setal pit found ventrally on XIV. Setae all fine, none specialized, those on anterior ventral short and close, both dorsal and ventral breaks rather slight. Setae number 33 (III), 26–27 (V), 49 (VI), 48 (IX), 42 (XXV); 17 (V), 22 (VI), 24 (VII) between spermathecal pores, 12 between male pores, setal formula $aa = 1.1–1.5ab$, $zz = 1.1–1.2zy$ in front of, $zz = 1.5zy$ behind clitellum. Male pores 1 pair in XVII, about 1/3 of circumference apart ventrally, each represented by a minute porophore accompanied by 2 small, round-topped papillae on its median side. The surrounded region by circular ridges and somewhat raised. Spermathecal pores 2 pairs in 5/6/7, but rather segmental at anterior edge of VI and VII, simple, about 5/12 of their ventral circumference, unpaired papillae ventromedially on VI–VII, presetal, medium-sized and round. Female pore single in XIV, medioventral.

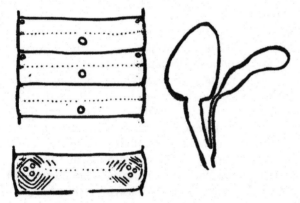

FIG. 178 *Amynthas tripunctus* (after Chen, 1946).

Internal Characters

Septa 8/9/10 absent, 6/7/8 thick, slightly muscular, 5/6 thin, 10/11/12 thinner than 7/8. Nephridial tufts on 5/6/7 thin. Pharyngeal gland moderately thick in front of 5/6. Gizzard moderate in size, onion-shaped, but smaller anteriorly and enlarged at posterior third. Intestine swelling in XV. Caeca simple and short, auricular in shape, extending anteriorly to XXIV. Vascular loops in IX and X asymmetrical and slender. Testis sacs large, both pairs separate and not communicated. Seminal vesicles large, warty in appearance, filling wholly segments XI and XII, dorsal lobes not apparent. Prostate glands well developed in XVI–XXI, its duct S-curved at middle and coiled ectally, enlarged at ental third. Accessory glands not clearly visible, a very thin velvety layer around base of each duct. Spermathecae 2 pairs, in VI and VII, fairly large. Ampulla heart-shaped, its duct longer, main pouch about 3 mm long. Diverticulum about 1/4 longer than main pouch, ental half enlarged, shaped like a cucumber, serving as seminal chamber. Accessory glands small, each as a small lobular mass connected with its respective papilla externally, sessile, its short cord under the nerve cord.

Color: Grey on dorsal side, greenish gray along dorsomedian line, pale ventrally. Clitellum brick red.
Distribution in China: Sichuan (Mt. Emei).
Etymology: The name of the species refers to the type locality.
Remarks: This species resembles *Amynthas triastriatus* (Chen, 1946) closely in the general shape of the body and the male region. However, it is distinct in the position of the spermathecal pores, the finer setae on anterior venter, the entirely separate testis sacs, and much less distinct in the seminal chamber of the diverticulum from ampulla duct.

182. *Amynthas zhangi* Qiu & Sun, 2009

Amynthas zhangi Qiu & Sun, 2009. *Revue Suisse De Zoologie*. 116(2):295–296.

External Characters

Length 124–200 mm, clitellum width 3.1–5.3 mm, segment number 186–206. Secondary annulations conspicuous in V–XIII. Prostomium combined prolobous and ? epilobous. First dorsal pore 12/13. Clitellum XIV–XVI, annular, thinly glandular, intersegmental furrows clear, 20–28 setae visible externally, evident only on ventrum. Setae number 50–64 (III), 60–76 (V), 64–70 (VIII), 60–70 (XX), 60–70 (XXV); 33–35 between spermathecal pores, 4–6 between male pores; $aa = 1.1–2ab$, $zz = 1.5–2zy$. Male pores 1 pair in XVIII, about 1/3 circumference apart ventrally, each on the top center of a slightly raised, conical porophore, with 2 unclear circular folds. Genital papillae large and round, paired in XVII and XIX, in line with male pores, 0.8–1.0 mm in diameter. Spermathecal pores 2 pairs in 5/6/7, eye-like, about 0.5 circumference apart. Genital markings absent. Female pore single, medioventral in XIV, in a small ovoid tubercle.

Internal Characters

Septa 8/9/10 absent, 5/6/7/8 thick and muscular, 10/11/12/13 slightly thickened. Gizzard in VIII–IX according to septum 7/8, long ball-shaped. Intestine enlarged distinctly from XVI. Intestinal caeca paired in XXVII, simple, slender, smooth, extending anteriorly to XXV. Esophageal hearts in X–XIII. Holandric, testis sacs 2 pairs in X–XI, undeveloped, second pair enclosing first pair of seminal vesicles. Seminal vesicles paired in XI–XII, small. Prostate glands small, extending from

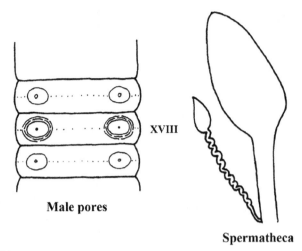

FIG. 179 *Amynthas zhangi* (after Qiu & Sun, 2009).

3/4 XVII–3/4 XIX, coarsely lobate. Prostatic duct S-shaped, slender. Accessory glands absent. Spermathecae 2 pairs in VI–VII, lanceolate, yellowish, about 1.9 mm long with a slender terminal duct about equally long. Diverticulum 3/4 length of main pouch, in zigzag fashion, terminal 1/7–1/6 dilated into pear-shaped chamber, with pointed tip, partially filled.

Color: Preserved specimens grayish on dorsum before clitellum, light brownish on dorsum after clitellum. Clitellum brownish.
Distribution in China: Hainan (Mt. Diaoluo).
Etymology: This species is named in honor of its collector, Xiaolong Zhang.
Deposition of types: Shanghai Natural History Museum and the Natural History Museum of Geneva.
Type locality: Hainan (Mt. Diaoluo).
Remarks: *A. zhangi* is similar to *A. sinuosus* (Chen, 1938) in having 2 pairs of spermathecal pores in 5/6/7, a rather large body, the diverticulum twisted in a zigzag fashion, and paired raised papillae in XVII and XIX in line with male pores. However, they differ markedly in that the male pores of *A. zhangi* do not have a lateral skin fold covering the pores, the diverticulum seminal chambers being oval, and the male organs being small and somewhat reduced. The male pores of *A. sinuosus* are partly covered by a lateral skin fold, the seminal chambers are not oval-shaped, and the prostate glands and other male organs are well developed.

pauxillulus Group

Diagnosis: *Amynthas* with spermathecal pores 3 pairs, at 4/5/6/7.
Global distribution: Oriental realm, ? Australasian realm, and ? introduced into the Oceanian realm. There is 1 species from China.

162. *Amynthas dignus* (Chen, 1946) (*v.ant.* 162)

Pheretima digna Chen, 1946. *J. West China. Border Res. Soc. (B).* 16:132–133, 160.
Amynthas dignus Sims & Easton, 1972. *Biol. J. Linn. Soc.* 4:236.
Amynthas dignus Xu & Xiao, 2011. *Terrestrial Earthworms of China.* 103.

pomellus Group

Diagnosis: *Amynthas* with spermathecal pores 2 pairs, at VII and VII.
Global distribution: Oriental realm, ? Australasian realm, and ? introduced into the Oceanian realm. There are 4 species from China.

TABLE 20 Key to Species of the *pomellus* Groups of the Genus *Amynthas* From China

1	Without papillae front clitellum .... *Amynthas biconcavus*
	With papillae front clitellum 2
2	Paired genital papillae on ventral side of XII and XIII .... *Amynthas pomellus*
	With papillae on ventral side of VII and VII 3
3	On VII and VII each with a small papilla situated posteriorly or a little ventrally .... *Amynthas sucatus*
	Male pores, 3 or more papillae usually occurring in this region, 1 in front and 2–3 behind setae, all surrounded by an incomplete glandular wall. 2 large ovoid ones placed ventrally on the same segment, closely paired in front of setae. 2 unpaired papillae, transverse-ovoid, placed midventrally and presetally on VII and VII. 2 smaller ones close to porophore on VII and 1 on left side on VII .... *Amynthas cupreae*

183. *Amynthas biconcavus* Quan & Zhong, 1989

Amynthas biconcavus Quan & Zhong, 1989. *Act Zootaxonomica Sinica.* 14(3):273–274.
Amynthas biconcavus Xu & Xiao, 2011. *Terrestrial Earthworms of China.* 85–86.

External Characters

Body rather small. Length 71–78 mm, width 2 mm, segment number 102–133. Prostomium 1/3 epilobous. First dorsal pore 12/13. Clitellum in XIV–XVI, annular, with or without setae. Setae fine; $aa = 1.2$–$1.5ab$, $zz = 1.5$–$2.0zy$; setae number 32–36 (III), 36–47 (V), 37–44 (VII), 38–45 (IX), 35–45 (XX); 4–5 (VII, VII) between spermathecal pores, no setae between male pores. Male pores 1 pair, on ventrolateral side of XVIII, about 1/6 of circumference apart ventrally, each with a small penis situated in a large elliptical depression. Spermathecal pores 2 pairs, on the anterior margin of VII and VII, close to ventromedian line, about 1/9 of ventral body circumference. Female pore single in XIV, medioventral.

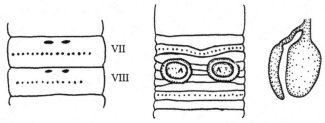

FIG. 180 *Amynthas biconcavus* (after Quan & Zhong, 1989).

Internal Characters

Septa 8/9/10 absent, 5/6/7 thickened, 10/11/12/13 thicker, membranous after 13/14. Gizzard pear shaped, smooth. Intestine swelling in XVI. Caeca simple, in XXVII, extending anteriorly to XXV. Hearts in X–XIII. Testis sacs unpaired, as transverse band. Seminal vesicles small, also transverse band, not dorsal; the first pair of seminal vesicles are contained in the second pair of testis sacs. Prostates in XVII–XXII or 1/2 XVI–XX, duct medially thickening, with accessory glands on the ectal end of the prostatic duct. Spermathecae 2 pairs in VII–VIII, ampulla elliptical, 1–1.2 mm long, seminal chamber long eggplant-shaped.

> Color: Preserved specimens gray-brown, slightly lighter on ventrum. Clitellum light brown to dark chestnut.
> Distribution in China: Hainan (Ledong).
> Etymology: The specific name indicates the large elliptical depression in male pore region.
> Type locality: Hainan (Jianfen Ridge, Ledong County).
> Deposition of types: Institute of Tropical Forestry, Chinese Academy of Forestry Science, Guangzhou.
> Remarks: This species is similar to *A. pomellus* (Gates, 1935b) but differs from the latter by its small penis in a large elliptical depression and the character of the seminal vesicles.

184. *Amynthas cupreae* (Chen, 1946)

Pheretima cupreae Chen, 1946. *J. West China. Border Res. Soc. (B).* 16:117–119, 140, 154.
Pheretima cupreae Sims & Easton, 1972. *Biol. J. Linn. Soc.* 4:224.
Amynthas cupreae Xu & Xiao, 2011. *Terrestrial Earthworms of China.* 99–100.

External Characters

Large-sized worm. Length 120 mm, width 6.5 mm, segment number 95. 3–4 annuli recognizable behind clitellum. Prostomium (?) destroyed. First dorsal pore 11/12. Clitellum XIV–XVI, annular, swollen, without setae. Setae evenly distributed, more noticeable but not enlarged at anterior ventral side, closer ventrally; setae

number 46 (III), 52 (VII), 50 (XXV); 25 (VII), 24 (VII) between spermathecal pores, 16 between male pores; *aa* = 1.2 *ab*, *zz* = 1.5–2*zy*. Male pores 1 pair in XVIII, about 1/3 of circumference apart ventrally, 3 or more papillae usually occurring in this region, 1 in front and 2–3 behind setae, all surrounded by an incomplete glandular wall. 2 large ovoid ones placed ventrally on the same segment, closely paired in front of setae. Spermathecal pores 2 pairs on VII and VII, in front of setae, much close to setae than the furrow, represented by a raised porophore, about 3/7 of ventral circumference. 2 unpaired papillae, transverse-ovoid, placed midventrally and presetally on VII and VII. 2 smaller ones close to porophore on VII and 1 on left side on VII. Female pore single in XIV, medioventral.

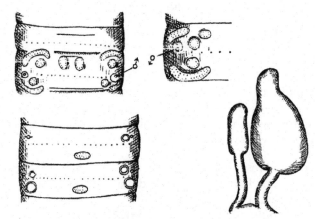

FIG. 181 *Amynthas cupreae* (after Chen, 1946).

Internal Characters

Septa 8/9/10 absent, 5/6/7/8, 10/11/12 moderately thickened, 12/13 less so. Nephridial tufts found on anterior faces of 5/6/7 thick. Gizzard barrel-shaped, anterior part slightly narrower, posterior part larger and abruptly terminated, with crop-like dilation in front. Intestine swelling in XV. Caeca not observed (destroyed when collected). Vascular loops in IX moderately large, 1 limb less developed, those in X absent. Testis sacs widely placed, not communicated, each sac about 2 mm, in diameter. Vas deferens small. Seminal vesicles in XI and XII, moderate in size, dorsal lobes not well developed, smooth on surface. Prostate glands well developed in XVII–XXII, with coarse lobules, its duct in a deep U-curve, lower limb thicker, its exit passing through a pore-bearing papilla, slightly posterior to setae. Accessory glands sessile on body wall, small. Spermathecae 2 pairs, in VII and VII, both pairs equally developed. Ampulla spatulate, about 2.5 mm long; its duct about equally long. Diverticulum shorter than main pouch; seminal chamber subspherical, well distended, with a characteristic color not unlike a dilute solution of copper sulphate. Accessory gland in round patches, largest about 2 mm in diameter, whitish and sessile.

Color: Pale generally except fleshy on dorsal side, deeper along dorsomedian line. Clitellum grayish flesh. Distribution in China: Hubei, Chongqing (Mt. Jinfo), Guizhou (Mt. Fanjing).

Remarks: This species is closely related to *A. mucrorimus* (Chen, 1946) but should be distinguished from the latter by possessing only 2 pairs of spermathecal pores which are situated more dorsally, the less numerous ventral setae, and the different character of genital papillae on the ventromedian side.

185. *Amynthas pomellus* (Gates, 1935)

Pheretima pomella Gates, 1935a. *Smithsonian Mus. Coll.* 93(3):14–15.

Pheretima pomella Chen, 1936. *Contr. Biol. Lab. Sci. Soc. China (Zool.).* 11:301–302.

Pheretima pomella Gates, 1939. *Proc. U.S. Natr. Mus.* 85:469–470.

Amynthas pomellus Xu & Xiao, 2011. *Terrestrial Earthworms of China.* 169–170.

External Characters

Length 87 mm, width 5 mm, segment number 92. Prostomium 1/2 epilobous. First dorsal pore 10/11. Clitellum entire, in XIV–XVI, annular, without setae. Setae all fine and evenly distributed; lacking dorsally on II. *aa* = 1.5*ab*, *zz* = 2*yz*; setae numbers 47 (VI), 48 (VII), 48 (XXV); 21 (VII) between spermathecal pores, 14 between male pore. Male pores 1 pair in XVIII, superficial, about 1/3 of circumference apart ventrally, each with 1 papilla on its anterior and posterior sides. Genital markings paired, presetal on XII, XIII and XVI, postsetal on XVIII; markings on XVIII in line with male pores, on

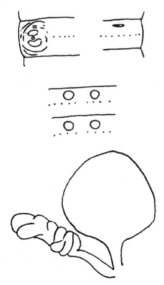

FIG. 182 *Amynthas pomellus* (after Chen, 1936).

XII–XIII absent in *ab*. Spermathecal pores 2 pairs, as transverse eye-like slits on VII and VIII, minute and superficial, slightly nearer to the intersegmental furrows than the setal circles, about 1/3 of circumference apart ventrally. Paired genital papillae present on ventral side of XII and XIII. Female pore single in XIV, medioventral.

Internal Characters

Septa 8/9/10 absent, anterior septa thick. Intestine swelling in XVI. Intestinal caeca simple, in XXVII, extending anteriorly to XXIV. Last hearts in XIII. Testis sacs of X and XI unpaired and united ventrally. Seminal vesicles very large, dorsal lobe not apparent. Prostate compact and thick, in XVII–XIX, with a long duct, about equal thickness throughout the length. Spermathecal ampulla large with a short and thin duct; diverticulum with a short, muscular stalk and an elongated tubular seminal chamber, the latter twisted into a ball-shaped mass of loops; seminal mass found in the coiled portion.

Color: Preserved specimens grayish dorsally.
Type locality: Chongqing (Xufu).
Distribution in China: Chongqing (Yibin, Xufu).
Deposition of types: Smithsonian Institution.
Remarks: Distinguished from *A. planata* by the posterior location of the spermathecal pore, the absence of copulatory chambers, and the locations of the genital markings.

186. *Amynthas sucatus* (Chen, 1946)

Pheretima sucata Chen, 1946. *J. West China. Border Res. Soc. (B).* 16:88–90, 138, 142.

Amynthas sucatus Sims & Easton, 1972. *Biol. J. Linn. Soc.*4:236.

Amynthas sucatus Xu & Xiao, 2011. *Terrestrial Earthworms of China.* 106–107.

External Characters

Large-sized worm. Length 110 mm, width 6.2 mm, segment number 109. Segment strongly built, anterior segment long and thick, somewhat like *Amynthas aspergillum*. Prostomium 1/3 epilobous, tongue narrow. First dorsal pore 11/12. Clitellum XIV–XVI, annular, glandular and roughened, with setal pits on the ventral side, setae absent. Setae rather large but all short and much closer ventrally. Anterior ventral ones short and close. *aa* = 1.1–1.2 *ab*, *zz* = 1.5–1.2*zy* in front of =2.5*zy* behind clitellum. Setae number 43 (III), 65 (VI), 58 (IX), 42 (XXV); 19 (VII), 20 (VII) between spermathecal pores, 12 between male pores. Male pores 1 pair in XVII, about 1/3 of circumference apart ventrally, each on lateral side of a large flat-shaped papilla, the latter surrounded by a deep groove and outward by a few circular ridges. No copulatory pouch

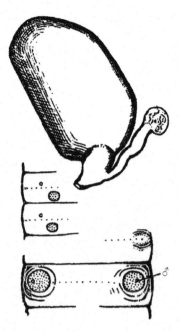

FIG. 183   *Amynthas sucatus* (after Chen, 1946).

nor any depression. Spermathecal pores 2 pairs on VII and VIII, between anterior and middle annuli, in a shallow furrow, also about 1/3 ventral circumference apart, each with a small papilla situated posteriorly or a little ventrally. Female pore single in XIV, medioventral.

Internal Characters

Septa 8/9/10 absent, 5/6/7/8 comparatively thick, but not particularly muscular, 11/12/13 thicker. Nephridial tufts on 5/6/7 not thick. Pharyngeal gland moderately thick and compact in front of 5/6. Gizzard large, globular in outline. Intestine swelling in XV. Caeca simple, pointed, in XXVII, extending anteriorly to XXIV. Vascular loops in IX and X stout. Testis sacs 2 pairs, broadly united, as transverse bands. Supernumerary vesicles found on posterior face of 12/13 rather large. Seminal vesicles fully filled in XI and XII, no dorsal lobes. Prostate glands lobular and compact, in XVII–XIX, its duct short and thick in a tight coil. Accessory glands in a large round mass, cord-like ductules not found. Spermathecae 2 pairs, in VII and VII, enormously large. Ampulla sac-like, about 4 mm in length, its duct very thick and short, not well marked off from the former. Diverticulum very short, ended with a disc-like enlargement not clearly distinct from its duct.

Color: Dark purplish gray on dorsal side, pale ventrally. Clitellum brick red.
Distribution in China: Sichuan (Mt. Emei).
Remarks: The present species is closely related to *A. pomellus*. As the aspects of the male pore region and the diverticulum are considerably different, it does not seem wise to include *A. sucatus* with this species.

Minor differences are also found, such as the size of *A. pomellus* is smaller, genital papillae ventromedially on XI & XIII instead of VII & VII, the spermathecal pores nearer the intersegmental furrow not as in the present species rather close to the setal circle, and its segments less strongly built and less pigmented.

rimosus Group

Diagnosis: *Amynthas* with spermathecal pores 4 pairs, at VI, VII, VII, IX.

Global distribution: Oriental realm, ? Australasian realm, and ? introduced into the Oceanian realm. There are 2 species from China.

TABLE 21   Key to the Species Groups of the Genus *Amynthas* From China

1   Male pores on large alate porophores in shallow invagination. Spermathecal pores on VI–IX dorsal, intrasegmental on small indistinct porophores in presetal annulus Male pores are round, minute; The paired genital markings are slightly eroded so that an exact description is impossible. Each marking appears to be thickly crescent-shaped, anteroposteriorly placed on a slightly protuberant region of XVIII; intersegmental furrows 17/18 and 18/19 slightly displaced anteriorly or posteriorly by the markings. Towards the outer margin of the marking and in line with the setal circle is the male pore. On each marking there are 2 diagonally placed fissures, 1 anterior and 1 posterior to the male pore; the midsegmental ends of the fissures pointing towards the male pore, the distal ends of the fissures pointing towards the midventral line. Spermathecal pores on VI, VII, VII and IX, on the anterior margins of the segments, very close to the intersegmental furrows; each pore a minute, transverse slit; the pores of a pair widely separated .... *Amynthas rimosus*
Male pores 1 pair in XVIII, on large alate porophores in shallow invagination. Spermathecal pores on VI–IX dorsal, intrasegmental on small indistinct porophores in presetal annulus of triannulate segments .... *Amynthas chaishanensis*

187. Amynthas chaishanensis James, Shih & Chang, 2005

Amynthas chaishanensis James, Shih & Chang, 2005. *Jour. Nat. His.* 39(14):1021–1022.
Amynthas chaishanensis Xu & Xiao, 2011. *Terrestrial Earthworms of China.* 92.

External Characters

Length 203–228 mm, width 8–11 mm, segment number 112–137. Prostomium epilobous, with tongue open. First dorsal pore in 12/13. Clitellum XIV–XVI, annular, setae invisible externally. Setae regularly distributed around segmental equators, size and distance regular; setal formula *aa: ab: yz: zz* = 2:1:1:3 at XXV, setae number 130–150 (VII), 126–138 (X), 104–20 (XXV); 20 between male pores. Male pores 1 pair in XVIII, on large alate porophores in shallow invagination, about 1/3 of body circumference apart. Spermathecal pores 4 pairs,

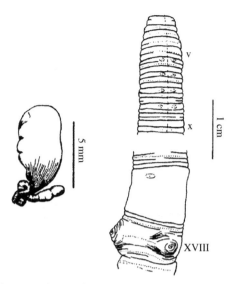

FIG. 184 *Amynthas chaishanensis* (after James et al., 2005).

on VI–IX dorsal, intrasegmental on small indistinct porophores in presetal annulus of triannulate segments, about 1/10 circumference apart. Female pore single in XIV, medioventral.

Internal Characters

Septa 9/10 absent, 8/9 present ventrally, 5/6/7/8 very thick and muscular, 10/11–13/14 very thick, diminishing in muscularity posteriorly. Gizzard in VII. Intestine swelling in XV. Caeca simple, long slender, in XXVII, extending anteriorly to XXV, XXIII, no incisions. Hearts 2 pairs in XII–XIII; X, XI lacking, commissural vessels VII, IX lateral, VII to gizzard. Testes, funnels in ventrally joined sacs in X; seminal vesicles large in XI, with dorsal lobe, enclosed within sac containing all segmental contents. Prostate glands in XVIII, numerous incised main lobes covering XVII–XX, duct short, thick, muscular, straight; numerous ductulets from glandular portion form undivided fan-shaped array, ental portion of prostatic duct with approximately 40 very small lumens, 1 larger lumen probably sperm duct; vasa deferentia join at duct-glandular portion junction; vasa deferentia nonmuscular. Spermathecae 4 pairs, in VI–IX, ampulla large ovate sac, duct stout but flaccid, half the length of ampulla, diverticulum stalk long convoluted kinks enclosed within membrane, chamber terminal ovate knob; diverticulum axis shorter than ampulla axis; no nephridia on spermathecal duct.

Color: Preserved specimens dark purple-brown pigment on dorsalmost third.
Type locality: Taiwan (Chaishan, near National Sun Yat-sen University, Kaohsiung City).
Distribution in China: Taiwan (Gaoxiong).
Etymology: This species is named after the locality of Chaishan, Kaohsiung City, Taiwan.

Deposition of types: National Museum of Natural Science, Taichung, Taiwan.
Remarks: In contrast to *A. formosae*, *A. chaishanensis* has numerous prostatic ductulets of equal size joining the large prostatic ducts at the same point, creating a single undivided fan of ductulets, all of which gather into small bundles prior to joining the main prostatic duct.

A. chaishanensis is 1 of 2 known proandric *Amynthas* species with dorsal intrasegmental pores on VI–IX, the other being *A. formosae*. It is further distinguished by having hearts only in XII and XIII. However, Gates (1959) noted the presence of hearts in X–XIII in his material and did not note this as a distinction between his material and *A. formosae*. Other differences from *A. formosae* are many more setae in the anterior segments, fewer setae between the male pores, slightly more widely spaced (from middorsal line) spermathecal pores, caeca originating in XXVII, lack of pseudovesicles in XII and XIII, and prostate glands divided into numerous main lobes, rather than only 2.

188. *Amynthas rimosus* (Gates, 1931)

Pheretimarimosa Gates, 1932. *Rec. Ind. Mus.* 34(4):408–411.
Amynthas rimosus Xu & Xiao, 2011. *Terrestrial Earthworms of China.* 173–174.

External Characters

Length 100–122 mm, width 4.5–5 mm, segment number 111–119. Prostomium 1/2 epilobous. Segments VI–XIII each have 2 secondary furrows; segments IX–XIII have in addition tertiary furrows. First dorsal pore in 12/13, there are functional pores in 13/14 or 16/17. Clitellum XIV–XVI, annular; without intersegmental furrows and dorsal pores. The setae begin on II, but can be recognized only on the ventral side of that segment. Dorsal or ventral break absent. The ventral setae of segments II–XIII and the dorsal setae of III–IX are slightly larger than elsewhere; setae number 44 (XX); 12 (VI), 13 (VII), 13 (VII) between spermathecal pores; 7 between genital markings on XVIII. Male pores 1 pair in XVIII, round, minute apertures in the setal circle. Paired genital markings appear to be thickly crescent-shaped, grayish, anteroposteriorly placed on a slightly protuberant region of XVIII; intersegmental furrows 17/18 and 18/19 slightly displaced anteriorly or posteriorly by the markings. On each marking there are 2 diagonally placed fissures, 1 anterior to and 1 posterior to the male pore; the midsegmental ends of the fissures pointing towards the male pore, the distal ends of the fissures pointing towards the midventral line. Spermathecal pores 4 pairs, on VI, VII, VII and IX, on the anterior margins of the segments, very close to the intersegmental furrows; each pore a minute, transverse slit; the pores of a pair widely separated. Female pore single in XIV, medioventral.

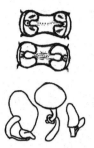

FIG. 185　*Amynthas rimosus* (after Gates, 1932).

Internal Characters

Septa 4/5/–8/9 are all present, membranous, 8/9 is attached to the dorsal blood vessel immediately behind the vessels passing from the dorsal trunk to the gizzard; 9/10 is apparently present ventrally only; 10/11 is present but very delicate; 11/12 and 12/13 are slightly thickened but membranous. Gizzard elongate. Intestine swelling in XV. Caeca in XXVII, extending anteriorly to XXV; pass under the intestine, long enough to reach into XXIII. The ventral margins of the caeca are slightly lobed. Last pair hearts in XIII. Testis sac 2 pairs in X and XI, rather large, first pair has a bilobed anterior margin. Seminal vesicles, large; the anterior pair pushes 10/11 and 9/10 anteriorly so that the vesicles appear to be in contact with the posterior portions of the sides of the gizzard. The posterior pair pushes 12/13 back into contact with 13/14.

Prostates, large, extending from XV to XIX on the right side and XVI to XXI on the left side. Each gland composed of a number of lobes, subdivided into smaller lobules; duct confined to XVIII; it is moderately thick throughout and looped. There are no copulatory chambers. Spermathecae 4 pairs, in VI, VII, VII; the last 2 pairs in VII; the posteriormost pair opening posteriorly. The ampullae are saccular, duct slender and of about the same length as the ampulla. Diverticulum arises from the anterior face of the bent duct which is shorter than the ampulla, elongate but not equal to the combined lengths of the ampulla and duct. The ectal portion is narrowly tubular, the ental portion slightly thicker.

Color: Preserved specimens dorsally light reddish, ventrally whitish. Clitellum yellowish brown.
Type locality: Myanmar (Kengtung State).
Distribution in China: Yunnan (Lancang).
Deposition of types: Indian Museum.

sieboldi Group

Diagnosis: *Amynthas* spermathecal pores 3 pairs, at 6/7/8/9.

Global distribution: Oriental realm, ? Australasian realm, and ? introduced into the Oceanian realm. There are 35 species from China.

1. Clitellum on XIV–XXII .... Glossoscolecidae	
Clitellum on XIII–XX 6	

TABLE 22　Key to Species of the *sieboldi* Group of the Genus *Amynthas* From China

1. With papillae front and behind clitellum 2
   With papillae front and/or behind clitellum 14

2. Papillae 2, or fewer than 2 in the male pore region 3
   Papillae more than 2 in the male pore region 9

3. Male pores on a round tubercle; a large long-round flat-topped papilla, which extends anteriorly to XVII and posteriorly to XIX, placed ventrally close to each pore .... *Amynthas daulis fanjinmontis*
   Genital marking not long-round 4

4. Genital marking round 5
   Genital marking not round 7

5. Male pores not visible on the surface, only an obscure whitish patch appearing vaguely under the skin; 1–2 round pits sometimes present median to each pore; paired in XVIII .... *Amynthas obscuritoporus*
   Genital marking flat and round or oval 6

6. Male pores on a flat and round porophore as large as a papilla which is usually placed at its median side, postsetal in position .... *Amynthas jaoi*
   Male pores round or oval, surface smooth, slightly convex, with or without a shallow horizontal slit (depression) in middle, surrounded by 2–3 circular folds .... *Amynthas proasacceus*

7. Male pores on a glandular papilla which encroaches over 1 segment, its rim not distinct, flat-topped. 1 large lens-shaped papilla placed medioventrally between setal lines of XVII and XVIII .... *Amynthas editus*
   Genital marking not lens-shaped 8

TABLE 22 Key to Species of the *sieboldi* Group of the Genus *Amynthas* From China—cont'd

8. Male pores on small knobs visible under hoods covering male pore openings .... *Amynthas huangi*
   Male pores represented by a papilla-like glandular area. 1 genital papilla constantly present, close to its posterior side on segment
   XVIII .... *Amynthas pingi chungkingensis*

9. Male pores on a round porophore closely connected before and behind with a pear-shaped glandular pad .... *Amynthas contingens*
   Genital marking between XVII–XX 10

10. Male pores on a round, white porophore, surrounded by 1–3 shallow skin folds. Genital papillae large, round, usually 2 pairs located slightly
    medially to the porophore in presetal XVIII and XIX, and an additional pair in XX, occasionally 1 pair only in presetal XVIII or XIX, or 2 pairs in
    XVIII with 1 pair anteromedial and the other posteromedial to male porophores, or 2 pairs in XVIII together with 1 pair in
    presetal XIX .... *Amynthas hongyehensis*
    Genital marking between XVII and/or XVIII 11

11. Male pores on a slightly raised papilla-like tubercle which is flat-topped. Genital papillae 2 pairs, always present between intersegmental furrows
    17/18 & 18/19, in line to or a little medial to male tubercle, often symmetrically paired. Each slightly transverse oval in outline, highly elevated,
    with a thick margin and a glandular depressed center, 6 or more times larger than male tubercle; intersegmental furrows around papillae often
    obliterated .... *Amynthas hupeiensis*
    Genital marking on XVIII 12

12. Male pores on round or ellipsoid tubercle at ventrolateral side of XVIII, 2 glandular pads placed in front and behind of porophore. A pair of
    papilla postsetally on XVII, or sometimes absent .... *Amynthas daulisdaulis*
    Genital marking flat-topped 13

13. Male pores on a small papilla on lateroventral side, surrounded by a few concentric ridges often not marked ventrally; 1 small papilla smooth but
    glandular and flat-topped placed medial to each pore, 2 on ventromedial side, present, slightly elevated, or variously arranged ..... *Amynthas
    dactilicus*
    Male pores situated in a small papilla. 3–4 ridges occurring exterior to male pore and a row of flat-topped papillae (7–8) presenting interiorly and
    presetally to each pore .... *Amynthas carnosus lichuanensis*

14. With papillae front clitellum 15
    With papillae front and behind clitellum 16

15. Male pores on a round porophore, surrounded by 3–4 skin folds. Paired papillae presetally on VII, VII .... *Amynthas loti*
    Male pores on ventral of VII, lacking associated genital markings. Genital marking paired, presetal VII–IX between third and fourth setal lines
    Amynthas monsoonus

16. Genital markings between XVII–XX and between VI–XII 17
    Genital markings in XVIII and between VI–IX 26

17. Some male pores with a small, less distinctly marked genital papilla pad adjacent laterally, genital papillae presetal, closely paired in midventral
    portion in XVIII–XX, in XVIII, 1 pair or 1 on left, in XIX, 1 pair or 1 on right or absent, in XX, 1 pair, each papilla small, round, center concave, size
    and arrangement similar to those in the preclitellar region. Genital papillae presetal, closely paired in midventral regions in IX–XII, number
    variable: in IX and X, absent, 1 on left or 1 pair, in XI, absent, 1 on right, 1 on left or 1 pair, in XII, absent or 1 on left, each papilla small, round, flat-
    topped or slightly concave .... *Amynthas tantulus*
    Genital markings between XVII–XX and between VI–X 18

18. Genital markings in XVII–XX and between VI–X 19
    Genital markings between XVII–XIX and between VI–X 20

19. Small papillae found on XVII–XX. Or pore-bearing papillae and pore apparently absent but instead found with as many as 7 or 9 small papillae on
    each side. With 1 rather large papilla placed medioventrally on VII, presetal. 2 series of small papillae from VI–X present slightly ventral to pores
    on each side. Medioventral papilla on VII absent .... *Amynthas moniliatus*
    Male pores surrounded by 3 circular papillae, 1 medial, 1 lateroanterior, 1 lateroposterior, the 3 papillae surrounded by 2–3 circular folds,
    additional postsetal papillae paired in XVII, paired or single in XIX and XX, similar in size and structure to those in the preclitellar regions. Genital
    papillae paired in VI–VII, medioanteriorly adjacent to the spermathecal pore, some of the papillae embedded partially in the pores, for some
    specimens, 2 additional pairs of presetal and postsetal papillae in VII, slightly medial to the spermathecal pores, each papilla circular, flat with a
    slightly concave center, surrounded by 1–2 circular folds .... *Amynthas tayalis*

20. Genital markings in XVII, XVIII and XIX; and between VI–X 21
    Genital markings in XVII, XVIII and/or XIX; and between VI–IX 24

21. Male pores absent or not detectable (softened), or present on XIX; each pore in a small depression surrounded by several concentric
    ridges. Similar papillae 2–3 in each group on each side of XVII and XVIII. Spermathecal pores in an eye-like depression. 2 stumpy cushions
    laterally projected on ventral side of X, on each of which are found from 5–13 small papillae. The glandular cushions are continuous
    ventromedially .... *Amynthas cruratus*
    Genital markings in XVII, XVIII and XIX; and between VI–IX 22

22. Male pores on a long-round porophore with 2 small papillae nearby, surrounded by concentric ridges. There are paired papillae on ventral rides
    of XVII, XVIII and XIX, sometimes lacking. Paired papillae on ventral sides of VII, VII and IX, or VII and VII, or VII only .... *Amynthas saccatus*

Continued

2. SYSTEMATICS

TABLE 22   Key to Species of the *sieboldi* Group of the Genus *Amynthas* From China—cont'd

Male pores on a small flat-topped papilla where the pore is marked by a whitish spot near the lateral side, 2 similar papillae situated posterior and anterior to the pore respectively, either in XVIII or XVII and XIX. 2 large horseshoe-shaped papillae, medial to the pore, each slightly raised and with a glandular top, usually placed between 17/18 and 18/19, in some cases, both confined to XVIII only, medial to the large papillae 1–2 round flat-topped papillae often present; sometimes, 4 such large papillae placed side by side occupying XVII–XIX inclusive with another small papilla on lateral side forming a large but shallow depression where the male pore is situated. The skin around the papillae somewhat swollen and glandular in character, sometimes with wrinkles particularly pronounced on lateral side. Spermathecal pores not easily visible externally unless there is a genital marking on skin. The skin around the pore swollen to form anterior and posterior lip-like projections, the pores of both sides fairly visible from dorsal aspect. In some cases, not so dorsal in position, from 1/2–6/11 apart ventrally, situated in a pale-colored region, the pores not visible from both dorsal and ventral aspects. Transverse grooves in middle or setal zones of segments VI–IX in the neighborhood of spermathecal pores, as very distinct genital markings. The pore, if visible, marked by a whitish spot just in the furrow, without a depression, sometimes surrounded by a whitish glandular patch of skin on 2 adjacent segments .... *Amynthas szechuanensis*

23. Male pores superficial, on a pore-bearing papilla on setal line, another and more distinct one immediately posterior to each porophore, 3 similar ones ventromedian and presetal; or 3 similar ones present, presetal on XIX. 5 small papillae present on ventral side of VII, presetal .... *Amynthas nubilus*
Genital markings in XVII and XVIII; and VII and/or IX 24

24. 1–2 round papillae adjacent laterally to each male pore. Additional papillae absent or present, if present, large, widely paired, presetal, immediately posterior to 17/18, and/or postsetal on medioventrum, closely paired or missing 1 of them. Each papilla round, center concave. Genital papillae presetal, widely paired in IX, close to intersegmental furrow. Each papilla large, round, concave .... *Amynthas fenestrus*
Genital markings in XVII and XVIII; and VII or IX 25

25. Genital markings unpaired and median, postsetal on XVII and XVIII; Genital papillae are median, unpaired, circular tubercles, each with a definite rim and a grayish-translucent center; presetal on VII, slightly nearer to the setae than 7/8; medioventral .... *Amynthas flexilis*
Male pores simple, on top of a papilla-like porophore surrounded by 7–8 circular folds, a genital papilla immediately anterior to setal line in the medial position between the male pores in XVIII, an additional papilla at the same location in XVII for some specimens, each papilla round, with a slightly concave center, about half of the distance between setal line and anterior intersegmental furrow. Genital papillae 2 transverse rows presetal in VII and IX, numbering 4–11 for each row, an additional pair of postsetal papillae present at the ventrolateral part of VII in some specimens .... *Amynthas tungpuensis*

26. Genital markings in XVIII and VI–VII 27
Genital markings in XVIII and between VII–IX 28

27. Male pores, number and structure of genital papillae variable: if contracted, male pores with 2 papillae, anterior papilla sandal-shaped, posterior papilla round, surrounded by several circular folds; if protruded, male pores located at the top of a corn-shaped porophore, associated closely with the genital papillae and surrounded by several circular folds; each papilla flat with slightly concave center. 3 pairs of large, oval, disc-like genital pads present in VI–VII in some specimens, each immediately medioanterior to spermathecal pores .... *Amynthas sexpectatus*
Male pores simple, round, papilla-like, surrounded by 2–3 incomplete circular folds, genital papillae presetal 5–11 or 18 and postsetal 1–4 or 8 in XVIII. Spermathecal pores invisible. Patches of small, round presetal genital papillae present ventromedially 0 or 2 in VI, 0, 5 or 11 in VII, 0–12 or 15 in VII .... *Amynthas tessellates paucus*

28. Genital markings in XVIII and VII–IX 29
Genital markings in XVIII and VII–VII or VII–IX 30

29. Male pores on the tip of large wart-like porophores whose base includes almost whole length of segment XVIII. Spermathecal pores are rather large cross slits on somewhat raised glandular areas. The genital marking, in the form of very small circular weakly raised papillae, in the region of the male pores as well as of the spermatheca pores. The former is limited to 3–4 papillae on the median side of each porophore, partly in front of and partly behind the setal zone, which stretches rather far to the median side of the porophore. In a regular pattern, apparently normal, the papillae mark the corners of a square which is symmetrically divided by the setae zone. The papillae of the spermathecal region usually are not found in groups but singly, and in 2 systems: partly in front of the setal zone, a small distance median from the line of spermatheca apertures, or partly behind the setal zone, a greater distance from the spermatheca apertures. Usually the spermathecal papillae are limited to segments VII and VII, therefore in fullest development on VII. Mostly some regular papillae are lacking. Rarely there appear a few papillae, or behind the setal zone on segment IX. Only 1 regular papilla was represented by 2–3 closely grouped papillae .... *Amynthas pecteniferus*
Male pores on a transverse ridge with 1 papilla in front and another behind in shape of a cross, on median wall of a crescent-shaped chamber; a rather raised skin pad medial to the chamber, usually with 1 very large papilla behind setae, or with 2–4 smaller ones distributed before and behind setal zone. Spermathecal pores preceded by a large papilla or several small ones and surrounded by glandular and wrinkled skin, sometimes greatly swollen forming a wide and deeply sunken slit; in many cases another glandular patch occurring in front of setal circle on VII, VII, IX, a papilla sometimes found in this patch; or spermathecal papillae less frequently present and surrounding glandular skin not marked. Genital papillae beside pore region generally present on ventral side of VII and VII, 1 pair in front and 1 behind setal circle of each segment. Very rarely 1 pair present in front of setal circle of segment IX. Those in front of setae more constantly present than those behind; those on VII again more constantly present than on VII. Or these papillae entirely absent. Each papilla rounded and flat-topped, perforated or slightly raised in center, without elevated margin. Its structure not different from those in male pore region .... *Amynthas yamadai*

30. Genital markings in XVIII and VII–VII or VII 31
Genital markings in XVIII and VII–IX 32

31. Male pores on a roundish papilla, or sometimes a large swelling, the pore marked as a pit in its center on setal line, without a slit though this part often sunken down due to contraction. 2 small nipple-like papillae anterior and posterior to, but a little ventral to each male papilla, sometimes 1

TABLE 22 Key to Species of the *sieboldi* Group of the Genus *Amynthas* From China—cont'd

placed still ventrally on setal line and opposite to male papilla. Each rounded and elevated, swollen or slightly depressed at its center. This group of papillae always surrounded by 3 or more circular ridges to form a transversely oval glandular ridge which extends anteriorly and posteriorly to intersegmental furrows. Sometimes a similar papilla present ventromedially just behind setal line of XVIII. Spermathecal pores in an elliptical pit, intersegmental but slightly close to posterior margin of anterior segment, fully visible when these segments well stretched, sometimes a glandular patch of skin paler in color around each pore. Similar round and raised papillae often paired medioventrally behind setae of segments VII and VII; an additional 1 present before or another 1 behind setae of VII .... *Amynthas leucocircus*

Male pores region with 3 large, flat, disc-like papillae surrounded by several circular folds: a lateral triangular papilla, a medioanterior sandal-shaped papilla, and a medioposterior sandal-shaped papilla. Each papilla with a slightly concave center. A pair of large, round, disc-like papillae in VII, occupying the entire space between setal line and 7/8 furrow. Each papilla round and flat .... *Amynthas binoculatus*

32. Male pores situated on the center of a flat-topped porophore, a large, round genital papilla situated medial to the male pore, the porophores and papillae surrounded by several oval folds. Genital papillae present or absent, if present, 1–2, small, round, on the posterior margin of the posterior 1–2 spermathecal pores .... *Amynthas taipeiensis*

Male pores situated on papilla-like porophore 33

33. Male pores on a cone with small papilla nearby, surrounded by concentric ridges somewhat raised; with small papillae, its front edge close to the pores in VII, IX .... *Amynthas heterogens*

Male pores on a papilla-like porophore surrounded by 3 genital papillae: 1 anterior, 1 posterior and 1 medial. Each papilla round, center depressed. The male porophore together with the genital papillae are surrounded by 3 diamond-shaped skin folds. Genital papillae 3 in presetal VII or widely paired in presetal VII; in IX, papillae absent, single median, or same number and arrangement as in VII. Each papilla round .... *Amynthas wuhumontis*

189. *Amynthas apapillatus* Zhao & Qiu, 2013

Amynthas apapillatus Zhao and Qiu, 2013. *J. Nat. Hist.* 47(33–36):2187–2191.

External Characters

Length 120 mm, width 4 mm, segment number 83. Prostomium epilobous. First dorsal pore in 11/12. Clitellum in XIV–XVI, ring-shaped, markedly glandular; ventral setae invisible but dorsal pore visible. Setae intensive, their numbers: 36 (III), 32 (V), 44 (VII), 48 (XX), 44 (XXV); 12 (V), 16 (VI), 16 (VII) between spermathecal pores, 12(XVIII) between male pores; setal formula $aa = 1.2–1.3ab$, $zz = 1.2zy$. Male pores paired in XVIII about 0.4 body circumference apart, ventrally placed, each on top of a round porophore. Genital papillae absent. Spermathecal pores 3 pairs in 6/7/8/9, ventral, each eye-shaped and obvious, distance between paired pores about 0.4 body circumference apart. Genital markings absent. Female pore single in the ventral center of XIV, ovoid, brown.

Internal Characters

Septa 8/9/10 absent; 5/6/7/8 comparatively thick and muscular; 10/11/12 thin. Gizzard ball-shaped, in IX–X. Intestine enlarged from XVI. Intestinal caeca simple, originating in XXV, extending anterior to XXVII, dark brown, with 3 distinct teeth-shaped diverticula dorsally, smooth ventrally. Hearts 4 pairs, in X–XIII, the last 3 pairs developed. Holandric, Testis sacs paired in X and XI. Seminal vesicles paired in XI and XII, well developed, both separate ventrally. Prostate gland in XVII–XVIII, small, coarsely lobated, with a stout U-shaped duct. Spermathecae 3 pairs, in VII–IX; ampulla elongated ovoid, duct stout, about 0.5 of ampulla, distinctly separated; diverticulum about 0.67 of main pouch and coiled, ental 0.25 enlarged, serving as the seminal chamber, silvery white in appearance inside the chamber.

Color: Preserved specimens brown dorsally and light brown ventrally. Clitellum light brown.
Distribution in China: Hainan (Limushan Nature Reserve).
Deposition of types: Shanghai Natural Museum.
Habitat: The specimens were collected in tropical rainforest on Quling Mountain in Limushan Nature Reserve at an elevation of 739 m.

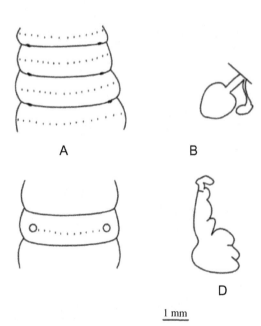

A   B

D

1 mm

FIG. 186 *Amynthas apapillatus* (after Zhao et al., 2013).

Remarks: Compared to the *Amynthas* species reported from China with 3 pairs of spermathecal pores in 6/7/8/9 (Chen 1931), *A. apapillatus* is most similar to *A. pingi chungkingensis* (Chen, 1936). They share some similar characters: 3 spermathecal pores, heart-shaped ampulla, short diverticulum, and ovoid seminal chamber. However, there are neither genital papillae nor genital markings, nor accessory glands in *A. apapillatus*. The prostate of *A. apapillatus* is small, while that of *A. pingi chungkingensis* is well developed.

190. *Amynthas binoculatus* Tsai, Shen & Tsai, 1999

Amynthas binoculatus Tsai, Shen and Tsai, 1999. *Jour. Nat. Taiwan Museum* 52(2):41.
Amynthas binoculatus Tsai, Shen and Tsai, 2000. *Zoological Studies* 39(4):286.
Amynthas binoculatus Tsai, Shen and Tsai, 2009. *Zootaxa* 2133:38.
Amynthas binoculatus Tsai, Shen and Tsai, 2009. *Earthworm Fauna of Taiwan.* 28–29.

External Characters

Length 196 mm, clitellum width 6.3 mm, segment number 113. No secondary segmentation. Prostomium epilobous. Clitellum XIV–XVI, annular, setae absent, dorsal pore absent. Setae 52 (VII), 73 (XX), 16 between male pores. Male pores 1 pair in XVIII, ventrolateral, 0.41 circumference apart ventrally. Each male pore region with 3 large, flat, disc-like papillae surrounded by several circular folds: a lateral triangular papilla, a medioanterior sandal-shaped papilla, and a medioposterior sandal-shaped papilla. Each papilla with a slightly concave center. Spermathecal pores 3 pairs in 6/7/8/9, lateral, 0.53 circumference apart ventrally, with light yellowish creamy white, swollen edge. A pair of large, round, disc-like papillae in VII, occupying the entire space between setal line and 7/8 furrow, distance between the papillae 0.21 body circumference apart ventrally. Each papilla round, flat, 1.5 mm in diameter. Female pore single, medioventral in XIV.

Internal Characters

Septa 8/9/10 absent, 10/11/–13/14 thickened. Gizzard in IX and X, round. Intestine from XV. Intestinal caeca paired in XXVII, smooth, wrinkled, extending anteriorly to XXIII. Lateral hearts in XI–XIII. Testis sacs 2 pairs in XI, small, medioventral. Seminal vesicles paired in XI and XII, each with a small dorsal lobe. Prostate glands paired in XVIII, large, folliculated, extending anteriorly to XV and posteriorly to XX. Prostatic duct n-shaped, proximal half enlarged. No accessory glands associated with

genital papillae in both spermathecal pore and male pore regions. Spermathecae 3 pairs in VII–IX, ampulla large, round, or peach-shaped, 2.7 mm long, stalk short, stout; diverticulum with a seminal chamber with folliculated surface, and a straight, stout stalk about 1.2 mm long.

Color: Live specimens yellow. Preserved specimens light yellow on head, dark brown on clitellum, light gray on body.
Type locality: Taiwan (Wufong, Taichung County).
Etymology: The species was named with reference to a pair of large, disc-like genital papillae in segment VII.
Deposition of types: Taiwan Endemic Species Research Institute, Jiji, Nantou County, Taiwan.
Distribution in China: Taiwan (Taichung).

191. *Amynthas carnosus lichuanensis* Wang & Qiu, 2005

Amynthas carnosus lichuanensis Wang and Qiu, 2005. *Journal of Shanghai Jiaotong University (Agricultural Science)* 23(1):24–25, 28.
Amynthas carnosus lichuanensis Xu & Xiao, 2011. *Terrestrial Earthworms of China.* 90–91.

External Characters

Length 300–405 mm, width 10–12 mm, segment number 157–163. Prostomium 1/2 epilobous. First dorsal pore in 12/13. Clitellum on segment XIV–XVI, annular, ventral glandular is not obvious shown conspicuously. Setae perichaetin, enlarged ventrally before segment XII; setal number 32–36 (III), 38–44 (V), 50–54 (VII), 120–126 (XX), 100–118 (XXV); 22–24 (VII), 22–26 (VII), between

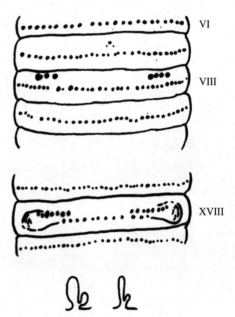

FIG. 187   *Amynthas carnosus lichuanensis* (after Wang and Qiu, 2005).

spermathecal pores, 34–38 between male pores. Male pores paired in XVIII, about 2/5 of circumferences apart ventrally, each situated in a small papillae of the segment XVIII. 3–4 ridges occurring to male pore and a row of 7–8 flat-topped papillae presenting interiorly and presetally to each pore. Spermathecal pores 3 pairs in 6/7/8/9, about 2/5 of circumferences apart ventrally, glandular part ventrally not obvious. Female pore single, medioventral in XIV.

Internal Characters

Septa 8/9/10 absent, 4/5/6/7/8, 10/11/12/13 muscular and thickened, 13/14/15 thick. Gizzard in VII–IX, barrel-shaped. Intestine swelling in XVI. Caeca paired in XXVII, simple, extending anteriorly to XXIII, serrated on ventral side. Hearts 4 pairs. Testis sacs 2 pairs in X and XI, a bit developed. Seminal vesicles paired in XI–XII, a pair of pseudo seminal vesicles smaller than the anterior pair in XIII. Prostate glands in XVIII, small, its duct smaller too; only duct found in some cases. Spermathecae 3 pairs, in VII–IX, in the forms of buds.

Color: Preserved specimens brown throughout.
Type locality: Hubei (Lichuan).
Etymology: The name of this species refers to the type locality.
Deposition of types: Laboratory of Environmental Biology, School of Agriculture and Biology, Shanghai Jiaotong University.
Distribution in China: Hubei (Lichuan).
Remarks: The present subspecies differs from *A. carnosus carnosus* and *A. carnosa chungkingensis* by its larger size, the different arrangement of papillae in male pore region, presence of a pair of pseudo seminal vesicles in XIII, and the vestigial prostate and spermathecae, which may indicate a trend of parthenogenesis.

192. *Amynthas contingens* (Zhong & Ma, 1979)

Pheretima contingens Zhong & Ma, 1979. *Acta Zootax. Sinica.* 4(3):229.
Amynthas contingens Xu & Xiao, 2011. *Terrestrial Earthworms of China.* 95.

External Characters

Length 40–80 mm, width 2.5–3.0 mm, segment number 74–108. Prostomium 1/2 epilobous. First dorsal pore 11/12. Clitellum in XIV–XVI, entire, annular, with ventral setal pits only of XV and XVI. Setae fine, close, even on anterior ventrum, none thick, setae number 37–42 (III), 47–57 (V), 46–56 (VII), 58 (IX), 40–55 (XX); 22 (VI), 24 (VII) between spermathecal pores, 10–15 between male pores; $aa = 1.2\ ab$, $zz = 2zy$. Male pores 1 pair on ventrolateral side of XVIII, about 1/3 of circumference apart ventrally,

each on a round porophore closely connected before and behind with a pear-shaped glandular pad. Spermathecal pores 3 pairs, intersegmental in 6/7/8/9, about 1/2 of ventral body circumference, no papillae. Female pore single in XIV, medioventral.

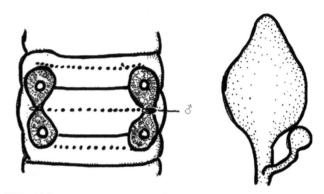

FIG. 188   *Amynthas contingens* (after Zhong & Ma, 1979).

Internal Characters

Septa 8/9/10 absent. Caeca palmate, in XXVII, extending anteriorly to XXIV, with 6–7 pointed pouches directed dorsally, ventral 3 longest. Testis sacs small, ventral, each pair close but not communicated, V-shaped. Seminal vesicles dorsal lobes invisible. Prostates in XVI–XX, its duct short, ectal end slight stout. Accessory glands on body wall. Spermathecae 3 pairs, ampulla around or heart-shaped, about 1.5 mm long; diverticulum shorter than main pouch with a round seminal chamber.

Color: Preserved specimens grayish dorsally.
Distribution in China: Chongqing (Shapingba), Sichuan (Mt. Emei).
Deposition of types: Department of Biology, Sichuan University.
Remarks: This species is similar to *A. szechuanensis szechuanensis* (Chen, 1931), but setae on anterior ventrum fine and close, seminal chamber round, and size smaller.

193. *Amynthas cruratus* (Chen, 1946)

Pheretima crurata Chen, 1946. *J. West China. Border Res. Soc. (B).* 16:126–127, 140, 150.
Amynthas cruratus Sims & Easton, 1972. *Biol. J. Linn. Soc.* 4:237.
Amynthas cruratus Xu & Xiao, 2011. *Terrestrial Earthworms of China.* 97–98.

External Characters

Medium-sized worm. Length 90 mm, width 3.5 mm, segment number 101. Prostomium 1/2 epilobous, tongue open behind, much wider than length of II. First dorsal pore 11/12. Clitellum XIV–XVI, entire, annular, rough on ventral side, or smooth, longer than intermediate post-clitellar segments, setae absent. Setae moderate in size and rather wider in intervals, slightly wider in space on anterior ventral side; *a*, *b*, or *c* larger. Setae number 26 (III), 30–32 (VI), 29–40 (IX), 35 (XXV); 12–14 (VI), 13–17 (IX) between spermathecal pores, 10 between male pores; $aa = 1.2\ ab$, $zz = 1.5zy$. Male pores absent or not detectable (softened), or present on XIX; if present, about 1/3 circumference apart ventrally, each pore in a small depression surrounded by several concentric ridges. Similar papillae 2–3 in each group on each side of XVII and XVIII. Spermathecal pores 3 pairs in 6/7/8/9, about 1/2 circumference apart, each in an eye-like depression. 2 stumpy cushions laterally projected on ventral side of X, on each of which are found 5–13 small papillae. The glandular cushions are continuous ventromedially, somewhat like that *Amynthas omeimontis* (Chen, 1931). Female pore single, medioventral in XIV.

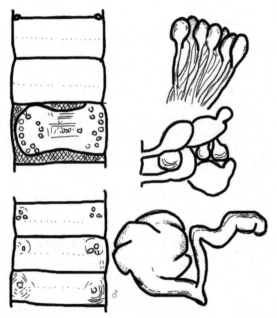

FIG. 189   *Amynthas cruratus* (after Chen, 1946).

Internal Characters

Septa 8/9/10 absent, 6/7/8 slightly muscular, 10/11/12 membranous. Gizzard large, narrower anteriorly in 1/2 IX and X. Intestine swelling in XV. Pharyngeal glands not particularly large, extending to V. Caeca simple, in XXVII, smooth on both sides, extending anteriorly to XXIV. Vascular loops in IX asymmetrically developed, those in X absent. Nephridial tufts fairly thick anteriorly on 5/6/7. Testis sacs in X widely placed but connected and communicated, posterior pair more so but communicated. Seminal vesicles in XI and XII moderately large, dorsal lobes distinct. Prostate glands well developed, in coarse lobules, its duct straight, curved at both ends, most

middle portion large, more so entally. Accessory glands stalked, each corresponding to respective papilla externally, gland conical or spherical. Spermathecae 3 pairs, large. Ampulla sac-like, duct very thick, more so entally, connected but not well marked off from the ampulla. Diverticulum nearly as long as main pouch, ental third slightly enlarged as seminal chamber, usually crooked. No seminal contents.

Color: Light chocolate on dorsum, pale on ventrum. Clitellum cinnamon-red.
Distribution in China: Chongqing (Beipei).
Etymology: The specific name *"cruratus"* probably refers to the structure of cushions, each of which is formed with 5–13 small papillae on ventral side of X.
Deposition of types: Previously deposited in the Institute of Zoology, Academic Sinica, Chongqing, the types were destroyed during the war in 1945–1949.
Remarks: The species appears like *A. omeimontis* (Chen, 1931) in the characters of the glandular cushions and the stalked glands. However, it is distinct. Regarding the male pore aspect, the 2 specimens all show abnormality, it being either totally absent or shifted to XIX.

194. *Amynthas dactilicus* (Chen, 1946)

Pheretima dactilica Chen, 1946. *J. West China Border Res. Soc. (B).* 16:106–108, 139, 150.
Amynthas dactilicus Sims & Easton, 1972. *Biol. J. Linn. Soc.* 4:237.
Amynthas dactilicus Xu & Xiao, 2011. *Terrestrial Earthworms of China.* 100–101.

External Characters

Length 30–70 mm, width 2.5–4 mm, segment number 65–108. First 2 segments always smaller. Prostomium 1/3 epilobous. First dorsal pore 4/5. Clitellum XIV–XVI, annular, smooth, well marked, not swollen, no setae. Setae on II invisible, on III distinct. Setal chain close both dorsally and ventrally; setae number 35 (IX), 34 (XIX), 35 (XXV); 18 (VII), 17 (VIII), 19 (IX) between spermathecal pores, 10–12 between male pores; $aa = 1.2$–$1.5ab$, $zz = 1.2$–$2zy$. Male pores paired in XVIII, superficial, about 1/3 circumference apart ventrally, each on a small papilla on lateroventral side, surrounded by a few concentric ridges often not marked ventrally; a small papilla smooth but glandular and flat-topped placed medial to each pore, 2 on ventromedial side, slightly elevated, or variously arranged (25 out of 60 specimens papillae found close to porophore; 11 of them entirely lacking in this region). Spermathecal pores 3 pairs in 6/7/8/9, about 1/2 (occasionally less than 1/2) circumference apart ventrally, intersegmental, each appearing as elliptical elevation, no swollen nor other genital markings in this region. Female pore single, medioventral in XIV.

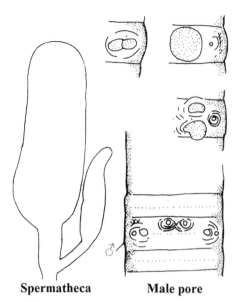

Spermatheca   **Male pore**

FIG. 190   *Amynthas dactilicus* (after Chen, 1946).

Internal Characters

Septa 8/9/10 absent, 9/10 traceable ventrally. Membranous on the other. Pharyngeal glands moderately developed. Gizzard onion-shaped, in VIII, IX. Intestine swelling in XV. Caeca simple, in XXVII, extending anteriorly to XXIV or XXIII. Vascular loops in X and IX as stout as hearts following those in IX, also distinct. Testis sacs anterior pair V-shaped and communicated, posterior pair broadly united. Seminal vesicles well developed, extending between the segments X and XIV or XV, meeting dorsally. Prostate gland massive, in XVI–XXIV, its duct of uniform thickness slender in a V-shaped curve. Accessory glands sessile, each corresponding to respective papilla externally. Spermathecae 3 pairs, in VII–IX, ampulla about 2 mm long, sac-like, with a short stout duct. Diverticulum short, seminal chamber dactyloid, pointed entally, as seminal chamber, its duct short and slender.

Color: Preserved specimens pale both dorsally and ventrally. Clitellum light fleshy.
Distribution in China: Hubei (Lichuan), Chongqing (Mt. Jinfo), Sichuan (Mt. Emei), Guizhou (Mt. Fanjing).
Etymology: The specific name *"dactilicus"* means dactyloid in Latin and refers to the structure of the seminal chamber.
Deposition of types: Previously deposited in the Institute of Zoology, Academic Sinica, Chongqing, the types were destroyed during the war in 1945–1949.
Remarks: This species is characterized by several important features: (1) anteriorly situated dorsal pore, (2) characteristic papillae usually found in the male pore region, (3) more laterally placed spermathecal pores, and (4) the smaller size. This species seems to prefer cold places. It is found between 1828 and 2438 m.

195. *Amynthas daulis daulis* (Zhong & Ma, 1979)

Pheretima daulis Zhong & Ma, 1979. *Acta Zootax. Sinica*, 4(3):229–230, 232.
Amynthas daulis daulis Xu & Xiao, 2011. *Terrestrial Earthworms of China*. 101.

External Characters

Length 49–93 mm, width 2.5–3.5 mm, segment number 61–105. Prostomium 1/2 epilobous. First dorsal pore often 4/5, occasionally 11/12 or 12/13. Clitellum entire, XIV–XVI, annular, with setae pit ventrally of XV, XVI, setae absent. Setae fine and close, even on anterior ventrum, setae number 29–34 (III), 34–40 (V), 36–48 (VII), 38–50 (IX), 40–44 (XX); 21–24 (VI), 22–26 (VII), 22–28 (VIII) between spermathecal pores, 11–14 between male pores; $aa = 1.0$–$1.2ab$, $zz = 1.5$–$2zy$. Male pores on ventrolateral side of XVIII, about 1/3 circumference apart ventrally, each on round or ellipsoid tubercle at ventrolateral side of XVIII, 2 glandular pads placed in front and behind of porophore. A pair of papillae postsetally on XVII, or sometimes absent. Spermathecal pores 3 pairs, intersegmental in 6/7/8/9, about 1/2 circumference apart. Female pore single, medioventral in XIV.

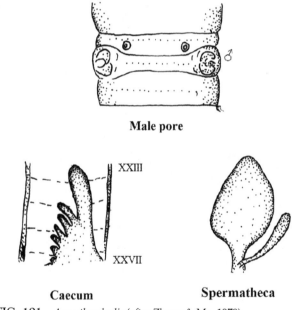

Male pore

Caecum   **Spermatheca**

FIG. 191   *Amynthas daulis* (after Zhong & Ma, 1979).

Internal Characters

Septa 9/10 absent, 8/9 thin. Caeca palmate, in XXVII, extending anteriorly to XXIII, with 4–5 small finger-shaped pouches on posterior dorsal half. Testis sacs as transverse bands, anterior pair slightly V-shaped. Seminal vesicles large, rough on surface, not dorsal. Prostate glands and accessory glands as usual. Spermathecal ampulla about 2 mm long. Diverticulum shorter, with ental club-shaped chamber.

Color: Preserved specimens light brownish.
Distribution in China: Hubei (Lichuan), Shichuan (Mt. Emei).
Etymology: The specific name probably refers to the structure of the male pores with 2 glandular pads in front and behind of porophore.
Deposition of types: Department of Biology, Sichuan University, Chengdu, Sichuan.
Remarks: This species is similar in many respects to *A. szechuanensis szechuanensis* (Chen, 1931), but differs from the latter by its first dorsal pore in 4/5, 2 glandular pads in front and behind of porophore, and also in the shape of the caeca.

196. *Amynthas daulis fanjinmontis* Qiu, 1992

Amynthas daulis fanjinmontis Qiu, 1992. *Sichauan Journal of Zoology*. 11(1):1–3, 2.
Amynthas daulis fanjinmontis Xu & Xiao, 2011. *Terrestrial Earthworms of China*. 101–102.

External Characters

Length 34–44 mm, width 2.0–2.5 mm, segment number 52–78. Prostomium 2/3 epilobous. First dorsal pore 3/4 or 4/5. Clitellum XIV–XVI, annular, no setae. Setae fine;

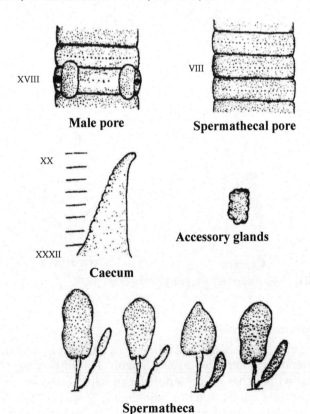

Male pore   **Spermathecal pore**

Accessory glands

Caecum

Spermatheca

FIG. 192 *Amynthas daulis fanjinmontis* (after Qiu, 1992).

setae number 30–42 (III), 38–56 (V), 52–72 (VIII), 41–64 (XX), 48–62 (XXV); 17–27 (VII), 20–28 (VIII) between spermathecal pores, 0–8 between male pores. Male pores 1 pair on ventral side of XVIII, about 1/3 circumference apart ventrally; each on a round tubercle; a large long-round flat-topped papilla, which extends anteriorly to XVII and posteriorly to XIX, placed ventrally close to each pore. Spermathecal pores 3 pairs in 6/7/8/9, about 2/5 circumference apart ventrally. No genital markings in this region. Female pore single, medioventral in XIV.

Internal Characters

Septa 8/9/10 absent, other thin. Gizzard long barrel-shaped. Intestine swelling in XV; caeca simple, very long, in XXVII, extending anteriorly to XX, with small tooth-like breach. Last hearts XIII. Testis sacs 2 pairs, ovoid, in X–XI, communicated. Seminal vesicles rather small, rough on the surface, no dorsal lobe. Prostate glands well developed, in XVII–XXI. Accessory glands round, attaching to body wall. Spermathecae 3 pairs in VII, VIII, IX, ampulla hearted or pyriform or like a long sac in shape, about 1.5–2.0 mm long; its wall thick; diverticulum shorter than main pouch, with terminal club-shaped chamber. No accessory glands in this region.

Color: Preserved specimens light brownish. Clitellum red-brown.
Distribution in China: Guizhou (Mt. Fanjing).
Etymology: The name of this species refers to the type locality.
Type locality: Guizhou (Mt. Fanjing).
Deposition of types: Institute of Biology, Guizhou Academy of Sciences, Guiyang.
Remarks: This subspecies differs from the nominate subspecies by: (1) rather small body; (2) large long-round flat-toped papillae, which are located close ventrally to male pores and extended anteriorly to XVII and posteriorly to XIX; (3) first dorsal pore in 3/4 or 4/5; (4) the absence of a septum 8/9; (5) simple caeca, and (6) conspicuous bounds between seminal chamber and its duct.

197. *Amynthas editus* (Chen, 1946)

Pheretima edita Chen, 1946. *J. West China Border Res. Soc. (B)*. 16:94–95, 138, 146.
Amynthas editus Sims & Easton, 1972. *Biol. J. Linn. Soc.* 4:237.
Amynthas editus Xu & Xiao, 2011. *Terrestrial Earthworms of China*. 108.

External Characters

Medium sized worm. Length 45 mm, width 4 mm, segment number 72. All Segments short, 3–7 annuli from VII posteriorly, VIII–X longer, abruptly shortened after XI, length of IX, X equal to 3 postclitellar ones. Prostomium epilobous, but inconspicuous. First dorsal pore 11/12 (?). Clitellum XIV–XVI, annular, smooth. Setae rather rough on anterior ventral side, equally distributed, slightly closer and shorter on dorsal side. Those behind 23rd segment scarcely visible. Setae number 23 (III), 38 (IX), 34 (XIX), 36 (XXV); 12 (VI, VII), 11 (VIII), 16 (IX) between spermathecal pores, 12 between male pores; $aa = 1.2ab$, $zz = 1.2$–$1.5zy$. Male pores paired in XVIII, about 1/3 circumference apart ventrally, each on a glandular papilla which encroaches over 1 segment, its rim not distinct, flat-topped. 1 large lens-shaped papilla placed medioventrally between setal lines of XVII and XVIII. Spermathecal pores 3 pairs in 6/7/8/9, last pair about 4/9 circumference apart ventrally, first 2 pairs about 1 seta interval ventrad. No genital papilla nor other markings. Female pore single, medioventral in XIV, in a small ovoid tubercle.

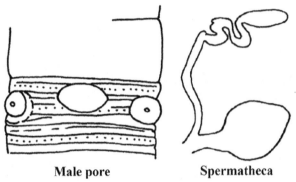

Male pore   **Spermatheca**

FIG. 193   *Amynthas editus* (after Chen, 1946).

Internal Characters

Septa 9/10 absent, 4/5 thin, 5/6/7 feebly muscular, with small nephridial tufts, 7/8 thin, 8/9 membranous, 10/11/12/13 thick and musculated, gradually thinning posteriorly. Pharyngeal gland moderately developed extending to V. Gizzard in IX and X, larger at both ends. Intestine swelling in XVI. Caeca simple and short, extending anteriorly to XXIV, smooth on both sides. Vascular loops in IX developed, those in X absent. Testis sacs in X about 4 mm high, 0.9 mm wide, extending to dorsal side but not meeting, those in IX larger, united dorsally, connected with a short bridge ventrally. Seminal vesicles small, about 1.5 mm in vertical length, warty or mulberry on surface, dorsal lobe two-thirds the size, also finely roughened. First pair in XI enclosed in testis sacs. Prostate glands with only 4–5 large lobes in XVI & XXI, its duct moderately long, in loose U-curve, ectal end stout. Accessory glands velvety and sessile, without definite outline. Spermathecae 3 pairs, in VII–IX, posterior to their respective septa. Ampulla small, heart-shaped, about 0.5 mm long. Diverticulum very long, with an elongate terminal seminal chamber and ectal portion of duct coiled.

Color: Pale both dorsally and ventrally, mediodorsal line dark, very narrow. Clitellum brownish.

Distribution in China: Sichuan (Mt. Emei).

Etymology: The specific name *"editus"* means elevated, high, or lofty in Latin and refers to the structures of the testis sacs.

Deposition of types: Previously deposited in the Institute of Zoology, Academic Sinica, Chongqing, the types were destroyed during the war in 1945–1949.

Remarks: This species appears superficially like *Desmogaster* (Rosa, 1895) in terms of prostomium, segmentation, and obsolete dorsal pores. Its identifying characteristics are the setal peculiarities, the long diverticulum, and the testis sacs.

198. *Amynthas fenestrus* Shen, Tsai & Tsai, 2003

Amynthas fenestrus Shen, Tsai & Tsai, 2003. *Zoological Studies.* 42(4):479–490.
Amynthas fenestrus Chang, Shen & Chen, 2009. *Earthworm Fauna of Taiwan.* 44–45.

External Characters

Length 60–73 mm, clitellum width 2.6–2.9 mm, segment number 83–103. Number of incomplete annuli (secondary segmentation) per segment 2–3 in IX–XIII, XVII, and XVIII. Prostomium epilobous. First dorsal pore 5/6 or 6/7. Clitellum XIV–XVI, annular, 2.2–3.4 mm long, setae absent, dorsal pore absent. Setae number 30–36 (VII), 33–43 (XX), 8–10 between male pores. Male pores paired in XVIII, lateroventral, 0.26–0.3 circumference apart. 1–2 round papillae adjacent laterally to each male pore. Additional papillae absent or present, if present, large, widely paired, presetal, immediately posterior to 17/18, and/or postsetal on medioventrum, closely paired or missing 1 papilla. Each papilla round, center concave, about 0.3 mm in diameter. Spermathecal pores 3 pairs in 6/7/8/9, ventrolateral in depressed furrow, 0.31–0.34 circumference apart ventrally. Genital papillae presetal, widely paired in IX, close to intersegmental furrow, V or VI intersetal distances apart. Each papilla large, round, concave, about 0.3 mm in diameter. Female pore single, medioventral in XIV.

Internal Characters

Septa 5/6/7/8, 10/11/12/13 thickened, 8/9/10 absent. Gizzard round in IX–X. Intestine from XV. Intestinal caeca paired in XXVII, simple, short, stocky, surface slightly wrinkled, extending anteriorly to XXV or XXIV. Esophageal hearts in XI–XIII. Testis sacs 2 pairs in XI, round. Seminal vesicles large, extending between XI and XIII, surface wrinkled, follicular, each with a large dorsal lobe with folliculated surface. Prostate glands paired in XVIII, wrinkled, extending anteriorly to XVI and posteriorly to XX, prostatic duct C-shaped. Accessory glands round, sessile or with stalk. Spermathecae 3 pairs in VII–IX, ampulla peach-shaped, about 1.0 mm long and 0.9 mm wide, stalk stout, 0.4–0.5 mm long, diverticulum stalk slightly bent or curved, 0.5–0.7 mm long, seminal chamber oval-shaped, 0.4 mm long. Accessory glands large, oval-shaped or divided into 3–4 round lobes, each lobe 0.3–0.5 mm long, stalk short, corresponding to external genital papillae.

Color: Preserved specimens brownish gray on dorsum, light gray on ventrum, light brownish gray around clitellum.
Distribution in China: Taiwan (Nantou).
Etymology: The specific name *"fenestrus"* means windows in Latin and refers to the window-like paired genital papillae in the preclitellar region.
Type locality: Rueyen Creek Nature Reserve (elevation 2300 m), Nantou County, Taiwan.
Deposition of types: Taiwan Endemic Species Research Institute, Jiji, Nantou County, Taiwan.
Remarks: *A. fenestrus* (Shen, Tsai & Tsai, 2003) has 3 pairs of spermathecal pores in 6/7/8/9 and thus belongs to the *sieboldi* group (Sims and Easton, 1972). This species is fairly similar to *A. tantulus* (Shen, Tsai & Tsai, 2003) except for the size and arrangements of the genital papillae and the structure of the accessory glands. *A. fenestrus* is large, and has comparatively higher segment and setal numbers, larger preclitellar and postclitellar genital papillae, and larger accessory glands than *A. tantulus*.

199. *Amynthas flexilis* (Gates, 1935)

Pheretima flexilis Gates, 1935a. *Smithsonian Mus. Coll.* 93(3):7–9.
Pheretima flexilis Chen, 1936. *Contr. Biol. Lab. Sci. Soc. China (Zool.).* 11:295–296.
Pheretima flexilis Gates, 1939. *Proc.U.S. Natr. Mus.* 85:433–434.
Amynthas flexilis Sims & Easton, 1972. *Biol. J. Linn. Soc.* 4:237.
Amynthas flexilis Xu & Xiao, 2011. *Terrestrial Earthworms of China.* 114.

External Characters

Length 40 mm, clitellum width 2 mm, segment number 67. Prostomium 1/3 epilobous, width of dorsal lobe broader (about twice) than length of II. First dorsal pore 13/14. Clitellum XIV–XVI, annular; intersegmental furrows, dorsal pores and setae absent. Setae evenly distributed all over the body, none particularly enlarged; both dorsal and ventral breaks slight; setae number 31 (III), 38 (VI), 36 (VIII), 36 (XII), 30 (XXV); 15 (VI), 16 (VIII)

between spermathecal pores, 8 between male pores. Male pores superficial, at the center of tiny, transversely oval areas, each area surrounded by several concentric furrows; about 1/3 circumference apart ventrally. Spermathecal pores minute and superficial, 3 pairs, in 6/7/8/9, about 3/5 circumference apart ventrally. Genital markings unpaired and median, presetal on VIII, slightly nearer to the setae than to 7/8, postsetal on XVII and XVIII, circular tubercles, each with a definite rim and a grayish-translucent center. Female pore single, medioventral in XIV.

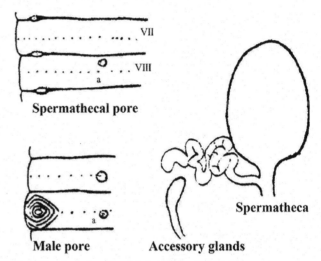

FIG. 194   *Amynthas flexilis* (after Gates, 1935a).

Internal Characters

Septa not particularly thickened. Gizzard large in VIII and IX. Intestine swelling in XV. Intestinal caeca elongate, in XXVII, extending anteriorly to XXIV, smooth. First pair of hearts in X, not observed, last pair of hearts in XIII. Anterior pair of testis sacs connected ventrally and also meeting dorsally, on X, horseshoe-shaped. 1 pair of vertical testis sacs in XI included within the posterior testis sacs. Seminal vesicles paired in XI destroyed, in XII very large. Prostate glands large, in XVII–XXII, nearly meeting dorsally, its duct 1 mm long, glistening, erect in the coelom, practically straight expect for a very short, slender, ental portion. Accessory glands with a short stalk. Spermatheca 3 pairs, ampulla distended and smooth, about 1.1 mm long, its duct extremely short; diverticulum long, with a short, muscular stalk, but irregularly coiled, and an elongate tubular seminal chamber, the latter variously bent, twisted or looped; ental half filled sperm masses. Genital marking glands with long, coelomic stalks.

Color: Preserved specimens grayish.
Distribution in China: Sichuan (Maogong, Dawei).
Etymology: The specific name *"flexilis"* means bent or curved in Latin and probably refers to the structure of the spermathecal diverticulum.

Type locality: Sichuan (from between Xiaojin County and Dawei).

Deposition of types: Smithsonian Institution.

Remarks: Distinguished from *A. hupeiensis* (Michaelsen, 1895) by the absence of septa 8/9/10 and from *A. leucocirca* (Chen, 1933) by the characteristics of the testis sacs and the included seminal vesicles and hearts.

The body wall is so transparent in places that exact enumeration of the setae is difficult.

200. *Amynthas heterogens* (Chen, Hsü, Yang & Fong, 1975)

Pheretima heterogens Chen, Hsü, Yang & Fong, 1975. *Acta Zool. Sinica*. 21(1):90–91.
Amynthas heterogens Xu & Xiao, 2011. *Terrestrial Earthworms of China*. 123–124.

External Characters

Length 63.5–157 mm, width 3.0–4.5 mm, segment number 64–113. Prostomium 2/3 epilobous. First dorsal pore 11/12. Clitellum XIV–XVI, annular, setae absent. Setae fine, *a–d* on II–VII stout, widely spaced; Setae number 40–58 (VIII), 34–53 (XVIII). Male pores in ventral region of XVIII, about 1/3 circumference apart ventrally, each on a cone with a small papilla nearby, surrounded by concentric ridges somewhat raised. Spermathecal pores 3 pairs in intersegmental furrows of 6/7/8/9, shaped like the eye of a needle, almost 1/2 circumference apart; anterior border of VIII and IX which immediately behind the pore with a small papilla respectively, which is absent in VII.

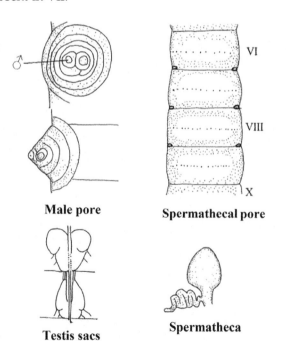

Male pore   **Spermathecal pore**

Testis sacs   **Spermatheca**

FIG. 195   *Amynthas heterogens* (after Chen et al., 1975).

Internal Characters

Septa 8/9/10 absent, 5/6/7/8 and 10/11/12 slightly thick, 12/13 and succeeding septa thin. Gizzard round, smooth, in IX. Intestine swelling in XVI; caeca simple, round and smooth, in XXVII, extending anteriorly to XXIV, incisions on dorsal and ventral margins. Last hearts in XIII. Testis sacs X–XI, 2 pairs, anterior pair in a thick V shape, not narrowly connected; posterior pair widely connected and communicated, in large sac including seminal vesicles. Seminal vesicles 2 pairs in XI and XII, large, rough, surrounded by dorsal lobes; posterior pair larger, extending to posterior segment. Prostate glands large and divided into several rough lobes, in XVI–XXI, prostatic duct stout, U-shaped, the base with accessory glands. Spermathecal ampulla heart-shaped and distally slightly pointed, its duct short and rich in muscle fibers, clearly demarcated from ampulla; diverticulum as long as main pouch, deeply zigzagged with short duct joining the main duct near body wall. No accessory glands.

Color: Preserved specimens gray-brown, setal ring light in color. Clitellum red.
Distribution in China: Fujian (Fuzhou), Guizhou (Mt. Fanjing).
Type locality: Fujian (Fuzhou).
Remarks: This species is similar to *A. diffringens* (Baird, 1869), but can be easily distinguished by the following points: (1) this species is rather small, no wider than 4.5 mm; (2) this species with much more setae, but without extraordinarily long setae on ventrum; (3) spermathecae 3 pairs; (4) prostate glands well developed; not cord-shaped or lump-shaped but kidney-shaped; (5) spermathecal diverticulum with 3–4 bend, seminal chamber no obvious.

77. *Amynthas hongyehensis* Tsai, Shen & Tsai, 2010 (*v.ant.* 77)

201. *Amynthas huangi* James, Shih & Chang, 2005

Amynthas huangi James, Shih & Chang, 2005. *Jour. Nat. His.* 39(14):1014–1015.
Amynthas huangi Xu & Xiao, 2011. *Terrestrial Earthworms of China*. 126–127.

External Characters

Length 70 mm, width 3.5 mm, segment number 101. Prostomium epilobous, with tongue open. First dorsal pore 12/13. Clitellum XIV–XVI, annular, setae invisible externally. Setae regularly distributed around segmental equators, size and distance regular; setae number 38 (VII), 48 (XXV); 10 between male pores. Male pores on small

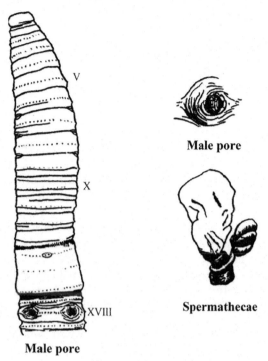

Male pore

Spermathecae

Male pore

FIG. 196   *Amynthas huangi* (after James, Shih & Chang, 2005).

knobs visible under hoods covering male pore openings. Spermathecal pores 3 pairs, on 6/7/8/9 lateral, deep slits. Genital markings not visible externally. Female pore single, medioventral in XIV.

Internal Characters

Septa 8/9/10 absent, 6/7/8 and 10/11–13/14 thinly muscular. Gizzard in VIII–X. Intestine beginning to swell from XVI. Caeca simple, in XXVII, extending anteriorly to XXIV. Hearts 4 pairs in X–XIII. Testes, funnels in ventral paired sacs in X, XI. Seminal vesicles small in XI, XII, without dorsal lobe. Prostate glands large in XVIII, deeply lobed; ducts thick, muscular, short; vasa deferens join duct at duct-glandular portion junction; vasa deferens nonmuscular; prostatic duct flanked by large sessile glandular masses on body wall. Spermathecae 3 pairs, in VII–IX, ampulla ovoid, large; diverticulum large flat ovate mass composed of tightly folded tubular chamber, short slender straight stalk; no nephridia on spermathecal ducts; genital marking glands with long stalks meeting body wall in VI–VIII next to spermathecal ducts.

Color: Preserved specimens pale brown dorsal pigmentation.
Distribution in China: Taiwan (Pingdong).
Etymology: This species is named after Mr. Chung-Chi Huang who helped extensively in the collection work.
Type locality: Taiwan (Shihwen, Pingdong County).
Deposition of types: National Museum of Natural Science, Taichung, Taiwan.

Remarks: *A. huangi* belongs to the *aelianus* species, it is similar to *A. taipeiensis* (Tsai, 1964). However, there are many differences as: smaller size than *A. taipeiensis*, fewer setae, no setal enlargement ventroanteriorly, male pore area different, hood or flap over male pores present, color different, intestinal origin in XVI not XV, seminal vesicles lack dorsal lobes, prostatic ducts short and straight, not coiled or bent, diverticulum chamber coiled with straight stalk versus stalk kinked in *A. taipeiensis*, and no genital marking glands in *A. taipeiensis*. Note that genital marking glands are present in XVIII even though no externally visible genital markings are present. This suggests that genital markings are hidden under the hoods partially obscuring the male pores. Furthermore, the spermathecal segment genital markings must be deep in the pore slits or even within the pores themselves, out of view.

202. Amynthas hupeiensis (Michaelsen, 1895)

Pheretima hupeiensis Chen, 1931. *Chin. Zool.* 7(3):122–123.
Pheretima hupeiensis Gates, 1935a. *Smithsonian Mus. Coll.* 93(3):11.
Pheretima hupeiensis Gates, 1939. *Proc. U. S. Natr. Mus.* 85:448–450.
Pheretima hupeiensis Kobaÿashi, 1938. *Sci. Rep. Tohokuo Univ.*, 13:152–153.
Pheretima hupeiensis Kobayashi, 1940. *Sci. Rep. Tohokuo Univ.* 15:273.
Pheretima hupeiensis Chen, 1946. *J. West China Border Res. Soc. (B).* 16:136.
Amynthas hupeiensis Sims & Easton, 1972. *Biol. J. Linn. Soc.* 4:237.
Amynthas hupeiensis Xu & Xiao, 2011. *Terrestrial Earthworms of China.* 127–128.

External Characters

Length 70–222 mm, clitellum width 3–6 mm, segment number 110–138. Prostomium epilobous. First dorsal pore 11/12 or 12/13. Clitellum XIV–XVI, annular, smooth, dorsal pore absent, setae generally present on ventrum. Setae very fine and numerous on each segment; more numerous on preclitellar; setae number 60–78 (II), 60–100 (III), 74–120 (VIII), 62–100 (XII), 66–88 (XXV); 14–22 (VIII) between spermathecal pores, 10–16 between male pores. Male pores paired on ventral side of XVIII, rather closely approximated, about 1/6 circumference apart ventrally, each on a slightly raised papilla-like tubercle which is flat-topped. Genital papillae 2 pairs, always present between intersegmental furrows 17/18/19, in line to or a little medial to male tubercle, often symmetrically paired. Each slightly transverse oval in outline,

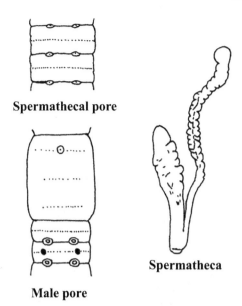

Spermathecal pore

Spermatheca

Male pore

FIG. 197 *Amynthas hupeiensis* (after Chen, 1933).

highly elevated, with a thick margin and a glandular depressed center, 6 or more times larger than male tubercle; intersegmental furrows around papillae often obliterated. Spermathecal pores 3 pairs in 6/7/8/9, ventral, near the line of male pores. Female pore single in XIV.

Internal Characters

Septa 8/9/10 present, 5/6/7/8 thickened. Gizzard VIII and IX, small, round. Intestine swelling in XV or XVI. Intestinal caeca paired in XXVII, simple, conical-shaped, smooth, extending anteriorly to XXIV. Hearts 4 pairs, in X–XIII. Seminal vesicle, testis, and seminal funnel all enclosed in a large but thin sac over dorsal side to communicate with that on opposite sides and also ventrally. The membranous sac taking its origin from anterior surface of each septum close to parietal wall and covering the seminal vesicles anteriorly to attach itself less intimately to posterior central part of preceding septum and then turning back close to the esophagus. Testis sacs large, first pair sometimes as large as vesicles. Testis very large, tufted, about 0.7 mm in diameter, situated in usual position close to nerve cord. Seminal vesicles small, narrow and elongate, each with a distinct dorsal lobe, enclosed in membranous sac. Prostate glands usually well developed, in XVI–XX, with large lobules of gland part and a long duct which is stouter distally and looped in an S shape. Accessory glands present corresponding to papillae outside, each in a roundish compact mass. Spermathecae 3 pairs, first pair in VII, last 2 pairs in VIII, or sometimes last pair IX; ampulla not sharply demarcated from its duct, spatulate in shaped, usually wrinkled on surface, its duct wide, about 1/2 or 1/3 of total length. Ampulla and duct about 1–3 mm long. Each diverticulum slender and very long, at least twice as long as main part

or 5 or more times as long, its ental portion constricted along surface appearing like compact loops and rounded at blind end, its ectal smooth portion serving as a duct, as long as, or longer than, main duct.

Color: Live specimens grass greenish on dorsum, greenish buff on anterior dorsum, light green on posterior dorsum, purplish green along mediodorsal line, lighter around setal zone; grayish pale ventrally. Clitellum milky or light chocolate. Preserved specimens dark green on dorsum, light green on ventrum. Or very faded, uniformly greyish at middle region of body, a purplish line visible along the dorsal pores; anterior 12 segments hardened and fleshy pale all over, posterior end a little less so; middle body softened. Global distribution: China, Japan, Korea, Taiwan, North America, and New Zealand.

Distribution in China: Beijing (Haidian, Huairou), Hebei, Liaoning (Shenyang, Dashiqiao, Dalian, Dandong), Jilin, Heilongjiang (Harbin), Shanghai, Jiangsu (Nanjing, Suzhou, Wuxi, Xuzhou), Zhejiang (Zhoushan, Ningbo, Shengxian, Lanxi, Tonglu, Fenshui), Anhui (Anqing, Chuzhou, Guichi), Fujian (Fuzhou, Xiamen), Taiwan (Taibei), Shandong (Yantai, Weihai), Jiangxi (Nanchang, Jiujiang, Dean), Shaaxi (Yuncheng, Xnjiang, Houma, Fushan, Linfen, Qixian), Henan (Xinxiang, Jiaozuo, Qinyang, Boai, Wenxian, Anyang, Luoyang, Kaifeng, Yexian, Xishan, Xiayi, Luoshan, Xinyang), Hubei (Wuchang, Qianjiang), Chongqing (Shapingba, Nanchuan, Beipei, Peiling, Jiangbei), Sichuan (Chengdu, Luzhou, Leshan), Guizhou (Mt. Fanjing), Yunnan, Shanxi, Gansu, Ningxia (Zhongning, Zhongwei).

Etymology: The name of this species refers to its type locality.

Type locality: Hupei (Shihuiyao near Wuchang).

Deposition of types: Hamburg Museum, Germany. Some specimens previously deposited in the Museum of Biological Laboratory of the Science Society of China, Nanjing. The types were destroyed during the war in 1937.

Remarks: In considering the geographical distribution of this species, it has only so far been reported from Fuzhou in China in spite of its frequent distribution in the provinces along the Yangtze River. It is also found in the central part of Japan. Its occurrence in Japan is probably a matter of human transference. It is perhaps also proper to note here that there is a single specimen of this species on the campus of the University of Pennsylvania, Philadelphia.

203. *Amynthas jaoi* (Chen, 1946)

Pheretima jaoi Chen, 1946. *J. West China Border Res. Soc. (B)*. 16:112–114, 140, 152.

Amynthas jaoi Sims & Easton, 1972. *Biol. J. Linn. Soc.* 4:237.

Amynthas jaoi Xu & Xiao, 2011. *Terrestrial Earthworms of China*. 130.

External Characters

Very small worm. Length 38 mm, width 1.5–2.0 mm, segment number 69. Segments after clitellum rather short. Prostomium 1/3 epilobous, its tongue as wide as length of II, open behind. Dorsal pore minute, first in 12/13. Clitellum annular, not well marked, shown in XIV–XVI by glandular epidermis and degenerated setae. Setae all very fine, scarcely visible under ordinary magnification, closer ventrally; those on anterior venter closer and numerous; Setae number 28–32 (III), 42–48 (VI), 48–52 (IX), 42–48 (XXV); 12–18 (VI), 14–18 (VII), 14–17 (VIII), 16–20 (IX) between spermathecal pores, 14–22 between male pores; $aa = 1.2ab$, $zz = 2.0$–$2.5zy$. Male pores paired in XVIII, about 1/3 circumference apart ventrally, each on a flat and round porophore as large as a papilla and which is usually placed at its median side, postsetal in position. Spermathecal pores 3 pairs in 6/7/8/9, intersegmental, about 1/3 circumference apart ventrally, no genital papillae nor other markings around each pore. Female pore single, medioventral in XIV.

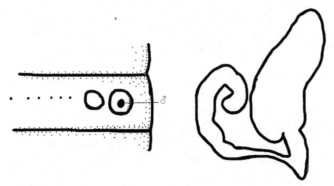

FIG. 198   *Amynthas jaoi* (after Chen, 1946).

Internal Characters

Septa 8/9/10 absent, all septa thin and little musculated. Nephridial tufts thin on anterior faces of 5/6/7. Pharyngeal gland always developed, greenish pale extending to VI. Gizzard elongate and smooth, in 1/2 IX and X. Intestine swelling in XVI. Caeca simple, in XXVII, extending anteriorly to XXIV or XXIII. Vascular loops in IX and X present but very small. Testis sacs, anterior pair more widely placed, posterior pair united. Seminal vesicles enormously developed, filling up several segments VIII–XIV (about 4 mm in anterioposterior length) dorsal lobes small; lateral surface smooth, median side usually subdivided irregularly. Prostate glands as large patch in XVII–XX, lobules indivisible,

its duct long and thin in a double U-curve. Accessory glands inconspicuous and sessile. Spermathecae 3 pairs, in VII–IX; ampulla elongate, thumb-shaped, main duct fairly short; diverticulum longer than main pouch if fully extended, ental two-thirds serving as seminal chamber but no seminal content inside; or observed, short and club-shaped probably in early stage of development.

Color: Preserved specimens light brick red on dorsum, purplish red on anterior half, a deep purple line along mediodorsal line extending anteriorly to II, reddish at anterior venter, pale posterior to clitellum. The latter dark red.

Distribution in China: Chongqing (Mt. Jinfo).

Etymology: The species was named after Dr. Rao Qinzhi (1900–1998), a great Chinese algologist.

Type locality: Chongqing.

Deposition of types: Previously deposited in the Institute of Zoology, Academic Sinica, Chongqing, the types were destroyed during the war in 1945–1949.

Habitats: This species is peculiar in its small size and its arboreal habit. All the specimens were collected from under mosses on bark as high as several meters. They are probably fond of the highly acid medium of the mosses. Fully mature specimens have not been obtained, yet in its full development of both male and female organs, it sufficiently warrants a valid description.

Remarks: It differs from *Amynthas acidophilus* (Chen, 1946) in its smaller size, the number of spermathecal pairs which are more ventrally situated, the greater setal number per segment, the different male pore aspect, and its possession of a sausage-shaped diverticulum instead of an ovoid one.

204. Amynthas leucocircus (Chen, 1933)

Pheretima leucocirca Chen, 1933. *Contr. Biol. Lab. Sci. Soc. China (Zool.).* 9:262–267.

Amynthas leucocircus Sims & Easton, 1972. *Biol. J. Linn. Soc.* 4:237.

Amynthas leucocircus Xu & Xiao, 2011. *Terrestrial Earthworms of China.* 136–137.

External Characters

Length 140–212 mm, width 4–7 mm, segment number 100–122. Prostomium 1/2 epilobous, as broad as, or broader than length of II, open behind. First dorsal pore 11/12. Clitellum XIV–XVI, annular, its length approximately equal to 3 immediately following postclitellar segments, without setae and furrows, dorsal pores not visible. Setae small and rather inconspicuous. Both dorsal and ventral breaks very slight; $aa = 1.2$–$1.5ab$, sometimes not evident posterior to clitellum, $zz = 1.2$–$2.0yz$, wider on

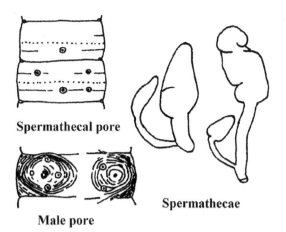

Spermathecal pore

Spermathecae

Male pore

FIG. 199　*Amynthas leucocircus* (after Chen, 1933)

preclitellar dorsum, sometimes more than 2 *yz*; setae number 21–25 (III), 32–35 (VI), 32–48 (VIII), 38–54 (XII), 50–58 (XXV); 16–18 between spermathecal pores, 8–12 between male pores. Male pores situated ventrolateral side of XVIII, about 1/3 circumference apart; each on a roundish papilla, or sometimes a large swelling, the pore marked as a pit at its center on setal line, without a slit though this part often sunken down due to contraction. 2 small nipple-like papillae anterior and posterior to, and also a little ventral to each male papilla, sometimes another placed ventrally on setal line and opposite to male papilla. Each one rounded and elevated, swollen, or slightly depressed at its center. This group of papillae always surrounded by 3 or more circular ridges to form a transversely oval glandular ridge which extends anteriorly and posteriorly to intersegmental furrows. Sometimes a similar papilla present ventromedially just behind setal line of XVIII. Spermathecal pores 3 pairs, in 6/7/8/9, about 1/3 circumference apart; each in an elliptical pit, intersegmental but slightly close to posterior margin of anterior segment, fully visible when these segments well stretched, sometimes a glandular patch of skin paler in color around each pore, no other genital markings. Similar round and raised papillae often paired medioventrally behind setae of VII and VIII; an additional 1 present before or another 1 behind setae of VIII (less constantly present), their distance to pore about one-third of distance between the latter and medioventral line. Female pore single, medioventral in XIV.

Internal Characters

Septa 8/9/10 absent, but 8/9 sometimes thin on ventral side only; 5/6/7/8 and 10/11/12/13 comparatively thick but not musculated, 13/14 less thickened; rest exceedingly thin and membranous. Gizzard large, barrel-shaped, in IX and X, larger posteriorly, dull pale and not smooth on surface. Intestine swelling in XV, or sometimes in XIV, where a deep constriction behind

septum 13/14 is marked. Intestinal caeca simple, elongate, in XXVII, extending anteriorly to XXIV, or sometimes to XXIII. Hearts 4 pairs, in X–XIII, all large in caliber, first pair sometimes very stout. Testis sacs very large, roundish and flattened, in X and XI, often a little larger than main vesicle; testis about 1.2 mm in diameter, occupying two-thirds of the sac, seminal funnel irregularly folded. Vasa deferens of each side meeting near or in front of septum 12/13, rather stout. Seminal vesicles always small, short and broad, inner surface with a thick elongate ridge in middle connecting large dorsal lobe and testis sac, outer surface pale and smooth; each with a large round or elongate oval dorsal lobe, about one-third or sometimes half the size of main vesicle. Prostate glands well developed, in XV–XXI, or sometimes XVI–XX, thick and expanded, whitish and smooth on surface, divided into 2–3 large lobes and again subdivided into small pieces; its duct short but stout, stouter at distal half, with single weak curve, pale or dull on surface. Accessory glands in this region well marked in XVIII around distal end of prostatic duct; each papilla associated with 1 large gland, round or kidney-shaped, whitish and compact, about 1 mm in diameter, with a short but stout stalk, usually concealed under gland part, or several glands sometimes associated with single papilla, smaller in size and crowded together as a thick cushion around distal end of prostatic duct. Spermathecae 3 pairs, first 2 pairs in VII, with their diverticula of first pair in VI or VII; last pair usually lying in VIII. Ampulla well distended, ovoid or pear-shaped, sometimes with an apical knob, its surface dull pale and smooth often irregularly constricted so as to form secondary lobes or dilations; ampullar duct shorter or longer than ampulla, one-half or one-third the width of the latter, or sometimes very slender and long in comparison with a large ampulla, sharply marked off from ampulla. Diverticulum shorter, or sometimes longer than main pouch, with an elongate ovoid, or heart-shaped, or horn-shaped, or cylindrical seminal chamber, often short, sometimes very long, about one-half or one-third its own duct which is slender, sometimes longer than main duct. Accessory glands occurring in this region very similar to those in prostate region, gland part large.

Color: Preserved specimens deep chestnut with setal zones creamy white dorsally, deep purplish or chocolate on preclitellar dorsum extending to ventrolateral side; whitish line along setal zones distinct but very narrow, slightly interrupted at dorsomedian line; pale ventrally. Clitellum chocolate brown.

Distribution in China: Jiangsu (Yixing), Zhejiang (Linhai).

Etymology: The specific name "*leucocircus*" is a compound word, "*leuco*" meaning white and "*circus*" meaning around or about in Latin. It probably refers to the color of this species.

Type locality: Jiangsu (Nanjing, Yixing), Zhejiang (Linhai, Chuxian).

Deposition of types: Previously deposited in the Zoological Museum of National Central University, Nanjing. The types were probably destroyed by the war in 1937.

Habitats: The specimens were first collected from a bamboo thicket on a hill at Qingshan in Yixing. The soil was humus. The worm when dug out of the earth was quite sluggish, relaxed, and immobile as if it were only half-awake.

Remarks: In some respects it is similar to *A. pingi* (Stephenson, 1925) in the shape of the spermathecae and male pore region, but they are by no means closely related.

205. *Amynthas loti* (Chen, Hsü, Yang & Fong, 1975)

Pheretima loti Chen, Hsü, Yang & Fong, 1975. *Acta Zool. Sinica.* 21(1):93–94.
Amynthas loti Xu & Xiao, 2011. *Terrestrial Earthworms of China.* 141–142.

External Characters

Small worm, width 2 mm. Prostomium 1/3 epilobous. First dorsal pore 10/11. Clitellum immature, setae absent. Setae scanty, wider spaced on ventrum than dorsum, not extraordinarily stout or wide. Medioventral gap in preclitellar region, $aa = 1.0$–$1.5ab$, $zz = 1.5$–$2.5zy$, slightly wide on postclitellar segments; Setae number 15 (III), 24 (VI), 30 (VIII), 26 (XVIII), 30 (XXV). Male pores in XVIII, about 1/3 circumference apart ventrally, each on a round porophore, surrounded by 3–4 skin folds. Spermathecal pores 3 pairs in 6/7/8/9; about 1/2 circumference apart. Paired papillae presetal on VII, VIII. Female pore single, medioventral in XIV.

Internal Characters

Septa 8/9/10 membranous, 5/6/7/8 and 10/11 slight thick, 11/12 and succeeding segments membranous. Gizzard long, in IX and X. Intestine caeca simple, in XXVII, extending anteriorly to XXIII. Last hearts in XIII. Testis sacs ventral, 2 pairs; anterior pair asymmetrically developed (the left is twice as large as the right), narrowly connected on ventrum; posterior pair developed, connected together, each with a posterior appendage. Seminal vesicles very large, the anterior pair not enclosed in testis sac, the posterior pair projected into XIII, dorsal lobe small. Prostate glands absent, prostatic duct U-shaped, ectal stout, ental on left nipple-shaped enlarged, maybe the trace of gland. Spermathecae 3 pairs, in VII–IX, ampulla

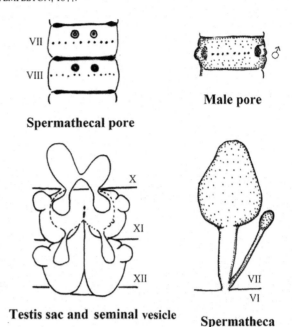

Spermathecal pore   Male pore

Testis sac and seminal vesicle   Spermatheca

FIG. 200   *Amynthas loti* (after Chen et al., 1975)

heart-shaped, its duct stout; diverticulum shorter, entally with ovoid seminal chamber. Accessory glands round-shaped, with short cord-like ducts which are communicated with each papilla.

Color: Preserved specimens red-gray on dorsum, white on ventrum, setal ring slightly whitish.
Distribution in China: Anhui (Huangshan), Sichuan (Mt. Emei).
Etymology: The specific name *"loti"* means lotus in Latin and "Lianhua" in Chinese means the same. It probably refers to the type locality of Lianhuagou, Mt. Huang.
Type locality: Anhui (Lianhuagou, Mt. Huang).
Remarks: This species is similar to *A. jaoi* (Chen, 1946), but membranous septa 8/9/10 present, it being absent in *A. jaoi*.

206. *Amynthas moniliatus* (Chen, 1946)

Pheretima moniliata Chen, 1946. *J. West China Border Res. Soc. (B).* 16:105–106, 139, 150.
Amynthas moniliatus Sims & Easton, 1972. *Biol. J. Linn. Soc.* 4:237.
Amynthas moniliatus Xu & Xiao, 2011. *Terrestrial Earthworms of China.* 150.

External Characters

Small-sized worm. Length 22–60 mm, width 2–3 mm, segment number 71–82. Prostomium 4/5 epilobous, its width wider than length II. Ventral side of I with a complete incursion. First dorsal pore 12/13 hardly visible,

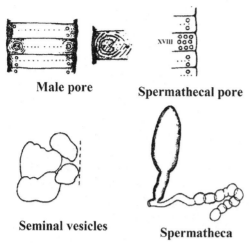

Male pore   **Spermathecal pore**

Seminal vesicles   **Spermatheca**

FIG. 201   *Amynthas moniliatus* (after Chen, 1946)

other also minute. Clitellum XIV–XVI, short, smooth all around, setae visible ventrally on XVI. Setae conspicuous, roughened on anterior ventral side; setal chain rather closed. Setae number 28–41 (III), 38–44 (VI), 33–47 (IX), 38–50 (XXV); 14–15 (VI), 16–18 (IX) between spermathecal pores, 12–13 between male pores; $aa = 1.8ab$, $zz = 1.5zy$. Male pores paired in XVIII superficial, about 1/3 circumference apart ventrally, each as a wart in a large depressed (or 2 coalesced) papilla, the region surrounded by a small ridge. Small papillae found on XVII–XX. Or pore-bearing papilla and pore apparently absent but instead are found as many as 7 or 9 small papillae on each side. Spermathecal pores 3 pairs in 6/7/8/9, about 1/3 circumference apart ventrally; or not present nor small papillae in this region, with 1 rather large papilla placed medioventrally on VII, presetal. 2 series of small papillae from VI–X present slightly ventral to pores on each side. Medioventral one on VII absent. Female pore single, medioventral in XIV.

Internal Characters

Septa 8/9/10 absent, 4/5–7/8 thin, or 10/11 not found, 11/12 membranous, 12/13–15/16 thicker and lightly musculated. Thick nephridial tufts looped anteriorly on 5/6/7. Micronephridial, both parietal and septal ones few but large. Pharyngeal glands large and soft, extending to VI, in front of septum 5/6. Gizzard large, globular (or very large, about 1.8 mm long, 3 mm in transverse diameter, obliquely placed between IX and X). Intestine beginning to swell from XVI. Caeca simple, in XXVII, extending anteriorly to XXIV or XXIII, smooth on both sides. Vascular loops in IX asymmetrically developed, those in X well developed. Anterior pairs of testis sacs extending forward and upward, narrowly connected but communicated ventrally, placed in 1/2XI–1/2XII, posterior pair also separate but communicated, in XII–1/2XIII. No supernumerary pair in next segment.

Seminal vesicles well developed, first pair closely applied to the anterior pairs of testis sacs (on account of absence of septum therein), second pair filling whole XII. Prostate glands absent, its duct as loose a S shape. Or both glands and its duct absent, or absent on left, well developed on right, gland portion as large lobules, in XVII–XXI, 5 mm in anterior-posterior length, duct long in a loose S-curve. Accessory glands stalked about 1 mm long, gland ovoid. Spermathecae 3 pairs, well developed, or absent. Ampulla elongate, its duct longer; diverticulum twice as long, zigzag constricted moniliform appearance, transparent; or spermatheca very rudimentary, about 0.5 mm long, diverticulum minute, bud-like. Stalked accessory glands similar to those in the male pore region.

Color: Preserved specimens pale generally, light purplish along dorsomedian line, reddish on both anterior and posterior ends. Clitellum reddish.
Distribution in China: Hubei (Lichuan), Congqing (Nanchuan, Beipei), Sichuan (Mt. Emei), Guizhou (Mt. Fanjing).
Etymology: The name of the species refers to the structure of the spermathecal diverticulum, which has a zigzag, constricted, and moniliform appearance.
Type locality: Anhui (Mt. Huang).
Deposition of types: Previously deposited in the Institute of Zoology, Academic Sinica, Chongqing, the types were destroyed during the war in 1945–1949.
Remarks: The reason why an athecal specimen was chosen as the type of the species is twofold: its male pore region is normal in appearance for the genus, and its initial discovery at Emei should mark the type locality of this interesting species.

207. Amynthas monsoonus James, Shih & Chang, 2005

Amynthas monsoonus James, Shih & Chang, 2005. *Jour. Nat. His.* 39 (14):1012–1014.
Amynthas monsoonus Xu & Xiao, 2011. *Terrestrial Earthworms of China.* 151.

External Characters

Length 102 mm, width 4 mm, segment number 83. Prostomium epilobous, with tongue open. First dorsal pore 12/13. Clitellum XIV–XVI, annular, setae invisible externally. Setae regularly distributed around segmental equators, size and distance regular; setae number 38 (VII), 42 (XXV); 16 between male pores. Male pores on ventral region of XVIII, lacking associated genital markings. Spermathecal pores 3 pairs, lateral in 6/7/8/9, genital marking paired, presetal VII–IX between third and fourth setal lines. Female pore single, medioventral in XIV.

Internal Characters

Septa 8/9/10 absent, 6/7/8 thickly muscular, 10/11–13/14 thinly muscular. Gizzard in VIII–X. Intestine beginning to swell from XVI. Caeca simple, in XXVII, extending anteriorly to XXIV, small incisions on ventral margin. Hearts 2 pairs in XII–XIII; X, XI lacking, commissural vessels VII, IX lateral, VIII to gizzard. Testes, funnels in paired ventral sacs in X, XI. Seminal vesicles large in XI, XII, with dorsal lobes. Prostate glands small in XVIII, 2 main lobes, ducts thick, muscular, vasa deferens join at duct-glandular portion junction; vasa deferens nonmuscular. Spermathecae 3 pairs, in VII–IX; ampulla warty spherical, duct very short, diverticulum small ovate, stalk muscular, straight; no nephridia on spermathecal duct; paired sessile genital marking glands in VII–IX.

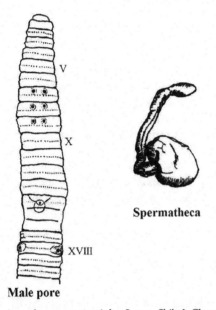

Spermatheca

Male pore

FIG. 202 *Amynthas monsoonus* (after James, Shih & Chang, 2005).

Color: Preserved specimens brown anterior dorsal pigmentation with unpigmented setal zones, pigment diminishing posteriorly but present to end.
Distribution in China: Taiwan (Pingdong).
Etymology: This species is named after the tropical monsoon forest in Nanrenshan, which is unusual for such a northern latitude.
Type locality: Taiwan (Nanrenshan, Kending, Pingdong County).
Deposition of types: National Museum of Natural Science, Taichung, Taiwan.
Remarks: *Amynthas monsoonus* belongs to the *aelianus* group. This group should also include 6 recently described species from Taiwan: *Amynthas binoculatus* (Tsai, Shen & Tsai, 1999), *Amynthas fenestrus* (Shen, Tsai & Tsai, 2003), *Amynthas sepectatus* (Tsai, Shen & Tsai, 1999), *Amynthas tayalis* (Tsai, Shen & Tsai, 1999),

Amynthas tenuis (Shen, Tsai & Tsai, 2003), and *Amynthas tungpuensis* (Tsai, Shen & Tsai, 1999). *A. monsoonus* differs from them all in lacking the anterior 2 pairs of hearts, like *A. nanrenensis* (James et al., 2005). It also has a different genital marking pattern from its Taiwanese sexthecal congeners. The missing hearts suggest that it is more closely related to *A. nanreaensis* than to the *aelianus* group members. No one has tested the hypothesis that the spermathecal battery is evolutionarily more conservative than details of the circulatory system, and there is evidence to the contrary. The locations of hearts are widely conserved among *Amyhthas*, across great variation in other characters, particularly the numbers and locations of spermathecae. *A. monsoonus* and the previous 2 species are quite unusual in having lost the hearts of X, or X and XI. These 3 are very similar with respect to other somatic characters and spermathecal morphology.

A. monsoonus is very similar to *A. carnosus* (Goto and Hatai, 1899), but the latter discounts the more anterior location of the three or four pairs of spermatecae, in the latter, as well as its possession of genital markings in XVIII and XIX, greater numbers of setae per segment, lack of genital marking glands, and the very different spermathecal morphology. Blakemore's (2003) diagnosis of *A. carnosus* and subsequent remarks all place its first pair of spermathecal pores in 5/6.

208. *Amynthas nubilus* (Chen, 1946)

Pheretima nubila Chen, 1946. *J. West China Border Res. Soc. (B).* 16:125–126, 141, 158, 159.
Amynthas nubilus Sims & Easton, 1972. *Biol. J. Linn. Soc.* 4:237.
Amynthas nubilus Xu & Xiao, 2011. *Terrestrial Earthworms of China.* 157.

External Characters

Length ?–90 mm, width 3.0–3.8 mm, segment number 107–?. Segments moderately long, about 12 segments immediately posterior to clitellum shortened. Prostomium 1/2 epilobous, tongue as wide as length of II. First dorsal pore 11/12. Clitellum XIV–XVI, annular, smooth, setae absent. Setae rather few per segment, wider in space on anterior ventral side. Setae number 20–23 (III), 31–32 (VI), 32–35 (IX), 36–45 (XXV); 9–10 (VI), 10–11 (IX) between spermathecal pores, 10–14 between male pores; $aa = 1.0$–$1.2ab$, $zz = 1.0$–$1.5zy$. Male pores superficial, paired in XVIII, about 1/3 circumference apart ventrally, each on a pore-bearing papilla on setal line, another and more distinct one immediately posterior to each porophore, 3 similar ones ventromedian and presetal; or 3 similar ones present, presetal on XIX. Spermathecal pores 3 pairs, intersegmental in 6/7/8/9, about 1/3

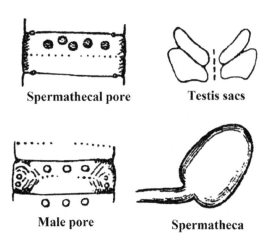

FIG. 203   *Amynthas nubilus* (after Chen, 1946).

circumference apart ventrally, 5 small papillae present on ventral side of VIII, presetal. Female pore single, medioventral in XIV.

Internal Characters

Septa 8/9/10 absent, anterior ones thin, 10/11/12/13 slightly muscular and stronger. Nephridial tufts on 5/6/7 not thick. Pharyngeal glands moderately thick. Gizzard elongate, barrel-shaped, narrower anteriorly, not shining, in 1/2 IX–1/2 XI, occupying 2 segments. Intestine beginning to swell from 1/2XV. Caeca simple, smooth on both sides, pointed, in XXVII, extending anteriorly to XXIV. Vascular loops in IX asymmetrical, those in X also asymmetrical and small. Paired lymph glands along dorsal vessel appearing behind XVI. Both pairs of testis sacs widely placed, each connected by membranous sac, anterior testis sac about 1 mm in diameter. Seminal vesicles in XI and XII small, about 1.2 mm in dorsoventral length, flattened, dorsal lobe small on dorsal side. Prostate glands either vestigial or very rudimentary in the 3 specimens observed, its duct always developed, in a long S-shaped curve, lower limb in a deep coil, equally thickened throughout. Accessory glands sessile, a patch of lobular glands corresponding to 1 papilla externally. Spermathecae 3 pairs, in VII–IX, diverticulum not observed. Ampulla and its duct about 2 mm long, normally developed; ampulla heart-shaped, its duct slightly shorter broadly connected with the latter. Accessory glands similar to those in the prostate region.

Color: Preserved specimens greyish brown to chocolate dorsally, setal circles pale, purplish on anterior ventral end, pale on ventral and lateral sides. Clitellum chocolate-brown.
Distribution in China: Chongqing (Beipei), Sichuan (Mt. Emei).
Etymology: The specific name *"nubilus"* means cloud in Latin, and "yun" in Chinese means the same. It probably refers to the type locality of Mt. Jinyun.

Type locality: Chongqing (Mt. Jinyun, Beipei). Deposition of types: Previously deposited in the Institute of Zoology, Academic Sinica, Chongqing, the types were destroyed during the war in 1945–1949.

209. *Amynthas obscuritoporus* (Chen, 1930)

Pheretima obscuritopora Chen, 1931. *Chin. Zool.* 7 (3):117, 119.
Pheretima obscuritopora Gates, 1935a. *Smithsonian Mus. Coll.* 93(3):12.
Amynthas obscuritoporus Sims & Easton, 1972. *Biol. J. Linn. Soc.* 4:237.
Amynthas obscuritoporus Xu & Xiao, 2011. *Terrestrial Earthworms of China.* 158.

External Characters

Length 180–210 mm, width 4–6 mm, segment number 130–172. Prostomium 1/2–1/3 epilobous, tongue wide with a longitudinal groove in the middle. First dorsal pore 12/13. Clitellum not present in all cases, with setae all around clitellar region not different from the neighboring segments. Setae slightly longer on anterior segments and shorter at the middle of the body, a little longer at the ventral side. Ventral setae $aa = 1.2ab$ preclitellar, 1.2–1.5ab postclitellar; dorsal setae $zz = 1.5yz$ preclitellar, 2.3–3yz postclitellar. Setae number 32–40 (III), 44–50 (VI), 48–54 (VIII), 48–54 (XII), 50–70 (XXV); 18–24 between male pores. Copulatory chamber absent. Male pores not visible on the surface, only an obscure whitish patch appearing vaguely under the skin. In some other cases, a very small crescent-shaped groove as if a shallow copulatory chamber were enclosed, (still indistinct unless viewed with a high-power magnifier), or a very small ridge around the pore. 1–2 round pits sometimes present median to each pore, paired in XVIII, more than 1/3 circumference apart ventrally. Spermathecal pores 3 pairs in 6/7/8/9, usually invisible externally, or if these segments are

Spermathecae

FIG. 204   *Amynthas obscuritoporus* (after Chen, 1930)

perfectly stretched, about 1/3 circumference apart ventrally, there appears a papilla-like area. No papilla occurs in this region. Female pore single, medioventral in XIV.

Internal Characters

Septa 8/9/10 absent, 5/6/7 and 10/11/12/13 thick, 7/8 and 13/14 less thickened. Gizzard in IX and X, globular or bell-shaped, rather large, with a dilated sac-like crop in front of it. Intestine beginning to swell from XVI or posterior part of XV. Caeca simple anteriorly in XXVII–XXIII or XXVII–XXII. Last hearts in XIII. Testis sacs 2 pairs. Testis sacs large, out of proportion to the small seminal vesicles, lying under septa 10/11, 11/12 respectively, the anterior pair not projecting into X. 2 sacs of each pair communicating or united into a transverse band. Testes very large, about 1 mm in diameter. Seminal vesicle 2 pairs, generally small and elongate at the posterior faces of septa 10/11, 11/12. 2 pairs of ovoid bodies present occupying similar position on the succeeding septa. The pair on the posterior surface of septum 12/13 usually very large and oval in outline, a half or one-third as the seminal vesicles and resembling them in appearance. However, they may be reduced to small strands. That on septum 13/14 also large or may be absent. Prostate glands very small, in segment XVIII only in a whitish compact mass or divided into fine lobules, sometimes extended to the next segment; prostatic duct large, with a loop medially placed. No accessory glands associated. Spermathecae 3 pairs in VII, VIII, IX, very small, about 1 mm in its total length; ampulla whitish, heart-shaped or ovoid, pointed at the end, the main duct very short and stout as if there were no duct. Diverticulum short at the median base of the main pouch or about half the length of the total spermatheca with a bulb-like dilated sac at the ental end. In the specimens from Suzhou, the spermathecal duct is usually shorter than main pouch. No stalked gland presents in this region.

Color: Preserved specimens chocolate on dorsum, deeper in color anteriorly; pale on ventrum.

Distribution in China: Jiangsu (Nanjing, Suzhou, Wuxi, Yixing, Puzhen), Zhejiang (Tonglu, Fenshui), Anhui (Chuzhou), Sichuan (Guanxian).

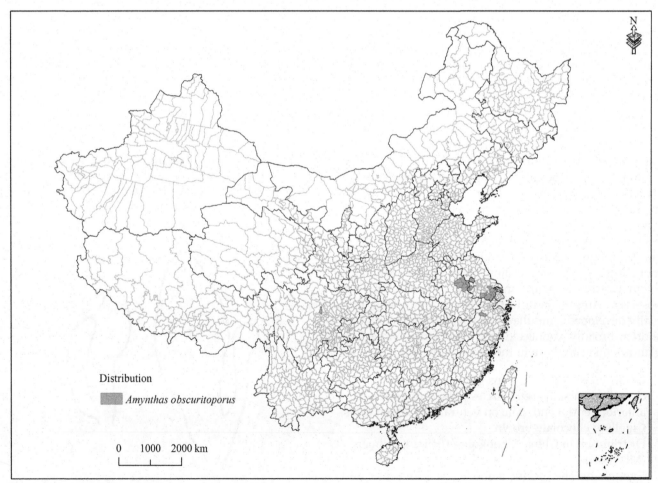

FIG. 205   Distribution in China of *Amynthas obscuritoporus*.

Etymology: The specific name *"obscuritoporus"* is a compound word with *"obscurit-"* meaning obscure in Latin and *"porus"* meaning pore in Latin. The word refers to the structure of the male pores.

Type locality: Jiangsu (Nanjing, Suzhou).

Deposition of types: Previously deposited in the Museum of the Biological Laboratory of the Science Society of China, Nanjing. The types were destroyed by the war in 1937.

210. *Amynthas pecteniferus* (Michaelsen, 1931)

Pheretima pectenifera Michaelsen, 1931. *Peking Natural History Bulletin.* 5(2):15–17.

Pheretima pectenifera Gates, 1935a. *Smithsonian Mus. Coll.* 93(3):13–14.

Pheretima pectenifera Gates, 1939. *Proc. U. S. Natr. Mus.* 85:460–465.

Pheretima pectenifera Chen, 1959. *Fauna atlas of China: Annelida.* 10.

Amynthas pecteniferus Sims & Easton, 1972. *Biol. J. Linn. Soc.* 4:237.

Amynthas pecteniferus Xu & Xiao, 2011. *Terrestrial Earthworms of China.* 164–165.

External Characters

Length 140–210 mm, width 8–10 mm, segment number 64–106. First dorsal pore 12/13. Prostomium epilobous. Clitellum XIV–XVI, annular, without setae. Setae delicate, dorsal rows of setae somewhat wider apart than the ventral, dorsal median partly interrupted, ventral continuous; setae number 60 (V), 70 (VIII), 72 (XII), 82 (XXVI); Male pores paired in XVIII, ventrolateral, about 1/3 circumference apart ventrally; each on the tip of large wart-like porophores whose base includes almost the whole length of XVIII. Spermathecal pores 3 pairs, ventrolateral on intersegments 6/7/8/9, about 3/7 circumference apart ventrally, they are rather large cross-like slits on somewhat raised glandular areas. The genital marking, in the form of very small circular and weakly raised papillae, in the region of the male pores as well as of the spermathecal pores. The former is limited to 3–4 papillae on the median side of each male porophore, partly in front of and partly behind the setae zone, which stretches rather far to the median side of the porophore. In a regular pattern, apparently normal, the papillae mark the corners of a square which is symmetrically divided by the setae zone. The papillae of the spermatheca region usually are not found in groups but singly, and in 2 systems: partly in front of the setae zone, a small distance median from the line of spermathecal apertures, and partly behind the setae zone, a greater distance from the spermathecal apertures. Usually the spermathecal papillae are limited to VII and VIII, therefore fullest development present on VIII.

Mostly some regular papillae are lacking. Rarely there appear a few papillae, or 1 pair before or behind the setae zone on segment IX. Very rarely 1 regular papilla was represented by 2–3 closely grouped papillae. Female pore single, medioventral in XIV.

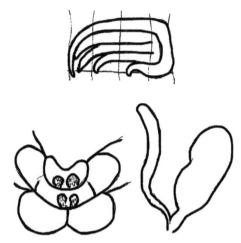

FIG. 206   *Amynthas pecteniferus* (after Chen, 1959).

Internal Characters

Septa 8/9/10 absent, 5/6 and 7/8 much thickened, the following delicate, scarcely visibly thickened. Gizzard large, behind septum 7/8. Intestine swelling in XV; Intestinal caeca strongly developed, 5–6 close together in segment XXVII, and extending forward with the forward end bent under. The largest and topmost extends to XXIV, the smallest to XXVI. Hearts in XI–XIII. Testes sacs 2 pairs, fused with each other for the entire breadth in X and XI. The anterior testis sacs on X are externally simple thick sausage-like U-shaped structures with rounded ends. The 2 sperm funnels lie rather close together in the median part of the semicircular sac. This large double testis sac connects with a pair of large seminal vesicles in XI, which extend around the intestine at its sides and grow together above the intestine. The seminal vesicles are compact, superficially net-like. The testis sacs of the posterior pair differ from the anterior pair in that they are not closed, but extend over the anterior pair for their entire breadth. Prostate glands from XVII–XX, divided into broad folds by several (4–6) deep grooves; in general reticular, with rough edges. Duct bent into a large S-shaped coil, proximal end thin and forming a small hook, distal end increasing in thickness and muscularity. Spermathecae 3 pairs in VII–IX, ampulla in normal position, long and sac-like, but many times pressed into various shapes. Duct sharply demarcated from the ampulla, in general cylindrical, somewhat narrowed at the distal opening, about 1/4 as long and 1/4 as thick as the normal ampulla. In the distal end of the duct there opens a blind hose-like diverticulum which is somewhat longer than

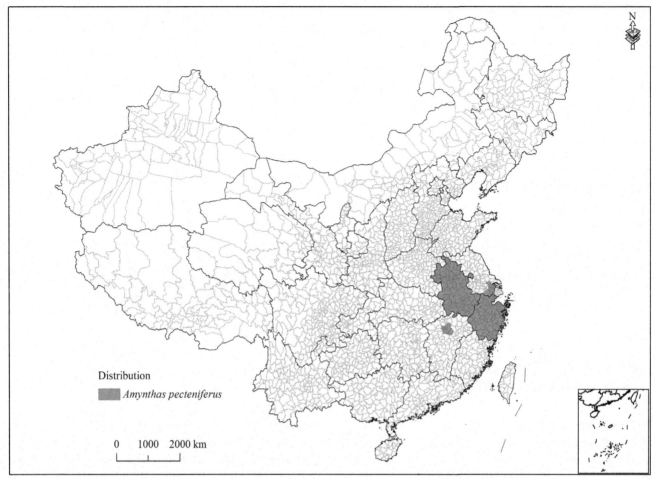

FIG. 207   Distribution in China of *Amynthas pecteniferus*.

the normally extended main pouch (ampulla plus duct). Only a small distal part of the diverticulum can be considered as the diverticulum stalk, scarcely as long as the duct of the main pouch. Somewhat proximal from the middle of the length of the diverticulum, the diverticulum reaches its largest size, the thickness slightly exceeding the thickness of the ampulla duct. The gradually decreasing coils here come to an end entirely; the thick proximal third of the diverticulum is entirely smooth.

Color: Preserved specimens dorsally almost uniform blue-gray, laterally shading gradually over into the reddish yellow of ventrum, setae zones scarcely noticeably lighter, intersegmental furrows outlined as very fine dark rings against the generally unpigmented ventrum. Clitellum brown.
Distribution in China: Jiangsu (Suzhou), Zhejiang, Anhui, Jiangxi (Nanchang).

Etymology: The specific name "*pecteniferus*" is a compound word, "*pecten*" meaning comb or rake in Latin. It refers to the structure of the intestinal caeca.

Type locality: Jiangsu (Suzhou).
Deposition of types: Hamburg Museum, Germany.
Remarks: The male pore region of this species is so remarkably like that of *A. yamadai* (Hatai, 1930), that the former may be, in reality, a synonym of the latter. Hatai's species is not, however, adequately characterized, and in as much as types or specimens of *A. yamadai* have not been available for examination, Michaelsen's species is retained. If Michaelsen's species is in fact composite and actually in part a synonym of *A. yamadai*, the name *pecteniferus* will have to be retained for the remaining part.

211. *Amynthas pingi chungkingensis* (Chen, 1936)

Pheretima pingi chungkingensis Chen, 1936. *Contr. Biol. Lab. Sci. Soc. China (Zool.).* 11:274–275.
Amynthas pingi chungkingensis Sims & Easton, 1972. *Biol. J. Linn. Soc.* 4:237.
Amynthas pingi chungkingensis Xu & Xiao, 2011. *Terrestrial Earthworms of China.* 165–166.

External Characters

Length 140–170 mm, width 5.5–7.0 mm, segment number 136–140. Prostomium 2/3 epilobous. First dorsal pore 12/13. Clitellum entire, in XIV–XVI, annular, without setae and other markings. Setae enlarged on III–VII or VIII, ventral ones more conspicuous, $aa = 1.2ab$, $zz = 2yz$. Setae number 22–26 (III), 26–31 (VI), 38–45 (VIII), 51–62 (XXV); 9–9 (VI), 10–11 (VII), 15 (VIII), 16 (IX) between spermathecal pores, 12–14 between male pores. Male pores 1 pair on ventrolateral side of XVIII, each represented by a papilla-like glandular area. No skin folding. 1 genital papilla constantly present, close to its posterior side on XVIII. No other genital papillae. Spermathecal pores 3 pairs, in intersegmental furrows 6/7/8/9, about 1/3 circumference apart ventrally, each with a large genital papilla closely behind. Not found near spermathecal region nor on ventral side of those segments. Female pore single, medioventral in XIV.

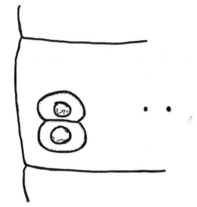

FIG. 208   *Amynthas pingi chungkingensis* (after Chen, 1936).

Internal Characters

Septa 8/9/10 absent, but 8/9 traceable ventrally. First pair hearts not found, probably absent. Seminal vesicles very large, in XI and XII; pseudovesicles on septa 12/13 & 13/14 very large and elongate. Prostate glands large, in XVI–XX. Spermathecae 3 pairs in VII–IX, rather small; ental third of each diverticulum club-shaped and whitish. Other characters as in typical form.

Color: Preserved specimens dark chestnut or purplish chocolate dorsally, darker at anterior dorsum, uniformly pigmented except slightly whitish around setal circles, pale or greyish pale ventrally. Clitellum brownish.
Distribution in China: Chongqing, Sichuan.
Etymology: The subspecific name indicates its type locality of Chongqing.
Type locality: Chongqing.

Deposition of types: Previously deposited in the Zoological Museum of National Central University, Nanjing. The types were probably destroyed by the war in 1937.
Remarks: This variety is perfectly identical to the typical form of this species except for 2 characters: (1) only 3 pairs of spermathecae present, and (2) genital papillae fewer in both spermathecal and male pore regions. Since these are constant in the specimens, it is considered that this may be a genuine variety of the typical form. Regarding the number of spermathecal pores, it is close to *A. leucocircus* (Chen, 1933). However, the latter species was found in its living condition to be quite different from *A. pingi pingi* (Stephenson, 1925) by its sluggish behavior. Chenyi was reluctant to combine these into a single variety. Very probably *A. pingi pingi* is identical with *A. carnosua* (Goto & Hatai, 1899) which is found in Tokyo, Japan. As it was characterized by these authors, these 2 species are similar except only 3 pairs of spermathecal pores in 5/6/7/8 found in the later. The synonymy of this species cannot be adequately dealt with unless a reexamination of the Japanese form is made.

128. *Amynthas proasacceus* Tsai, Shen & Tsai, 2001 (*v.ant.* 128)

212. *Amynthas saccatus* Qiu, Wang & Wang, 1993

Amynthas saccatus Qiu, Wang & Wang, 1993. *Guizhou Science*. 11(4):3–6.
Amynthas saccatus Xu & Xiao, 2011. *Terrestrial Earthworms of China*. 177–178.

External Characters

Length 101–124 mm, width 3–4 mm, segment number 64–92. Prostomium 1/2 epilobous. First dorsal pore 11/12. Clitellum XIV–XVI, annular, with some setae on ventral side of XVI. Setae enlarge and widely distributed before VIII; setae numbers 18–26 (III), 24–28 (V), 32–35 (VIII), 34–44 (XX), 31–46 (XXV); 8–14 (VII), 10–14(VIII) between spermathecal pores, 5–10 between male pores. Male pores paired on ventrolateral side of XVIII, about 1/3 circumference apart ventrally; each situated on a long-round porophore with 2 small papillae nearby, surrounded by concentric ridges. There are paired papillae on ventral sides of XVII, XVIII, and XIX, sometimes lacking. Spermathecal pores 3 pairs in 6/7/8/9, about 2/5 circumference apart ventrally. Paired papillae on ventral sides of VII, VIII and IX or VII and VIII or only VIII. Female pore single, medioventral in XIV.

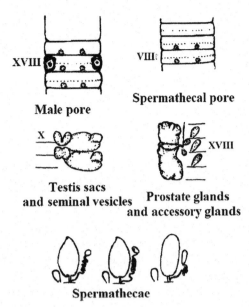

FIG. 209   *Amynthas saccatus* (after Qiu et al., 1993a).

Internal Characters

Septa 9/10 absent, 8/9 present and very thin. Gizzard developed, barrel-like. Intestine beginning to swell from XV. Caeca simple, in XXVII, extending anteriorly to XXIV. Testis sacs 2 pairs, in X and XI, rather developed, communicated. Seminal vesicles well developed. Dorsal lobes conspicuous. Prostate glands in XVII–XX. Accessory glands long-sac-shaped with smooth surface, well developed, arranged as the papillae outside. Spermathecae 3 pairs, in VII–IX. Ampulla heart-shaped or long-round with small and short duct. Diverticulum shorter than main pouch, twisted heavily on its middle part, and with round or long-round seminal chamber. Accessory glands as those on prostate region.

> Color: Clitellum red-brown.
> Distribution in China: Guizhou (Mt. Taiyang, Mt. Yueliang, Congjiang).
> Etymology: The specific name *"saccatus"* means shaped like a sac in Latin and refers to the saced accessory glands of the prostate and spermathecae.
> Type locality: Guizhou (Mt. Taiyang, Jiaya Village, Congjiang County).
> Deposition of types: Institute of Biology, Guizhou Academy of Science, Guiyang.
> Remarks: This species is similar to *A. flexilis* (Gates, 1935a), but differs from the latter by the male pore, which is situated on a long-round porophore with small papillae nearby, surrounded by concentric ridges; diverticulum with a round or long-round seminal chamber, well-developed seminal vesicles, which are not enclosed in testis sac and long second accessory gland with smooth surface. Meanwhile, this

species can be separated from *A. leucocircus* (Chen, 1933) by the male pore, the ampulla with short and small duct; long, small, and twisted diverticulum with a small long-round or round seminal chamber, and the large and long second accessory glands.

213. *Amynthas sexpectatus* Tsai, Shen & Tsai, 1999

Amynthas sexpectatus Tsai, Shen & Tsai, 1999. *Journal of the National Taiwan Museum.* 52(2):33–46.
Amynthas sexpectatus Chang, Shen & Chen, 2009. *Earthworm Fauna of Taiwan.* 80–81.

External Characters

Length 193–258, segment number 102–140. Number of annuli (secondary segmentation) per segment 3 in IX–XIII. Prostomium epilobous. First dorsal pore 12/13. Clitellum XIV–XVI, annular, 6.2–8.6 mm long, setae absent, dorsal pore absent. Setae number 56–62 (VII), 66–94 (XX); 17–24 between male pores. Male pores paired in XVIII, lateroventral, 0.25–0.30 circumference apart ventrally, number and structure of genital papillae variable: if contracted, male pores with 2 papillae, anterior sandal-shaped, posterior round, surrounded by several circular folds; if protruded, male pores located at the top of a corn-shaped porophore, associated closely with the genital papillae and surrounded by several circular folds; each papilla flat with slightly concave center. Spermathecal pores 3 pairs in 6/7/8/9, lateral, 0.4–0.5 circumference apart ventrally. 3 pairs of large, oval, disc-like genital pads present in VI–VIII in some specimens, each immediately medioanterior to spermathecal pores, 1.2 mm in diameter. Female pore single in XIV.

Internal Characters

Septa 9/10 absent, 7/8/9 and 10/11–13/14 thickened. Gizzard in IX and X, round, with a small pharyngeal crop in VIII. Intestine from XV. Intestinal caeca paired in XXVII, smooth, extending anteriorly to XXIV. Lateral hearts in X–XIII. Testis sacs 2 pairs in XI and XII, medioventral, small, smooth. Seminal vesicles paired in XI and XII, smooth, with a large dorsal lobe. Prostate glands paired in XVIII, large, lobulated, extending anteriorly to XVII and posteriorly to XIX, prostatic duct n-shaped, diameter gradually increased from the distal end to the proximal end. No accessory glands associated with external genital papillae in the spermathecal pore and male pore regions. Spermathecae 3 pairs in VII–IX, each with a large, peach-shaped ampulla and a short, slightly curved stalk, diverticulum with a slender, slightly curved stalk and a granulated seminal chamber. A pair of small, round nephridia on the dorsal side of intestine in each segment, nephridia tufted in V and VI. Ovaries paired in XIII, medioventral.

Color: Live specimens yellow or greenish brown. Preserved specimens brown on dorsum, light gray on ventrum, dark brown on clitellum.

Distribution in China: Taiwan (Taichung, Nantou).

Etymology: The name was given with reference to the 3 pairs of large oval, disc-like pads in the spermathecal pore region of this species.

Type locality: Guoshing, Nantou County, Taiwan.

Deposition of types: Taiwan Endemic Species Research Institute, Jiji, Nantou County, Taiwan.

214. *Amynthas szechuanensis szechuanensis* (Chen, 1931)

Pheretima szechuanensis Chen, 1931. *Chin. Zool. 7* (3):160–167.

Pheretima szechuanensis Gates, 1939. *Proc. U. S. Natr. Mus.* 85:485–488.

Pheretima szechuanensis Chen, 1946. *J. West China Border Res. Soc. (B).* 16:136.

Amynthas szechuanensis Sims & Easton, 1972. *Biol. J. Linn. Soc.* 4:237.

Amynthas szechuanensis szechuanensis Xu & Xiao, 2011. *Terrestrial Earthworms of China.* 183–184.

External Characters

Length 85–220 mm, width 4–9 mm, segment number 60–135. Secondary annulation very inconspicuous, triannular from VIII or IX backward, but not visible on ventral side. Prostomium 2/3 epilobous, open behind. First dorsal pore 12/13. Clitellum extensive, in XIV–XVI, annular, smooth without setae nor intersegmental furrows but the segments sometimes distinguished, its glandular part not constricted nor particularly raised; dorsal pores absent. Setae small, closer ventrolaterally, those on ventral region of II–IX slightly or noticeably longer and widely spaced; setae number 31–38 (III), 37–49 (VI), 45–52 (VIII), 49–66 (XII), 56–65 (XXV); 20–32 between spermathecal pores, 15–22 between male pores. Male pores paired in XVIII, about 1/3 circumference apart ventrally, each on a small flat-topped papilla where the pore is marked by a whitish spot near the lateral side, 2 similar papillae situated posterior and anterior to the pore respectively, either in XVIII or XVII and XIX. 2 large horseshoe-shaped papillae, medial to the pore, each slightly raised and with a glandular top, usually placed between 17/18 and 18/19, in some cases, both confined to XVIII only, medial to the large papillae 1–2 round flat-topped papillae often present; sometimes 4 such large papillae placed side by side occupying XVII–XIX inclusive with another small papilla on lateral side forming a large but shallow depression where the male pore is situated. No copulatory chamber. The skin around the papillae somewhat swollen and glandular in character, sometimes with wrinkles

particularly pronounced on lateral side. Setae absent from the inner and outer areas of this region. Spermathecal pores 3 pairs, in 6/7/8/9, not easily visible externally unless there is a genital marking of the skin. The skin around the pore swollen to form anterior and posterior lip-like projections, about 3/5 circumference apart ventrally, the pores of both sides fairly visible from dorsal aspect. The pore, if visible, marked by a whitish spot just in the furrow, without a depression, sometimes surrounded by a whitish glandular patch of skin of 2 adjacent segments. No genital papillae visible around there nor on ventral side of the region. Female pore single, medioventral in XIV.

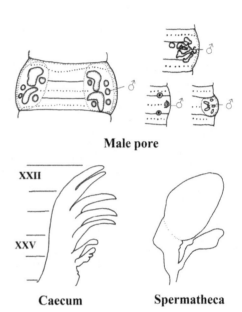

Male pore

XXII

XXV

Caecum   **Spermatheca**

FIG. 210   *Amynthas szechuanensis szechuanensis* (after Chen, 1931).

Internal Characters

Septa 9/10 absent, 8/9 present but very thin, covering posterior side of gizzard, 10/11 exceedingly thin only on ventral side; 4/5–7/8, 11/12–15/16 all equally thin, membranous, transparent, 6/7, 11/12 comparatively thick. Grape-like glands sometimes present in V and VI. Very thick nephridial tufts on anterior faces of septa 5/6/7, non on 4/5, very little on 7/8. Gizzard large, barrel-shaped, narrower at both ends, or larger posteriorly, in IX and X, surface usually dull white, not smooth. Intestine swelling from middle part of XVIII. The commencement of intestine varying, in XVI, XVII, or even XX but more often in XVII. Caeca lobulated, with 9 elongate diverticula, finger-shaped, extending from XXVII forward to XX or XIX, whitish in color, lying on dorsolateral side of intestine and overlapping dorsally. Hearts 4 pairs in X–XIII, first pair rather large in caliber in front of septum 10/11. Testis sacs comparatively small, in XI and XII under the vesicles. Seminal vesicles 2 pairs, very large rhombic, without dorsal lobe, meeting dorsally from both sides. Prostate glands very large, meeting dorsally, divided into large lobules, whitish and smooth on surface, in XVII–XX or sometimes in XVI–XXII. Each with a very long duct forming a deep loop at its middle, of which the proximal arm is slender and coiled while the distal arm straight and stout becoming slender within body wall and turning spirally therein to its pore on the lateral side of a small papilla outside. A large patch of diffusely scattered glands in this region appearing like the velvet tufts close to body wall. Spermathecae 3 pairs, in VII, VIII, and IX, ampulla pear-shaped, smooth and greenish grey on surface, well distended, its size varied even in the same individual, much stouter at ental end where it may be enlarged as part of the ampulla. Diverticulum with its ental half or third enlarged into an elongate oval seminal chamber, much larger than the remaining portion. In some cases, the enlargement gradually increasing and with some constrictions therein, the ental end round or pointed, its entire length about 3 mm, seminal chamber occupying half of it and 1 mm in diameter, its duct very slender at about 3 mm in diameter, joining to the main duct a little away from the body wall.

Color: Preserved specimens from greyish brown to dark brown or dark purplish on anterior dorsum, rich brown or chestnut brown posteriorly; greyish at anterior ventral end, greyish pale on the rest of ventrum; the brownish dark color of the dorsal side extending just to the middle of lateral side of body. Clitellum light chocolate, greyish pale, or cream.

Distribution in China: Chongqing (Mt. Qingcheng), Sichuan (Chengdou, Mt. Emei, Guanxian), Guizhou (Mt. Fanjing).

Etymology: The name of this species refers to the type locality.

Type locality: Chongqing (Mt. Qingcheng, Mt. Emei).

Deposition of types: Previously deposited in the Museum of the Biological Laboratory of the Science Society of China, Nanjing. The types were destroyed by the war in 1937.

Remarks: This species is found at rather high elevations on Mt. Qingcheng in Dujiangyan. Few have the clitellum distinct and genital organs well developed. Those without clitellum generally have smaller spermathecae, prostate glands, and small seminal vesicles. 2 large flat-topped papillae often present in 17/18/19 while the small ones usually absent. The color of the specimens from Emei is rich brown or reddish brown and may be due to different fixations. The purplish or dark greyish brown as found in specimens from Dujiangyan should be considered as more natural.

215. *Amynthas taipeiensis* (Tsai, 1964)

Pheretima taipeiensis Tsai, 1964. *Quarterly Journal of the Taiwan Museum.* 17(1&2),12–13.
Amynthas taipeiensis Sims & Easton, 1972. *Biol. J. Linn. Soc.* 4:237.
Amynthas taipeiensis Chang, Shen & Chen, 2009. *Earthworm Fauna of Taiwan.* 86–87.
Amynthas taipeiensis Xu & Xiao, 2011. *Terrestrial Earthworms of China.* 185–186.

External Characters

Length 132–136 mm, clitellum width 4.0–4.5 mm, segment number 107–116. First dorsal pore 11/12. Clitellum XIV–XVI, annular, setae absent, dorsal pore absent. Setae number 25 (V), 46–52 (VIII), 61–63 (XX), 9–10 between male pores. Male pores paired in XVIII, ventrolateral, situated on the center of a flat-topped porophore, a large, round genital papilla situated medial to the male pore, the porophores and papillae surrounded by several oval folds. Spermathecal pores 3 pairs in 6/7/8/9, lateroventral. Genital papillae present or absent, if present, 1–2, small, round, on the posterior margin of the posterior 1 or 2 spermathecal pores. Female pore single, medioventral in XIV.

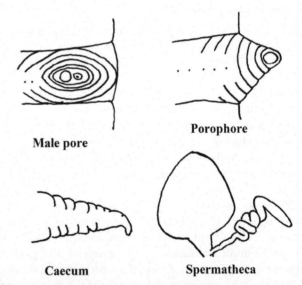

Male pore

Porophore

Caecum

Spermatheca

FIG. 211 *Amynthas taipeiensis* (after Tsai, 1964).

Internal Characters

Septa 8/9/10 absent, 5/6/7/8, 10/11/12/13 thickened. Gizzard in IX and X, cone-shaped. Intestine from XVI. Intestinal caeca simple in XXVI, extending anteriorly to XXIII, serrated ventrally and dorsally. Last hearts in XIII. Spermathecae 3 pairs in VI–IX, ampulla round or heart-shaped with a short stout stalk, diverticulum slender, long, with elongated heart-shaped seminal chamber and coiled stalk. Ovaries paired in XIII. Testis sacs 2 pairs in X and XI, posterior pair larger. Seminal vesicles paired in XI and XII. Prostate glands paired in XVIII, extending anteriorly to XVI and posteriorly to XXI, each divided into 4–6 lobules, prostatic ducts coiled.

Color: Preserved specimens purplish green on dorsum, light green or gray on ventrum and setal lines, brown on clitellum.
Distribution in China: Taiwan (Taibei).
Etymology: The name of the species refers to its type locality of Taibei.
Type locality: Chon-ho Village, Taibei, Taiwan.
Deposition of types: Types are missing.

216. *Amynthas tantulus* Shen, Tsai & Tsai, 2003

Amynthas tantulus Shen, Tsai & Tsai, 2003. *Zoological Studies*. 42(4),479–490.
Amynthas tantulus Chang, Shen & Chen, 2009. *Earthworm Fauna of Taiwan*. 88–89.
Amynthas tantulus Xu & Xiao, 2011. *Terrestrial Earthworms of China*. 186–187.

External Characters

Length 32–58 mm, clitellum width 1.7–2.2 mm, segment number 65–90. Number of incomplete annuli (secondary segmentation) per segment 2–3 in VIII–XI. Prostomium epilobous. First dorsal pore 5/6. Clitellum XIV–XVI, annular, setae absent, dorsal pores present or absent. Setae number 26–35 (VII), 28–35 (XX), 8–10 between male pores. Male pores paired in XVIII, lateroventral, 0.26–0.31 circumference apart ventrally, male aperture minute in a round porophore of about 0.3 mm in diameter, surrounded by 3–4 slight circular or diamond-shaped folds, some male pores with a small, less distinctly marked genital papilla pad adjacent laterally, genital papillae presetal, closely paired in medioventral portion in XVIII–XX, in XVIII, 1 pair or 1 on left, in XIX, 1 pair or 1 on right or absent, in XX, 1 pair, each papilla small, round, center concave, size and arrangement similar to those in the preclitellar region. Spermathecal pores 3 pairs in 6/7/8/9, ventrolateral in depressed furrow, 0.3–0.36 circumferences apart ventrally. Genital papillae presetal, closely paired in medioventral regions in IX–XII, number variable among specimens and segments: in IX and X, absent, 1 on left or 1 pair, in XI, absent, 1 on right, 1 on left or 1 pair, in XII, absent or 1 on left, each papilla small, round, flat-topped or slightly concave, about 0.2 mm in diameter. Female pore single, medioventral in XIV.

Internal Characters

Septa 5/6/7/8, 10/11/12/13 thickened, 8/9/10 absent. Gizzard round in IX and X. Intestine from XV. Intestinal caeca paired in XXVII, simple, short, stocky, surface slightly wrinkled, extending anteriorly to XXV. Esophageal hearts in XI–XIII. Testis sacs 2 pairs, first pair large in X, connected with posterior pair of lesser enlargement in XII. Seminal vesicles large, extending between XI and XIII or XIV, surface wrinkled, follicular. Prostate glands paired in XVIII, wrinkled, extending anteriorly to XVI and posteriorly to XX, prostatic duct C-shaped. Accessory glands similar in size and structure to those in the spermathecal region. Spermathecae 3 pairs in VII–IX, ampulla peach-shaped, the posterior 2 pairs larger than the anterior pair, stalk 0.4–0.5 mm long, diverticulum stalk straight or slightly bent, 0.4–0.7 mm long, seminal chamber oval. Accessory glands round, flat, with extremely short stalk, corresponding to external genital papillae.

Color: Preserved specimens dark reddish brown on dorsum, light gray on ventrum, light orange brown around clitellum.
Distribution in China: Taiwan (Nantou).
Etymology: The specific name "*tantulus*" mean "so small" in Latin and refers to the small size of this species.
Type locality: Rueyen Creek Nature Reserve, Nantou County, Taiwan.
Deposition of types: Taiwan Endemic Species Research Institute, Jiji, Nantou County, Taiwan.
Holotype: coll. no. 1999-24-Shen. Paratypes: same collection data as for holotype.
Remarks: *A. tantulus* is sexthecal and belongs to the *sieboldi* group (Sims & Easton, 1972). Its small size, large seminal vesicles, segment and setal numbers, and the position of the first dorsal pore are fairly similar to those of *A. dactilicus* (Chen, 1946) of Sichuan. However, *A. tantulus* has short-stalked accessory

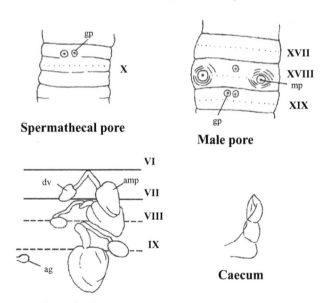

Spermathecal pore

Male pore

Caecum

Spermathecae

FIG. 212   *Amynthas tantulus* (after Shen, Tsai & Tsai, 2003).

glands and preclitellar genital papillae, whereas *A. daclitellar* has sessile accessory glands and no preclitellar genital papillae. The 2 species also have a different spermathecal structure.

217. *Amynthas tayalis* Tsai, Shen & Tsai, 1999

Amynthas tayalis Tsai, Shen & Tsai, 1999. *Journal of the National Taiwan Museum*. 52(2),33–46.
Amynthas tayalis Chang, Shen & Chen, 2009. *Earthworm Fauna of Taiwan*. 90–91.

External Characters

Length 120–125 mm, width 3.4–3.7 mm (clitellum), segment number 85–98. No secondary segmentation. Prostomium epilobous. First dorsal pore 11/12. Clitellum XIV–XVI, annular, setae absent, dorsal pores absent. Setae number 40–41 (VII), 56 (XX); 11–13 between male pores. Male pores paired in XVIII, ventrolateral, 0.34 circumference apart ventrally, each male opening surrounded by 3 circular papillae, 1 medial, 1 lateroanterior, 1 lateroposterior, the 3 papillae surrounded by 2–3 circular folds, additional postsetal papillae paired in XVII, paired or single in XIX and XX, similar in size and structure to those in the preclitellar regions. Spermathecal pores 3 pairs in 6/7/8/9, ventrolateral, 0.3 circumference apart ventrally. Genital papillae paired in VI–VIII, medioanteriorly adjacent to the spermathecal pore, some of the papillae embedded partially in the pores, for some specimens, 2 additional pairs of presetal and postsetal papillae in VIII, slightly medial to the spermathecal pores, each papilla circular, flat with a slightly concave center, surrounded by 1–2 circular folds, around 0.3 mm in diameter. Female pore single, medioventral in XIV.

Internal Characters

Septa 6/7 thickened, 7/8 thin, 8/9/10 absent, 10/11–13/14 slightly thickened. Gizzard in IX and X, peach-shaped. Intestine from XV. Intestinal caeca paired in XXVII, simple, with smooth border, extending anteriorly to XXV. Lateral hearts paired in XI–XIII. Testis sacs 2 pairs in X and XI, medioventral, small, oval. Seminal vesicles paired in XI and XII, large, surface wrinkled, each with a small smooth, oval dorsal lobe. Prostate glands paired in XVIII, large, folliculated, extending anteriorly to XIV and posteriorly to XXI, with a thick, slightly curved prostatic duct. Accessory glands present for genital papillae in both preclitellar and postclitellar regions, each gland round, mushroom-like, with a very short stalk. Spermathecae 3 pairs in VII–IX, each with a smooth, peach-shaped ampulla of about 1.8 mm long and a short, stout stalk of about 1.1 mm, diverticulum with a long, slightly pointed seminal chamber and a slender, slightly curved stalk. Ovaries closely paired, medioventral in XIII.

Color: Preserved specimens dark brown on dorsum and around clitellum, light gray on ventrum and lateral part of head, yellowish white on setal lines; the dark brown dorsal and white setal line colorations giving the worm a striped appearance.
Distribution in China: Taiwan (Taibei).
Etymology: The species was named after the Tayal tribe, a minority aboriginal tribe in northern Taiwan.
Type locality: Hsiju, Taibei County, Taiwan.
Deposition of types: Taiwan Endemic Species Research Institute, Jiji, Nantou County, Taiwan.
Holotype: coll. no. 1998-60. Paratypes: same collection data as for holotype.

218. *Amynthas tessellatus paucus* Shen, Tsai & Tsai, 2002

Amynthas tessellatus paucus Shen, Tsai & Tsai, 2002. *The Raffles Bulletin of Zoology*. 50(1),1–8.
Amynthas tessellatus paucus Chang, Shen & Chen, 2009. *Earthworm Fauna of Taiwan*. 94–95.
Amynthas tessellatus paucus Xu & Xiao, 2011. *Terrestrial Earthworms of China*. 188–189.

External Characters

Length 42–109 mm, clitellum width about 3 mm, segment number 68–107. Prostomium epilobous. First dorsal pore 11/12. Clitellum XIV–XVI, annular, setae absent. Setae number 31–40 (VII), 36–46 (XX); 9–11 between male pores. Male pores paired in XVIII, porophore simple, round, papilla-like, surrounded by 2–3 incomplete circular folds, about 0.28 circumference apart ventrally, genital papillae presetal 5–11 or 18 and postsetal 1–4 or 8 in XVIII. Spermathecal pores invisible. Patches of small, round presetal genital papillae present ventromedially 0 or 2 in VI, 0, 5 or 11 in VII, 0–12 or 15 in VIII. Female pore single, medioventral in XIV.

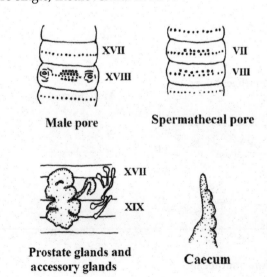

Male pore   Spermathecal pore

Prostate glands and accessory glands   Caecum

FIG. 213   *Amynthas tessellatus paucus* (after Shen, Tsai & Tsai, 2002).

Internal Characters

Septa 5/6/7/8 thickened, 8/9/10 absent, 10/11–13/14 greatly thickened. Gizzard IX–X, round. Intestine from XVI. Dorsal typhlosole from XXVII, with height about one-third of that of intestine. Intestinal caeca paired in XVII, simple, extending anteriorly to XXIII. Esophageal hearts in XI–XIII. Testis sacs 2 pairs in XI. Seminal vesicles paired in XI and XII, each with a round dorsal lobe. Prostate glands paired in XVIII, large, follicular, extending anteriorly to XVI and posteriorly to XX, prostatic duct U-shaped with the slender proximal end connecting to prostate gland and the enlarged distal end connecting to male pore. Accessory glands present in the spermathecal and male pore regions, each gland round, with a slender stalk 0.7–1.4 mm long, corresponding to each external genital papilla. Spermathecae 3 pairs in VI–VIII, ampulla oval, diverticulum with an oval seminal chamber and a slender stalk.

Color: Preserved specimens whitish olive on dorsum, olive-gray on ventrum, dark brown on clitellum.
Distribution in China: Taiwan (Nantou).
Etymology: The name "paucus" was given to this subspecies with reference to its "lesser degree" of mosaic formation in the patch of genital papillae compared with that of *A. tessellatus* (Shen, Tsai & Tsai, 2002).
Type locality: A mountain slope along Nanshan Creek (elevation 800–900 m), Jenay, Nantou County, Taiwan.
Deposition of types: Taiwan Endemic Species Research Institute, Jiji, Nantou County, Taiwan.
Holotype: coll. no. 1999-20-Shen. Paratypes: same collection data as for holotype.
Remarks: *A. tessellatus paucus* differs from *A. tessellatus tessellatus* (Shen, Tsai & Tsai, 2002) by having patches of both presetal and postsetal genital papillae only in XVIII in the postclitellar region, and by having accessory glands with long, slender stalks.

219. *Amynthas tungpuensis* Tsai, Shen & Tsai, 1999

Amynthas tungpuensis Tsai, Shen & Tsai, 1999. *Journal of the National Taiwan Museum.* 52(2),33–46.
Amynthas tungpuensis Chang, Shen & Chen, 2009. *Earthworm Fauna of Taiwan.* 96–97.

External Characters

Length 142–160 mm, clitellum width 4.5–5.3 mm, segment number 119–128. No secondary segmentation. Prostomium epilobous. First dorsal pore 11/12, 12/13 or 13/14. Clitellum XIV–XVI, annular, setae absent, dorsal pores present or absent. Setae number 40–47 (VII), 54–59 (XX); 12–16 between male pores. Male pores paired in XVIII, lateroventral, 0.26–0.29 circumference apart

ventrally, each pore simple, on top of a papilla-like porophore surrounded by 7–8 circular folds, a genital papilla immediately anterior to setal line in the medial position between the male pores in XVIII, an additional papilla at the same location in XVII for some specimens, each papilla round, with a slightly concave center, about half of the distance between setal line and anterior intersegmental furrow. Spermathecal pores 3 pairs in 6/7/8/9, ventrolateral, 0.26–0.29 circumference apart ventrally, each in a depressed furrow with a tiny projection. Genital papillae 2 transverse rows presetal in VIII and IX, numbering 4–11 for each row, an additional pair of postsetal papillae present at the ventrolateral part of VIII in some specimens. Female pore single, medioventral in XIV.

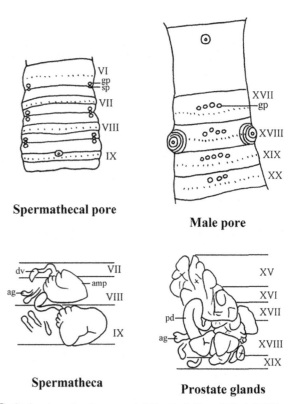

Spermathecal pore

Male pore

Spermatheca

Prostate glands

FIG. 214   *Amynthas tungpuensis* (after Tsai, Shen & Tsai, 1999).

Internal Characters

Septa 8/9/10 absent, 6/7/8 thin, 10/11–14/15 slightly thickened. Gizzard in IX and X, round. Intestine beginning to swell from XV. Intestinal caeca paired in XXVII, simple, smooth, extending anteriorly to XXIII, with wide basal and slender distal portions. Lateral hearts in X–XIII. Testis sacs 2 pairs in X and XI, smooth, medioventral. Seminal vesicles paired in XI and XII, surface follicular, each with a small, dorsal lobe. Prostate glands paired in XVIII, racemose, large, extending anteriorly to XIV and posteriorly to XIX, prostatic duct n-shaped, proximal half enlarged. No accessory glands associating with preclitellar and postclitellar genital papillae. Spermathecae 3 pairs in VII–IX, ampulla large, round, peach-shaped, with a short, stout stalk, diverticulum with a small, peach-shaped seminal chamber and a straight, slender stalk. Nephridial batteries large, paired in V and VI. Ovaries closely paired, medioventral in XIII.

Color: Preserved specimens violet-brown on dorsum with a dark violet longitudinal dorsal line, light gray on ventrum, brown on clitellum, light yellowish white on setal line.
Distribution in China: Taiwan (Nantou).
Etymology: The name of the species refers to the type locality, Tungpu, a small village wita h hot spring in the Central Mountain Range.
Type locality: Tungpu, Nantou County, Taiwan.
Deposition of types: Taiwan Endemic Species Research Institute, Jiji, Nantou County, Taiwan.
Holotype: coll. no. 1998-61. Paratypes: Same collection data as for holotype.
Remarks: The number of genital papillae in this species shows great individual variation. That might be the reason for *A. monsoonus* (James, Shih & Chang, 2005) being regarded as a distinct species from *A. Tungpuensis* by James et al. (2005), but after examining type specimens, Tsai et al. (2009) argued that the former is a junior synonym of the latter. This dispute is probably a matter of opinion on morphological variations, with the solution relying on new data in the future (S. James, personal communication).

220. *Amynthas wuhumontis* Shen, Chang, Li, Chih & Chen, 2013

Amynthas wuhumontis Shen, Chang, Li, Chih & Chen, 2013. *Zootaxa*. 3599 (5):475–477.

External Characters

Small to medium; length 65–134 mm, width 2.69–4.08 mm (clitellum), segment number 78–112. Prostomium epilobous. First dorsal pore 11/12 or 12/13. Clitellum XIV–XVI, setae and dorsal pores absent. Setae number 38–45 (VII), 49–61 (XX); 11–12 between male pores. Male pores paired in XVIII, 0.21–0.29 circumference apart ventrally, each on a papilla-like porophore surrounded by 3 genital papillae: 1 anterior, 1 posterior and 1 medial. Each papilla round, center depressed, 0.35–0.6 mm in diameter. The male porophore together with the genital papillae are surrounded by 3–4 diamond-shaped skin folds. Spermathecal pores 3 pairs in intersegmental furrows of 6/7/8/9; distance between paired pores 0.19–0.25 circumference apart ventrally. Genital papillae 3 in presetal VIII or widely paired in presetal VIII; in IX, papillae absent, single median, or same number and arrangement as in

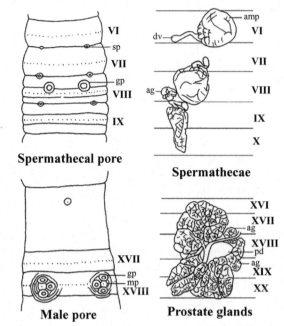

FIG. 215 *Amynthas wuhumontis* (after Shen et al., 2013).

VIII. Each papilla round, center depressed, 0.4–0.6 mm in diameter. Female pore single, medioventral in XIV.

Internal Characters

Septa 8/9/10 absent, 5/6/7/8 thick, 10/11–13/14 muscular. Nephridial tufts on anterior faces of 5/6/7. Gizzard large in VIII–X. Intestine from XV or XVI. Intestinal caeca paired in XXVII, extending anteriorly to XXIII–XXV, each simple, distal end either straight or bent. Esophageal hearts in XI–XIII. Holandric, testis sacs oval, 2 pairs in X and XII. Seminal vesicles large, smooth, 2 pairs in XI and XII, each vesicle with a large, round dorsal lobe. Prostate glands large, occupying 4–5 segments in XVI–XXI, divided into several lobules by grooves. Prostatic duct stout, straight, enlarged in the middle. Accessory glands large, sessile, 0.45–1.05 mm long and 0.4–0.85 mm wide, corresponding to external genital papillae. Spermathecae 3 pairs, in VI, VII, and VIII or in VI, VIII, and IX, ampulla round or elongated oval-shaped, surface wrinkled, spermathecal stalk short, stout. Diverticulum with a slender stalk 0.4–0.8 mm long and an elongated oval seminal chamber 0.3–0.62 mm in length. Accessory glands large, pad-like or flower-like, with shallow grooves on the surface, each corresponding to external genital papilla.

Color: Preserved specimens brown to grayish brown. Clitellum dark to light brown.
Distribution in China: Fujian (Jinmen, Quanzhou).
Etymology: This name of this species refers to the type locality of Mt. Wuhu.
Type locality: Fujian (Jinmen).

Deposition of types: The Invertebrate Zoology and Cell Biology Lab, Department of Life Sciences, National Taiwan, Taibei, Taiwan.

Habitats: *Amynthas wuhumontis* (Shen, Chang, Li, Chih & Chen, 2013) is distributed only in areas around Mt. Wuhu and Mt. Taiwu in east Jinmen.

Remarks: It is fairly similar to *A. leucocircus* (Chen, 1933) from the coastal provinces of central China. Both species are sexthecal with spermathecal pores in 6/7/8/9 and have similar number of setae and similar papillae arrangement in the male pore region. However, *A. leucocircus* is larger with a body size of 140–212 mm long and 4–7 mm wide; it has papillae postsetal in VII–VIII, 4 pairs of hearts in X–XIII, small seminal vesicles, and stalked accessory glands. *A. wuhumontis* is smaller (65–134 mm long and 2.69–4.08 mm wide), and has papillae in presetal VIII–IX, 3 pairs of hearts in XI–XIII, large seminal vesicles, and sessile accessory glands. Furthermore, the ventral distance between paired spermathecal pores is wider in *Amynthas leucocircus* (about 1/3 circumference) than that in *A. wuhumontis* (0.19–0.25 circumference).

221. *Amynthas yamadai* (Hatai, 1930)

Pheretima yamadai Chen, 1933. *Contr. Biol. Lab. Sci. Soc. China (Zool.).* 9:255–261.

Amynthas yamadai Sims & Easton, 1972. *Biol. J. Linn. Soc.* 4:237.

Amynthas yamadai Xu & Xiao, 2011. *Terrestrial Earthworms of China.* 199–200.

External Characters

Length 100–150 mm, width 5–9 mm, segment number 95–108. Prostomium 1/2–1/3 epilobous. First dorsal pore 12/13. Clitellum XIV–XVI, annular, setae absent, glandular part somewhat swollen. Setae conspicuous; setae number 34–65 (III), 48–85 (VI), 52–82 (VIII), 54–88 (XII), 62–88 (XXV); 28–34 between spermathecal pores. Male pores on ventrolateral side of XVIII, each on a transverse ridge with 1 papilla in front and 1 behind in shape of a cross, on median wall of a crescent-shaped chamber; a rather raised skin pad medial to the chamber, usually with 1 very large papilla behind setae, or with 2–4 smaller ones distributed before and behind setal zone. Spermathecal pores 3 pairs in 6/7/8/9, about 3/7–9/20 circumference apart ventrally, each pore intersegmental but slightly anterior to furrow, preceded by a large papilla or several small ones and surrounded by a glandular and wrinkled skin, sometimes greatly swollen forming a wide and deeply sunken slit; in many cases another glandular patch occurring in front of setal circle on VII, VIII, IX, a papilla sometimes found in this patch; or spermathecal papillae less frequently present and

surrounding glandular skin not marked. Genital papillae beside pore region generally present on ventral side of VII and VIII, 1 pair in front and 1 behind setal circle of each segment. Very rarely another pair present in front of setal circle of IX. Or these papillae entirely absent. Each papilla rounded and flat-topped, perforated or slightly raised in center, without elevated margin. Female pore single, medioventral in XIV.

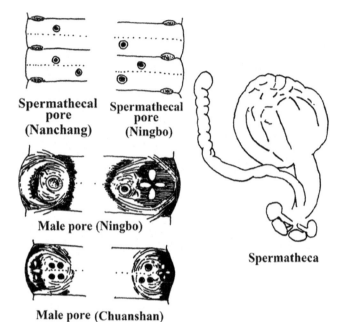

Spermathecal pore (Nanchang) **Spermathecal pore (Ningbo)**

Male pore (Ningbo) **Spermatheca**

Male pore (Chuanshan)

FIG. 216 *Amynthas yamadai* (after Chen, 1933).

Internal Characters

Septa 8/9/10 absent, 5/6/7 thick, or sometimes not much musculated, 7/8 less thickened, 11/12/13/14 also thick and muscular, 10/11 less so, the rest thin and membranous. Gizzard large and rounded, in IX and X. Intestine from XV. Intestinal caeca large and lobulated, in XXVII, extending anteriorly to XXIV or XXII, various degrees of lobulation found on ventral edge, posterior ones usually longer. Hearts 4 pairs, in X–XIII. Testis sacs moderate in size, anterior pair thick U-shaped, each sac elongate and rounded anteriorly, partly in X, their median connection rather thick; posterior pair partly under anterior, in XI, broadly united medially. Testes rather small, about 0.8 mm in diameter, seminal funnels large. Vasa deferens meeting near second sac. Seminal vesicles enormously developed, fully occupied in XI and XII and projected in XIII and part of X, rounded or divided into 2–3 small lobes, with a very small and usually inconspicuous dorsal lobe, round and knob-like. Prostate glands well developed, in XVI–XXI, small, or in XVII–XIX or XX, transversely divided into lobes and again subdivided into small pieces; prostatic duct short and stout, usually U-curved, proximal part rather slender, sometimes longer, with several coils proximally. Accessory glands occurring medial to the duct, very fine and velvety or villiform, close to or partly embedded in body wall, in roundish or elongate patches, granular on surface. Spermathecae 3 pairs in VII, VIII, IX. Ampulla round sac-like, usually flattened and not well-distended, sometimes heart-shaped, seldom elongate oval, frequently transverse-ovoid. Ampullar duct slightly shorter than, or as long as ampulla, clearly marked off from the latter, ental half a little enlarged. Diverticulum elongate, tubular, its length varying from 2–8 mm straight or slightly curved or twisted, about half the caliber of main duct, the whole length serving as the seminal chamber, not distinctly marked off from its duct, the latter as long as, or longer than, ampullar duct, joining the latter near body wall. Accessory glands found anteriorly to main duct and ventral side corresponding to papillae outside. Gland part bulb-like, with a stout duct, well-developed, only 1 large or from 2–5 glands associated with 1 papilla outside, more so in those anterior to spermathecae, but sometimes villiform in appearance, similar to those in prostate region.

Color: Live specimens rather dark chocolate or reddish brown on anterior dorsum, chestnut on posterior half, clitellum dark fleshy red and gizzard region reddish, pale ventrally, Or fleshy red or brownish fleshy anteriorly, greenish or grayish fleshy posteriorly, greenish pale ventrally. Preserved specimens uniformly purplish black or purplish on anterior dorsum, dark brownish on posterior dorsum, pale or buff ventrally, greyish at anterior end of ventrum. Distribution in China: Jiangsu (Nanjing, Wuxi, Yixing,), Zhejiang (Zhoushan, Ningbo, Chuanshan, Xindeng), Anhui (Chuzhou), Jiangxi (Nanchang), Sichuan.

Deposition of types: Mus. Zool. Cent. Univ. Nanking. Remarks: This species was described by Michaelsen in 1931 as *Amynthas pectiniferus* (Michaelsen, 1931) from the Suzhou specimens. But Chen described as *Amynthas yamadaifrom* from Lower Yangtze Valley, Hatai from Central Japan. There are 3 characters which are more often present constant from Chuanshan: (1) 3–4 small papillae on the medial side of the male porophore, (2) genital papillae on ventral side of VII–IX less constantly not always present (lacking in Japanese form), and (3) caeca generally finger-like.

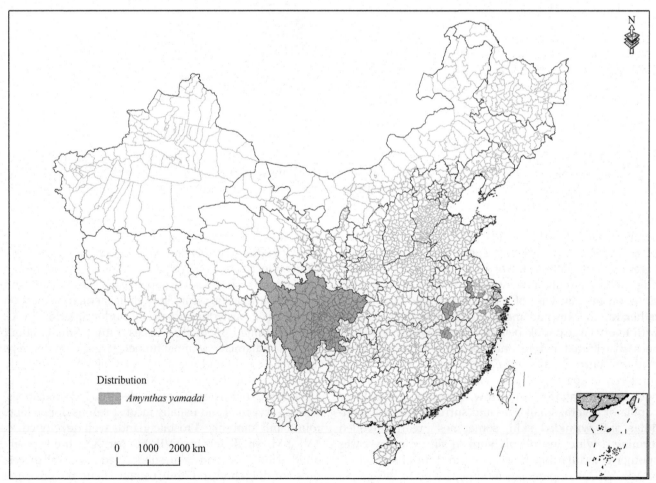

FIG. 217 Distribution in China of *Amynthas yamadai*.

supuensis Group

Diagnosis: *Amynthas* with spermathecal pores 1 pair, at 8/9.

There are 2 species from China.

222. *Amynthas antefixus* (Gates, 1935)

Pheretima antefixa Gates, 1935a. *Smithsonian Mus. Coll.* 93 (3):6.
Pheretima antefixa Chen, 1936. *Contr. Biol. Lab. Sci. Soc. China (Zool.).* 11:293–294.
Pheretima abdita Gates, 1939. *Proc. U. S. Natr. Mus.* 85:418–420.
Pheretima antefixa Chen, 1946. *J. West China Border Res. Soc. (B).* 16:136.
Amynthas antefixus Sims & Easton, 1972. *Biol. J. Linn. Soc.* 4:237.
Amynthas antefixus Xu & Xiao, 2011. *Terrestrial Earthworms of China.* 80.

External Characters

Length 80–132 mm, width 3.2–5.2 mm, segment number 88–112. Prostomium about 1/3 epilobous. First dorsal pore 12/13. Clitellum well marked, in XIV–XVI, annular, usually with setae on ventral side. Setae prominent, those on ventral side of III–X (especially on V–VIII) stouter and more widely spaced; both dorsal and ventral breaks slight; $aa = 1.2ab$, $zz = 1.5$–$3.0yz$; setae number 15–20 (III), 19–23 (VI), 20–24 (VIII), 27–25 (XII), 32–40 (XXV); 10–12 (VIII) between spermathecal pores, 8–12 between male pores. Male pores on ventrolateral corners of XVIII, superficial, each on a minute tubercle surrounded by 5–8 concentric ridges which are interrupted at both sides of setal line; toward lateral margins of short, transverse ridges. A small papilla found on median side near the margin of the circular ridge. Spermathecal pores 1 pair, in 8/9, a little nearer medioventral than mediodorsal line, superficial, or close to posterior edge of VIII, each on a round tubercle intersegmentally situated, skin near orifice swollen and longitudinally wrinkled. Genital papillae moderate in size, median, presetal on III, IV, and V. Female pore single, medioventral in XIV.

Internal Characters

Septa 8/9/10 absent, 5/6–7/8 thick but not particularly so. Gizzard round, surface smooth and glittering. Intestine beginning to swell from XV. Intestinal caeca simple, in XXVII, extending anteriorly to XXV, with ventral shallow saw-like edge. First pair hearts in X absent. Testis sacs of X and XI unpaired and ventral, connected medioventrally. Seminal vesicles very large, with dorsal lobes on anterodorsal side, similar to *Amynthas hawayanus* (Rosa, 1891). Prostate glands large in XVII–XXII, coarsely lobular, ental 2/3 of its duct twice as stout as ectal, moderately long. Accessory glands several connected to 1 papilla, stalked. Spermathecae 1 pair, ampulla long and sac-like, about 4 mm long, 1.5 mm wide, distinct from a short duct; diverticulum if extended about as long as main pouch, with a short, muscular stalk and an elongately tubular seminal chamber, the later nearly straight, twisted, or looped. Genital markings stalked and coelomic.

Color: Preserved specimens greyish white, clitellum light brown.
Distribution in China: Hubei (Lichuan, Qianjiang) Chongqing (Peiling, Beipei, Xufu), Sichuan (Chengdou, Yibin, Xufu).
Etymology: The specific name "*antefixus*" is a compound word, "*ante-*" meaning anterior and "*fixus*" meaning immovable in Latin. It refers to the location of the spermathecal pores.
Type locality: Sichuan (Yibin).
Deposition of types: Smithsonian Institution.
Remarks: *A. antefixus* is distinguished from all other bithecal species of *Amynthas* by the spermathecal pore on 8/9, and by the unpaired, presetal, median genital markings and their anterior location.

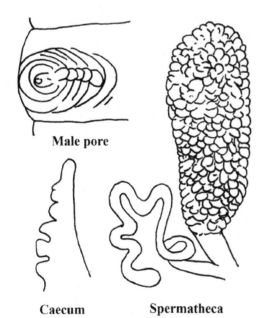

Male pore

Caecum   **Spermatheca**

FIG. 218   *Amynthas antefixa* (after Chen, 1936).

223. *Amynthas liaoi* Zhang, Li, Fu & Qiu, 2006

Amynthas liaoi Zhang, Li, Fu & Qiu, 2006. *Annales Zoologici.* 56 (2):251–252.

External Characters

Length about 55 mm, clitellum width 1.1–2.0 mm, segment number 105–131, secondary annuli present in XI–XIII. Prostomium epilobous. First dorsal pore 11/12.

Clitellum XIV–XVI, annular, intersegmental furrow and dorsal pore present or absent, setae invisible externally. Setae begin on II, regularly distributed around segmental equators; setae number 39–45 (III), 48 (V), 48–51 (VIII), 39–45 (XX), 42 (XXV); 12 (VIII), 14 (IX) between spermathecal pores, 8 between male pores; $aa = 2.2ab$, $zz = 1.1zy$. Male pores 1 pair, about 0.5 circumference apart, each in the center of a slightly convex porophore in XVIII, companied by 3 papillae, with 2–3 slightly circular folds. 2 horizontal rows of 3–8 presetal and postsetal papillae present in XVIII. Spermathecal pores 1 pair, in 8/9, ventral, sometimes inconspicuous. 1 horizontal row of 5 postsetal papillae present in IX. Female pore single, medioventral in XIV.

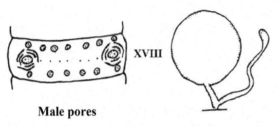

Male pores

Spermatheca

FIG. 219   *Amynthas liaoi* (after Zhang et al., 2006).

Internal Characters

Septa 8/9/10 absent, 6/7/8, 10/11/12/13 comparatively thick. Gizzard in VIII–X, short pumpkin-shaped, moderately large, with vertical fiber. Intestine beginning to swell from XVI. Intestinal caeca simple, small, paired in XXVII, extending anteriorly to XXVI. Esophageal hearts moderately large in X–XIII. Testis sacs 2 pairs, in X and XI, large, separated from each other. Seminal vesicles 2 pairs, in XI and XII, small, with large dorsal lobes. Prostate glands paired in XVIII, moderately large, finger-like, extending anteriorly to XVI and posteriorly to XX. Prostatic duct slender, C-shaped. No accessory glands observed. Spermathecae 1 pair in IX, ampulla ball-shaped, white, about 0.45 mm long, 0.43 mm wide, with a short slender stalk about 0.15 mm long. Diverticulum a little thicker than the duct, slightly shorter than the main pouch, sinuously curved and passed into the duct at the base. No accessory glands observed.

Color: Preserved specimens grayish dorsum, whitish ventrum. Clitellum brown.
Distribution in China: Guangdong (Mt. Dinghu).
Etymology: The species is named after Prof. Chonghui Liao, who has worked in the field of soil fauna biology and ecology for 20 years and has made a great contribution to earthworm study in China.
Type locality: Guangdong (Mt. Dinghu).

Deposition of types: Laboratory of Environmental Biology, School of Agriculture and Biology, Shanghai Jiaotong University, Shanghai, China.
Remarks: *Amynthas liaoi* (Zhang, Li, Fu & Qiu, 2006) appears to be related to *A. antefixus* (Gates, 1935a) and *A. supuensis* (Michaelsen, 1896). They both have 1 pair of spermathecal pores in 8/9. However, *A. liaoi* is characterized by the constant presence of 2 horizontal rows of 3–8 presetal and postsetal papillae in XVIII, and 1 horizontal row of about 5 postsetal papillae in IX. In addition, the first dorsal pore position and the ampulla shape and size are markedly different compared with the other 2 species above.

swanus Group

Diagnosis: First spermathecal pores at 4/5, 2 thecal segments.
There is 1 species from China.

224. *Amynthas swanus* (Tsai, 1964)

Amynthas swanus Tsai, 1964. *Quarterly Journal of the Taiwan Museum.* 17(1&2),13–17.
Amynthas swanus Chang, Shen & Chen, 2009. *Earthworm Fauna of Taiwan.* 84–85.
Amynthas nanshanensis Xu & Xiao, 2011. *Terrestrial Earthworms of China.* 182–183.

External Characters

Length 141–182 mm, clitellum width 5.0–5.5 mm, segment number 154–175. Number of annuli (secondary segmentation) per segment 3 in VII, 5 after middle body. First dorsal pore 12/13. Clitellum XIV–XVI, annular, setae absent, dorsal pores absent. Setae number 98 (V), 103–104 (VIII), 76–82 (XX); 16–18 between male pores. Male pores paired in XVIII, situated within a chamber with tuberculated surface covered by lateral skin wall and exteriorly surrounded by several circular folds. Spermathecal pores invisible externally. Preclitellar papillae absent. Female pore single, medioventral in XIV.

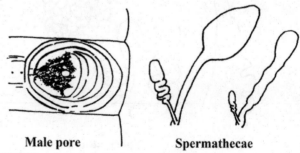

Male pore   **Spermathecae**

FIG. 220   *Amynthas swanus* (after Tsai, 1964).

Internal Characters

Septa 8/9/10 absent, 5/6/7/8 greatly thickened, 10/11/12/13 thickened. Gizzard in X, round. Intestine from XVI. Intestinal caeca paired in XXVII, simple, small, wrinkled with smooth border, extending anteriorly to XXV or XXVI. Last hearts in XIII. Testis sacs 2 pairs in X and XI. Seminal vesicles paired in XI and XII, the anterior pair enveloped within testis sacs. Prostate glands paired in XVIII, small, extending anteriorly to XVII and posteriorly to XIX, with 2 separate lobes connected by 2 narrow, short branches of prostatic ducts, prostatic ducts C-shaped. Spermathecae 2 pairs in V and VI or 2 pairs plus a left rudimentary spermatheca in VII, ampulla peach-shaped with a slender, straight stalk, diverticulum small, with a heart-shaped seminal chamber and a coiled stalk. Ovaries paired in XIII.

> Color: Preserved specimens grayish brown on clitellum, white in other parts of the body.
> Distribution in China: Taiwan (Taibei).
> Etymology: The specific name "*swanus*" derives from "swan" and indicates the white coloration of the species.
> Type locality: Main Campus of National Taiwan University, Taibei City, Taiwan.
> Deposition of types: The types are missing.
> Remarks: This species is the endemic earthworm species recorded in the plain regions in Taiwan. It is rare and endangered due to urbanization. It is extinct from its type locality, the main campus of National Taiwan University, and there have only been a few records in the past decade.

tokioensis Group

Diagnosis: *Amynthas* with spermathecal pores 2 pairs, at 6/7/8.

There are 6 species from China.

TABLE 23 Key to Species of the *tokioensis* Group of the Genus *Amynthas* From China

1. Spermathecal pores on dorsum 2
   Spermathecal pores on ventrum 3

2. Male pores on a rectangular glandular patch, surrounded by several concentric ridges, minute papillae found among the ridges; a small papilla found presetally on VII .... *Amynthas vividus*
   Male pores in a semilunar skin fold, a large transversely ovate papilla placed closely to its median side; occasionally papilla ventrally on VII .... *Amynthas acidophilus*

3. Male pores paired in XVIII, between 1 presetal and 1 postsetal papilla in the same segment. Genital papillae paired in VI–VIII, presetal .... *Amynthas candidus*
   Without papillae preclitellar 4

Continued

TABLE 23 Key to Species of the *tokioensis* Group of the Genus *Amynthas* From China—cont'd

4. Male pores on the top of 1 pair slightly large papillae, this papilla is at the top of a circular protuberance which is high and flat .... *Amynthas zhongi*
   Papillae between XVII–XX 5

5. Male pores on a raised conical porophore genital papillae minute, pimple-like, 10 on right side, 8 on left, confined to segments XVII–XX .... *Amynthas axillis*
   Male pores on a glandular pad of 1/3 XVII–1/3 XIX; pad rectangular, sharply marked off from surrounding skin; each male pore on a small conical papilla. 2 pairs large, truncated additional papillae between 2 male pores .... *Amynthas quadrapulvinatus*

225. Amynthas acidophilus (Chen, 1946)

Pheretima acidophila Chen, 1946. *J. West China Border Res. Soc. (B)*. 16:111–112, 140.
Amynthas acidophilus Sims & Easton, 1972. *Biol. J. Linn. Soc.* 4:237.
Amynthas acidophilus Xu & Xiao, 2011. *Terrestrial Earthworms of China*. 78–79.

External Characters

Medium-sized worm. Length 60 mm, width 1.8–3.0 mm, segment number 118. Segments shortened after clitellum. Prostomium 1/2 epilobous, its width the same as length of II. First dorsal pore 11/12. Clitellum annular, not well marked, glandular in XIV–XVI, setae disappearing intersegmental furrows rather indistinct. Setae all fine, evenly distributed, slightly closer ventrally; setae number 16–26 (III), 27–33 (VI), 29–36 (IX), 34 (XXV); 17–24 (VI), 19–24 (VII), 16–26 (VIII) between spermathecal pores, ?–6 between 2 papillae of male pores; $aa = 1.1ab$, $zz = 1.1$–$1.5zy$. Male pores paired in XVIII, about 1/3 circumference apart ventrally, each in a semilunar skin fold, a large transversely ovate papilla placed closely to its median side. Spermathecal pores 2 pairs, intersegmental in 6/7/8, about 6/11 circumference apart ventrally. No genital markings therein; occasionally papilla ventrally on VIII. Female pore single, medioventral in XIV.

Internal Characters

Septa 8/9/10 absent, all rest thin and membranous, 10/11/12/13 thin, due to pressure from seminal vesicles, 13/14/15 appearing thicker but not musculated. Nephridial tufts on anterior faces of 5/6/7 thick. Pharyngeal glands moderately developed. Gizzard in IX and X, round, smooth but not shining on surface, a crop-like dilation in front. Intestine actually beginning to swell from XVII, caeca simple, short, extending anteriorly to XXIV. Vascular loops in IX slender equally developed,

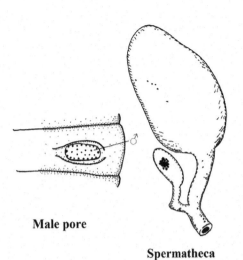

Male pore

Spermatheca

FIG. 221   *Amynthas acidophilus* (after Chen, 1946).

those in X absent. Testis sacs moderately large, those in X narrowly connected and communicated; in XI widely placed, also communicated. Seminal vesicles very well developed, dorsal lobes dorsal in position, vesicle subdivided into small lobes, both pairs occupying segments 1/2X–XIV. Prostate glands well developed in XVII–XIX, its duct S-curved middle portion larger, about 0.5 mm thick, ectal end coiled. Accessory glands small and sessile. Spermathecae 2 pairs, in VII and VIII, ampulla large, heart-shaped, duct moderately long. Diverticulum about half as long as main pouch, seminal chamber ovoid. No accessory glands found in this region.

Color: Preserved specimens reddish grey on middle and posterior, reddish purple on anterior dorsum, purplish red at anterior ventrum (ca. 10 segments), pale ventrally. Clitellum dark purplish red.
Distribution in China: Chongqing (Mt. Jinfo).
Etymology: The specific name indicates the habitat of its type locality of the underside of mosses that may secrete acids.
Type locality: Chongqing (Mt. Jinfo).
Deposition of types: Previously deposited in the Institute of Zoology, Academic Sinica, Chongqing, the types were destroyed during the war in 1945–1949.
Habitats: This species is usually found under mosses, not on the ground but on trees, occasionally on dead trees. They usually make their burrows between crevices in the bark. When the mosses are stripped off, they still crawl between pieces of bark and will readily fall with any disturbance. They are agile and very active creatures. It has not been found in the soil nearby. It may be a habit of the species to crawl up the tree as the humidity there is exceedingly high.

226. *Amynthas axillis* (Chen, 1946)

Pheretima axillis Chen, 1946. *J. West China Border Res. Soc. (B).* 16:122–123, 140, 158.
Amynthas axillis Sims & Easton, 1972. *Biol. J. Linn. Soc.* 4:237.
Amynthas axillis Xu & Xiao, 2011. *Terrestrial Earthworms of China.* 82–83.

External Characters

Small-sized worm. Length 45 mm, width 2.8 mm, segment number 80. 3 annuli found on VII–VIII. Prostomium 1/3 epilobous. First dorsal pore 11/12. Clitellum XIV–XVI, annular, entire, and smooth. Setae all fine, none enlarged, evenly distributed. Setal chain nearly closed; setae number 34 (III), 36 (VI), 40 (IX), 40 (XXV); 20 (VII) between spermathecal pores, 12 between male pores; $aa = 1.2ab$, $zz = 1.2$–$1.5zy$. Male pores paired in XVIII, about 1/3 circumference apart ventrally, each on a raised conical porophore, genital papillae minute, pimple-like, 10 on right side, 8 on left, confined to XVII–XX. Spermathecal pores 2 pairs in 6/7/8, not visible externally, about 1/3 circumference apart ventrally, No genital papillae. Female pore single, medioventral in XIV.

Internal Characters

Septa 8/9/10 absent, all rest very thin. Nephridial tufts in front of 5/6/7 thick. Gizzard round, onion shaped, in IX and X. Intestine beginning to swell from XV. Caeca simple, in XXVII, extending anteriorly to XXIII. Vascular loops in IX asymmetrical, those in X large. Testis sacs in X and XI separate. Seminal vesicles in XI and XII, and a part in XIII, each subdivided into several lobes. Prostate glands entirely absent, nor its duct visible. Vasa deferens leading directly to its exit on XVIII. Accessory glands very small, each with spherical glandular portion and a long duct (gland about 0.2 mm in diameter, duct 0.6 mm long). Each corresponding to its respective papilla externally. Spermathecae 2 pairs, in VII and VIII, very small, ampulla date-shaped, with a long duct. Diverticulum represented only by a bud-like structure at ectal 1/3 of main duct.

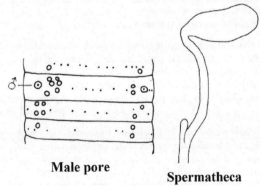

Male pore   **Spermatheca**

FIG. 222   *Amynthas axillis* (after Chen, 1946).

Color: Greyish ventrally, reddish on anterior end, reddish gray on dorsum, reddish along mediodorsal line. Clitellum cinnamon brown.

Distribution in China: Chongqing (Nanchuan), Sichuan (Leshan).

Etymology: The specific name *"axillis"* means axilla in Latin and refers to the structure of the spermathecal diverticulum.

Type locality: Chongqing (Nanchuan).

Deposition of types: Previously deposited in the Institute of Zoology, Academic Sinica, Chongqing, the types were destroyed during the war in 1945–1949.

Remarks: Differs from *Amynthas moniliatus* (Chen, 1946) in terms of the spermatheca and diverticulum and only having 2 pairs of spermathecal pores. As a result the present form is described as a separate species.

227. *Amynthas candidus* (Goto & Hatai, 1898)

Perichaeta candida Goto & Hatai, 1898. *Annot. Zool. Jap.* 2:77–78.

Amynthas candidus Chang, Shen & Chen, 2009. *Earthworm Fauna of Taiwan.* 30.

Amynthas candidus Xu & Xiao, 2011. *Terrestrial Earthworms of China.* 89.

External Characters

Length 150 mm, clitellum width 6 mm, segment number 95. Setae number 44 (VII), 44 (XVIII), 12 between male pores. First dorsal pore 13/14. Clitellum XIV–XVI, annular, setae absent. Male pores paired in XVIII, between 1 presetal and 1 postsetal papillae in the same segment. Spermathecal pores 2 pairs in 6/7/8. Genital papillae paired in VI–VIII, presetal. Female pore single, medioventral in XIV.

Internal Characters

Septa 9/10/11 absent, 6/7/8/9, 10/11–13/14 thickened. Gizzard in IX–X. Intestine from XV. Intestinal caeca paired in XXVII, extending anteriorly to XXV. Last hearts in XIII. Spermathecae 2 pairs in VII and VIII, diverticulum long, about 3 times the length of the main part. Ovaries absent (?). Testis sacs 2 pairs in X and XI. Seminal

vesicles paired in XI and XII. Prostate glands paired in XVIII, large, lobular, extending anteriorly to XVII and posteriorly to XXII.

Color: Dark brown on dorsum, light gray on ventrum, with metallic luster.

Distribution in China: Taiwan (Taibei).

Etymology: The specific name *"candidus"* literally means "shining white" in Latin.

Type locality: Taiwan (Taibei).

Deposition of types: The type is missing.

Remarks: The single missing type specimen is the only known record of this species to date.

228. *Amynthas quadrapulvinatus* Wu & Sun, 1997

Amynthas quadrapulvinatus Wu & Sun, 1997. *Sichauan Journal of Zoology.* 16(1):3–5.

Amynthas quadrapulvinatus Xu & Xiao, 2011. *Terrestrial Earthworms of China.* 172–173.

External Characters

Length 49–77 mm, width 2.7–4.1 mm, segment number 102–106. First dorsal pore 10/11. Clitellum XIV–XVI, annular, setae absent, with dorsal pores. Setae stout, widely spaced; setae number 28–36 (III), 50–58 (VI), 58–68(VIII), 34–36 (XVIII), 68–72 (XXV); 12–14 (VII) between spermathecal pores, 0 between male pores. Male pores 1 pair, on ventrolateral region of XVIII, about 1/6 circumference apart ventrally, both on a glandular pad of 1/3 XVII–1/3 XIX; pad rectangular, sharply marked off from surrounding skin; each male pore on a small conical papilla. 2 pairs large, truncated additional papillae between 2 male pores. Spermathecal pores 2 pairs in 6/7/8, a little eye-like, about

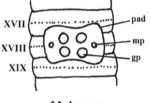

Male pore

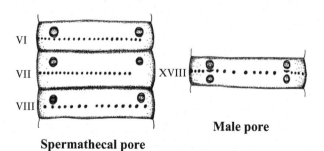

Spermathecal pore   **Male pore**

FIG. 223 *Amynthas candidus* (after Goto & Hatai, 1898).

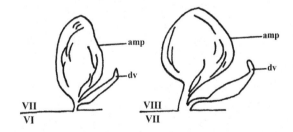

Spermathecae (anterior and posterior)

FIG. 224 *Amynthas quadrapulvinatus* (after Wu & Sun, 1997).

1/5 circumference apart ventrally. No genital papillae in this region. Female pore single, medioventral in XIV.

Internal Characters

Septa 9/10 absent, 5/6/7/8 as thick as 10/11/12/13. Gizzard in VIII–X. Intestine beginning to swell from XV. Caeca simple, in XXVII, extending anteriorly to XXIII. Hearts 4 pairs, last pairs in XIII. Testis sacs 2 pairs, both sides separate, in X and XI. Seminal vesicles well developed, in XI and XII. Prostate glands XVII–1/2XXI, leaf-shaped. No other accessory glands. Spermathecae 2 pairs, in VII and VIII, posterior pair larger than anterior pair. Anterior ampula long-round shaped, its ducts short. Diverticulum shorter than main part, duct straight, seminal chamber club-shaped.

Color: Preserved specimens yellowish brown. Clitelum dark gray.
Distribution in China: Yunnan (Xishuangbanna).
Etymology: The specific name indicates the structure of the rectangular pad in the male pore region.
Type locality: Yunnan (Xishuangbanna).
Deposition of types: Department of Biology, Hangzhou Normal College, Hangzhou.
Remarks: This species is similar to *Amynthas bisemicircularis* (Ding, 1985) in having a glandular pad in the male pore region, but possesses the following difference: (1) spermathecal pores 2 pairs; (2) glandular pad rectangular; (3) no semicircular groove.

143. Amynthas vividus (Chen, 1946) (v.ant. 143)

229. Amynthas zhongi Qiu, Wang & Wang, 1991

Amynthas zhongi Qiu, Wang & Wang, 1991. *Guizhou Science.* 9 (4):301–304.
Amynthas zhongi Xu & Xiao, 2011. *Terrestrial Earthworms of China.* 201–202.

External Characters

Small-sized worm. Length 37–41 mm, width 0.8–1.2 mm (clitellum), segment number 90–114. Prostomium proepilobous. First dorsal pore 11/12. Clitellum XIV–XVI, annular, setae absent. Setae all fine, evenly distributed, setae number 24–34 (III), 31–42 (V), 35–52 (VIII), 32–38 (XX), 34–40 (XXV); 12–15 (VII) between spermathecal pores, 6–7 (XVIII) between male pores. Male pores paired in XVIII, each on the top of a rather large papilla, about 2/3 circumference apart ventrally; this papilla is at the top of a circular protuberance which is high and flat. Genital markings absent in this region. Female pore single, medioventral in XIV.

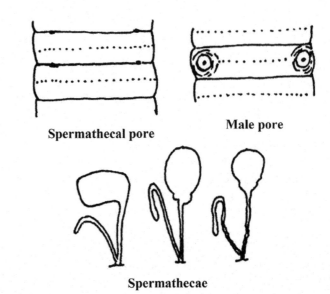

Spermathecal pore   **Male pore**

Spermathecae

FIG. 225   *Amynthas zhongi* (after Qiu et al., 1991a).

Internal Characters

Septa 8/9/10 absent, 5/6/7/8 rather thickened and rest thin. Gizzard long-round shaped, slightly developed, in IX–X. Intestine from XVI; caeca simple, smooth, in XXVII, extending anteriorly to XXIV. Hearts 4 pairs, in X–XIII, developed. Testis sacs 2 pairs, in X–XI, slightly small, ovoid-shaped, separated. Seminal vesicles paired in XI–XII, undeveloped, dorsal lobe small, round. Prostate glands developed, in XVI–XX, chunk-like lobate. Prostatic duct S-shaped, slender. Accessory glands absent. Spermathecae 2 pairs in VI–VII; ampulla long-round shaped, slightly small; ampulla duct stout and long. Accessory glands absent.

Color: Clitellum red-brown.
Distribution in China: Guizhou (Liping, Jinping).
Etymology: This species was named after Prof. Zhong Yuanhui, a brilliant Chinese earthworm taxonomist.
Type locality: Guizhou (Liping, Jinping).
Deposition of types: Institute of Biology, Guizhou Academy of Science, Guiyang.
Remarks: This species is similar to *Amynthas vividus* (Chen, 1946) in having 2 pairs of spermathecal pores (in 6/7/8), and in the shape of spermathecae and diverticulum. However, this species can easily be distinguished from *Amynthas vividus* by (1) body size is small; (2) spermathecal pores ventrally; (3) male pore on the central of large round papillae; (4) septum 8/9 absent; (5) both testis sacs and seminal vesicles are small and undeveloped; (6) spermathecae front septa 6/7 and 7/8, in VI and VII; (7) diverticulum shorter than main pouch, etc.

youngi Group

Diagnosis: *Amynthas* with spermathecal pores 2 pairs, at VI and 5/6.

There is 1 species from China.

230. *Amynthas youngi* (Gates, 1932)

Pheretima youngi Gates, 1932. *Rec. Ind. Mus.* 34 (4): 406–408.

Amynthas youngi Sims & Easton, 1972. *Biol. J. Linn. Soc.* 4:237.

Amynthas youngi Xu & Xiao, 2011. *Terrestrial Earthworms of China.* 200–201.

External Characters

Length 77–90 mm, width 4 mm, segment number 71. Prostomium epilobous (?). First dorsal pore 10/11. Clitellum XIV–XVI, annular, setae unrecognizable externally, without intersegmental furrows, and positions of the dorsal pores indicated by nonfunctional pore-like markings. Setae number 29–33 (V), 40–48 (XX); 5–9 (VI) between spermathecal pores (dorsally), 4–7 between male pores; Male pores small, in XVIII, anteroposterior slits, on the genital markings (both pores are near the median margins of the genital markings or near the lateral margins of the markings). 1 pair of genital markings 1 pair on XVIII, each marking anteroposteriorly elongated with bluntly rounded ends and an even lateral margin, but with a single slight notch or indentation of the median margin about in line with the setae of XVIII, extending slightly anterior to 17/18 and slightly posterior to 18/19. The marking is sunk into the parietes; around the depression there is a slightly protuberant and glistening whitish rim. The surface of the genital marking is not level but rises slightly into a small, not sharply demarcated, somewhat conical protuberance on which is the male pore. The markings are about 6–8 intersetal distances wide transversely. Spermathecal pores 2 pairs,

dorsal, very small, transverse slits; 1 pair in the setal circle of VI, the other pair in 6/7, the anterior pores about 2 intersetal distances lateral to the posterior pores. The pores are on tiny, whitish, conical protuberances which are readily visible to the unaided eye against the pigmentation of the dorsum. (In the cotypes, the eye against the pigmentation of the dorsum of posterior pores). Female pore single, medioventral in XIV.

Internal Characters

All septa are present from 4/5 posteriorly; none are particularly thickened. Intestine from XV. Caeca simple, in XXVII, extending anteriorly to XXV or XXIV. The last hearts in XIII. Testis sacs unpaired, ventral. Seminal vesicles large, covering the dorsal blood vessel in XI and XII; the anterior vesicles pushing 10/11 forward into contact with 9/10, the posterior vesicles pushing 12/13 into contact with 13/14. Prostate glands extend through XVII–XX, displacing slightly 16/17 and 20/21; its ducts bent into a short, U-shaped loop, the closed end of the loop towards the nerve cord, the ectal limb posterior to the ental limb. Spermathecae 2 pairs, ampulla duct is shorter than the ampulla, the diverticulum longer than combined lengths of duct and ampulla. The ental portion of the diverticulum is looped, the loops pressed together and against the anterior face of the spermatheca, the looping may or may not be in a regular zigzag pattern. The entalmost portion of the diverticulum is widened. The diverticulum passes into the anterior face of the duct within the parietes; the duct ectal to this junction abruptly narrowed. In the body wall dorsal to the genital marking there is a tough mass of whitish tissue which does not project into the coelom.

Color: Preserved specimens reddish brownish anteriorly to light yellowish brown on dorsum. Clitellum greyish (or brownish).
Distribution in China: Yunnan (Mengmeng).
Etymology: The specific name was named after Mr. H. Young who collected the specimens.
Type locality: Myanmar (PangWo).
Deposition of types: The types are missing.
Remarks: This species somewhat resembles *A. sulcatus* (Gates, 1932) from the Tenasserin division of the province. *Amynthas youngi* and *Amynthas sulcatus* are the only Burmese species of the genus with dorsal spermathecal pores.

zebrus Group

Diagnosis: *Amynthas* with spermathecal pores 1 pair, at 7/8.

There are 5 species from China.

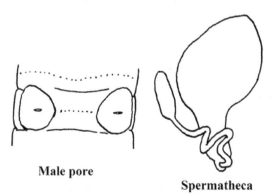

Male pore

Spermatheca

FIG. 226   *Amynthas youngi* (after Gates, 1932).

TABLE 24 Key to Species the of *zebrus* Group of the Genus *Amynthas* From China

1. Male pores on the lateral part of a large papilla which extends to 3/4XVII and 1/4XIX; sometimes, a large round flat-top papilla on anterior of medioventral setal ring of XX .... *Amynthas megapapillatus*
   With papillae front clitellum 2

2. With or without papillae front and behind VIII 3
   Papillae between IX–XIII 4

3. 1–2 large round flat-topped papillae on anterior of medioventral setal ring of VIII, sometimes absent; 1 large round flat-topped papilla on anterior of medioventral setal ring of X occasionally .... *Amynthas fasciculus*
   Male pores on a large flat-topped papilla, surrounded by a few concentric ridges; large papillae situated on VII and VIII occasionally .... *Amynthas palmosus*

4. 1 pair papillae front each setal ring of IX–XII or X–XI ventral *Amynthas eleganus*
   1 pair papillae on anterior and posterior of each setal ring of IX–XIII ventrally, and 1 pair papillae on anterior of setal ring of X .... *Amynthas xingdoumontis*

231. *Amynthas eleganus* Qiu, Wang & Wang, 1991

Amynthas eleganus Qiu, Wang & Wang, 1991. *Guizhou Science.* 9 (3):220–221.
Amynthas eleganis Qiu & Wang, 1992. *Acta Zootaxonomica Sinica.* 17(3):262–264.
Amynthas domosus Xu & Xiao, 2011. *Terrestrial Earthworms of China.* 109–110.

External Characters

Length 75–123 mm, width 3.5–4.0 mm, segment number 81–107. Prostomium 1/3 epilobous. First dorsal pore 11/12. Clitellum XIV–XVI, annular, with setae on ventral side. Setae fine, evenly distributed; setae number 22–34 (III), 33–42 (V), 32–46 (VIII), 42–48 (XX), 42–50 (XXV); 16–19

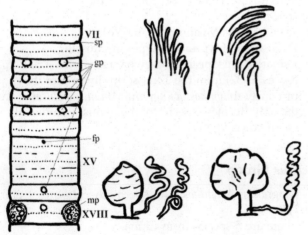

FIG. 227 *Amynthas eleganus* (after Qiu & Wang, 1992).

(VIII) between spermathecal pores, 7–10 between male pores. Male pores 1 pair, in ventrolateral side of XVIII, each on a small round papilla, other 3 genital papillae close to porophore; about 1/3 circumference apart ventrally. Spermathecal pores 1 pair in 7/8, about 4/9 circumference apart ventrally; 2 pairs papillae on ventral side of X and XI or 4 pairs on IX–XII; papillae about 1/3 circumference apart ventrally. Female pore single, medioventral in XIV.

Internal Characters

Septa 8/9/10 absent, rest rather thin. Gizzard barrel-like, slightly developed. Intestine from XV. Caeca compound, with 5–9 long finger-shaped pouches on ventral side, in XXVII, extending anteriorly to XXII. Testis sacs 2 pairs in X and XI, communicated. Seminal vesicles paired in XI and XII, well developed. Prostate glands large in XVII–XX, a group of small accessory glands in this region, each with a small stalk. Spermathecae 1 pair in VIII, ampulla heart-shaped; diverticulum longer than the main part, terminal half twisted and enlarged. Accessory glands in this region.

Color: Preserved specimens dark brown on dorsum, white gray on ventrum. Clitellum red-brown.
Distribution in China: Hubei (Lichuan), Guizhou (Mt. Fanjing).
Etymology: The specific name probably refers to the elegant posture of this species.
Type locality: Guizhou (Mt. Fanjing).
Deposition of types: Institute of Biology, Guizhou Academy of Science, Guiyang.
Remarks: This species is similar to *A. palmosus* (Chen, 1946), but differs from the latter in the following characters: (1) male pore on a small papilla with other 3 papillae close to it; (2) caeca compound, but not palm-shaped; (3) diverticulum longer than the main part, terminal half sausage-shaped or closed in a zigzag fashion; (4) accessory glands well developed, each with a small stalk.

232. *Amynthas fasciculus* Qiu, Wang & Wang, 1993

Amynthas fasciculus Qiu, Wang & Wang, 1993. *Acta Zootaxonomica Sinica.* 18(4):407–408.
Amynthas fasciculus Xu & Xiao, 2011. *Terrestrial Earthworms of China.* 112–113.

External Characters

Length 134–152 mm, width 3.5–5.0 mm, segment number 120–132. Prostomium 1/2 epilobous. First dorsal pore 12/13. Clitellum XIV–XVI, annular, with some setae on ventral side of XVI. Setae fine, evenly distributed; setae number 30–34 (III), 36–38 (V), 38–42 (VIII), 43–51 (XX), 52–56 (XXV); 13–14 (VII), 15–16(VIII) between

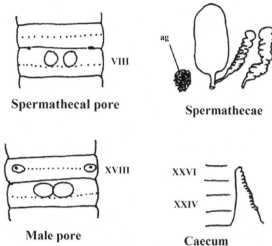

FIG. 228 *Amynthas fasciculus* (after Qiu et al., 1993b).

spermathecal pores, 11–14 between male pores. Male pores 1 pair, each situated on the center of a rather large long-round papilla in ventrolateral side of XVIII, or on a small round papilla, about 1/3 circumference apart ventrally; a pair large long-round and flat-topped papillae on anterior of medioventral setal ring of XIX. Spermathecal pores 1 pair in 7/8, about 2/5 circumference apart ventrally; each situated on a round papilla with both anterior and posterior margins glandular. 1–2 large round flat-topped papillae on anterior of medioventral setal ring of VIII, sometimes absent; 1 large round flat-topped papilla on anterior of medioventral setal ring of X occasionally.

Internal Characters

Septa 8/9/10 absent, 5/6/7/8 rather thickened, others thin. Gizzard developed, barrel-shaped, in IX–X. Intestine from XV. Caeca simple, with tooth-like breach on ventrum, in XXVII, extending anteriorly to XXIV. Testis sacs 2 pairs in X and XI, round, rather small, connected but not communicated. Seminal vesicles 2 pairs in XI and XII, small, its dorsal lobes small too. Prostate glands all absent of or rudimentary. Accessory glands fascicled in XVIII–XIX. Spermathecae 1 pair in VIII, ampulla rather large, long-sac-shaped, about 4 mm long, duct short and clearly demarcated from ampulla; diverticulum shorter than the main part, accessory glands absent.

> Color: Preserved specimens dark brown on dorsum, white-gray on ventrum. Clitellum red-brown.
> Distribution in China: Guizhou (Mt. Leigong).
> Etymology: The specific name indicates the structure of the fascicled accessory glands.
> Type locality: Guizhou (Mt. Leigong).
> Deposition of types: Institute of Biology, Guizhou Academy of Science, Guiyang.

Remarks: This species is similar to *A. palmosus* (Chen, 1946), but differs from the latter in the simple caeca and the fascicled accessory glands.

233. *Amynthas megapapillatus* Qiu, Wang & Wang, 1991

Amynthas megapapillatus Qiu, Wang & Wang, 1991. *Guizhou Science.* 9(3):221–222.
Amynthas magnipapillata Qiu & Wang, 1992. *Acta Zootaxonomica Sinica.* 17(3):264–265.
Amynthas megapapillatus Xu & Xiao, 2011. *Terrestrial Earthworms of China.* 147–148.

External Characters

Length 84–167 mm, width 3.5–4.0 mm, segment number 76–103. Prostomium 1/3 epilobous. First dorsal pore 12/13. Clitellum XIV–XVI, annular, setae absent. Setae fine, evenly distributed; setae number 34–36 (III), 32–42 (V), 42–46 (VIII), 48–54 (XX), 43–52 (XXV); 10–14(VIII) between spermathecal pores, 9–12 between male pores. Male pores 1 pair, in ventrolateral side of XVIII, each on the lateral part of a large papilla which extends to 3/4XVII and 1/4XIX, about 2/5 circumference apart ventrally. Sometimes a large round flat-topped papilla on anterior of medioventral setal ring of XX. Spermathecal pores 1 pair in 7/8, about 3/8 circumference apart ventrally; Female pore single, medioventral in XIV.

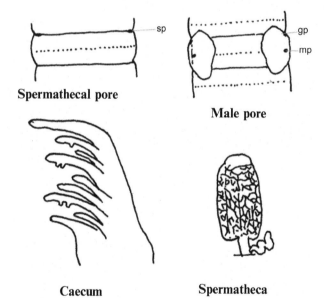

FIG. 229 *Amynthas megapapillatus* (after Qiu et al., 1991b).

Internal Characters

Septa 8/9/10 absent, other thin. Gizzard long round-shaped, developed. Intestine beginning to swell from XVI. Caeca compound, with 8–9 long finger-shaped pouches on ventrum, in XXVII, extending anteriorly to XXII. Testis sacs 2 pairs in X and XI, communicated. Seminal vesicles paired in XI and XII, well developed. Prostate glands very large in XVII–XX. Accessory glands in XVII–XIX, sessile. Spermathecae 1 pair in VIII–X, ampulla long sac-shaped, about 8 mm long; diverticulum short and stout, about 1/6 of the main part, terminal ends enlarged. Accessory glands absent.

Color: Preserved specimens dark brown on dorsum, greyish white on ventrum. Clitellum red-brown.
Distribution in China: Guizhou (Mt. Fanjing).
Etymology: The specific name refers to the large round flat-topped papilla in the male pore region.
Type locality: Guizhou (Mt. Fanjing).
Deposition of types: Institute of Biology, Guizhou Academy of Science, Guiyang.
Remarks: This species is similar to *A. palmosus* (Chen, 1946), but differs from the latter in the following characters: (1) male pore on the lateral part of a very large papilla which extends to 3/4XVII and 1/4XIX; (2) caeca compound, not palm-shaped, with 8–9 long finger-shaped pouches on the ventral side.

234. *Amynthas palmosus* (Chen, 1946)

Pheretima palmosa Chen, 1946. *J. West China Border Res. Soc. (B).* 16:116–117, 140, 144.
Amynthas palmosus Sims & Easton, 1972. *Biol. J. Linn. Soc.* 4:237.
Amynthas palmosus Xu & Xiao, 2011. *Terrestrial Earthworms of China.* 161.

External Characters

Medium-sized worm. Length 65–160 mm, width 4.0–4.5 mm. Segments rather long, VIII–XIII longest, shorter in I–IV. Prostomium 1/2 epilobous, tongue twice as long as length of II. First dorsal pore 12/13. Clitellum

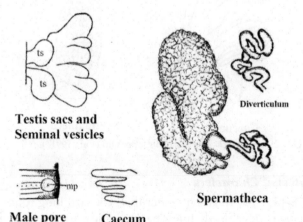

Testis sacs and Seminal vesicles

Male pore   **Caecum**

Diverticulum

Spermatheca

FIG. 230   *Amynthas palmosus* (after Chen, 1946).

XIV–XVI, annular, elongate, and swollen, setae absent. Its length equal to 5 immediately posterior segments. Setae all fine, not very distinct in II–IV, slightly closer ventrally; setae number 28 (III), 34–36 (VII), 33–34 (IX); 20–23 (VII) between spermathecal pores, 8 between male pores; $aa = 1.5ab$, $zz = 1.5$–$2zy$ or $2.5zy$. Male pores paired in XVIII, about 1/3 circumference apart ventrally, each on a large flat-topped papilla, surrounded by a few concentric ridges. Spermathecal pores 1 pair in 7/8, about 6/13 circumference apart ventrally, intersegmental, in an eye-like area. 2 large papillae situated on VII and VIII, ventrally to each pore. Female pore single, medioventral in XIV.

Internal Characters

Septa 8/9/10 absent, all rest thin and membranous, 10/11 comparatively thick. Nephridial tufts on 5/6/7 thick. Pharyngeal glands well developed in III–VI. Gizzard small in X, elongate and shining, crop not marked. Intestine from XV. Caeca compound, in XXVII, palmate in outline, about 4–5 diverticula directed anteriorly to XXIV, topmost longest and whitish in color. Vascular loops in X absent. Testis sacs united, anterior pair in X, V-shaped, posterior pair in XI, U-shaped, 2 limbs directed backward. Vasa deferens of each side very large up to XVIII, nearly as large as nerve cord. No supernumerary pair on 12/13. Seminal vesicles well developed, in X–XIV, dorsal lobe mediodorsal in position. Prostate glands well developed, in XVI–XXI, with large lobes, subdivided; its duct long, S-curved, middle portion straight, and thickest, slender towards both ends. Accessory glands as small patches embedded in the connective tissues. Spermathecae in sexually mature specimens well developed, constricted at middle, extending backward under the seminal vesicles, usually corrugated on surface. Diverticulum comparatively small, confined to the side of main duct; ental 2/3 serving as seminal chamber, closely coiled, about 6 whorls or more, its duct shorter than main duct. Accessory glands as small lobules, attached to body wall.

Color: Reddish purple on dorsum, pale on ventrum except first 5 segments where it is purplish. Clitellum cinnamon brown.
Distribution in China: Chongqing (Mt. Jinfo).
Etymology: The specific name indicates the structure of the palm-shaped caeca.
Type locality: Chongqing (Mt. Jinfo).
Deposition of types: Previously deposited in the Institute of Zoology, Academic Sinica, Chongqing. The types were destroyed during the war in 1945–1949.
Remarks: The peculiarity of this species lies in the single pair of spermathecal pores which are more dorsally situated, and the palmate shape of the caeca.

235. *Amynthas xingdoumontis* Wang & Qiu, 2005

Amynthas xingdoumontis Wang & Qiu, 2005. *Jour. Shanghai Jiaotong Univ. (Agri. Sci.).* 23(1):25–26.
Amynthas xingdoumontis Xu & Xiao, 2011. *Terrestrial Earthworms of China.* 198–199.

External Characters

Length 85–170 mm, width 3–6 mm, segment number 66–95. Prostomium 4/5 epilobous. First dorsal pore 12/13. Clitellum XIV–XVI, annular, setae absent, dorsal pore visible. Setae perichaetin; setae number 30–40 (III), 40–50 (V), 42–50 (VIII), 50 (XX), 50 (XXV); 26–28 (VII) between spermathecal pores, 18 between male pores. Male pores paired, ventral in XVIII, each situated at the center of 1 flat-topped papilla, the same size papillae interior to the former; 3–4 skin folds surrounding the 2 papillae; about 2/5 circumference apart ventrally. Spermathecal pores 1 pair in 7/8, in the center of papillae which transversely in intersegmental groove; 1 pair papillae behind setal ring of IX–XIII, about 1/15 circumference apart ventrally; and the front setal ring of X similarly, about 1/13 of circumference apart ventrally. Female pore single, medioventral in XIV.

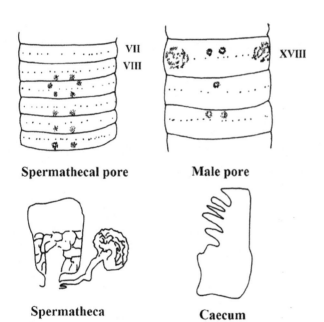

Spermathecal pore   **Male pore**

Spermatheca   **Caecum**

FIG. 231   *Amynthas xingdoumontis* (after Wang & Qiu, 2005).

Internal Characters

Septa 8/9/10 absent, the rest thin. Gizzard in VIII–IX, barrel-shaped. Intestine from XV. Caeca paired in XXVII, compound, extending anteriorly to XXV. Hearts 4 pairs in X–XIII, not developed. Testis sacs 2 pairs in X and XI, oval, covered with continuous membrane. Seminal vesicles paired in XI and XII, meeting dorsally; dorsal lobes well developed, whitish. Prostate glands in XVII–XIX, developed, racemose; its duct short, middle part a little thickened. Accessory glands with stalk. Spermathecae 1 pair, in VIII, ampulla 4.0 mm long, connective tissue found on surface; its duct short and stout, clearly demarcated from ampulla. Diverticulum slim and long, twisted to something like a massive ball, its seminal chamber not swollen. Accessory glands invisible nearby, but many sessile accessory glands distributed on both sides of ventral vessels on IX–XIII.

Color: Preserved specimens light brown on dorsum, grey on ventrally. Clitellum grey.
Distribution in China: Hubei (Lichuan).
Etymology: The name of the species refers to its type locality of Mt. Xingdou.
Type locality: Hubei (Lichuan).
Deposition of types: Laboratory of Environmental Biology, School of Agriculture and Biology, Shanghai Jiaotong University.
Remarks: The species appears to be closely related to *A. eleganus* (Qiu, Wang & Wang), but differs mainly from the latter in such aspects as: (1) reticular connective tissue distributed on spermathecal ampulla surface; (2) diverticulum slender and twisted to something like a massive clew, its diameter the same as the width of 1 segment, and its end not swollen.

II. *BEGEMIUS* EASTON, 1982

Begemius Easton, 1982. *Aust. J. Zool.* 30:717.

Type species: *Begemius jiamisoni* Easton, 1982.
Diagnosis: Megascolecidae of varying sizes with cylindrical bodies. Setae numerous, regularly arranged around each segment. Clitellum annular, XIV–XVI; rarely XIII–XVII. Male pores paired discharging directly onto the surface of XVIII (rarely XIX), without copulatory pouches opening onto XVIII. Female pore single, rarely 2, XIV. Spermathecal pore small or large, usually paired, rarely singular or absent, between 4/5–8/9.

Gizzard present between 7/8 and 8/9. Esophageal pouches absent. Intestinal caeca present originating in XXV. Testes in X and XI, or XI. Ovaries paired in XIII. Spermathecae usually paired, rarely singular or absent. Meronephridial, nephridia rarely present on spermathecal ducts.

Global distribution: Oriental realm, Australasian realm.

There are 5 species from China (Guangdong, Guangxi, and Taiwan).

236. *Begemius dinghumontis* Zhang, Li, Fu & Qiu, 2006

Begemius dinghumontis Zhang, Li, Fu & Qiu, 2006
Annales Zoologici. 56(2):250–251.

External Characters

Length 13–60 mm, width 0.6–2.0 mm, segment number 72–114, secondary annuli inconspicuous. Prostomium epilobous. First dorsal pore 12/13. Clitellum XIV–XVI, annular, intersegmental furrow and dorsal pore present or absent, setae invisible externally. Setae minute, setae number 24–36 (III), 33–42 (V), 27–36 (VIII), 36–39 (XX), 33–34 (XXV); about 7 (VIII) and 9 (IX) between spermathecal pores, 0–2 between male pores. Male pores on ventrolateral region of XVIII or XVII, about 0.17–0.25 circumference apart ventrally, each in the center of a slightly convex ellipse-shaped porophore in XVIII or XVII, no surrounding circular folds. Male region papillae absent. Spermathecal pores 1 pair, in 8/9, about 0.25 circumference apart ventrally, whitish glandular swelling at anterior and posterior sides of each pore. Genital markings absent. Female pore single, medioventral in XIV.

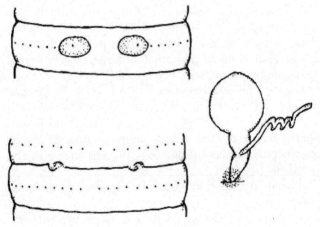

FIG. 232 *Begemius dinghumontis* (after Zhang et al., 2006).

Internal Characters

Septa 8/9/10 absent, 5/6/7/8 comparatively thick, 10/11/12/13 slightly thick. Gizzard in IX–X, short pot like, moderately large, with whitish vertical fiber. Intestine swelling from XIV or XV. Intestinal caeca simple, in XXV, extending anteriorly to XXIII or XXIV. Esophageal hearts moderately large in X–XIII. Testis sacs in X and XI, the right and left testis sacs have blended into a single body, like a whitish long transverse moon. Seminal vesicles in XI and XII, small, oval-shaped, dorsal lobes conspicuous. Prostate glands paired in XVIII, moderately large or small or even vestigial, extending anteriorly to XVI and posteriorly to XX or XXII. Prostatic duct slender or thick, usually straight and long. Accessory glands absent. 1 pair in IX, ampulla ball-shaped or pear-like, yellow, reddish yellow or white, 0.25–0.46 mm long, 0.26–0.42 mm wide, with a short, stout stalk about 0.26 mm long. A slight twist at midpoint of the spermathecal duct where it is joined by the zigzag-shaped bent slender diverticulum. Diverticulum a little shorter than the main pouch, distal seminal chamber not enlarged. Whitish ball-like accessory glands present or absent at the base of the spermathecal duct.

Color: Preserved specimens grayish white or white body. Clitellum brown, yellow, or white.
Deposition of types: Laboratory of Environmental Biology, School of Agriculture and Biology, Shanghai Jiaotong University, China.
Etymology: The name of the species refers to the type locality.
Type locality: Guangdon (Mt. Dinghu).
Distribution in China: Guangdon (Mt. Dinghu).

237. *Begemius heshanensis* (Zhang, Li & Qiu, 2006)

Amynthas heshanensis Zhang, Li & Qiu, 2006. *J. Nat. Hist.* 40(7–8):396–398.
Begemius heshanensis Xu & Xiao, 2011. *Terrestrial Earthworms of China.* 203.

External Characters

Length 110 mm, width 2–2.5 mm (preclitellum), segment number 94–108. Annulets conspicuous in segments V–XIII. Prostomium proepilobous. First dorsal pore 11/12. Clitellum XIV–XVI, annular, setae visible externally. Setae number 30 (III), 39–42 (V), 42–51 (VIII), 48–54 (XX), 39–60 (XXV); 13 between male pores; setal formula $aa = 1.1$–$1.5ab$, $zz = 1.2zy$. Male pores 1 pair in XVIII, 0.4 circumference apart ventrally, each on the top of a capsula-like porophore surrounded by several circular folds. Genital papillae absent. Spermathecal pores 2 pairs in 6/7/8, ventrally, inconspicuous externally. Genital papillae absent. Female pore single, medioventral in XIV.

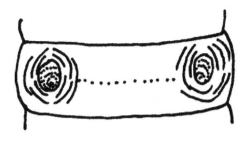

FIG. 233   *Begemius heshanensis* (after Zhang et al., 2006a).

Internal Characters

Septa 8/9/10 absent, 6/7/8, 10/11/12 comparatively thickened. Gizzard short pachyrhizus-shaped, moderately developed, in VIII–X. Intestine swelling from XV. Intestinal caeca paired in XXV, simple, smooth, extending anteriorly to XXII. Lateral hearts in X–XIII, whitish. Holandric, Testis sacs 2 pairs, small, in X–XI, separated from each other. Seminal vesicles paired in XI–XII, well developed with large conspicuous dorsal lobes. Prostate glands developed, extending from XVII–XXII. Prostatic duct slender, slightly curved at distal part. Accessory glands absent. Spermathecae 2 pairs in VII–VIII, ampulla oval-shaped, about 2 mm long; the duct is so short as to be inconspicuous. Diverticulum curved, a little longer than the ampulla plus duct, terminal one-quarter dilated into a long chamber. Accessory glands absent.

Color: Preserved specimens greyish on dorsum, whitish on ventrum. Clitellum pinkish.
Deposition of types: Laboratory of Environmental Biology, School of Agriculture and Biology, Shanghai Jiaotong University, China.
Etymology: The name of the species refers to the type locality (Heshan county-level city).
Type locality: Guangdong (Heshan station).
Distribution in China: Guangdong (Heshan).
Habitat: Collected from vegetable plots near an orchard in Heshan County, Guangdong.
Remarks: *Begemius heshanensis* (Zhang, Li & Qiu, 2006) is somewhat similar to *Amynthas zhangi* (Qiu et al., 1991), *Amynthas sanchongensis* (Hong and James, 2001), *Amynthas paiki* (Hong et al., 2001) and *Amynthas*

taebaekensis (Hong and James, 2001). However, it is easy to distinguish *Begemius heshanensis* from *A. zhongi* by its large body size, the much shorter spermathecal duct, and the caeca originating in XXV, from *Amynthas sanchongensis*, *Amynthas paiki*, and *Amynthas taebaekensis* by the simple structure of male pores, the absent genital markings and papillae, and the simple caeca originating in XXV. *Amynthas sanchongensis* has a distinct lateral crescentic groove on the apex of the porophore. *Amynthas paiki* has 2 presetal pairs of genital papillae in XVIII, presetal paired sets of 2 genital markings in segments VII and VIII, and its caecum is manicate, each consisting of 6–7 finger-shaped sacs, originating from XXVII. *Amynthas taebaekensis* has only 1 pair of postsetal genital papillae in XVIII.

238. *Begemius jiangmenensis* (Zhang, Li & Qiu, 2006)

Amynthas jiangmenensis Zhang, Li & Qiu, 2006. *J. Nat. Hist.* 40(7–8):398–399.
Begemius jiangmenensis Xu & Xiao, 2011. *Terrestrial Earthworms of China.* 204.

External Characters

Length 65 mm, width 1.7 mm (preclitellum), segment number 110. Annulets inconspicuous. Prostomium proepilobous. First dorsal pore 11/12. Clitellum XIV–XVI,

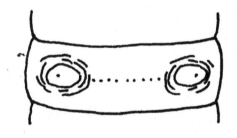

FIG. 234   *Begemius jiangmenensis* (after Zhang et al., 2006a).

annular, setae invisible externally. Setae slightly long, thick, and comparatively sparse at III–VIII, setae number 27 (III), 33 (V), 48 (VIII), 42 (XX), 39 (XXV); 16 (VII) between spermathecal pores, 9 between male pores; setal formula $aa=1.6ab$, $zz=1.5\,zy$. Male pores 1 pair in XVIII, on the center of a slightly raised, transverse, ellipse-like porophore, 0.3 circumference apart, surrounded by several circular folds. Genital papillae absent. Spermathecal pores 2 pairs in 6/7/8, ventral, inconspicuous externally. Genital papillae absent. Female pore single, medioventral in XIV.

Internal Characters

Septa 8/9/10 absent, 5/6/7/8, 10/11/12 comparatively thickened. Gizzard short pot-shaped, moderately developed, in VIII–X. Intestine swelling from XIV. Intestinal caeca paired in XXV, simple, smooth, with some indentations on 1 edge, extending anteriorly to XXII. Lateral hearts in X–XIII. Holandric. Testis sacs 2 pairs, small, in X–XI, separated from each other. Seminal vesicles paired in XI–XII, small. Prostate gland well developed, the duct much thickened, enlarged at midlength. Accessory glands absent. Spermathecae 2 pairs in VII–VIII, ampulla long, narrow oval-shaped, whitish, about 1.5 mm long; not marked off from the duct. The duct is slender, about 0.6 mm long. Diverticulum is as long as the duct, with an enlarged chamber. Accessory glands absent.

> Color: Preserved specimens greyish brown on dorsum. Whitish on ventrum, whitish setae line is conspicuous. Clitellum pinkish.
>
> Deposition of types: Laboratory of Environmental Biology, School of Agriculture and Biology, Shanghai Jiaotong University, China.
>
> Etymology: The name of the species refers to the type locality (Heshan County-level City is located in Jiangmen City).
>
> Type locality: Guangdong (Heshan station).
>
> Distribution in China: Guangdong (Mt. Dinghu).
>
> Habitat: Collected from an abandoned nursery in an orchard in Heshan County, Guangdong.
>
> Remarks: *Begemius jiangmenensis* is closely related to *Begemius heshanensis* (Zhang, Li & Qiu, 2006). However, they differ markedly in the structures of the male porophore and spermathecae. In addition, the seminal vesicles of *Begemius jiangmenensis* were smaller than those of *Begemius heshanensis*.

239. *Begemius paraglandularis* (Fang, 1929)

Pheretima paraglandularis Fang, 1929. *Sinensia*. 1(2):15–24.
Amynthas paraglandularis Xu & Xiao, 2011. *Terrestrial Earthworms of China*. 163–164.

External Characters

Length 115–236 mm, width 6–10.5 mm, segment number 116–146. Number of annuli per segment 2 in VI–VIII, 1 anterior and 1 posterior to setal zone; 3 in IX–X, XIII, 2 anterior and 1 posterior; 4 in XI and XII, 2 anterior and 2 posterior. Body anterior to clitellum somewhat flattened dorsoventrally; ventral surface of segments XVIII–XIX also flattened. Prostomium 1/2 epilobous. First dorsal pore 12/13. Clitellum XIV–XVI, annular, seta and dorsal pore absent. Setae beginning on II, in rings along the whitish band on remaining segments, slightly longer at anterior segments and shorter at posterior, closer and slightly longer at ventral side; ventral break indistinguishable, very slight if present; dorsal break before clitellum, $zz=1.5yz$, behind clitellum $zz=3\,yz$; setae number 34 (III), 42 (VI), 55 (XII), 68 (XVII), 70 (XXV), about 85 in middle of the body, about 78 at the posterior end; 14 between male pores. Male pores on segments XVIII, about 2/5 circumference apart ventrally, a large transverse slit with its neighborhood swollen up to form an oval ridge with several somewhat elliptical grooves on it, medial side of the ridge narrower and pointed; copulatory pouch moderate with an elongate genital papilla in it; more than 20 minute papillae on the ridged region, which are the indications of the openings of the stalk glands; about 10 setae between the oval ridges. Spermathecal pores 2 pairs, lateral in furrow 7/8/9, about 7/15 circumference apart ventrally; probably there are

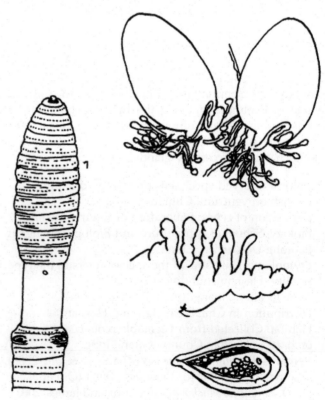

FIG. 235   *Begemius paraglandularis* (after Fang, 1929).

a few minute papillae for opening of the stalk glands, a tubercle protruded about the pore. Female pore single, medioventral in XIV.

Internal Characters

Septa 9/10 absent, 6/7/8 considerable thickened; 8/9 less so; 10/11/12/13 also thickened; 13/14 slightly thicker than those following. Gizzard somewhat bell-shaped, narrower anteriorly and broader posteriorly, situated between septa 8/9, and 10/11 (2 segments) more than 10 small longitudinal blood vessels with minute lateral branches around and on the surface of gizzard. Intestine swelling from XV. Intestinal caeca originating in segment XXV, flattened and only extending 1 segment long lateroventrally, each with 5–6 or more diverticula, each diverticulum usually with lobules. Last hearts in XIII. Testis sacs 2 pairs, in X and XI, close to the anterior side of septa 10/11, 11/12 respectively; first pair somewhat V-shaped, separate in their dorsal view, conjoined posteromedially at ventral side; second pair conjoined medially, narrower than the first pair, more or less oblong with an anteromedial deep and narrow notch, the conjoined paired testis sacs connected to each other with a slender tube at the posterior end (slightly dorsal in position). Seminal vesicles 2 pairs, large, in XI and XII, each with 2–3 large distinct dorsal lobes, the lobes in each segment meeting each other in the middle line; the ventral side of the first pair of vesicles also lobed. Prostates occupying segments XVI–XXI, much cut into lobes, and these into lobules; the duct forming a loop pointing forwards and inwards, its ental limb longer and wider than ectal; more than 20 stalk glands or capsulogenous glands near the male pore. Spermathecae 2 pairs, in VIII, IX; ampulla large, the posterior one nearly occupying IX and X, somewhat pear-shaped with its apical region narrower and rounded, more or less laterally flattened, alveolar in appearance, fully filled with contents; its duct short; diverticulum single, tubular, twisted, and slightly enlarged at its distal portion, communicated with spermathecal duct at the ectal end, slightly over 1/2 the length of the ampulla when extended; about 13–25 stalk glands on the inner wall at the region near the opening of spermathecal duct; the stalk glands generally fewer in the first spermathecal region.

Color: Preserved specimens banded with chestnut and whitish rings alternately; chestnut band intersegmental, beginning at anterior end and on dorsal side, and at furrow 19/20 on ventral, much broader than whitish one (about 3 times that of whitish band on dorsal side in the middle of body), darker and broader on dorsal than ventral; the chestnut color of the bands after clitellum slightly protruding into anteriorly and posteriorly at dorsal pore region to form an interrupted dorsomedial stripe; whitish band at setal zone; clitellum dark chestnut all round.

Deposition of types: Previously deposited in the Metropolitan Museum of Natural History, Beipei, Chongqing. The types were destroyed during the war in 1949.

Etymology: The specific name "*paraglandularis*" comes from the Japanese *Metaphire glandularis* (Goto & Hati). *Begemius paraglandularis* is somewhat related to *Metaphire glandularis* in having numerous stalk glands on the spermathecal and spermiducal segments.

Type locality: Guangxi (Lingyun).

Distribution in China: Guangxi (Lingyun).

240. *Begemius yuhsi* (Tsai, 1964)

Pheretima yuhsi Tsai, 1964. *Quar. Jou. Taiwan Mus.* 17(1&2):5–8.

Metaphire yuhsii Chang, Shen & Chen, 2009. *Earthworm Fauna of Taiwan.* 138–139.

Begemius yuhsi Xu & Xiao, 2011. *Terrestrial Earthworms of China.* 204–205.

External Characters

Length 177–318 mm, clitellum width 11 mm, segment number 80–163. Number of annuli per segment 3 after V. Prostomium epilobous. Setae 77 (IV), 103 (VIII), 123 (XX), 14–34 between male pores. First dorsal pore 13/14. Clitellum XIV–XVI, annular, dorsal pore absent, setae absent. Male pores paired, situated on setal line close to lateral border of XVIII, on large conical porophores, surrounded by several circular folds, about 1/3 circumference apart ventrally. Spermathecal pores 4 pairs in VI–IX, intrasegmental, mediodorsal, situated at anterior edge of each segment, about 0.95–0.97 circumference apart ventrally. No genital papillae in the preclitellar region. Female pore single, medioventral in XIV.

Internal Characters

Septa 5/6/7/8 thickened, 8/9/10 absent, 10/11–13/14 greatly thickened. Gizzard in VIII–X. Intestine from XV. Intestinal caeca paired in XXVII, simple, extending anteriorly to XXIV. Lateral hearts in X–XIII. Nephridia tufted, attached to the postsegmental septa, surrounding the segmental chambers in V and VI. Ovaries paired in XIII, medioventral, close to the 12/13 septum. Testis sacs 1 pair in X, oval, smooth, medioventral in front of 10/11. Seminal vesicles paired in XI, large, each one with a folliculate dorsal lobe. Prostate glands paired in XVIII, large, separated into 2 main lobes, extending anteriorly to XV and posteriorly to XX. Spermathecae 4 pairs in VI–IX, mediodorsal, ampulla large, peach-shaped, with a slender stalk about half the ampulla length, diverticulum with a small oval seminal chamber on the tip and tightly coiled stalk.

FIG. 236   *Begemius yuhsi* (after Tsai, 1964).

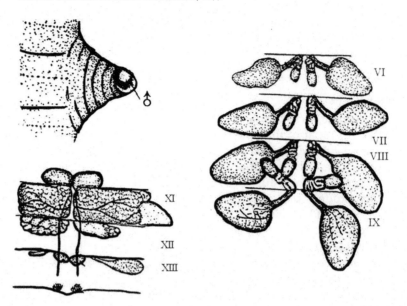

Color: Preserved specimens dark purplish blue on dorsum and clitellum, light grey on ventrum.

Deposition of types: The type is missing.

Etymology: The species epithet *"yuhsi"* was given in remembrance of the Taiwanese zoologist Dr. Yu-Hsi Wang, the former chief of the Department of Zoology, National Taiwan University, Taiwan.

Type locality: Yuantung Temple, Taibei, Taiwan.

Distribution in China: Taiwan (Taibei City, Hsinchu County).

Remarks: *Begemius yuhsi* (Tsai, 1964) is an anecic species, having permanent vertical burrows. It is active around the upper layer of soil or on the ground at night, but stays 30 cm or more below the ground in the soil during the day.

III. LAPTIO KINBERG, 1866

Megascolex Stephenson, 1930. *The Oligochaeta.* 837.
Laptio Gates, 1972. *Trans. Am. Phil. Soc.* 62(7):130–132.

Type species: *Laptio mauritii* Kinberg, 1866.

Diagnosis: Setae, the fewest in the middle and hinder parts of the body, numerous (more than 8) in each segment. Spermathecal pores usually 1–5 pairs, between segments IV and IX (the exceptions are constituted by the few cases where the pores are fused in the middle line, or where they are numerous on each side in each segment).

1 Gizzard in V, VI, or VII. Micronephridia throughout the body and meganephridia in addition in all the postclitellar segments. Prostates with branched system of ducts.

Global distribution: Sri Lanka and India; Australia (Queensland, New South Wales, Victoria, Tasmania, South Australia, and South-West Australia); New Caledonia; Norfolk Island; Vietnam. *L. mauritii* is peregrine all over the coasts and islands of the Indian Ocean, and across South-West Asia and the Malay Archipelago.

About 117 species.

There is 1 species from China.

241. *Laptio mauritii* (Kinberg, 1866)

Megascolex mauritii Gates, 1931. *Rec. Ind. Mus.* 33:361.
Megascolex mauritii Gates, 1932. *Rec. Ind. Mus.* 34(4): 374.
Megascolex mauritii Chen, 1938. *Contr. Biol. Lab. Sci. Soc. China (Zool).* 12(10):381–382.
Laptio mauritii Gates, 1972. *Trans. Am. Phil. Soc.* 62(7): 133–135.
Laptio mauritii Reynolds, 1994. *Megadrilogica.* 5(4):37.
Laptio mauritii Xu & Xiao, 2011. *Terrestrial Earthworms of China.* 206.

External Characters

Length 95–155 mm, width 3–6 mm. Segment number 157–201. Prostomium prolobous or 1/2 epilobous. First dorsal pore in region 10/11–12/13. Clitellum setae retained, from 13/14 or a postsetal portion of XIII to 17/18. Setal circles are present on all the clitellar segments. There is always a wide, midventral gap in each setal circle. On the preclitellar segments there is almost always a fairly wide, middorsal gap. On the postclitellar segments the middorsal gap is definite, usually fairly wide though slightly variable in width from segment to segment. Few specimen has very slight middorsal breaks on the preclitellar segments and almost no middorsal breaks on the postclitellar segments. Seta *b* on XIV and XIX is either just median to the male pore line or actually on the male pore line. Setae number 26–29(III), 40–51

(VIII), 38–50 (XII), 30–43 (XX). Male pores tiny, transverse slits on large, nearly circular, slightly protuberant poro-phores that dislocate 17/18 and 18/19 slightly. Genital markings absent. Spermathecal pores 3 pairs in 6/7/8/9, pores larger than female apertures. No genital papil-lae in this region. Female pores 1 pair, ventral in XIV.

Internal Characters

Gizzard in V. Intestine from XV, Typhlosole, quite insignificant. Seminal vesicles in IX and XII. Prostate glands confined to XVIII, ducts straight, about 2 mm long. Spermathecae 3 pairs in VI–VIII, ampulla is 2–4 times as long as the duct and is narrowed ectally. Spermathecal duct with transversely slit-like lumen, barrel-shaped, bul-bous entally, abruptly narrowed prior to entrance into the parietes; diverticulum paired, into the latter and median faces of the narrowed, coelomic portion of the duct, each diverticulum shorter than the duct and with a very short, slender stalk and ovoidal seminal chamber.

Color: Preserved specimen grayish, yellowish, brownish.
Deposition of types: Naturhistoriska Riksmuseet, Stockholm.
Etymology: The name of the species refers to the type locality.
Type locality: Mauritius.
Distribution in China: Hainan (Shamojue, Baoping), Hong Kong (Jiulong).

IV. METAPHIRE (SIMS AND EASTON, 1972)

Rhodopis Kinberg, 1867. *Annulata Nova*. 102.
Amyntas (part) Beddard, 1900. Proc. *Zool. Soc. London*. 1900:612.
Pheretima (*Pheretima*) (part) Michaelsen, 1928a. *Arkiv For Zoologl*. 20(2):8.
Metaphire Sims & Easton, 1972. *Biol. J. Linn. Soc*. 4:215.
Metaphire Xu & Xiao, 2011. *Terrestrial Earthworms of China*. 206–211.

Type species: *Metaphir javanica* Kinberg, 1867
Diagnosis: Megascolecidae with cylindrical bodies. Setae numerous, regularly arranged around each seg-ment. Clitellum annular, XIV–XVI. Male pores paired within copulatory pouches on XVIII, rarely XIX or XX. Female pore single, rarely paired, XIV spermathecal pores large, transverse slits, seldom small; paired, occa-sionally single or multiple, rarely single, between 4/5 and 9/10. Gizzard present between septa 7/8 and 9/10. Esophageal pouches absent. Intestinal caeca pre-sent, originating in or near XXVII. Testes holandric, rarely proandric or metandric. Prostate glands racemose. Cop-ulatory pouches present, often with stalked glands;

secretory diverticula absent. Ovaries paired in XIII. Sper-mathecae paired, rarely single or numerous. Meronephri-dial, nephridia absent from the spermathecal ducts.

Global distribution: Oriental realm from Japan south-wards through the Indo-Australasian archipelago to the rainforests of Australasia eastwards through the Ocea-nian realm.

There are 73 species from China.

Remarks: Species of this genus are separable from those of *Amynthas* by the presence of copulatory pouches and they differ from species of *Pheretima* by the absence of nephridia from the spermathecal ducts.

TABLE 26 Key to the Species Groups of the Genus *Metaphire* From China

1. Spermathecal pores at segment 2	
Spermathecal pores at intersegment 5	
2. Spermathecal pores 1 pair, at V .... *biforatum* group	
Spermathecal pores 3 pairs or more than 3 pairs 3	
3. Spermathecal pores 3 pairs, at V–VII .... *thecodorsata* group	
Spermathecal pores more than 3 pairs 4	
4. Spermathecal pores 4 pairs, at V–VIII .... *posthuma* group	
Spermathecal pores multitheca, at VI–VIII .... *multitheca* group	
5. Spermathecal pores 1 pair 6	
Spermathecal pores 2 pairs or more than 2 pairs 7	
6. Spermathecal pores 1 pair, at 5/6 .... *plesiopora* group	
Spermathecal pores 1 pair, at 7/8 .... *densipapillata* group	
7. Spermathecal pores 2 pairs 8	
Spermathecal pores more than 2 pairs 10	
8. Spermathecal pores 2 pairs, at 5/6/7 .... *exilis* group	
Spermathecal pores 2 pairs, behind 5/6/7 9	
9. Spermathecal pores 2 pairs, at 6/7/8 .... *glandularis* group	
Spermathecal pores 2 pairs, at 7/8/9 .... *schmardae* group	
10. Spermathecal pores 3 pairs 11	
Spermathecal pores 4 pairs, at 5/6/7/8/9 .... *bucculenta* group	
11. Spermathecal pores 3 pairs, at 4/5/6/7 .... *flavarundoida* group	
Spermathecal pores 3 pairs, behind 4/5/6/7 12	
12. Spermathecal pores 3 pairs, at 5/6/7/8 .... *anomola* group	
Spermathecal pores 3 pairs, at 6/7/8/9 .... *houlleti* group	

anomola Group

Diagnosis: *Metaphire* with spermathecal pores 3 pairs, in 5/6/7/8.

Global distribution: Oriental realm from Japan south-wards through the Indo-Australasian archipelago to the rainforests of Australasia eastwards through the Oceanian realm.

There are 3 species from China.

Remarks: Species of this genus are separable from those of *Amynthas* by the presence of copulatory pouches and differ from species of *Pheretima* by the absence of nephridia from the spermathecal ducts.

TABLE 27   Key to Species of the *anomola* Group From China

1. Spermathecal pores dorsal in position .... *Metaphire abdita*
    Spermathecal pores ventral in position 2

2. No genital papillae in male pore region .... *Metaphire birmanica*
    Genital markings, small, paired, disc-like, in XVII–XVIII or
    XIX .... *Metaphire anomala*

242. Metaphire abdita (Gates, 1935)

Pheretima abdita Gates, 1935a. *Smithsonian Mus. Coll.*
93(3):5–6.
Pheretima abdita Chen, 1936. *Contr. Biol. Lab. Sci. Soc.*
China (Zool). 11:292–293.
Pheretima abdita Gates, 1939. *Proc. U.S. Natn. Mus.*
85:415–418.
Pheretima abdita Chen, 1946. *West China. Border Res. Soc.*
(B). 16:136.
Metaphire abdita Sims & Easton, 1972. *Biol. J. Soc.* 4:237.
Metaphire abdita Xu & Xiao, 2011. *Terrestrial Earthworms*
of China. 211–212

External Characters

Length 80–140mm, width 3.5–6.0mm. Segment number 119–122. Prostomium proepilobous, small. First dorsal pore 12/13. Clitellum not occupying 3 full segments, beginning and ending a little distance from intersegmental furrows 14/15 and 16/17 respectively. No visible clitellar setae but with setal pits. Setae numerous and evenly distributed, ventral break indistinct, dorsal break about 1.5 *yz*. Setae number 51–55 (III), 60–68 (VI), 64–64 (VII), 60–68 (VIII), 48–60 (XXV); 34–38 (VI), 36–40 (VIII) between spermathecal pores, 12–16 between male pores. Male pores in a small lateral parietal invagination, enclosing medially a fleshy pad surrounded by several folds, a penis (0.8mm long) projecting out from the semilunar

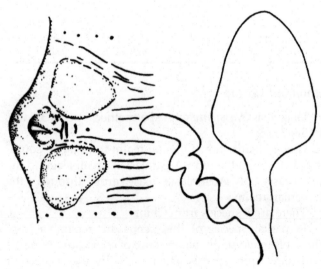

FIG. 237   *Metaphire abdita* (after Chen, 1936).

pouch. Genital papillae 2, paired in 17/18 and 18/19 respectively, slightly medial to male openings, each roundish and flattened, very large in size, about 1.2–1.5 mm in diameter. These very constantly present. Spermathecal pores rather small, 3 pairs in 5/6/7/8, rather dorsal in position, about 3/5 circumference apart ventrally. No genital papillae in this region. Genital markings paired, presetal on XVIII and XIX. Female pore single, medioventral in XIV.

Internal Characters

Septa 5/6–9/10 thickly muscular; 10/11/12/13 also muscular, especially 10/11, but not as thick as the anterior septa. Gizzard small in front of 8/9. Intestine swelling from XVI. Intestinal caeca simple, long, slender, with smooth margins, in XXVII, extending anteriorly to XXVI. First pair of hearts not observed. Testis sacs extremely large, testis of X annular, testis sacs of XI U-shaped or annular; not connected dorsally. Seminal vesicles of XI squeezed outward at lower side of intestine and included in the posterior testis sac; posterior pair of vesicles very large, in XII and XIII, without distinct dorsal lobes. Prostate glands paired in XVI–XIX. Accessory gland sessile, close to body wall. Spermathecae 3 pairs in VI–VIII, ampulla about 1.8mm wide, diverticulum with a short, muscular stalk, a middle portion more or less regularly bent back and forth in a zigzag fashion, and a terminal, ovoidal seminal chamber. Genital marking glands sessile, sometimes slightly protuberant into the coelom.

> Color: Preserved specimens grayish pale all over the body. Clitellum chocolate brown.
> Deposition of types: Smithsonian Institution
> Etymology: The specific name "*abdita*" means hidden or concealed in Latin and probably refers to the structure of the male pores.
> Type locality: Sichuan (Yibin).
> Distribution in China: Chongqing (Shapingba), Sichuan (Mt. Emei, Yibin).
> Remarks: *Metaphire abdita* is close to *Amynthas indicus* and *Metaphire gemella* (Gates, 1931) but is distinguished from both by the restriction of the male invaginations to the parietes, by the muscularity of septa 8/9/10, by the genital markings on XVIII and XIX, and by the 3 pairs of spermathecae.

243. Metaphire anomala (Michaelsen, 1907)

Pheretima anomala Michaelsen, 1907. *Mitt. Mus.*
Hamburg. 24:167.
Pheretima anomala Gates, 1972. *Trans. Am. Phil. Soc.*
62(7):166–168.
Metaphire anomala Xu & Xiao, 2011. *Terrestrial*
Earthworms of China. 214.

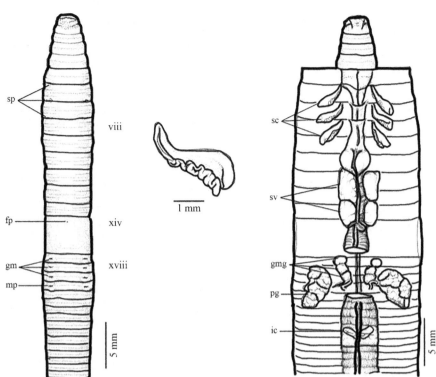

FIG. 238 *Metaphire anomala* (after Bantao-wong et al., 2011).

External Characters

Length 80–200 mm, width 3–7 mm. Segment number 119–130. Prostomium epilobous. First dorsal pore 12/13. Clitellum XIV–XVI. Setae small, closely spaced, setae number 60–68 (III), 90–96 (VIII), 78–95 (XII), 81–90 (XIII); 17–22 (VI), 17–23 (VII) between spermathecal pores, 16–18 (XIX), 15–21 (XX) between male pores. Male pores in XX, minute, each at center of a small disc in a parietal invagination with transversely slit-like equatorial aperture, the invagination eversible to a columnar protuberance with male porophore at its distal end. Genital markings, small, paired, disc like, each with obvious central pore and in a parietal invagination with transversely slit-like equatorial aperture, invaginations eversible to columnar protuberance with discs at distal ends, in XVII–XIX. Spermathecal pores 3 pairs in 5/6/7/8, each pore minute, at center of a small disc in a small parietal invagination with a transversely slit-like aperture. Female pore single, medioventral in XIV.

Internal Characters

Septa 8/9/10 aborted, none very thickly muscular. Intestine from XV. Intestinal caeca simple, margins with septal incisions, in XXVII, extending anteriorly to XXI. Lateral hearts in X–XII. Holandric. Testis sacs unpaired and ventral. Seminal vesicles, large, in XI and XII. Prostate glands large, extending through some or all of XVI–XVIII, ducts 4–7 mm long, muscular, looped. Spermathecae 3 pairs in VI–VIII, ampulla duct slender and nearly as long as the indistinctly demarcated ampulla, diverticulum from anterior face of duct in parietes, longer than axis, with short and slender stalk, looped or twisted middle region in which lumen is gradually widened, and a club-shaped seminal chamber. Genital marking glands, mushroom-shaped, erect in coelom, with soft head and straight, spindle-shaped, muscular stalks.

Color: Preserved specimen red, faint on dorsum, or lacking behind the clitellum.
Type locality: Botanical Gardens, Sibpur, Bengal.
Deposition of types: Indian Museum and Hamburg Museum.
Distribution in China: Yunnan (Tengyue).

244. *Metaphire birmanica* (Rosa, 1888)

Pheretima birmanica Gates, 1972. *Trans. Am. Phil. Soc.* 62(7):172–173.
Metaphire birmanica Bantaowong, Chanabun, Tongkerd et al., 2011. *Tropical Natural History.* 11(1):55–69.
Metaphire birmanca Xu & Xiao, 2011. *Terrestrial Earthworms of China.* 218.

External Characters

Length 100–160 mm, width 4–7 mm. Segment number 112. Prostomium epilobous, tongue open. First dorsal pore 12/13. Clitellum XIV–XVI, annular, without externally recognizable setae. Setae about 70 per segment.

FIG. 239 *Metaphire birmanica* (after Bantaowong, Chanabun, Tongkerd, et al., 2011).

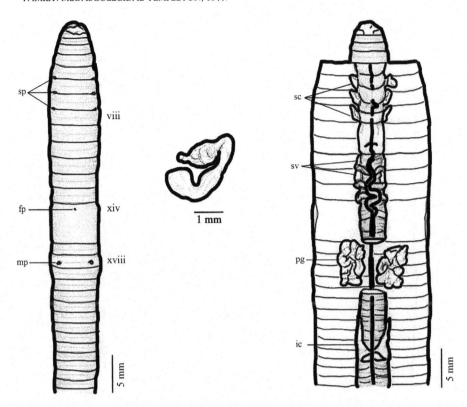

Male pores in XVIII, small, each in a longitudinal groove on a conical penial body in a small copulatory pouch with an equatorial aperture. Spermathecal pores 3 pairs in 5/6/7/8, each pore minute, transverse slits, on a tubercle on the top of a small parietal invagination, about 1/2 circumference apart ventrally. No genital papillae in male pore region and spermathecal pore region. Female pore single, medioventral in XIV.

Internal Characters

Septa 8/9/10 aborted, 5/6/7 thickened. Intestine from XV. Intestinal caeca manicate, dorsalmost of 3–6 secondary caeca the longest, in XXVII, extending anteriorly to XXIII. Last hearts in XIII. Holandric. Ovaries paired in XIII. Testis sacs paired and ventral. Seminal vesicles small, in XI and XII. Prostate glands in XVI–XX, ducts each in a U-shaped loop. Spermathecae 3 pairs in VI–VIII, ampulla duct seemingly widened and thickened within the parietes but not distinguishable externally from the parietal invagination, with narrow lumen ental to diverticular junction, diverticulum from duct in parietes, longer than the main axis, looped in a more or less regularly zigzag manner.

Color: Preserved specimen brownish to slate on dorsum.
Deposition of types: Genoa Museum, Genoa, Italy.
Etymology: The name of the species refers to the type locality.

Habitats: Soils, red, dark, rich, gravelly, in fields, manured gardens, ridges of paddy fields, open areas under grass, bamboo, bush, and hill jungles. Mud. Manure.
Type locality: Burma (Bhamo).
Distribution in China: Yunnan (Lancang).
Remarks: *M. birmanica* distinguish with *Amynthas defecta* (Gates, 1930) as: differentiation of a discoidal porophore around each spermathecal pore; retraction of each porophore into a small parietal invagination; invagination of the male porophore into the coelomic cavity as an eversible copulatory chamber, and elevation from the top of copulatory chamber of the small porophore into a conical penial body.

biforatum Group

Diagnosis: *Metaphire* with spermathecal pores 1 pair, in V.

Global distribution: Oriental realm from Japan southwards through the Indo-Australasian archipelago to the rainforests of Australasia eastwards through the Oceanian realm.

There is 1 species from China.

Remarks: Species of this genus are separable from those of *Amynthas* by the presence of copulatory pouches and differ from species of *Pheretima* by the absence of nephridia from the spermathecal ducts.

245. *Metaphire biforatum* Tan & Zhong, 1987

Metaphire biforatum Tan & Zhong, 1987. *Act Zootaxonomica Sinica.* 12(2):128–129.
Metaphire biforatum Xu & Xiao, 2011. *Terrestrial Earthworms of China.* 217–218.

External Characters

Length 121–142 mm, width 5–6.2 mm, segment number 97–115. Prostomium 1/3–1/4 proepilobous. First dorsal pore 12/13. Clitellum XIV–XVI, annular, with setae on ventral side, with dorsal pores. Setae, closely spaced on both dorsum and ventrum, setae number 77–85 (III), 86–101 (V), 97–116 (VII), 95–123 (IX), 85–90 (XIII), 80–85 (XX); 24–27 (V) between spermathecal pores, 8–11 between male pores. Male pores in XVIII, ventrally, within copulatory pouches which are surrounded by folds. Porophores small, round, the base covered by the skin folds. Spermathecal pores 1 pair, dorsally and on posterior edge of V, about 1/4–2/7 circumference apart ventrally. Female pore single, medioventral in XIV.

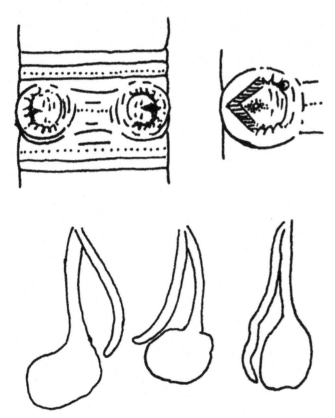

FIG. 240 *Metaphire biforatum* (after Tan & Zhong, 1987).

Internal Characters

Septa 5/6–9/10 very thickly muscular, 10/11 also muscular, but not as thick as anterior septa. Intestine swelling from XVI. Intestinal caeca simple, long sac-shaped, in XXVII, extending anteriorly to XXIII, smooth. Testis sacs 2 pairs, the anterior pair in X, developed, upward coiled and united on dorsum of intestine, covered by the septum 10/11, slightly kidney-shaped, an incision located about 1/3 of the way from anterior edge to distal end; the posterior pair in XI, slightly petal-shaped, the first pair of seminal vesicles included in it; each pair connected but not directly communicated on ventral side. Seminal vesicles 2 pairs, in XI and XII, lump-like, the junction with testis sacs slender, stalk-like, dorsal lobe conspicuous, slightly small, slightly round. Prostate glands developed, in XVII–XIX, or 1/2 XVI–1/2 XX or XVI–XX; each consists of 2 lobes, each lobe separated into numerous lobules. Prostatic duct, U-shaped, the ectal half thicker than the ental half, accessory glands disappearing from the position of prostate glands passing into the parietes. Spermathecae 1 pair in VI, ampulla oblate or long sac-shaped or irregular, duct thick, without particularly enlarged seminal chamber.

Color: Preserved specimens brownish, no difference between dorsal and ventral sides; a black brown stripe along dorsomedian line to the end of body behind XII, width about 0.8 mm. Clitellum dark brown.
Deposition of types: Hunan Agricultural University.
Etymology: The specific name *"biforatum"* is a compound word with *"bi"* meaning twice or double in Latin and *"foratum"* meaning holes in Latin. This word refers to the number of spermathecal pores.
Type locality: Hunan (NanXian, Yiyang City).
Distribution in China: Hunan (NanXian, Yiyang City).
Remarks: This species is similar to *Metaphire exiloides* (Chen, 1936) but differs from it in having 1 pair of spermathecal pores, dorsal in position, on the posterior edge of segment V and having no Z-shaped diverticulum.

bucculenta Group

Diagnosis: *Metaphire* with spermathecal pores 4 pairs, in 5/6/7/8/9.
Global distribution: Oriental realm from Japan southwards through the Indo-Australasian archipelago to the rainforests of Australasia eastwards through the Oceanian realm.
There are 18 species from China.
Remarks: Species of this genus are separable from those of *Amynthas* by the presence of copulatory pouches, and they differ from species of *Pheretima* by the absence of nephridia from the spermathecal ducts.

TABLE 28 Key to the *bucculenta* Group From China

1. Spermathecal pores dorsal in position 1
   Spermathecal pores ventral in position 3

2. Spermathecal pores, above the lateral midline, about 0.55 circumference apart ventrally. No genital papillae in the spermathecal region .... *Metaphireta hanmonta*
   Spermathecal pores about 0.63 circumference apart ventrally. Genital papillae on laterodorsum, 0.1–0.2 mm dorsal to spermathecal pore adjacent to intersegmental furrows, paired in 2 longitudinal rows in VII–IX with varying numbers and locations, each papilla circular, flat-topped with concave center, about 0.2 mm in diameter, surrounded by a circular fold .... *Metaphire yeni*

3. With a pair of genital papillae in front of the setal line of VI, VII, VIII and IX .... *Metaphire heteropoda*
   No genital papillae around spermathecal pores region 4

4. With genital papillae in male regions 5
   No genital papillae in male regions 10

5. Genital papillae more than 2 6
   Genital papillae 2 7

6. Male pores surrounded by 1–3 shallow skin folds, with 1–8 genital papillae at medioanterior and/or medioposterior portion outside the skin fold. Postclitellar papillae similar in shape and size to those in the preclitellar region, also in transverse patches between male pores; presetal papillae 20–61 in XVIII, 8–59 in XIX, and 0–1 in XX; postsetal papillae 0–8 in XVII, 0–5 in XVIII, and 0–3 in XX .... *Metaphire pavimentus*
   Genital papillae paired in XIX and XX, single in XXI, arranged longitudinally, each papilla on setal line, large, occupying nearly entire segment, with a slightly concave center and surrounded by a circular fold .... *Metaphire puyuma*

7. Genital markings paired, presetal on XVIII ..... *Metaphire bucculenta*
   Genital markings paired, embedded in the copulatory pouch, or ventral to aperture of the pouch 8

8. A small genital pad present in the front of the opening of the copulatory pouch .... *Metaphire feijani*
   Male pores on semilunar sac 9

9. Male pores on a small tubercle situated at the bottom of a lateral parietal invagination, its aperture semilunar in shape; a large and round papilla constantly present medial to each pore tubercle usually exposed in part, its diameter occupying three-fifths of the segmental length .... *Metaphire fangi*
   Genital markings paired ventral to aperture of the pouch 10

10. Male pores not on semilunar aperture 11
    A pair of genital papillae situated just in setal line and slightly medial to male pores in segment xvii and xix respectively, each papilla round in shape with depressed top, slightly larger than male papilla .... *Metaphire posthuma*

11. A large and round papilla immediately medial to the pore area .... *Metaphire bipapillata*

246. *Metaphire bipapillata* (Chen, 1936)

Pheretima bipapillata Chen, 1936. *Contr. Biol. Lab. Sci. Soc. China (Zool).* 11:286–288.
Pheretima bipapillata Chen, 1946. *J. West China. Border Res. Soc.* 16:136.

Metaphire bipapillata Sims & Eason, 1972. *Biol. J. Soc.* 4(3):238.
Metaphire bipapillata Xu & Xiao, 2011. *Terrestrial Earthworms of China.* 218.

External Characters

Length 110–210 mm, width 3.5–5.0 mm, segment number 100–141. Prostomium 1/3 epilobous, with longitudinal furrow in middle. First dorsal pore 12/13. Clitellum XIV–XVI, annular and glandular, setae absent. Setae on VI–VII more prominent, slightly longer in length and wider in intervals, those on other parts of body fine, slightly closer ventrally, $aa = 1.2\ ab$, $zz = 2\ yz$; setae number 24–32 (III), 38–44 (VI), 38–46 (VIII), 49–66 (XXV); 13–1 (VI), 12–14 (VIII) between spermathecal pores; 10–12 between male pores. Male pores 1 pair, represented by 2 shallow semilunar pouches on ventrolateral sides of XVIII, each pore on a glandular area at bottom of the pouch, with a large and round papilla immediately medial to the pore area. The papilla largely exposed and ventral to aperture of the pouch; its diameter about the length of semilunar slit of the pouch. Distance between 2 pores about 1/3 circumference apart ventrally but that between 2 papillae about 1/6 circumference of segment. Spermathecal pores 4 pairs, in 5/6–8/9, superficial and intersegmental, about 2/7 circumference apart ventrally. No genital papillae in this region. Female pore single, medioventral in XIV.

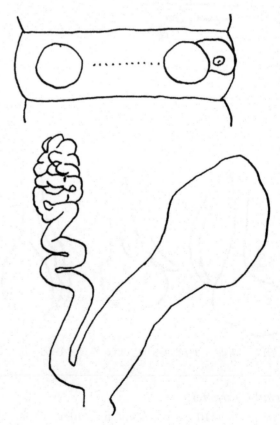

FIG. 241 *Metaphire bipapillata* (after Chen, 1936).

Internal Characters

Septa 8/9/10 absent, 5/6/7/8 muscular and equally thickened, 10/11/12 less thickened, 12/13 and succeeding septa thin and slightly muscular. Gizzard in IX large, sometimes elongate. Intestine swelling from XVI. Intestinal caeca simple, weakly constricted on both sides, extending anteriorly to a small part of XXIII. Lateral hearts 4 pairs, in X–XIII, first pair not stout. Testis sacs in X well-distended, connected medially, in XI broadly united. Seminal vesicles in XI very large, with a small dorsal lobe, anterodorsal in position, those in XII smaller, about 1/3 of the former but equally large in cotypes. Prostate glands large, in XV–XXI, with slender and long lobules, about 4 mm long; each with a long duct, about 9 mm long, coiled at middle. A round patch of accessory gland near its base, which is sessile and smooth on surface, opening externally through the large papilla referred to above. Spermathecae 4 pairs, in VI–IX, gradually enlarged posteriorly; ampulla heart-shaped and rather small, about 1 mm wide, its duct about 2 mm long, 0.8 mm wide; ampulla fairly large, about 3 mm long, 1.5 mm wide, sharply marked off from its duct; diverticulum long, its ental portion compactly coiled, ectal portion about 1 mm long, straight, serving as a duct entering the main duct close to body wall; the portion between entrance of diverticular duct and body wall suddenly becoming slender.

Color: Preserved specimens chocolate brown or dark slaty on dorsum, pale on ventrum.

Deposition of types: Previously deposited in the Museum of the Biological Laboratory of the Science Society of China, Nanjing. The types were destroyed during the war in 1937.

Etymology: The species name "bipapillata" was given to describe the 2 papillae medial to the male pouch that are constantly present in this species ("bi" means 2 in Latin).

Type locality: Chongqing.

Distribution in China: Chongqing (Shapingba, Baixi, Beibei).

Remarks: This species is distinct. It probably lives in sandy soil. Most specimens in the present collection were obtained from the sandy bank of the Jialing River at Chongqing.

247. Metaphire bucculenta (Gates, 1935)

Pheretima bucculenta Gates, 1935a. *Smithsonian Mus. Coll.* 93(3):7.
Pheretima bucculenta Chen, 1936. *Contr. Biol. Lab. Sci. Soc. China (Zool).* 11:294–295.
Pheretima bucculenta Gates, 1939. *Proc. U.S. Natn. Mus.* 85:425–427.
Pheretima bucculenta Chen, 1946. *J. West China. Border Res. Soc. (B).* 16:136.

Pheretima bucculenta Chen, 1959. *Fauna Atlas of China: Annelida.* 14.
Metaphire bucculenta Sims & Easton, 1972. *Biol. J. Soc.* 4:238.
Metaphire bucculenta Xu & Xiao, 2011. *Terrestrial Earthworms of China.* 220–221.

External Characters

Length 110–210 mm, width 3.5–7 mm, segment number 127–141. Prostumium 2/3 epilobous. First dorsal pore 12/13. Clitellum XIV–XVI, annular, setae absent. Setae on anterior ventral side of III–VIII larger and more widely spaced; $aa=1.2$ ab, $zz=1.5$ yz; setae number 40 (III), 55 (VI), 64 (VIII), 60 (XII), 77 (XXV); 14–23 (VIII) between spermathecal pores, 11–17 between male pores. Male pores on tiny, conical tubercles in the dorsalmost portions of deep parietal invaginations with longitudinally slit-like apertures. Genital markings paired, presetal on XVIII. Spermathecal pores minute and superficial, 4 pairs, in 5/6–8/9, intersegmental, each with a small tubercle closely placed on posterior side. No genital papillae around spermathecal region. Female pore single, medioventral in XIV.

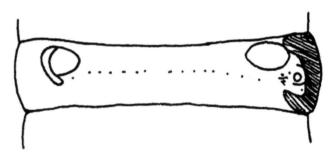

FIG. 242 *Metaphire bucculenta* (after Chen, 1936).

Internal Characters

Septa 8/9/10 absent, 6/7/8 and 10/11/12 thickly muscular, 12/13 muscular. Intestine from XV. Intestinal caeca elongate, simple, with 6–8 very definite but short and stumpy, rather broad, lobes on the ventral margin, length of lobes less than dorsoventral diameter of the main portion of the sac. Testis sacs of X and XI unpaired and ventral. Seminal vesicles of XI and XII are medium-sized vertical bodies, in contact transversely over the dorsal blood vessel. The prostate glands extend through XVII–XVIII. Prostatic duct about 4 mm in length, bent into a U shape, the ectal half thicker than the ental half. Genital marking glands sessile but protuberant through the parietes into the coelom. Accessory glands in prostate region in a large compact mass (about 1.5 mm in diameter), with very short ducts, each about 1.5 mm in diameter. Spermathecae and its diverticulum comparatively small, without seminal content; 4 pairs in VI–IX, grape-like nodules found at inner surface of body wall, also tubercles found on 1 spermathecal ampulla.

Color: Preserved specimens puce on dorsum, grey on ventrum, slightly taupe on anterior end.

Deposition of types: Smithsonian Institution.

Etymology: The specific name "*bucculenta*" means full cheeked in Latin and probably refers to the shape of the male pore region.

Type locality: Sichuan.

Distribution in China: Chongqing (Beipei, Shapingba), Sichuan (Yibin), Gansu.

Remarks: Distinguished from other octothecal (having 8 spermathecae) Chinese species of *Amynthas* by the combination of superficial spermathecal pores and deeply invaginated male pores.

The male parietal invaginations are very similar to those of *Metaphire tschiliensis* (Michaelsen, 1928a), *Metaphire praepinguis* (Gates, 1935a), and *Metaphire paeta* (Gates, 1935a).

Metaphire fangi (Chen, 1936) is distinguished from *Metaphire bucculenta*, according to Chen, by the larger size of the genital marking in the male pore invagination, the larger size of the male invagination, the stout hearts of X, and the coiling of the spermathecal diverticulum. Slight differences in size of genital markings or of hearts, in depth of the male pore invaginations (even if they exist) as well as coiling of spermathecal diverticulum are not acceptable criteria of specific distinctness in the genus *Metaphire*.

248. *Metaphire bununa* Tsai, Shen & Tsai, 2000

Metaphire bununa Tsai, Tsai & Liaw, 2000. *J. Nat. Hist.* 34:1736–1738.
Metaphire bununa Chang, Shen & Chen, 2009. *Earthworm Fauna of Taiwan.* 104–105.
Metaphire bununa Xu & Xiao, 2011. *Terrestrial Earthworms of China.* 221.

External Characters

Length 255–352 mm, clitellum width around 10.6 mm, segment number 189–221. Number of annuli per segment 3 in IV–VI, 5–7 in VIII–XIII, and 5 in body segments after XVII. Prostomium prolobous. First dorsal pore 12/13. Clitellum XIV–XVI, annular, length around 10.0 mm, dorsal pore absent, setae absent. Setae number 103–111 (V), 114–158 (VII), 119–145 (XX), 19–29 between male pores. Male pores paired in XVIII, lateroventral, about 0.25 circumference apart ventrally, each pore C-shaped with swollen and folliculated edge surrounded with circular folds, extending to the intersegmental furrows of 17/18 and 18/19, an oval pad situated behind the setal line of XVII, close to the anterior end of the male pore area, linked to the male aperture through a seminal groove. Genital papillae absent in the male pore area. Spermathecal pores 4 pairs in 5/6–8/9, 0.36 circumference apart ventrally. No genital papillae in the preclitellar region. Female pore single, medioventral in XIV.

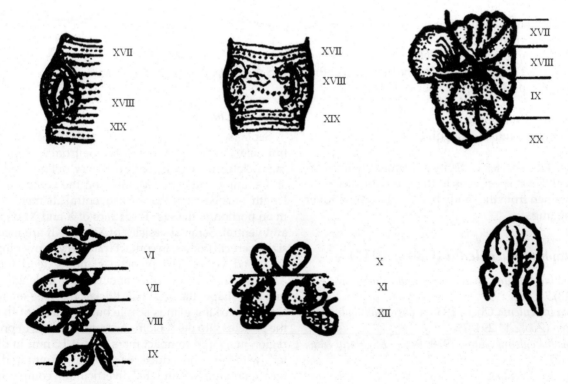

FIG. 243   *Metaphire bununa* (after Tsai et al., 2000a).

Internal Characters

Septa 8/9/10 absent, 5/6/7/8 thickened, 10/11/12/13 greatly thickened. Gizzard in VIII–X. Intestine swelling from XV. Intestinal caeca paired in XXVII, simple, extending anteriorly to XXVI. Lateral hearts in IX, XII and XIII. Spermathecae 4 pairs in VI–IX, ampulla elongated, peach-shaped, with a slender stalk, diverticulum with a long stalk, with straight proximal portion and coiled or twisted distal portion and a small seminal chamber. Nephridia tufted, attached to the postsegmental septa, surrounding the segmental chambers in V and VI. Ovaries paired in XIII, medioventral, close to the 12/13 septum. Testis sacs 1 pair in X, oval, smooth, medioventral in front of 10/11. Seminal vesicles paired in XI, large, each with a folliculate dorsal lobe. Prostate glands paired in XVIII, large, lobular, extending anteriorly to XVII and posteriorly to XX.

Color: Preserved specimen dark purplish blue on dorsum, light grey on ventrum, grayish brown on clitellum.
Deposition of types: Taiwan Endemic Species Research Institute, Jiji, Nantou, Taiwan. Holotype: coll. no. 1998–32.
Etymology: The species name *"bununa"* was given with reference to the Bunun Tribe of the aboriginal people of Nantou, Taiwan.
Type locality: Mt. Jilong between Jiji and Zhongliao, Nantou County, Taiwan.
Distribution in China: Taiwan (Nantou).

249. *Metaphire fangi* (Chen, 1936)

Pheretima fangi Chen, 1936. *Contr. Biol. Lab. Sci. Soc. China (Zool).* 11:275–278.
Pheretima fangi Chen, 1946. *J. West China. Border Res. Soc. (B).* 16:136.
Metaphire fangi Sims & Easton 1972. *Biol. J. Soc.,* 4:238.
Metaphire fangi Xu & Xiao, 2011. *Terrestrial Earthworms of China.* 226–227.

External Characters

Length 100–135 mm, width 4–5 mm, segment number 112–139. Postclitellar segments considerably shorter, about half of longest preclitellar segment; segment XIV and those thereafter (except XVIII) becoming short. Clitellum equal to 2.5 preclitellar and 4 postclitellar segments. Prostomium 1/2 epilobous. First dorsal pore 11/12, minute, 12/13 very distinct. Clitellum in XIV–XVI, annular, smooth and setae absent, glandular portion not swollen. Setae on II–VIII or IX longer, those on III–VI wider in intervals, shedding in some places; no difference between dorsal and ventral ones; those on postclitellar segments more densely spaced; $aa = 1.5–2$ ab, $zz = 2–2.5$ yz; those on postclitellar segments more numerous; setae number

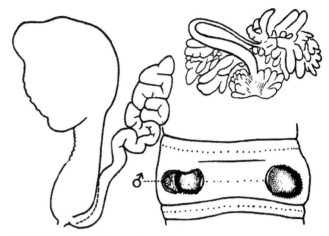

FIG. 244　*Metaphire fangi* (after Chen, 1936).

24–28 (III), 34–37 (VI), 34–40 (VIII), 48–56 (XXV); 11–12 (V), 12–13 (VI), 13–14 (VII), 14–16 (VIII) between spermathecal pores; 8–16 between male pores. Male pores 1 pair, in XVIII ventrally, about 1/3 of circumference apart ventrally; each on a small tubercle situated at the bottom of a lateral parietal invagination, its aperture semilunar in shape; a large and round papilla constantly present medial to each pore tubercle usually exposed in part, its diameter occupying three-fifths of segmental length; No other papillae in other parts of body. Spermathecal pores 4 pairs, in 5/6–8/9, intersegmental in position, superficial, appearing eye-like, about 2/7 circumference apart ventrally. Female pore single, medioventral in XIV.

Internal Characters

Septa 8/9/10 absent, 4/5 thin, 5/6/7/8 thickened, 10/11/12 very thin and membranous. Gizzard elongate barrel-shaped, in 1/2 IX and X. Intestine swelling in XVI. Intestinal caeca simple, slender and pointed, in XXVII, extending anteriorly to XXII, ventral edge with low teeth. Lateral hearts 4 pairs, first pair in X stout. Testis sacs in X communicated, in a transverse sac, or thick V shape, posterior pair in XI united. Seminal vesicles very large, in XI and XII, granular on surface, about 3 mm in anteroposterior length, each with a small dorsal lobe. Prostate glands in XVI–XX, with elongate lobules, each with a long duct, ental half much more slender, about 8 mm long. Accessory glands sessile in 2 closely approximate patches at median side of the base of prostatic duct, in XVII– XIX, about 4 mm in anteroposterior length. Spermathecae 4 pairs, in VI–IX; ampulla heart-shaped or subspherical, distinct from its broad duct which is about equal in length to the ampulla, enlarged at ectal end; diverticulum shorter (rarely longer) than main pouch (ampulla & duct), ental half closely twisted in a zigzag fashion, coiled on same plane, its limbs inseparable, about 7–11 coils; its duct nearly as long as ampullar duct. Ampulla 2mm wide, 2.2mm long; ampullar duct 2.2mm long and 1mm at its widest part. Diverticular duct about 0.26mm in width.

Color: Preserved specimens faint, light chocolate on dorsum.

Deposition of types: Previously deposited in the Museum of the Biological Laboratory of the Science Society of China, Nanjing. The types were destroyed during the war in 1937.

Etymology: This species was named after Dr. Fang Bingwen who collected the specimens and was also a brilliant animal taxonomist.

Type locality: Sichuan (Xufu).

Distribution in China: Chongqing (Shapingba, Baixi, Beipei), Sichuan (Xufu).

Remarks: This species is closely related to *M. bucculenta* (Gates, 1935a). However, it differs mainly in having a large-sized papilla medial to each male pore and a greatly twisted diverticulum. In the type of *M. bucculenta*, many characters are found that are similar to *Amynthas pingi* (Stephenson, 1925), such as the spermathecal papillae, the diverticulum, and the setae (except the male pore region). In that single specimen, there are numerous grape-like structures attached to the inner surface of the body wall, which are also found on a spermathecal ampulla. This shows that the particular specimen is highly infected with parasites. Of course, the present species is distinct from both *Metaphire bucculenta* and *Amynthas pingi* in its coiled diverticulum, stout hearts in X, larger male pouch with a larger papilla, and many other characters.

250. *Metaphire feijani* Chang & Chen, 2004

Metaphire feijani Chang & Chen, 2004. *Taiwania.* 49(4):219–224.
Metaphire feijani Chang, Shen & Chen, 2009. *Earthworm Fauna of Taiwan.* 108–109.

Metaphire feijani Xu & Xiao, 2011. *Terrestrial Earthworms of China.* 227–228.

External Characters

Length 215–310mm, clitellum width 8.0–12mm, segment number 95–140. Number of annuli per segment 3 in VI–IX, 5 in X–XIII, 3 in body segments behind XVII. Prostomium prolobous. First dorsal pore 12/13. Clitellum XIV–XVI, annular, length 10.2–11.0mm, dorsal pore absent, setae absent. Setae number 76–96 (VII), 101–104 (XX), 20–22 between male pores. Male pores small, paired, situated on setal line close to lateral border of XVIII, about 0.26 circumference apart ventrally, each copulatory pouch compressed to the outer body wall, surrounded by 2–6 circular folds, laterally bordered by a thick skin lip, the opening of copulatory pouch forming a split parallel to the body axis, facing the medioventral line, male aperture inconspicuous, embedded in the copulatory pouch, a small genital pad present in front of the opening of the copulatory pouch. Spermathecal pores 4 pairs in 5/6–8/9, invisible from outside, ventral, 0.27–0.35 circumference apart ventrally. No genital papillae in the preclitellar region. Female pore single, medioventral in XIV.

Internal Characters

Septa 9/10 absent, 5/6/7/8 thickened, 8/9 thin, 10/11–13/14 greatly thickened. Gizzard in VIII–X. Intestine swelling from XV. Intestinal caeca paired in XXVII, simple, extending anteriorly to XXVI. Lateral Hearts in X–XIII. Testis sacs 1 pair in X, oval, smooth, medioventral in front of 10/11. Seminal vesicles paired in XI, large, each with a folliculate dorsal lobe. Prostate glands paired in XVIII, large, lobular, extending anteriorly to XVII and posteriorly to XIX. Spermathecae 4 pairs in VI–IX; ampulla large, about 2.8–4.2mm in length, with a stalk about 1.4–2.4mm in length; diverticulum short, around the middle of spermathecae, with a small oval seminal chamber on the tip, stalk long, tightly coiled, forming a short and thick appearance.

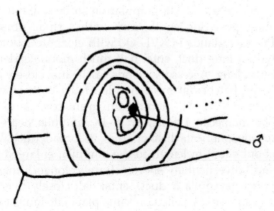

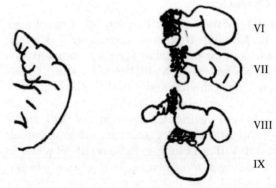

VI

VII

VIII

IX

FIG. 245   *Metaphire feijani* (after Chang & Chen, 2004).

Color: Live specimens bluish brown with metallic luster on dorsum, light reddish brown on ventrum. Preserved specimens purplish brown on dorsum, light brown on ventrum.

Deposition of types: Institute of Zoology, National Taiwan University, Taibei, Taiwan.

Etymology: The species epithet "*feijani*" was given in remembrance of the Taiwanese evolutionary biologist Dr. Fei-Jan Lin.

Type locality: Taiwan (Majia, Pingdong).

Distribution in China: Taiwan (Pingdong).

Remarks: *M. feijani* is a member of the *M. formosae* species group.

251. *Metaphire glareosa* Tsai, Shen & Tsai, 2000

Metaphire glareosa Tsai, Tsai & Liaw, 2000. *J. Nat. Hist.* 34:1738–1740.
Metaphire glareosa Chang, Shen & Chen, 2009. *Earthworm Fauna of Taiwan.* 112–113.
Metaphire bununa glareosa Xu & Xiao, 2011. *Terrestrial Earthworms of China.* 221–222.

External Characters

Length 204–330mm, clitellum width around 10.0mm, segment number 124–155. Prostomium epilobous. Setae number 46 (V), 61–90 (VII), 81–91 (XX), 6–27 between male pores. First dorsal pore 12/13. Clitellum XIV–XVI, annular, length around 10.0mm, dorsal pore absent, setae absent. Male pores paired in XVIII, lateroventral, each pore C-shaped with swollen and folliculated edge surrounded by circular folds, extending to the intersegmental furrows of 17/18 and 18/19, an oval pad situated behind the setal line of XVII, close to the anterior end of the male pore area, linked to the male aperture through a seminal groove. Spermathecal pores 4 pairs in 5/6–8/9. No genital papillae in the preclitellar region. Female pore single, medioventral in XIV.

Internal Characters

Septa 8/9/10 absent, 5/6/7/8 thickened, 10/11/12/13 greatly thickened. Gizzard in VIII–X. Intestine swelling from XV. Intestinal caeca paired in XXVII, simple, extending anteriorly to XXV. Lateral hearts in X–XIII. Nephridia tufted, attached to the postsegmental septa, surrounding the segmental chambers in V and VI. Ovaries paired in XIII, medioventral, close to the 12/13 septum. Testis sacs 1 pair in X, oval, smooth, medioventral in front of 10/11. Seminal vesicles paired in XI, large, each with a folliculate dorsal lobe. Prostate glands paired in XVIII, large, lobular, extending anteriorly to XVII and posteriorly to XIX. Spermathecae 4 pairs in VI–IX; ampulla elongated, peach-shaped, with a slender stalk; diverticulum with a long stalk, with straight proximal

portion and coiled or twisted distal portion and a small white seminal chamber.

Color: Preserved specimen purplish blue on dorsum, light grey on ventrum.

Deposition of types: Previously deposited in Taiwan Endemic Species Research Institute, Jiji, Nantou, Taiwan. The types were destroyed in a strong earthquake in 1999.

Etymology: The name "*glareosa*" is Latin adjective meaning "gravelly". It was given to indicate the habitat of the type locality.

Type locality: Jhihben Forest Recreation Park, Jhihben, Taidong, Taiwan.

Distribution in China: Taiwan (Taidong).

Remarks: *M. glareosa* is a member of the *M. formosae* species group. It lives in the mountains where the vegetation is broadleaf forest and can be found in both virgin and secondary forests. *M. Glareosa* is an anecic species, having permanent vertical burrows. It is active around the upper layer of soil or on the ground at night, but stays 30 cm or more below the ground in the soil during the day.

252. *Metaphire heteropoda* (Goto & Hatai, 1898)

Perichaeta heteropoda Goto & Hatai, 1898. *Annot. Zool. Jap.* 2:69–70.
Metaphire heteropoda Sims & Easton, 1972. *Biol. J. Soc.* 4:237.
Metaphire heteropoda Xu & Xiao, 2011. *Terrestrial Earthworms of China.* 232.

External Characters

Length 100mm, width 4mm. Segment number 72. All the segments except the first and the last of the same width. First dorsal pore 10/11. Clitellum XIV–XVI, setae absent. Setae of segments II–XIII thicker and longer; 32 in the spermathecal segments; 12 between male pores. Male pores in XVIII, margin of the pores slightly elevated; without genital markings in male pore region. Spermathecal pores 4 pairs in 5/6–8/9, with a pair of genital papillae in front of the setal line. Female pore single, medioventral in XIV.

Internal Characters

Septa 8/9/10 absent, 5/6/7/8 and 10/11–15/16 thickened. Gizzard in VIII–IX. Intestine swelling from XVII. Intestinal caeca simple, in XXVI, extending anteriorly to XXIII. Last heart in XIII. Testes in X, XI. Seminal vesicles in XI, XII. Prostate gland absent; terminal bulb present situated a little in front of the external male pore. Spermathecae 4 pairs in VI–VIII, with diverticula which are not convoluted but with the blind end simply enlarged.

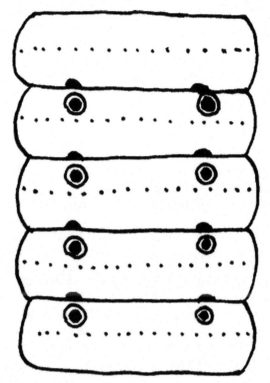

FIG. 246　*Metaphire heteropoda* (after Goto & Hatai, 1898).

Clitellum annular, not appearing, with setae all around the segments XIV–XVI. Setae small and more or less evenly distributed; setae number 37 (VI), 39 (VIII), 41 (XII), 49 (XXV); 16 (VI), 16 (VIII), between spermathecal pores, 10 between male pores. The male pore are tiny slits on the roof and toward the median side of transversely slit-like depressions. The margin of the slit is smooth and glistening; external to the smooth circumferential lip there are several concentric circumferential furrows. No genital markings. Spermathecal pores in parietal invaginations with transversely slit-like apertures, 4 pairs, in 5/6–8/9. Female pore single, medioventral in XIV.

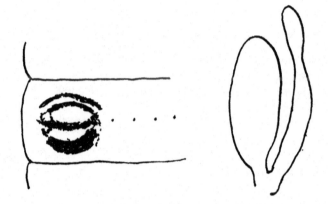

FIG. 247　*Metaphire ignobilis* (after Chen, 1936).

Color: Preserved specimen brown, yellowish on clitellum.
Deposition of types: Smithsonian Institution.
Etymology: The specific name "*heteropoda*" is a compound word with "*hetero-*" meaning other or different in Latin and "*-poda*" meaning feet or foot-like parts. It probably refers to the setae with different lengths.
Type locality: Japan (Tokyo, Tokorosawa, Kamakura).
Distribution in China: Taiwan.

253. *Metaphire ignobilis* (Gates, 1935)

Pheretima ignobilis Gates, 1935a. *Smithsonian Mus. Coll.* 93(3):11.
Pheretima ignobilis Chen, 1936. *Contr. Biol. Lab. Sci. Soc. China (Zool).* 11:299.
Pheretima ignobilis Gates, 1939. *Proc. U.S. Natn. Mus.* 85:450–451.
Metaphire ignobilis Sims & Easton, 1972. *Biol. J. Soc.* 4:238.
Metaphire ignobilis Xu & Xiao, 2011. *Terrestrial Earthworms of China.* 235.

External Characters

Length 50–55 mm, width 3–4 mm, segment number 97. Prostomium 3/4 epilobous. First dorsal pore 11/12.

Internal Characters

Septa 8/9/10 absent. Intestine swelling from XV. Intestinal caeca simple, in XXVII, extending anteriorly to XIII, but from the ventral margin of the caecum there protrude ventrally several short, stumpy, finger-like lobes, the dorsoventral length of these lobes is less than the dorsoventral diameter of the main portion of the caecum. First pair of hearts in X present. Testis sacs of X and XI paired and ventral. Seminal vesicles are vertical bodies, each with a primary ampulla reaching to the dorsal blood vessel. There is a pair of relatively large pseudovesicles in XIII. Prostate gland small, not developed, confined to XVII–XIX, with a slender duct, bent into a C shape, the ectal portion thicker than the ental portion. Spermathecae 4 pairs, in VI–IX, juvenile. The coelomic portion of the duct is of about the same thickness as the ampulla and of about same length, but within the parietes the duct becomes thicker and its lumen wider. The diverticulum, which passes into the anterior face of the duct in the parietes and is as long as or slightly longer than the main pouch (duct and ampulla), is slender and tubular with just a slight suggestion of a spheroidal widening of the ental end.

Deposition of types: Smithsonian Institution.
Type locality: Sichuan (Xichang).

Distribution in China: Sichuan (Xichang).

Remarks: *Metaphire ignobilis* cannot be adequately characterized at present. Distinguished from all octothecal Chinese species of *Pheretima* by the presence of the spermathecal pore (parietal only?) invaginations, with large, transversely slit-like, secondary apertures.

254. Metaphire paiwanna liliumfordi Tsai, Shen & Tsai, 2000

Metaphire paiwanna liliumfordi Tsai, Tsai & Liaw, 2000. *J. Nat. Hist.* 34, 1734–1736.
Metaphire paiwanna liliumfordi Chang, Shen & Chen, 2009. *Earthworm Fauna of Taiwan.* 122–123.
Metaphire paiwanna liliumfordi Xu & Xiao, 2011. *Terrestrial Earthworms of China.* 244–245.

External Characters

Length 345–356 mm, clitellum width 9–16 mm, segment number 182–183. Number of annuli per segment 3 in V–VIII, 5 in IX–XIII, and 3 in body segments behind XVII. Prostomium epilobous. First dorsal pore 12/13. Clitellum XIV–XVI, annular, length 8.5–9.7 mm, dorsal pore absent, setae absent. Setae 115–135 (VII), 119–138 (XX), 30–32 between male pores. Male pores paired, situated on setal line close to lateral border of XVIII, each male pore area enlarged, slightly C-shaped, with the opening of the C facing the ventral setal line, and length about twice the length of XVIII, extending to the setal line of XVII and XIX, bordered by a thick skin wall and surrounded by circular folds; the male aperture situated on the end of the ventral setal line, on the middle or slightly posterior to the middle of the male pore area, a horizontal ridge extending from the setal line to the male aperture, an oval pad situated between the male aperture and the anterior end of the male pore area, linked to the male aperture through a ditch-like structure. Spermathecal pores 4 pairs in 5/6–8/9. No genital papillae in the preclitellar region. Female pore single, medioventral in XIV.

Internal Characters

Septa 8/9/10 absent, 5/6/7/8 slightly thickened, 10/11–13/14 greatly thickened. Gizzard in VIII–X. Intestine swelling from XV. Intestinal caeca paired in XXVII, simple, extending anteriorly to XXV. Lateral hearts in XI–XIII. Testis sacs 1 pair in X, oval, smooth, medioventral in front of 10/11. Seminal vesicles paired in XI, large, each with a folliculate dorsal lobe. Prostate glands paired in XVIII, large, lobular, extending anteriorly to XVII and posteriorly to XIX. Spermathecae 4 pairs in VI–IX, each with a slightly tuberculate ampulla, and a long stalk, about half ampulla length, diverticulum small, with an oval seminal chamber, and a short, twisted stalk, reaching the base of ampulla.

Color: Preserved specimens bluish or purplish brown on dorsum, light gray on ventrum.
Deposition of types: Taiwan Endemic Species Research Institute, Jiji, Nantou, Taiwan.
Etymology: The name "*liliumfordi*" was given to indicate the type locality of a daylily plantation.
Type locality: Taiwan (Mt. Setou in the middle of the Coastal Mountain Range, Hualien County).
Distribution in China: Taiwan (Hualien)
Remarks: *M. paiwanna liliumfordi* is a member of the *M. formosae* species group.

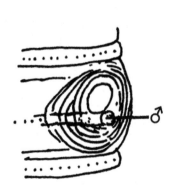 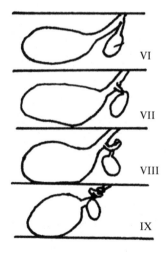 

FIG. 248 *Metaphire liliumfordi* (after Tsai et al., 2000a).

255. *Metaphire nanaoensis* Chang & Chen, 2005

Metaphire nanaoensis Chang & Chen, 2005. *J. Nat. Hist.* 39(18),1469–1482.
Metaphire nanaoensis Chang, Shen & Chen, 2009. *Earthworm Fauna of Taiwan.* 116–117.
Metaphire nanaoensis Xu & Xiao, 2011. *Terrestrial Earthworms of China.* 243.

External Characters

Length 335–429 mm, clitellum width 10.1–14.9 mm, segment number 132–177. Number of annuli per segment 3 in V–IX, 5 in X–XIII, 3 in body segments behind XVII. Prostomium prolobous. First dorsal pore 12/13. Clitellum XIV–XVI, annular, length 13.8–14.4 mm, dorsal pore absent, setae absent. Setae number 103–114 (VII), 120–131 (XX), 19–23 between male pores. Male pores paired, situated on setal line close to lateral border of XVIII, each male pore area C-shaped, enlarged, with the opening of the C facing the ventral setal line, and length about twice the length of XVIII, extending to the setal line of XVII and XIX, laterally bordered by a thick skin wall, with several folds on lateral side; male apertures situated at the ends of the ventral setal line, partially covered by the skin wall bordering the male pore area; the region covered by the skin wall swollen, forming a smooth appearance; a very small pad present on the posterior end of the male pore area. Spermathecal pores 4 pairs in 5/6–8/9, about 0.4 circumference apart ventrally. No genital papillae in the spermathecal region. Female pore single, medioventral in XIV.

Internal Characters

Septa 9/10 absent, 8/9 thin, 5/6/7/8 thickened, 10/11–13/14 greatly thickened. Gizzard in VIII–X. Intestine swelling from XV. Intestinal caeca paired in XXVII, simple, extending anteriorly to XXIII. Lateral hearts in X–XIII. Nephridia tufted, attached to the postsegmental septa, surrounding the segmental chambers anterior to the 6/7 septum. Ovaries paired in XIII, medioventral, close to the 12/13 septum. Testis sacs 1 pair in X, oval, smooth, medioventral in front of 10/11. Seminal vesicles paired in XI, filling the space between septa, a pair of very small vestiges of seminal vesicles present in XII in some specimens. Prostate glands paired in XVIII, large, lobular, extending anteriorly to XVII. Spermathecae 4 pairs in VI–IX; ampulla large, about 3.5–5.5 mm in length, with a stalk about 0.5–1.4 mm in length; diverticulum short, beyond the middle of spermathecae, with a small oval seminal chamber on the tip.

Color: Live specimens bluish brown or dark purplish gray with metallic luster on dorsum, reddish brown on ventral. Preserved specimens purplish brown on dorsum, light grayish brown on ventrum.
Deposition of types: Institute of Zoology, National Taiwan University, Taibei, Taiwan.

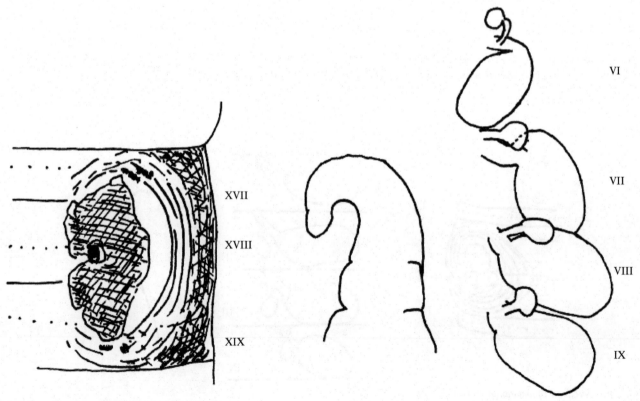

FIG. 249   *Metaphire nanaoensis* (after Chang & Chen).

Etymology: The species name "*nanaoensis*" refers to Nanao in Ilan County in Taiwan, where the species was first collected.

Type locality: Nanao, Ilan County, Taiwan.

Distribution in China: Taiwan (Ilan).

Remarks: *M. nanaoensis* is a member of the *M. formosae* species group. The male pore area of *M. nanaoensis* is C-shaped, and with respect to body size, color, number of spermathecae, and position and shape of prostate gland and intestinal caeca, *M. nanaoensis* resembles *M. yuanpowa* (Chang & Chen, 2005), *M. paiwanna paiwanna* (Tsai, Shen & Tsai, 2000), and *M. bununa* (Tsai, Shen & Tsai, 2000). However, the male pore area of *M. yuanpowa* has an oval pad on each side of the male aperture, and the width between the paired spermathecal pores of *M. yuanowa*, which is about 0.5 body circumference apart ventrally, is wider than that of *M. naniensis*. The major difference the 2 species is that *M. yuanpowa* is holandric, while *M. nanaoensis* is protandric. *M. paiwanna paiwanna* and *M. bununa* are also protandric species, but the male pore areas of these 2 species are different from that of *M. nanaoensis*. The male pore area of *M. paiwanna paiwanna* and *M. bununa* has an oval pad between the male aperture and the anterior end of the male pore area, and the male pore area of *M. paiwanna* also has a horizontal ridge. These differences are easily seen under a dissection microscope, or even with the naked eye.

256. Metaphire paiwanna paiwanna Tsai, Shen & Tsai, 2000

Metaphire paiwanna paiwanna Tsai, Tsai & Liaw, 2000. *J. Nat. Hist.* 34:1732–1736.

Metaphire paiwanna paiwanna Chang, Shen & Chen, 2009. *Earthworm Fauna of Taiwan.* 118–119.

Metaphire paiwanna paiwanna Xu & Xiao, 2011. *Terrestrial Earthworms of China.* 245–246.

External Characters

Length 170–300 mm, clitellum width 6.8–11.0 mm, segment number 132–177. Number of annuli per segment 3 in V–IX, 5 in X–XIII, 3 in body segments behind XVII. Prostomium prolobous. First dorsal pore 12/13. Clitellum XIV–XVI, annular, length 8.5–9.7 mm, dorsal pore absent, setae absent or rarely 2 setae remaining on ventrum XVI. Setae minute, setal number 108–170 (VII), 126–170 (X), 100–103 (XX), 104–126 (XXV), 22–40 between male pores. Male pores paired, ventrolateral in XVIII, the distance between the pores 0.25–0.30 circumference apart ventrally. each pore area C-shaped or slightly S-shaped, bordered laterally by a thick skin wall and medially by a male disc, which has a horizontal ridge, extending from the setal line to the middle of the pore, along with tubercles and folds. The male pore area surrounded by circular

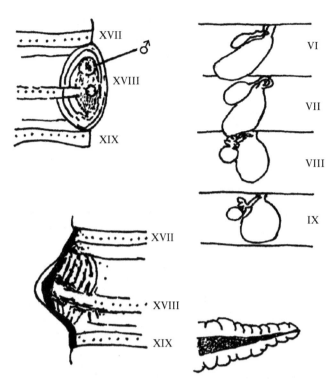

FIG. 250 *Metaphire paiwanna paiwanna* (after Tsai et al., 2000a).

folds, which extend anteriorly to the postsetal annulet of XVII and posteriorly to the presetal annulet of XIX. Male aperture located at the distal end (tip) of the horizontal ridge, coiling into the copulatory pouch. Genital papillae absent in the male pore and spermathecal regions. Spermathecal pores 4 pairs in 5/6–8/9, distance between the paired pores about 0.5 body circumference apart, edge of each pore slightly swollen, wrinkled, and pale in color. Female pore single, medioventral in XIV.

Internal Characters

Septa 8/9/10 absent, 5/6/7/8 slightly thickened, 10/11–13/14 greatly thickened. Gizzard in VIII–X, large, yellowish white, cylindrical, with a slight constriction at 9/10 that divides the Gizzard into a long anterior part in VIII and IX and a short posterior part in X. Intestine swelling from XV. Intestinal caeca paired in XXVII, simple, extending anteriorly to XXIV. Lateral hearts in X–XIII. Testis sacs 1 pair in X, oval, smooth, medioventral in front of 10/11. Seminal vesicles paired in XI, each with a folliculate dorsal lobe. Prostate glands paired in XVIII, large, racemose, lobular, extending anteriorly to XVII; prostatic duct short, stout, slightly curved. Spermathecae 4 pairs in VI–IX, each with a slight tuberculate ampulla, about 7 mm long, and a very short stalk, about 1/3 of ampulla length; diverticulum small, with an oval seminal chamber, and a short, twisted stalk, reaching to or beyond the middle of ampulla.

Color: Preserved specimens bluish or purplish brown on dorsum, light gray on ventrum.

Deposition of types: Taiwan Endemic Species Research Institute, Jiji, Nantou, Taiwan.

Etymology: The specific name *"paiwanna"* refers to the Taiwanese aboriginal Paiwan tribe.

Type locality: Taiwan (Mudan, Pingdong).

Distribution in China: Taiwan (Pingdong).

Remarks: *M. paiwanna* has a pair of C-shaped or S-shaped male pores, each with a specific male disc with a horizontal ridge coiling laterally into the copulatory pouch. This character resembles those of *Begemius yuhsi* (Tsai, 1964) and *M. aggera* (Kobayashi, 1934). In the original descriptions there was no mention of a copulatory pouch in *Amynthas swanus* (Tsai, 1964), but a mention of "a shallow copulatory pouch" in *M. aggera* (Kobayashi, 1934). Sims and Easton assigned *Begemius yuhsi* to the genus *Begemius*, and *Metaphire aggera* to the genus *Metaphire*.

Amynthas swanus has a male disc only slightly coiled into the copulatory pouch, whereas in *M. aggera* the male disc is completely embedded within the copulatory pouch and covered by the lateral wall. In this respect, one may consider that *A. swanus* is a primitive form, *M. aggera* is a specialized form, and *M. paiwanna* is an intermediate from between the 2 nominal species.

93. Metaphire pavimentus Tsai & Shen, 2010 (*v.ant.* 93)

257. Metaphire posthuma (Vaillant, 1868)

Pheretima posthuma Gates, 1959a. *Amer. Mus. Novitates.* 141:15–16.

Pheretima posthuma Tsai, 1964. *Quer. Jou. Taiwan Mus.* 17:4–5.

Metaphire posthuma Sims & Easton 1972. *Biol. J. Soc.,* 4:239.

Metaphire posthuma Chang, Shen & Chen, 2009. *Earthworm Fauna of Taiwan.* 124–125.

Metaphire posthuma Xu & Xiao, 2011. *Terrestrial Earthworms of China.* 247–248.

External Characters

Length 60–210, clitellum width 3–8 mm, segment number 91–124. Prostomium epilobous or tanylobous. Setae number 111–130 (VIII), 71–83 (XX), 18–19 between male pores. First dorsal pore 12/13. Clitellum XIV–XVI, annular, setae absent, dorsal pore absent. Male pores paired on lateroventral side of segment XVIII, within copulatory pouches, each pore situated on the center of a comparatively large and round papilla with a depressed center, occasionally surrounded by a few circular ridges, in the setal line. A pair of genital papillae situated just in setal line and slightly medial to male pores in segment XVII and XIX respectively, separated by 15–17 setae. Each papilla round in shape with depressed top, slightly larger than male papilla. Spermathecal pores 4 pairs in 5/6–8/9, ventrolateral, 0.3 circumference apart ventrally. No genital papillae are present in this region. Female pore single, medioventral in XIV.

Internal Characters

Septa 9/10 absent, 5/6–8/9, 10/11/12 slightly thickened, membranous. Gizzard round in VIII. Intestine

FIG. 251   *Metaphire posthuma* (after Vaillant, 1868).

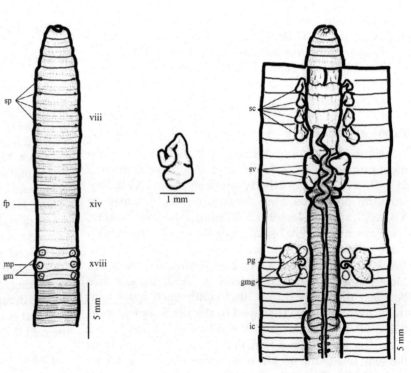

swelling from XV. Intestinal caeca paired in XXVII, simple, extending anteriorly to XXV or XXIV. Last hearts in XIII. Ovaries paired in XIII. Testis sacs 2 pairs in X and XI, anterior pair oval in shape, connected widely on ventral ends, attaching to anterior surface of 10/11 septum; posterior pair much larger, connected widely on their ventral ends. Vas deferens meeting in XII, but not fused to each other throughout. Seminal vesicles paired in XI and XII, the anterior pair a little smaller than the posterior ones, included in posterior testis sacs besides their anterior middle portions. Each vesicle divided into small pieces on its surface by irregular deep grooves, with a large round ental dorsal lobe whose surface is granular, and which is light yellow in color. Prostate glands well developed, occupying through 3–4 segments, from XVII to XIX, and separated radially into lobules and then subdivided into small pieces on surface by the shallow groove. Prostatic ducts looped at terminal half part, hooked and enlarged. Accessory glands in XVII and XIX, corresponding to external papillae, each large, round and of compact mass. Spermathecae 4 pairs in VI–IX or VI–VIII, if in VI–VIII, last segment containing last 2 pairs; ampulla round or heart-shaped with short, stout stalk, diverticulum elongated, heart-shaped with a short and slender stalk.

Color: Live specimens greenish brown or greenish gray. Preserved specimens light brown on dorsum, light gray on ventrum; occasionally, purplish green on dorsum and light blue on ventrum. Brown on clitellum.
Deposition of types: Paris Museum.
Type locality: Java.
Distribution in China: Taiwan (Taibei, Pingdong), Yunnan (Tengyuan).
Remarks: *Metaphire posthuma* (Vaillant, 1868), *Amynthas hupeiensis* (Michaelsen, 1892), and *Metaphire peguana* (Rosa, 1890), have a definite pattern of genital markings location. The markings in *posthuma* always develop in the setal rings and are symmetrically placed in adults with reference to metameric equators. In *hupeiensis*, which may well have been mistaken for *posthuma* on several occasions, the markings never are in the setal ring and usually appear to be intersegmental, as they are in *peguana*.

258. *Metaphire puyuma* Tsai, Shen & Tsai, 1999

Metaphire puyuma Tsai, Shen & Tsai, 1999. *Journal of the National Taiwan Museum.* 52(2):33–46.
Metaphire puyuma Chang, Shen & Chen, 2009. *Earthworm Fauna of Taiwan.* 126–127.

External Characters

Length 62 mm, clitellum width 3.1 mm, segment number 113. Number of annuli per segment 3 in VI–IX, 5 in X–XIII. Prostomium epilobous. Setae number 115 (VII), 73 (XX), 16 between male pores. First dorsal pore 12/13. Clitellum XIV–XVI, annular, 2.6 mm long, setae absent, dorsal pore absent. Male pores paired in XVIII, ventrolateral, copulatory pouches with C-shaped openings surrounded by oval swelling area, genital papillae paired in XIX and XX, single in XXI, arranged longitudinally, each papilla on setal line, large, occupying nearly entire segment, with a slightly concave center and surrounded by a circular fold. Spermathecal pores 4 pairs in 5/6–8/9, ventrolateral, 0.29 circumference apart ventrally. No genital papillae in the preclitellar region. Female pore single, medioventral in XIV.

Internal Characters

Septa 9/10 absent, 8/9 and 10/11 thickened. Gizzard in IX and X, round. Intestine swelling from XV. Intestinal caeca paired in XXVII, small, simple, surface wrinkled, extending anteriorly to XXV. Lateral hearts in XI–XIII. Testis sacs 2 pairs in X and XI, medioventral, small. Seminal vesicles paired in XI and XII, large, folliculated, with a large dorsal lobe. Prostate glands paired in XVIII, large, racemose, extending anteriorly to XVI and posteriorly to XIX, prostatic duct coiled. Spermathecae 4 pairs in VI–IX, each with a small, round or peach-shaped ampulla, and a short, stout stalk, diverticulum with a large, peach-shaped seminal chamber and a short, straight, stout stalk.

Color: Preserved specimens light brown on head, brown on clitellum, light gray on body.
Deposition of types: Taiwan Endemic Species Research Institute, Jiji, Nantou County, Taiwan.
Etymology: The name of this species refers to the Puyuma tribe, the aboriginal people living in the north part of Taidong County.
Type locality: The shore of Monhuan Lake, south of Taidong City, Taidong County, Taiwan.
Distribution in China: Taiwan (Taidong).

259. *Metaphire tahanmonta* Chang & Chen, 2005

Metaphire tahanmonta Chang & Chen, 2005. *J. Nat. Hist.* 39(18),1475–1481.
Metaphire tahanmonta Chang, Shen & Chen, 2009. *Earthworm Fauna of Taiwan.* 130–131.
Metaphire tahanmonta Xu & Xiao, 2011. *Terrestrial Earthworms of China.* 252–253.

External Characters

Length 291–408 mm, clitellum width 12.9–14.7 mm, segment number 122–191. Number of annuli per segment 3 in V–IX, 5 in X–XIII, and 3 in body segments behind XVII. Prostomium prolobous. First dorsal pore 12/13. Clitellum XIV–XVI, saddle-shaped, length 12.8–15.5 mm,

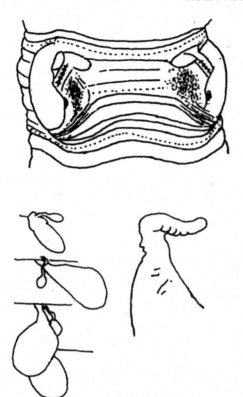

FIG. 252   *Metaphire tahanmonta* (after Chang & Chen, 2005).

dorsal pore absent, setae absent. Setae number 122–144 (VII), 134–156 (XX), 24–30 between male pores. Male pores paired, on setal line close to lateral border of XVIII, each male pore area enlarged, slightly C-shaped, with the opening of the C facing the ventral setal line and a length about twice the length of XVIII, extending to the setal line of XVII and XIX, bordered by a thick skin wall and surrounded by circular folds, with the appearance of the anterior part of the skin wall tubercular; male apertures on the extended line of the ventral setal line, slightly posterior to the middle of the male pore area, a horizontal ridge extending from the setal line, backward to the male aperture, an oval pad situated behind the setal line of XVII, close to the anterior end of the male pore area, linked to the male aperture through a seminal groove. Spermathecal pores 4 pairs in 5/6–8/9, lateral, above the lateral midline, about 0.55 circumference apart ventrally. No genital papillae in the spermathecal region. Female pore single, medioventral in XIV.

Internal Characters

Septa 8/9/10 absent, 5/6/7/8 thickened, 10/11–13/14 greatly thickened. Gizzard in VIII–X. Intestine swelling from XV. Intestinal caeca paired in XXVI, simple, extending anteriorly to XXIII. Lateral hearts in X–XIII. Nephridia tufted, attached to the postsegmental

septa, surrounding the segmental chambers anterior to the 6/7 septum. Ovaries paired in XIII, medioventral, close to the 12/13 septum. Testis sacs 2 pairs in X and XI, the anterior pair oval, smooth, medioventral in front of 10/11, the posterior pair much larger than the anterior pair, filling the space between septa. Seminal vesicles paired in XI and XII, the anterior pair included in the posterior testis sac, both pairs moderate in size. Prostate glands paired in XVIII, large, lobular, extending anteriorly to XVII. Spermathecae 4 pairs in VI–IX, ampulla large, about 3.6–5.6 mm in length, with a stalk about 1.2–1.9 mm in length, diverticulum short, usually shorter than one-third of spermathecae length, with a small oval seminal chamber on the tip.

Color: Live specimens dark purplish gray with metallic luster on dorsum, reddish brown on ventrum. Preserved specimens purplish brown on dorsum, light grayish brown on ventrum.
Deposition of types: Institute of Zoology, National Taiwan University, Taibei, Taiwan.
Etymology: The name *"tahanmonta"* refers to Mt. Tahan, where this species was collected.
Type locality: Jinshuiying Nature Reserve, Pingdong, Taiwan.
Distribution in China: Taiwan (Pingdong).
Remarks: *M. tahanmonta* (Chang & Chen, 2005) has a length greater than 300 mm, a width over 10 mm, 4 pairs of spermathecae, and C-shaped male pores with an oval pad and horizontal ridge. These characters are shared with *M. paiwanna* (Tsai, Shen & Tsai, 2000). However, *M. paiwanna* is protandric, while *M. tahanmonta* is holandric. In pheretimoid earthworms, most species have a holandric condition, while rare species have a protandric condition. Accordingly, it is reasonable to regard the holandric specimens as a different species.

Regarding body size, position of seminal vesicles, number of spermathecae, and holandry, *M. tahanmonta* shares characters with *M. yuanpowa*. However, the male pores of the 2 species are very different. The male pore of *M. tahanmonta* has a horizontal ridge and only 1 oval pad, while that of *M. yuanpowa* has no horizontal ridge and 2 oval pads on each side of the male aperture. Therefore, they are easily distinguished by their external characters.

260. *Metaphire taiwanensis* Tsai, Shen & Tsai, 2004

Metaphire taiwanensis Tsai, Tsai & Shen, 2004. *J. Nat. Hist.* 38(7),877–887.
Metaphire taiwanensis Chang, Shen & Chen, 2009. *Earthworm Fauna of Taiwan.* 132–133.

Metaphire taiwanensis Xu & Xiao, 2011. *Terrestrial Earthworms of China.* 253–254.

External Characters

Length 637–655 mm, clitellum width 16.1–17.2 mm, segment number 183–228. Number of annuli per segment 3 in V–XIII and on the ventrolateral portion of some body segments. Prostomium prolobous. First dorsal pore 12/13 or 13/14. Clitellum XIV–XVI, annular, length 16.5–19.9 mm, setae absent, dorsal pore absent. Setae number 159–188 (VII), 135–145 (XX), 15–24 between male pores. Male pores paired in XVIII, small, lateroventral, 0.27–0.29 circumference apart ventrally, each pore C-shaped or L-shaped, bordered laterally by a thick skin surrounded with circular folds, male aperture inconspicuous, porophore circular, top flat, tuberculate, an oval pad situated anteriorly to the porophore, slightly larger than the porophore, porophore and oval pad surrounded by circular folds. Spermathecal pores 4 pairs in 5/6–8/9, minute, invisible externally. No genital papillae in the preclitellar region. Female pore single, medioventral in XIV.

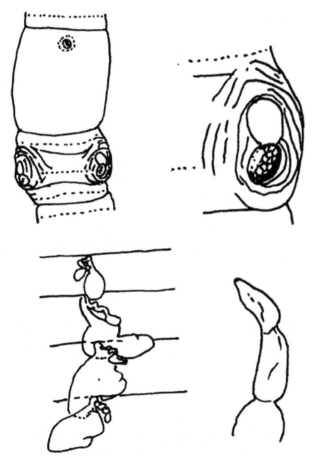

FIG. 253 *Metaphire taiwanensis* (after Tsai, Tsai & Shen, 2004).

Internal Characters

Septa 8/9/10 absent, 7/8 and 10/11–12/13 thickened. Gizzard in IX and X. Intestine swelling from XV. Intestinal caeca paired in XXVII, simple, long, surface slightly wrinkled, extending anteriorly to XIX. Lateral hearts in X–XIII. Testis sacs 1 pair in X, oval, smooth, medioventral in front of 10/11. Seminal vesicles paired in XI and XII, large in XI, surface tuberculate, with a large dorsal lobe, vestigial in XII, surface tuberculate. Prostate glands paired in XVIII, large, oval, extending anteriorly to XVII and posteriorly to XIX. Spermathecae 4 pairs in VI–IX, ampulla oval, with a short stalk, diverticulum small, with an oval, smooth seminal chamber and a short coiled or twisted stalk.

Color: Preserved specimens bluish brown on dorsum, light grayish brown on ventrum, clitellum dark bluish brown on dorsum, greyish brown on ventrum.
Deposition of types: Taiwan Endemic Species Research Institute, Jiji, Nantou, Taiwan.
Etymology: The species name "*taiwanensis*" refers to "Taiwan", to which this species is endemic .
Type locality: Mt. Beidongyan, Nantou County, Taiwan.
Distribution in China: Taiwan (Nantou).
Remarks: *Metaphir taiwanensis* (Tsai, Shen & Tsai, 2004) is a member of the *M. formosae* species group.
Habitats: This is the largest species among Taiwanese earthworms. It lives in the mountains where the vegetation is broadleaf forest and can be found in both virgin and secondary forests. *M. taiwanensis* is an anecic species, having permanent vertical burrows. It is active around the upper layer of soil or on the ground at night, but stays 30 cm or more below the ground in the soil during the day.

261. *Metaphire yeni* Tsai, Shen & Tsai, 2000

Metaphire yeni Tsai, Shen & Tsai, 2000. *J. Natl. Taiwan Mus.* 53:7–14.
Metaphire yeni Chang, Shen & Chen, 2009. *Earthworm Fauna of Taiwan.* 136–137.

External Characters

Length 53–62 mm, clitellum width 2.3–2.6 mm, segment number 59–75; Number of annuli 3 in preclitellar segments, 5–7 in postclitellar segments. Prostomium epilobous. First dorsal pore 11/12. Clitellum XIV–XVI, annular, setae and dorsal pore absent. Setae number 37–40 (VII), 41–43 (XX); 7–9 between male pores. Male pores paired in XVIII, ventral, 0.26 circumference apart ventrally, contracted into copulatory pouches or semiprotruded, if contracted, each male pore with a large oval opening with wrinkled edge surrounded by 3–4 circular folds, if semiprotruded, each male pore with a large, stout porophore with a flat top and 3 large papillae around the male aperture, with the base of porophore surrounded by

3–4 circular folds, each papilla with concave center, often surrounded by a fold like that in the preclitellar region. Spermathecal pores 4 pairs in 5/6–8/9, lateral, each on a tubercle in intersegmental furrow, highly visible, 0.63 circumference apart ventrally. Genital papillae on laterodorsum, 0.1–0.2 mm dorsal to spermathecal pore adjacent to intersegmental furrows, paired in 2 longitudinal rows in VII–IX with varying numbers and locations, each papilla circular, flat top with concave center, about 0.2 mm in diameter, surrounded by a circular fold, occupying entire annuli. Female pore single, medioventral in XIV.

Internal Characters

Septa 8/9/10 absent, 6/7/8 and 10/11/12/13 thickened. Gizzard in XI and X, large, round. Intestine swelling from XV. Intestinal caeca simple, paired in XXVII, surface wrinkled with a pointed end, extending anteriorly to XXIV. Lateral hearts in XI–XIII. Testis sac 2 pairs in XI and XII, small, oval, smooth. Seminal vesicles paired in XI–XII, large, irregular, follicular, with a granulated dorsal lobe. Prostate glands large and lobulated, small and wrinkled, or vestigial, prostatic ducts U-shaped or S-shaped. Accessory glands aggregated in a circular mass around the base of each prostatic duct. Spermathecae 4 pairs in VI–IX, each with an oval or peach-shaped ampulla and a short, straight, stout stalk, diverticulum with an oval seminal chamber and a straight or slightly bent stalk, with length varying from slightly shorter to slightly longer than that of spermathecal stalk. Accessory glands solitary, round, surface follicular, stalk very short or absent.

> Color: Preserved specimens dark brown on dorsum, light yellowish white on ventrum and laterum where spermathecal pores and genital papillae present, dark brown around clitellum, light yellowish white on setal line.
> Type locality: Taiwan (Pingdong).
> Deposition of types: Taiwan Endemic Species Research Institute Jiji, Nantou, Taiwan.
> Distribution in China: Taiwan (Pingdong).
> Etymology: This species name "*yeni*" was given to this species in honor of former Director Jen-Teh Yen for his endeavors in establishing the Taiwan Endemic Species Research Institute.

262. Metaphire yuanpowa Chang & Chen, 2005

Metaphire yuanpowa Chang & Chen, 2005. *Jour. Nat. His.* 39(18):1470–1473.
Metaphire yuanpowa Xu & Xiao, 2011. *Terrestrial Earthworms of China.* 261–262.

External Characters

Length 215–425 mm, clitellum width 13.9–15.6 mm, segment number 129–189. Prostomium epilobous. Number of annuli per segment 3 in V–IX, 5 in X–XIII and 3 in body segments behind XVII. First dorsal pore 12/13. Clitellum XIV–XVI, smooth, saddle-shaped, length 9.8–14.1 mm, dorsal pore absent, setae absent. Setae number 114–122 (VII), 126–162 (XX); 30–35 between male pores. Male pores paired, situated on setal line close to lateral border of XVIII. Each male pore area is C-shaped, with the opening of the C facing the ventral setal line, bordered by a thick skin wall, which has several folds on its lateral side. The male pore area is enlarged, with a length about twice the length of XVIII, extending to the setal line of XVII and XIX. The male aperture is situated on the end of the ventral setal line, with an oval pad on each side. The 2 oval pads are a vertical bar-shaped structure extending from the male aperture. These structures are sometimes partially covered by the skin wall bordering the male pore area. Genital papillae absent in the male pore area. Spermathecal pores 4 pairs in 5/6–8/9, lateral, distance between the paired pores about 0.5 circumference. No genital papillae in the spermathecal region. Female pore single, medioventral in XIV.

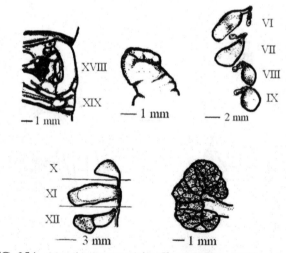

FIG. 254   *Metaphire yuanpao* (after Chang & Chen, 2005).

Internal Characters

Septa 9/10 absent, 8/9 thin, 5/6/7/8 thickened, 10/11–13/14 greatly thickened. Gizzard large, round in VIII–X. Intestine swelling from XV. Intestinal caeca paired in XXVII, simple, extending anteriorly to XXIII. Lateral hearts in X–XIII. Testis sacs 2 pairs, medioventral in X and XI, the anterior pair oval, smooth, medioventral in front of 10/11, the posterior pair is much larger than the anterior pair, filling the space between septa. Vas deferens connected in XII. Seminal vesicles 2 pairs in XI and XII, the anterior pair included in the posterior testis sac, both pairs moderate in size. Prostate glands paired in XVIII, large, lobular, extending to XVII. Spermathecae 4 pairs in VI–IX, ampulla large, about 3.5–6.2 mm in length, with a stalk about 0.5–1.3 mm in length. Diverticulum short, beyond the middle of spermathecae.

Color: Live specimens bluish brown or dark purplish gray with metallic luster on dorsum, reddish brown on ventrum. Preserved specimens purplish brown on dorsum, lightly grayish brown on ventrum.
Distribution in China: Taiwan (Taibei).
Etymology: The name "*yuanpowa*" refers to the male pore area of this species. The male pore area of this species is like Chinese "yuanpau", which is a kind of money made by silver in the past in China.
Deposition of types: Department of Life Science, National Taiwan University, Taibei, Taiwan.
Remarks: *Metaphire yuanpowa* (Chang & Chen, 2005) resembles the other 2 sympatric species, *Amynthas aspergillum* (Perrier, 1872) and *Amynthas formosae* (Michaelsen, 1922), in body size, shape and color, but it is easily distinguished in the field by external characters. The male pore area of *M. yuanpowa* is enlarged, with an obvious oval pad on each side of the male pore, while the other 2 species do not have these characters. The spermathecal pores of *A. formosae* are mediodorsal in 5/6–8/9, and the pores are evident to the naked eye through the lighter color of the pore borders.
A. aspergillum has many papillae on the male pore area. Furthermore, *A. aspergillum* was recognized as a peregrine species in Taiwan and is usually found in cultivated soil, especially in the grassland of public parks, while *M. yuanpowa* and *A. formosae* are often found on mountain slopes, where the soil is less disturbed.

M. yuanpowa shares most of its characters with *Metaphire paiwanna* (Tsai, Shen & Tsai, 2000) and *Metaphire bununa* (Tsai, Shen & Tsai, 2000), including body size, color, number of spermathecae, and the position and morphology of prostate gland and intestinal caeca. However, the male pore area of *M. paiwanna* and *M. bununa* has only 1 oval pad between the male pore aperture and the anterior end of the male pore area, and the male pore area of *M. yuanpowa* also has a horizontal ridge. Moreover, *M. yuanpowa* has testis sacs paired in X and XI, while both *M. paiwanna* and *M. bununa* have testis sacs paired only in X.

Except for number of spermathecae, *M. yuanpowa* is similar to *M. trutina* (Tsai, Chen, Tsai & Shen, 2003), which has only 3 pairs of spermathecae and belongs to the *houlleti* group. It is not yet known if the 2 species are distantly related due to the difference in the number of spermathecae, or are more closely related due to the similarity of other characters. According to field surveys, *M. yuanpowa* is restricted to the western edge of the Shei-Shan Mountain Range, and *M. trutina* is found only in the eastern edge of the Shei-Shan Mountain Range. Although they seem to be allopatric, the detailed distribution pattern of the 2 species is still unknown.

densipapillata Group

Diagnosis: *Metaphire* with spermathecal pores 1 pair, at 7/8

Global distribution: Oriental realm from Japan southwards through the Indo-Australasian archipelago to the rainforests of the Australasia realm eastwards through the Oceanian realm.

There are 2 species from China.

Remarks: Species of this genus are separable from those of *Amynthas* by the presence of copulatory pouches. They differ from species of *Pheretima* by the absence of nephridia from the spermathecal ducts.

TABLE 29 Key to the *densipapillata* Group From China

1. Three or two large gland areas on the ventral sides of VIII and IX .... *Metaphire coacervata*   Large raised ovoid papillae paired presetally on VIII and IX, in front of setae b, c, those on VII postsetal, between c, d, and e .... *Metaphire tecta*

263. *Metaphire coacervata* Qiu, 1993

Metaphire coacervata Qiu, 1993. *Sichuan Journal of Zoology*. 12(4):1–4.
Metaphire coacervata Xu & Xiao, 2011. *Terrestrial Earthworms of China*. 223–224.

External Characters

Length 75–76 mm, width 2.5–3.0 mm, segment number 91–103. Prostomium 2/3 epilobous. First dorsal pore 11/12. Clitellum XIV–XVI, annular, setae absent. Setae all fine; Setae number 27–28 (III), 31–33 (V), 40–46 (VIII), 44–45 (XX), 47–49 (XXV); 24–25 (VIII) between spermathecal pores, 9–11 between male pores. Male pores 1 pair, situated in small and shallow copulatory pouches that are on the ventrolateral side of XVIII; about 1/3 circumference apart ventrally; the copulatory pouches' aperture appears as a vertical slit with a large wrinkle of skin on the lateral side of the pouches, which cover the apertures. Wide gland area around the copulatory pouch extending anteriorly to 1/2 XVII and posteriorly to 1/2 XIX, no genital papilla in or out of the chambers. Spermathecal pores 1 pair in 7/8, about 5/9 circumference apart ventrally; 2–3 large gland areas on the ventral sides of VIII and IX. Female pore single, medioventral in XIV.

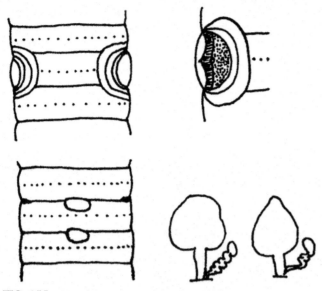

FIG. 255   *Metaphire coacervata* (after Qiu, 1993).

Internal Characters

Septa 8/9/10 ventral present, very thin. Gizzard well developed, bar-like. Intestine swelling from XVI. Intestinal caeca simple, originating in XXVI and extending anteriorly to XXIV. Last hearts in XIII. Testis sac 2 pairs in X–XI, small, the first pair united in a V shape, those in XI united broadly. Seminal vesicles paired in XI–XII, developed, their dorsal lobe conspicuous. Seminal chamber long-round, rather small. Accessory glands in ventral side of VIII and IX, well developed. Prostate glands in XVI–XXI, well-developed. Accessory glands well developed, in the ventral side to the prostate glands. Spermathecae 1 pair in VIII; ampulla oblate or heart-shaped, about 2.5 mm long and 2.0 mm wide, its wall thin; spermathecal duct thick and long and there is no membrane around it. Diverticulum is shorter than main pouch and twisted in a zigzag fashion or wave-shaped.

> Color: Preserved specimens red-brown on dorsum, gray-white on ventrum. Red-brown on clitellum.
> Distribution in China: Guizhou (Suiyang).
> Etymology: The specific name *"coacervata"* probably indicates the structure of the male pore regions of this species.
> Type locality: Guizhou (Mt. Taiyang, Kuankuoshui forest region, Suiyang County).
> Deposition of types: Guizhou Institute of Biology, Guiyang.
> Remarks: This species is similar to *Metaphire tecta* (Chen, 1946) in its 1 pair of spermathecal pores in 7/8, the short and twisted diverticulum, well-developed accessory glands, prostate glands and seminal

vesicles, and in having a copulatory pouch, but differs from it by the small and shallow copulatory pouch, the simple caeca, and the presence of septa 8/9/10. Both ampullar and diverticular ducts are not invested thickly with muscles growing with the peritoneum.

264. *Metaphire tecta* (Chen, 1946)

Pheretima tecta Chen, 1946. *J. West China. Border Res. Soc. (B).* 16:120–122.
Metaphire tecta Sims & Easton, 1972. *Biol. J. Linn. Soc.* 4:239.
Metaphire tecta Xu & Xiao, 2011. *Terrestrial Earthworms of China.* 254–255.

External Characters

Large-sized worm. Length 145 mm, width 5 mm, segment number 137. Prostomium 1/3 epilobous, its width as length of segment II. Annulation pronounced in preclitellar segments, number of annulets per segment 5–7 in VII–XIII; length of VI or VIII equal to XIX. First dorsal pore 12/13. Clitellum XIV–XVI, annular, not well shown but setae shorter and slightly glandular in appearance. Setae noticeable in preclitellar segments, more so on ventral side, but none modified or lengthened, more widely spaced on anteroventral side; those of postclitellar segments shorter, slightly closer on dorsal side. Setae number 29 (III), 36 (VI), 38 (IX), 46 (XXV); 23 (VII) between spermathecal pores, 10 between male pores; $aa = 1.2$–$1.5ab$, $zz = 1.5\ zy$. Male pores paired in XVIII, rather lateral in position, each in a copulatory pouch of parietal invagination, pore found on the lower limb of a large raised papilla which has its ventral side round and broader and dorsal side elongate and narrower, its dorsoventral length about 2.2 mm, partly exposed on ventral side. Another smaller papilla situated deeper on posterior side entirely covered by pouch wall. A pair of large ovoid papillae found ventrally on 17/18. Spermathecal pores 1 pair in 7/8, about 5/8 circumference apart ventrally, large raised ovoid papillae paired presetally on VIII and IX, in front of setae *b, c*, those on VII postsetal, between *c, d*, and *e*. Female pore single, medioventral in XIV.

Internal Characters

Septa 8/9/10 absent, anterior septa thicker but not muscular, 10/11/12/13 less thickened, thin after 14/15. Nephridial tufts very thick in front of septa 5/6/7. Gizzard bell-shaped, posterior end wider. Intestine swelling from XV. Intestinal caeca small, about 5–7 finger-like diverticula. Testis sacs large, united; anterior pair U-shaped, narrowly connected, posterior pair wedge-shaped, with broad connection. Seminal vesicles well developed, in IX and XIII, smooth on surface, dorsal lobes

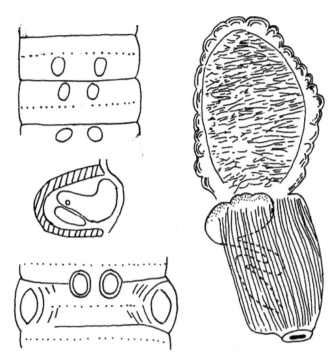

FIG. 256 *Metaphire tecta* (after Chen, 1946).

moderate in size, on anterior dorsal side. Prostate glands separated into anterior and posterior masses, each finely lobate, about 4.5 mm in length, 2 branches of the prostatic duct long. Prostatic duct S-curved, enlarged at middle and tapering toward both ends. Accessory glands in large thick masses, each papilla connected with 4–5 indistinctly separable lobules. Spermathecae very large in VIII, extending posteriorly to X. Ampulla on left side 3 mm, on right 5 mm long, wrinkled on surface. Diverticulum as short cap-like sac, invariably whitish, transversely placed ental to ampullar duct. Both ampullar and diverticular ducts invested thickly with muscles growing with the peritoneum from the body wall, diverticular duct twisted within the investment. Accessory glands large, 5 or more lobules connected with short cords, large lobule larger than seminal chamber of diverticulum.

Color: Pale on ventrum, grayish on dorsum, brownish purple on clitellum. Setal zone pale.
Distribution in China: Hubei (Lichuan), Chongqing (Nanchuan).
Deposition of types: Guizhou Institute of Biology, Guiyang.
Remarks: Although the clitellum in this species has not yet been shown clearly, the sexual organs have reached their full development. In the diverticulum, there is a seminal mass (glistening or shining on surface) and thus it shows that they have undergone copulation. The seminal vesicles and testis sacs are all fully developed.

exilis Group

Diagnosis: *Metaphire* with spermathecal pores 2 pairs, at 5/6/7.

Global distribution: Oriental realm from Japan southwards through the Indo-Australasian archipelago to the rainforests of the Australasia realm and eastwards through the Oceanian realm.

There are 2 species from China.

Remarks: Species of this genus are separable from those of *Amynthas* by the presence of copulatory pouches and differ from species of *Pheretima* by the absence of nephridia from the spermathecal ducts.

TABLE 30 Key to the *exilis* Group From China

1.	Genital markings paired, presetal on XVII and XIX, probably postsetal .... *Metaphire exilis*
	No genital papillae in XVII and XIX .... *Metaphire exiloides*

265. *Metaphire exilis* (Gates, 1935)

Pheretima exilis Gates, 1935a. *Smithsonian Mus. Coll.* 93(3):7.
Pheretima exilis Chen, 1936. *Contr. Biol. Lab. Sci. Soc. China (Zool).* 11:295.
Pheretima exilis Gates, 1939. *Proc. U.S. Natn. Mus.* 85:431–432.
Metaphire exilis Sims & Easton, 1972. *Biol. J. Linn. Soc.* 4:238.
Metaphire exilis Xu & Xiao, 2011. *Terrestrial Earthworms of China.* 224–225.

External Characters

Length 68–85 mm, width 2.0–2.5 mm, segment number 129. Prostomium 1/3 epilobous. First dorsal pore 12/13. Clitellum entire, XIV–XVI, annular, dorsal pores and intersegmental furrows absent, with setae ventrally on XVI. Setae evenly distributed and numerous; setae number 54 (III), 78 (VI), 76 (VIII), 50 (XX), 40 (XXV); 37 (VI) between spermathecal pores, 8 between male pores. Male pores minute, on a smooth, glistening, indistinctly demarcated area. Genital markings paired on XVII and XIX, probably postsetal. Spermathecal pores minute and superficial, 2 pairs, in 5/6/7 or on the posteriormost margin of V and VI, about 3/5 circumference apart ventrally. Female pore single, medioventral in XIV.

Internal Characters

Septa 9/10 absent, 5/6/7/8 thickly muscular, 8/9 represented only by a thin ventral rudiment, 10/11/12/13 membranous but slightly strengthened. Intestine swelling from XV. Intestinal caeca simple, smooth on both sides, in XXVII, extending anteriorly to XXV. First pair of hearts in X, last pair of hearts in XIII. There is a pair

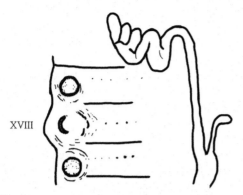

FIG. 257   *Metaphire exilis* (after Chen, 1936).

of testis sacs on the anterior face of 10/11; no transverse connection between the sacs noted. The testis sac or sacs of XI extend dorsally at the sides of the esophagus to the dorsal blood vessel and contain the hearts of XI as well as the seminal vesicles of that segment. The seminal vesicles, paired in XI and XII, are small vertical bodies. Prostate glands in XVI–XXII, rather compact, with a slender and long duct. Genital marking glands sessile but slightly protuberant through the parietes into the coelom. Spermathecal diverticulum long (about 2.4 mm), its ental end closely coiled, with a terminal globular seminal chamber, whitish. All spermathecal ampulla vestigial in size. Accessory glands compact, without stalks.

Distribution in China: Chongqing, Sichuan (Chengdu, Yibin).
Etymology: The specific name "*exilis*" means thin or small in Latin and probably refers to the small size of the body.
Type locality: Chongqing (Yibin).
Deposition of types: Smithsonian Institution.
Remarks: The types are almost certainly abnormal. Examination of normal specimens may enable recognition of further abnormalities in the types.
M. exilis (Gates, 1935a) can be distinguished from other quadrithecal Chinese species of *Metaphire* with spermathecal pores in 5/6/7, by the inclusion of the seminal vesicles of XI within the posterior testis sacs.

266. *Metaphire exiloides* (Chen, 1936)

Pheretima exiloides Chen, 1936. *Contr. Biol. Lab. Sci. Soc. China (Zool).* 11:288–291.
Metaphire exiloides Sims & Easton, 1972. *Biol. J. Linn. Soc.* 4:238.
Metaphire exiloides Xu & Xiao, 2011. *Terrestrial Earthworms of China.* 225–226.

External Characters

Length 55–88 mm, width 2.5–3.0 mm, segment number 94–112. Prostomium 1/4 proepilobous, as a small dorsal

process, completely cut off behind. First dorsal pore 12/13. Clitellum entire, setae absent, XIV–XVI, annular; or very long and swollen. Setae minute and numerous, none specially enlarged, slightly closer on dorsomedian side; both dorsal and ventral breaks very slightly noticeable; setae number 54–58 (III), 70–72 (VI), 60–74 (VIII), 62 (XII), 62–64 (XXV); 30–35 (V), 38–44 (VI), (according to posterior pair), 35–45 (VII) between spermathecal pores; 14–16 between male pores. Male pores 1 pair, situated ventrolaterally, each in a lateral crescent-shaped pouch, the region medial to it wrinkled but not glandular. The pore either represented by a tubercle or a minute penis-like process. No genital papillae therein and none in neighboring segments. Spermathecal pores 2 pairs, in 5/6/7, superficial and intersegmental, each with a very minute tubercle. Posterior pairs visible from dorsal aspect, about 4/7 circumference apart ventrally; anterior

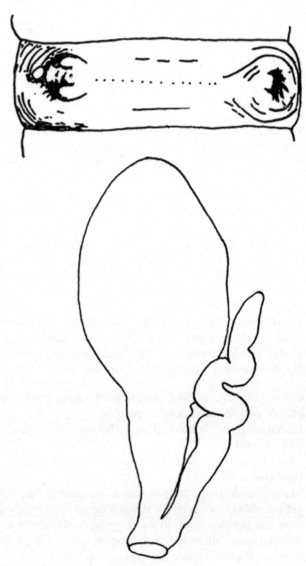

FIG. 258   *Metaphire exiloides* (after Chen, 1936).

pair more ventrally situated, about 4 or 5 setae ventral. No genital papillae in or around the pores. Female pore single, medioventral in XIV.

Internal Characters

Septa 6/7–9/10 very thick, 10/11 less so, 5/6 thin, 11/12/13 only slightly thicker than the following; rest septa thin and membranous. Gizzard very small, slightly larger than diameter of esophagus, soft and not very muscular, in front of septum 8/9. Intestine swelling in XV. Intestinal caeca simple, in XXVII, extending anteriorly to a small part of XXIV, moderate in size, very slightly wrinkled on both sides. Lateral hearts 4 pairs, in X–XIII, first pair fairly stout. Testis sacs in X V-shaped and connected medially, posterior pair also united ventrally. Seminal vesicles in XI, XII, very large, filling up the segments, each with a very minute dorsal lobe. Seminal vesicles of XI probably included within the posterior testis sacs. Prostate glands in XVII–XIX, with large and irregular lobes, its duct equal in caliber, U-shaped. Spermathecae 2 pairs, in VI, VII. Posterior pair large. Ampulla ovoid, indistinctly or clearly marked off from its duct which is shorter and broad; diverticulum shorter than main pouch, ental two-thirds loosely twisted, no particularly enlarged seminal chamber which is probably not yet differentiated, ectal one-third straighter, probably as its duct. Ampulla 1.1mm long, 0.72mm wide; diverticulum about 0.2mm wide; or diverticulum as long as ampullar duct, zigzag coiled and little swollen.

Color: Preserved specimens greenish on dorsum, grass green along dorsomedian line, pale on ventrum. Clitellum cinnamon brown.
Deposition of types: Previously deposited in the Museum of the Biological Laboratory of the Science Society of China, Nanjing. The types were destroyed during the war in 1937.
Etymology: The specific name "exiloides" is a compound word with "exil" probably from the word "axil" and "-oides" used to form genus names from other genus names and which means "resembling". It probably indicates the structures of the spermathecae.
Type locality: Chongqing.
Distribution in China: Chongqing (Beipei), Sichuan.
Remarks: This species resembles Metaphire exilis (Gates, 1935a) in several characters, such as body size, setal character, and the number of spermathecal pores. However, it differs in the prostomium being proepilobous, the first dorsal pore in 11/12, and the genital papillae around the male pores being absent. It was not noted whether the anterior 2 pairs of spermathecal pores were situated on the same level. The important character of the presence of the gizzard septa in M. exilis was neglected by either Gates or Chen. These are probably not absent. It is also a

question as to whether the vestigial ampulla of M. exilis is a specific character. It is fully developed in all specimens of this species.

flavarundoida Group

Diagnosis: Metaphire with spermathecal pores 3 pairs, at 4/5/6/7.
Global distribution: Oriental realm from Japan southwards through the Indo-Australasian archipelago to the rainforests of Australasia and eastwards through the Oceanian realm.
There is 1 species from China.
Remarks: Species of this genus are separable from those of Amynthas by the presence of copulatory pouches and they differ from species of Pheretima by the absence of nephridia from the spermathecal ducts.

267. Metaphire flavarundoida (Chen, 1935)

Pheretima flavarundoida Chen, 1935b. Bull. Fan. Mem. Inst. Biol. 6:51–56.
Metaphire flavarundoida Sims & Easton, 1972. Biol. J. Linn. Soc. 4:238.
Metaphire flavarundoida Xu & Xiao, 2011. Terrestrial Earthworms of China. 228.

External Characters

Length 95–145mm, width 6–8mm, segment number 138–178. Prostomium prolobous. Form rather short and robust; preclitellar segments relatively long but postclitellar ones short, about 20 segments at posterior end extremely short, weak annulation on IV and V, the number of annuli 3 in VI–VIII, 5–7 in IX–XIII, 5 in most segment of middle body, no annuli at posterior end. First dorsal pore in 11/12. Clitellum XIV–XVI, highly glandular and irregularly grooved, but smooth; setae generally traceable on ventral side, about 12 on XVI, 4–6 on each of the preceding ones; dorsal pores absent. Setae small and numerous; setae number 72–76 (III), 86–90 (IV), 104–106 (VI), 100–112 (VIII), 85–102 (XII), 82–92 (XVII), 74–85 (XXV); 42–45 (IV), 44–48 (V), 48–50 (VI) between spermathecal pores; 16 between male pores. Male pores 1 pair, on lateral one-third of a transverse ridge, about 1/3 circumference apart ventrally, laterally covered by a shallow crescent pouch; with minute papillae ventrally but obliquely placed on anterior and posterior sides of ridge, surrounded by a few circular ridges. 2 patches of minute papillae (about 40 each) on ventrolateral sides of XVII. Spermathecal pores 3 pairs, intersegmental in 4/5/6/7, situated rather dorsally, about 6/11 circumference apart ventrally; a skin flap which seem to be an outgrowth of the preceding segment usually situated in intersegmental furrow, glandular in character, either continuous or discontinuous; each pore found near lateroposterior side of the flap; small papillae

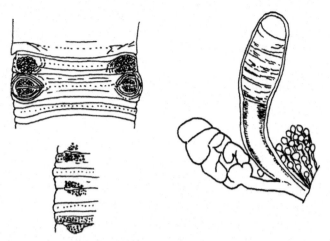

FIG. 259   *Metaphire flavarundoida* (after Chen, 1935b).

whether numerous (about 20) or few always found in front of and behind the flap, usually more on posterior side. No genital markings found outside this region. Female pore single, medioventral in XIV.

Internal Characters

Septa 8/9/10 absent; 4/5 very thin enclosed upon pharyngeal mass, 5/6/7/8 highly muscular and considerably thickened, nearly as thick as body wall but normally inserted; 10/11/12/13 (13/14) slightly muscular and thin. Gizzard in IX and X, moderate in size, round and dull, invested with large longitudinal vessels. Intestine swelling from XV, usually large in caliber. Intestinal caeca 1 pair, simple and very small, in XXVII, extending anteriorly to XXVI. Lateral hearts 4 pairs, in X–XIII, first 2 pairs about equal in caliber enclosed together with seminal vesicles, testis sacs and seminal funnels in a membranous sac, posterior 2 pairs stout. Commissural vessels in IX totally absent, leaving part of dorsal vessel paler, those in X asymmetrical, 1 limb extending down about half its way. Dorsal vessel large in caliber, each pair simple. Testis sacs usually absent, or common membranous sac communicating both dorsally and ventrally, in which the seminal products are stored (serving as testis sac). That in X larger on ventral side but narrower on dorsal side where the seminal products of both sides do not meet; that in XI completely communicating dorsally and well filled with seminal products. Testis rosette and large, in usual position. Seminal funnel iridescent and large. Sperm ducts of each side meeting in XII. Seminal vesicles 2 pairs, first pair in XI small, enclosed together with testes; second pair in XII larger, about 1.4mm in dorsoventral length, dorsal lobe large, whitish and granular on surface, about 1mm in

dorsoventral length. Prostate glands always compact and elevated, confined to XVII–XIX, divided into about 5 or 6 lobes, about 3mm in anteroposterior length; prostatic duct short and thick, about 2mm in length, with a U-shaped curve. Accessory glands small and closely clustered; each with a smooth ampulla-like glandular portion clearly distinguished from a long and slender duct leading out to respective papilla externally. Spermathecae 3 pairs, in V–VII; first 2 pairs similar in size, last pair slightly larger; ampulla generally small, spatulate or heart-shaped, about 1.5mm in length, not distinguishable from its duct which is broad and as long as, or longer than ampulla; diverticulum usually longer than ampulla, always compactly coiled on 1 plane, its coils indivisible, thick, and distended, in sections, appearing like compartments of a beehive, each filled with granules about 4 μm in diameter (probably mucin secretions) which take up an eosin stain; color on outer surface uniformly grayish buff, about one-third or one-fourth of the thickness of ampullar duct; diverticulum and its duct generally shorter than ampulla and its duct.

Color: Preserved specimens yellowish on both dorsum and ventrum, a little grayish on postclitellar region (due to intestinal content). Clitellum light chocolate brown.
Distribution in China: Hong Kong (Huangzhukeng).
Etymology: The name of the species refers to its locality ("*flava*" means yellow in Latin).
Type locality: Hong Kong (Huangzhukeng).
Deposition of types: Previously deposited in the Fan Memorial Institute of Biology, Beijing. The types were destroyed during the war in 1941–45.
Remarks: This is a very unique species in respect of its forward position of spermathecae, prolobous prostomium, enormous thickness of the anterior septa, membranous sac enclosing testis, seminal vesicle, seminal funnel and heart, blood sinuses along dorsal vessel, etc.

glandularis Group

Diagnosis: *Metaphire* with spermathecal pores 2 pairs, at 6/7/8.
Global distribution: Oriental realm from Japan southwards through the Indo-Australasian archipelago to the rainforests of Australasia and eastwards through the Oceanian realm.
There are 3 species from China.
Remarks: Species of this genus are separable from those of *Amynthas* by the presence of copulatory pouches and they differ from species of *Pheretima* by the absence of nephridia from the spermathecal ducts.

TABLE 31   Key to Species of the *glandularis* Group From China

1. Without papillae in spermathecal pore region   *Metaphire jianfengensis*
   With papillae in spermathecal pore region 2

2. 1 pair of presetal ventral papillae in VIII .... *Metaphire dadingmontis*
   1 pair of flat-topped postsetal ventral papillae in VII ...... *Metaphire nanlingmontis*

268. *Metaphire dadingmontis* Zhang, Li, Fu & Qiu, 2006

Metaphire dadingmontis Zhang, Li, Fu & Qiu, 2006, *Annales Zoologici*. 56(2):253.

External Characters

Length 80–85 mm, clitellum width 2.9–3.4 mm, segment number 82–90, secondary annuli not present. Prostomium epilobous. First dorsal pore 12/13. Clitellum in XIV–XVI, annular, dorsal pores present, setae invisible externally. Setae minute, setae number 32–40 (III), 30–36 (V), 38–42 (VIII), 40–44 (XX), 43–46 (XXV); 7–9 between male pores. Male pores 1 pair, in XVIII, about 0.4 of circumference apart ventrally, on the surface of a small protuberance situated in a large copulatory chamber in XVIII, surrounded by longitudinal epidermis folds. No other male region papillae. Spermathecal pores 1 pair, in 6/7/8, ventral, about 0.33 circumference apart ventrally; 1 pair of presetal ventral papillae in VIII. Female pore single, medioventral in XIV.

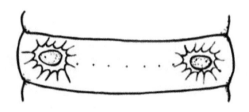

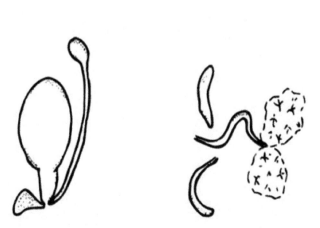

FIG. 260   *Metaphire dadingmontis* (after Zhang et al., 2006).

Internal Characters

Septa 8/9/10 absent. Gizzard in VIII–X, small drum-like. Intestine enlarged from about XVI. Intestinal caeca simple, paired in XXVII, extending anteriorly to XXIV. Esophageal hearts not developed well in X–XIII. Testis sacs in X and XI, small, ball-like, connected on ventrum. Seminal vesicles in XI and XII, well developed, with inconspicuous dorsal lobes. Prostate glands paired in XVIII, moderately large. Prostatic duct simple curved. 2 finger-shaped accessory glands present. Spermathecae 2 pairs in VII–VIII, ampulla oval-shaped, about 2 mm long, 0.8 mm wide, with a straight stout stalk about 0.4 mm long. Diverticulum passed into the duct at the base. Diverticulum at least as long as the whole spermatheca, distal seminal chamber enlarged. Nephridia absent on the spermathecal ducts. Accessory glands mushroom-like.

Color: Preserved specimens light yellowish brown on dorsum, grayish white on ventrum. Clitellum slightly yellowish white.
Distribution in China: Guangdong (Nanling).
Etymology: The species is named for its type locality of Dadingshan station in Nanling National Natural Reserve.
Type locality: Guangdong (Nanling).
Deposition of types: Laboratory of Environmental Biology, School of Agriculture and Biology, Shanghai Jiaotong University, China.
Remarks: *Metaphire dadingmontis* appears to be closely related to *Metaphire nanlingmontis* Zhang, Li, Fu & Qiu, 2006. They both have the same spermathecal pore number and position, and similar male pore copulatory pouch. However, *M. dadingmontis* is much smaller in body size. In addition, it is also easy to distinguish them by their different diverticulum shape and accessory glands in prostatic region.

269. *Metaphire jianfengensis* (Quan, 1985)

Pheretima jianfengensis Quan, 1985. *Acta Zootaxonomica Sinica*. 10(1):18–20.
Metaphire jianfengensis Zhong & Qiu, 1992. *Guizhou Science*, 10(4):41.
Metaphire jianfengensis Xu & Xiao, 2011. *Terrestrial Earthworms of China*. 236.

External Characters

Length 160–250 mm, width 6–10 mm, segment number 131–173. Prostomium 1/2 epilobous. First dorsal pore 12/13. Clitellum XIV–XVI, annular, setae absent. Setae evenly distributed; setae number 52–56 (III), 58–60 (V), 65–66 (VIII), 76–78 (XII), 74–76 (XVIII), 82–84 (XXV); 12–15 (VII), 14–15 (VIII) between spermathecal pores, 4–6 between male pores. Male pores in ventrum of XVIII,

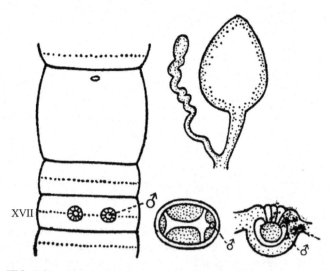

FIG. 261   *Metaphire jianfengensis* (after Quan, 1985).

about 1/4 circumference apart ventrally. The male pore is an invagination formed by the extrusion of 2 round papillae and 3 small skin pads on the inner wall of pouch. Spermathecal pores 2 pairs in 6/7/8, about 1/4 circumference apart ventrally. Female pore single, medioventral in XIV.

Internal Characters

Septa 8/9/10 absent, 4/5–7/8 thin membrane, 10/11/12/13 very thick, 13/14 slightly thin. Gizzard bead-shaped, in IX–X. Intestine swelling from XV. Intestinal caeca simple, in XXVII, extending anteriorly to XXV. Lateral hearts 4 pairs, in X–XIII. Testis sacs 2 pairs, anterior pair slightly small, behind ventral side of septum 10/11 in X, a hemisphere-like swelling, slightly broad at base, but not connected; posterior pair not developed. Seminal vesicles well developed, anterior pair with triangular-shaped dorsal lobes, only occupy sac 1/4–1/3, the left lobes and the right lobes separated on dorsum but united on dorsal vessels;posterior pair with larger dorsal lobes, lobate or long and triangular, the left lobes and the right lobes coiled together but not connected or communicated. The first pair of seminal vesicles encloses the second pair of seminal vesicles. Prostate glands well developed, in XVII–XIX, fan-shaped, with many interlaced lobules; prostatic duct C-shaped, thick and large, but ental ends slender. Accessory glands absent. Spermathecae 2 pairs, in VII–VIII, ampulla oval-shaped, distal end slightly blunt, diverticulum shorter than main pouch, usually curved, distal end serving as seminal chamber, passing into the main duct close to the body wall.

Color: Live specimens reddish, slightly transparent. Clitellum red-brown.
Distribution in China: Hainan (Jianfengling).
Etymology: The name of the species refers to the type locality.
Type locality: Hainan (Jianfengling).

Deposition of types: Institute of Tropical Forestry, Chinese Academy of Forestry Science, Guangzhou.
Remarks: This species is similar to *Metaphire multitheca* (Chen, 1938), in the size of body and presence of the special copulatory pouch, but differs in having 2 pairs instead of many pairs of spermathecae and spermathecal pores, the spermathecae larger, and ovoid instead of pot-like in shape instead of much smaller and pot-like, and the spermathecal pores being situated intersegmenatlly in 6/7 and 7/8 instead of on posterior border of VI, VII, and VIII.

270. *Metaphire nanlingmontis* Zhang, Li, Fu & Qiu, 2006

Metaphire nanlingmontis Zhang, Li, Fu & Qiu, 2006. *Annales Zoologici.* 56 (2):252–253.

External Characters

Length 110–150 mm, width 3–4 mm, segment number 101–150, secondary annuli not present. Prostomium epilobous. First dorsal pore 12/13. Clitellum XIV–XVI, annular, smooth. Setae tiny, setae number 36–45 (III), 33–54 (V), 45–60 (VIII), 45–66 (XX), 39–72 (XXV); 10–14 (VII) between Spermathecal pores, 9–10 between male pores. Male pores on ventrolateral region of XVIII, about 0.33 circumference apart ventrally, on the surface of a small protuberance which is situated in a large copulatory pouch, surrounded by longitudinal epidermis folds. No other male region papillae. Spermathecal pores 2 pairs, in 6/7/8 ventral, about 0.33 circumference apart ventrally, 1 pair of flat-topped postsetal ventral papillae in VII. Female pore single, medioventral in XIV.

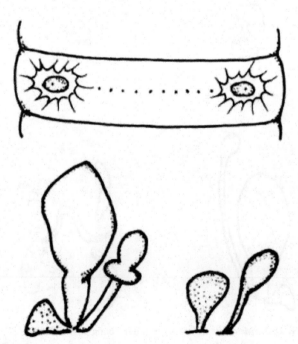

FIG. 262   *Metaphire nanlingmontis* (after Zhang et al., 2006).

Internal Characters

Septa 8/9/10 absent, 5/6, 11/12/13 comparatively thick. Gizzard in VIII–X, small drum-like. Intestine starts from XV but distinctly enlarged from XXII. Intestinal caeca simple, paired in XXVII, smooth, extending anteriorly to XXIII. Lateral hearts moderately large in X–XIII. Testis sacs in X and XI, small, ball-like, connected on ventrum. Seminal vesicles in XI and XII, well developed, small dorsal lobes with coarse surface due to presence of many small grains. Prostate glands paired in XVIII, well developed, extending anteriorly to XVI and posteriorly to XXI. Prostatic duct U-shaped, thick at lateral parts. Accessory glands mushroom-like or small sac-like with short stalk. Spermathecae 2 pair in VII–VIII. Ampulla broad oval-shaped, about 2.2 mm long, 0.9 mm wide, with a short, straight stout stalk about 0.5 mm long. Diverticulum passed into the spermathecal duct at the base. Diverticulum stalk a little shorter than the main pouch, distal seminal chamber enlarged, in addition, a special ring-like swelling at the border between the duct and chamber. Nephridia absent on the spermathecal ducts. Accessory glands mushroom-like.

Color: Preserved specimens light yellowish brown on dorsum, grayish white on ventrum. Clitellum slightly yellowish white.
Distribution in China: Guangdong (Nanling).
Etymology: The name of the species refers to the type locality.

Type locality: Guangdong (Nanling).
Deposition of types: Laboratory of Environmental Biology, School of Agriculture and Biology, Shanghai Jiaotong University, China.
Remarks: M. nanlingmontis (Zhang, Li, Fu & Qiu, 2006) appears to be related to M. jianfengensis (Quan, 1985). They have the same spermathecal pore number and position, and similar male pore copulatory pouches. However, M. nanlingmontis is much smaller in body size than M. jianfengensis. In addition, there is a special ring-like swelling at the border between the duct and chamber in the diverticulum, and a mushroom-like accessory gland present beside the duct.

houlleti Group

Diagnosis: Metaphire with spermathecal pores 3 pairs, at 6/7/8/9.

Global distribution: Oriental realm from Japan southwards through the Indo-Australasian archipelago to the rainforests of Australasia and eastwards through the Oceanian realm.
There are 20 species from China.
Remarks: Species of this genus are separable from those of Amynthas by the presence of copulatory pouches and they differ from species of Pheretima by the absence of nephridia from the spermathecal ducts.

TABLE 32 Key to species of the *houlleti* Group From China

1. Spermathecal pores 3 pairs in 6/7/8/9 dorsally .... *Metaphire hunanensis*
   Spermathecal pores 3 pairs in 6/7/8/9 ventrally 2

2. Genital papillae located in male pore area and spermathecal pore area 3
   Genital papillae located neither in male pore area nor spermathecal pore area 5

3. Copulatory pouches much deeper, situated rather laterally, with a large (round or elongate) button-shaped papilla at the medial side and 2 smaller ones at the bottom; another pointed papilla at the bottom of the chamber with a male pore on its tip, often protruding laterally. 2 or more large button-shaped papillae situated posterior and anterior to each spermathecal pore .... *Metaphire kiangsuensis*
   Papillae size similar in male pore area 4

4. Male pores wrinkled laterally including a crescent groove which indicates the opening of the chamber. It is eversible with a nipple-like bulb on the top and below which are several circular grooves which mark off bulb from the large base, a small flat-topped papilla occasionally present at the anteromedian side of the chamber. There are 2 small papillae posteriorly and both papillae may be absent from spermathecal cavity ....
   Metaphire vulgaris Agricola
   Male pores in ventrum of XVIII each on small copulatory pouches with a small opening and surrounded by skin folds, 1–4 small papillae in pouch (sometimes 1–2, sometimes outside). Spermathecal pores each bordered anteriorly and posteriorly by glandular lips, 1–2 small papillae in anterior lips .... *Metaphire kokoan*

5. Genital papillae located anterior clitellum 6
   No genital papillae located anterior clitellum 13

6. Genital papillae begin in VI 7
   Genital papillae begin posterior VI 9

7. Papillae located anterior and posterior spermathecal pores 8
   Spermathecal pores on tiny, conical protuberances into large club-shaped spermathecal chambers, apertures of spermathecal chambers large, transversely slit-like, 1 fairly large papilla situated on anterior side .... *Metaphire graham*

Continued

TABLE 32   Key to species of the *houlleti* Group From China—cont'd

8. The spermathecal depression anteriorly or posteriorly or both, 1–2 very small simple swelling-like genital papillae are often found in VI–IX or on some of them. As such, the papillae of the present species are clearly different in size and in position from *Metaphire asiatica* (Michaelsen, 1900) *Metaphire aggera*
Spermathecal aperture in a not obviously lip-shaped transverse slit .... *Metaphire tschiliensis*

9. Segment VII, VIII and IX with papillae located in front or behind the setae
Spermathecal pores each with a wide slit as the outlet of the "spermathecal chamber", wrinkled at both anterior and posterior margins, better seen at the anterior margin. The cone-shaped structure as the terminal part of the spermathecal duct ordinarily inside of the chamber is protruded with 2 papillae visible externally .... *Metaphire vulgaris*

10. Genital papillae located on the ventral region of spermathecal pores
Spermathecal pores each on a tubercle in intersegmental furrow, sometimes anterior and posterior of apertures lip-shaped swelling; transversely broad slit-like, with shallow spermathecal chamber; ventral side of segment VII, VIII, and IX with 1 pair small papillae respectively, 1 specimen with 1 pair of papillae on presetal XIII adjacent to intersegmental furrow or ventral region VII .... *Metaphire viridis*

11. Spermathecal pores in a not obviously lip-shaped transverse slit .... *Metaphire tschiliensis*
Genital papillae 1 or more pairs 12

12. Spermathecal pores usually invisible externally. 5 pairs of genital papillae on the ventral side of segment VII, VIII, and IX .... *Metaphire youyangensis*
Spermathecal pores on tiny tubercles located in parietal invaginations with transversely slit-like apertures, 1 circular genital marking on the anterior wall of each invagination .... *Metaphire praepinguis*

13. Genital papillae located male pores area 14
No genital papillae located in front of male pore and Clitellum

14. Papillae within copulatory pouch 15
Papillae outside copulatory pouch 17

15. The male pore in a shallow and slitlike copulatory pouch, On these lobes or between the lobes there are setae, usually 2–3 for each invagination *Metaphire guillelmir*
Many papillae within copulatory pouch 16

16. Male pores very large, nearly circular holes with thick well-developed lips, the edge of these holes is formed of a prepuce-like fold, which holds a thick and short pistil-like penis. This short thick penis shows in the middle point of its tip the minute invisible primary male aperture, and near this, some small circular gland papillae, which are not always clearly recognizable. There is a short broad, penis pouch whose wall surrounds the short broad penis like a prepuce .... *Metaphire tibetana*
A round porophore protruded from the male pore and surrounded by its papillae border .... *Metaphire ichangensis*

17. Male pores large, C-shaped, reaching 17/18 and 18/19 segmental furrows; porophore, a small tubercle with a small male aperture at the laterocentral end of male disc, a narrow longitudinal ridge extending from setal annulus of XVIII. A pair of large, nearly equally sized round-shaped genital papillae located on anterior and posterior ends of the male disc. Each of the genital papillae with a slightly concave center. Lateral wall of the copulatory pouch smooth externally (laterally) but slightly folliculated internally .... *Metaphire trutina*
Male pores within shallow copulatory pouches with C-shaped opening and surrounded by skin folds, 2 setae situated between papillae in each pouch; usually with 1 small flat papilla on opening, entirely or half outside; 1 specimen with 2 flat papillae, line-shaped, 1 inside; posterior margin of flat papillae with 4–5 bar-shaped protuberances; the papilla at the bottom of the chamber with male pore on its tip .... *Metaphire lanzhouensis*

18. Male pores situated in a rather shallow copulatory pouch, no papilla in or out of the copulatory pouch .... *Metaphire ptychosiphona*
Male pores in copulatory pouch with C-shaped opening (slit), surrounded by a round swollen area with numerous transverse ridges .... *Metaphire houlleti*

271. *Metaphire aggera* (Kobayashi, 1934)

Pheretima aggera Kobayashi, 1938b. *Sci. Rep. Tohoku Univ.* 13:153–155.
Pheretima aggera Kobayashi, 1940. *Sci. Rep. Tohoku Univ.* 15:273–277.
Metaphire aggera Sims & Easton, 1972. *Biol. J. Linn. Soc.* 4:238.
Metaphire aggera Xu & Xiao, 2011. *Terrestrial Earthworms of China.* 213–214.

External Characters

Length 175–298 mm, width 5.5–10.0 mm. Segment number 150–171. First dorsal pore 12/13, distinct and functional; an indistinct and nonfunctional pore-like marking may be rarely found in 11/12. Clitellum entire, in XIV–XVI, annular, setae absent. Setae moderate in size; setae number 34–47 (IV), 52–67 (XII); 15–22 (VI), 16–23 (VII), 16–24 (VIII) between spermathecal pores, 16–23 between male pores; 3–4 of the male pore setae are always found on the medial part of male disc, and may be invisible when latter are withdrawn into the coelom. Male pores in XVIII, crescent-shaped, with moderate elevation; the body wall lateral to the pore is light-colored, prominently elevated and lacks the setae; within the shallow copulatory pouch is found a wrinkled and glandulated male disc, on which 3–4 setae planted; on the disc, at the extreme lateral side, which is slightly lower than the medial part or often slightly sunken into the coelom, is found a very small, transversely club-shaped male porophore. The porophore is anteriorly and posteriorly provided with 2 very small oval brownish yellow papilla-like swellings which are half buried into the ground tissue. Male pore opens as a minute aperture at the lateral part of the club-shaped porophore. These porophore and papilla-like swellings form a small but firm, laterally directed arrowhead-like body; the body may be variable in form corresponding to the degree of contraction of the specimens. Spermathecal pores 3 pairs in 5/6/7/8, each situated on a small papilla-like swelling which is found within a large, eye-shaped depression. Both anterior and posterior borders of the depression are moderately swollen. When the depression is closed, the pore is invisible externally. To the spermathecal depression anteriorly or posteriorly or both, 1–2 very small simple swelling-like genital papillae are often found in VI–IX or on some of them. The papillae of the present species are clearly different in size and in position from *Metaphire asiatica* (Michaelsen, 1900). Female pore single, medioventral in XIV.

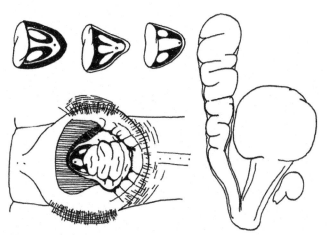

FIG. 263 *Metaphire aggera* (after Kobayashi, 1941).

Internal Characters

Septum 9/10 absent, 8/9 ventrally present but thin. Intestine swelling from XV. Intestinal caeca manicate, large and long, in XXVII, finger-shaped with broad basal portion, each ventrally with several serriformed outgrowths, extending anteriorly to XXII. Lateral hearts in X–XIII, fairly large; those in X subequal in caliber to those

in XI–XIII. Testis sacs ventral in position and moderate in size; the anterior pair forming a low or a very low U-shaped sac and the posterior pair a quadrate sac. Seminal vesicles moderate in size or often rather small, circular or oval in shape, vesicular on surface, each with a moderately-sized, distinctly constricted, ovoidal, smooth dorsal lobe; usually the anterior pair of vesicles are smaller than the posterior. Prostate gland small, occupying only 2 or more segments, in XVIII (or 1/2 XVII)–XIX, divided into many finger-shaped pieces. Prostatic duct usually curved in a U-like manner, the ental half is slender being of nearly equal thickness while the ectal half is very thick and musculated, being about 5 or more times as thick as the former. With the narrowed ental end, the duct enters, through a cushion-like, moderately-sized glandular tissue of the male disc, into the arrowhead-like body, and opens at the tip of the porophore within the shallow copulatory pouch; on both sides of the ectalmost of the duct, 2, very small, oval-shaped, apparently soft accessory glands are recognizable embedded in the tissue without a stalk. Spermathecae large, 3 pairs in VI–VIII, ampulla rounded, sometimes unevenly or irregularly wrinkled on surface. Spermathecal duct thick and musculated, slightly narrowed in the ectal part, nearly equal in length to the ampulla, sharply marked off from the latter; diverticulum arising from the ectal end of the ampullar duct, always longer than the main portion; the ectal portion which serves as duct is short and subequal in length to the ampullar duct, which is thick-walled and nearly straight, while the remaining longerental portion is thin-walled, slightly dilated, coiled into several compact loops running in zigzag manner, and is enclosed within a delicate sheath. Small stalked accessory glands are found inside corresponding to genital papillae externally present.

> Color: Live specimens dark brown on dorsum and concentrated on the preclitellar segments, dark gray on ventrum, clitellum reddish chocolate. Preserved specimens dark brown on dorsum, concentrated middorsally and on preclitellar segements, light brown or sometimes rather pale on ventrum. Clitellum chocolate.
> Deposition of types:
> Type locality: Chongqing (Shapingba, Xufu).
> Distribution in China: Liaoning (Jinzhou, Huludao, Dashiqiao, Dalian), Inner Mongolia (Chifeng).
> Remarks: This species is similar to *M. asiatica* (Michaelsen, 1900), but it can easily be distinguished by the following points: (1) the minute structure of the terminalia of the male gentialia; (2) the size, shape and relative position of genital papillae found in the spermathecal region; (3) the size and shape of both seminal vesicle and prostate (with duct).

272. Metaphire asiatica (Michaelsen, 1900)

Amynthas asiaticus Michaelsen, 1900. *Das Tierreich, Section* 10.

Pheretima tschiliensis Gates, 1939. *Proc. U.S. Natn. Mus.* 85:488–494.

Pheretima tschiliensis Kobayashi, 1940. *Sci. Rep. Tohoku Univ.* 15:277–282.

Pheretima asiatica Chen, 1958. *The Commercial Press.* 96–103.

Metaphire asiatica Sims & Easton, 1972. *Biol. J. Linn. Soc.* 4:236.

Metaphire asiatica Xu & Xiao, 2011. *Terrestrial Earthworms of China.* 215.

External Characters

Length 154–197 mm, width 7–8 mm, segment number 140–145. Prostomium 1/2 epilobous, moderate in length and width. First dorsal pore 12/13. Clitellum XIV–XVI, annular; intersegmental furrows, setae and dorsal pores absent. Setae moderate in size, usually those on II–IX slightly enlarged. Middorsal breaks very slight if present, and midventral break slight; setae number 38–43 (III), 50–54 (V), 54–56 (VI), 53–57 (VIII), 57–60 (IX), 59–61 (XII), 57–58 (XIII), 63–69 (XXV); 20–22 (VI), 21–24 (VII), 22–26 (VIII), 23–28 (IX) between spermathecal pores; 21–24 between male pores. Male pores paired in XVIII, crescent-shaped, the concave side of the crescent facing midventrally. The lateral wall of the invagination is thin and without setae; the ventral margin of this wall forming a crescentic, lateral lip of the aperture of the invagination. Spermathecal pores, minute, widely separated; 3 pairs, in 6/7/8/9. Each pore is at the center of a tiny, circular to transversely oval, smooth area, the margins of which are usually not clearly demarcated. The spermathecal pores are readily recognizable though minute owing to the presence of a tiny whitish rim around the margin of a pore. A spermathecal pore tubercle may be very slightly depressed into the body wall. Female pore single, medioventral in XIV.

Internal Characters

Septa 8/9/10 absent, 5/6/7/8, 10/11/12 much thickened, 12/13 less thickened. Gizzard behind 7/8, globular in shape. Intestine swelling from XV. Intestinal caeca simple, large and long, horn-shaped, in XXVII, extending anteriorly to XXII or XXIII, each ventrally with several serriformed outgrowths. Lateral hearts in X–XIII; those in X are large, but sometimes they are bound to the anterior face of 10/11. Testis sacs ventral in position, massive and large; the anterior pair forming a U shape and the posterior pair a transverse sac. Seminal vesicles, vesicular, voluminous, fairly large, occupying all the interior of the respective segment and dorsally meeting with each other; each with a large dorsal lobe. Dorsal lobe of the

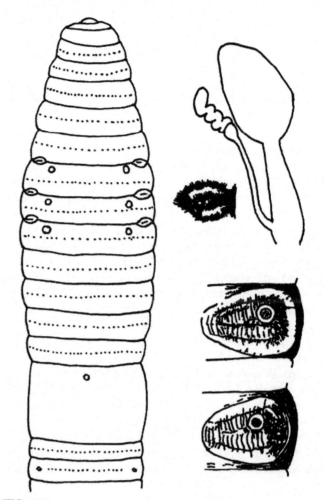

FIG. 264 *Metaphire asiatica* (after Kobayashi, 1940).

anterior pair may sometimes be subdivided into 2–3 smaller lobes. Prostate gland large, usually in XVI–XXI, divided into many finger-shaped pieces; duct long, looped in a hairpin-shape. Accessory glands with solid stalks are found close to the ectal end of the prostatic duct corresponding externally to the smaller papillae. A similar gland but of larger size, corresponding to the larger papilla is found near these. Spermathecae fairly large, 3 pairs in VII–IX, ampulla large, ovoidal, with smooth surface; ampullar duct thick, slightly shorter than the diameter of ampulla, not sharply marked off from the latter, slightly enlarged at the ectal part; diverticulum subequal to (or a little shorter than) the main portion; the ectal half slender but thick-walled, entering into the ampullar duct at its ectal end close to the parietes, and the ental half thin-walled, slightly dilated, coiled into several compact loops running in a zigzag manner; a delicate sheath present around these but rather indistinct. No accessory glands are found close to the ectal end of the ampullar duct. Slightly posteriorly to each spermatheca is found a large accessory gland with stalk; its glandular portion is often divided into 2 or more numbers of smaller lobes.

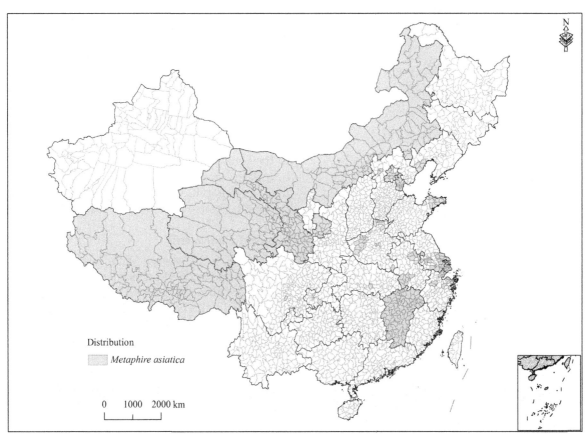

FIG. 265 Distribution in China of *Metaphire asiatica*.

Color: Live specimens dorsally brownish blue anteriorly, purplish brown posteriorly, ventrally dusty gray or pale. Clitellum dusty chocolate. In preserved specimens the bluish color has faded away.
Etymology: The name of the species refers to the type locality.
Type locality: Tianjin.
Distribution in China: Beijing (Xicheng, Dongcheng, Haidian, Tongzhou, Changping, Shunyi, Huairou, Fangshan, Fengtai, Shijingshan, Mentoyugou, Yanqing, Pinggu), Tianjin, Hebei (Xuanhua), Liaoning (Huludao), Shanghai, Jiangsu (Nanjing, Zhenjiang, Suzhou, Wuxi, Yixing, Nantong, Puzhen), Zhejiang (Zhoushan, Ningbo, Tonglu, Xindeng, Fenshui, Shengxian, Jiaxing), An'hui (An'qing, Chuzhou), Shandong (Yantai, Weihai), Jiangxi, Hubei (Qiangjiang), Henan (Boai, Anyang, Luoyang, Xishan, Yanling, Xuchang, Shangcheng), Chongqing (Beipei, Nanchuan, Shapingba), Sichuan (Mt. E'mei), Xizang, Neimenggu, Gansu, Shaanxi (Jiushan, Daixian), Qinghai.
Remarks: *M. asiaticus* was first recorded in 1900 by Michaelsen, and then mentioned in the paper about earthworms of Tibet in 1902. In *The Oligochaeta of China* published by Michaelsen in 1931, the records of *"Pheretima (Ph.) asiatica* (Mich.) (= *Amyntas asiaticus* Mich) Prov. Hopei, Tientsin (Michaelsen, 1900). -*Ph. asiatica* is an endemic species." and *"Pheretima (Ph.) tschiliensis* Mich. Prov. Hopei, 50 Chinese miles east of Hsanhua-hsin (Michaelsen, 1928a) -Endemic." indicated the relationship between the 2 species. *Pheretima asiatica* also mentioned in *Notes on a Small Collection of Earthworms from Ichang, Hupeh*, which was published by Fang Bingwen in 1933, but just recorded that "setae 50/V" and "male pore with round papilla".

Pheretima tschiliensis recorded by Gates (1939), was based on the descriptions of *Pheretima asiatica* in 1903, *Pheretima tschiliensis* in 1928, *Pheretima kiangsuensis* in 1930, and *Pheretima tibetana* in 1931. However, the record was obviously different from Michaelsen's description of *Pheretima tschiliensis* in 1928, especially in body color, spermathecae shape, diverticulum shape, and caeca shape. Therefore, unlike the past records that regarded *Pheretima tschiliensis* as *Metaphire asiatica*, *Pheretima tschiliensis* and *Pheretima asiatica* were recorded as 2 species in Xu & Xiao (2011). If these 2 forms belong to the same species, it should be named *Metaphire asiatica* (Michaelsen, 1900) instead of *M. tschiliensis* by principle of preference of time. Based on the above content, *M. asiatica*, which was regarded as a subspecies of *M. tschiliensis*, is a single species.

273. *Metaphire grahami* (Gates, 1935)

Pheretima grahami Gates, 1935a. *Smithsonian Mus. Coll.* 93(3):9–10.

Pheretima grahami Chen, 1936. *Contr. Biol. Lab. Sci. Soc. China (Zool).* 11:298–299.

Pheretima grahami Gates, 1939. *Proc. U.S. Natn. Mus.* 85:437–439.

Metaphire graham Sims & Easton, 1972. *Biol. J. Linn. Soc.* 4:238.

Metaphire graham Xu & Xiao, 2011. *Terrestrial Earthworms of China.* 228–229.

External Characters

Length 230–285 mm, width 11–15 mm, segment number 116–131. Prostomium 1/2 epilobous, width of dorsal lobe about 2/3 of length of II. First dorsal pore 12/13 or 13/14. Clitellum XIV–XVI, annular. Setae number 42–44 (III), 65–66 (VI), 69–80 (VIII), 76–82 (XII), 90–115 (XXV); 24–25 (VI), 24–27 (VIII) between spermathecal pores; 12–19 between male pores. Male pores on broadly conical

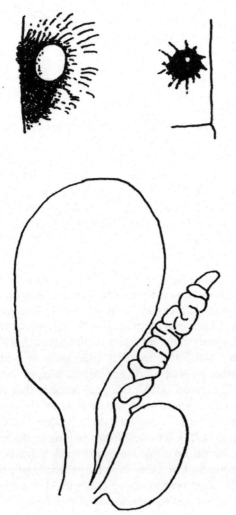

FIG. 266   *Metaphire grahami* (after Chen, 1936).

tubercles in large copulatory pouches with apertures approximating to transversely slit-like, about 1/3 circumference apart ventrally. External genital markings lacking; 5–6 circular or oval, flat-surfaced markings in each copulatory pouch; 1 large oval marking on the dorsal wall of each spermathecal chamber. Spermathecal pores on tiny, conical protuberances into large club-shaped spermathecal chambers, the latter invaginated posteriorly and deeply into the coelom and bound by connective tissue to the ventral parietes; apertures of spermathecal chambers large, transversely slit-like, 3 pairs, in 6/7/8/9. 1 fairly large papilla situated on anterior side but concealed in the slit. Female pore single, medioventral in XIV.

Internal Characters

Septa 8/9/10 absent, 5/6/7/8 and 10/11/12/13 thickly muscular, 13/14 muscular. Intestine swelling from XV. Intestinal caeca elongate, in XXVII, extending anteriorly to XXII, with ventral tooth-like projections. First pair of hearts in X present. Testis sacs of X and XI unpaired and ventral. Seminal vesicles of XI and XII are firm vertical bodies filling their segment, and in contact transversely above the dorsal blood vessel; with distinct dorsal lobes. Prostate glands extend through XVII–XIX, prostatic duct is about 10 mm long, bent into a C shape, ectal two-thirds much thickened. On the floor of XVIII on each side and just median to the ectal end of the prostatic duct is a large glandular mass, which can be separated with care into several discrete glands, from each of which a bundle of cords or ducts passes to a genital marking. Spermathecal ampulla sac-like, about 6 mm long, with a rather slender duct; diverticulum slightly shorter than main pouch; with a horn, muscular stalk and an elongate tubular seminal chamber, the latter often looped in a regular zigzag fashion. Accessory gland large and globular, about 2 mm in diameter, closely placed anterior to ampullar duct.

Distribution in China: Sichuan (Ya'an, Panzhihua, Xichang, Daxiangguanling).
Etymology: The species was named after Dr. David Crockett Graham (1884–1961), who found many of the specimens of this species in Sichuan.
Type locality: Sichuan (Daxiangling).
Deposition of types: Smithsonian Institution.
Remarks: *Metaphire graham* (Gates, 1935a) is distinguished from *Metaphire vulgaris* (Chen, 1930) by the ventral unpaired testis sacs of X and XI, the larger size of the spermathecal chamber, the posterior direction of the chamber, the attachment of the chamber to the ventral parietes, and the single large genital marking within the chamber. It is noteworthy that *M. graham* can be easily distinguished from *M. asiatica* (Michaelsen, 1900) by the primary spermathecal pores: those of the former are contained

within an invagination so large that it not only passes through the parietes into the coelomic cavity but extends posteriorly on the ventral parietes well toward the septum next behind, while those of the latter are superficial.

274. *Metaphire guillelmi* (Michaelsen, 1895)

Pheretima guillelmi Gates, 1935a. *Smithsonian Mus. Coll.* 93(3):10.

Pheretima guillelmi Chen, 1936. *Contr. Biol. Lab. Sci. Soc. China (Zool).* 11:270.

Pheretima guillelmi Gates, 1939. *Proc. U.S. Natn. Mus.* 85:440–445.

Metaphire guillelmi Sims & Easton, 1972. *Biol. J. Linn. Soc.* 4:238.

Metaphire guillelmi Xu & Xiao, 2011. *Terrestrial Earthworms of China.* 229–230.

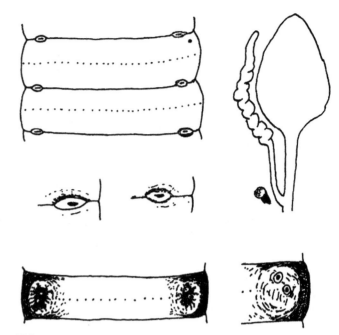

FIG. 267   *Metaphire guillelmi* (after Chen, 1959).

External Characters

Length 96–150 mm, width 5–8 mm, segment number 88–156. Prostomium epilobous. First dorsal pore 12/13 or 13/14. Clitellum XIV–XVI, annular, setae and dorsal pore absent. Setae begin on II. Middorsal and midventral breaks may be entirely lacking in the setal circles; the middorsal breaks, when present, variable in width; a midventral break, when present, very slight. Setae number 12–19 (VI), 13–22 (VIII), 14–22 (XVII), 14–21 (XVIII), 16–22 (XIX), 44–64 (XX). Male pores paired in XVIII, the apertures of the male pore invaginations are crescentic, the concave side facing midventrally. In many specimens the apertures gape open, disclosing more or less of the medial wall of the invagination. The invaginations are rather shallow. The lateral lip or wall is thin and lacks setae. The median wall of the invagination is firmer than the lateral wall and is ridged, the ridge in line with the male setae and cut up, as a rule, by short furrows into fine lobes. On these lobes or between the lobes there are setae, usually 2–3 for each invagination, which have been included in the male setae in the preceding table. On the ventral face of the tubercle is the minute male pore. There are no definite genital markings or papillae within the invagination aside from the male pore tubercle. The lobulations of the median ridge sometimes look much like genital markings or tubercles, especially in 1 worm where 1 of the lobulations in each invagination has been crowded anteriorly into a position just in front of the main portion of the ridge. No pores have been found on these lobulations or demarcation into rims and central areas as on genital markings associated with glands. Spermathecal pores 3 pairs, in 6/7/8/9, each pore is a transverse slit opening into a deep pit. Female pore single, medioventral in XIV.

Internal Characters

Septa 8/9/10 absent, 5/6/7/8 and 10/11/12/13 thickly muscular, 13/14 muscular, 14/15 slightly muscular. Gizzard large in VIII and IX. Intestine swelling from XV. Intestinal caeca in XXVII, extending anteriorly to XXII, simple, but the ventral margins, especially posteriorly, are slightly incised in such a way as to produce an appearance of a row of very short but definite lobulations. There is a small, whitish, occasionally lobed, glandular collar on the esophagus just behind the gizzard. All hearts of IX–XIII pass into the ventral vessel. Testis sacs of X and XI are ventral and unpaired. The seminal vesicles of XI and XII fill their segments and reach into contact transversely above the dorsal vessel. The prostate glands extend through XVI or XVII to XIX, XX, or XXI; the prostatic ducts are 6–10 mm long, each duct usually bent into a hairpin shape, with the ectal limb much thicker than the ental limb; or C-shaped or U-shaped. No stalked glands or glandular masses can be found on the parietes or within the parietes in the vicinity of the prostatic ducts. Spermathecae 3 pairs, in VII, VIII and IX, ampulla saclike. Spermathecal duct is smooth, the coelomic portion about the same diameter throughout and about equal in length to the ampulla; the diverticulum passes into the anterior face of the duct close to the parietes, ectal to this junction the duct is much narrowed. The diverticular stalk is slender, smooth, and firm, about equal in length to the coelomic portion of the spermathecal duct or slightly shorter, but always shorter than the seminal chamber. The latter is wider than the stalk, thin-walled, and looped in a zigzag manner, apparently within a

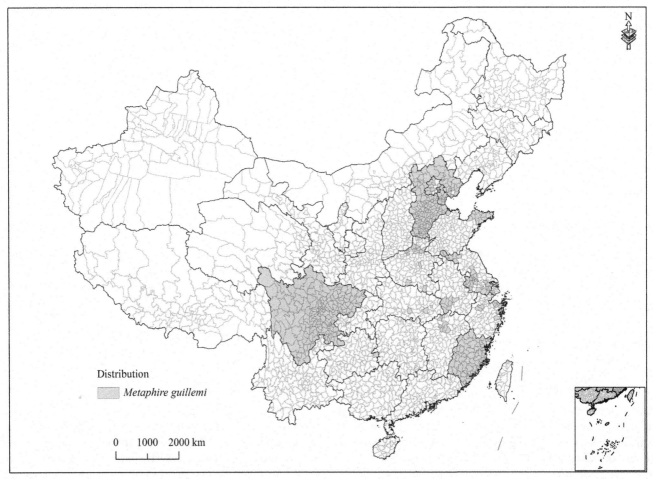

FIG. 268   Distribution in China of *Metaphire guillelmi*.

delicate, transparent, connective tissue sac or investment. The limbs of the loops are very short and in contact.

Color: Bluish yellow or pale on dorsum, dark bluish on mediodorsal line.

Distribution in China: Beijing (Xicheng, Haidian, Tongzhou, Changping, Huairou, Fangshan, Yanqing, Pinggu), Tianjin, Hebei, Shanghai (Jinshan, Songjiang), Jiangsu (Nanjing, Zhenjiang, Suzhou, Wuxi, Yixing, Xuzhou, Yangzhou), Zhejiang (Ningbo, Zhoushan, Chuanshan, Linhai, Tonglu, Xindeng), Anhui (Anqing), Fujian, Shandong (Yantai, Weihai), Jiangxi (Nanchang, Jiujiang, De'an), Henan (Xinxiang, Boai, Qinyang, Wenxian, Tanghe, Luoshan, Yanling), Hubei (Wuchang, Huangzhou, Qianjiang), Chongqing (Jiangbei), Sichuan.

Type locality: Hubei (Shihuiyao near Wuchang).

Deposition of types: Hamburg Museum.

Remarks: The worms are characterized by the presence of setae in the male pore invaginations and by the posterior location of the spermathecal stalked glands. They differ from other specimens of *M. guillelmi* by the presence of glandular material on the parietes just median to the ectal ends of the prostatic ducts. No definite genital markings were noted in the male pore invaginations. The first functional dorsal pore is on 12/13 on each of the 4 specimens, but on 2 specimens there is a pore-like marking on 11/12.

The specimens of *M. ichangensis* have been compared side by side with Stephenson's specimens of *M. houlleti* (= *M. guillelmi*) and with the specimens of *M. guillelmi*. The only difference found was the presence in both specimens of *M. ichangensis* (Fang, 1933), in XVIII median to the prostatic duct, of a stalked gland opening to the exterior by a pore on a rather indefinite genital marking in the male pore invagination. Retention of *M. ichangensis* on the basis of such an unimportant characteristic can scarcely be justified.

M. guillelmi is distinguished from *M. houlleti* (with which it has been confused) by the restriction of the male pore invaginations to the parietes, the conformation of the male porophore, and the presence of setae within the male pore invagination.

275. Metaphire houlleti (Perrier, 1872)

Metaphire houlleti Sims & Easton, 1972. *Biol. J. Linn. Soc.* 4:238.

Metaphire houlleti Shen, Tsai & Tsai, 2005. *Taiwania,* 50(1):11–21.

Amynthas huangi James, Shih & Chang, 2005. *J. Nat. Hist.* 39(14):1007–1028.

Metaphire houlleti Chang, Shen & Chen, 2009. *Earthworm Fauna of Taiwan.* 114–115.

Metaphire houlleti Xu & Xiao, 2011. *Terrestrial Earthworms of China.* 232–233.

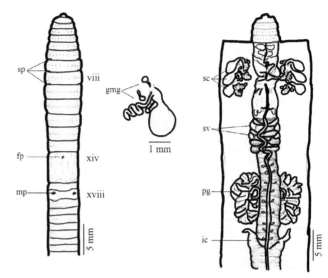

FIG. 269   *Metaphire houlleti* (after Shen, Tsai & Tsai, 2005).

External Characters

Length 107–118 mm, clitellum width 2.4–3.6 mm, segment number 86–102. Number of incomplete annuli (secondary segmentation) per segment 3 in VI–XIII. Prostomium epilobous. First dorsal pore 9/10. Clitellum XIV–XVI, annular, 3.2–4.4 mm long, dorsal pore absent, each segment with about 40 setal pits. Setae number 30–38 (VII), 50–52 (XX), 9–10 between male pores. Male pores paired in XVIII, about 0.28 circumference apart ventrally, each in copulatory pouch with C-shaped opening (slit), surrounded by a round swollen area with numerous transverse ridges. Genital papilla absent in both preclitellar and postclitellar regions. Spermathecal pores 3 pairs in 6/7/8/9, ventrolateral, slit-like, wrinkled at both anterior and posterior margins, buried deeply in intersegmental furrow, 0.3–0.32 circumference apart ventrally. Female pore single, medioventral in XIV.

Internal Characters

Septa 8/9/10 absent, 5/6/7/8 and 10/11/12/13 thickened. Gizzard large in IX–X. Intestine from XV. Intestinal caeca paired in XXVII, simple, stocky, wrinkled, extending anteriorly to XXIV. Lateral hearts in X–XIII. Testis sacs 2 pairs in X and XI, round, second pair vestigial. Seminal vesicles paired in XI and XII, small, folliculate, posterior pair larger, each with a round or oval dorsal lobe. Prostate glands paired in XVIII, large, lobed, smooth, extending anteriorly to XVI and posteriorly to XXII or XXIII, prostatic duct U-shaped, slender at proximal half and enlarged at distal half. White patches of accessory glands immediately anterior to the bulge at the base of the prostatic duct. Spermathecae 3 pairs in VII–IX, ampulla oval, large, surface wrinkled, 1.7–2.6 mm long, 1.1–1.9 mm wide, duct long, stout, 0.8–1.9 mm long, with a swollen basal portion, diverticulum originating from below the swollen portion of the spermathecal duct, stalk slender at the proximal end, 0.4–0.6 mm long, enlarged and greatly coiled toward distal end. Accessory glands stalked, stalk length

0.4–1.1 mm, with a round or slightly lobed head, connecting to the swollen basal portion of the spermathecal duct.

Color: Preserved specimens black on dorsum, grayish on ventrum, dark brown around clitellum.

Deposition of types: Museum National d'Histoire Naturelle, Paris, France.

Etymology: The species was named after Mr. Houllet.

Type locality: Calcutta, India.

Distribution in China: Tianjin, Jiangsu (Nanjing), Taiwan (Pingdong).

Remarks: *Metaphire houlleti* complex consists of different morphs distinguishable by somatic and genital characters (Gates, 1972). Its number of spermathecae ranges from 3 pairs of the normal, original morph to 0 in the athecal (without spermathecae) morph. According to Gates (1972), the differences in the reproductive organs might be attributable to parthenogenetically induced modifications, and moreover, an amphimictic ancestral population could have differentiated into different geographical races.

Michaelsen (1895) described *M. guillelmi* (Michaelsen, 1895) as a new species from Hubei, China, but later he considered it synonymous with *M. houlleti* (Michaelsen 1897, 1900). Michaelsen (1931) concluded that *M. guillelmi* was a special variety of *M. houlleti* after further examination of the original material. Chen (1933) considered these 2 species to be specifically different in view of the lower numbers of setae and the absence of spermathecal chamber in *M. houlleti*, and their different aspect of the male pore region. Gates (1935a) distinguished the 2 species by their different male pore invaginations. The geographical distribution also varies for the 2 species. *M. guillelmi* is found in central China

(Michaelsen, 1895; Chen, 1933), while *M. houlleti* is widely distributed in Southeast Asia (Gates, 1972). Accordingly, this study retained *M. guillelmi* as a valid species.

276. Metaphire hunanensis Tan & Zhong, 1986

Metaphire hunanensis Tan & Zhong, 1986. *Act Zootaxonomica Sinica.* 11(2):144–146.
Metaphire hunanensis Xu & Xiao, 2011. *Terrestrial Earthworms of China.* 233–234.

External Characters

Length 80–85 mm, width 3.0–3.5 mm, segment number 84–102. Prostomium 2/3 epilobous. First dorsal pore 12/13. Clitellum XIV–XVI, annular, with setae on ventral side of XVI. Setae small, evenly distributed, setae number 39–53 (III), 40–53 (V), 46–60 (VII), 45–58 (XI), 42–50 (XX); 15–21 (VI), 11–16 (VII), 3–8 (VIII) dorsally between spermathecal pores, 5–8 between male pores. Male pores situated deep in copulatory pouches which are on ventrolateral sides of XVIII, about 1/2 circumference apart ventrally. Each copulatory pouch aperture as a large longitudinal slit. No intersegmental furrows between the apertures, but with 7–10 longitudinal grooves on the skin. Spermathecal pores 3 pairs in 6/7/8/9 dorsally; first pair about 1/3 circumference apart dorsally, the third pair near to dorsomedian line, about 1/10 circumference apart dorsally, these 3 pairs of spermathecal pores arranged in an inverted trapezoid on the back of the body. Female pore single, medioventral in XIV.

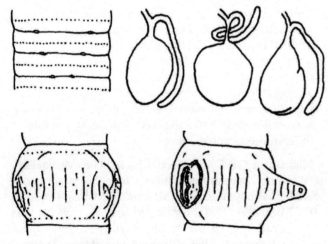

FIG. 270   *Metaphire hunanensis* (after Tan & Zhong, 1986).

Internal Characters

Septa 8/9/10 absent, 5/6/7 thickened and slightly transparent, and 10/11/12/13 thick and muscular, rest membranous. Gizzard slightly ball-shaped, in VIII–X. Intestine swelling from XVI. Intestinal caeca simple, in XXVI, extending anteriorly to XXIV or XXIII, smooth. Testis sacs 2 pairs in X and XI, meeting on dorsal side of esophagus and communicated ventrally. Seminal vesicles 2 pairs, in XI and XII, dorsal lobe conspicuous, the first pair contained in the second pair of testis sacs. Prostate glands developed, in XVI–XIX, or XVI–XX or XVI–XXI; with 12–20 finger-like lobules. Prostatic duct slender entally, thickened at distal 2/3 and entering into ental top of columnar copulatory pouch, about 1.5–2.0 mm long, 1.0–1.2 mm wide. Spermathecae 3 pairs in VII–IX, ampulla elliptical, about 1.8–2.0 mm long and 1.2–1.5 mm wide, or round with a diameter of about 1.5 mm, or pear-like with diameter of 1.0 mm and length of 1.2 mm. Diverticulum shorter than main pouch or as long as it. Seminal chamber tube-shaped not differentiated from its duct.

Color: Preserved specimens brownish, preclitellar segments light in color. Clitellum chestnut.
Deposition of types: Hunan Agricultural University.
Etymology: The name of the species refers to the type locality.
Type locality: Hunan (Changsha).
Distribution in China: Hunan (Changsha).
Remarks: This species is similar to *M. thecodorsata* (Chen, 1933) in having 3 pairs of spermathecal pores dorsally and deep copulatory pouches. It differs from the latter in having spermathecal pores in 6/7/8/9 arranged in an inversed ladder shape dorsally, no septa 8/9/10, and the tube-shaped seminal chamber not differentiated from its duct.

277. Metaphire ichangensis (Fang, 1933)

Pheretima ichangensis Fang, 1933. *Sinensia.* 3(7):180–184.
Metaphire ichangensis Sims & Easton, 1972. *Biol. J. Linn. Soc.* 4:238.
Metaphire ichangensis Xu & Xiao, 2011. *Terrestrial Earthworms of China.* 234–235

External Characters

Length 103–185 mm, width 6–8 mm, segment number 99–107; those anterior to clitellum generally each with 2 annuli, 1 anterior and 1 posterior to setae zone except II–IV or V, which with a single posterior annulus only. Setae zone generally ridged. Prostomium 2/3 epilobous, tongue open behind. First dorsal pore 12/13. Clitellum XIV–XVI, annular, setae and dorsal pore absent. Setae beginning on segment II, *zz* subequal to *zy* before, and about 0.8 *zy* behind clitellum; weak on segments IX–XIII, less so on segments before IX, strong on segments behind clitellum; setae number 34 (V), 38 (VIII), 50 (XIII), 52 (XIX), 56 (XXV), about 28 on segments near posterior

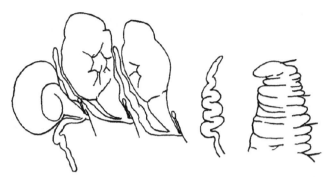

FIG. 271   *Metaphire ichangensis* (after Fang, 1933).

end; 12 between male pores. Male pores paired in XVIII, about 0.25 circumference apart ventrally, a round porophore protruded from the male pore and surrounded by its papillae border. Spermathecal pores 3 pairs in 6/7/8/9, ventrolateral. Female pore single, medioventral in XIV.

Internal Characters

Septa 9/10 absent, 5/6/7/8 and 10/11/12/13 considerably thickened, 8/9 only a small thin membrane developed at the ventral side. Gizzard large, in VIII–X. Intestine swelling from XIII. Intestinal caeca paired in XXVI, extending anteriorly to XXIV, tip pointed and curved, gradually expanded toward the base. Last heart in segment XIII. Testis sacs 2 pairs in X and XI, medioventral; the anterior pair of sacs broader than long, much narrower posteriorly, somewhat concave in anterior side, having 2 short rounded anterolateral horns; they are ventrally communicated with each other completely along the median edge to form a single capsule. The posterior pair much narrower than the anterior pair, also broader than long, truncate in front, somewhat rounded at sides and behind, they are also communicated to form a single capsule, but without demarcations both on dorsal and ventral sides. Seminal vesicles 2 pairs, large, fully occupying segments XI and XII, dorsally meeting each other slightly right of the median line; the outer surface fairly smooth; a topographically median lobe (actually lateral) present in each left vesical; the inner surface well lobulated. Prostate glands large, occupying segments 1/2 XVI–XX, consisting of a small dorsomedian and 2 large ventrolateral lobes. The latter subdividing into many lobules. The duct forming loops pointing inward and slightly hindward. The ectal duct V-shaped, its principal part nearly straight, thick and expanded at its greater middle part; the ental part also V-shaped, with the rami or V differing in length and pointing outward, much shorter and thinner than the principal part of the ectal duct. About 7 accessory glands near or on the rounded elevation. Spermathecae 3 pairs in VII–IX, ampulla generally anteroposteriorly compressed, first (3.2 × 3 mm)

much narrower than the second (3.3 × 3 mm), both somewhat heart-shaped with tips toward distal, alveolations present at the basal middle part; third corn-shaped or kidney-shaped, broader than long (2.4 × 3.2 mm), with a semiovoid and somewhat elevated area near its basal portion. Spermathecal duct differentiated from ampulla, much broader proximally, about 1/2 or less as long as ampulla. Diverticulum linear, emptying into spermathecal duct at its ectal end, narrower and somewhat straight at basal 1/5, where the diverticulum bent, broader, and twisting in more or less zigzag manner over its remaining distal 4/5; shorter than main part in situ, subequal or slightly shorter or longer when extended, Usually a single stalked gland about half the length of the duct of spermatheca, setting very close to opening of spermathecal duct and at the opposite side of the diverticulum, viz. lateral in position.

Color: Live specimens bluish. Preserved specimens fleshy, clitellum dilute dark brown.
Deposition of types: Previously deposited in the Metropolitan Museum of Natural History, Beipei, Chongqing. The types were destroyed during the war in 1949.
Etymology: The name of the species refers to the type locality.
Type locality: Hubei (Yichang).
Distribution in China: Hubei (Yichang, Qianjiang).
Remarks: This species is related to *M. asiatica* (Michaelsen, 1900) and *M. tibetana* (Michaelsen, 1931), but it is quite easily differentiated from them by the presence of stalked and accessory glands near spermathecae and prostatic duct. This earthworm is thick, short, and bluish when dug out from the soil in winter. It differs markedly from the formalin preserved extended specimens in both habitat and coloration.

278. *Metaphire kiangsuensis* (Chen, 1930)

Pheretima kiangsuensis Chen, 1930. *Sci. Rep. Natn. Cent. Univ. Nanking.* 1:11–37.
Pheretima kiangsuensis Chen, 1931. *Chin. Zool.* 7(3):119–122.
Metaphire ichangensis Sims & Easton, 1972. *Biol. J. Linn. Soc.* 4:238.
Metaphire kiangsuensis Xu & Xiao, 2011. *Terrestrial Earthworms of China.* 236–237.

External Characters

Length 200–350 mm, width 6–12 mm, segment number 87–188. Prostomium 2/3 epilobous. First dorsal pore 12/13. Clitellum XIV–XVI, annular, greatly swollen, setae and dorsal pores absent. Setae generally brittle and projecting out with tips charcoal black, noticeably

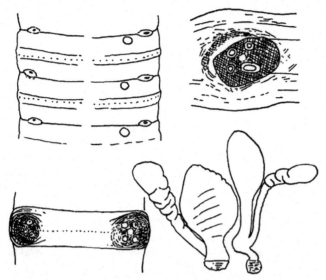

FIG. 272 *Metaphire kiangsuensis* (after Chen, 1930).

longer at anterior ventral side. Setae number 32–40 (III), 50–58 (VI), 52–60 (VIII), 62–78 (XXV); 18–20 (VIII) between spermathecal pores, 12–24 between male pores. Male pores in XVIII, about 1/3 circumference apart ventrally. Copulatory pouches much deeper, situated rather laterally, with a large (round or elongate) button-shaped papilla at the medial side and 2 smaller ones at the bottom; another pointed papilla at the bottom of the chamber with a male pore on its tip, often protruding laterally. The part between the male pore and the medial larger papilla is not musculated but with very soft tissue in connection with glandular tissue inside; 5–10 setae borne on the medial wall of the chambers; inner surface of the chamber wrinkled. Spermathecal pores 3 pairs in 6/7/8/9, about 1/3 circumference apart ventrally, skin around spermathecal pores usually strongly swollen and glandular in appearance, pale in color, often wrinkled around the pores; genital markings also indicated as transverse slits along setal zones on the ventrolateral surface of segments VI, VII, VIII, or IX. 2 or more large button-shaped papillae situated posterior and anterior to each pore. Female pore single, medioventral in XIV.

Internal Characters

Septa 8/9/10 absent, 5/6/7/8, 10/11/12 greatly thickened, 12/13 less thickened. Gizzard in IX, X, globular or sometimes slightly elongate, with a small rim at posterior side, the portion on anterior side connected with the esophagus is a little dilated. Intestine swelling from XV. Intestinal caeca simple, in XXVII, extending anteriorly to XXII, large and elongate, with distinct indentations along the ventral edge. Lateral hearts 4 pairs; first pair in X sometimes growing into the septum. Testis sacs 2 pairs, the posterior pair united into an elongated transverse band while the anterior pair in the form of 2

globular sacs projecting into the segment X but also communicating with each other. Seminal vesicles very large, fully filling up segments XI and XII, with a small dorsal lobe, 2 pairs of ovoid bodies attached to posterior faces of the 2 following septa; Sperm ducts of each side approaching nearer each other in segment XIV but remaining separate throughout. Prostate gland very large, in XVI–XXI, divided into large elongate lobules, smooth on surface; proximal third of the duct slender, middle part straight about 5–6 times as stout, looped at both ends. Central portion of the gland mass filled with numerous slender, coiled, and filiform structures, each of them not tubular but solid and somewhat perforated in its middle, being glandular in character. Similar compact glands also present in spermathecal region, about 4 or 5 mm in diameter, pale in color, surrounding the basal portion of spermathecal duct. Spermathecae ampulla sometimes very large, about 9 mm in width, sac-like, with a stout duct, but its diverticulum not proportionally enlarged, but looped in a zigzag manner, passing into the main duct close to the body wall.

Color: Live specimens from chocolate to grayish or purplish violet or bluish purple (grape-color) on anterior dorsum; light chocolate on posterior dorsum, buff on lateral region of the body; yellowish pale or light fleshy or buff on ventrum. Clitellum light purplish dusty to bluish ashy or dusty pale. Preserved specimens from purplish brown to grape violet or dark slaty violet. Clitellum cinnamon brown.

Deposition of types: Previously deposited in the Museum of the Biological Laboratory of the Science Society of China, Nanjing. The types were destroyed during the war in 1937.

Etymology: The name of the species refers to the type locality.

Type locality: Jiangsu (Nanjing, Suzhou).

Distribution in China: Jiangsu (Nanjing, Suzhou), Sichuan (Mt. Emei, Guanxian).

Remarks: The color of the species is generally dark purplish, compared with other species which Chen collected from the hilly regions in Nanjing. The color of those from the plain regions is generally brownish or chocolate. These specimens were picked up while they were straying during hot weather in August on Mt. E'mei. This habit again agrees with that of the Nanjing forms which often come out of their burrows following showers during periods of hot weather.

279. *Metaphire kokoan* (Chen & Fong, 1975)

Pheretima tschilliensis kokoan Chen, Hsü, Yang & Fong, 1975. *Acta Zool. Sinica.* 21(1):94–95.
Metaphire kokoan Xu & Xiao, 2011. *Terrestrial Earthworms of China.* 237–238.

External Characters

Length 107–160 mm, width 5.5–8.0 mm, segment number 66–120. Prostomium 1/2 epilobous. First dorsal pore 12/13. Clitellum XIV–XVI, annular, short, setae absent. Setae normal in ventral region of II–IX, sometimes slightly wide, but not stout and not long; setae number 34–54 (III), 47–68 (VIII), 52–73 (XII), 40–61 (XVIII), 52–66 (XXI); 19–32 (VIII) between spermathecal pores, 13–20 (XVIII) between male pores. Male pores in ventral region of XVIII, each on small copulatory pouches with a small opening and surrounding by skin folds, 1–4 small papillae in pouch (sometimes 1–2, sometimes outside), about 1/3 circumference apart ventrally. Spermathecal pores 3 pairs in intersegmental furrows of 6/7/8/9, about 1/2 circumference apart ventrally, each bordered anteriorly and posteriorly by glandular lips, 1–2 small papillae in anterior lips. Female pore single, medioventral in XIV.

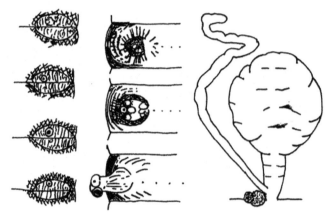

FIG. 273   *Metaphire kokoan* (after Chen, et al., 1975).

Internal Characters

Septa 8/9/10 absent, 5/6/7/8 thickly muscular, some or all of septa 10/11–13/14 thickly muscular. Gizzard behind 7/8, globular in shape. Intestine swelling from XV. Intestinal caeca simple, large and long, horn-shaped, in XXVII, extending anteriorly to XXII or XXIII, each ventrally with several serriformed outgrowths. Testis sacs ventral in position, massive and large; the anterior pair forming a U shape and the posterior a transverse sac. Seminal vesicles, vesicular, voluminous, fairly large, occupying all the interior of the respective segment and dorsally meeting with each other; each with a large dorsal lobe. Dorsal lobe of the anterior pair may sometimes be subdivided into 2–3 smaller ones. Prostate gland large, usually in XVI–XXI, divided into many finger-shaped pieces; duct long, looped and hairpin-shaped; entally rather thin and ectalwards gradually increasing in thickness and muscularity. Ectal end of the duct becomes thinner and enters, through a fairly large cushion-like glandular tissue, into a firmly formed body situated at the bottom of the copulatory pouch, and opened at the tip of the porophore. Accessory glands with solid stalks are found close to the ectal end of the prostatic duct corresponding externally to the smaller papillae. A similar gland but of larger size, corresponding to the larger papilla, is found near these. Spermathecae 3 pairs, in VII–IX; ampulla oval-shaped, its duct about 2/3 length of ampulla; diverticulum longer than main pouch, straight, rarely in a weak zigzag fashion.

Color: Preserved specimens gray-brown on dorsum.
Etymology: The name of the species refers to the type locality.
Type locality: Gansu (Mt. Xinglong), Qinghai (Xining).
Distribution in China: Qinghai (Xining), Gansu (Lanzhou, Mt. Xinglong).
Remarks: This species has 4 characteristics: (1) small or moderate in body size, similar to *typica* but differs from *M. grahami* (Gates, 1935a); (2) copulatory pouches small and superficial, about 1/3 the segment, rarely in *typica*; (3) chamber in spermathecal pore absent, but glandular lips on anteroposterior apertures, rarely in *typica* and *M. grahami* (Gates, 1935a); (4) the spermathecal diverticulum long and straight, rarely in a weak zigzag fashion.

280. *Metaphire lanzhouensis* (Feng, 1984)

Pheretima tschiliensis lanzhouensis Feng, 1984. *Zoological Research*. 5(1):47–50.
Metaphire tschiliensis lanzhouensis Zhong & Qiu, 1992. *Guizhou Science*. 10(4):41.
Metaphire lanzhouensis Xu & Xiao, 2011. *Terrestrial Earthworms of China*. 238–239.

External Characters

Length 245–310 mm, width 6–7 mm, segment number 111–149. Prostomium 1/2 epilobous. First dorsal pore 12/13. Clitellum XIV–XVI, annular, setae absent. Setae fine, closely arranged; setae number 32–40 (III), 50 (V), 35–46 (VI), 46–55 (VIII). Male pores paired in XVIII ventrally, within shallow copulatory pouches occupying 1/3 XVIII with C-shaped opening and surrounding by skin folds, 2 setae situated between papillae in each pouch; usually with 1 small flat papilla on opening, entirely or half outside; 1 specimen with 2 flat papillae, line-shaped, 1 inside; posterior margin of flat papillae with 4–5 bar-shaped protuberances; the papilla at the bottom of the chamber with male pore on its tip; about 1/3 circumference apart ventrally. Spermathecal pores 3 paired, ventrally, inconspicuously in 6/7/8/9; pore small, hardly visible from outside. Spermathecal chamber absent, about 1/3 circumference apart ventrally. Genital papillae absent. Female pore single, medioventral in XIV.

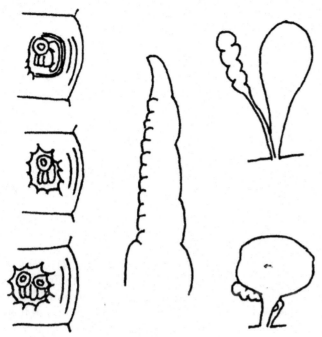

FIG. 274 *Metaphire lanzhouensis* (after Feng, 1984).

Internal Characters

Septa 9/10 absent, 8/9 traceable ventrally and membranous, 5/6/7/8 thickened, 10/11 backward membranous. Gizzard in IX and X, bell-shaped. Intestine swelling from XV. Intestinal caeca complex, in XXVII, extending anteriorly to XXIII, with distinct ventral indentations. Last pair of hearts in XIII. Testis sacs paired in X and XI. Seminal vesicles in XI and XII. Prostate gland small, in XVII–XIX, prostatic duct V-shaped. Spermathecae 3 pairs in VII, VIII, IX; ampulla spherical, surface wrinkled into 4–5 transverse lines; diverticulum equal in length to main pouch, ental 2/3 closely twisted in a zigzag fashion.

Color: Preserved specimens yellowish brown, dark gray on middorsal line.
Etymology: The name of the species refers to the type locality.
Type locality: Gansu (Lanzhou).
Distribution in China: Shaanxi (Fusnan), Gansu (Lanzhou).
Remarks: This species has 5 characteristics: (1) similar to *Metaphire graham* (Gates, 1935a) in large body size, differs from *Metaphire typica* and *M. kokoan* (Chen & Fong, 1975) which have small body size; (2) similar to *M. grahami* (Gates, 1935a) and *M. kokoan* in the absence of flat papillae on anterior margin of spermathecal pores and ventrum of XII, VIII and IX, but rarely in *M. typica*; (3) similar to *M. kokoan* with shallow copulatory pouches surrounded by skin folds, occupying 1/3 of the segment, but differs from *M. graham* which has deep copulatory pouches with round-shaped openings; (4) similar to *M. kokoan* in that both sides of spermathecae with shallow transverse lines, however, others always smooth; (5) the septa traceable ventrally and membranous in this species, but absent in *M. grahami*, *M. typica*, and *M. kokoan*. Above all, the Lanzhou specimens should not be regarded as a subspecies but as a single species.

281. *Metaphire praepinguis jiangkouensis* Qiu & Zhong, 1993

Metaphire praepinguis jiangkouensis Qiu & Zhong, 1993.
Guizhou Science. 11(1):40–41.
Metaphire praepinguis jiangkouensis Xu & Xiao, 2011.
Terrestrial Earthworms of China. 248–249.

External Characters

Length 146–278mm, width 6.0–8.5mm, segment number 89–131. Prostomium 1/2 epilobous. First dorsal pore 12/13. Clitellum XIV–XVI, annular, with some setae on ventral side. Setae thick before the segment of XVIII, close on the dorsal side and scattered on the ventral side; setae fine and close behind the segment of XVIII, evenly distributed; setae number 35–38 (III), 42–48 (V), 52–62 (VIII), 72–78 (XX), 75–82 (XXV); 20–23 (VII), 21–24 (VIII) between spermathecal pores, 21–24 (XVIII) between male pores. Male pores 1 pair, each situated on a small papilla in the copulatory pouch on the ventrolateral side of XVIII; about 1/3 circumference apart ventrally. There are 1–3 round or nephroid papillae in or out of the copulatory chamber close to the male papilla and 2 rows (about 23–29) of papillae on the ventral side of XVIII. Spermathecal pores 3 pairs, in 6/7/8/9, each bordered anteriorly and posteriorly by glandular lips, about 2/5 circumference apart ventrally. There are about 14 small papillae on the ventral side of VII–IX or 48 small papillae on the ventral sides of VII–XII, sometimes no papilla in this region. Female pore single, medioventral in XIV.

Internal Characters

Septa 9/10 absent, 8/9 rather thick, 5/6/7/8 thickened and musculated, 10/11–14/15 slight thick, others thin. Gizzard slightly developed, barrel-like. Intestine swelling from XVI. Intestinal caeca simple, finger-like sac, with small tooth-shaped protuberances on ventral side, in XXVII, extending anteriorly to XXIII. Last pair hearts in XIII. Testis sacs 2 pairs, in X and XI, ovoid. Seminal vesicles 2 pairs, in XI and XII, well developed, meeting each other on dorsal side; dorsal lobes apparent, anterior pair larger, posterior pair small, ovoid. Prostate glands well developed, in XVI–XX, lump-like or bar-like lobulated;prostatic duct short and stout, U-shaped. Accessory glands developed, lump-like, with short

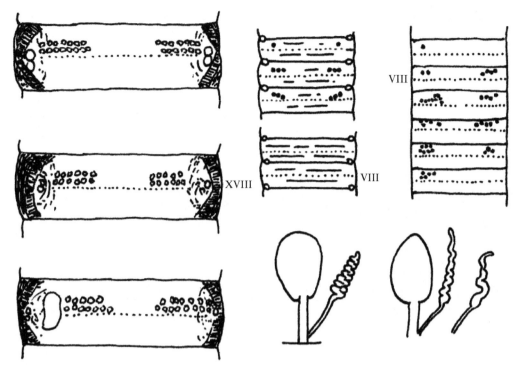

FIG. 275 *Metaphire praepinguis jiangkouensis* (after Qiu & Zhong, 1993).

cord-like ducts. Spermathecae 3 pairs, in VII, VIII, IX; ampulla long-rounded, about 3.5 mm long; ampulla duct stout and long, demarcation between ampulla sac and ampulla duct distinct; diverticulum shorter than main part, distal end of 1/2–1/3 twisted in a zigzag fashion. Accessory glands lump-shaped, connected with cord-like ducts.

Color: Preserved specimens grayish brown on dorsum, grayish white on ventrum. Clitellum red-brown.

Deposition of types: Guizhou Institute of Biology, Guiyang.

Etymology: The name of the species refers to the type locality of "Jiangkou" County.

Type locality: Guizhou (Mt. Fanjing, Jiangkou).

Distribution in China: Guizhou (Jiangkou).

Remarks: This subspecies can be distinguished from the nominate subspecies *Metaphire praepinguis praepinguis* (Gates, 1935a) by the presence of septa 8/9, the papillae in the copulatory chamber and on the surface of the body, and the lack of pseudovesicles.

282. *Metaphire praepinguis praepinguis* (Gates, 1935)

Pheretima praepinguis Gates, 1935a. *Smithsonian Mus. Coll.* 93(3):15.
Pheretima praepinguis Chen, 1936. *Contr. Biol. Lab. Sci. Soc. China (Zool).* 11:302.

Pheretima praepinguis Gates, 1939. *Proc. U.S. Natn. Mus.* 85:471–473.
Metaphire praepinguis Sims & Easton, 1972. *Biol. J. Linn. Soc.* 4:238.
Metaphire praepinguis praepinguis Xu & Xiao, 2011. *Terrestrial Earthworms of China.* 249.

External Characters

Length 207 mm, width 16 mm. First dorsal pore 12/13. Clitellum XIV–XVI, annular, only very slight traces of intersegmental furrows and dorsal pores visible; circles of setal pits present on XIV–XVI but no setae visible. Setae begin on II, on which segment there is a circle; there is no definite midventral gap in the setal circles; the middorsal breaks of variable width; setae number 23 (VII), 24 (VIII), 20 (XVII), 9/13? (XVIII), 22 (XIX), 93 (XX). Male pores on tubercles in the lateral-most portions of deep parietal invaginations with crescentic apertures, lateral walls of the invaginations thin and nonsetigerous. Just median to each male pore tubercle a single genital marking; on the median wall of the male invagination transversely oval presetal genital markings. External genital markings paired, presetal on VII, VIII, and IX. Spermathecal pores on tiny tubercles located in parietal invaginations with transversely slit-like apertures, 3 pairs, in 6/7/8/9. 1 circular genital marking on the anterior wall of each invagination. Female pore single, medioventral in XIV.

Internal Characters

Septa 8/9/10 absent, 5/6/7/8 and 10/11/12/13 thickly muscular; 13/14 muscular. Intestine swelling from XV on the right side, from XVI on the left side. Intestinal caeca simple. Last pair of hearts in XIII. Testis sacs of X and XI unpaired and ventral. Seminal vesicles of XI and XII are vertical bodies reaching into contact with the dorsal blood vessel. Prostate glands are relatively rather small; the right prostate glands confined to XVIII though 17/18 and 18/19 are dislocated anteriorly or posteriorly; a small lobe of the left prostate glands extending into XVII. The prostatic duct is about 12 mm long, bent into a hairpin-shaped loop, the ectal limb of the loop thicker than the ental. Spermathecae 3 pairs in VI–VIII; ampulla is about as long as or slightly longer than the duct. The latter is stout and with a rather bulbous appearance as it passes into the parietes. Within the body wall the duct is very abruptly and very considerably narrowed; the very short, slender portion of the duct opening to the exterior by a minute pore on the ventral face of a tiny, smooth, rather conical tubercle on the top of the spermathecal invagination. The spermathecal invagination, transversely slit-like in section, is confined to the outer half of the rather thick body wall. On the anterior face of the spermathecal duct, close to the parietes is a spheroidal, glandular mass. A bundle of stalks or ducts from this gland passes through the parietes on the anterior face of the spermathecal duct to a circular genital marking located on the anterior wall of the spermathecal invagination. The junction of the diverticular stalk with the spermathecal duct close to the parietes is concealed from view by the anterior gland. Spermathecal diverticulum comprises a smooth, glistening stalk and a much wider seminal chamber, the latter with 2–3 slight constrictions.

Color: Live specimens sky blue. Preserved specimens dark blue-gray.
Deposition of types: Smithsonian Institution.
Type locality: Sichuan (Mt. Emei).
Distribution in China: Sichuan (Mt. Emei, Xikang).
Remarks: Distinguished from *Metaphire asiatica* (Michaelsen, 1900) by the spermathecal invaginations and the genital markings therein. Stalked glands or glandular masses in connection with the presetal genital markings were not found. No setae were observed on the median wall of the male parietal invagination but unusually small or very deeply retracted setae may have been overlooked.

Metaphire praepinguis (Gates, 1935a) is closely related to *Metaphire tschiliensis* from which it may be distinguished at present by the location of the primary spermathecal pores in the parietal invagination.

Chen (1936) maintains that *praepinguis* is a synonym of *tschiliensis* and that the type of the former is not only "perfectly identical" with some of Chen's earlier specimens from Sichuan but also with the types of *grahami*. *M. graham* is quite clearly distinguished from *praepinguis* by the unusual spermathecal chambers and the presence of copulatory chambers. What Chen's earlier specimens from Sichuan actually are cannot be determined from his description. *M. praepinguis* (Gates, 1935a) is, as was stated above, close to *tschiliensis* and, of course, cannot be satisfactorily characterized from a single specimen. The spermathecal pores of 1 species are superficial but in the other species are within definite, or what appear to be definite, parietal invaginations with large transversely slit-like apertures, which would normally be shut so as to conceal the primary pore from sight externally. Such an invagination appears to be of sufficient importance to distinguish *praepinguis* (Gates, 1935a) from *tschiliensis*, in view of the lack of intraspecific variation with regard to the spermathecal pore. If the 2 forms are really specifically distinct, examination of additional specimens of each should disclose further distinguishing characteristics.

283. Metaphire ptychosiphona Qiu & Zhong, 1993

Metaphire ptychosiphona Qiu & Zhong, 1993. *Guizhou Science.* 11(1):38–40.
Metaphire ptychosiphona Xu & Xiao, 2011. *Terrestrial Earthworms of China.* 250–251.

External Characters

Length 196–295 mm, width 6.0–9.0 mm, segment number 80–151. Prostomium 1/2 epilobous. First dorsal pore 11/12. Clitellum XIV–XVI, annular. Setae fine and close, evenly distributed; setae number 57–61 (III), 64–72 (V), 64–79 (VIII), 87–117 (XX), 85–124 (XXV); 27–31 (VII), 28–34 (VIII) between spermathecal pores, 15–27 (XVIII) between male pores. Male pores 1 pair, each situated in a rather shallow copulatory pouch on the ventrolateral side of XVIII, about 1/3 circumference apart ventrally; no papilla in or out of the copulatory pouch. Spermathecal pores 3 pairs, in 6/7/8/9, each bordered anteriorly and posteriorly by glandular lips, nearly 2/5 circumference apart ventrally. No papilla in this region. Female pore single, medioventral in XIV, round.

Internal Characters

Septa 9/10 absent, 8/9 membranous, 5/6/7/8 thick and muscular, others thin. Gizzard ball-shaped, slightly small. Intestine swelling from XV. Intestinal caeca simple, finger-sac shaped, in XXVII, extending anteriorly to XXIII. Last pair of hearts in XIII. Testis sacs 2 pairs, in X and XI, rather small, separated widely; seminal vesicles 2 pairs, in XI and XII, meeting dorsally. Prostate glands

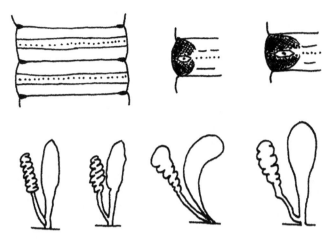

FIG. 276 *Metaphire ptychosiphona* (after Qiu & Zhong, 1993).

small, divided into 12–15 small long-bar-shaped lobules, in XVI–XVIII or XVII–XIX; prostatic duct stout, U-shaped, distal end passing into a semicircular sac-shaped structure formed by the copulatory pouch invaginating into body cavity. Accessory glands absent. Spermathecae 3 pairs, in VII, VIII IX; ampulla wooden club-like or curved crescent-shaped, about 3 mm long, 1.0–1.5 mm wide; diverticulum shorter than the main part, ental 3/5 twisted in a tight zigzag fashion. Accessory glands absent.

> Color: Preserved specimens grayish brown. Clitellum red-brown.
> Deposition of types: Guizhou Institute of Biology, Guiyang.
> Type locality: Guizhou (Kuan Kuo Shui forest region, Suiyang).
> Distribution in China: Guizhou (Suiyang).
> Remarks: This species is similar to *Metaphire grahami* (Gates, 1935a), but differs from the latter by the present of septum 8/9, the widely separated small testis sacs, the shallow and small copulatory pouch (in or out of which there is no papilla), and no papilla in spermathecal pore region.

284. *Metaphire tibetana* (Michaelsen, 1931)

Pheretima tibetana Michaelsen, 1931. *Peking Natural History Bulletin.* 5(2):13–15.
Metaphire tibetana Sims & Easton, 1972. *Biol. J. Linn. Soc.* 4:238.
Metaphire tibetana Xu & Xiao, 2011. *Terrestrial Earthworms of China.* 255–256.

External Characters

Length 75–110 mm, width 5–7 mm, segment number 95–104. Clitellum XIV–XVI, annular, setae absent. Setae are somewhat enlarged in the middle of the body in front of the clitellum and the ventral ones are somewhat larger

than the dorsal; setae number 44 (V), 50 (VIII), 60 (XIII), 68 (XIX); Male pores paired in XVIII ventrally, very large, nearly circular holes with thick well-developed lips, its middle points about 1/3 circumference apart ventrally; the edge of these holes is formed of a prepuce-like fold, which holds a thick and short pistil-like penis. This short thick penis shows in the middle point of its tip the minute invisible primary male aperture, and near this, some small circular gland papillae, which are not always clearly recognizable. There is no copulatory pouch, but a short broad, penis pouch whose wall surrounds the short broad penis like a prepuce. Spermathecal pores almost invisible cross slits, 3 pairs, ventrolateral on intersegments 6/7/8/9, those of 1 pair almost 1/2 circumference apart ventrally. Female pore single, medioventral in XIV.

Internal Characters

Septa 8/9/10 absent, 5/6/7/8, 10/11/12 moderately thickened. Gizzard large, behind septum 7/8. Intestinal

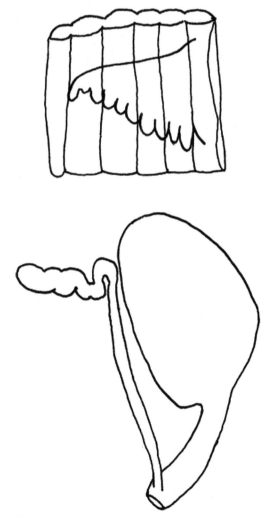

FIG. 277 *Metaphire tibetana* (after Michaelsen, 1931).

caeca manicate, paired in XXVII, extending anteriorly to XXIII, in regular manner; these regular digestive caeca are broad, 3-sided, at the base almost as wide as long. The caeca are not always so regularly formed. Testes 2 pairs, rather large, ventral in X and XI, the pairs ventromedially fused with each other through its entire width, they together form a single capsule lying transversely and somewhat rounded at the sides. The tufted testes extend from the lateral part of the anterior wall into the lumen of the capsule, which otherwise is completely filled with the large seminal funnels. There are no sorts of excrescences of the testis sac. Each testis sac is connected by a short thin tube with a large wide seminal vesicle in the following segment, XI or XII. The broad sac-like seminal vesicles embrace the alimentary canal, touching each other ventrally as well as dorsally, without fusing together. On the dorsal edge of the seminal vesicles externally separated from the main part only by a slight furrow there formed a complete portion, buried in the main part. Prostate glands with large broad glandular part extending into 4–5 segments, split into many rounded right-angled projections which are 2–3 times as long as they are broad; prostatic duct bent into an irregular screw-shape; proximal half thin, opaque, somewhat thick compared with the strongly muscular shining, spindle-shaped, swollen distal half. The 2 sperm ducts of the 2 sides open together into the thin proximal end of the prostate duct and push themselves distally inside its wall. Spermathecae 3 pairs in VII–IX, ampulla egg-shaped; ampulla duct somewhat sharply distinguished from the ampulla, almost half as long, proximal part thin, swollen in the distal part, with a somewhat metallic luster. In the distal end of the ampulla duct there opens a blind hose-shaped diverticulum which is somewhat longer than the main pouch (ampulla and duct). The stalk-forming distal part of the diverticulum is thin and snake-like, only weakly and irregularly bent; the lumen-forming proximal third presses against the proximal blind end rather forcefully, and shows several not very widely extending twists, through alternately placed bends.

Color: Preserved specimens even yellow to gray-brown.
Deposition of types: Hamburg Museum.
Etymology: The name of the species refers to the type locality.
Type locality: Xizang (River Dracü)
Distribution in China: Xizang.

285. *Metaphire trutina* Tsai, Chen, Tsai & Shen, 2003

Metaphire trutina Tsai, Chen, Tsai and Shen, 2003.
Endemic Species Res. 5:83–88.

Metaphire trutina Chang, Shen & Chen, 2009.
Earthworm Fauna of Taiwan. 134–135.
Metaphire trutina Xu & Xiao, 2011. *Terrestrial Earthworms of China.* 256–257.

External Characters

Length 215–425 mm, clitellum width 11.0–15.6 mm, segment number 96–189. Prostomium prolobous. Number of annuli per segment 3 in VI–IX and 5 in X–XIII, 3 in segments behind XVII. Setae slightly wider than the adjacent annuli. First dorsal pore 12/13. Clitellum XIV–XVI, annular, length 7.89–14.1 mm, dorsal pores absent, setae absent in XIV and XV, but 5 setae on medioventrum of XVI. Setae number 109–118 (VII); 108–128 (XX); 22–35 between male pores. Male pores paired in XVIII, ventrolateral, large, C-shaped, 2.87 mm in longitudinal length, reaching 17/18 and 18/19 segmental furrows; porophore, a small tubercle with a small male aperture at the laterocentral end of male disc, a narrow longitudinal ridge extending from setal annulus of XVIII. No setae on the male disc. A pair of large, nearly equally sized round-shaped genital papillae located on anterior and posterior ends of the male disc. Each of the genital papillae 1.0–1.1 mm in diameter with a slightly concave center. Lateral wall of the copulatory pouch smooth externally (laterally) but slightly folliculated internally. Spermathecal pores invisible externally due to heavy secondary segmentation, 3 pairs, ventrolateral in 6/7/8/9, but detected from inside after dissection. Genital papillae absent. Female pore single, medioventral in XIV.

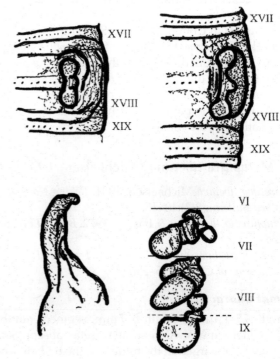

FIG. 278   *Metaphire trutina* (after Tsai et al., 2003).

Internal Characters

Septa 8/9/10 absent, 10/11–13/14 very thick. Gizzard large, round in IX–X. Intestine swelling from XIV. Intestinal caeca paired in XXVII, extending anteriorly to XXII, each simple with a pointed end, surface wrinkled with vertical lines. Lateral hearts in XI–XIII. Testes 2 pairs, medioventral in X and XI, each round, white in color, with smooth surface, sperm ducts connected in XIII. Seminal vesicles 2 pairs in XI and XII, anterior pair large, pale in color, surface wrinkled, each with a round brownish, granulated dorsal lobe, filling whole segmental cavity; posterior pair smaller in size compared with the anterior ones, granulated surface, each with a large, brown colored, granulated dorsal lobe. Prostate glands paired in XVIII, each lobed, surface folliculated. Prostatic duct stout and straight. No accessory glands or visible structure associated with the genital papillae in copulatory pouches. Spermathecae 3 pairs in VII–IX, each with an oval or round ampulla with heavily pressed wrinkles at proximal end, and a very short stout stalk. Diverticulum with silver-colored and oval-shaped seminal chamber and a heavily twisted stalk. Ovaries paired in XIII.

Color: Live specimens bluish brown or dark purplish gray with metallic luster on dorsum, reddish brown on ventrum. Preserved specimens purplish brown on dorsum, light grayish brown on ventrum.
Deposition of types: Taiwan Endemic Species Research Institute, Jiji, Nantou, Taiwan.
Etymology: The name "*trutina*" is given to this species with reference to an earthworm possessing a pair of large genital papillae, each with a slightly concave center and connected by a narrow male disc with a porophore at the mediolateral center, like a balance, in each shallow copulatory chamber.
Distribution in China: Taiwan (Ilan)
Habitat: The specimens were collected from a roadside ditch covered with about 10 cm of sand without vegetation at an elevation of 150 m.
Remarks: *M. trutina* (Tsai, Chen, Tsai & Shen, 2003) is sexthecal (3 pairs of spermathecae) with spermathecal pores in 6/7/8/9, and has no genital papillae in 17/18 and 18/19. It belongs to the *houlleti* group. It is a large earthworm with a pair of shallow copulatory chambers with C-shaped openings, the characters closely related to *M. vulgaris* (Chen, 1930) of Nanjing, China, *M. viridis* (Feng & Ma, 1987) from Gansu, northwest China, *M. praepinguis* (Gates, 1935a) of Sichuan (Gates 1935a, 1939), *M. aggera* (Kobayashi, 1934) of Korea and northeast China (Kobayashi 1934, 1940), and *M. tschiliensis* (Michaelsen, 1928a) of China. The last species consists of 3 subspecies: (1) *Metaphire t. tschiliensis* (Michaelsen, 1928a) of North and Central China (=*M. kiangsuensis* Chen, 1930 (Chen 1930, 1931, 1933)),

(2) *Metaphire t. grahami* (Gates, 1935a) from Sichuan (Gates 1935a, 1939), and (3) *Metaphire t. kokoana* (Chen & Fong, 1975) of Qinghai and Gansu (Chen et al., 1975).

M. trutina is easily distinguished from the above related species and subspecies of China by possessing a higher number of setae, absence of preclitellar genital papillae, absence of wrinkled, swollen lips or crescent ridges bordered anteriorly and posteriorly to each of the spermathecal pores, and the presence of 2 large genital papillae on both anterior and posterior ends of male disc in each shallow copulatory chamber.

286. *Metaphire tschiliensis* (Michaelsen, 1928)

Pheretima tschiliensis Michaelsen, 1928a. *Arkiv For Zoologl.* 20(2):13–15.
Metaphire tschiliensis Xu & Xiao, 2011. *Terrestrial Earthworms of China*. 257.

External Characters

Length 190–210 mm, width 6.5–7 mm, segment number 200. Prostomium 1/4 epilobous. First dorsal pore 11/12. Clitellum XIV–XVI, annular; setae, dorsal pore, and intersegmental furrow absent. Setae evenly distributed, stout and strong, slightly widely spaced on ventrum, almost continuous on dorsum;Setae number 50 (V), 55 (IX), 72 (XIII), 76 (XXV). Male pores paired in XVIII, situated ventrolaterally at the copulatory chamber, about 2/5 circumference apart ventrally; aperture on the

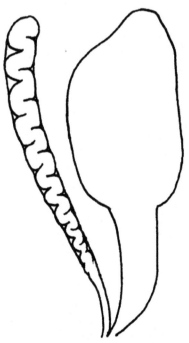

FIG. 279   *Metaphire tschiliensis* (after Michaelsen, 1928a).

curve of a crescent-shaped slit which extends to the center; slit with concave center and protruding edge, the penis concealed by the glandular protuberances around the slit. Spermathecal pores 3 pairs in 6/7/8/9, ventral, about 1/3 circumference apart ventrally; aperture in a not obviously lip-shaped transverse slit. Female pore single, medioventral in XIV.

Internal Characters

Septa 8/9 absent, if 9/10 present, represented only by a thin ventral rudiment, 5/6/7/8 thick, 10/11/12 uniform thickness, 12/13/14 slightly thickened. Gizzard behind septum 7/8, fairly large. Intestine swelling from XV. Intestinal caeca simple, originating in XVII and extending anteriorly to XXIII, with diaphragm constriction. Last pair of hearts in XIII. Testis sac 2 pairs, well developed, in X and XI; separated totally by septum. Seminal vesicles 2 pairs in XI and XII. Spermathecae 3 pairs in VII–IX, ampulla rectangular and pocket-shaped, ental end of ampulla duct stout, ectal end abruptly thinned; diverticulum fairly twisted, equal in thickness with ampulla duct.

Color: Preserved specimens yellowish brown. Clitellum dark brown.
Deposition of types: Stockholm Museum.
Etymology: The name of the species refers to the type locality.
Type locality: Hebei (Xuanhua).
Distribution in China: Hebei (Xuanhua).
Remarks: This species has the following characteristics: spermathecal ampulla rectangular and pocket-shaped, ental end stout, ectal end abruptly thinned, diverticulum fairly twisted and equal in thickness with ampulla duct.

287. Metaphire viridis Feng & Ma, 1987

Metaphire viridis Feng & Ma, 1987. A*cta Zootax. Sinica,* 12(3):248–250.
Metaphire viridis Xu & Xiao, 2011. *Terrestrial Earthworms of China.* 257–258.

External Characters

Length 192–230 mm, width 9.5–10 mm, segment number 124–128. Prostomium 1/2 epilobous. First dorsal pore 12/13. Clitellum XIV–XVI, annular, setae absent, and with trace of dorsal pore. Setae fine and close, middorsal and midventral breaks (setal breaks) continuous; setae number 47–50 (III), 49–54 (V), 60–65 (VI), 64–67 (VII), 64–70 (VIII), 92–95 (XXV). Male pores paired in XVIII, ventrolaterally, within the large and deep copulatory pouches with C-shaped openings and surrounded by skin folds, about 1/2 circumference apart ventrally; a presetal flat-topped papilla situated on the opening, a large conical papilla at the bottom of the pouch with male pore

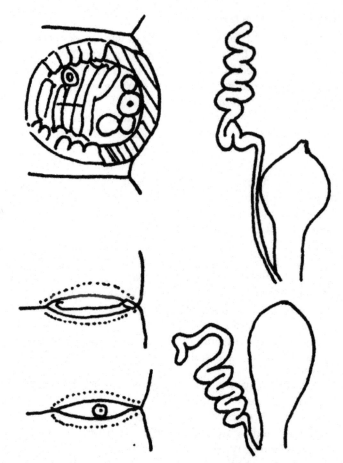

FIG. 280   *Metaphire viridis* (after Feng & Qiu, 1987).

on its tip, a papilla in front of porophore and 2 papillae behind the porophore, the pouch about 4/5 segment length. Spermathecal pores 3 pairs in 6/7/8/9, lateral, each on a tubercle in intersegmental furrow, sometimes anterior and posterior of apertures lip-shaped and swollen; transversely broad slit-like, with shallow spermathecal chamber, about 1/2 circumference apart; ventral side of segment VII, VIII, and IX with 1 pair small papillae respectively, 1 specimen with 1 pair of papillae on presetal XIII adjacent intersegmental furrow or ventral VII. Female pore single, medioventral in XIV.

Internal Characters

Septa 9/10 absent, 8/9 ventrally present. 5/6/7/8 thickly muscular, 10/11–13/14 slightly thickened, 14/15 and succeeding septa thin and membranous. Gizzard ball-shaped, in VIII and IX. Intestine swelling from XV. Intestinal caeca simple, in XXVI, with distinct teeth-like incisions along ventral edge, extending anteriorly to XXII. Last pair of hearts in XIII. Testis sac 2 pairs, the left pair communicates with the right pair, anterior pair rectangular, with center concave on anterior edge, width 4 mm, length 2 mm; posterior pair butterfly-shaped, larger than anterior pair, length 2 mm, width 4.3 mm. Seminal vesicles 2 pairs, in XI–XII, equal in size; dorsal side of seminal vesicles almost connect dorsal vessels, dorsal lobes obvious; the first pair of seminal vesicles not in testis sacs. Prostate glands mass-like, in XV–XX or XVII–XIX, prostatic duct C-shaped, ectal end slightly stout, with 1 mass-like accessory gland on the base. Spermathecae 3 pairs in VII, VIII, and IX, ampulla ball-shaped, anterior end slightly sharp, or oval, ampulla duct stout; diverticulum slender, longer than main part or sometimes of equal length, ental end 1/2 with 5–7 appressed curves; diverticulum inserted though antero-medial of ampulla duct or inside of body parietes, with 1 mass-like accessory gland on the base.

Color: Live specimens dark green on dorsum, greenish on ventrum.
Deposition of types: Department of Biology, Lanzhou University School of Medicine, Lanzhou, Gansu Province.
Etymology: The specific name "*viridis*" means green in Latin and indicates the body color of this species.
Type locality: Gansu (Lanzhou).
Distribution in China: Gansu (Lanzhou).
Remarks: This species is allied to *M. tschiliensis tschiliensis* (Michaelsen, 1928a) and *M. praepinguis praepinguis* (Gates, 1935a). It differs from *M. tschiliensis tschiliensis* by: (1) body smaller; (2) color dark grass green on dorsal side, greenish on ventral; (3) spermathecal pores 3 pairs, in 6/7/8/9, about 1/2 circumference apart; (4) clitellum with traces of dorsal pores; (5) the male pouch present, with 4 papillae in it; (6) caeca simple, with distinct indentations along ventral edge. This species differs from *M. praepinguis praepinguis* by: (1) body smaller; (2) color dark grass green on dorsal side, greenish on ventral side; (3) setae numerous; (4) with shallow "spermathecal chamber", without papillae at both anterior and posterior margins, papillae usually on VII, VIII, IX.

288. *Metaphire vulgaris agricola* (Chen, 1930)

Pheretima vulgaris agricola Chen, 1930. *Sci. Rep. Natn. Cent. Univ. Nanking*, 1:18–23, 34–36.
Pheretima vulgaris Gates, 1939. *Proc. U.S. Natn. Mus.* 85:497–502.
Metaphire vulgaris agricola Sims & Easton, 1972. *Biol. J. Linn. Soc.* 4: 238.
Metaphire vulgaris agricola Xu & Xiao, 2011. *Terrestrial Earthworms of China*. 258–259.

External Characters

Length 154–240 mm, width 6–9 mm, segment number 80–122. Annuli less obvious than in *Metaphire vulgaris vulgaris* (Chen, 1930), triannuli begin to appear after VIII or

IX backward. Prostomium 2/3 proepilobous. First dorsal pore 12/13. Clitellum present, not raised up, sometimes with intersegmental furrows, in XIV–XVI, annular, without setae, but sometimes with the setal pits. In young forms, the setae are present but without bifid. Setae on II–IX very slightly longer, their arrangement regular and equal in length. Both dorsal and ventral breaks very small, preclitellar region indistinct. Setae number 33–37 (III), 39–45 (VI), 45–56 (IX), 52–62 (XII), 44–60 (XVII), 58–62 (XXV); 19 (VIII) between the spermathecal orifices. 16–20 (those in the copulatory chamber included) between male pores. Male pores paired in XVIII, copulatory chamber present, ventrolaterally placed at XVIII, wrinkled laterally including a crescent-shaped groove which indicates the opening of the chamber. Inner surface is smooth and not swollen at the external lateral body wall where the setae may be occasionally present. At the median side, close to the margin of the aperture, there is large elevation on which 2–3 setae are present. It is eversible with a nipple-like bulb on the top below which are several circular grooves which mark off the bulb from the large base, a small flat-topped papilla occasionally present at the anteromedian side of the chamber. Sometimes the top of the bulb is chestnut in color with some minute papillae and small whitish spit which marks the pore of the prostate gland. Spermathecal pores 3 pairs in 6/7/8/9, each with a slit similar to that of *Metaphire vulgaris vulgaris* (Chen, 1930), wrinkled at both anterior and posterior edges with flat-topped papilla at the posterior median side and a papilla-like ridge at the anterolateral side. Sometimes, there are 2 small papillae posteriorly and, in many cases, both papillae may be absent. Or both anterior and posterior edges not wrinkled but raised as if greatly swollen. No other glands outside the aperture. No

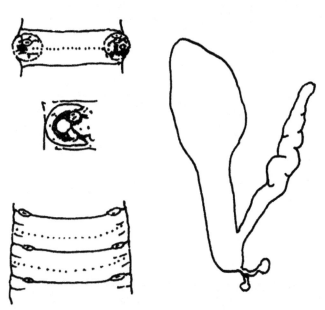

FIG. 281 *Metaphire vulgaris agricola* (after Chen, 1930).

such internally raised "spermathecal chamber" as in *Metaphire vulgaris vulgaris* (Chen, 1930). Female pore single, pore on a 4-cornered depression or a slight elevation in the setal circle of XIV.

Internal Characters

Septa 8/9/10 absent, but 8/9 with a thin membrane at the ventral side, 5/6/7/8 very thick and muscular; 10/11/12 also very thick, 12/13 less thickened, 13/14 still thicker than the subsequent segment. Gizzard small, in XI–X, globular shape, with a posterior rim. Intestinal caeca simple, originating in XXVII and extending anteriorly to XXIII, including 5 or 4.5 segments, with dorsal and ventral constrictions of the edge. Last pair of hearts in XIII. Testis sac 2 pairs in X, XI; the anterior pair globular, connected medially under the septum of 10/11, their 2 round sacs projecting into the segment X; while the posterior pair, V-shaped or transversely united at the posteromedian side of the first pair of seminal vesicles. Vas deferens at each side running close together after XIV but remaining separate throughout the whole length. Seminal vesicles large, 2 pairs in XI, XII, with large dorsal lobes overlapping each other over the dorsal side. Prostate glands in XVI–XX, often very large, lobate, much cut up into small lobules occupying XVI–XXI, its long duct with deep loop directed outward or upward, its ectal limb very long and stout; the gland closely lying upon the body wall, not noticeable unless dissected out, with a bundle of strands leading to the flat-topped papilla referred to above. Also a small velvet-like patch around the base of the duct; it is glandular in appearance. Spermathecae 3 pairs, in VII–IX; the last pair partly under the gizzard. Ampulla about 2 mm wide and 3 mm long, heart-shaped or oval in outline, whitish and smooth on the surface, not flattened; its duct long, about 1 mm wide and 2 mm long, in some cases, longer than the ampulla, about equally stout ectally and entally. Diverticulum shorter or slightly longer than main pouch made of 5 or more indistinct loops, their arms are in close contact with a straight elongate tip: its duct shorter than the main duct joining to anteromedian side of the latter, a short distance away from the body wall where the common duct becomes gradually smaller at the ectal end opening to the anterolateral side of the slit. The large lumen of the main duct is lost after joining the diverticular duct.

Color: Preserved specimens from grayish slaty to greenish gray on dorsum, grayish buff or greenish orange on ventrum, dark grayish along dorsomedian line. Clitellum grayish or greenish pale.
Deposition of types: Previously deposited in the Museum of the Biological Laboratory of the Science Society of China, Nanjing. The types were destroyed during the war in 1937.

Etymology: The name of the species refers to the type locality.
Type locality: Beijing, Jiangsu (Nanjing, Yangzhou, Xuzhou), Zhejiang (Taizhou).
Distribution in China: Jiangsu (Nanjing, Yangzhou, Xuzhou), Zhejiang (Taizhou), Beijing.
Remarks: This species is distinguished from either *M. houlleti* (Perrier, 1872) or *M. companulata* (Rosa, 1890) in having no modified clitella setae and penial setae as described by Professor Gates.

289. *Metaphire vulgaris vulgaris* (Chen, 1930)

Pheretima vulgaris Chen, 1930. *Sci. Rep. Natn. Cent. Univ. Nanking.* 1:12–18, 34.
Pheretima vulgaris Fang, 1933. *Sinensia*, 3(7):179–180.
Pheretima vulgaris Gates, 1935a. *Smithsonian Mus. Coll.* 93(3):19.
Metaphire vulgaris vulgaris Xu & Xiao, 2011. *Terrestrial Earthworms of China.* 259–260.

External Characters

Length 120–215 mm, width 5–8 mm, segment number 90–124. Triannuli from VI backward. Prostomium 2/3 epilobous and slightly prolobous, no transverse groove behind the tongue. First dorsal pore 11/12, the first is smaller and usually indistinct or absent. Clitellum generally not present, with setae all around, shorter but not modified in shape. Clitellum if it is present, in XIV–XVI, annular, without setae when it is swollen up as ring. Setae number 44–58 (III), 60–75 (VIII), 60–75 (XXV); 22 between spermathecal pores, 12–22 between male pores. Male pores paired in XVIII, situated ventrolaterally at the bottom of the copulatory chamber. The openings of the copulatory chamber ordinarily closed, about 1/4 circumference of the segment wrinkled around the aperture. They are protractile and may be easily everted as a penis which is vertical and slightly median. A part of the outer margin of the copulatory chamber is turned out with saccular small wrinkles at the tip and a papilla at the median posterior corner looking like half-opened blossom or it may be totally turned inside out. Copulatory chamber present, swollen out at the lateral body wall of the segment. Inner surface all rough with irregular ridges like the internal surface of the stomach of higher mammals and an elongate ovoid pad at the median side of the chamber on which there appear 10 or more transverse grooves or transversely elongated pits separated from each other by transverse ridges or broken up into small tubercles. Around the sides of the pad to a deep furrow. 2 moderately large flat-topped papillae situated deep at the lower side of the pad and lower than these is the round-topped papilla on which opens the duct of the prostate gland. Near the posterior median corner of

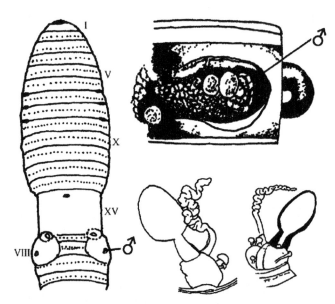

FIG. 282 *Metaphire vulgaris vulgaris* (after Chen, 1930).

but testes and funnels are not so large. Seminal vesicles 2 pairs in XI, XII respectively very large with 2 large dorsal lobes, meeting each other dorsally. Prostate glands large, finger-shaped expanded into segments XVII–XXI or XVI–XXII with many elongate lobes, usually cut up into smaller pieces. The duct has a long loop directed forward and inward at the side of intestine; the ectal limb stout, turning 2–3 times in the body wall before opening on to the exterior through a round-topped papilla at the median lower part of the elongate ovoid pad in the copulatory chamber. Small glands, whitish, 2–3 patches, similar to those in the spermathecal regions, are situated around the ental surface of the body wall near the base of prostate gland. Their stalks are embedded in the body wall opening on to the exterior through the flattened papillae referred to above.

Color: Live specimens variable, different combinations of orange, green and gray; from light olive green, yellowish light olive to ashy on the dorsal side, a little deeper anteriorly, light greenish orange on the posterior dorsolateral sides; a dark olive line along the dorsal pores, fleshy olive green in the young specimens; from light fleshy olive to light orange on the ventral side; Clitellum fleshy or pale to light chocolate or brownish. Preserved specimens greenish ashy on the dorsum with a dark olive streak along the dorsomedian line of dorsal pores; light greenish fleshy on the ventrum. Clitellum light brownish or light chocolate.

Deposition of types: Previously deposited in the Museum of the Biological Laboratory of the Science Society of China, Nanjing. The types were destroyed during the war in 1937.

Type locality: Jiangsu (Nanjing, Zhenjiang, Suzhou).

Distribution in China: Beijing (Xicheng, Dongcheng, Haidian, Tongzhou, Changping, Daxing, Fangshan, Chaoyang), Tianjin, Jiangsu (Nanjing, Zhenjiang, Suzhou), Zhejiang (Ningbo, Taizhou), Shandong, Hubei (Lichuan, Yichang, Qiangjiang), Hunan.

Remarks: This earthworm is similar to *M. houlleti* (Perrier, 1872) in having a severable copulatory chamber, a pouch similar to a "spermathecal chamber," and in general colocation. However, it differs from it in having no modified clitellar setae, and more than 1 stalked gland.

the aperture of the chamber is another flat-topped papilla which is occasionally seen externally. Spermathecal pores 3 pairs in 6/7/8/9, each with a wide slit as the outlet of the "spermathecal chamber", wrinkled at both anterior and posterior margins, better seen at the anterior margin. In many cases, the cone-shaped structure as the terminal part of the spermathecal duct ordinarily inside of the chamber is protruded by 2 papillae visible externally: the outer, as the outlet of the stalked gland situated at the lateral side, is flat-topped and the inner one, as the opening of the spermathecal duct, is round-topped. Another larger papilla, the outlet for the stalked gland at the median side, is covered by the posterior wall of the chamber situated at the posterior median corner. Female pore single, medioventral in XIV.

Internal Characters

Septa 8/9/10 absent, 8/9 very thin, only at the ventral side between second and third pairs of the spermathecae, 5/6/7 very thick, 7/8 less thickened, 10/11/12/13 also thickened, 13/14 less thickened. Gizzard in XI–X, globular shape with a posterior rim. Intestine swelling from XV. Intestinal caeca simple, originating in XXVII and extending anteriorly to XXIII, including 4–5 segments pointed at the anterior end and with irregularly wrinkled ventral edge. Last heart in XIII. Testis sac 2 pairs in X, XI; The anterior pair elongate ovoid, projected through the septum 10/11 into the posterior part of segment X, or suspended at the anterior lower face of the septum with a round sac directed downward. The posterior pair round or somewhat elongate situated posterioventral to the first pair of seminal vesicles. Both pairs have a median constriction communicating with each other. Testis sacs large

The account above (normal forms) differs somewhat from that of Chen, especially with respect to the copulatory chambers, testis sacs, and male deferent ducts. The everted copulatory chambers are club-shaped but with the narrowed portion of the everted body nearest the parietes. The testis sacs of X and XI of the Hamburg specimens are U-shaped. Chen's description of the testis sacs is, however, not clear, so that an adequate basis for

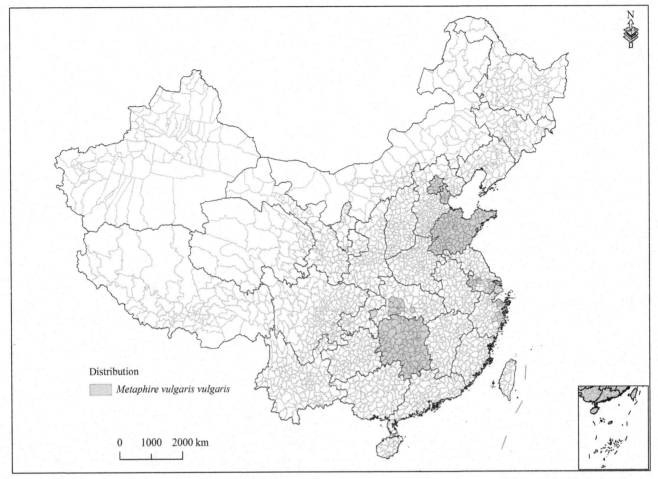

FIG. 283 Distribution in China of *Metaphire vulgaris vulgaris*.

comparison is not available. In the Hamburg specimens the 2 male deferent ducts of a side come into contact in segment XII, whereas in Chen's specimens the vasa deferentia of a side pass posteriorly into XVIII independently of each other. The differences just mentioned appear to be rather unimportant and insufficient justification for the erection of a new species, especially in view of the similarities of the copulatory and spermathecal chamber. The Hamburg specimens have accordingly been referred of *M. vularis*.

Fang (1933) refers to three worms "apparently without clitellum" from Nan-hu to "*Metaphire vulgaris* Chen. Yichang, Hubei, 1929." The specimen is quite obviously aclitellate. The worm is characterized by large, club-shaped, copulatory chambers, a U-shaped testis sac belonging to X, and spermathecal invaginations into the coelomic cavities as in *Metaphire vulgaris*. Other sex organs are more or less rudimentary.

290. *Metaphire youyangensis* (Zhong, Xu & Wang, 1984)

Pheretima youyangensis Zhong, Xu & Wang, 1984. *Act Zootaxonomica Sinica*. 9(4):356–359.

Metaphire youyangensis Zhong & Qiu 1992. *Guizhou Science*. 10(4):40.
Metaphire youyangensis Xu & Xiao, 2011. *Terrestrial Earthworms of China*. 260–261.

External Characters

Length 45–100 mm, width 3.2–4.6 mm, segment number 74–119. First dorsal pore 12/13. Clitellum XIV–XVI, annular, setae absent. Setae fine, enlarged in ventral surface of preclitellar segment, setae number 30–44 (III), 40–54 (V), 41–60 (XI), 46–66 (XX); 13–18 between male pores. Male pores paired in XVIII, ventrolateral, each on a small round papilla, situated in copulatory chamber. 1 small papilla lying close to porophore, other or 4 genital papillae on each side of XVIII, 2 of them posterior and 1–2 anterior to the setal ring. The lateral papilla in front of the setal ring, if present, concealed in the copulatory chamber. Spermathecal pores 3 pairs in 6/7/8/9, usually invisible externally. 5 pairs of genital papillae on the ventral side of segment VII, VIII, and IX. Female pore single, medioventral in XIV.

Internal Characters

Septa 8/9/10 absent, 5/6/7/8 thick, 10/11–13/14 slightly thickened, 14/15 backward membranous.

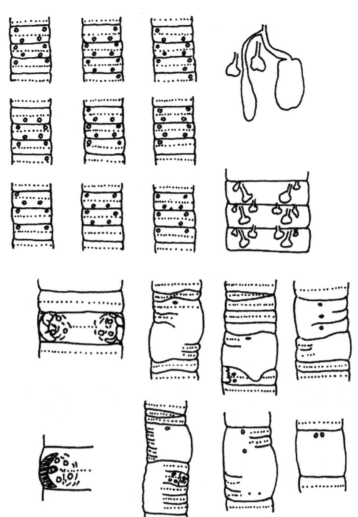

FIG. 284 *Metaphire youyangensis* (after Zhong, Xu & Wang, 1984).

Gizzard pear-shaped. Intestine swelling from XV. Intestinal caeca simple, in XXVII, extending anteriorly to XX. Last pair of hearts in XIII. Testis sac rounded, in X and XI, meeting ventrally. Seminal vesicles as transverse bands, no dorsal lobe; 3–4 stalked glands around the prostate duct. Prostate glands in XVI–XXI, usually divided into 2 large lobes, and then divided into many lobules. Spermathecae 3 pairs in VII–IX, ampulla oval, with a tube-like or long sac-shaped seminal chamber. Stalked glands in this region very conspicuous, the arrangement and number identical to the genital papillae of segments VII, VIII, and IX.

Color: Live specimens puce on dorsum, livid on middorsal line, dark orange on clitellum.
Deposition of types: Department of Biology, Sichuan University.
Etymology: The name of the species refers to the type locality.
Type locality: Sichuan (Youyang, Xiushan).

Distribution in China: Sichuan (Youyang, Xiushan).
Remarks: This species is similar to *M. houlleti* (Perrier, 1872) in general morphology and in reproductive organ polymorphism, but differs from the latter in having no setae on clitellum, no convoluted diverticulum and long-stalked gland connecting with spermathecae.

There are 6 specimens with abnormal clitella on which develop1–3 female pores, lying in XIV, XIV–XV, or XIV–XVI respectively. There are also 2 pairs of ovaries. Other genital organs are either rudimentary or lacking.

multitheca Group

Diagnosis: *Metaphire* with spermathecal pores 3 pairs, at VI–VIII.
Global distribution: Oriental realm from Japan southwards through the Indo-Australasian archipelago to the

rainforests of Australasia and eastwards through the Oceanian realm.

There is 1 species from China.

Remarks: Species of this genus are separable from those of *Amynthas* by the presence of copulatory pouches and differ from species of *Pheretima* by the absence of nephridia from the spermathecal ducts.

291. *Metaphire multitheca* (Chen, 1938)

Pheretima (Pheretima) multitheca Chen, 1938. *Contr. Biol. Lab. Sci. Soc. China (Zool).* 12(10):383–385.

Metaphire multitheca Sims & Easton, 1972. *Biol. J. Linn. Soc.* 4:239.

Metaphire multitheca Xu & Xiao, 2011. *Terrestrial Earthworms of China.* 241–242.

External Characters

Length 155 mm, width 7 mm, segment number 95. Prostomium 1/3 epilobous. First dorsal pore 12/13. Clitellum XIV–XVI, annular, setae absent. Ventral setae of VII–X (*a–e*) slightly enlarged, ventral break distinct, dorsal break in front of clitellum not apparent; $aa = 2.0$–$2.5\,ab$, $zz = 1.2$–$2.0\,yz$; setae number 33–36 (III), 50–53 (VI), 48–60 (XIX), 50–60 (XXV); 4 between male pores. Male pores about 1/4 circumference apart ventrally, each in a laterally retracted semilunar pouch, male pore on top of a median skin pad, 2 large papillae placed 1 in front of, and another behind each pore. Male pore portion with 2 papillae appearing nipple-like when everted. Spermathecal pores numerous, irregularly placed on posterior border of VI, VII, and VIII, each minute and ampulla-like, nearly as large as setal pit, about 12 on VI, 10–12 on VII, 8 on VIII. Female pore single, medioventral in XIV.

Internal Characters

Septa 8/9 present, very thin (membranous), 9/10 absent, 7/8 muscular but thin, 5/6/7 thick and muscular, 10/11/12/13 also thick. Gizzard rounded, in IX–1/2 X. Intestine swelling from XV. Intestinal caeca simple, small, in XXVII, extending anteriorly to XXV. Lateral hearts 4 pairs, first pair in X stout. Testis sacs round; both broadly united on ventral side. Seminal vesicles paired in XI and XII, enormously developed, rather smooth on surface, each with a distinct dorsal lobe. Prostate glands small and compact, in XVII–XIX, about 5 mm in anteroposterior length, its duct short and weakly curved, ental 1/4 slender and ectal portion greatly thickened. Spermathecae in VII–IX; 10 in VII, 11 in VIII and 8 in IX, main pouch consisting of a spatulate ampulla and an indistinctly marked duct about 1/3 of whole length (whole length about 2 mm). Diverticulum small, with a bulb-like whitish seminal chamber and a slender long duct joining ectal end of main duct, its whole length as long as main pouch.

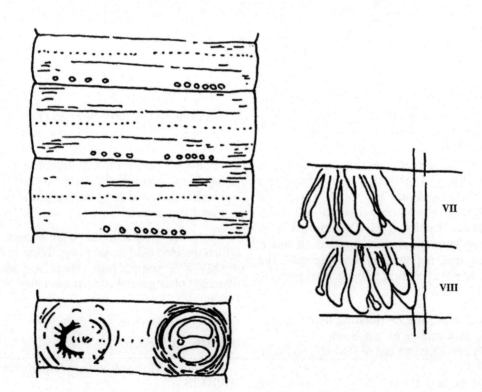

FIG. 285 *Metaphire multitheca* (after Chen, 1938).

Color: Preserved specimens dark grayish on dorsum, pale on ventrum.

Deposition of types: Previously deposited in the Metropolitan Museum of Natural History, Beipei, Chongqing. The types were destroyed during the war in 1949.

Etymology: The name *"multitheca"* indicates the number of spermathecae of this species.

Distribution in China: Hainan (Sanya).

plesiopora Group

Diagnosis: *Metaphire* with spermathecal pores 1 pair, at 5/6.

Global distribution: Oriental realm from Japan southwards through the Indo-Australasian archipelago to the rainforests of Australasia and eastwards through the Oceanian realm.

There is 1 species from China.

Remarks: Species of this genus are separable from those of *Amynthas* by the presence of copulatory pouches and differ from species of *Pheretima* by the absence of nephridia from the spermathecal ducts.

292. *Metaphire plesiopora* (Qiu, 1988)

Pheretima plesiopora Qiu, 1988b. *Sichuan Journal of Zoology.* 7(1):1–2.
*Metaphire plestopora.*Zhong & Qiu 1992. *Guizhou Science.* 10(4):41.
Metaphire plesiopora Xu & Xiao, 2011. *Terrestrial Earthworms of China.* 246–247.

External Characters

Length 44–80 mm, width 2–3 mm, segment number 70–116. Prostomium proepilobous. First dorsal pore 11/12. Clitellum XIV–XVI, annular, with setae on ventral side. Setae fine, darkish, numerous and close together; setae number 74–93 (III), 96–117 (V), 101–127 (VIII), 104–116 (XX), 97–123 (XXV); 7–8 (V), 8–9 (VI) between spermathecal pores, 2–3 rows setae on ventral region of XVII, and XIX–XXII or XXVI, but one row on dorsal side of these segments. Male pores 1 pair, in XVIII ventrally, each in a small and shallow copulatory chamber, ectal margin of copulatory chamber covered by a large skin fold and ental margin swelling, protuberance large and tall, the 2 margins almost fused together, apertures at the bottom of the copulatory chamber, about 1/4 circumference apart ventrally, papillae absent. Spermathecal pores 1 pair, intersegmental in 5/6, apertures on a shuttle–shaped papilla, anterior and posterior body wall swelling, about 1/15 circumference apart ventrally, without papillae around it. Female pore single, medioventral in XIV.

Internal Characters

Septa 4/5–9/10 very thick and muscular, not transparent; 10/11/12/13 less thickened, transparent, the rest thin. Gizzard in VIII, small, ball-shaped. Intestine swelling from XV; Intestinal caeca simple, in XXVII, finger-like, extending anteriorly to XXIV, smooth on both edges. Last pair of hearts in XIII. Testis sacs 2 pairs, in X and XI; the anterior pair smaller, both pairs long round-shaped. Seminal vesicles 2 pairs, anterior pair slightly small, the first pair of seminal vesicles enclosed in the second pair of testis sacs; posterior pair slightly large, dorsal lobe small, round-shaped or triangular, anteroposition. Prostate glands small, in XVII–XVIII, prostatic duct stout and long, U-shaped. Accessory glands absent. Spermathecae 1 pair, in VI; ampulla ball-shaped or ellipse-shaped, duct slightly curved; diverticulum of equal length to or slightly shorter than main part; twisted in a Z-shape in the middle part, and its seminal chamber not enlarged.

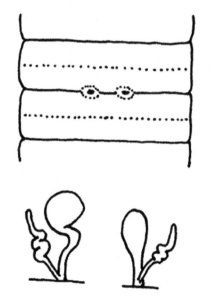
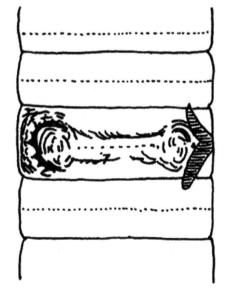

FIG. 286   *Metaphire plestopora* (after Qiu, 1988b).

Color: Preserved specimens gray-white, clitellum orange-brown.

Deposition of types: Guizhou Institute of Biology, Guiyang.

Etymology: The species name indicates the short distance between spermathecae.

Type locality: Guizhou (Chishui).

Distribution in China: Guizhou (Chishui).

Remarks: This species is similar to *Amynthas limellus* (Gate, 1935a), but can distinguished from it by the spermathecae close to medioventral line, 2–3 rows setae on ventral region of XVII, XIX–XXII, or XXVI and the Z-shaped twist in the middle part of the spermathecal diverticulum.

schmardae group

Diagnosis: *Metaphire* with spermathecal pores 2 pairs, at 7/8/9

Global distribution: Oriental realm from Japan southwards through the Indo-Australasian archipelago to the rainforests of Australasia and eastwards through the Oceanian realm.

There are 18 species from China.

Remarks: Species of this genus are separable from those of *Amynthas* by the presence of copulatory pouches and differ from species of *Pheretima* by the absence of nephridia from the spermathecal ducts.

TABLE 33   Key to the *schmardae* Group From China

1. Spermathecal pores 2 pairs in 7/8/9 dorsally 2
   Spermathecal pores 2 pairs in 7/8/9 ventrally 5

2. Genital papillae located anterior Clitellum 3
   No genital papillae located anterior Clitellum 4

3. There is a genital papilla on anterior site of setal ring near the spermathecal pore, sometimes another is present on ventrolateral side of VIII .... *Metaphire aduncata*
   Each intersegmental, appearing as minute tubercle at posterior edge of segments VII and VIII, encircled anteriorly and laterally by 3–6 minute ampulla-like papillae, a glandular depression usually posterior to each pore on VIII and IX. Skin weakly glandular around this area. No other genital papillae .... *Metaphire magna*

4. Posterior pair spermathecal pores closer between setae *a* and *b*; anterior pair more lateral, in line with *c* and *d*. Each as elliptical slit, quite wide *Metaphire longipenis*
   Anterior pair more laterally situated, in line to 4th or 5th seta, posterior pair to 2nd seta. Each represented by an eye-like swelling bearing a central pore .... *Metaphire prava*

5. Genital papillae located anterior Clitellum 6
   No genital papillae located anterior Clitellum 7

6. Fewer than 5 papillae 16
   Greater than 5 papillae 17

7. Spermathecal pores usually invisible externally. 1 pair of papillae on ventral side of segments VIII and IX respectively ...... *Metaphire brevipenis*
   Spermathecal pores on circular to oval areas within deep invaginations with transversely slit-like apertures, A large papilla close to anterior edge of VIII and IX. Male pores on the dorsal wall of large copulatory chambers conspicuously protuberant into the coelom. On the median wall of the copulatory chamber a presetal, transversely oval genital marking .... *Metaphire paeta*

8. Spermathecal pore large and obvious, transversely fissured, 2–3 oblong papillae on anterior and posterior margin ...... *Metaphire guizhouense*

9. Genital papillae located posterior Clitellum 3
   No genital papillae located posterior Clitellum 4

10. Fewer than 5 papillae 16
    Greater than 5 papillae 17

11. Male pores situated in a deep copulatory pouch, openings of the latter as 2 transverse slits, often widely opened and each wrinkled around its margin, Cuticle extending to inner side of wrinkled margin. Prostatic duct entering from top of the chamber through a roundish papilla. 2 shallow pits on anterior and posterior sides of the chamber in which the lateral glands open .... *Metaphire schmardae*
    Male pores in deep pouches, lateroventral in position, each pouch rather deep, its median border wrinkled and usually with few papillae exposed, 2–3 in front of, 1–2 behind setal line .... *Metaphire leonoris*

12. Male pores minute, each in a very small tubercle often at end of a columnar protuberance from the top of an eversible, deep, transversely slit-like parietal invagination .... *Metaphire californica*
    Male pores each situated in a shallow copulatory pouch, as in *Metaphire californica* (Kinberg, 1867), the openings transverse slit-like when invaginated, with annular ridges and longitudinal slits around .... *Metaphire myriosetosa*

13. Male pores in a shallow copulatory chamber, a small fleshy pad on medial side of the chamber, male pore situated on lateral lower side of the pad. The pad in outline resembling the tip of a nipple when it is wholly everted .... *Metaphire hesperidum*
    No pad in copulatory chamber 14

14. Male pores 1 pair, situated on the ventrolateral side of XVIII. Spermathecal pores 2 pairs in 7/8/9 .... *Metaphire capensis*
    Male pores are large, and are not situated on papillae. Spermathecal pores 2 pair, in 7/8/9 .... *Metaphire browni*

293. *Metaphire aduncata* Zhong, 1987

Metaphire aduncata Zhong, 1987. Journal of Sichuan University Natural Science Edition. 24(3):336–339.
Metaphire aduncata Xu & Xiao, 2011. Terrestrial Earthworms of China. 212–213.

External Characters

Length 54–111 mm, width 4.0–5.0 mm. Segment number 63–95. Prostomium 2/3 epilobous. First dorsal pore 12/13. Clitellum entire, XIV–XVI, annular, with setae on ventral side. Setae small, setae number 29–33 (III), 27–30 (V), 39–42 (VII), 43–54 (IX), 44–51 (IX), 50–54 (XIII), 47–51 (XX); 6–8 (VIII), 7–9 (IX) between spermathecal pores, 11–24 between male pores. Male pores situated in deep copulatory chambers which are ϕ-shaped on ventrolateral side of XVIII, about 2/5 circumference apart ventrally. Each copulatory chamber aperture as a transverse slit with longitudinal grooves on the margin and the surface of the skin. Spermathecal pores 2 pairs, intersegmental, dorsally in 7/8/9, about 1/5 circumference dorsally apart, arranged in a rectangle. There is a genital papilla on anterior site of setal ring near the spermathecal pore, sometimes another is present on ventrolateral side of VIII. Female pore single, medioventral in XIV.

Internal Characters

Septa 8/9/10 absent, 5/6/7 slightly thick, 7/8 thin, 10/11–13/14 muscular. Gizzard garlic-shaped, smooth on surface. Intestine from XV. Intestinal caeca manicate, in XXVII, extending anteriorly to 1/2 XXIV or XXIV, with 4–5 finger-like sacs. Last pair of hearts in XIII. Testis sacs holandric, first pair united in a V shape, second pair in XI, united portion broad. Seminal vesicles in XI and XII, dorsal lobe conspicuous. Prostate glands in XVI–XX or XVI–XXI or XVII–XIX, its duct enters into ental top of copulatory chamber and a thin membrane covers the surface of both duct and chamber. Copulatory chamber columnar or rounded bulb-like with inner surface having no genital papillae but strongly wrinkled. Spermathecae 2 pairs, in VIII and IX, ampulla oblate about 3.0 mm in diameter or heart-shaped with surface strongly wrinkled; diverticulum is shorter than main pouch and closely twisted in a zigzag manner or W-shaped; seminal chamber not differentiated from its duct conspicuously.

Color: Preserved specimens brown on dorsum, slightly gray on postclitellar ventrum. Setal circles on VII or IX–XIII and XVII–XXI or XVII–XIX unpigmented, form lighter color rings. A black stripe on mediodorsal line from clitellum to posterior end of body. Clitellum dark maroon.
Deposition of types: Department of Biology, Sichuan University, Chengdu, Sichuan.
Etymology: The specific name *"aduncata"* means hooked in Latin and probably indicates the aduncate ectal end of the prostate glands.
Type locality: Chongqing (Wan County).
Distribution in China: Chongqing (Wanzhou).
Remarks: This species is similar to *M. longipenis* (Chen, 1946) in having 2 pairs of spermathecal pores dorsally in 7/8/9 and a copulatory chamber present, but differs from it in having spermathecal pores arranged in a rectangle and the absence of a spermathecal chamber; caeca manicate and with stalked gland in VIII. This species also differs from *Metaphire prava* (Chen, 1946) by the caeca being manicate, the copulatory chamber visible in the body cavity, and no large papilla placed deep in the pouch; setae fewer.

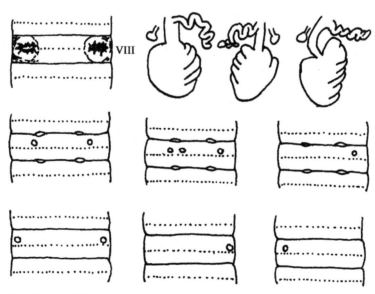

FIG. 287 *Metaphire aduncata* (after Zhong, 1987).

294. *Metaphire brevipenis* (Qiu & Wen, 1988)

Pheretima brevipenis Qiu & Wen, 1988. *Act Zootaxonomica Sinica*. 13(4):340–342.
Metaphire brevipenis Zhong & Qiu, 1992. *Guizhou Science*. 10(4):41.
Metaphire brevipenis Xu & Xiao, 2011. *Terrestrial Earthworms of China*. 219.

External Characters

Body rather small. Length 55–95 mm, width 3–4 mm, segment number 78–117. Prostomium 1/2 epilobous. First dorsal pore 12/13. Clitellum XIV–XVI, annular, without setae. Setae fine; setal number 29–36 (III), 32–43 (V), 37–50 (VIII), 48–63 (XX), 46–66 (XXV); 17–21 (VIII) between spermathecal pores, 9–14 (XVIII) between male pores. Male pores 1 pair, situated in ventrolateral side of XVIII, about 1/3 circumference apart ventrally; each situated in the bottom of a small copulatory chamber. Spermathecal pores 2 pairs in 7/8/9, about 2/5 circumference apart ventrally, usually invisible externally. 1 pair of papillae on ventral side of segments VIII and IX respectively. Female pore single, medioventral in XIV.

Internal Characters

Septa 8/9/10 absent, 5/6/7/8 slightly thickened. Gizzard developed, rugby. Intestine swelling from XVI. Intestinal caeca simple, originating in XXVII and extending anteriorly to XXIV. Lateral hearts 4 pairs, in X–XIII. Testis sac 2 pairs in X–XI, developed, ball-shaped, communicated ventrally. Seminal vesicles paired in XI–XII, not developed, but their dorsal lobe conspicuous. Prostate glands in XVI–XXI, developed. Accessory glands mass-like, with cord-like ducts, in front of the prostatic ducts. Spermathecae 2 pairs in VIII and IX; ampulla elliptical, about 4 mm long, and 3 mm wide, or round with a diameter of about 3 mm. Diverticulum convoluted, longer than the main part. Stalk glands in this region very conspicuous, the arrangement and number identical to the genital papillae of segments VIII and IX.

Color: Preserved specimen unpigmented. Clitellum red-brown.
Deposition of types: Guizhou Institute of Biology, Guiyang.
Etymology: The name *"brevipenis"* indicates the short penis of this species.
Type locality: Guizhou (Kuankuoshui forest region, Suiyang).
Distribution in China: Guizhou (Suiyang).
Remarks: This species is similar to *M. leonoris* (Chen, 1946), but differs from it in having no spermathecal chamber, a longer diverticulum, a long-stalk gland connecting with the spermathecae, and a rather small copulatory pouch. Also, this species is distinguished from *M. pedunculata* (Chen & Hsü, 1977) by the absence of septa 8/9, the shape of spermathecae pore, the longer and twisted diverticulum, and the accessory gland with cord-like ducts.

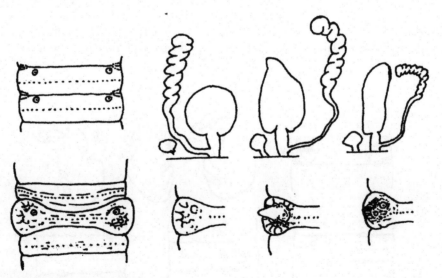

FIG. 288   *Metaphire brevipenis* (after Qiu & Wen, 1988).

295. *Metaphire browni* (Stephenson, 1912)

Pheretima browni Stephenson, 1912. *Rec. Indian Mus.* 7:273–274.
Pheretima browni Gates, 1931. *Rec. Ind. Mus.* 33: 372–373.
Metaphire browni Sims & Easton, 1972. *Biol. J. Linn. Soc.* 4:238.
Metaphire browni Xu & Xiao, 2011. *Terrestrial Earthworms of China.* 220.

External Characters

Length 102 mm, width 3 mm, segment number 108. Prostomium small, prolobous. First dorsal pore 11/12. Clitellum XIV–XVI, annular, setae absent. Setae form a ring which is closed ventrally, and almost closed dorsally; the setae are a little closer together ventrally than laterally and dorsally, and those of segments IV–IX are enlarged; setae number 23 (V), 34 (IX); 12 between male pores. Male pores 1 pair, situated in ventrolateral side of XVIII, about 1/3 circumference apart ventrally; the apertures are large, and are not situated on papillae. Spermathecal pores 2 pairs, in 7/8/9. No genital papillae or other special marks around spermathecal region. Female pore single, medioventral in XIV.

Internal Characters

Septa 5/6/7/8 thickened, 8/9 is represented by a ventral fragment only, 9/10 absent. Gizzard in VIII and IX. Intestine swelling from XV. Intestinal caeca in XXVI, extending anteriorly to XXIII; elongated, conical, without secondary projections. Last pair of hearts in XIII. The seminal funnels in X and XI, enclosed in small testicular sacs; the sacs of each pair are separate, not conjoined across the middle line. The seminal vesicles are paired, of moderate size, in segments XI and XII. Prostate glands are of moderate size, and flattened against the body wall; each consists of 2 principal lobes, 1 anterior, the other posterior to the origin of the duct; both lobes are divided up into numerous lobules. Spermathecae 2 pairs, in VIII and IX, possess an irregularly shaped, roughly ovoid ampulla, with a broad short duct. Diverticulum arises from near the distal end of the duct; it is variable, often coiled, thin and narrow for the most part, and dilated at its internal end; when uncoiled it is about equal in length to the ampulla or somewhat shorter.

FIG. 289 *Metaphire browni* (after Gates, 1931).

Color: Preserved specimens dark brown, often with a purple tinge.
Deposition of types: Indian and British Mus.
Etymology: The species was named after Mr. J. Coggin Brown (1884–1962), who collected many of the specimens for the Geological Survey of India.
Type locality: Yunnan (Tengyuan).
Distribution in China: Yunnan (Tengyue, Yiliang, Kunming, Lancang).
Remarks: According to the records by Stephenson and Gates, the specimens were all in a poor condition for detailed study.

296. *Metaphire californica* (Kinberg, 1867)

Pheretima californica Chen, 1936. *Contr. Biol. Lab. Sci. Soc. China (Zool).* 11:270.
Pheretima californica Gates, 1939. *Proc. U.S. Natn. Mus.* 85:427–429.
Pheretima californica Chen, 1946. *J. West China. Border Res. Soc. (B).* 16:135, 142.
Pheretima californica Chen, 1959. *Fauna Atlas of China: Annelida.* 10.

Pheretima californica Gates, 1959a. *Amer. Mus. Novitates*, 141:5–6.

Pheretima californica Tsai, 1964. *Quer. Jou. Taiwan Mus.* 17:23–25.

Pheretima californica Gates, 1972. *Trans. Amer. Philos. Soc.* 62(7):174–175.

Metaphire californica Sims & Easton, 1972. *Biol. J. Linn. Soc.* 4:238.

Metaphire californica James, Shih & Chang, 2005. *Jour. N. His.* 39(14):1026.

Metaphire californica Chang, Shen & Chen, 2009. *Earthworm Fauna of Taiwan*. 106–107.

Metaphire californica Xu & Xiao, 2011. *Terrestrial Earthworms of China*. 222.

External Characters

Length 50–156 mm, clitellum width 3–5 mm, segment number 55–115. Prostomium epilobous. Setae number 38–41 (VII), 55 (XX), 14–20 between male pores. First dorsal pore 11/12. Clitellum XIV–XVI, annular, dorsal pore and setae absent. Male pores in XVIII, minute, each in a very small tubercle often at end of a columnar protuberance from the top of an eversible, deep, transversely slitlike parietal invagination. Spermathecal pores 2 pairs in 7/8/9, 0.3–0.5 circumference apart ventrally. No genital papillae in the preclitellar region. Male pores paired in XVIII, about 0.4 circumference apart ventrally, in lateral slits of retracted copulatory pouches. Female pore single, medioventral in XIV.

Internal Characters

Septa 8/9/10 absent, 6/7/8, 10/11/12/13 thickened. Gizzard in VIII–IX. Intestine from XIV, XV or XVI. Intestinal caeca paired in XXVI or XXVII, simple, extending anteriorly to XXV–XXIII. Lateral hearts in X–XIII. Testis sacs 2 pairs in X and XI. Seminal vesicles paired in XI and XII. Prostate glands paired in XVIII, extending anteriorly to XVII and posteriorly to XX or XXI, racemose. Spermathecae 2 pairs in VIII and IX, ampulla large, heart-shaped, stalk short and stout, diverticulum tubular, slender, and coiled. Ovaries paired in XIII.

> Color: Live specimens dark red-brown on dorsum, white on clitellum.
> Deposition of types: Stockholm Museum, Sweden.
> Etymology: The name of the species refers to the type locality of California.
> Type locality: Sausalito Bay, California, USA.

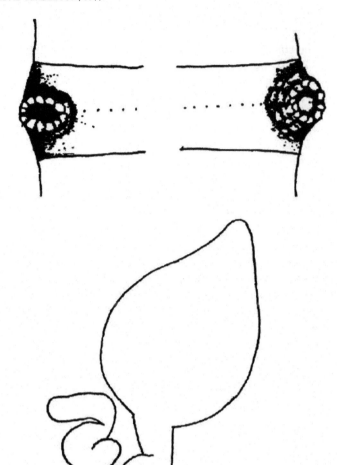

FIG. 290 *Metaphire californica* (after Chen, 1959).

Distribution in China: Jiangsu, Zhejiang, Anhui, Fujian, Taiwan (Taibei, Pingdong), Jiangxi, Hubei (Lichuan), Hunan, Chongqing (Nanchuan, Shapingba, Beipei, Peiling, Jiangbei), Sichuan (Chengdu, Luzhou, Leshan), Guizhou (Mt. Fanjing), Yunnan.

Remarks: There is nothing whatsoever in Beddard's account of his *M. sandvicensis* (Beddard, 1896) to indicate specific distinction from *M. californica* (types and the Hong Kong specimens). If Beddard's specimens cannot be found, *M. sandvicensis* will have to be regarded as a synonym of *M. californica*.

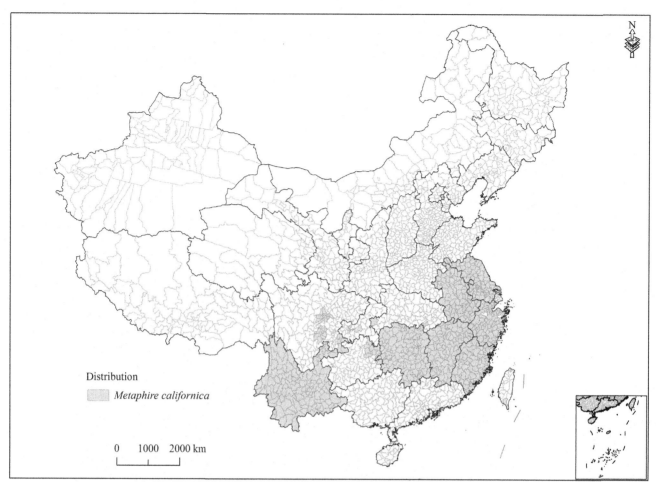

FIG. 291   Distribution in China of *Metaphire californica*.

297. *Metaphire capensis* (Horst, 1883)

Megascolex capensis Horst, 1883. *Notes from the Leyden Museum.* 5:195.
Amyntas capensis Beddard, 1900c. *Proc. Zool. Soc. London.* 1900:617–618.
Metaphire capensis Xu & Xiao, 2011. *Terrestrial Earthworms of China.* 223.

External Characters

Length 130 mm, segment number 110. First dorsal pore 5/6. Clitellum XIV–XVI, annular. Setae number 40 (VIII), 60 (XXV); 12 between male pores. Male pores 1 pair, situated on the ventrolateral side of XVIII. Spermathecal pores 2 pairs in 7/8/9, ventral. Female pore single, medioventral in XIV.

Internal Characters

A large prostatic gland, consisting of numerous slender lobes. In XV–XXIV. Spermathecae 2 pairs in VIII and IX; each spermathecae consists of 2 parts, a large round vesicle, and a thick, cylindrical tube, more than twice as long as the vesicle and usually bent around it;

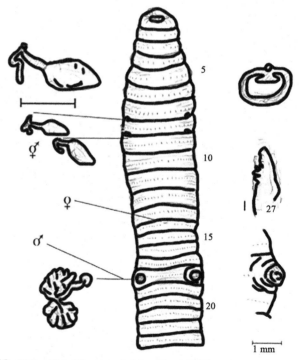

FIG. 292   *Metaphire capensis* (after Horst, 1883).

over its whole length the tube has the same diameter, but near its free end it suddenly becomes narrower and ends in a small oval sac. Diverticulum very long and often coiled like a sheep's horns.

Etymology: The name of the species refers to the type locality.
Type locality: Africa (Cape of Good Hope).
Distribution in China: Hong Kong.

298. *Metaphire extraopaillae* Wang & Qiu, 2005

Metaphire extraopaillae Wang & Qiu, 2005. *J. Shanghai Jiaotong Univ. (Agri. Sci.).* 23(1):26–27.
Metaphire extraopaillae Xu & Xiao, 2011. *Terrestrial Earthworms of China.* 226.

External Characters

Length 70–122 mm, width 3–4 mm, segment number about 92. Prostomium 1/3 epilobous. First dorsal pore 12/13. Clitellum XIV–XVI, annular, setae absent. Setae perichaetine, setae number 40–42 (III), 50–52 (V), 58–62 (VIII), 52–56 (XX), 52–56 (XXV); 24–28 between spermathecal pores, 15–18 between male pores. Male pores paired, ventral in XVIII, located in the bottom of the smaller copulatory chamber; a depressed pit present anteriorly and postsetally to each copulatory chamber, in which many glandular papillae occurring; about 1/3 circumference apart ventrally. Spermathecal pores 2 pairs in 7/8/9, situated on a slightly small papilla, glandular tumescence on anterior edges of spermathecae pores, approaching setal zone; about 2/5 circumference apart ventrally. Female pore single, medioventral in XIV.

Internal Characters

Septa 8/9 absent, 9/10 present ventrally, rest thin. Gizzard in VIII–IX, barrel-shaped, developed. Intestine

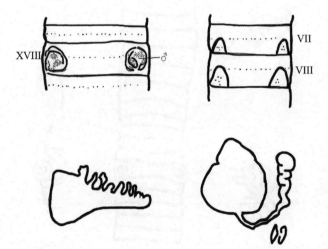

FIG. 293   *Metaphire extraopaillae* (after Wang & Qiu, 2005).

swelling from XV. Intestinal caeca paired in XXVII, manicate, extending anteriorly to XXIV, about 8 finger-like sacs. Testis sacs 2 pairs in X and XI, well developed, united. Seminal vesicles paired in XI and XII, developed, meeting on dorsal side. Prostate glands in XVII–XXI, developed, its duct thickened, accessory glands stick-shaped. Spermathecae 2 pairs, in VIII–IX; ampulla round, heart-shaped or ovoid, diverticulum longer than main pouch, ectal half twisted and swollen. Accessory gland similar in character to those near prostate gland.

Color: Preserved specimens violet-brown on dorsum, gray-brown on ventrum, setal ring white.
Deposition of types: Laboratory of Environmental Biology, School of Agriculture and Biology, Shanghai Jiaotong University.
Etymology: The specific name probably indicates the structure of the spermathecal pores of this species.
Type locality: Hubei (Mt. Xingdou, Lichuan).
Distribution in China: Hubei (Lichuan).
Remarks: This species is similar to *Metaphire aduncata* (Zhong, 1987), but differs from the latter with the respect to 3 aspects: (1) spermathecal pores are about 2/5 circumference apart ventrally; (2) there is a postsetal depressed region with many papillae near the copulatory pouch; (3) there are many accessory glands close to the prostate gland.

299. *Metaphire guizhouense* Qiu, Wang & Wang, 1991

Metaphire guizhouense Qiu, Wang & Wang, 1991. *Guizhou Science.* 9(3):222–223.
Metaphire guizhouense Xu & Xiao, 2011. *Terrestrial Earthworms of China.* 230–231.

External Characters

Length 132–282 mm, width 6–8 mm, segment number 98–145; Prostomium 1/3 epilobous. First dorsal pore 12/13. Clitellum XIV–XVI, annular, setae absent. Setae fine; setae number 34–42 (III), 40–54 (V), 50–62 (VIII), 60–72 (XX), 66–72 (XXV); 24–29 (VIII) between spermathecal pores, 6–10 (XVIII) between male pores. Male pores 1 pair in XVIII, ventrolateral about 1/3 circumference apart ventrally; each situated in a large and deep copulatory pouch; Spermathecal pores 2 pairs in 7/8/9, large and obvious, transversely fissured, about 1/2 circumference apart; 2–3 oblong papillae on anterior and posterior margin, rarely, copulatory pouch invisible, without other papillae. Female pore single, medioventral in XIV.

Internal Characters

Septa 9/10 absent, 8/9 present ventrally, thin; 5/6/7 thick, 7/8 thick and muscular. Gizzard long ball-shaped,

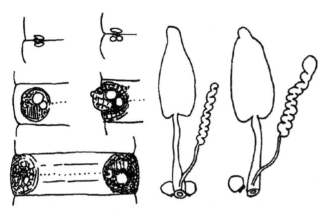

FIG. 294   *Metaphire guizhouense* (after Qiu et al., 1991b).

in IX–X. Intestine swelling from XV. Intestinal caeca simple, finger-sac-like, in XXVII, extending anteriorly to XXIII. Testis sacs 2 pairs in X and XI, ovoid, communicated. Seminal vesicles large, in XI and XII. Prostate glands developed, in XVI–XXI. Spermathecae 2 pairs in VIII and IX, ampulla large, heart-shaped, about 7–8 mm long; ampulla duct stout and long, diverticulum slightly shorter than main part, half or 2/3 of terminal part in a twisted Z shape. Accessory glands sessile.

Color: Preserved specimens brown on dorsum, slightly gray on ventrum. Clitellum red-brown.
Deposition of types: Guizhou Institute of Biology, Guiyang.
Distribution in China: Guizhou (Leishan).
Etymology: The name of the species refers to the type locality.
Type locality: Guizhou (Mt. Leigong).
Remarks: This species is similar to *Metaphire leonoris* (Chen, 1946), but differs from the latter by the presence of septum 8/9, the rather large body, and the shape of the copulatory pouch.

300. *Metaphire hesperidum* (Beddard, 1892)

Perichaeta hesperidum Beddard, 1892. *P. Z. S*: 69.
Amynthas hesperidum Beddard, 1900c. *P. Z. S*: 633.
Pheretima hesperidum Chen, 1931. *Contr. Biol. Lab. Sci. Soc. Chin. Zool.* 7(3):137–142.
Pheretima hesperidum Chen, 1933. *Contr. Biol. Lab. Sci. Soc. China (Zool).* 9:275–277.
Metaphire hesperidum Sims & Easton, 1972. *Biol. J. Soc.,* 4:238.
Metaphire hesperidum Xu & Xiao, 2011. *Terrestrial Earthworms of China.* 231–232.

External Characters

Length 80–150 mm, width 3.5–5.0 mm, segment number 76–112; preclitellar segments thick in musculature. Prostomium 1/2 epilobous, tongue open behind. First

dorsal pore 11/12. Clitellum as long as it is broad, annular, equal to or a little longer than 3 following postclitellar segments, in XIV–XVI, dorsal pore absent, without setae and intersegmental furrows, glandular part not particularly swollen nor constricted. Setae in continuous rings, both dorsal and ventral breaks very slight, the former wider than the latter, preclitellar segments smaller; setae number 20–30 (III), 28–40 (VI), 32–48 (VIII), 40–52 (XII), 42–54 (XXV); 16–17 between spermathecal pores, 12–16 between male pores. Male pores in a shallow copulatory chamber on ventrolateral side of XVIII, about 5/13 circumference apart ventrally; a small fleshy pad on medial side of the chamber, male pore situated on lateral lower side of the pad. The pad in outline resembling the tip of a nipple when it is wholly everted. The everted part sometimes high and pointed, or low and broadly rounded with radiating grooves and sometimes with a few transverse grooves perpendicular to the former, pointed ventrally or laterally. The chamber with a round opening wrinkled radially around its margin, without circular ridges and deep furrows. Genital papillae absent in this region. Spermathecal pores 2 pairs, in intersegmental furrows 7/8/9, about 6/13 circumference apart ventrally, its surrounding part depressed, transversely elliptical or eye-like, with a small opening in the middle appearing as a whitish spot. Genital papillae absent in this region. Female pore single, medioventral in XIV.

Internal Characters

Septa 8/9/10 absent, or 8/9 ventrally present, 5/6/7/8, 10/11–13/14 thick, particularly thickened in 5/6/7, 10/11/12, of which 13/14 thinnest; silk-shining on surface. Gizzard short and rounded or sometimes vase-shaped, surface smooth and glittering whitish, lying in IX and X. Intestine swelling from XV. Intestinal caeca simple, moderate in size, in XXVI, extending anteriorly to XXIV or XXIII, with ventral edge wrinkled. Lateral hearts 4 pairs in X–XIII. Testis sacs very large but testis and seminal funnel rather small. Anterior pair prominently projected out before septum 10/11 in X, more or less elongated or rounded, in a thick V shape, narrowly connected medially. Posterior pair round, in XI, broadly connected with each other. Each about 2 mm in diameter. 2 sperm ducts of each side meeting in middle part of XII. 2 pairs of vestigial ovoid bodies present on posterior faces of septa 12/13/14. Seminal vesicles large, fully occupying segments XI and XII, each irregularly indented or grooved on its anterodorsal side, nodular or granular, rarely smooth on surface, with a small dorsal lobe, round or flat, distinctly constricted off from the main lobe, sometimes easily distinguished from the latter by its grayish color. Prostate glands moderate in size, usually directed a little backward in XVII–XXI, occasionally in XVI–XX or XIX, kidney-shaped, divided into transverse lobules which are again subdivided or sometimes finely granulated.

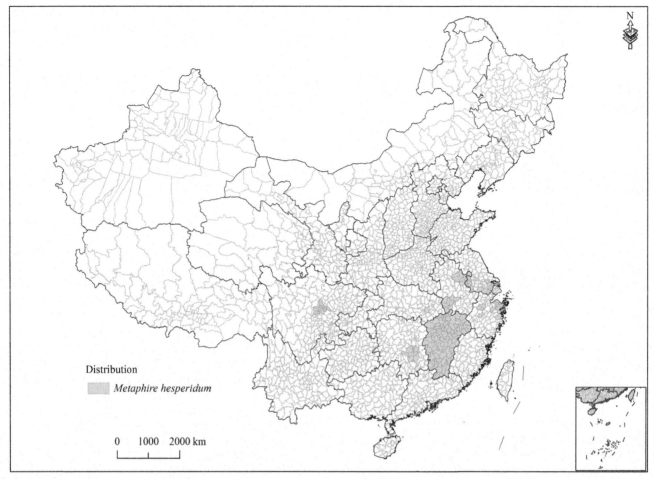

FIG. 295 Distribution in China of *Metaphire hesperidum*.

Prostatic duct long, whitish and glittering on surface, looped at both ends, ental loop deep, distal end spirally coiled, its middle part straight or sometimes slightly S-curved with a comparatively large caliber. Spermathecae 2 pairs, in VII and IX, diverticulum of the latter in front of the rudimentary septum 8/9; ampulla heart-shaped or oval in outline, about 3 mm long, yellowish or greenish pale, smooth, usually with an apical knob, with a very short duct, about 3 mm long, sharply marked off from the ampulla. Diverticulum tube-like, slender and often coiled in a zigzag fashion or in a round mass, as long as, or longer than, the main part if it is fully extended, about 3/4 or 2/3 of the whole tube serving as seminal chamber which is sometimes wholly distended, or distended merely at its middle portion or ental end; diverticular duct longer than the main duct, its connection with the latter very close to body wall.

Color: Preserved specimens grayish brown or faint brown on dorsum, dark azure gray at postclitellar region, grayish at posterior dorsum; pale or grayish on ventrum. Clitellum reddish brown or orange-brown sharply distinguished from the rest of body.

Deposition of types: Previously deposited in the Museum of the Biological Laboratory of the Science Society of China, Nanjing. The types were destroyed during the war in 1937.

Etymology: The specific name probably indicates the habitat of its type locality.

Distribution in China: Jiangsu (Nanjing, Suzhou, Yixing), Zhejiang (Lanxi, Tonglu, Fenshui, Xindeng, Ningbo, Zhoushan, Chuanshan, Fenghua, Linhai, Shengxian), Anhui (Anqing, Chuzhou), Jiangxi, Hong Kong, Hubei (Huangzhou, Wuchang), Hunan (Hengyang), Sichuan (Chengdu).

Remarks: Beddard in his original description of this species found an "extremely small terminal sac" but in his revision he put this form under the category "no bulbus at end of spermiducal gland-duct". Apparently his so-called terminal sac is not a true sac. While in Chen's specimens from Szechuan a cushion-like whitish connective tissue is constantly present around the terminal end of the prostatic duct, it is questionable whether it is what Beddard once found.

301. *Metaphire leonoris* (Chen, 1946)

Pheretima leonoris Chen, 1946. *J. West China Border Res. Soc. (B).* 16:114–116, 140, 154.

Metaphire leonoris Sims & Easton, 1972. *Biol. J. Linn. Soc.* 4:238.

Metaphire leonoris Xu & Xiao, 2011. *Terrestrial Earthworms of China.* 239–240.

External Characters

Length 100–170 mm, width 5–7 mm, segment number 78–134. Prostomium 1/3 epilobous, tongue open behind, its width as length of II. First dorsal pore 12/13. Clitellum XIV–XVI, annular, setae absent, its length equal to 8 segments posterior to male pore region. Setae number 28–32 (III), 40–46 (VI), 41–52 (IX), 54–70 (XXV); 18–23 (VII), 20–23 (VIII), 20–25 (IX) between spermathecal pores, 8–16 between male pores, $aa = 1.2$–$2.0\ ab$, $zz = 1.2$–$1.5\ zy$. Male pores in deep pouches, lateroventral in position, paired in XVIII, about 1/3 circumference apart ventrally, each pouch rather deep, its median border wrinkled and usually with few papillae exposed, 2–3 in front of, 1–2 behind setal line; a pointed papilla situated laterally on setal line, which bears the male pore. No papillae found outside of each pouch. Spermathecal pores 2 pairs in 7/8/9, about 1/3 to 1/2 circumference apart ventrally, each in an ellipsoid glandular depression, which is usually sunk between the segments, genital papillae lacking, sometimes present on VII, VIII (rarely on VI and IX) ventral but close to each pore region. Female pore single, medioventral in XIV.

Internal Characters

Septa 8/9/10 absent, 6/7/8, 10/11/12 very thick and well musculated, 12/13 as thick as 5/6. Nephridial tufts found anteriorly on septa 5/6/7. Micronephridial. Gizzard barrel-shaped, in IX and X, without a crop. Intestine swelling from XV. Intestinal caeca simple, in XXVII, extending anteriorly to XXII, smooth on both edges. Lateral hearts in IX asymmetrical, in X large. Testis sacs communicated. Seminal vesicles small, in XI and XII (exceedingly large, extending between IX and XIV, visible externally through the nonpigmented body wall in the juvenile specimens), no dorsal lobes. Supernumerary pair on 12/13 not found. Prostate gland always developed, in XVI–XXI or XXII, finely lobate, its duct S-curved, middle part straight and large (ca. 1 mm thick) becoming slender close to body wall. Accessory glands in large masses, sessile. Spermathecae 2 pairs, in VIII and IX, ampulla either heart-shaped or sac-like, about 3.5 mm long (largest about 5 mm long), duct shorter. Diverticulum shorter than main pouch, its ental 2/3 twisted in a zigzag manner, with a round terminal knob. Accessory glands present close to ampullar duct, as white lobules, sessile; similar glands on ventral side if the papillae are present.

Color: Preserved specimens grayish purple on dorsum, chocolate on anterior dorsum, pale ventrally, darker on clitellum. In juvenile specimens pale and lucid both dorsally and ventrally; seminal vesicles visible as whitish ring.

Deposition of types: Previously deposited in the Institute of Zoology, Academic Sinica, Chongqing. The types were destroyed during the war in 1945–1949.

Etymology: The name of the species refers to the type locality.

Type locality: Chongqing (Mt. Jinfo).

Distribution in China: Chongqing (Mt. Jinfo).

Remarks: This species is widely distributed on the top of Mount Jinfo. It is found either in the clay or cultivated soil. The present species is very close to *Metaphire paeta* (Gates, 1935a), which was known only in Songpan, but differs from it chiefly in the intestinal caeca. In the former, they are definitely simple, ventral edge either smooth or with weak wrinkles. In the latter, they are decidedly compound in view of Gates' original description "The intestinal caeca are compound, with 8–11 anteriorly directed, secondary caeca, the dorsal most the longest. The dorsal most caecum may have several tertiary caeca

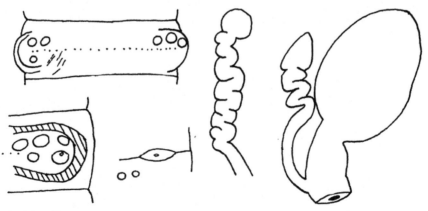

FIG. 296 *Metaphire leonoris* (after Chen, 1946).

on its ventral margin, these usually ventrally directed [. . .]. It is therefore a distinct species."

302. *Metaphire longipenis* (Chen, 1946)

Pheretima longipenis Chen, 1946. *J. West China Border Res. Soc. (B).* 16:91–93, 138, 144.
Metaphire longipenis Sims & Easton, 1972. *Biol. J. Linn. Soc.* 4:238.
Metaphire longipenis Xu & Xiao, 2011. *Terrestrial Earthworms of China.* 240.

External Characters

Length 45–60 mm, width 3.0–3.2 mm, segment number 78–81. Segments relatively long, longest preclitellar about 1.5 mm. Prostomium 1/3 epilobous, its width equal to length of II. First dorsal pore 10/11. Clitellum XIV–XVI, annular, smooth, not well marked off from neighboring segments. Setae all alike, slightly close ventrally, setal breaks wider on dorsal side. Setae number 24–27 (III), 28–34 (IX), 26–32 (XIX), 30–32 (XXV); 8–10 between male pores; $aa=2$ ab, $zz=3$–$4zy$. Male pores represented by 2 deep pouches, pouch orifices about 1/3 circumference apart ventrally. Penis slender and long, pointed at its free end, sometimes eversible. Inner surface circularly wrinkled. Spermathecal pores 2 pairs in 7/8/9, close to mediodorsal line. Posterior pair closer between setae a and b; anterior pair more lateral, in line with c and d. Each as elliptical slit, quite wide. No genital markings. Female pore single, medioventral in XIV.

Internal Characters

Septa 8/9/10 absent, anterior septa extremely thin, posterior septa behind Gizzard also thin, membranous. Pharyngeal glands thick, extending to VII. Gizzard in X, subspherical. Intestine swelling from XV. Intestinal caeca simple, extending anteriorly to XXIII. Vascular loops in IX feebly developed, those in X also slender. Testis sacs in X either widely placed as 2 wings or closely united as large transverse band, communicated; posterior pair usually adjoined and united. Seminal vesicles well developed in all cases observed, extending anteriorly to IX and posteriorly to XIV, dorsal lobe at anterodorsal side of each vesicle. Prostate glands well developed, in XVII–XXI, its duct about 6 mm long, twisted, emerging from the top of an everted bell-like bulb enclosing a copulatory chamber. Coelomic surface of the chamber smooth and slightly shining. Accessory glands absent. Spermathecae very large, posterior pair larger still, 2 pairs. Ampulla usually kidney-shaped, 2.5 mm wide, with very short duct. Diverticulum shorter than main pouch, with a globular or date-shaped seminal chamber which may be relatively long but never twisted in a zigzag manner in the cases observed. Both ducts of ampulla and diverticulum meeting close to spermathecal chamber. Spermathecal chamber about 1.8 mm in height, smaller in tube type, surface shining. Its internal lumen not well defined, filled with trabecular tissues. The common duct partly inserted in the chamber but attached to the parietes of the chamber. Accessory glands absent.

Color: Dark purplish chestnut dorsally, setal lines whitish, pale ventrally. Clitellum dark chocolate.
Deposition of types: Previously deposited in the Institute of Zoology, Academic Sinica, Chongqing. The types were destroyed during the war in 1945–1949.
Etymology: The specific name indicates the long penis of this species.
Type locality: Sichuan (Mt. Emei).
Distribution in China: Sichuan (Mt. Emei).
Remarks: This species is characterized by the presence of spermathecal and copulatory chamber, and a long penis.

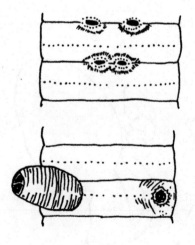

FIG. 297   *Metaphire longipenis* (after Chen, 1946).

303. *Metaphire magna* (Chen, 1938)

Pheretima (Pheretima) magna Chen, 1938. *Contr. Biol. Lab. Sci. Soc. China (Zool).* 12(10):416–419.
Metaphire magna Sims & Easton, 1972. *Biol. J. Linn. Soc.* 4:238
Metaphire magna Xu & Xiao, 2011. *Terrestrial Earthworms of China.* 241.

External Characters

Length 680 mm, width 22 mm, segment number 178. Prostomium proepilobous. First dorsal pore 12/13. Clitellum well marked, XIV–XVI, annular, setae and intersegmental furrows absent. Setae similarly built on dorsal and ventral sides, except preclitellar ones which are slightly large, ventral ones of anterior segments similarly spaced. Both dorsal and ventral breaks indistinct; $aa = 1.1–1.2\ ab$, $zz = 1.2\ yz$ before and $=2.5\ yz$ behind clitellum. Setae number 76 (III), 106 (VII), 110 (IX), 132 (IX), 140 (XXV); 30 (VII), 29–31 (VIII), 32 (IX) between spermathecal pores, 19–20 between male pores. Male pores 1 pair in XVIII, about 1/3 circumference apart ventrally, each in a small and shallow slit, surrounded by several concentric ridges which extend to small parts of neighboring segments. A small round papilla usually present medially among these ridges, on setal line. Spermathecal pores 2 pairs in 7/8/9, about 1/3 circumference apart ventrally. Each intersegmental, appearing as minute tubercle at posterior edge of segments VII and VIII, encircled anteriorly and laterally by 3–6 minute ampulla-like papillae, a glandular depression usually posterior to each pore on VIII and IX. Skin weakly glandular around this area. No other genital papillae. Female pore single, medioventral in XIV.

Internal Characters

Septa 9/10 absent, 8/9 membranous, wrapping over gizzard, 4/5 membranous, 5/6/7/8 thickened, each with many strong ligaments attached posteriorly to body wall, 10/11/12 about 6 times thicker than, 12/13 about 3 times, 13/14 as thick as septum 6/7, remaining septa membranous. Gizzard large, in IX–X. Intestine swelling from 1/2 XV. Intestinal caeca in XXVII, extending anteriorly to XXIII, elongate and pointed, with distinct ventral teeth-shaped diverticula (12 or more, about 1–2 mm deep). Lateral hearts in X–XIII, thick-walled and muscular, first pair as stout as following ones. Testis sacs in X broadly connected ventromedially with connective tissue, not communicated; those in XI enclosed in a common membranous sac with seminal vesicles and hearts of that segment. Seminal vesicles in XI less than half the size of those in XII; the latter about 4 mm wide, 10 mm high including dorsal lobe which is about half the size of the vesicle. Ovoid bodies found on posterior side of septa 12/13/14, posterior pair many times larger. Prostate gland portion large, granular in

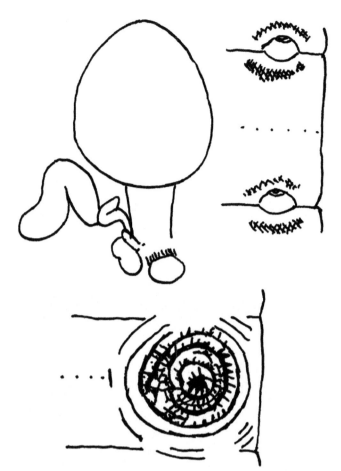

FIG. 298 *Metaphire magna* (after Chen, 1938).

appearance, in XVII–XIX, its duct about 10 mm long, uniform in caliber, slender only at ectal end. 3 distinct accessory glands, each roundish and enclosed in peritoneum, close to base of prostatic duct. Median gland considerably large in proportion to its small-sized papilla as referred to above. Duct cord-like, fairly thick. Spermathecae 2 pairs, in VIII and IX, relatively large. Ampulla large sac-like, thin-walled, about 5 mm in diameter, its duct about 1.5 mm wide, less than half the length of ampulla. Diverticulum if extended longer than main pouch. Seminal chamber thumb-shaped, weakly U-curved, about 5 mm long, 1 mm wide; its duct slender, generally coiled, as long as chamber. Accessory glands 1 batch associated with each small papilla externally, each consisting of a gland mass and a cord-like duct leading to each papilla.

Color: Preserved specimens dark slaty on dorsal and lateral sides, grayish on ventral side.
Deposition of types: Previously deposited in the Fan Memorial Institute of Biology, Beijing. The types were destroyed by the war in 1941–45.

Etymology: The specific name indicates the large-sized body.
Type locality: Hainan (Wanning, Sha-mom-chiu).
Distribution in China: Hainan (Wanning, Sha-mom-chiu).
Remarks: This is the largest species ever recorded from the region. *Metaphire musica* (Horst, 1883) distributed in Java is about 15 mm wide, up to 570 mm long. (*Amynthas jampeanus?*) *Pheretima jampeja* is measured at about 15–20 mm in thickness. *Archipheretima ophiodes* (Michaelsen, 1930) found in the Philippines reaches 20 mm in diameter of the body. This species appears to be the first record of the largest size in *Metaphire*.

304. *Metaphire magnaminuscula* Qiu & Zhao, 2013

Metaphire magnaminuscula Zhao, Jiang, Sun & Qiu, 2013. *Zootaxa*. 3619(3):388–390.

External Characters

Length ?–297 mm, width 10–12 mm, segment number ?–137; body cylindrical in cross section, gradually tapered towards head and tail. Prostomium epilobous. First dorsal pore 11/12. Clitellum XIV–XVI, annular, markedly glandular, setae cannot be seen externally. Setae number 36–46 (III), 44–46 (V), 52–56 (VIII), 40–60 (XX), 52–78 (XXV); 12–20 between spermathecal pores, 0 between male pores, $aa = 1.2–1.5$ ab, $zz = 1–2$ zy. Male pores paired in XVIII surrounded by 3–4 circular folds, a little longer than 0.3 circumference apart ventrally, each in a copulatory pouch. Genital papilla absent. Spermathecal pores 2 pairs in 7/8/9, ventral, eye-like and obvious, about 0.4 circumference apart ventrally. Genial marking absent. Female pore single, medioventral in XIV.

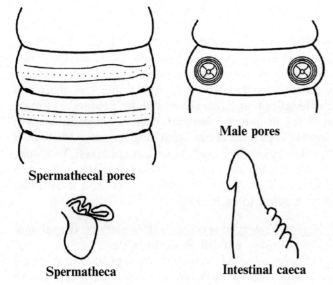

Spermathecal pores

Male pores

Spermatheca

Intestinal caeca

FIG. 299 *Metaphire magnaminuscula* (after Qiu & Zhao, 2013).

Internal Characters

Septa 9/10 absent, 5/6/7/8 comparatively thick and muscular, 10/11/12/13 thin, 8/9 thin, membranous. Gizzard in XI–X, ball-shaped. Intestine swelling from XVI. Intestinal caeca simple, originating in XXVII and extending anteriorly to XXIV, with distinct teeth-shaped incision on dorsum, smooth ventrally. Dorsal blood vessel single, continuous onto pharynx; Lateral hearts in X–XIII, the first pair is small. Holandric, Testis sacs 2 pairs in X–XI. Seminal vesicles paired in XI–XII, developed, ventrally separated. Prostate glands in 2/3 XVII–2/3 XIX, small but stout, closely adhering to body wall; prostatic duct U-shaped. Accessory glands absent. Spermathecae 2 pairs in VIII–IX, 3 mm long; ampulla heart-shaped, about 2.5 mm long, spermathecal duct as long as 0.2 main pouch; diverticulum a little longer than main pouch by 0.2 mm, weakly bent in a zigzag fashion or U-curved, terminal 0.67 enlarged as a clavate seminal chamber; no nephridia on spermathecal ducts.

Color: Preserved specimen dark brown on dorsum, light brown on ventrum. Clitellum dark brown.
Deposition of types: Station Biologiqué, Université de Rennes 1, France. (Holotype). Shanghai natural History Museum, Shanghai, China (Paratype).
Etymology: The specific name indicates that it is smaller than *Metaphire magnamagna* (Chen, 1938).
Type locality: A rubber plantation in Longmen Town, Tunchang City, Hainan province.
Distribution in China: Hainan (Tunchang).
Remarks: This specimen is similar to *Metaphire magna* (Chen, 1938) in its body color dark, setal formula, 2 pairs of spermathecal pores in 7/8/9, septa in 8/9 membranous, distinct teeth-shaped intestinal caeca, and type of spermathecae. However, this specimen is much smaller than described for *M. Magna*, setae are fewer, and there is no accessory gland. In view of the slight but nonetheless clear cut differences, Qiu and Zhao gave subspecies status to this specimen.

305. *Metaphire myriosetosa* (Chen & Hsü, 1977)

Pheretima myriosetosa Chen & Hsü, 1977. *Acta Zool. Sinica*, 23(2):175–176.
Metaphire myriosetosa Xu & Xiao, 2011. *Terrestrial Earthworms of China*. 242–243.

External Characters

Length 111–167 mm, width 7–8 mm, segment number 63–145. Prostomium 1/2 epilobous. First dorsal pore 12/13. Clitellum XIV–XVI, annular, setae absent. Setae fine; setal number 80–95 (III), 100–110 (VI), 98–114 (VIII), 88–100 (XVIII), 92–114 (XXV); 45–56 (VIII) between spermathecal pores, 27–30 (XVIII) between male pores. Male pores each situated in shallow copulatory pouches,

FIG. 300 *Amynthas myriosetosa* (after Chen & Hsü, 1977).

large dorsal lobe, triangular, almost occupying 2/3–1/2 of whole vesicles, the left and the right lobes meeting and coiled on dorsal vessels, but not communicated. Prostate glands developed, long finger-like lobes, its duct about 4 mm long, middle stout, and both ends thin. Spermathecae 2 pairs, well developed, ampulla ball-shaped or ovoid-like, 2.5–3 mm in diameter, faint transverse stripes visible on surface; its duct stout, demarcation indistinct, about 1 mm long, ectal end thin, connected with the swollen portion (the shallow spermathecal pouch) which is equal length with the duct. Diverticulum longer than main pouch, slender or loosely twisted, its seminal chamber enlarged, joining the main spermathecal strium.

Color: Preserved specimens dark brown on middorsal line, anterior ventrum brown, the rest taupe. Clitellum red-brown.

Etymology: The specific name indicates the numerous setae of this species.

Type locality: Guangdong (Fengkai).

Distribution in China: Guangdong (Fengkai).

Remarks: This species is similar to *Metaphire californica* (Kinberg, 1867) in general morphology, but it can easily be distinguished from the latter by the shallow spermathecal pouch, male pores situated at the bottom of copulatory pouches, the well-developed seminal vesicles, ectal end of prostatic dust without fiber heap, spermathecal ampulla not heart-shaped, and diverticulum longer than main pouch.

306. *Metaphire paeta* (Gates, 1935)

Pheretima paeta Gates, 1935a. *Smithsonian Mus. Coll.* 93(3):13.

Pheretima paeta Chen, 1936. *Contr. Biol. Lab. Sci. Soc. Chin.(Zool).* 11:300.

Pheretima paeta Gates, 1939. *Proc. U.S. Natn. Mus.* 85:456–459.

Metaphire paeta Sims & Easton, 1972. *Biol. J. Linn. Soc.* 4:239.

Metaphire paeta Xu & Xiao, 2011. *Terrestrial earthworms of China.* 244.

External Characters

Length 75–136 mm, width 5–7 mm, segment number 71–86. Prostomium 2/3 epilobous. First dorsal pore 11/12 or 12/13. Clitellum XIV–XVI, annular. Ventral setae on III–IX longer and more widely spaced (especially on IV–VII), black tips like in *Amynthas pingi* (Stephenson, 1925), those on dorsal side also marked; ventral break not distinct, dorsal break $zz = 1.5$–3.0 yz behind clitellum; setae number 32–33 (III), 39–43 (VI), 42–48 (VIII), 50–57 (XII), 64 (XXV); 21–23 (VIII) between spermathecal pores, 14–16 between male pores. Male pores 1 pair, in XVIII, on the dorsal wall of large copulatory chambers

as in *Metaphire californica* (Kinberg, 1867), the openings transverse slit-like when invaginated, with annular ridges and longitudinal slits around them, about 2/5 circumference apart ventrally. Spermathecal pores 2 pairs, in 7/8/9, intersegmental each bordered anteriorly and posteriorly by glandular lips, nearly 1/2 circumference apart. Female pore single, medioventral in XIV.

Internal Characters

Septa 8/9 traceable ventrally, 9/10 absent, 5/6/7/8 thick, 10/11/12/13 slightly thick, 13/14 and succeeding septa thin and membranous. Gizzard round barrel-shaped. Intestine swelling from XVI. Intestinal caeca simple, in XXVII, extending anteriorly to XXIV. Last pair of hearts in XIII. Anterior pair testis sacs well developed. First pair seminal vesicles not found in second pair of testis sacs. Both pairs of seminal vesicles all well developed,

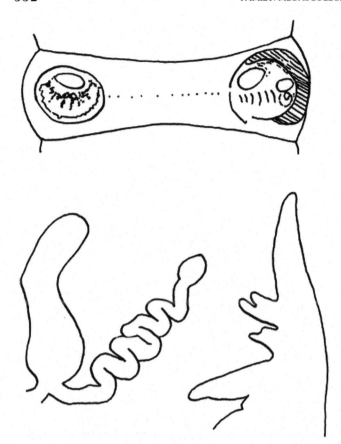

FIG. 301   *Metaphire paeta* (after Chen, 1936).

conspicuously protuberant into the coelom. On the median wall of the copulatory chamber a presetal, transversely oval genital marking, occasionally also a postsetal marking; on the top of the chamber 1–2 further markings of variable shape and size. Spermathecal pores on circular to oval areas within deep invaginations with transversely slit-like apertures, 2 pairs, in 7/8/9, about 1/2 circumference apart. A large papilla close to anterior edge of VIII and IX. External genital markings paired, on the posteriormost margins of VII and VIII, each marking 1–3 intersetal intervals median to the aperture of the spermathecal invagination. Female pore single, medioventral in XIV.

Internal Characters

Septa 8/9/10 absent, 5/6/7/8 slight thickened, 10/11–12/13 muscular. Intestine swelling from XV. Intestinal caeca manicate, in XXVII, with 3–4 rather long diverticula, dorsalmost secondary caecum are the longest, extending anteriorly to XXIII. First pair of hearts present. Testis sacs of X and XI unpaired and ventral, ventrally connected. Seminal vesicles very large, each with a dorsal lobe on anterodorsal side. Prostate glands extend through XVII–XIX or XX; the prostatic duct 3–6 mm long, bent into a U shape or C shape, the ectal

portion thickened. Accessory glands sessile. Spermathecae 2 pairs, in VII and IX, they are obviously juvenile even in the mature specimens, ampulla elongate, 3 mm long, with a short duct; diverticulum slightly longer, with a muscular stalk and an elongate tubular seminal chamber, the latter looped back and forth in a regular zigzag manner.

Deposition of types: Smithsonian Institution.
Type locality: Sichuan (Songpan).
Distribution in China: Hubei (Lichuan), Sichuan (Songpan).
Remarks: *Metaphire paeta* distinguished from *Amynthas omeimontis* (Chen, 1931) by the copulatory chambers and even in young aclitellate specimens by the parietal male pore invaginations. From *M. schmardae* (Horst, 1883), *M. paeta* is distinguished by the invaginated spermathecal pore, the median location of the copulatory chamber glands, and the presence of preclitellar genital markings. A satisfactory diagnosis of *M. paeta* cannot, of course, be given until fully clitellate specimens have been made available for study.

307. Metaphire pedunclata (Chen & Hsü, 1977)

Pheretima pedunclata Chen & Hsü, 1977. *Acta Zool. Sinica*, 23(2):176–177.
Metaphire pedunclata Xu & Xiao, 2011. *Terrestrial Earthworms of China*. 246.

External Characters

Length 94–139 mm, width 4–5 mm, segment number 103–127. Prostomium 1/3 proepilobous. First dorsal pore 12/13. Clitellum XIV–XVI, annular, setae absent. Setae on II–IX ventrally stout, widely spaced; postclitellar setae fine and closely spaced, setae on dorsum fine and indistinct;setal number 17–23 (III), 31–41 (VIII), 25–38 (XVIII), 48–63 (XXV); 13–18 (VIII) between spermathecal pores, 0–7 (XVIII) between male pores. Male pores about 1/3 circumference apart ventrally, each situated in a shallow copulatory chamber, with 1–4 small papillae, a larger papilla was easily mistaken for a porophore;3–4 annular grooves outside of the chamber, extending anteriorly to 1/3 XVII and posteriorly to 1/3 XIX. Spermathecal pores 2 pairs, in intersegmental furrows 7/8/9, in eye-like nest, shaped like the eye of a needle, with 1–4 genital papillae around, irregularly arranged, usually 2 inside and 1 outside of the postintersegmental pore, 1 inside the preintersegmental pore; a little more than 1/2 circumference apart. Female pore single, medioventral in XIV.

Internal Characters

Septa 8/9 traceable ventrally, 9/10 absent, 5/6/7 and 10/11/12 thick, 12/13/14 slightly thick, 14/15 and

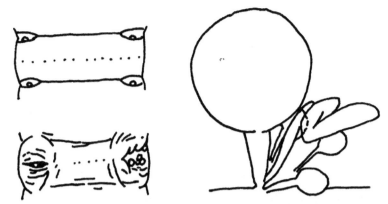

FIG. 302 *Metaphire pedunclata* (after Chen & Hsü, 1977).

succeeding septa thin and membranous. Gizzard ball-like. Intestine swelling from XV. Intestinal caeca simple, in XXVII, extending anteriorly to XXIII, 2/3 of base swelling, teeth-shaped sacs on ventrolateral side, yellow sand inside. Last pair of hearts stout. Testis sacs in X, V-shaped, narrowly connected on ventral side and communicated, posterior pair slightly smaller. Seminal vesicles well developed, the left and the right vesicles meeting on dorsal side, with large dorsal lobes. Prostate glands in XVII–XXI or XXII, lobules after lobes, its duct 3 mm long, S-shaped, ental end slight thick. Accessory glands each with a long stalk. Spermathecae 2 pairs, in VIII and IX; ampulla irregular sac-like, round-shaped, heart-shaped or transverse oval-shaped, demarcation with ducts obvious, main part about 3 mm long, ducts occupying 1/3. Diverticulum longer than main part, with 3–4 curves, or twisted, distal end finger-like, expanding and serving as seminal receptacle, passing inside by the parietes, accessory glands each with a long stalk.

> Color: Preserved specimens red-brown on dorsum, gray-white to slightly brownish on ventrum. Clitellum brown.
> Etymology: The specific name indicates the long-stalked accessory glands of this species.
> Type locality: Yunnan (Luosuo River, Menglun).
> Distribution in China: Yunnan (Xishuangbanna).
> Remarks: This species is similar to *Metaphire longipenis* (Chen, 1946), but differs from the latter by the habits, male pores at the bottom of copulatory pouch, anterior and posterior of spermathecal pores without glandular lips, septum 8/9 traceable ventrally, and the diverticulum with fewer curves and longer than the main part. It is also similar to *Amynthas robustus* (Perrier, 1872) in morphology of spermathecae, accessory glands with long stalk, the position and the number of spermathecal pores, and the distribution of sexual characteristics, but it can easily be distinguished as in *A. robustus* the copulatory pouch is absent.

308. *Metaphire prava* (Chen, 1946)

Pheretima prava Chen, 1946. *J. West China. Border Res. Soc. (B)*. 16:90–91, 138, 144.
Metaphire prava Sims & Easton, 1972. *Biol. J. Linn. Soc.* 4:238.
Metaphire prava Xu & Xiao, 2011. *Terrestrial Earthworms of China.* 249–250.

External Characters

Length 45–50 mm, width 2.5 mm, segment number 80–119. Prostomium 1/2 epilobous, its tongue as wide as length of II. First dorsal pore 11/12. Clitellum in 3 full segments, XIV–XVI, annular, smooth all over. Setae all very short, barely visible on dorsal side; none of them enlarged and modified. Setae number 20–25 (III), 35–37 (IX), 30–31 (XXV); 8–10 between male pores; $aa = 1.5\ ab$, $zz = 2\ zy$. Male pores about or slightly over 1/3 circumference apart ventrally; each in small pouch of parietal invagination and on lateral side of a large papilla placed deep in the pouch. No penis-like structure or other genital marking. Each pouch orifice surrounded by a few circular ridges. Spermathecal pores 2 pairs, intersegmental,

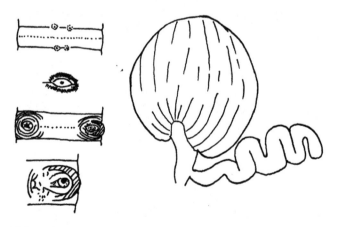

FIG. 303 *Metaphire prava* (after Chen, 1946).

dorsally on 7/8/9, close to dorsomedian line. Anterior pair more laterally situated, in line to 4th or 5th seta, posterior pair to 2nd seta. Each represented by an eye-like swelling bearing a central pore. No other genital marking. Female pore single, medioventral in XIV.

Internal Characters

Septa very thin, scarcely musculated, 8/9/10 absent, 10/11 and others equally membranous. Nephridial tufts on 5/6/7 rather thick. Pharyngeal thick, extending to V. Gizzard in IX, elongate, round. Intestine swelling from XV. Intestinal caeca simple, elongate, extending anteriorly to XXI. Lateral hearts in IX asymmetrically developed, in X moderate in size. Testis sacs in X very large, extending to lateral sides of Gizzard, narrowly connected; those in XI smaller, broadly united, constricted at middle. Seminal vesicles large, filling more than 2 segments in X–XIII, dorsal lobe on anterodorsal side of each vesicle, spotted in appearance. Prostate glands with coarse lobules, in XVII–XXII, no copulatory pouch visible in the body cavity. Prostatic duct slender in its ectal and ental portions, about 0.8 mm wide, middle portion straight, very stout, about 2.2 mm wide. Accessory glands diffuse, sessile, surrounding ectal end of prostatic duct. Spermathecae very large, 2 pairs. Ampulla rounded, with a comparatively short duct; diverticulum twisted in a zig-zag manner on the same plane, its ental 4/5 serving as a seminal chamber; diverticular duct short and not well marked. No trace of spermathecal chamber. Accessory glands absent.

Color: Chestnut brown on dorsum, setal circles paler, pale on ventrum. Clitellum chocolate.
Deposition of types: Previously deposited in the Institute of Zoology, Academic Sinica, Chongqing. The types were destroyed during the war in 1945–1949.
Etymology: The specific name "prava" means crooked or deformed in Latin and probably indicates the shape of spermathecal diverticulum.
Type locality: Sichuan (Mt. Emei).
Distribution in China: Sichuan (Mt. Emei).
Remarks: This species is similar to M. longipenis (Chen, 1946) in respect of (1) the coloration, (2) the dorsal position of the spermathecal pores, (3) the presence of a shallow male pouch, and (4) the characters of the septa, caeca etc. However, on account of lacking a large and deeply invaginated copulatory pouch, a long penis, a conspicuous spermathecal chamber, and, most important of all, possessing a twisted diverticulum, it clearly distinguishes itself as an independent species. The male apparatus is fundamentally different

between these 2 species. The spermathecae pore in this species is a minute orifice, while in M. longipenis it is a wide slit probably for receiving the peculiar penis during coitus.

309. *Metaphire schmardae* (Horst, 1883)

Pheretima (Pheretima) Schmardae Chen, 1931. *Chin. Zool.* 7(3):125–131.
Pheretima schmardae Gates, 1939. *Proc. U.S. Natn. Mus.* 85:482–485.
Pheretima schmardae Chen, 1959. *Fauna Atlas of China: Annelida.* 12.
Metaphire schmardae Chang, Shen & Chen, 2009. *Earthworm Fauna of Taiwan.* 128–129.
Metaphire schmardae Xu & Xiao, 2011. *Terrestrial Earthworms of China.* 251–252.

External Characters

Length 68–125 mm, clitellum width 2–5 mm, segment number 76–96. Prostomium 2/3–3/4 epilobous. First dorsal pore 12/13, rarely 11/12. Clitellum XIV–XVI, annular, setae and dorsal pore absent; about 3 mm long as well as wide. Setae small and delicate, setae number 24–30 (III), 22–28 (VI), 49–53 (VIII), 44–57 (XII), 50–54 (XXV); 30–34 between spermathecal pores, 14–20 between male pores. Male pores paired in XVIII, each pore situated in a deep copulatory pouch, openings of the latter as 2 transverse slits, often widely opened and each wrinkled around its margin, about 2/5 circumference apart ventrally, never everted in cases observed. Cuticle extending to inner side of wrinkled margin. A number of irregular skin folds or fimbriae enclosed by the margin in the chamber and lined with a highly glandular epithelium which is different from that of the outer body wall bearing a few papillae connected to lateral glands inside. Prostatic duct entering upon the top of the chamber through a roundish papilla. 2 shallow pits on anterior and posterior sides of the chamber in which the lateral glands open. Spermathecal pores 2 pairs in 7/8/9, ventrolateral, about 3/5 or 5/9 circumference apart ventrally, completely visible from dorsal aspect, eye-like, about 0.5 mm wide, each marked by a whitish spot which is the terminal end of the spermathecal duct, often with a hair-like process (the cuticular border of the duct) brownish in color, its exposed part about 0.3 mm long; the elliptical or eye-like depression seemingly free from the body wall, so the ampullar duct very easily broken off from the body wall. Genital papillae absent. Female pore single, medioventral in XIV.

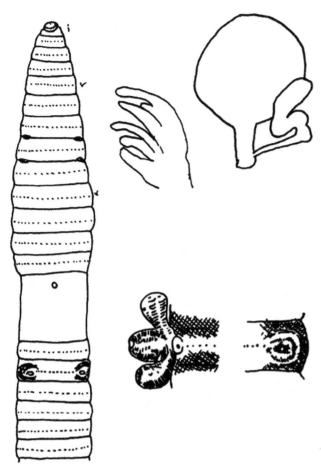

FIG. 304   *Metaphire schmardae* (after Chen, 1959).

Internal Characters

Septa 8/9/10 absent, but 8/9 ventrally present, 5/6/7/8 and 10/11–13/14 thickened, but not very musculated, 7/8 and 10/11 nearly as thick as 14/15. Gizzard large in IX, X and XI; goblet-shaped, anterior end narrower. Intestine swelling in XV. Intestinal caeca in XXVII, manicate or complex, extending anteriorly to XXIV, each with 4–7 diverticula or lobules. Last pair of hearts in XIII. Testis sacs moderate in size, 2 pairs in X and XI, anterior pair smaller than posterior pair. Seminal vesicles paired fully occupying XI and XII, very large. Prostate glands paired in XVIII, racemose, extending anteriorly to XVII or XVI, and posteriorly to XX or XXI, divided into 15–20 lobes; prostatic duct stout at middle, looped at both ends, smooth and glittering on surface, inserted on the coelomic surface of the top of the chamber and terminated with a small papilla. Accessory glands small, often present on the coelomic surface of the chamber. Spermathecae 2 pairs in VII or VIII and IX, ampulla large, round, stalk short and stout, diverticulum with 3 or more

bends at its ental two-thirds and this portion comparatively thick, as long as the main part, much longer if extended. Its ental end dilated into an elongate oval seminal chamber, often whitish in color.

Color: Preserved specimens chestnut brown on dorsal side anteriorly, light grayish brown posteriorly, the brown color extending downward nearly to ventrolateral side, dark brown streak along the entire dorsomedian line visible at the clitellum, somewhat fading before the first dorsal pore, whitish around setal zones, each as broad as, or broader than, the chestnut intersegmental zone, but interrupted at dorsomedian line. Clitellum chocolate brown on dorsal side and orange-brown ventrally, but the whitish zones still apparent, buff or light brown on preclitellar ventrum, grayish or pale on postclitellar ventrum.

Deposition of types: Leidenand Vienna Museums.

Type locality: Hawaii.

Distribution in China: Tianjin, Zhejiang (Fenghua, Lanxi, Fenshui, Linhai), Fujian (Quanzhou, Jinmen), Taiwan (Taibei), Jiangxi, Hubei (Lichuan, Huangzhou, Qianjiang), Hong Kong (Jiulong), Aomen, Chongqing (Beipei), Sichuan (Jiading, Leshan), Guizhou (Mt, Fanjing), Yunnan (Lancang).

Remarks: Michaelsen's *M. schmardae macrochaeta* from Kowloon, Guangzhou, is the only Chinese record of this species which agrees very well with Chen's specimens in the character of setae. Beddard's *M. trityphla* from Barbados Island is doubtlessly the same species, as Beddard himself and Michaelsen have already considered it to be synonymous with *M. schmardae*. The Japanese form, *Metaphire vesiculata*, as recorded by Goto & Hatai, is most probably the same species judging from its general structures with the exception of the position of spermathecae in 6/7/8 instead of 7/8/9, *Metaphire schmardae*, has a similar distribution as *M. hesperilum* or *A. hawayanus* between China and the West Indies.

Beddard's *sumatrana*, 1892, is in all probability, either *californica* or *shmardae*. The presence of large copulatory chambers rules out *californica*. Similarly, compound intestinal caeca would rule out *shmardae* were it not for the fact that all but the dorsalmost secondary caeca may be overlooked unless the gut (in a specimen dissected from the dorsal side as is usual) is rolled well over to one side or the other. The dorsalmost secondary caecum, failing this precaution, would then appear to be a simple caecum.

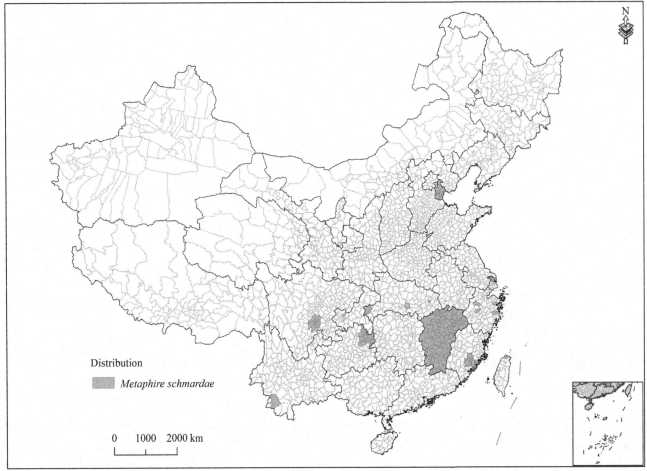

FIG. 305 Distribution in China of *Metaphire schmardae*.

310. *Metaphire wuzhimontis* Qiu & Sun, 2013

Metaphire wuzhimontis Zhao, Jiang, Sun & Qiu, 2013. *Zootaxa*. 3619(3):387–388.

External Characters

Length 163 mm, width 5.5 mm, segment number 94; body cylindrical in cross section, gradually tapered towards head and tail. Prostomium epilobous. First dorsal pore 9/10. Clitellum XIV–XVI, annular, markedly glandular, setae cannot be seen externally. Setae number 20 (III), 24 (V), 40 (VIII), 56 (XX), 40 (XXV); 13–16 between spermathecal pores, 13 between male pores, $aa = 1$–2 ab, $zz = 2$ zy. Male pores paired in XVIII, less than 0.5 body circumference apart ventrally, each in a copulatory pouch, surrounded by 5–6 circular folds. Genital papillae absent. Spermathecal pores 2 pairs in 7/8/9, ventral, eye-like and conspicuous, less than 0.5 body circumference apart. Each pore with a small round marking placed between the pore and the setae circle of VII and VIII, another marking of the same size placed ventromedially on VIII. Female pore single, medioventral in XIV.

Internal Characters

Septa 8/9/10 absent, 5/6/7/8 comparatively thick, 10/11/12 slightly thickened. Gizzard in IX–X, ball-shaped. Intestine swelling from XV. Intestinal caeca simple, originating in XXVI and extending anteriorly to XXIV, 3 distinct incisions on dorsal margins, smooth ventrally. Dorsal blood vessel single, continuous onto pharynx; hearts in X–XIII, the first pair is small. Holandric. Testis sacs 2 pairs in X–XI. Seminal vesicles paired in XI–XII. Prostate glands in XV–XIX, well developed, closely adhering to body wall; prostatic duct stout, U-shaped. Accessory glands absent. Spermathecae 2 pairs in VIII–IX, large; ampulla peach-shaped, spermathecal duct short, as long as 0.14 main pouch; diverticulum equal to main pouch, straight, terminal 0.5 enlarged, ectal 0.33 served as an elongated seminal chamber. Accessory glands present.

Color: Preserved specimens dark brown on dorsum, gradually lighter after clitellum, pale on ventrum. Clitellum dark brown.
Deposition of types: Station Biologiqué, Université de Rennes 1, France.

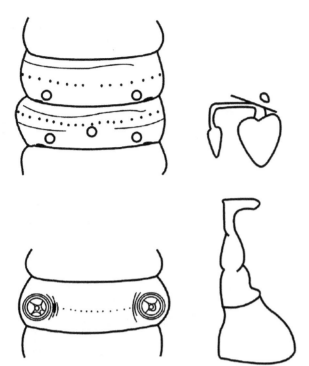

FIG. 306   *Metaphire wuzhimontis* (after Zhao et al., 2013).

Etymology: The name of the species refers to the type location.

Type locality: Hainan (Wuzhishan National Nature Reserve).

Distribution in China: Hainan (Wuzhishan National Nature Reserve).

Remarks: *M. wuzhimontis* (Qiu & Sun, 2013) keys to the *javanica* group according to Sims and Easton (1972). There are only 5 species in the *javanica* group in China and Southeast Asia. They are *M. californica* (Kinberg, 1867), *M. javanica* (Kinberg, 1867), *M. longipenis* (Chen, 1946), *M. magnamagna* (Chen, 1938) (including *M. magna minuscula* Qiu & Zhao, 2013), and *M. prava* (Chen, 1946). *M. wuzhimontis* is a moderately large species, obviously 2–4 times smaller than *M. magna magna*, but nearly 3 times larger than *M. longipenis* and *M. prava*. Its first dorsal pore is in 9/10, different from the others. Only *M. wuzhimontis* and *M. magna magna* have genital markings. However, 3–6 markings in *M. magnamagna* encircle the spermathecal pores, while in *M. wuzhimontis* 1 pair of markings is placed between the pore and the setae circle of VII and VIII, and another ventromedially in IX.

thecodorsata Group

Diagnosis: *Metaphire* with spermathecal pores 3 pairs at V–VII.

Global distribution: Oriental realm from Japan southwards through the Indo-Australasian archipelago to the rainforests of Australasia and eastwards through the Oceanian realm.

There are 3 species from China.

Remarks: Species of this genus are separable from those of *Amynthas* by the presence of copulatory pouches and differ from species of *Pheretima* by the absence of nephridia from the spermathecal ducts.

TABLE 34   Key to the *thecodorsata* Group From China

1.	Spermathecal pores 3 pairs in V, VI, VII ventrally *Metaphire bifoliolare*
	Spermathecal pores 3 pairs, in V, VI, VII dorsally 2
2.	Spermathecal pores in eye-like shallow nest, transverse slit-like, between *xy* or *yx* .... *Metaphire cruroides*
	Spermathecal pores on posterior edge as large transverse slits, with anterior and posterior lips, situated very near to middorsal line *Metaphire thecodorsata*

311. Metaphire bifoliolare Tan & Zhong, 1987

Metaphire bifoliolare Tan & Zhong, 1987. *Act Zootaxonomica Sinica.* 12(2):129–131.
Metaphire bifoliolare Xu & Xiao, 2011. *Terrestrial Earthworms of China.* 215–216.

External Characters

Length 89–103 mm, diameter 3.5–4.0 mm, segment number 94–101. Prostomium 1/3 epilobous. First dorsal pore 12/13. Clitellum XIV–XVI, annular, with setae on ventrum and with dorsal pores. Setae fine; setae number 46–52 (III), 48–52 (V), 51–60 (VII), 52–54 (IX), 48–56 (XIII),

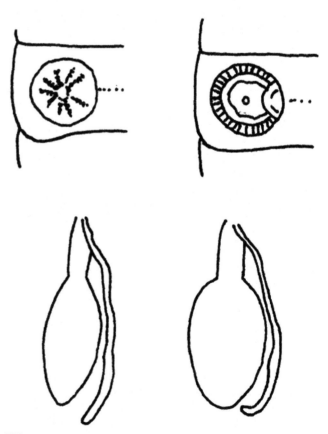

FIG. 307   *Metaphire bifoliolare* (after Tan & Zhong, 1987).

46–54 (XX); 20–28 (V), 21–31 (VI), 26–30 (VII) between spermathecal pores, 9–10 between male pores. Male pores 1 pair, ventrolateral XVIII, within copulatory chambers which are deep and formed by parietal invagination, the opening large, irregular, body wall round with radial grooves; inner surface with few longitudinal streaks from the opening to the bottom; the bottom flat and with a small porophore in the center. Anterior, posterior, and ectal parts of the bottom swollen and forming a semicircular ridge of differing width and a circular groove with the wall. Spermathecal pores 3 pairs in V, VI, VII ventrally, transverse slit-like, on the posterior edge, situated 1/3 between intersegmental furrow and setal circle, about 1/2 circumference apart. Female pore single, medioventral in XIV.

Internal Characters

Septa 5/6–10/11 thickened and muscular. 5/6/7 with well-developed nephridium, 11/12 slightly thickened but thinner than the above-mentioned, 12/13 and succeeding septa thin and membranous. Gizzard slightly long, round-shaped, in VIII–X. Intestine swelling from XV. Intestinal caeca paired in XXVII, extending anteriorly to XXIV or 1/2 XXIII, simple, long sac-like, free end wider than base. Lateral hearts in X–XIII, anterior 2 pairs more slender than posterior 2 pairs. Testis sacs 2 pairs in X and XI, first pair hemispheric, anterior margin swelling and posterior margin flat, narrowly connected; second pair covers first pair of seminal vesicles, broadly united and forming a transverse strip. The second pair of seminal vesicles well developed, lump-shaped, dorsal lobes conical. Prostate glands large, divided into 2 lobes, in XVI–XX or XVII–XX, the anterior and posterior lobes separated obviously in XVIII, each lobe divided into many lobules; prostatic duct divided into 2 ducts, the anterior shorter and the posterior longer, these 2 ducts then conjoined and forming the thick prostatic duct, it passing into porophore through the central portion of the copulatory chamber protrudes into coelomic. Spermathecae 3 pairs in V–VII, third pair is larger than first and second pairs; first pair ampulla long sac-like, third pair ampulla long oblate-shaped; its duct slightly stout. Diverticulum slightly longer than main pouch, 1/3 distal end slightly thick, serving as the unapparent seminal chamber; seminal chamber sausage-shaped, distal end straight or slightly bent.

Color: Preserved specimens black-brown on dorsum, and brown on ventrum. Clitellum slightly black-brown.
Deposition of types: Hunan Agricultural College.
Etymology: The specific name "biforatum" is a compound word with "bi" meaning twice or double in Latin and "foratum" meaning holes in Latin. It refers to the number of spermathecal pores and probably indicates the structure of the prostate glands.

Type locality: Hunan (Changde, Anxiang, Nanxian).
Distribution in China: Hunan (Changde, Anxiang, Nanxian).
Remarks: This species is similar to *Metaphire guillelmi* (Michaelsen, 1895), but differs from the latter by the spermathecal pores on the posterior edge of V, VI, VII, septa 8/9/10 present, and the sausage-shaped seminal chamber.

312. *Metaphire cruroides* (Chen & Hsu, 1975)

Pheretima cruroides Chen, Hsü, Yang & Fong, 1975. *Acta Zool. Sinica*. 21(1):93.
Metaphire cruroides Xu & Xiao, 2011. *Terrestrial Earthworms of China*. 224.

External Characters

Length 91 mm, width 4 mm, segment number 90. Prostomium prolobous. First dorsal pore 12/13. Clitellum XIV–XVI, annular, smooth, generally with a few setae on ventral region of XVI, about 4–5 postclitellar segments in length. Setae fine and close; setal number, 58 (III), 67 (VI), 64 (VIII), 39 (XVIII), 50 (XXV); 11 between male pores. Male pores in ventral region of XVIII, about 2/5 circumference apart ventrally, there are 2 foot-like penises, 4 mm length and 1.2 mm width on base, 14–15 circular grooves on the surface. Spermathecal pores 3 pairs, dorsally and posteriorly on V–VII, in eye-like shallow nest, transverse slit-like, between xy or yx. No other genital markings. Female pore single, medioventral in XIV.

Internal Characters

Septa 8/9/10 thick, 5/6/7/8 greatly thickened, and behind 13/14 membranous. Gizzard ball-like, in 1/2 VIII–IX. Intestine swelling from XV. Intestinal caeca in

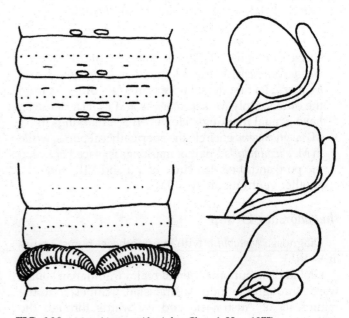

FIG. 308   *Metaphire cruroides* (after Chen & Hsu, 1975).

XXVII, extending anteriorly to XXV or XXVI. Last pair of hearts in XIII. Testis sacs in X V-shaped, meeting dorsally, first pair seminal vesicles contained in second pair of testis sacs. Prostate glands well developed, in XVI–XX, prostatic duct thin and short, U-shaped. Spermathecae 3 pairs, in VI–VIII, ampulla oval-shaped, its duct connected with ampulla thicker, serving as inner parietes of ampullar. Diverticulum slightly longer than main pouch, with long duct, ental 1/2 or 1/4 expanded as seminal chamber.

Color: Preserved specimens puce-green on mediodorsal line, the rest part of body flesh-brown. Clitellum chocolate.

Etymology: The specific name *"cruroides"* is a compound word with *"crur"* from the Latin word *"cruris"* meaning leg, and *"-oides"* which is used to form genus names from other genus names and means "resembling." It indicates the shape of the penis of this species.

Type locality: Hubei (Yichang).

Distribution in China: Hubei (Yichang, Qianjiang).

Remarks: This species is similar to *Metaphire longipenis* (Chen, 1946), but can distinguished from it by the position and number of pairs of spermathecal pores, very big penis, numerous setae, and testis sacs meeting dorsally.

313. *Metaphire thecodorsata* (Chen, 1933)

Pheretima thecodorsata Chen, 1933. *Contr. Biol. Lab. Sci. Soc. China (Zool)*. 9:244–249.
Metaphire thecodorsata Xu & Xiao, 2011. *Terrestrial Earthworms of China*. 255.

External Characters

Length 98–110 mm, width 3.5–5.0 mm, segment number 88–90. Prostomium 1/3 epilobous. First dorsal pore 12/13.

Clitellum entire, XIV–XVI, annular, highly glandulate, smooth on surface, its length equal to either immediate preclitellar or postclitellar 4 segments; setae invisible but setal pits present on ventral side of XIV–XVI. Setae very inconspicuous, evenly distributed all over body. Not particularly long nor unevenly spaced in any part of body, those on segments near posterior end as numerous as on those of middle body. Both dorsal and ventral breaks invisible, somewhat wider on anterior ventrum. Setae number 45–46 (III), 54–60 (VI), 55–60 (VIII), 49–55 (XII), 50–56 (XXV); 6 (VI) between spermathecal pores, 11–12 between male pores. Male pores paired in very deep copulatory pouches on ventrolateral side of XVIII, about 1/3 circumference apart ventrally; each secondary aperture wrinkled around its margin, with a very small opening. Each copulatory pouch very deep and spacious, with its inner surface transversely wrinkled into skin folds and longitudinally grooved from porophore along its median side to opening. Porophore as a large fleshy pad with a secondary rim, situated at bottom of chamber. No other papillae found in or out of chamber. Spermathecal pores 3 pairs, as large transverse slits, with anterior and posterior lips, situated very near to middorsal line, on posterior edge of V, VI, and VII, about 1/4 or 1/5 nearer to furrow than to setae of the segment, about 1/9 circumference dorsally apart. No other genital markings. Female pore single, medioventral in XIV.

Internal Characters

Septa 8/9/10 normally present, 4/5 thin, 5/6/7/8 much thickened, 6/7/8 very musculated, 8/9 thicker than 7/8, 9/10 twice thicker than 7/8, 10/11 thicker than 8/9, 11/12 as thick as 8/9, 12/13 thin and membranous. Gizzard moderate in size, rounded, in IX and X, onion shining on surface. Intestine swelling from XV. Intestinal caeca simple, long and narrow, in XXVII–XXIV, very smooth on both sides, except slightly constricted along insertion

FIG. 309 *Metaphire thecodorsata* (after Chen, 1933).

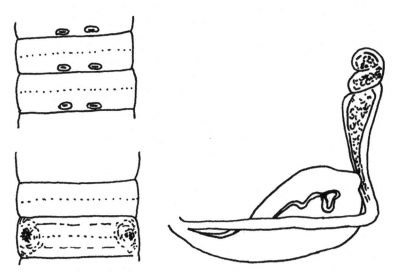

of septa. Lateral hearts 4 pairs, in X–XIII, last pair larger. Testis sacs 2 pairs in X and XI, enormously developed, each pair filling up whole segment and meeting but not communicated on dorsal side, narrowly connected ventrally, both pairs similar in size and structure. Testis large, situated at anterior inner corner of each sac. Seminal vesicles 1 pair, very large, in XII but extending back to XV, tubercular on surface, without a dorsal lobe. Prostate glands fairly large, in XVII–XXI, with rather few finger-like lobules, divisible into anterior and posterior main lobes. Prostatic duct slender but very long, with a deep loop, about 3–4 mm long, 0.2 mm wide. Vas deferens entering into its proximal end; its distal end not enlarged entering into ental top of a thumb-shaped copulatory chamber and through a porophore into the chamber. Spermathecae 3 pairs, in VI, VII, and VIII, medium-sized. Ampulla not differentiated from its duct, longer than the latter, slightly narrower or with knob-like apex, the whole spatulate in shape. Diverticulum twice longer than main pouch, its duct slender, about one-third longer than the latter; its seminal chamber enlarged but not constricted off from its duct; its ental end round but often contracted or twisted as long as ampulla and about 1/3 of its width. Diverticular duct joining the main duct near body wall.

Color: Preserved specimens faint, grayish green near and along middorsal line, cinnamon brown on clitellum, pale on rest of body.
Distribution in China: Jiangxi (Jiujiang).
Etymology: The specific name indicates the spermathecal pores on the dorsum.
Deposition of types: Previously deposited in the Zoological Museum of National Central University, Nanjing. The types were probably destroyed during the war in 1937.
Remarks: Very few species have been recorded with their spermathecal pore so dorsally situated. *Metaphire michaelseni* (Ude, 1925) has only 1 unpaired pore in furrow 8/9 along the middorsal line. *Metaphire dorsalis* (Michaelsen, 1928a) from Borneo has 2 pairs in 7/8/9 near the middorsal line, about 0.9 mm apart. This is the distinct character of the species.

V. PERIONYX PERRIER, 1872

Octochaetinae Stephenson, 1930. The Oligochaeta. 841.

Type species: *Perionyx excavatus* Perrier, 1872.
Diagnosis: Setae numerous per segment, in rings which are often almost closed. Male pores often approximated to greater or less degree, and may be very close to the middle line. Female pore unpaired (always ?). Spermathecal pores like the male pores often very near the middle line, the last pair in 7/8 or 8/9.

Gizzard very frequently more or less vestigial, in V or VI. Meganephridial. 2 pairs of testes and funnels. Prostate glands with branched system of ducts.

Global distribution: India and Sri Lanka, Australia, Auckland Islands. Several species are peregrine.

There is 1 species from China.

314. *Perionyx excavatus* (Perrier, 1872)

Perionyx excavatus Kobayashi, 1938c. *Sci. Rep. Tohoku Univ.* 13:201–203.
Perionyx excavatus Chang, Shen & Chen, 2009. *Earthworm Fauna of Taiwan.* 140–141.
Perionyx excavatus Xu & Xiao, 2011. *Terrestrial Earthworms of China.* 262–263.

External Characters

Length 50–180 mm, clitellum width 2.5–5 mm, segment number 115–178. No secondary segmentation. Prostomium epilobous. Body dorsoventrally flattened. Setae perichaetine, 44 (XII), 40–54 (XX) with narrow middorsal breaks, no setae between male pores but tips of black penial setae present around male pores. First dorsal pore 4/5 or 5/6. Clitellum in XIII–XVII, annular, setae present. Spermathecal pores 2 pairs in 7/8/9, large, very obvious, closely paired, same width apart as male pores. Male pores closely paired in XVIII, in deep clefts in a common depressed but tumid field. Female pore single, medioventral in XIV, anterior to setal line.

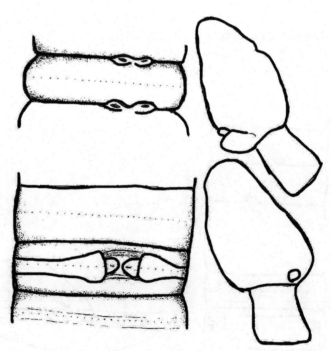

FIG. 310 *Perionyx excavatus* (after Kobayashi, 1938c).

Internal Characters

Septa 7/8/9 slightly thickened. Gizzard absent or vestigial in VI. Intestine swelling from XVI, XVII, or XVIII. Intestinal caeca absent. Lateral hearts in IX–XII, XIII. Nephridia holonic. Ovaries paired in XIII, large. Testes 2 pairs in X and XI. Seminal vesicles in IX–XII. Prostate glands paired in XVIII, large, round, racemose. Spermathecae 2 pairs in VIII and IX, ampulla large, stalk short and stout, diverticulum short.

Color: Live specimens dark red or violet-red.
Deposition of types: Paris Museum.
Etymology: The specific name *"excavatus"* literally means "to make hollow" in Latin.
Type locality: Ho Chi Minh City, Vietnam.
Distribution in China: Taiwan (Taibei, Xinzhu).
Remarks: This species is frequently cultured with *Eisenia andrei* (Bouché, 1972) in earthworm farms and sold with the latter as "red earthworms" for fishing in Taiwan.

VI. *PITHEMERA* KINBERG, 1867

Amyntas (part) Beddard, 1900c. *Proc. Zool. Soc. London.* 1900:612.
Pheretima (*Pheretima*) (part) Michaelsen, 1928a. *Arkiv For Zoologl.* 20(2):8.
Pithemera Sims & Easton, 1972. *Biol. J. Linn. Soc.* 4:202.

Type species: *Pithemera bicincta* Perrier, 1875.
Diagnosis: Small- to medium-sized Megascolecidae with cylindrical bodies usually under 130 mm in length. Setae numerous, regularly arranged around each segment. Clitellum annular, XIV–XVI. Male pores paired, discharging directly onto surface of XVIII. Female pores single or closely paired on XIV. Spermathecal pores small, 3, 4, or 5 pairs between, 4/5 and 8/9.
Gizzard present between 7/8/9/10. Esophageal pouches absent. Intestinal caeca present originating in or near XXII, rarely XXIV, paired laterally or single midventrally. Ovaries paired in XIII. Testes holandric or metandric. Prostate glands racemose. Copulatory pouches absent. Spermathecae 3, 4, or 5 pairs in V to IX. Meronephridial, no nephridia in spermathecal ducts.
Global distribution: Solomon Islands, New Britain, Fiji, Samoa, 1 species peregrine.
There are 4 species from Yunnan and Taiwan, China.
Remarks: Species of this genus are separable from those of allied genera by the occurrence of intestinal caeca in or near segment XXII.

TABLE 35   Key to the *Pithemera* From China

1.	Spermathecal pores 5 pairs 2
	Spermathecal pores fewer than 5 pairs 3
2.	Spermathecal pores 5 pairs in 4/5–8/9, Male pores on porophores with variable shapes, minute .... *Pithemera bicincta*
	Spermathecal pores 5 pairs in 4/5–8/9, No genital papillae in the spermathecal region. Male porophores round or oval, smooth, slightly elevated, surrounded by several slight circular folds, male aperture not visible; genital papillae present in XIX, XX–XXI, paired in XX, occupying nearly entire presetal part of XX between setal line and 19/20 intersegmental furrow .... *Pithemera lanyuensis*
3.	Spermathecal pores 2 pairs in 7/8/9, widely separated, not definitely recognizable .... *Pithemera bicincta*
	Male pores paired in XVIII. Genital markings absent. Spermathecal pore single, middle at 4/5–6/7 .... *Pheretima tao*

315. *Pithemera doliaria* (Gates, 1931)

Pheretima doliaria Gates, 1931. *Rec. Ind. Mus.* 33:374–378.
Pheretima doliaria Gates, 1972. *Trans. Am. Phil. Soc.* 62(7):180–181.
Pheretima doliarius doliarius Sims & Easton 1972. *Biol. J. Soc.* 4:234.
Pheretima doliarius Zhong & Qiu 1992. *Guizhou Science.* 10(4):40.
Pithemera doliarius Xu & Xiao, 2011. *Terrestrial Earthworms of China.* 263–264.

External Characters

Length 81–129 mm, width 4–6 mm, segment number 109–135. Prostomium epilobous, tongue open. First dorsal pore 12/13. Clitellum XIV–XVI, annular; intersegmental furrows and dorsal pores absent. The setae begin on II, the setal circles appear to be complete, without dorsal or ventral breaks, the setae regularly spaced and nearly the same distance apart from each other dorsally as ventrally; setae number 50–68 (XX); 18–26 (VIII) between spermathecal pores, 10–24 (XVII), 3–17 (XVIII), 11–20 (XIX) between male pores. Male pores paired in XVIII, minute, diagonal slits in line with the setae, porophore of variable size and shape. The male genital marking consists of a pair of elongate oval, diagonally placed areas on XVIII, separated by 4 setae, with the anterior ends directed towards the midventral line and the posterior ends directed laterally; each area 1.5 mm long and 0.5 mm wide, and surrounded by a narrow but deep and completely circumferential furrow. Outside of this furrow are incompletely circumferential furrows; the posterior third of the oval area separated from the anterior two-thirds by a furrow and bears the male pore. Spermathecal pores 2 pairs in 7/8/9, widely separated, not definitely recognizable. Female pore single, medioventral in XIV.

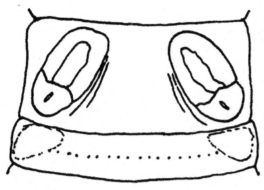

FIG. 311   *Pithemera doliaria* (after Gates, 1931).

Internal Characters

Septa 8/9/10 absent, 4/5–7/8 membranous; 10/11 also appears absent, 11/12 and succeeding septa are present, membranous. Gizzard elongate. Intestine swelling from XV. Intestinal caeca simple, in XXVII, extending anteriorly to XIX. Lateral hearts in XIII, the heart commissures of IX–XIII all pass into the ventral vessel. Testis sacs 2 pairs in X and XI. Seminal vesicles paired in XI and XII. Prostate glands large, extending from XVI to XXIII on the left side and from XVI to XXIII on the right side; the prostate glands broken up into a number of small lobes; the prostatic ducts long, the ectal portion of each duct greatly thickened and muscular; the ducts extend through segments XVIII–XX; no trace of a copulatory pouch; the vasa deferentia small but readily traced. The ovaries and oviduct funnels in XIII. Spermathecae 4 pairs, in VII and VIII but pass into the parietes posteriorly. The duct very short and sunk into the ventral face of the ampulla and not visible until the spermatheca pulled off from the parietes. Diverticulum short, with a narrow ectal portion and enlarged ental part which is more or less ovoid.

> Color: Preserved specimens dorsally reddish to dark reddish gray, ventrally whitish. Clitellum yellowish brown.
> Etymology: The specific name "*doliaria*" literally means "doliarious" in Latin.
> Type locality: Burma (Mong Ko, Kentung).
> Distribution in China: Yunnan (Mengmeng).
> Remarks: Dehiscence of setae, sometimes with abortion of follicles, was noted on the ventrum of VIII, in circles of XVII and XIX just in front of and just behind male porophores, but not in II (even in the dorsum) of any specimens.

316. *Pithemera bicincta* (Perrier, 1875)

Pheretima bicincta Gates, 1959a. *Amer. Mus. Novitates.* 141:4–5.

Pheretima bicincta Easton, 1982. *Aust. J. Zool.* 30:711–735.
Pithemera bicincta Shen & Tsai, 2002. *Endemic Species Research.* 4(2):1–8.
Pithemera bicincta Chang, Shen & Chen, 2009. *Earthworm Fauna of Taiwan.* 142–143.
Pithemera bicincta Xu & Xiao, 2011. *Terrestrial Earthworms of China.* 264.

External Characters

Length 33–80 mm, clitellum width 2–3 mm, segment number 77–125. Prostomium epilobous. First dorsal pore 11/12 or 12/13. Clitellum XIV–XVI, annular. Male pores paired in XVIII, 0.2 circumference apart ventrally, on porophores with variable shapes, minute. Genital markings absent or present, if present, paired smooth areas extending from XVIII to XIX. Spermathecal pores 5 pairs in 4/5–8/9, 0.26 circumference apart ventrally. Female pores closely paired, ventral in XIV.

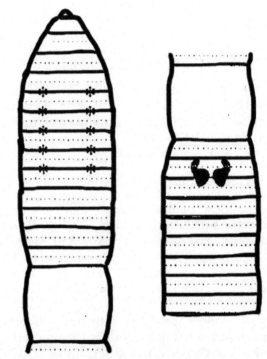

FIG. 312   *Pithemera bicincta* (after Easton, 1982).

Internal Characters

Septa 5/6–7/8 thickened, 8/9 absent, 9/10–12/13 greatly thickened. Gizzard in VIII–IX. Intestine swelling from XV. Intestinal caeca paired in XXII, simple, small, extending anteriorly to XXI. Lateral hearts in X, XI–XII. Nephridia meroic. Ovaries paired in XIII. Testis sacs 2 pairs in X and XI. Seminal vesicles paired in XI and XII. Prostate glands paired in XVIII, racemose. Spermathecae 5 pairs in V–IX, duct short, diverticulum slender, shorter than duct, with an oval seminal chamber.

Color: Reddish on dorsum, red on clitellum.
Deposition of types: Paris Museum.
Etymology: The specific name "*bicincta*" literally means "two girdles" in Latin.
Type locality: Philippines.
Distribution in China: Taiwan (Taibei).
Remarks: Pigment seen by Gates only in specimens that were quite preserved. GM glands, lacking in Mexican and Taiwanese worms, sometimes present and presumably with external pores even though the epidermis is not sufficiently modified for genital markings to be recognizable.

317. *Pithemera lanyuensis* Shen & Tsai, 2002

Pithemera lanyuensis Shen & Tsai, 2002. *Journal of the National Taiwan Museum.* 55(2):1–7.
Pithemera lanyuensis Chang, Shen & Chen, 2009. *Earthworm Fauna of Taiwan.*144–145.

External Characters

Length 37–46 mm, clitellum width 1.6–2.0 mm, segment number 79–90. No secondary segmentation. Prostomium epilobous or zygolobous. Setae number 52–57 (VII), 46–52 (XX); 9–12 between male pores. First dorsal pore 12/13. Clitellum XIV–XVI, annular, 0.9–1.5 mm long, dorsal pore and setae absent. Male porophores paired in XVIII, lateroventral, about 0.23 circumferences apart ventrally, round or oval, smooth, slightly elevated, surrounded by several slight circular folds, male aperture not visible; genital papillae present in XIX, XX–XXI, paired in XX, occupying nearly entire presetal part of XX between setal line and 19/20 intersegmental furrow, each papilla round, center concave, about 0.4 mm in diameter, 7–12 intersetal distances apart from each other, paired in XIX if present, single in right XXI. Spermathecal pores 5 pairs in 4/5–8/9, ventrolateral, 0.25–0.33 circumference apart ventrally. No genital papillae in the spermathecal region. Female pores closely paired, medioventral in XIV.

Internal Characters

Septa 8/9 absent, 9/10–12/13 thick. Gizzard round in IX. Intestinal caeca paired in XXII, small, short, bent. Esophageal hearts in X–XII. Testes 2 pairs in X and XI, small, round. Seminal vesicles paired in XI and XII, rudimentary. Prostate glands paired in XVIII, large, wrinkled, extending anteriorly to XVI and posteriorly to XX, with C-shaped ducts. Accessory glands in XX, round, sessile, pad-like, about 0.2 mm in diameter, each corresponding to external genital papillae. Spermathecae 5 pairs in V–IX, small, ampulla peach-shaped, 0.4–0.5 mm long, about 0.2 mm wide, stalk straight about 0.2 mm long, diverticulum small or vestigial, stalk short, straight to slightly bent, seminal chamber rudimentary or absent.

Color: Preserved specimens dark brown on dorsum, dark gray on ventrum, brown around clitellum.
Deposition of types: Taiwan Endemic Species Research Institute, Jiji, Nantou, Taiwan. Etymology: The name "*lanyuensis*" was given referring to the Lanyu Island, where the species was first collected.
Type locality: Southeastern Lanyu, Taidong County, Taiwan.
Distribution in China: Taiwan (Lanyu Island, Taidong County).
Remarks: *Pithemera lanyuensis* (Shen & Tsai, 2002) is the only endemic species of *Pithemera* in Taiwan.

318. *Pithemera tao* Wang & Shih, 2010

Pheretima tao Wang & Shih, 2010. *Zootaxa.* 2341:56–58.

External Characters

Length 27–43 mm, clitellum width 2.1–2.5 mm, segment number 55–90. Prostomium epilobous. First dorsal pore 12/13. Clitellum XIV–1/2 XVI, annular, setae on XIV and XV. *aa: ab: yy: yz*=1.5: 1: 3: 1.75 on XXV. Male pores paired in XVIII. Genital markings absent. Spermathecal pore single, middle at 4/5–6/7. Female pores paired in XIV.

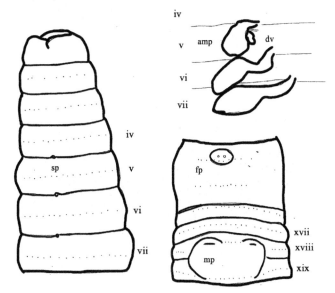

FIG. 313   *Pithemera tao* (after Wang & Shih, 2010).

Internal Characters

Septa 8/9/10 absent, 4/5–7/8 thin, 10/11–13/14 thick. Gizzard in VIII–X or IX–X. Intestine swelling from XIV, typhlosole simple 1/4 lumen diameter from XXI. Intestinal caeca in XXI, simple and short, extending anteriorly to XIX. Esophageal hearts absent. Holandric, testis sacs in X and XI. Seminal vesicles in XI and XII. Prostate glands in XVIII, small, round-shaped, extending XVII to

XIX. Spermatheca single in V–VII, ampulla ovoid, adiverticulate or diverticulate.

Deposition of types: Zoological Collections of the Department of Life Sciences, National Chung Hsing University.

Etymology: The name of the species refers to the aboriginal Tao Tribe, who dwell on the island of Lanyu. The name is used as a noun in apposition.

Type locality: Taiwan, Lanyu Island.

Distribution in China: Taiwan (Lanyu Island).

Remarks: *Pithemera tao* (Wang & Shih, 2010) is similar to *Pithemera bicincta* (Perrier, 1875), *P. palaoensis* (Ohfuchi, 1941), *P. philippinensis* (James, Hong & Kim, 2004), *P. sedgwicki sedgwicki* (Benham, 1897), and *P. sempoensis* (Kobayashi, 1938c), but can be distinguished by the spermathecal pore and spermatheca. *P. bicincta* and *P. philippinensis* have 5 pairs of spermathecal pores at 4/5–8/9. *P. sedgwicki sedgwicki* and *P. sempoensis* have 3 pairs of spermathecal pores at 5/6–7/8. *P. palaoensis* has 3 pairs of spermathecal pores at 4/5–6/7. All of the above species have paired spermathecal pores and spermathecae with diverticula. *Pithemera tao* has only a single spermathecal pore at 4/5–6/7.

VII. PLANAPHERETIMA MICHAELSEN, 1934

Planapheretima Sims & Easton, 1972. *Biol. J. Linn. Soc.* 4:208–209.
Planapheretima Easton, 1979. *British Museum (Natural History).* 35(1):64.

Type species: *Pheretima moultoni* Michaelsen, 1914.

Diagnosis: Small- to medium-sized Megascolecidae with depressed bodies under 120mm in length. Setae numerous, crowded ventrally on each segment. Clitellum annular XIII, XIV–XVI, XVII. Male pores paired XVIII, usually discharging directly onto the body surface but occasionally within copulatory pouches. Female pore single, rarely paired, XIV. Spermathecal pores small, occasionally large transverse slits, 1–5 pairs between 4/5 and 8/9.

Gizzard present between 7/8 and 9/10. Intestinal caeca usually absent, if present rudimentary, restricted to XXVII. Esophageal pouches absent. Meronephridial, nephridia absent from the spermathecal ducts. Ovaries paired in XIII. Testes holandric. Prostate glands racemose. Copulatory pouches usually absent. Spermathecae paired.

Global distribution: China (Sichuan, Chongqing, Guizhou), Indonesia (Sumatra, Celebes, Borneo), New Guinea, New Hebrides.

There are 4 species from Sichuan, Chongqing and Guizhou of China

Remarks: Species of this genus are unique among pheretimoid worms in having depressed bodies and the ventral setae crowded together.

TABLE 36   Key to Species of the Genus *Planapheretima* From China

1.	Spermathecal pores 2 pairs in 7/8/9 ...... *Planapheretima bambophila*
	Spermathecal pores 3–4 pairs 2
2.	Spermathecal pores 3 pairs in 5/6/7/8 (or 4/5/6/7) *Planapheretima continens*
	Spermathecal pores 4 pairs, in furrow 5/6/7/8/9 3
3.	Posterior end somewhat tapering like a "tail" .... *Planapheretima lacertina*
	Posterior end not tapering like a "tail" ..... *Planapheretima tenebrica*

319. *Planapheretima bambophila* (Chen, 1946)

Pheretima bambophila Chen, 1946. *J. West China. Border Res. Soc. (B).* 16:86–88, 138, 143.
Planapheretima bambophila Sims & Easton, 1972. *Biol. J. Linn. Soc.* 4:233.
Planapheretima bambophila Xu & Xiao, 2011. *Terrestrial Earthworms of China.* 265–266.

External Characters

Length 40–55mm, width 3–4mm, segment number 88–94. Body pointed anteriorly, more so posteriorly, invariably flattened and grooved ventromedially. A glandular zone, whitish, about 1mm wide and 12–14 setae intervening, appearing behind male pore region to posterior end of body. Prostomium 2/3 epilobous. First dorsal pore 10/11, distinct. Clitellum feebly developed, setae on XIV–XVI slightly shorter, not glandular. Setal chain close, *aa* slightly wider than *ab*, *zz* wider, equal to 1/2 *zy* at anterior and posterior ends, about 2/3 *zy* at middle. Ventral ones generally closer, those on glandular zone more so, widely spaced on ventrolateral and dorsal sides of middle of body, *ab*=1.5 *zy* at both anterior and posterior ends, *ab*=2–3 *zy* on middle of body. None of

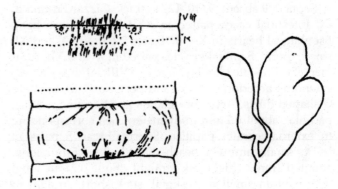

FIG. 314   *Planapheretima bambophila* (after Chen, 1946).

them enlarged or modified. Setae number 84 (X), 54 (XIX), 52 (XXV), 44 (XI). Male pores on XVIII, about 0.25 circumference apart ventrally, visible as minute pores, each on a highly glandular area which extends around each pore and between the pores as whitish glandular skin; setae absent therein, a space of about 14 setae. Male pores in younger specimens slightly wider, about 1/3 circumference apart ventrally, with visible 24/28 setae intervening when the glandular area not well shown. Spermathecal pores 2 pairs in 7/8/9, hardly visible externally, about 0.25 circumference apart ventrally, 14 setae at interval. Skin in their vicinity pale and highly glandular. Female pore unpaired, ventrally in XIV.

Internal Characters

Glandular tissues soft and whitish, hulky in II–V wholly filling these segments. No septa particularly thickened. Gizzard septa very thin, membranous. Gizzard large in VIII, barrel-shaped. Intestine swelling from XV. Intestinal caeca absent, intestinal wall thicker and glandular, distinctly pouched in XXII–XXIX, less conspicuous up to XXXVIII. Hearts in IX stout, in X also developed. Testis sacs in X broadly united, in XI as a thick V shape, communicated. Seminal vesicles unusually developed extending to several segments (between IX and XVII), dorsal lobes indistinctly marked. Prostate glands in XVII–XX digitate, its duct swollen on distal third, very slender. Spermathecae 2 pairs, in VIII and IX, closely under gizzard, small, posterior pair larger. Ampulla about 1.5 mm long. Diverticulum club-shaped with short and narrow duct, close to ectal end of ampullar duct, as long as main pouch.

> Color: Deep purplish or dark chestnut, iridescent on dorsum, pale ventrally, paler along ventral glandular zone.
> Deposition of types: Previously deposited in the Institute of Zoology, Academic Sinica, Chongqing. The types were destroyed during the war in 1945–1949.
> Etymology: The name "*bambophila*" indicates the habitat of the type locality.
> Type locality: Sichuan (Jiulaodong, Mt. Emei).
> Distribution in China: Sichuan (Mt. Emei).
> Remarks: Chen found this species generally inhabits small bushy bamboos. Its coloration vividly mimics the stems and branches of the purplish bamboos growing in the region. The glandular zone along ventromedian side and closer setae therein evidently help a firmer grip on the smooth surface of the plant. Upon examining its intestinal contents, we have found some fibrous structures of dead bamboo leaves and fragments of algae, mosses etc., probably taken from the stems and branches of the plant. It is an agile creature.

320. *Planapheretima continens* (Chen, 1946)

Pheretima continens Chen, 1946. *J. West China. Border Res. Soc. (B).* 16:95–97, 139, 146.
Planapheretima continens Sims & Easton, 1972. *Biol. J. Linn. Soc.* 4:233.
Planapheretima continens Xu & Xiao, 2011. *Terrestrial Earthworms of China.* 266.

External Characters

Small-sized worm. Length 33–38 mm, width 2 mm, segment number 94–102. Anterior segments II/XII about equally long, those behind clitellum very short. Length of 2 preclitellar segments equal to 4 postclitellar segments. Prostomium 1/3 epilobous. First dorsal pore 11/12. Clitellum in XIV–XVI, annular, smooth, extending nearly to setal line of either XIII or XVII respectively. Setae uniformly built, shorter on preclitellar segments, slightly closer ventrally. Setae number 30–40 (III), 54–58 (VIII), 52–55 (XXV); 22 (VI), 21 (VII), 22 (VIII) between spermathecal pores, 10–14 between male pores; $aa = 1.2\ ab$, $zz = 1.5$–$1.8\ zy$. Male pores paired in XVIII, about 1/3 circumference apart ventrally, each in a flat papilla, surrounded by concentric ridges, a pore-bearing minute tube exposed slightly longer than a seta, usually visible. Spermathecal pores 3 pairs in 5/6/7/8 (or 4/5/6/7). No papilla near the pore but a little swollen. Female pore single, medioventral in XIV.

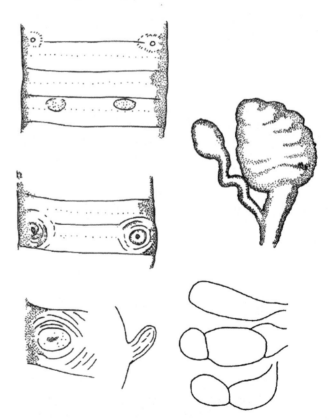

FIG. 315   *Planapheretima continens* (after Chen, 1946).

Internal Characters

Septa 8/9/10 absent, 5/6/7/8 relatively thick, 10/11 very thin, 11/12–14/15 equally thickened. Nephridial tufts on 5/6/7 thick. Gizzard onion-shaped, dull pale. Intestine swelling from XVI. Intestinal caeca very rudimentary, stumpy, about 1 segment long, intestinal wall after caecal region glandular. Hearts in IX and X rudimentary. Testis sacs large, anterior pair extending to dorsal side but not meeting. Seminal vesicles small, each with a large dorsal lobe. First pair situated dorsally and medially to posterior pair of testis sacs, closely connected with and enclosed in the latter. Prostate glands large, in XVI–XXI, mulberry in appearance, its duct S-shaped, enlarged in most of middle portion. Accessory glands sessile, each in a round patch, cord-like ductules found only in the body wall. Spermathecae 3 pairs, in VI–VIII. Ampulla heart-shaped, with a moderately long duct. Diverticulum longer than main pouch, with a globular or date-shaped seminal chamber.

> Color: Pale generally, slightly gray on dorsomedian side. Clitellum light chocolate red.
> Deposition of types: Previously deposited in the Institute of Zoology, Academic Sinica, Chongqing. The types were destroyed during the war in 1945–1949.
> Etymology: The specific name "*continens*" literally means "bordering, neighboring, connected, or continuous" in Latin.
> Distribution in China: Chongqing (Shapingba, Beipei), Sichuan (Mt. Emei).
> Remarks: Specimens are identical in several important points, e. g., clitellum extending more than 3 segments, the male pore with a minute tube, first pair of seminal vesicles enclosed in the testis sacs, and the characters of septa, setae etc. However, the Chongqing specimens differ in having 2 pairs of spermathecal pores (in 5/6/7, testis sacs more widely placed but connected, caeca a little longer, and possessing 1 unpaired papilla on VIII, IX respectively (found in 1 specimen only).

321. *Planapheretima lacertina* (Chen, 1946)

Pheretima lacertina Chen, 1946. *J. West China. Border Res. Soc. (B).* 16:109–111, 139, 152.
Planapheretima lacertina Sims & Easton, 1972. *Biol. J. Linn. Soc.* 4:233.
Planapheretima lacertina Xu & Xiao, 2011. *Terrestrial Earthworms of China.* 266–267.

External Characters

Length 81–82 mm, width 4.0–4.5 mm, segment number 90–100. Segments VII–XIII rather long, length of X–1/2 XIII = length of clitellum, length of X–XII = XIX–1/2 XXIII (4.5 segments), grooved along ventral side, posterior end somewhat tapering like a "tail". Prostomium 2/3 epilobous, with a median longitudinal groove,

extending to segment II, its tongue narrower than length of II. First dorsal pore 10/11. Clitellum XIV–XVI, annular, swollen and smooth, setae absent. Setae all fine, especially fine and closer on midventral side along the grooved zone; setal chain closed both dorsally and ventrally; those on anterior ventral very numerous and short, invisible unless highly magnified; setae number 72–78 (III), 78–86 (VI), 81–84 (IX), 72–79 (XXV); 33–44 (VI), 40 (IX) between spermathecal pores, 18 between male pores. Male pores paired in XVIII, about 1/2 circumference apart, each in a large glandular depression, its diameter slightly less than segmental length. No other genital markings. Segment XVIII somewhat expanded and swollen on lateral side of the pore region. Spermathecal pores 4 pairs in 5/6–8/9, about 4/9 circumference apart ventrally, about 4–5 setae ventral to ventrolateral mark. No genital markings. Female pore single, medioventral in XIV.

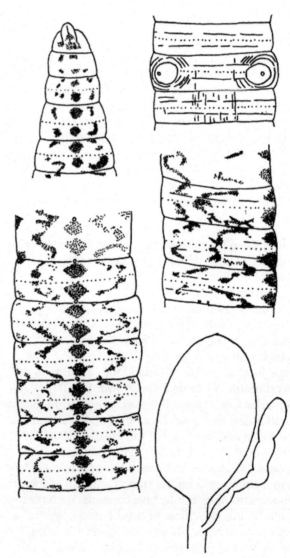

FIG. 316   *Planapheretima lacertian* (after Chen, 1946).

Internal Characters

Septa 8/9/10 present, 6/7/8/9 thicker, slightly musculated, 5/6, 9/10/11 thin, thinner posteriorly. Pharyngeal glands well developed, in III–VIII or IX. Micronephridial. Gizzard in VIII but pushed back in 1/2 IX, X, round but a little oblong, shiny on surface, with crop-like dilation in front. Intestine swelling from XV. Intestinal caeca rudimentary, ear-like projection confined to segment XXVI, smooth and pale in color. Side of intestine behind caecal region glandular. Hearts in IX stout, those in X absent. Testis sacs united, anterior pair rounded, with a broad bridge, posterior pair united. Seminal vesicles large, lying in XI–XIV, dorsal lobes indistinct. Prostate glands well developed, filling the segments XVI–XXIV, its duct evenly thickened, in a deep curve. Accessory glands velvety and whitish as cushion around the base of prostatic duct. Spermathecae 4 pairs, lying in VI–IX, ampulla heart-shaped, with a moderately long duct. Diverticulum as long as main pouch, ental 2/3 crooked and whitish serving as seminal chamber.

Color: Patterns on preserved specimens unique to this species, in general appearance like a beautiful *Eumeces* lizard, dark chestnut or brownish black with a bluish background. The pattern in each segment quite regular throughout the body; dorsomedian line forming a trapezium in front and a large rhomboid behind, setal circle, whole line from the first to the last segment as a chain of beads, darker than the rest; a zigzag dorsolateral line starting from the first segment, more distinct from the fifth, to the posterior end, not evenly pigmented; another zigzag ventrolateral line from the second segment posteriorly less pigmented and gradually disappearing toward posterior end. Similar patterns present but much vaguer on clitellum. Pale ventrally. Clitellum chocolate.

Deposition of types: Previously deposited in the Institute of Zoology, Academic Sinica, Nanjing. The types were destroyed during the war in 1945–1949.

Etymology: The specific name "*lacertina*" literally means "lacertine" in Latin and probably refers to the unique coloration of the species.

Type locality: Chongqing (Mt. Jinfo).

Distribution in China: Chongqing (Mt. Jinfo), Sichuan (Mt. Emei), Guizhou (Mt. Fanjing).

Remarks: This species is remarkable on account of its beautiful coloration. The specimens were collected from moss on the bark of large trees on the top of the mountain. They live on the moss debris. The beautiful coloration might be an adaptive feature in securing protection among the mosses. The flattened ventral side and numerous setae therein probably help its movement on the trees. In many structural respects this species resembles *Planapheretima tenebrica* (Chen, 1946), referred to below.

322. *Planapheretima tenebrica* (Chen, 1946)

Pheretima tenebrica Chen, 1946. *J. West China. Border Res. Soc. (B).* 16:93–94, 138, 146.
Planapheretima tenebrica Sims & Easton, 1972. *Biol. J. Linn. Soc.* 4:233.
Planapheretima tenebrica Xu & Xiao, 2011. *Terrestrial Earthworms of China.* 267–268.

External Characters

Small-sized worm. Length 35–60 mm, width 2.0–2.8 mm, segment number 82. Preclitellar segments moderately long, longest equal to 1.5 postclitellar segments. Prostomium 1/3 epilobous, with a median longitudinal furrow extending to a part of segment II. First dorsal pore 9/10. Clitellum XIV–XVI, usually encroaching a part of neighboring segments, i.e., extending over intersegmental furrows 13/14 and 16/17, smooth. Setae fine, uniformly distributed, closer ventrally. Ventral setae about 20–28 from II to about X particularly closely spaced; setae number 46–50 (III), 52–60 (IX), 50–52 (XIX), 50–52 (XXV); 25–38 between spermathecal pores, 16–22 between male pores; $aa = 1.0–1.1\ ab$, $zz = 2.0–2.5\ zy$. Male pores superficial, paired in XVIII, about 0.5 circumference apart, each on lateral side of a depressed area which is composed of 2 anteroposterior papillae. The region surrounded by a few circular furrows. Spermathecal pores 4 pairs in 5/6–8/9, intersegmental, about or more than 0.5 circumference apart ventrally. Genital papillae or other markings absent. Female pore single, medioventral in XIV.

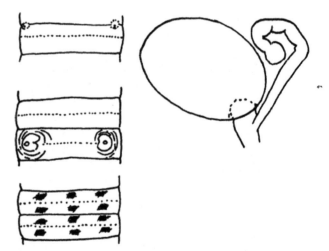

FIG. 317　*Planapheretima tenebrica* (after Chen, 1946).

Internal Characters

Septa all membranous, none musculated, ligaments attached to their posterior side strong, 8/9/10 present. Pharyngeal glands very large, lobular and delicate, extending to anterior part of gizzard. Gizzard elongate barrel-shaped. Intestine swelling from XV. Intestinal caeca absent, intestinal wall in XXVI–XXXVI thicker (especially in anterior 4 segments), pale in appearance. Hearts in IX and X present. Testis sacs in X united but not directly communicated (in 2 sacs inside); those in XI as a transverse thick band. Seminal vesicles exceedingly developed, occupying 5–7 segments, dorsal lobe not distinct, anteriormost division probably representing it. Prostate glands with large lobules, in XVI–XXI, its duct long, in deep V-curved ectal half about 3 times stouter. Accessory glands diffuse and velvety, sessile. Spermathecae 4 pairs, in VI–IX, ampulla egg-shaped, about 1mm long, ampullar duct 0.3mm long; diverticulum short, straight, with small spherical seminal chamber.

Color: Preserved specimens with 3 longitudinal reddish brown stripes, dorsomedian stripe deeper and 2 dorsolateral stripes lighter or with 3 indistinct but deeper stripes on a chocolate background. Clitellum from brick red to dark chocolate.

Deposition of types: Previously deposited in the Institute of Zoology, Academic Sinica, Chongqing. The types were destroyed during the war in 1945–1949.

Etymology: The specific name "*tenebrica*" derives from the Latin "*tenebricus*" and means darkened, indicating the coloration of the species.

Type locality: Chongqing (Mt. Jinfo).

Distribution in China: Chongqing (Nanchuan), Sichuan (Mt. Emei).

Remarks: In the characters of the pharyngeal glands, septa, seminal vesicles, testis sacs, the absence of caeca, and the striped coloration, this species is close to *Planapheretima lacertina* (Chen, 1946), an arboreal species found on Mt. Jinfo. *P. lacertina* is fluvial so differs in habitat from *P. tenebrica*, yet they are structurally identical. The differences between the specimens at those 2 localities are slight. In the Nanchuan specimen the coloration is more brilliant, with 3 brownish stripes on a bluish background; its setal number is greater: 58 (III), 60 (IX), 64 (XXV); 30 (VI), 6 (IX), 33 (XVII) between the pores, its nephridia are comparatively large, and its vascular loops in X are absent. It is an aclitellate specimen.

VIII. *POLYPHERETIMA* EASTON, 1979

Polypheretima Easton, 1979. *British Museum (Natural History).* 35(1):28–29.

Type species: *Perichaeta stelleri* Michaelsen, 1892.

Diagnosis: Megascolecidae with an esophageal gizzard in VIII and an intestine lacking caeca and gizzards. Setae perichaetine. Male pores on circular porophores, lacking associated crescentic markings, occasionally within copulatory pouches. Nephridia absent from the spermathecal ducts.

Global distribution: Autochthonous species recorded throughout the *Pheretima* group domain except for New

Britain, Solomon Islands, New Hebrides, Caroline Islands, Mariana Islands.

There is 1 species from China.

Remarks: This genus is readily recognizable among introduced pheretimoid populations, being the only acaecate genus to occur outside the *Pheretima* domain.

323. *Polypheretima elongata* (Perrier, 1872)

Pheretima elongata. Gates, 1930. *Rec.Ind. Mus.* 32:309–310.

Pheretima elongata. Gates, 1932. *Rec. Ind. Mus.* 34(4):391–392.

Pheretima elongata. Gates, 1936. *Rec. Ind. Mus.* 38:413–415.

Pheretima elongata. Gates, 1959a. *Amer. Mus. Novitates,* 141:9.

Pheretima elongata. Gates, 1972. *Trans. Amer. Philos. Soc.* 62(7):182–183.

Metapheretima elongata. Sims & Easton 1972. *Biol. J. Soc.* 4:233.

Polaypheretima elongata. Easton 1979. *Bull. Br. Mus. Nat. Hist. (Zool).* 35:53–54.

Polaypheretima elongata. Zhong & Qiu 1992. *Guizhou Science.* 10(4):39.

Polaypheretima elongata. James, Shih & Chang 2005. *Jour. N. His.* 39(14):1026.

Polypheretima elongate Chang, Shen & Chen, 2009. *Earthworm Fauna of Taiwan.* 146–147.

Polypheretima elongate Xu & Xiao, 2011. *Terrestrial Earthworms of China.* 268–269.

External Characters

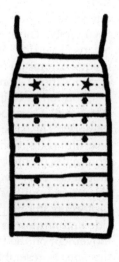

FIG. 318   *Polypheretima elongate* (after Easton, 1982).

Length 75–300 mm, clitellum width 3–6 mm, segment number 136–297. Prostomium prolobous or epilobous. Setae number 67–104 (VIII), 55–75 (XX). First dorsal pore 12/13 or 13/14. Clitellum XIV–XVI, annular, setae present ventrally or absent, dorsal pore absent. Male pores paired in XVIII, on raised porophores. Genital papillae paired in XIX–XXIII, XXIV, presetal, large, raised, with smooth center. Spermathecal pores absent. Female pore single, medioventral in XIV.

Internal Characters

Septa 4/5–7/8 greatly thickened, 8/9/10 absent. Gizzard in IX. Intestine swelling from XV. Intestinal caeca absent. Hearts in IX, X–XII, XIII. Nephridia meroic. Ovaries paired in XIII. Testis sacs 2 pairs in X and XI. Seminal vesicles in XI and XII. Prostate glands paired in XVIII, racemose, extending anteriorly to XVI and posteriorly to XXI. Spermathecae absent.

Color: Live specimens light grey with pink anterior.
Deposition of types: Paris Museum.
Etymology: This name of the species refers to its elongated body shape.
Type locality: Peru.
Global distribution: Taiwan (Kaohsiung).
Distribution in China: Taiwan (Gaoxiong).
Remarks: *Polypheretima elongata* (Perrier, 1872) forms a species complex with 4 other morphologically similar species from western Indonesia and Borneo. It is the only allochthonous member of the complex, often being reported from tropical regions throughout the world and, less frequently, from the warmer parts of temperate regions.

Family: Microchaetidae Beddard, 1895

Microchaetidae Gates, 1972. *Trans. Am. Phil. Soc.* 62(7):233.

Normal setae in 8 longitudinal rows. Copulatory setae if present are not of the grooved type. Male pores preclitellar or intraclitellar. Spermathecal pores usually altogether behind the testis segments. Esophageal gizzard, calciferous glands, and strengthening of the musculature at the beginning of the intestine may be present or absent, but there is no well-marked intestinal. Metagynous (having the ovaries only in segment XIII or a homoeotic segment); seminal vesicles usually short, not extending back for several segments by penetration of the septa.

Global distribution: South Africa; Madagascar, Tropical East Africa, North-East and Central Africa; The Republic of Congo, Cameroon, Togoland, Gambia, Central and South America.

GENUS: *GLYPHIDRILUS* HORST, 1889

Glyphidrilus Gates, 1972. *Trans. Amer. Philos. Soc.* 62(7):234–236.

Type species: *Glyphidrilus weberi* Horst, 1889.
Diagnosis: Setae widely paired anteriorly, more closely behind; in the hind part of the body *dd* is at most a little greater than *aa*. Male pores intraclitellar, situated a variable distance behind segment XVI, on a level area between a pair of long ridges (ridges of puberty). Spermathecae in front of the male pores. Esophagus with a gizzard in the region of segments VII–VIII, calciferous glands absent. One pair of nephridia per segment. Holandric and metagynous, no testis sacs, no copulatory sacs, prostates present or absent.

Global distribution: The genus *Glyphidrilus* has a widely distributed range from Tanzania in Africa to South Asia and Southeast Asia. In Asia many species have been recorded from Yunnan, southwest China, Hainan, Indochina, Malaysia, Laos, Thailand, Singapore, and some Sunda Shelf islands such as Sumatra, Java, Borneo, and Celebes. It occurs along river banks, lakes, canals, swamps, and in rice paddy systems.

There are 2 species from Hainan and Yunnan, China.

324. *Glyphidrilus papillatus* (Rosa, 1877)

Glyphidrilus papillatus Chen, 1938. *Contr. Biol. Lab. Sci. Soc. China (Zool).* 12(10):426.
Glyphidrilus papillatus Gates, 1972. *Trans. Amer. Phil. Soc.* 62(7):235–236.
Glyphidrilus papillatus Xu & Xiao, 2011. *Terrestrial Earthworms of China.* 283.

External Characters

Length 74–180 mm, width 3–6 mm, segment number 104–330. Prostomium zygolobous. Clitellum, annular, in XIII–XL (less frequently on XIII–XXXIII); intersegmental furrows not obliterated, setae retained (including all segments with reproductive apertures?). Wings on ventrolateral sides of XVIII–XXIII (or XXIV), lateral to *b*. Genital papillae paired in segments before and behind wings, on ventral region of XII (or XI)–XVII and XXIV–XXV or –XXVII, divided into 2 series: lateral series lateral to *b*, in line with wings, ventral series between *aa*, often occurring in XVII, occasionally on XII or XI.

Internal Characters

Gizzard in VII and VIII, small but distinct, fairly muscular. Intestine swelling from XIV, and an intestinal typhlosole but without caeca. Testes and funnels in X and XI, seminal vesicles in IX–XII, the last twice as large. Spermathecae sac-like, or globular, well distended, about 1 .. mm in diameter, sessile, in 4 transverse rows, in XIV–XVII, extending from midventral line to level of seta *d*, about 6–9 on each side, fewer in XIV and XVII respectively. Ovisacs in XIV and XV.

Color: Preserved specimens unpigmented.
Deposition of types: Genoa Museum.
Etymology: The specific name refers to the wings with numerous papillae.

Terrestrial Earthworms (Oligochaeta: Opisthopora) *of China*
https://doi.org/10.1016/B978-0-12-815587-5.00012-3

Type locality: Cobapo (Menglun).

Distribution in China: Hainan (Baopeng, Sha-mom-chiu).

Remarks: Chen's specimens from Hainan have some variations compared to those collected by other authors. In Chen's, the clitellum is not clearly marked off at anterior end, the glandular character appears ventrally on XIII but is not distinct until XVI or XVII. The ventral series of papillae occurs generally on XVII and less frequently on XI or XII. The number of spermathecae in each transverse row is much higher than that recorded by others.

325. *Glyphidrilus yunnanensis* Chen & Hsü, 1977

Glyphidrilus yunnanensis Chen & Hsü, 1977. *Acta Zool. Sinica* 23(2):178.

Glyphidrilus yunnanensis Xu & Xiao, 2011. *Terrestrial Earthworms of China*. 283–284.

External Characters

Length 123 mm, width 6 mm, segment number 139. Number of annuli per segment 7 or more in preclitellar segments, indistinct or absent in postclitellar segments. Body shape quadrangular from middle part in cross section, dorsal side slightly wide, anus longitudinal slit-like. Easily autotomized. Prostomium zygolobous. Dorsal pore absent. Clitellum saddle-shaped, flattened on ventrum, in XVIII–XXXVIII, demarcation indistinct at ends. Wing ventrolateral, in XXII–XXXII (left side) or in XXXIII (right side), between *b* and *c*, and closer to *b*, short ridge-like or horn-shaped. Genital papillae paired or asymmetrical, wings 4 on antelateral XVIII–XXI, 3 on XIX–XXI left, postlateral absent and 3 on XXXIII–XXXIV left; each papilla about 1–1.5 mm wide, margin wide with circular grooves between it and the central papillae; papilla rows absent on medioventral line. Setae 4 pairs per segment; in XII: $aa = 2ab = bc = 2cd = 5/6dd$; in L: $aa = 2ab = bc = 2cd = 2/3dd$. Nephridiopores in front of the setae *b*. No obvious male pores, female pores, and spermathecal pores.

Internal Characters

Septa 6/7–11/12 gradually thickened. Gizzard in VIII, undeveloped. Intestine swelling from XVI. Meganephridial. Heart 5 pairs, in VII–XI. Two pairs large funnels free in X and XI. Four pairs irregular seminal vesicles in IX–XII, last pair is well developed. Ovisacs in XIV. Testes, prostates and spermathecae undetected.

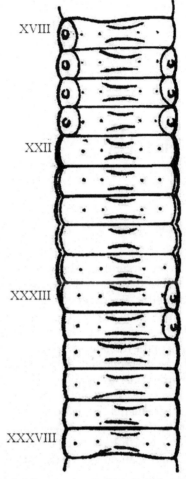

FIG. 319 *Glyphidrilus yunnanensis* (Chen & Hsü, 1977).

Color: Live specimens unpigmented, faint on clitellum. Preserved specimens brown on postclitellar segments, anterior end light brown, the rest of body light gray, intestine faintly visible.

Etymology: The name of the species refers to the type locality.

Type locality: Yunnan (Menglun).

Distribution in China: Yunnan (Menglun).

Remarks: This species is similar to *Glyphidrilus annandalei* (Michaelsen 1910) in that it lives at the water's edge, is easily autotomized, unpigmented, zygolobous, dorsal pore absent, quadrangular from middle part, 2 pairs of testis funnels free in X and XI, 4 pairs irregular seminal vesicles in IX–XII, and prostates undetected. However, its clitellum is saddle-shaped, wings not on XXV–XXVII, or XXVIII–XXXII, or XXXIII, but on XXII–XXXII or XXXIII; genital papillae on medioventral line absent, and spermathecae absent.

Family: Moniligastridae Claus, 1880

Moniligastridae. Gates, 1930. *Rec. Ind. Mus.* 32:264.
Moniligastridae Gates, 1972. *Trans. Am. Phil. Soc.* 62(7):238–240.

Type genus: ?

External Characters

Prostomium, prolobous but separated from I, protuberant from roof of buccal cavity behind level of 1/2. Clitellum 1 cell in thickness, extending over 3–6 segments, including those bearing the genital pores. Setae, sigmoid and single pointed, (penial and copulatory setal lacking), 4 pairs per segment. Dorsal pore none. Male pores, 1 or 2 pairs, in or near furrows 10/11/12, or 12/13. Spermathecal pores 1 or 2 pairs, in 7/8 or 8/9, or 7/8 and 8/9. Female pores 1 pair, in 11/12 or on XIII or XIV.

Internal Characters

Gizzard multiple, either in front of the testis segment or segments, or at the beginning of the intestine. Last heart 2 segments in front of the ovarian segment. Meganephridial. Testes and funnels 1 or 2 pairs, enclosed in 1 or 2 pairs of testis sacs which are suspended on the septa; vasa deferentia opening into prostate glands, or on the surface independently of them. One pair of ovaries; 1 pair of ovisacs extending backwards from the ovarian segment. One or two pairs of spermathecae, with long tubular ducts.

Global distribution: Cosmopolitan. Found in South India, Ceylon, Myanmar, the East Himalayas, India, Malay Archipelago, Philippine Islands, Japan, China, Caroline Islands, and the Bahamas. The Syngenodrilinae have been found only in tropical East Africa.

GENUS: *DESMOGASTER ROSA*, 1890

Desmogaster Gates, 1939. *Proc. U. S. Natn. Mus.* 85:406.
Desmogaster Gates, 1972. *Trans. Am. Phil. Soc.* 62(7):241.

Type specie: *Desmogaster doriae* Rosa, 1890.

External Characters

Setae single or closely paired, often invisible. Genital setae in some species. Meganephirdia from a few pregonadal segments to the distal end except for the gonadal segments, pores near ventral setae. Clitellum of a single layer of cell, mostly in the region of the genital pores. Spermathecal pores 1 or 2 pairs, in 7/8 or 8/9 and 8/9. Female pores anteriorly on XIV. Spermathecae without atrial dilatation or stalked glands at ectal end.

Internal Characters

Gizzards 3–10. Last heart in XI. 2 pairs of testes and funnels, enclosed in testis sacs on septal 10/11 and 11/12; 2 pairs of much elongated prostates; 2 pairs of male pores, in furrows 11/12 and 12/13. Ovaries in XIII; ovisacs extending back from septum 13/14.

There is only 1 species from China.

Global distribution: Burma, China, Sumatra.
Distribution in China: Jiangsu (Suzhou, Wuxi).

326. *Desmogaster sinensis* Gates, 1930

Desmogaster sinensis Chen, 1933. *Contr. Biol. Lab. Sci. Soc. China (Zool).* 9:180–189.
Desmogaster sinensis Gates, 1939. *Proc. U. S. Natn. Mus.* 85:406.
Desmogaster sinensis Xu & Xiao, 2011. *Terrestrial Earthworms of China.* 44–45.

External Characters

Length 290–540 mm, width 8–12 mm, segment number 360–588. Each segment (except first 3) with 3 circular furrows in anterior third, 2 in middle and 0 in posterior half of body. Prostomium epilobous, large and broad, transversely elongate. Dorsal pore absent. Clitellum not apparent, in X–XV; ventral surface somewhat deeper in color and slightly roughened on segments XI–XIV, rather smooth on XV. Setae absent even in posterior segments. Male pores 2 pairs, in intersegmental furrows 11/12/13, about 1/3 circumference apart ventrally, each pore

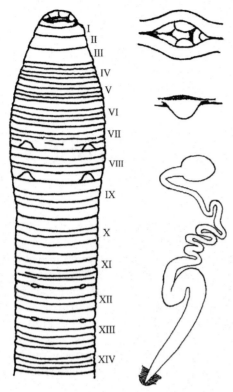

FIG. 320 *Desmogaster sinensis* (after Chen, 1933).

pairs, in VI–XI. Testis sacs 2 pairs, on septa 10/11/12, posterior pair much larger; vas deferens much coiled, penetrating the septa backward to inner side of each prostate, which is compact and tubercular on surface in XI and XII, closely applied to body wall. Spermathecae on posterior faces of septa 7/8/9; ampulla ovoid, whitish, its duct long and winding down to its opening. Ovaries as long strings on posterior face of septum 13/14 and a pair of long egg sacs on dorsal side of XIV and XV over anterior part of gizzards, connected to posterior face of septum 14/15.

Color: Live specimens unpigmented, slightly transparent, light yellowish at anterior end.
Distribution in China: Jiangsu (Nanjing, Suzhou, Wuxi).
Etymology: The name of the species refers to the type locality.
Type locality: Jiangsu (Suzhou).
Deposition of types: Smithsonian Institution.
Remarks: These worms are often found on low hills, whether rich in humus or hard soil, or even in cultivated land close to the hillside. Their burrows are not very deep and are usually indicated by their very stout, ill-defined castings. When alive, they appear sluggish and motionless as if they were only half awake. They will not move if they are left undisturbed for a long time. They then slowly protrude their prostomium and extend the anterior half of their body.

GENUS: *DRAWIDA* MICHAELSEN, 1900

Drawida Michaelsen, 1900. *Das tierreich, Berlin.* 10:114.
Drawida Gates, 1972. *Trans. Amer. Phil. Soc.* 62(7):244.

Type species: *Moniligasater barwelli* Beddard, 1886.

External Characters

Body moderate in size, or small. Clitellum, including the whole or greater part of X–XIII. Dorsal pores absent or present. Setae paired, 4 pairs per segment. Spermathecal pores, at 7/8. Male pores, at or near 10/11. Female pores, at or just behind 11/12.

Internal Characters

Septa, 5/6–9/10 strengthened (usually thickly muscular), parietal insertion of 9/10 dislocated posteriorly, 10/11/12 approximated. Gizzards 2–8, in region of XII–XXVII. Last connectives between extraesophageal and dorsal trunks on posterior face of 9/10, another pair associated with 8/9. Lateral hearts, in each of VIII–IX,

opening on top of a tubercle enclosed in center of a transverse and large slit or pouch; each male tubercle with its surrounding part wholly visible if both lips of pouch widely open, but entirely invisible if closed past a certain point; anterior and posterior walls of pouch sometimes swollen, glandular, or wrinkled as thick lips. Spermathecal pores 2 pairs, on posterior margin of VII and VIII, each on top of a teat-like protuberance, close to intersegmental furrows 7/8/9. Female pores 1 pair, on anterior annulus of XIV, usually in tertiary furrow, slightly medial to male pore or in line to inner corner of male pouch, in front of seta *b* if it were present; each pore visible in a small, transverse and depressed fissure, sometimes very inconspicuous. No glandular markings in this region.

Internal Characters

Septa 3/4/5 very thin and confined to segment V, 5/6 fairly thick, in a deep dome-shape enclosing the muscular pharynx; 6/7/8/9 enormously thickened and strongly musculated; 9/10 about two-thirds the thickness of preceding septa; all septa behind 10/11 to posterior end of body very thin and membranous. Gizzards 3, in XIV–XVI; moderate in size, very strong in musculature, appearing short and broad. Intestine thin-walled, swelling immediately behind last gizzard. Lateral hearts 6 after joining connectives from extraesophageal trunks unite mesially above gut and then communicate with dorsal trunk through a short vertical vessel in median plane. Spermathecae with or without atrial dilation at ectal, without glands in connation with ectal end of duct. Testes, in 9/10. Prostates, in X. Ovaries, in XI.

Global distribution: India, Burma, Malay Peninsula, Thailand, Indo-China, China, Korea, Manchuria, Siberia, Japan, Philippine Islands, Borneo. Sri Lanka (?). (Possibly also Sumatra and Java?)

There are 23 species from China.

TABLE 37 Key to Species of the Genus *Drawida* From China

1. First dorsal pore probably in 6/7, those before clitellum hardly visible, behind evident .... *Drawida glabella*
   Dorsal pore absent or nonfunctional 2

2. Dorsal pores visible after furrow 2/3 or 3/4, nonfunctional 3
   Dorsal pore absent 6

3. Penis conical, situated in a shallow pouch not raised in body cavity, the top pointed distally .... *Drawida anchingiana*
   Penis pointed distally, the top pointed distally 4

4. Penis a little pointed distally, large at base, situated in a deeply sunken penis pouch .... *Drawida gisti nanchangiana*
   Penis pointed distally and elongate 5

5. Clitellum in X–XIII or extending to XIV and a part of IX .... *Drawida gisti*
   Clitellum in 1/2 IX–XIV or 1/2 XIV .... *Drawida cheni*

6. Skin cavernous or with sieve-like small pits all over body .... *Drawida sinica*
   Skin not cavernous or with sieve-like small pits 7

7. With penis 8
   Without penis (or inconspicuous) 13

8. Male pores in posterior of segment X 9
   Male pores in the intersegmental furrow 10/11 10

9. Male pores projecting out from posterior edge of segment X. Length 30–200 mm .... *Drawida japonica*
   Male pores on small porophores seated on 10/11, intersegmental furrow 10/11 ending blindly against the lateral and median sides of the porophores, length 40–55 mm .... *Drawida japonica siemssens*

10. Male pores in the intersegmental furrow 10/11, penis short, pointed distally, usually pointing ventromedially when wholly everted *Drawida syring*
    Male pores in the intersegmental furrow 10/11, penis top not sharp 11

11. Penis large and transversely ovoid, gradually rounded (never pointed) distally, as a round tubercle-like protuberance. Length 37–58 mm *Drawida linhaiensis*
    Penis conical 12

12. Penis can project out of the aperture, and inner surface of the chamber visible; length 60–86 mm .... *Drawida sulcata*
    Conical penis usually not visible externally, length 65–120 mm .... *Drawida nemora*

13. Male pores on the porophore that located in posterior of segment X 14
    Male pores in posterior of segment X, or intersegmental furrow 10/11 16

14. No annulion in each segment .... *Drawida koreana*
    With annulion in part segments 15

15. Two to three annulion each segment in middle region of body. Length 87–119 mm .... *Drawida bimaculata*
    With annulion in posterior of segments X, 3 annulion in X. Length 52–66 mm .... *Drawida jeholensis*

16. Male pores are minute, transverse slits on X, nearer to *b* than to *c* and nearer to 10/11 than to the transverse setal line.
    Length 73–130 mm .... *Drawida propatula*
    Male pores in the intersegmental furrow 10/11 17

17. Male pores between *b* and *c*, pore rather on posterior edge of X, its anterior and posterior sides of neighboring 2 segments swollen as lips ...
    Male pores on small porophores seated on 10/11, the anterior margin of the porophore may be indicated by a slight transverse groove. Length 40–55 mm .... *Drawida grahami*

18. Length 40–60 mm .... *Drawida omeiana*
    Length 100–160 mm .... *Drawida changbaishanensis*

327. *Drawida anchingiana* (Chen, 1933)

Drawida gisti anchingiana Chen, 1933. *Contr. Biol. Lab. Sci. Soc. China (Zool).* 9:202.
Drawida anchingiana Kobayashi, 1936c. *Sci. Rep. Tohoku Univ.* 11:333–337.
Drawida anchingiana Xu & Xiao, 2011. *Terrestrial Earthworms of China.* 47

External Characters

Length 56–80 mm, width 3–5 mm, segment number 125–145. Prostomium prolobous. Similar large papillae generally present in VII–XI, but absent in some cases, with a large raised glandular area and dark center, similar to those of *Drawida gisti gisti* (Michaelsen, 1931). Dorsomedian line thinner, nonfunctional dorsal pores visible after furrow 2/3. Clitellum in X–XIII, indistinct. Setae may be stouter than *Drawida gisti gisti*, but their arrangement is similar to the latter. Those on II not weakly built but slightly smaller than the rest; no marked difference in size between anterior and posterior segments to the clitellum distinguishable; *ab*=*cd*, in general *aa* narrower than *bc*, but greater on II–IV or V, *dd*=about 4/7 circumference. Penis conical, situated in a shallow pouch not raised in body cavity similar to *Drawida gisti gisti*. Spermathecal pores 1 pair in 7/8, each represented as a large transverse slit, just medial to *c*. Genital papillae circular, with a pigmented glandulated center, distinctly elevated, usually surrounded by swollen skin. Female pores 1 pair, in line with *b*, rather distinct as transverse slits at the anterior border of XII.

Internal Characters

Septa 5/6 thick, 6/7/8/9 very much thickened, the rest thin and membranous; 9/10 pushed to middle part of X, 10/11 to anterior part of XII to meet 11/12 forming ovarian chamber, 12/13 to anterior part of XIII, the posterior septa resuming nearly normal attachment. Gizzards generally 3, in XII–XVI or XI–XVI. Lateral hearts 4 pairs in VI–IX, relatively stout. Male atrium slender, cylindrical, the largest about 2 mm long and 1 mm wide, smooth on surface, opening into a very shallow penis pouch. Ovarian chamber formed by septa 10/11 and 11/12 which meet only on dorsal side. Egg sacs slender and short, empty in both collections. Spermathecal atrium often short, or cylindrical, longest about 2 mm, invariably in front of septum 7/8, ampulla small with its long duct on posterior surface of septum, the duct entering into posterior ectal third of atrium. A large gland often associated with ectal end of atrium and also with penis pouch; these papillae always invisible externally.

Color: Preserved specimens dark greenish blue dorsally, light greenish ventrally, clitellum fleshy.
Type locality: Anhui (Anqing), Jiangsu (Pukou).
Distribution in China: Jiangsu (Pukou), Anhui (Anqing), Shaanxi (Jiexiu).
Etymology: The name of the species refers to the type locality.
Deposition of types: Previously deposited in the Museum of the Biological Laboratory of the Science Society of China, Nanjing. The types were destroyed during the war in 1937.

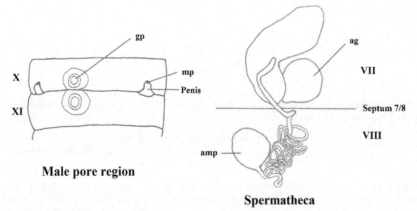

FIG. 321 *Drawida anchingiana* (after Kobayashi, 1936c).

Male pore region

Spermatheca

328. *Drawida bimaculata* Zhong, 1992

Drawida bimaculate Zhong, 1992. *Acta Zootax. Sinica,* 17(3):268–270.
Drawida bimaculate Xu & Xiao, 2011. *Terrestrial Earthworms of China.* 47–48.

External Characters

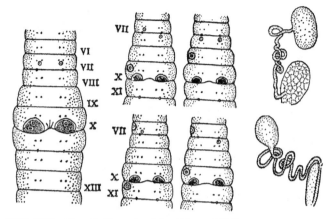

FIG. 322   *Drawida bimaculate* (after Zhong, 1992).

Length 87–119 mm, width 4.5–5.0 mm, segment number 174–212. Two to three annulion each segment in middle region of body. Prostomium prolobous. Dorsal pore absent. Clitellum in X–XIII. Setae 4 pairs per segment, fine, $aa = 0.8–1bc$, $ab = cd$, $aa = 5–6ab$, $dd = 1/2$ circumference. Male pores 1 pair, on X ventrally, slightly ectal of b, near to 10/11 intersegmental furrow. Porophores bun-like, not protruded from the body skin, present a large translucent glandular area. Spermathecal pores 1 pair, in 7/8, in b line. Genital markings, circular, usually on VII, VIII, and X, not elevated from body surface, each with a circular translucent area and a central pore; other markings, about 1 mm in diameter, occurring in some specimens, are elevated from body surface and located on lateral part of body. Female pores 1 pair, minute, in intersegmental furrow 11/12.

Internal Characters

Septa 2/3/4/5 rudimentary, 5/6–8/9 developed, well musculated, 9/10 slightly thin and testis sacs suspended on it, 10/11/12 united as ovarian chamber, 12/13 and succeeding septa membrane. Gizzards 3, in XII–XIV. Lateral hearts 4 pairs, in VI–XI, slightly large. Testis sacs kidney-shaped or ovoid, sometimes long sac-shaped. Vas deferens turns in by septum 9/10, bypasses the last pair of hearts and turns lower laterally, passing through the septa into prostate directly. Prostates pyriform and erect. Spermathecae ampulla elliptical or round, 1.2–1.5 mm long, 1.0 mm broad or 1.0–1.2 mm in diameter. Its duct 4.5–6.0 mm long, conducts the atrium near the ectal end. Spermathecal atrium, finger-shaped, 1.3–1.8 mm long, 0.3 mm in diameter.

Color: Preserved specimens light brownish. Clitellum reddish brown.
Distribution in China: Chongqing (Youyang).
Type locality: Chongqing (Youyang).
Deposition of types: Department of Biology, Sichuan University.
Remarks: This species differs from *Drawida grahami* (Gates, 1935) in having (1) spermathecal pores in 7/8, in b line; (2) porophores not elevated from body surface, with a large, translucent, bun-like glandular area, and (3) prostates pyriform and erect. This species also differs from *Drawida japonica* by the vas deferens which does not pass into the parietes.

329. *Drawida changbaishanensis* Wu & Sun, 1996

Drawida changbaishanensis Wu & Sun, 1996. *Sichauan Journal of Zoology.* 15(3):98–99.
Drawida changbaishanensis Xu & Xiao, 2011. *Terrestrial Earthworms of China.* 49–50.

External Characters

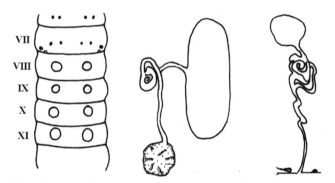

FIG. 323   *Drawida changbaishanensis* (after Wu & Sun, 1996).

Length 110–160 mm, width 3.2–6.1 mm, segment number 120–180. Prostomium prolobous. Clitellum in X–XIII, rarely in IX–XIV, glandular part slightly swollen, less glandulated ventrally. Setae paired, $dd < 1/2$ circumference, $ab = cd$, $aa = 1.0–1.2ab$, $aa = 1.2–1.8bc$. Male pores in intersegmental furrow 10/11, minute, each pore in a transverse slit possessing anterior and posterior elevated swollen lips, when everted, penis visible. Genital papillae usually paired situated on VIII–XI. Spermathecal pores 1 pair, in 7/8, between $b$and c, nearer to c, and 1 smaller papilla on each side of spermathecal pore at posterior edge of VII. Female pores 1 pair in 11/12, minute, not obvious appearances, parallel with ab.

Internal Characters

Septa 5/6–8/9 slightly thickened, 2/3/4/5 faintly visible, 9/10 greatly thickened, 10/11/12 united as ovarian chamber. Gizzards 4, in XIII–XVI, almost equal in size, parietes thick, with longitudinal muscular wrinkles on surface; surrounded by ovaries. Lateral hearts 4 pairs, in VI–IX, posterior 2 pairs slightly large. Testis sac 1 pair well developed, long round-shaped, about 10 mm long, 3–4 mm wide, in IX–XIV; vas deferens comes out through middle-upper part of sac, runs for 3 mm, twisted and coiled, joining to anterior prostate glands of X. Male atrium absent. Penis short and small, always within the male pore. Prostate glands oblate-like, clinging to body wall, mostly in XI, the remainder part extending into X, surface granulated. Spermathecae 1 pair, behind the septum 7/8, ampulla oval- or heart-shaped, always empty inside. Spermathecal duct comes out from ampulla, coiling as loose mass-shaped. The coiling duct is connected by connective tissue, distal end swollen, passing into parietes by spermathecal pore; without atrium, which is usually present in other known species. Accessory glands in VII–XI.

> Color: Preserved specimens black-gray, clitellum white; or red, clitellum ochre-red.
> Distribution in China: Jilin (Mt. Changbai).
> Etymology: The name of the species refers to the type locality.
> Deposition of types: Department of Biology, Hangzhou Normal College.
> Remarks: This species resembles *Drawida papillifer* (Stephenson, 1917), but differs by the genital papillae arranged in pairs in VIII–XI, 1 smaller papilla on each side of spermathecal pore; no male atrium; no spermathecal strium; testis sacs large and 4 gizzards in XIII–XVI.

330. *Drawida cheni* Gates, 1935

Drawida cheni Gates, 1935b. *Lingnan Sci. Journ.* 14(3):446–449.
Drawida cheni Xu & Xiao, 2011. *Terrestrial Earthworms of China.* 50.

External Characters

Length 122–157 mm, width 4.0 mm; or longest 177 mm. segment number 146–198. Prostomium prolobous, Dorsal pore. Clitellum in 1/2 IX–XIV, or to 14/15, or to 1/2 XV. The setae begin on II and are fine; on XX, $aa < bc$. The male pores are at narrowed tips of penes. The penes when fully protruded are dorsoventrally flattened and anteriorly directed, reaching almost to 9/10, pressed dorsally against the ventral parietes, almost triangular in outline with the broad base at 10/11 extending from b almost to c. On each penis there is a small, circular, translucent, grayish genital marking, located usually on the ventral

face but occasionally toward the median margin. In retraction the penes are drawn posteriorly within a chamber that appears to be formed by a posteriorly directed invagination of the anterior most margin of XI reaching in the parietes at least to the transverse setal line of XI. When the fully protruded penis is cut off close to the parietes a transversely oval aperture on the anterior margin of XI just behind 10/11 becomes apparent. The genital markings are small, circular, of a grayish translucence, often protuberant, comparable to the circular portion or the transversely oval markings of *Drawida propatula* (Gates, 1935b). The epidermis in the immediate vicinity of a marking may be slightly tumescent and finely wrinkled. The markings are presetal on VIII, IX, and X, in aa or bc, and postsetal on VII. Spermathecal pores are very tiny, transverse slits in 7/8, in cd or just median to c; The parietes in the immediate vicinity of these slits are slightly swollen due to the presence of parietal glands, the external markings of which may not be readily recognizable. Female pores are on 11/12 in ab.

Internal Characters

Gizzards 3, in XIII–XV. The last pair hearts in IX. On the posterior face of 7/8 but covered over by connective tissue is a pair of dorsoventral commissures. The vas deferens is 5–6 mm in length, twisted into several very short loops; falling ventrally to the floor of the coelom in X, it penetrates into the parietes and passes laterally under a small strand of the longitudinal musculature and then emerges into the coelom and at once passes into the extreme ental end of the prostate. The prostates are 10–12 mm in length or slightly longer if the portion of the prostatic duct within the penis is counted; variously bent or looped. An ental portion 6 mm long is noticeably thicker than the remainder of the gland and mainly as the result of an increase in the thickness of the granular layer. The central body is elongate, slender, and tubular and has, after the granular layer is scraped off, an iridescent appearance. Diagonal muscle fibers inserted in the parietes ventrally and laterally pass across the prostate binding it closely to the body wall. Ectally the prostate passes into a vaguely demarcated, not very conspicuous mass of connective tissue the size of which varies, apparently, with the extent of the protrusion of the penis. Within this mass of tissue the prostatic duct is bent into several short quirks. Also within the connective tissue and median to the prostatic duct is a pear-shaped, tough-walled, grayish-translucent gland with the narrowed portion passing into the penis and to the genital marking thereon. The ovarian chamber appears to be closed off dorsally but laterally cannot be separated from the parietes and is opened when the specimen is pinned out after a middorsal incision. The ovisacs extend into XIII, XIV, or XV. The spermathecal duct is 17–20 mm long, very slightly thickened ectally. The spermathecal atria are 4–5 mm long, club-shaped, narrowed ectally, erect in VIII and flattened against the posterior face of 7/8. The atrial wall is thick and ridged internally. The spermathecal duct passes into the posterior face of the atrium near the ental end, but is continued ectally within the atrial wall for some distance before opening into the atrial lumen. In the first part of its course within the atrium the duct can be easily dissected out, but ectally such dissection becomes more difficult. A parietal gland is usually in close contact with the ectal end of the atrium and this gland sometimes appears to be imbedded within the wall of the atrium but in each case the gland is connected with an external genital marking and can be dissected off from the atrium without opening the latter. The glands of the genital markings are tough-walled, spheroidal to ovoidal, sessile but conspicuously protuberant into the coelom.

Color: Preserved specimens show slight traces of a dark bluish pigmentation. Clitellar coloration dark reddish.
Distribution in China: Jiangsu (Suzhou).
Etymology: This species was named after Professor Chen Yi, who made great contributions to the earthworm systematics of China.
Deposition of types: Smithsonian Institution.
Remarks: *Drawida cheni* (Gates, 1935b) is distinguished from *Drawida gisti* which it most resembles by the larger, flattened, almost triangular penis, the posteriorly directed, parietal, penial chamber, the pear-shaped gland within the penis, the genital marking on the penis, and the absence of an urn-shaped gland in the wall of the spermathecal atrium. In *Drawida cheni* the vas deferens is shorter than the prostate.

The structures in *Drawida cheni* to which the term penis has been applied are not exactly comparable to the penis of *D. hehoensis* (Stephenson, 1924) or *Metaphire abdita* (Gates, 1935b). In both of these species the penis is slender, tubular, and contained with a penial or copulatory chamber which is conspicuously protuberant into the coelom. When the male terminalia are fully everted or protruded the penial or copulatory chamber is represented by a spheroidal or conical protuberance from the body wall on which is seated the penis. In *D. cheni* the penes are not slender and tubular and there is on the protruded terminalia no indication of demarcation of the penis from the everted chamber. The penis of *cheni* is probably more strictly comparable to the penial body produced by the eversion of the copulatory chambers of *Metaphire thecodorsata* (Chen, 1933).

331. *Drawida chenyii* Zhang, Li and Qiu, 2006

Drawida cheni Zhang, Li & Qiu, 2006. *J. Nat. Hist.* 40(7–8):399–401.
Drawida chenyii Xu & Xiao, 2011. *Terrestrial Earthworms of China.* 50–51.

External Characters

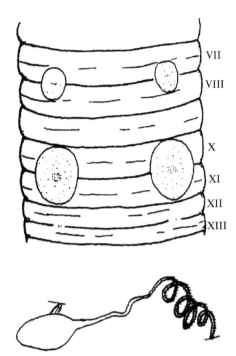

FIG. 324   *Drawida chenii* (after Zhang et al., 2006a).

Body large, smooth, slightly sharp at head and blunt at the caudal part. Length 110–185 mm, width 9–11 mm (X), segment number 165–174, segment XII is distinctly short. Annulets are conspicuous in IV–XXIX. Prostomium prolobous. Dorsal pore absent. Clitellum not observed. Setae invisible externally. Male pores 1 pair in 10/11 intersegmental furrow, 0.4 circumference ventrally apart, slightly swollen, each on the center of a large, longitudinally orientated, flat elliptical whitish glandular membrane patch. Genital papillae absent. Spermathecal pores 1 pair, in 7/8, intersegmental furrow, 0.4 circumference apart ventrally, each on the center of a longitudinally orientated, flat elliptical whitish glandular membrane patch. Genital markings not present. Female pores 1 pair in 12/13, 0.4 circumference apart ventrally.

Internal Characters

Septa 5/6/7/8/9 greatly thickened, muscular. Gizzards 5, in XII–XXII segments (in XII–XIV, XV–XVI, XVII–XVIII, XIX–XX, and XXI–XXII, respectively), same size, shining on the surface with whitish vertical fibers. Intestine enlarged distinctly at XXIV–XXVI, just behind the last gizzard. Esophageal hearts greatly thickened, black in VI–IX. Meganephridia present, beginning at least from VI, about 20–30 mm long, 1 pair in each segment close to the anterior septum. There are a few brownish black dots on the surface of segment anterior to VII. Testis sacs 1 pair, about 6 mm long, 3.5 mm wide, yellowish, each suspended in middle part of septum 9/10. Ovarian chambers are in XI–XII, about 3.5mm, vertical long pouch-shaped or palm-like. Spermathecae 1 pair in VIII, ampulla oval-shaped, yellowish, about 2.5mm long, 1.2mm wide, narrowly attached to the surface of septum 7/8 with a short connective tissue; from its lower side a duct arising, gradually narrowing and making a number of large coils from its median part, about 23mm in total length, spermathecal atrium inconspicuous or absent. Accessory glands absent.

Color: Body with yellowish pigment.
Distribution in China: Guangdong (Mt. Dinghu).
Etymology: This species was named after Professor Chen Yi, who made great contributions to the earthworm systematics of China.
Deposition of types: Laboratory of Environmental Biology, School of Agriculture and Biology, Shanghai Jiaotong University, China.
Habitat: Collected from ravine rainforest on Mt. Dinghu, Guangdong.
Remarks: *Drawida chenyii* (Zhang, Li & Qiu, 2006) is somewhat similar to *Drawida sulcata* (Zhong, 1986) from Yunnan. They share similar characters in their external appearance, large body size, absent dorsal pore and clitellum, oval-shaped testis sac and ampulla, and convoluted duct. However, *Drawida chenyii* is distinguished from *D. sulcata* and other species by the 5 gizzards which otherwise can only be found in *D. syringa* (Chen, 1933), the absent genital papillae, and the inconspicuous thumb-shaped spermathecal atrium, as in *Drawida nemorus* (Kobayashi, 1936c) whose spermathecal duct also terminates without any trace of atrial dilation. Furthermore, *Drawida chenyii* (Zhang, Li and Qiu, 2006) is characterized by its smooth body without setae, the intestine origin immediately behind the last gizzard, the female pore in 12/13, and the superficial male pore covered by a glandular membrane patch, with no atrium or penis present.

332. *Drawida dandongensis* Zhang & Sun, 2014

Drawida dandongensis Zhang & Sun, 2014. *Zoological Systematics*. 39(3):422–444.

External Characters

Body large, slightly sharp at head end and blunt at the caudal part. Length 80–164mm, width 3.5–4.2mm (X), segment number 147–198, without annulets. Prostomium prolobous. Clitellum X–XIV, 4.2–5.9mm wide, wider than other segments. Dorsal pores absent. Setae hardly visible before clitellum, black dots after clitellum, $aa = 1.5$–$2.0bc$, $ab = cd$, $dd = 3/5$ circumference. Male pores 1 pair in 10/11 intersegmental furrow, 1/4 body circumference apart from each other, between setae b and c,

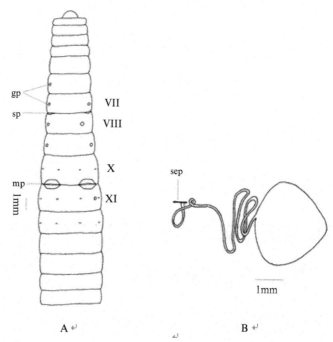

FIG. 325   *Drawida dandongensis* (after Zhang & Sun, 2014).

greatly swollen, each on the center of a large, latitudinally orientated, flat elliptical whitish glandular membrane patch. Spermathecal pores 1 pair in 7/8 intersegmental furrow, 2/5 body circumference apart ventrally, each on the center of a latitudinally orientated, slit-like depression. Female pore inconspicuous. Genital papillae present on VI–XI, sometimes absent, the arrangement seems to be irregular.

Internal Characters

Septa 5/6 – 8/9 comparatively thickened, muscular. Gizzards 4 in XII–XVII, 9mm long, 3.5mm wide, same size, generally shiny on the surface with whitish vertical fibers. Esophageal hearts 4 pairs, each pair the same size, greatly thickened, black in VI–IX. Testis sacs 1 pair, about 6mm long, 3.5mm wide, yellowish, each suspended in middle part of septum 9/10. Spermathecae 1 pair in VIII, spermathecal ampulla fairly large, heart-shaped, yellowish, about 3mm long, 2.5mm wide, narrowly attached to the septum 7/8 with connective tissue; from its lower side, a duct arising and beginning to make a number of large coils, about 15mm in total length, without atrial dilation. Ovarian chambers in XI–XII, about 3mm long, palm-shaped.

Color: Preserved specimens dorsally dark green and shiny with a blue tone present.
Type locality: Mt. Beidongyan, Nantou County, Taiwan.
Deposition of types: Department of Ecology, College of Resources and Environmental Sciences, China Agriculture University, Beijing.

Etymology: The name *"dandongensis"* is given to this species with reference to its type locality of Maokuishan Park, Dandong, China.

Distribution in China: Liaoning (Dandong).

Remarks: This species is somewhat similar to *D. nemora* (Kobayashi, 1936c), which occurs in China and North Korea. They share similar characters of spermathecal pore and spermathecal ampulla. However, this species is distinguished from *D. nemora* by having the spermathecal pore in 7/8 intersegmental furrow, while the spermathecal pore of *D. nemora* is found on the posterior edge of VIII. The spermathecal duct of *D. dandongensis* is longer and has more twists than that of *D. nemora*. Furthermore, this species is dark green on the dorsal surface and shiny with a blue tone present, distinguishing it from *D. Nemora*, which is dorsally dark bluish.

333. *Drawida gisti gisti* Michaelsen, 1931

Drawida gisti Michaelsen, 1931. *Peking Nat. Hist. Bull.* 5 (2):7–11.

Drawida gisti gisti Chen, 1933. *Contr. Biol. Lab. Sci. Soc. China (Zool).* 9:194.

Drawida gisti Gates, 1935a. *Smithsonian Mus. Coll.* iii (3):2–3.

Drawida gisti Kobayashi, 1938b. *Sci. Rep. Tohoku Univ.* 13:95–99.

Drawida gisti Gates, 1939. *Proc. U. S. Natn. Mus.* 85:406–408.

Drawida gisti Kobayashi, 1940. *Sci. Rep. Tohoku Univ.* 15:271–272.

Drawida gisti gisti Chen, 1946. *J. West China. Border Res. Soc. (B).* 16:135.

Drawida gisti gisti Xu & Xiao, 2011. *Terrestrial Earthworms of China.* 51.

External Characters

Length 78–122 mm, width 3.0–4.2 mm or rarely 6.0 mm, segment number 146–198, variable in size. Prostomium prolobous. Dorsal pore, not functional, indistinctly present, thinner and slightly grayish along line of dorsal pore after 2/3 backward. Clitellum in X–XIII or extending into XIV and a part of IX, ventral side of X and XI somewhat medial to male apertures and less glandulated. Setae 4 pairs per segment, $aa = cd$, aa about 1/5 narrower than bc, equal to bc in or before clitellum, greater before VII or VI, aa on II–IV about one-third greater; dd of dorsal distance a little greater than that of ventral distance, about 4/7 circumference apart dorsally. Male pores represented by 1 pair of large transverse slits as secondary apertures in 10/11, between bc, nearer to c, inconspicuous or marked by surrounding swollen skin, enclosing a fairly deep penis pouch which is filled up

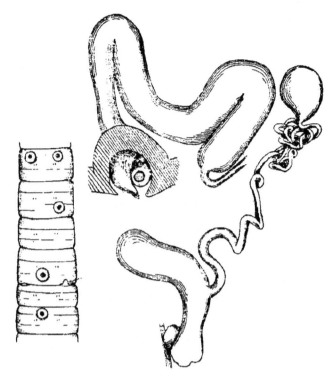

FIG. 326 *Drawida gisti gisti* (after Chen, 1933).

by a penis; the latter pointed distally and rounded at base, conical or elongate, with its free end thinner and anteroposteriorly compressed, largest about 1 mm in total length and 0.6 mm in width, with a small primary pore situated at its distal end; sometimes wholly concealed or slightly visible externally when in retracted condition, or sometimes wholly everted. Spermathecal pores 1 pair, in intersegmental furrow 7/8, each sunken down as a chamber, frequently indicated by anterior and posterior swollen lips, between cd, the pore large at bottom of the depression; small papillae sometimes found anterior to, or sometimes posterior to, each pore. Female pores 1 pair, invisible or merely visible, in intersegmental furrow 11/12, between ab, or slightly closer to b, no glandular skin around each pore.

Internal Characters

Septa 5/6 thick, 6/7/8/9 greatly thickened, the rest thin and membranous; 9/10/11 greatly pushed backward on dorsal side but slightly ventrally. Lateral hearts 4 pairs, in VI–XI. Nephridia similar to other species of the genus, not present in first 2 and last few segments. Gizzards 3, in XII–XIV, very strong. Testis sacs 1 pair, moderately large, ovoid or elongate kidney-shaped, suspended on septum 9/10. Maleatrium usually greatly elongated, cylindrical, with warty but solid surface, ental end usually larger and roundish, irregularly placed under intestine or ovarian chamber. Egg sacs long and pointed posteriorly pushed back from septum 11/12 on

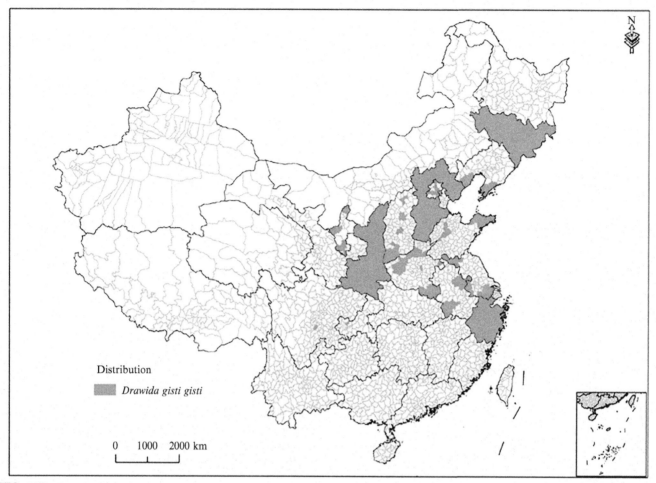

FIG. 327   Distribution in China of *Drawida gisti gisti*.

both sides of gizzards to XVI or XVII. Spermathecae 1 pair, with its atrium wholly lying upon posterior surface of septum 7/8, each with a thin-walled ampulla attached to septum, often empty inside, from its lower side arising a duct not quite sharply marked off, gradually narrowing and making a number of large coils at its proximal end, less so at its distal two-thirds. Spermathecal atrium elongate, cylindrical, pointed upward and remaining free in VIII.

Color: Preserved specimens greenish pale, grayish buff, or ashy gray on dorsum, always darker posteriorly and lighter or greenish at anterior end, pale ventrally; clitellum pinkish.
Type locality: Hebei, Shandong (Jinan), Beijing.
Distribution in China: Beijing (Dongcheng, Haidian, Tongzhou, Changping, Shunyi, Huairou, Fengtai, Miyun, Shijingshan), Hebei, Liaoning (Huludao, Dalian), Jilin, Jiangsu (Nanjing, Xuzhou, Pukou, Suzhou, Wuxi, Yixing), Zhejiang, Anhui (Anqing, Chuzhou), Shandong (Jinan, Yantai, Weihai), Shaanxi (Yuncheng, Wanrong, Jishan, Xinjiang, Houma, Quwo, Yicheng, Fushan, Linfen, Jiexiu, Pingyao, Qixian, Yangqu, Xinzhou, Yuanping, Shanyin,

Huairen), Henan (Xinxiang, jiaozuo, Boai, Wenxian, Luoyang, Xiayi, Xishan, Yanling, Xinyang), Chongqing (Beipei), Sichuan (Mt. Emei), Shanxi, Ningxia (Shzuishan, Zhomgwei, Guyuan, Xiji).
Deposition of types: Hamburg Museum.
Remarks: The species is probably widely distributed. Of course, it is not as abundant as *Drawida japonica* (Michaelsen, 1892). It seems to be a highly variable species judged from the arrangement of setae, the male and spermathecal atria, and other characters.

334. *Drawida gisti nanchangiana* Chen, 1933

Drawida gisti nanchangiana Chen, 1933. *Contr. Biol. Lab. Sci. Soc. China (Zool).* 9:200.
Drawida gisti nanchangiana Xu & Xiao, 2011. *Terrestrial Earthworms of China.* 52.

External Characters

Length 55–98 mm, width 2.2–3.0 mm, segment number 122–136. Clitellum in X–XIII, not marked. Setae 4 pairs per segment, slender and long, very prominent, each pair widely spaced, their exposed portion about one-third of segmental length; $aa = cd$, $aa = bc$, $aa = 3\ ab$ or

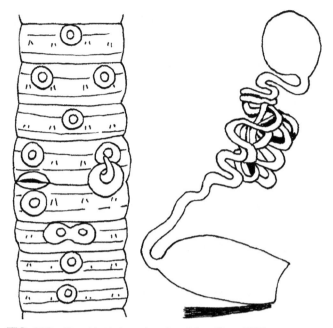

FIG. 328 *Drawida gisti nanchangiana* (after Chen, 1933).

Color: Preserved specimens generally unpigmented, or with irregular grayish patches, grayish dark posteriorly.

Type locality: Jiangxi (Nanchang, Jiujiang).

Distribution in China: Shandong (Yantai, Weihai), Jiangxi (Nanchang, Jiujiang).

Etymology: The subspecific name refers to the type locality.

Deposition of types: Previously deposited in the Museum of the Biological Laboratory of the Science Society of China, Nanjing. The types were destroyed by the war in 1937.

Remarks: *Drawida gisti nanchangiana* (Chen, 1933) is similar to *Drawida bahamensis* (Beddard, 1892). It is somewhat identical with that species in most points, however, Beddard's figures are concerned with 2 points, viz., lack of genital papillae as not shown in his figure, and smaller egg-shaped male atrium. In addition, the straighter spermathecal duct and smaller atrium as judged from his drawings are also slightly different. These differences may be due to the fact that the drawings are too diagrammatic and probably drawn from the reconstruction of his section-cutting specimen.

4*ab*; *dd* of dorsal distance about 3/5 or 5/8 circumference; setae on II and III shorter, those on II indistinct or absent. Male pores 1 pair; penis a little pointed distally, large at base, situated in a deeply sunken penis pouch, its length about half depth of pouch, or everted with a large circular base between *bc*, about one-third nearer to *c*. Genital papillae larger and more numerous, rarely absent, in VII–XIII or XIV, paired on segments near male and spermathecal apertures, sometimes about 4–5 on 1 segment.

Internal Characters

Gizzards 3, in XII–XV. Testis sacs usually very large, each with its larger part in X, ovoid, its duct taking a usual course, much convoluted in front of septum 9/10, passing around the last heart and penetrating through the septum into X to join proximal end of male atrium; Atrium slender, short, club-shaped, or enlarged entally, pear-shaped, about 2.0–2.5 mm long, its distal third slender as a duct, joining an internally raised penis pouch, a pear-shaped gland usually associated with the latter. Ovarian chamber formed by septa 10/11 and 11/12 which do not meet dorsally, that is, 10/11 on posterior edge of XI, 11/12 on middle of XII, about half a segmental length intervening; ovaries rosette at anteromedial corner of chamber; egg sacs slender, short, reaching about 3 segments posteriorly. Spermathecae ampulla round, small, with a much convoluted duct entering into proximal end or side of a short thumb-shaped atrium. Atrium often minute, bulb-like embedded in body wall, rarely elongate like other varieties.

335. *Drawida glabella* Chen, 1938

Drawida glabella Chen, 1938. C*ontr. Biol. Lab. Sci. Soc. China (Zool).* 12(10):377–379.
Drawida glabella Xu & Xiao, 2011. *Terrestrial Earthworms of China.* 53.

External Characters

Length 52 mm, width 2.0 mm, segment number 120. First dorsal pore probably in 6/7, those before clitellum hardly visible, behind evident. Clitellum in 1/2 IX–XIV, segment XIV and posterior portion of IX, also ventral side of X–XIII between ventral setae slightly glandular, swollen and glandular on dorsal and lateral sides. Lateral setae on X–XIII invisible. Setae 4 pairs per segment, closely paired; *aa* = *cd*; a few segments just before clitellum *aa* = 6*ab*, *bc* slightly wider than *aa*, *dd* of dorsal distance wider and about 2/3 circumference apart, behind clitellum *aa* = 9*ab*, *bc* = 8*ab*, *dd* about 4/9 circumference on dorsal side. Male pores 1 pair in 10/11, between *bc*, closer to *b*, about 1/4 circumference apart ventrally, each as transverse slit, about 0.07 mm in width, anterior and posterior lips inconspicuously swollen, posterior larger, neither genital papillae nor other markings found in this region. Spermathecal pores 1 pair, in 7/8, in line with setae *cd*, its neighborhood neither swollen nor glandular; genital markings absent. Female pores in 11/12, intersegmental, slightly lateral to male pores.

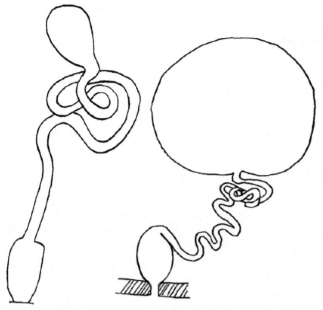

FIG. 329   *Drawida glabella* (after Chen, 1938).

Internal Characters

Septa 5/6–8/9 thicker (especially 6/7/8/9), 9/10 thin, 11/12/13 also thin, enclosing an ovarian chamber. Gizzards fairly muscular, shining on surface, separable into 4, about equal in size, in XII–XV. Lateral hearts large, in VI–IX. Testis sacs 1 pair, large, about 1 mm in diameter, suspended on septum 9/10, with a small portion projected out in front of the septum. Vas deferens with a few coils in front of septum and rather straight behind, joining ental end of an atrium. Male atrium roundish, about 0.5 mm long, rather smooth on surface, glandular portion probably enclosed by peritoneal sheet (not to be sectioned). Ovarian chamber in XI, formed by septa 10/11 and 11/12, these nearly meeting on dorsal side but widely separated on ventral side to enclose a large chamber. Ovisacs very large and long, extended posteriorly to segment XXI, constricted around every septum. No terminal appendages. Spermathecae attached to posterior side of septum 7/8, extending nearly to dorsal side; ampulla ovoid, about 0.16 mm in width, flattened, duct weakly curved, about 1.75 mm long entering ental end of spermathecal atrium, the latter thumb-shaped, about 0.2 mm in length, situated closely behind septum. Accessory gland absent.

Color: Preserved specimens pale throughout. Clitellum brownish.
Distribution in China: Hainan (Sha-mom-chiu).
Etymology: The subspecific name "glabella" refers to the smooth surface of the frontal bone lying between the superciliary arches. It originates from the word "glaber" which means smooth, in this species indicating its smooth surface without genital markings.

Deposition of types: Previously deposited in the Museum of the Biological Laboratory of the Science Society of China, Nanjing. The types were destroyed during the war in 1937.

Remarks: Only a single specimen was collected. The species is characterized by (1) its small size; (2) having no genital markings; (3) both spermathecal and male pores simple in character; and (4) male atrium small and not glandular in appearance.

336. Drawida grahami Gates, 1935

Drawida grahami Gates, 1935a. *Smithsonian Mus. Coll.* 93(3):3.
Drawida grahami Chen, 1936. *Contr. Biol. Lab. Sci. Soc. China (Zool).* 11:291–292.
Drawida grahami Gates, 1939. *Proc. U. S. Natn. Mus.* 85:408–411.
Drawida grahami Chen, 1946. *J. West China. Border Res. Soc. (B).* 16:134–135,142.
Drawida grahami Xu & Xiao, 2011. *Terrestrial Earthworms of China.* 53–54.

External Characters

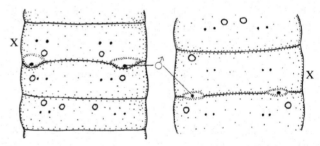

FIG. 330   *Drawida graham* (after Chen, 1946).

Length 40–55 mm, width 2.5–4.0 mm, segment number 88–123. Clitellum in X–XIII segments, also extending neighboring to IX and XIV; glandular. Setae short, *aa* about equal to or slightly greater than *bc*, *ab* = *cd*. Male pores in *bc*, nearer to *b* than *c*, on small porophores seated on 10/11, intersegmental furrow 10/11 ending blindly against the lateral and median sides of the porophores, the anterior margin of the porophore may be indicated by a slight transverse groove, which does not pass into the intersegmental furrow, the posterior margin of the porophore is marked off by a short transverse furrow that passes laterally and mesially into 10/11. Genital markings on VII–XIII, each marking with a firm, rounded gland projecting through the parietes into the coelom. Spermathecal pores 1 pair, in intersegmental furrow 7/8, midway between *b* and *c*. The female pore was not identified.

Internal Characters

Septa 5/6–8/9 thickened, 9/10 thin and displaced posteriorly. Gizzards 3 in XII–XIV. The last pair of hearts in IX. There is a band of opaque material on each side of the dorsal blood vessel. Testis sacs are usually flattened laterally and nearly fill the available space in IX and X. Vas deferens is short, rather thick relative to the size of the worm, and passes into the prostate mesially without first passing into the parietes. Prostates are flattened discs of circular outline, sessile on the parietes, central body tiny, ovoidal, the more pointed end within the parietes. Segment XI is reduced to a horseshoe-shaped and closed-off ovarian chamber. The ovisacs are laterally flattened and confined to XII in the clitellate specimen. In other worms, the ovisacs are more slender and also confined to XII. Spermathecae atrium finger-shaped, erect on the posterior face of 7/8. The spermathecal duct (7–8 mm long) passes into the atrium near the ental end of the latter but runs ventrally in the atrial wall for a short distance before opening into the atrial lumen.

Color: Preserved specimens clitellum reddish.
Type locality: Chongqing (Xufu).
Distribution in China: Chongqing (Beipei, Nanchuan), Sichuan (Yibin, Mt. Emei)
Deposition of types: Smithsonian Institution.
Remarks: *Drawida grahami* (Gates, 1935a) is distinguished from *Drawida japonica japonica* (Michaelsen, 1892) as follows: Location of the spermathecal pores in mid-*bc* rather in or just median to *c*; direct entrance of the vas deferens into the prostate (rather than first passing into the parietes); prostates disc-shaped and sessile on the parietes (rather than erect or vertical and columnar to club-shaped); the very small, ovoidal, central body of the prostate with pointed end buried in the parietes (rather than the elongate digitiform central body nearly 1 mm in length); absence of an elongate rod-like appendix on the ovisacs. The exact morphological location of the male pores was not determined, but the pores are in line with 10/11, though the latter is not recognizable across the male porophores. If the male pores are to be placed on 10/11 or the site of 10/11 this will be further distinction from *japonica*, in which the pores are quite definitely segmental, postsetal on X.

337. *Drawida japonica japonica* Michaelsen, 1892

Drawida japonica Michaelsen, 1931. *Peking Nat. Hist. Bull.* 5(2):7.
Drawida japonica Chen, 1933. *Contr. Biol. Lab. Sci. Soc. China (Zool).* 9:189.
Drawida japonica Gates, 1935a. *Smithsonian Mus. Coll.* 93 (3):3–4.

Drawida japonica Gates, 1939. *Proc. U. S. Natn. Mus.* 85:411–413.
Drawida japonica Kobayashi, 1940. *Sci. Rep. Tohoku Univ.* 15:263–264.
Drawida japonica Chen, 1946. *J. West China. Border Res. Soc. (B).* 16:135.
Drawida japonica Chang, Shen & Chen, 2009. *Earthworm Fauna of Taiwan.* 158.
Drawida japonica Xu & Xiao, 2011. *Terrestrial Earthworms of China.* 54–55.

External Characters

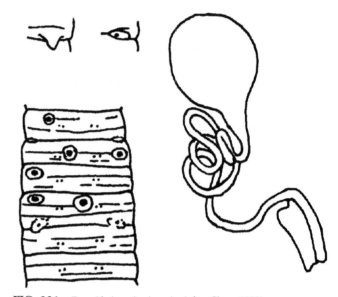

FIG. 331   *Drawida japonica japonica* (after Chen, 1933).

Length 30–200 mm, width 2.0–5.5 mm, segment number 165–195, variable in size. Prostomium prolobous. Dorsal pore absent. Clitellum in X–XIII, well marked, not glandular on ventral side of X and XI. Setae small and closely paired, 4 pairs per segment; $aa = cd$, $aa = 5$ or 6 ab; aa about one-third wider than bc in II–V, equal to or slightly wider than bc in VI–XI, equal to or slightly narrower than bc behind XI; dd of dorsal distance slightly greater than or equal to half of body circumference. Male pores on median side of a transversely raised ovoid tubercle between bc, projecting out from posterior edge of segment X; each pore as a transverse slit, often visible slightly on median side of a raised and transversely ovoid tubercle between setae bc, about 1/3 nearer to b. Spermathecal pores 1 pair, in intersegmental furrow 7/8, near c. Genital papillae absent or a few only, irregularly placed on VII–XIII, more frequently present in male and spermathecal regions, often pairs on VII, VIII, and X, unpaired on IX, XII, and XIII, that on XIII rarely present. Female pores paired in intersegmental furrow 11/12, in line with b.

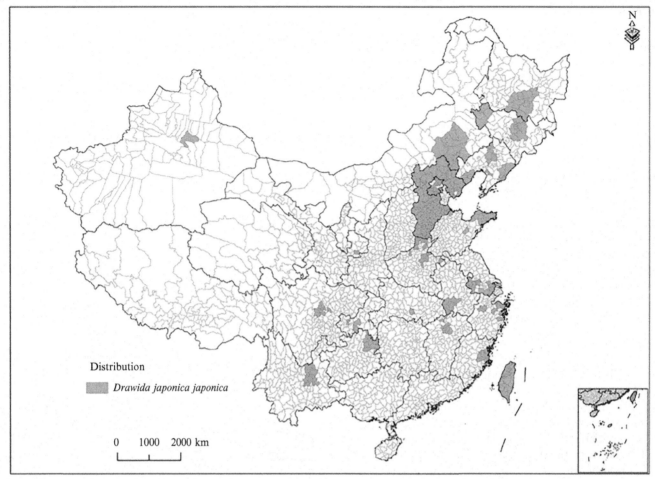

FIG. 332 Distribution in China of *Drawida japonica japonica*.

Internal Characters

Septa 5/6–8/9 thickened, 10/11/12 meeting dorsally to form ovarian chamber, Hearts 4 pairs, in VI–XI. Nephridia meganephridial. Gizzards 2–3, in XII–XIV. Testis sacs 1 pair, large, suspended on septum 9/10, sperm duct greatly coiled in front of the septum and in segment X. Atrium thumb-shaped, stouter at ental third, narrower ectally, or cylindrical throughout. Egg sacs long in XII–XV or XIX. Spermathecae attaching to posterior face of septum 7/8, ampulla oval, its duct loosely or compactly coiled, joining ental side or sometimes middle part of spermathecal atrium, which is slender and elongate, situated posteriorly to septum.

Color: Live specimens variable, light bluish gray or greenish violet dorsal side, grayish pale on ventral side, fleshy on clitellum. Preserved specimens unpigmented, with a very light grayish blue along dorsal side in young forms; deep grape or bluish violet dorsally, slight bluish gray toward dorsolateral side, light grayish pale on ventral side, bluish anteriorly; clitellum purplish red or fleshy.
Type locality: Japan.

Distribution in China: Beijing (Dongcheng, Haidian, Tongzhou, Changping, Huairou, Daxing, Fengtai, Miyun), Hebei, Liaoning (Shenyang, Dashiqiao, Huludao, Dandong), Jilin (Jilin, Tumen, Baicheng), Heilongjiang (Harbin), Jiangsu (Nanjing, Zhenjiang, Suzhou, Yixing), Zhejiang (Linhai, Ningbo, Chuanshan, Shengxian, Tonglu, Xindeng), Anhui (Anqing), Fujian (Fuzhou), Taiwan, Shandong (Yantai, Weihai), Jiangxi (Nanchang, Jiujiang, De'an), Henan (Xinxiang, Huixian, Anyang, Boai, Wenxian, Kaifeng, Yanling, Xiayi), Hubei (Qianjiang), Chongqing (Shapingba, Beipei, Nanchuan, Peiling), Sichuan (Chengdu), Yunnan (Kunming), Guizhou (Mt. Fanjing), Neimenggu (Chifeng), Gansu (Jingchuan), Ningxia (Haiyuan, Guyuan, Zhongning), Xinjiang (Wurumuqi).

Etymology: The name of the species refers to the type locality of Japan.
Deposition of types: Zoologisches Institut und Zoologisches Museum, Universitat Hamburg, Germany.

Remarks: The specimens are small (about 28 mm long, 3 mm wide in the case of the former, not longer than 60 mm in the latter) or large, some reaching about 200 mm in length.

338. *Drawida japonica siemenensis* Michaelsen, 1910

Drawida japonica siemenensis Michaelsen, 1931. *Peking Natural History Bulletin.* 5(2):7
Drawida japonica siemenensis Gates, 1939. *Proc. U. S. Natn. Mus.* 85:411–413.

External Characters

The genital marking and the male porophores are different from those of *Drawida japonica japonica*. The clitellar glandularity appears to be only partially developed.

A very questionable variety and perhaps only the result of a malformation.

Type locality: Fujian (Fuzhou).
Deposition of types: Hamburg Museum.
Remarks: The form *siemsseni* was erected on the basis of a single specimen that was distinguished from f. *typica* by the greater length, greater thickness, greater number of segments (said to be "sehr ungenau"), and "ungefhr" 6 gizzards. The internal organs were removed in the course of the original dissection and have been lost.

The type of f. *siemsseni* is quite clearly specifically distinct from *D. japonica*, but the species cannot be adequately characterized in the absence of the internal organs.

339. *Drawida jeholensis* Kobayashi, 1940

Drawida jeholensis Kobayashi, 1940. *Sci. Rep. Tohoku Univ.*, 15:268–271.
Drawida jeholensis Zhong & Qiu. 1992. *Guizhou Science.* 10(4):39.
Drawida jeholensis Xu & Xiao, 2011. *Terrestrial Earthworms of China.* 55–56.

External Characters

Length 52–66 mm, width 2.8–3.5 mm, segment number 153–160. The region of the preclitellar segments is somewhat conical in shape. Prostomium prolobous. Dorsal pore absent, epidermis along middorsal line not thin. Clitellum in IX–XIV, slightly swollen and slightly glandulated; glandularity of XIV is less developed. In most of the specimens, the ventrolateral parts of both segments X and XI are also slightly glandulated and projects more or less ventralwards. Setae beginning on II, closely paired, 4 pairs per segment; small and not conspicuously projected, apparently a little stouter anteriorly than

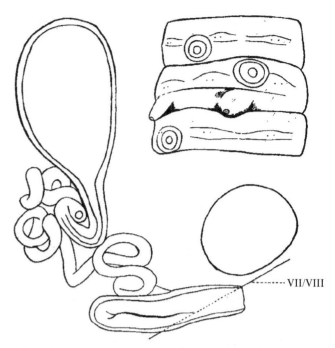

FIG. 333   *Drawida jeholensis* (after Kobayashi, 1940).

posteriorly; setal distance *ab* is equal to *cd; aa* is slightly larger than *bc* in the preclitellar region, but is smaller than the latter behind the clitellum; *dd* is subequal to, or slightly smaller than, 1/2 of the circumference. Male pores are minute, in the form of transverse slits, seated on the top of somewhat conical or mamma-like porophore with oval base; the base of the porophore is about 0.7–0.8 mm in diameter and occupies about the posterior 1/3–2/5 of X. The posterior margin of the porophore is always placed on the anterior most part of XI beyond the intersegmental furrow 10/11, and the pore is found in the position corresponding to 10/11, between *b* and *c* but nearer to *b*. No trace of penis is found. In the anterior side to the male porophores, the posterior furrow of the middle annulus of X (X with 3 annuli) is so deeply grooved, that it appears to be an intersegmental furrow. Spermathecal pores are minute, forming longitudinal slits on the posterior margin of VII, just medial to *c*; the region around the pore is slightly depressed in a crescent-shaped. Each genital papilla is spherical, forming a small and dully glistening tubercle surrounded by a whitish, smooth, thick, and slightly elevated glandular rim of the body skin. The papillae are found on VII–XI presetally or postsetally, between *b* and *c* or medial to *b* or midventrally; in most specimens, a single, small spermathecal papilla is found within each spermathecal depression, to be just lateral to the aperture. Female pores are also in the form of transverse slits, on anterior most edge of XII, in *ab* or *b* or just lateral to *b*, minute and not readily recognizable; the location of the pores can, however, be determined indirectly by the grayish translucent appearance of the epidermis surrounding the apertures.

Internal Characters

Septa 5/6–8/9 moderately thickened, the rest thin; 10/11 is dorsally displaced posteriorly into the anterior part of XII and ventrally into about 1/2 XI. Gizzards 2 or 3, in XII and XIII or in XI–XIII; large and globular in shape; the first one is, in most cases, well musculated even when 3 present, and slightly pushes anteriorwards the testis sacs. Last hearts in IX. Testis sacs 1 pair, of moderate size; each rather deeply constricted by 9/10, usually lying its smaller part in IX, 2–3.5 mm long and 1.3–2 mm broad. Sperm duct is short with a few loose loops and passes into parietes just medial to prostate. Prostates 1 pair, each rather small, thumb-shaped, usually slightly compressed, 1–2 mm long, narrowed ectally, slightly warty on surface. Central body slender and tubular and somewhat finger-shaped, about 0.6–1 mm long. Ovarian chamber in XI, formed by 10/11 and 11/12; both septa are entirely fused dorsally but widely separated ventrally. Ovisacs are slender and rod-like, and extend posteriorly into about XVI–XX. The ovisacs may possibly become finger-shaped when they attain their maturity and are filled with ripe ova. Spermathecae with atria found on the posterior face of 7/8. Ampulla ovoidal or rounded, about 0.7–1.2 mm in diameter. Spermathecal duct thin, moderate in length, about 5–7 mm long, looped almost in its entire length, and not sharply marked off from the ampulla; it passes into the lateral side of atrium at about ental third, and its lumen is fused with the ectal end of the latter. Atrium is columnar, nearly equal in length to the diameter of ampulla. Each accessory gland is tough-walled, ball-like in shape and with a very short duct. The gland corresponding to the spermathecal papilla is usually found just in front of the septum 7/8.

Color: Preserved specimens uniformly whitish gray.
Etymology: The name of the species refers to the type locality of Jehol, previously known as a former province of NE China, north of the Great Wall, divided between Hebei, Liaoning, and Inner Mongolia in 1955. Distribution in China: Neimenggu (Chifeng), Ningxia (Zhongning, Xiji).
Remarks: In its external appearance the present species closely resembles D. koreana. In its internal feature it closely resembles D. koreana, D. japonica japonica (Michaelsen, 1892), and D. propatula (Gates, 1935a). It is easily distinguishable from the last 2 mentioned by the large and rather conical male porophore and by its general body-shape. It is easily distinguishable from the first, which it most closely resembles, mainly by the distinctly longer and thinner spermathecal duct and also by the coloration of the body.

340. *Drawida koreana* Kobayashi, 1938

Drawida koreana Kobayashi, 1938. *Sci. Rep. Tohoku Imp. Univ.* 13(2):102–107.
Drawida koreana Kobayashi, 1940. *Sci. Rep. Tohoku Univ.* 15:268.
Drawida koreana Zhong & Qiu 1992. *Guizhou Science.* 10(4):39.
Drawida koreana Xu & Xiao, 2011. *Terrestrial Earthworms of China.* 56–57.

External Characters

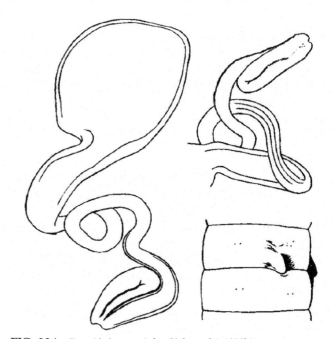

FIG. 334   *Drawida koreana* (after Kobayashi, 1938b).

Length 63–100 mm, width 3–4 mm, segment number 130–186. In shape of the body, I–IX rather cylindrical. Prostomium prolobous. Dorsal pore absent, and middorsal line not thinner than the rest. Clitellum in X–XIII, not distinctly swollen from the neighboring segments; but clearly distinguishable from the rest in glandulation and coloration. The ventral side of it less glandulated, and the midventral portion is rather nonglandulated and tends to become somewhat flattened, especially on X and XI, and besides, the medial sides of the male porophores are slightly concave. Setae 4 pairs per segment, short and closely paired; those on II and on clitella segments may be slightly smaller than the rest; setal distance $ab=cd$; anteriorly to about VI aa larger than bc, on about VII–XV aa nearly equal to bc, behind this region, especially behind XX aa much smaller than bc; dd nearly equal to, or a very little larger than 1/2 circumference. Just medial to c-lines and also just lateral to b-lines, the skin behind clitellar region thinner, and each appears to be a distinct pale line. Male pores 1 pair, each on top of a moderately sized nipple-like or sometimes conical poro-phore, which is always formed by markedly elevated glandular skin near the posterior about 1/3–1/2 X, and situated between *b* and *c*, much nearer to *b*. This poro-phore is always projects slightly beyond 10/11 into ante-rior part of XI; anterior border of XI facing each porophore may be also slightly elevated. Spermathecal pores 1 pair, in 7/8, in *c*-line or just medial to *c*; each on a small tip which projects from the posterior edge of the preceding segment and is moderately sunken into intersegmental furrow; mouth of this depression may be largely opened and its vicinity may be glandulated in mature specimens. Genital papillae resemble in shape those of *Drawida japonica*, but are much fewer in number, and not found infrequently; 1–4 in number if present, and irregularly placed on VII–XII, most frequently on VIII–X unpairly, and midventrally or in *ab*-line or between *b* and *c*. Female pores 1 pair, minute, slit-like, scarcely visible in line with *b*, on quite anteriormost border of XII; rather easily visible in macerated specimens.

Internal Characters

Septa 5/6 and 8/9 moderately, 6/7/8 much thickened and somewhat musculated, the rest thin; all septa expect 10/11 almost normally inserted. Dorsal part of septum 10/11 displaced posteriorly to meet dorsolaterally with 11/12 in intersegmental space of 11/12. Gizzards 3, each somewhat pot-shaped or barrel-shaped, small but strongly musculated and shining, in XII–XIV; when 3 giz-zards are present, the middle is usually slightly larger the others. Testis sacs 1 pair, moderate in size, each sus-pended by 9/10 with the larger half lying in X. Its lateral face is usually constricted moderately by the septum, so it appears to be kidney-shaped. Those in mature specimens are massive, yellowish white, and granular on surface. Sperm duct short, loosely twisted with its larger part placed in IX, and piercing 9/10 into X, and entering into parietal wall just at anteromedial side of ectal end of the prostate. Prostate thick and short, about 2 mm long and 1 mm wide, usually compressed and weakly constricted at its ventral side in about middle portion, somewhat resembling testis sac in shape, but rather thumb-shaped in immature and semimature specimens. Ovarian cham-ber in XI, formed by 10/11/12, both septa entirely fused dorsolaterally in intersegmental space of 11/12, but sep-arated by about 1 segment ventrally, also closed around esophagus in an inverted U shape. Ovaries fairly large, much tufted. Ovisacs situated dorsolaterally on alimen-tary tract, long and slender, moniliformed or distinctly constricted by each septum through its course, extending posteriorly into about XVIII, seldom behind this into XXII or XXIII and in front of this only into XIV or XV. Sper-mathecae 1 pair, with its atrium wholly lying upon posterior face of 7/8. Ampulla thin-walled, spherical (in immature specimens usually ovoid and empty); duct very short but very thick and slightly shining, not making a mass of coils in any parts, but loosely twisted, its ental portion much thicker than the rest, about 2–3 times of the latter. This much thicker and shorter portion resembles the ampulla duct and the following thinner and longer portion a diverticulum, in a spermatheca of the genus *Pheretima*. It is not sharply marked off from the ampulla; ectally joining with ental and dorsal surface of the atrium, but its lumen not fusing with that of the atrium until near the ectal end of the latter. Atrium small and short, sac-like, always shorter than diameter of the ampulla, its ental end rounded but not enlarged; when it is separated off from parietes, spermathecal opening may appear as a minute aperture at its ectal end. Accessory glands elastic, thickly discoidal, always found projecting through parietes into coelom, corresponding to the exter-nal genital papillae; similar ones may be found internally on a few occasions even when papillae are not found externally.

Color: Live specimens dark blue or dark reddish blue dorsally, lighter ventrally, clitellum reddish or pinkish. Distribution in China: Liaoning (Anshan, Shenyang). Etymology: The name of the species refers to the type locality. Remarks: This species closely resembles *D. japonica japonica* (Michaelsen, 1892) in many respects, but differs considerably in (1) general form of the body, (2) aspects of male porophore, and (3) length and thickness of spermathecal duct.

The region of I–IX is longer than that of *D. japonica japonica*; the shape of this region is cylindrical in the for-mer and is rather shortly ovoid, being weakly constricted in front of clitellum in the latter; anteroposterior length of each segment of this region is always, at least in equal-sized worms, longer than that of the latter.

Situation of male pores resembles each other. How-ever, the male porophore in this species is much larger and always somewhat nipple-like or rather penis-like in shape, while it is only a small oval tumescence in the lat-ter. Proximal position of the porophore is also different; in the former it is at about the posterior 1/3 or often almost 1/2 of X, while in the latter it is usually at about the posterior 1/5. Length of spermathecal duct shows slight variation in both species.

This species also resembles *D. propatula* (Gates, 1935a), but differs mainly in body size, aspects of male poro-phores, and length and thickness of spermathecal duct. The differences between this species and both *D. propatula* or *D. japonica japonica* appear to be more dis-tinct than those between the latter 2.

341. *Drawida linhaiensis* Chen, 1933

Drawida linhaiensis Chen, 1933. *Contr. Biol. Lab. Sci. Soc. China (Zool).* 9:211.
Drawida linhaiensis Gates, 1935a. *Smithsonian Mus. Coll.* 93(3):4.
Drawida linhaiensis Xu & Xiao, 2011. *Terrestrial Earthworms of China.* 58–59.

External Characters

Length 37–58mm, width 3–4mm, segment number 138–144. A short and robust worm, with its anterior end short and blunt; anterior half large and cylindrical. Prostomium prolobous. Dorsal pore absent, indistinctly thinner in pore spaces. Clitellum in X–XIII, these segments longer and more thickly glandular with a lighter color; less glandular ventrally on XII and XIII and also between male apertures; or traceable. Setae 4 pairs per segment, moderate in size, but not very conspicuous under a magnifier, *aa*=*cd*, *aa*=9 or 10*ab*, *aa* equal to *bc* or slightly narrower than *bc* by a distance of *ab* or 1/2*ab*; *dd* completely visible if viewed from ventromedial aspect, that of dorsal distance about 7/11–11/18 circumference. Male pores 1 pair of transverse slits in furrow 10/11, with a prominent penis, always partly protruding in preserved specimens, almost fully occupying penis pouch. Penis large and transversely ovoid, gradually rounded (never pointed) distally, as a round tubercle-like protuberance, situated ventrally to seta *d*; a small pore appearing on anterior side of its top. Spermathecal pores as 1 pair of large slits directed ventrally, in line or lateral to seta *c*. Genital papillae symmetrically paired, on ventral side of segments XIII–X, constantly present in specimens examined. 2 papillae on ventral side of VIII, in front of, but lateral to, seta *b*, or between *bc*, closer to *b*. Each papilla with a round glandular portion, smooth and slaty gray in color, its diameter about a distance of 2–3*ab*, often deeply sunken down. Female pores 1 pair, as very minute slit-like openings, in intersegmental furrow 11/12, close to anterior edge of XII, in line, or slightly lateral to seta *b*.

Internal Characters

Septa 5/6–8/9 well-musculated and very thick, in normal attachment to their respective segments, 9/10 thin attached dorsally to posterior third of X, 10/11 to intersegmental space of XI/XII, 11/12 to middle part of XII, both not meeting dorsally to form ovarian chamber. Hearts 4 pairs, in VI–XI. Nephridia meganephridial. Gizzards generally 3, in XIII–XVI, or 4 whereas the first smaller and less muscular, the posterior 3 thick-walled and very large in size, usually with longitudinal minute ridges. Intestine large, yellowish or fuscous in color. Testis sacs 1 pair, large, round or elongate with its midventral part constricted and suspended on septum 9/10. Testis rosette, near anterior side of funnel. Sperm duct taking usual course with more coils in front of septum 9/10, straight after piercing through the septum and passing from base of atrium to its ental end to join the atrial duct at its ental lateral side. Male atrium usually very large and long, irregularly placed under intestine and ovarian chamber, its ental 1/4 enlarged and rounded with thick glandular wall, its ectal 1/4 much slender and without glandular wall serving apparently as a duct. The large-sized penis all muscular in structure with its nonglandular epithelium continuous to inner surface of penis pouch. Ovaries very conspicuous, as long posteriorly directed lobules, almost occupying the whole spacious chamber. Egg sacs usual in shape and position, about 3mm in length, often distended slightly with their contents. Spermathecal ampulla and ampullar duct on posterior surface of septum 7/8, but atrium in front of septum. Ampulla small about 0.2mm in diameter, or sometimes a little larger than 0.8mm, situated to the middle part of intestine. Its duct slender and much coiled at beginning and loosely coiled at end; first part slender, distal part joining to lateral side of ental half of atrium, but remaining separate from lumen of atrium almost to the opening. Atrium usually small, bulb-like, its ental end large and round with its surface sometimes uneven, yellowish in color, its wall thick but musculated, its lumen very large with irregular and centrally projected ridges. Its ectal half

FIG. 335　*Drawida linhaiensis* (after Chen, 1933).

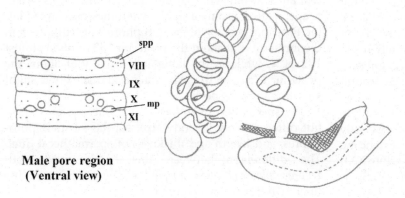

**Male pore region
(Ventral view)**

Spermatheca

closely coalesced with the septum and embedded in body wall. Genital glands similar to *Drawida japonica* (Michaelsen, 1892), globular in shape, large ones about 4 mm in diameter, or small and embedded among muscle fibers of body wall, each gland with a thick muscular capsule continuous with body wall and with a thick cord-like duct directly connected with its papilla.

Color: Preserved specimens light grayish or grayish pale with a bluish tinge, largely due to intestinal contents and iridescence of skin, irregular pale patches vaguely visible under skin also due to internal structures; yellowish or pale on first 7–8 segments, greyish pale on posterior third of body. Subneural vessel very noticeable along ventromedian line on anterior middle of body becoming whitish posteriorly; dorsal vessel also visible on anterior middle of body but not so distinct. Clitellar segments slightly paler.

Distribution in China: Zhejiang (Linhai).

Etymology: The name of the species refers to the type locality.

Deposition of types: Previously deposited in the Museum of the Biological Laboratory of the Science Society of China, Nanjing. The types were destroyed during the war in 1937.

Remarks: The species stands as a distinct species characterized by the following structures distinguishing it from *D. gisti nanchangiana* (Chen, 1933) and *D. syringa* (Chen, 1933); (1) penis more dorsal in position and tubercle-like in shape, (2) spermathecal atrium usually small and placed in front of the septum, (3) definite arrangement of the genital papillae, (4) widely separated septa 10/11 and 11/12, and some other minor differences such as the character of setae and blood vessels.

342. *Drawida nemora* Kobayashi, 1936

Drawida nemora Kobayashi, 1936b. *Sci. Rep. Tohoku Univ.* 11:141–146.

Drawida nemora Kobayashi, 1940. *Sci. Rep. Tohoku Univ.* 15:272–273.

Drawida nemora Xu & Xiao, 2011. *Terrestrial Earthworms of China*. 59–60.

External Characters

Length 65–120 mm, width 4–5 mm, segment number 165–200. Prostomium prolobous, narrow but long, separated from first segment by a groove and projecting into buccal cavity and extending inward to slightly over 1/2. Dorsal pore absent. Clitellum in X-XIII, distinct, glandular part appearing thicker and slightly swollen, clearly distinguished from its neighboring segments by its glandular skin and different coloration. Setae beginning on II; 4 pairs per segment, small and closely paired; *aa* usually

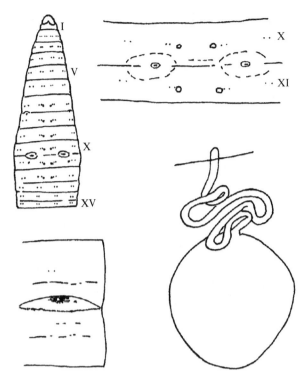

FIG. 336   *Drawida nemora* (after Kobayashi, 1936b).

slightly wider than *bc*, *ab* nearly equal to or slightly wider than *cd*, *dd* nearly equal to 4/7–5/8 circumference, on spermathecal region *aa* = 1.14–1.25 *bc*, *ab* = *cd*, *dd* = 0.625 circumference, on clitella region *aa* = 1.32–1.33*bc*, *aa* = *cd*, *dd* = 0.62 circumference, on middle portion of the body *aa* = 1.4–1.42*bc*, *ab* = *cd*, *dd* = 0.57 circumference; each seta sigmoid, curved at both ends in a weak S shape. Male pores 1 pair of very large transverse slits in 10/11, between *b* and *c*, representing the secondary male pore in which is situated a conical penis usually not visible externally; each slit largely opened on top of light-colored, not sharply conical protuberant tubercle of an epidermal thickening of both posterior and anterior about 1/3 (incomplete annulation) of X and XI, its base transversely oval but not distinctly demarcated and in most cases the intersegmental furrow faintly runs on each side. Spermathecal pores, 1 pair of minute, longitudinal slits, in 7/8, close to the posterior border of VII just medial to *c* or line with *c*. Genital papillae present on VI–XIII, often externally invisible, sometimes totally absent; their arrangement seems to be regular, having 1–3 pairs near the spermathecal pores and 1–2 pairs on every segment of or some of VI–XIII, either 1 pair just medial to *a* or to *c*, or 2 pairs combining in both of the 2 cases; each papilla usually rather indistinct, a small, whitish or pale, circular tubercle which is very slightly protuberant from the general surface, sometimes it is slightly sunken into the body wall, more frequently externally invisible, but, after dissection, oval glands are found internally where the papillae should be present externally; seldom, totally absent both externally and internally. Female pores 1 pair of transverse slits, minute but rather easily visible in well-preserved specimens under the lens, in 11/12, strictly on anterior border of XII in line with *b*.

Internal Characters

Septa 5/6–8/9 thickened, of these posterior 2 much thickened and the last in middle part of IX, the remaining septa all thin and membranous, 9/10 and 10/11 in posterior part of X and XI respectively, 11/12 fused with 10/11 dorsally forming ovarian chamber, posterior septa to this normal in attachment. Hearts 4 pairs, in VI–XI. Nephridia meganephridial. Gizzards brownish or golden in color, generally smooth and shiny on surface, the middle ones thick ring-shaped, and first and last ones somewhat cup-shaped. Testis sacs 1 pair, each suspended with its slightly smaller part in IX and larger part in X (sometimes, each part in IX and X nearly equal); about middle part neck-likely constricted, especially marked on the side facing esophagus, perhaps due to attachment of septum, nearly rectangular in shape as a whole, about 4–6 mm in anteroposterior length, 2–2.8 mm in width, anteroposterior length of part in IX about 1.8–2.2 mm, part in X 2–4 mm, yellowish white and granular on surface. Testis large, nearly circular, seminal funnel whitish and shiny just behind the testis on ventral side of each sac. Sperm duct long, moderately closely coiled, running in front of 9/10 around last heart, piercing the septum into X and touching nephridium of X to a certain degree and finally passing under the anterior side of prostate to connect directly with the anterior and basal portion of penis. Prostate thick oval or circular disc, sessile on parietes, occupying, mostly, about 1.67 segments in X and XI remaining for a short length anteriorly and posteriorly; on top of the penis a large transverse slit-like primary pore opens. Ovarian chamber in XI, formed by septa 10/11 and 11/12, both septa entirely fused dorsally but rather widely separated ventrally, also closed around esophagus in an inverted U shape. Spermathecae, 1 pair; ampulla and duct wholly lying upon posterior face of 7/8; ampulla thin-walled, large, round, of about 1–2 mm in diameter, internally filled with whitish coagulated sex products in all cases examined (preserved specimens); from its lower side rises a thin and long (5–8.5 mm) duct moderately sharply marked off, running with loose and irregular twists and finally becoming nearly straight to enter directly into the parietes without atrial dilation, its ectalmost portion buried in the parietes usually slightly extending beyond intersegmental furrow 7/8 into VII and is a trifle thicker than the rest but often without enlargement.

> Color: Preserved specimens dorsally dark bluish, ventrally yellowish gray. Clitellum similar to ventral surface.

Distribution in China: Liaoning (Shenyang), Jilin (Changcun, Tumen).

Etymology: The name of the species refers to the habitat of the type locality.

Deposition of types: Forest Experiment Station.

Remarks: The absence of a spermathecal atrium and circular disc-like prostate perhaps relate this species to *Drawida rara* (Gates, 1925), but by the same author's reports (spermathecal atrium traceably present (1926 and 1931) or its presence is not clear (1933)), their similarity is much decreased, and furthermore it differs from the latter in genital markings, position of the ectalmost part of the spermathecal duct, position of gizzards, and body size. It is also easily distinguished from the other Korean *Drawida* forms, by (1) external general form of the body, (2) large, transverse slit-like secondary male pore, situated on top of the tubercle which is not sharply conical and is formed by thickening of both posterior and anterior borders of X and XI, (3) disc-like prostate with smooth surface; vas deferens passing under the anterior side of prostate to connect with the anterior and basal portion of small penis which is deeply buried with its rather wide base into the ventral center of the prostate, (4) spermathecal pore situated on posterior border of VII (which is, in most cases, slightly depressed in a crescent shape), minute longitudinal slit in line with or just menial to *c*; no spermathecal atrium, and ectalmost portion of the spermathecal duct which is buried in the parietes, terminates anteriorly often just passing over the intersegmental furrow 7/8, and (5) mostly small, whitish or pale genital papillae, in most cases 1 pair, sometimes with additional ones, on every segment of or some of VI–XIII, normally just medial to either *a* or *c* and mostly concentrated in 1–3 pairs near the spermathecal pore.

343. *Drawida nepalensis* Michaelsen, 1907

Drawida nepalensis Gates, 1972. *Trans. Am. Phil. Soc.* 62(7):256–257.

External Characters

Length 78–130 mm, width 4–5 mm, segment number 129–180. Clitellum IX–XIV. Setae, $aa =$ or slightly > or $< bc$, $dd =$ or slightly $>1/2$ circumference. Male pores obvious, at or median to the line in middle of *bc*, each usually on or near end of a protuberant ventrally directed porophore apparently independent of both X and XI. Spermathecal pores small, transverse slits, just median to *c*, 1 pair, in 7/8, intersegmental. Genital markings, 1 small, circular translucent area on lateral or anterior face of each male porophore, a similar area in VII just anterior to each spermathecal pore, also grayish or whitened areas in which epidermis may or may not be thickened, 1 in front of and 1 behind each male porophore, others (when present) paired or unpaired and variously located in VII–XI.

Internal Characters

Gizzards 2–4, in XII–XX (XXIII?), often some or all not in consecutive segments. Intestinal origin, in XXVII (±1). Sperm ducts, a postseptal portion of each thickened and in a cluster of loops that may be larger than the testis sac, passing into ental end of prostate directly. Prostatic capsule, 2–4 mm long, slender club-shaped, slightly and gradually widened entally, glandular investment continued to parietes. Spermathecal diverticula, saccular, 3–5 mm long, an ectal portion of variable length stalk-like, in VII. GM glands of male porophores and in front of spermathecal pores solid, spheroidal, with translucent walls.

> Color: Preserved specimens unpigmented.
> Distribution in China: Yunnan (Mongmong).
> Etymology: The name of the species refers to the type locality.
> Type locality: Nepal (Gowchar, near Katmandu).
> Deposition of types: Indian Museum.
> Remarks: Relationships, as indicated by the genital marking glands and the spermathecal diverticula, seem to be with *Drawida papillifer* (Stephenson, 1917).

344. *Drawida omeiana* Chen, 1946

Drawida omeiana Chen, 1946. *J. West China. Border Res. Soc. (B)*. 16:84–85.
Drawida omeiana Xu & Xiao, 2011. *Terrestrial Earthworms of China*. 61.

External Characters

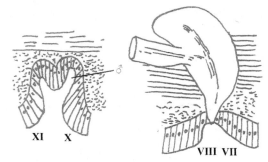

Male pore region   **Spermatheca and pariet**

FIG. 337   *Drawida omeiana* (after Chen, 1946).

Length 40–60 mm, width 2.5–3.3 mm, segment number 100–118. Clitellum X–XIII, glandular, also extending neighboring to segments of IX and XIV. Setae short, *aa* about 1/5 greater than *bc*; *ab*=*cd*, *dd* about 5/11

circumference ventrally. Male pores intersegmental in 10/11, midway between *b* and *c*, generally without porophore; pore rather on posterior edge of X, its anterior and posterior sides of neighboring 2 segments swollen as lips, sometimes a skin fold at posterior side of each pore, fold hardly visible externally. Very rarely the skin fold as large as in *Drawida grahami* (Gates, 1935a) but differing from the latter in not bearing a pore; no porophore as in *Drawida japonica japonica* (Michaelsen, 1892) or *Drawida grahami*. Genital papillae never found in this region. Spermathecal pores 1 pair, in 7/8, intersegmental, ventral to *c* by a distance of *cd*. No genital papillae; rarely with a single papilla posterior to 1 pore. Female pores 1 pair in 11/12, close but slightly lateral to *b*.

Internal Characters

Septa 5/6–8/9 thick and well-muscular, 9/10 thin. Gizzards 3 in all cases observed. Last hearts in IX. Testis large, somewhat kidney shaped, about 2 mm in diameter, suspended on septum 8/9. Vas deferens of each site running down on anterior surface in a fashion of screw wire, to the lower side piercing through to the posterior side in 1–2 coils, then entering the parietes and passing along the wall of male atrium, pouring into lumen of the latter at its ental end. Prostate finger-shaped, about 2 mm long, 0.2–0.3 mm thick, warty on surface. Ovisacs elongate, extending 7–8 segments behind the last gizzard, all filled with reproductive material, very slender and long in immature specimens. Spermathecal ampulla large, round, about 0.5–0.8 mm in diameter; duct greatly coiled on posterior site of septum 7/8, about 4 mm long if uncoiled. Atrium thumb-shaped, very inconspicuous, largely embedded in tissues of body wall, about 0.2 mm long, duct entering at ental lateral side.

> Color: Pale generally, grayish on dorsal side, paler on ventral side; clitellum reddish.
> Distribution in China: Sichuan (Mt. Emei).
> Etymology: The specific name indicates its type locality.
> Deposition of types: Previously deposited in the Institute of Zoology, Academic Sinica, Chongqing. The types were destroyed during the war in 1945–1949.
> Remarks: In appearance, this species is like *Drawida grahami* (Gates, 1935a), which is also common in China. However, its general body form is smaller and more slender. The male "porophore" of this new species when present resembles that of the latter species but differs in not bearing the male pore. The latter is generally round anterior to the fold or "porophore". The main distinguishing features are: (1) the prostate is not sessile but finger-shaped, (2) the spermathecal atrium is minute, and (3) the genital papillae are usually absent. It is common at lower

elevations on the mountain and is found as high as about 2000 m.

345. *Drawida propatula* Gates, 1935

Drawida propatula Gates, 1935b. *Lingnan Sci. J.* 14(3):449–450.
Drawida propatula Kobayashi, 1940. *Sci. Rep. Tohoku Univ.* 15:265–267.
Drawida propatula Xu & Xiao, 2011. *Terrestrial Earthworms of China.* 61–62.

External Characters

Length 73–130 mm, width 4.0–5.0 mm. segment number 149–179. Clitellum in IX–XIV. Setae begin on II, closely paired, small and not conspicuously protuberant, apparently stouter and darker anteriorly than posteriorly, especially *a* and *b* on II to X or XI; on XX, *aa* < *bc*. Male pores are minute, transverse slits on X, nearer to *b* than to *c* and nearer to 10/11 than to the transverse setal line; on rather indefinite, not sharply demarcated, whitish porophores approximating to transversely oval, very slightly tumescent, especially a central portion on which the pore is located. No trace of a penis. The genital markings are transversely oval with a central, circular area comprising a grayish, concave depression surrounded by a regular, sometimes slightly raised, whitish band external to which is an irregularly jagged and faintly grayish zone narrower than the whitish band. External to the grayish zone there is a smooth, whitened zone which may or may not be sharply demarcated peripherally. The markings are located as follows: presetal on VII in *bc*; postsetal on VII, in *ab*, in *bc*, or just lateral to the spermathecal pore; presetal on IX, in *bc*; postsetal on X in *ab* and in *bc*. Spermathecal pores are minute, longitudinal, diagonal, or transverse slits on 7/8 or the posterior most margin of VII, just median to *c*. Female pores are transverse slits on 11/12 in *ab*, *b* or just lateral to *b*, minute and not readily recognizable; the location of the pores can, however, be determined by the grayish translucent appearance of the epidermis immediately surrounding the apertures.

Internal Characters

Septa 5/6–8/9 much thickened, the remaining septa thin or membranous; 9/10 and 12/13 slightly displaced posteriorly, dorsal part of 10/11 displaced to the anterior part of XII, the others almost normally inserted. Gizzards 3, in XII–XIV. The last hearts in IX but there is on the posterior face of 9/10 a pair of dorsoventral commissures covered by transparent connective tissue. The testis sacs are constricted by 9/10. The vas deferens is short with a few loose loops and passes into the parietes from whence it emerges to pass into a groove on the prostate. In this groove it is partially concealed from view by the granulations, which are rather loose. The prostates are fairly large, 2–3 mm long, club-shaped, narrowed ectally, with the granulations extending to the parietes. The central body is slender, tubular and finger-shaped, narrowed ectally, with the granulations extending to the parietes. The central body is slender, tubular and finger-shaped, about 1.5 mm long. Ovarian chamber appears to be closed off dorsally but attached to the parietes laterally. The ovisacs are finger-shaped and extend into XVI–XVIII, and are provided with posterior, rod-like appendices. The spermathecal duct is 9–11 mm long, looped; it passes into the lateral side of the atrium just below the ental end. The atrium is columnar, 1.5–2.0 mm high, erect in VII against the posterior face of 7/8. In 1 specimen there are several tiny, spheroidal projections from the ental ends of the atria. The glands of the genital markings are tough-walled and sessile but conspicuously protuberant into the coelom.

Color: Preserved specimens unpigmented or uniformly light blue due to the content of the alimentary tract. Clitella reddish.
Type locality: Jiangxi (Jiujiang).
Distribution in China: Jilin (Yanji), Jiangxi (Jiujiang).
Deposition of types: Smithsonian Institution.
Remarks: This species is distinguished from *Drawida japonica japonica* (Michaelsen, 1892), which it most resembles, by the larger size and by the absence of the elongate, rod-like appendices of the ovisacs.

346. *Drawida sinica* Chen, 1933

Drawida sinica Chen, 1933. *Contr. Biol. Lab. Sci. Soc. China (Zool).* 9:205.
Drawida cheni Gates, 1935b. *Lingnan Sci. Journ,* 14(3):450–451.
Drawida sinica Gates, 1935a. *Smithsonian Mus. Coll.* 93(3):4.
Drawida sinica Gates, 1939. *Proc. U. S. Natn. Mus.* 85:414.
Drawida sinica Xu & Xiao, 2011. *Terrestrial Earthworms of China.* 62.

External Characters

Length 50–95 mm, width 2.2–3.2 mm, segment number 175–202. All medium-sized, rather elongate, anterior half of body comparatively stouter. Prostomium prolobous, its lobe swollen and projecting out and extending ventrally to about segment 1/2. Dorsal pore absent, a mid-dorsal line indistinctly appearing behind clitellum region. Skin smooth with our naked eye examination, but very cavernous or with sieve-like small pits all over body under a high-power magnifier. Each small pit is a gland cell, resembling the skin of *Desmogaster* (Rosa,

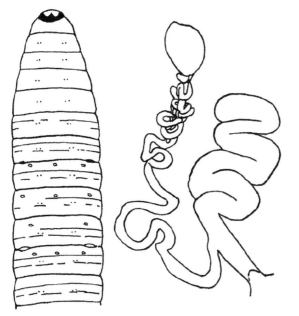

FIG. 338   *Drawida sinica* (after Chen, 1933).

1895). Clitellum not marked, slightly glandular on venter of segments X–XIII. Setae minute, hardly visible or appearing as pits under magnifier, traceable or invisible on II, absent on first 1 and last 2 segments, uniformly built on remaining segments, *aa* equal to *bc* or slightly narrower; *ab* = *cd*, distance of *ab* equal, at least, to 1/10*aa* or *bc*, *dd* of dorsal distance a little more than half of body circumference, about 3/5–5/9 of segmental circumference; *aa* of II–VII wider than *bc* (especially in V, VI). Each seta sigmoid, curved at both ends in a weak S shape, external free end often abruptly curved backward as a hook or sometimes gently curved. Male pores 1 pair of large transverse slits in 10/11, between *bc*, about 2/5 nearer *c*, representing the secondary male aperture in which is situated 1 conical penis usually not visible when retracted. The latter small, about 0.18 mm in length, enlarged at base and narrowing gradually toward top, its distal end round or transversely flattened (not pointed), often with a circular constriction between base of penis and body wall; not everted in the cases observed. Genital papillae small, on ventral side of VIII–XI, regular in arrangement, whether unpaired ventrolaterally between *aa* or paired ventrolaterally between *bc*. The ventromedial ones on VIII, IX, and XI frequently present, near anterior part of each segment with that on X absent in cases observed. The ventrolateral papillae on VIII and XI always closer to seta *b*, on anterior half of segments, those on XI sometimes absent; those on IX closer to *c*, on anterior half of segments; often another additional pair at posterior portion of segment X in line with or slightly medial to anterior pair. Each papilla minute, as a transversely oval depression with a pigmented spot in center. Female pores 1 pair, in 11/12, in line to *b*.

Internal Characters

Septa 5/6–8/9 well-thickened and all pushed a little behind intersegmental furrows, 2 rudimentary septa visible in front of 5/6; 9/10 on posterior third of X, 10/11 on posterior edge of XI, 11/12 on middle or posterior part of XII; 12/13–16/17 or 17/18 similarly pushed about 1 segment or less backward, inserted on posterior part of XIII–XVII or XVIII respectively. Lateral hearts 4 pairs, in VI–IX. Nephridia meganephridial. Gizzards usually 3, in XII–XVI, or 4, in XI–XVI. Testis sacs 1 pair, suspended on septum 9/10, triangular, with equal parts on both sides of the latter, usually elongate and pointed anteroposteriorly, convex dorsally and concave in form of a hilum ventrally, its sperm duct very long and greatly coiled before and behind septum 9/10 and suddenly enlarged to form a long atrial tube which forms a deep loop with 1 arm closely twisted with another, passing directly to the penis. Spermathecal atrium very characteristic in this species, as a large and long compactly looped zigzag tube on 1 plane, about 4 or more loops, free in body cavity just before septum 7/8, its lumen large and irregularly shaped with internally projected ridges at its ental main part, smaller and circular near its ectal end, its inner columnar epithelium thick and surrounded by a thick muscular layer. Ampulla thin-walled, usually empty, very small or moderately large, ranging from 1–2 mm in diameter, situated on posterior surface of septum 7/8, close to dorsolateral side of intestine, its duct slender, loosely or closely coiled, but less coiled ventrally, piercing through ventral side of septum to join lower ectal end of atrium, a short distance from body wall, and communicated with main lumen. Genital gland very small, embedded in body wall. Ovarian chamber similarly formed by septa 10/11 and 11/12, but wider dorsally and laterally, narrower ventromedially.

Color: Preserved specimens generally free from pigmentation, with a light bluish grey shimmer irregularly distributed on dorsal side largely due to intestinal contents and blood vessels. Anterior end pale, due to thickened septa internally; a deeper greenish line along dorsomedian line.
Type locality: Jiangxi (Nanchang).
Distribution in China: Jiangxi (Nanchang).
Etymology: The name of the species refers to the type locality.
Deposition of types: Previously deposited in the Museum of the Biological Laboratory of the Science Society of China, Nanjing. The types were destroyed during the war in 1937.
Remarks: *Drawida sinica* (Chen, 1933) was collected from only 1 place in Nanchang, a graveyard where the soil was hard, compact, and covered with grasses and shrubs in fair abundance. Chen believed they must occur in either Jiangxi or neighboring areas. It is

interesting to note that its male atrial duct and the spermathecal atrium are so different from other species recorded in this region that it stands as a distinct species. In general appearance, it looks more like *Desmogaster* than *Drawida*. In life, it is unpigmented and appears sluggish. Its skin is smooth and somewhat iridescent.

347. *Drawida sulcata* Zhong, 1986

Drawida sulcate Zhong, 1986a. *Act Zootaxonomica Sinica*. 11(1):28–31.
Drawida salcata. Zhong & Qiu, 1992. *Guizhou Science*. 10(4):39.
Drawida sulcate Xu & Xiao, 2011. *Terrestrial Earthworms of China*. 64–65.

External Characters

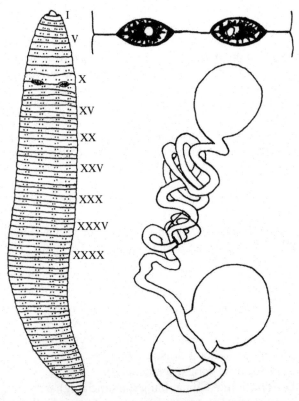

FIG. 339   *Drawida sulcata* (after Zhong, 1986a).

Length 60–86 mm, width 9–12.5 mm, segment number 71–108. Prostomium prolobous. The body stouter, pointed, and slender on anterior and posterior ends. Smooth on surface, without papilla. No secondary segmentation, III–XI slightly wider than other segments. The body wall fairly thick, and intersegment furrows deep. Dorsal pore absent. Setae 4 pairs per segment, fine; $aa = 1.2–1.5bc$, $ab = bc$, $aa = 9–12ab$, $dd = 2/3–3/4$ circumference. The male pores 1 pair, on copulatory pouch of

10/11; the aperture of copulatory pouch between *bc* and nearer to *c*, eye-shaped or oval, ental end smaller than ectal end if oval. Anteroposterior margins of aperture swollen, with slender longitudinal grooves; a conical penis situated in the chamber, when everted, the penis can project out of the aperture, and inner surface of the chamber visible. Spermathecal pores 1 pair, each situated in a deep and narrow spermathecal chamber formed by body wall invagination in intersegmental furrow 7/8; The fertilized sac is rounded, on a larger papilla on the back wall of chamber in VIII; the opening of the spermathecal chamber slit-like, located on the midline of *cd*, it is visible when the intersegmental furrow is opened. Female pores 1 pair, in intersegmental furrow 11/12, tiny slit-like, in line to middle of *ab*.

Internal Characters

Septa 2/3–4/5 rudimentary, visible; 5/6–9/10 slightly thick, and testis sacs suspended on septum 9/10, 10/11 and 11/12 united as ovarian chamber, 12/13 and succeeding septa membranous. Gizzards 3, in XII–XIV or XI–XIV. Lateral hearts 4 pairs, in VI–IX. Testis sacs long round-shaped or thick disc-like, suspended on septum 9/10, occupying half of IX and X; sperm duct arising from sac and coiled behind 9/10, running a little distance, piercing the septum 9/10 and coiling down around ental side of heart, then turning out near body wall and crossing the heart, piercing the septum 9/10 again, and passing into prostate glands in X. Prostate glands smooth on surface, compact, covered , "☒" shaped, passing through posterosuperior part into copulatory chamber. A semicircular protuberance projecting through the parietes into the coelom, penis projected out from posterosuperior part of parietes, the base slightly small and the tip usually curved ectally; inner surface of the chamber with many wrinkles and usually outside of the chamber opening. Spermathecal ampulla behind septum 7/8, ball-shaped, in mature specimens a milky spermatophore in a smaller ball in ampulla; diverticulum about 15–16 mm long, 0.2–0.3 mm in diameter, the enlarged portion slightly rough and cylindrical; a semicircular protuberance projecting through the parietes into the coelom.

Color: Preserved specimens brown, no difference between dorsal and ventral sides.
Type locality: Yunnan (Mt. Ailao).
Distribution in China: Yunnan (Mt. Ailao).
Etymology: The specific name "*sulcata*" is from the Latin word "sulcus" meaning furrow and refers to the intersegment furrows.
Deposition of types: Department of Biology, Sichuan University.
Remarks: This species is similar to *D. decourcyi* (Stephenson, 1914) but differs from the latter in (1) sperm duct short (10–12 mm) and coiled in a plane

instead of very long (640 mm) with loops in a cluster longer than the testis sac; (2) Gizzards 3, in XI–XIV, instead of 7–9, in XIV–XXVII.

348. *Drawida syringa* Chen, 1933

Drawida syringa Chen, 1933. *Contr. Biol. Lab. Sci. Soc. China (Zool).* 9:203–205.
Drawida syringa Gates, 1935a. *Smithsonian Mus. Coll.* 93(3):4.
Drawida syringa Xu & Xiao, 2011. *Terrestrial Earthworms of China.* 65.

External Characters

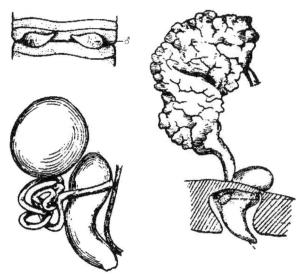

FIG. 340 *Drawida syring* (after Chen, 1933).

Length 50–100 mm, width 2.5–3.0 mm, about 80 mm long in most cases, segment number 124–172. Clitellum in X–XIII, glandular and very distinct, its purplish red color extending to a part of IX and XIV. Setae 4 pairs per segment, short, dorsal setae (*dd*) about half of body circumference, *aa* a trifle smaller than, or equal to, *bc*. Penis short, stouter at basal two-thirds and pointed distally, usually pointing ventromedially when wholly everted. Genital papillae on ventral side absent in most cases, or only 1

found present in IX without a large elevated base as in *Drawida gisti gisti* (Michaelsen, 1931). However, those situated anteriorly or posteriorly to spermathecal apertures and also anteriorly to penis usually present, all not visible outside but concealed within intersegmental grooves.

Internal Characters

Gizzards 4–5, very rarely 2–3. Testis sacs and sperm duct usual, but male atrium very short and broad, thick glandular layer on its ental two-thirds very loosely composed, highly warty and soft, or less so. Egg sac very conspicuous, fully filled with sex products, distended but deeply constricted around each septum, exceedingly long, extending far back to XXV (about 10–15 segments). Spermathecae ampulla round and very large (about 1 mm in diameter), rather low on posterior surface of septum, its duct entering into ental side of atrium. Spermathecal atrium short and stout, club-shaped, always behind septum 7/8. This latter character being unique to this species and not found in any example of *Drawida gisti gisti* (Michaelsen, 1931).

Color: Preserved specimens dark fleshy or greyish on dorsal side, clitellum purplish red.
Type locality: Zhejiang (Fenghua, Linhai).
Distribution in China: Zhejiang (Fenghua, Linhai).
Etymology: The subspecific name refers to the type locality.
Deposition of types: Zoological Central Museum of Nanjing University.
Remarks: *D. syringa* (Chen, 1933) was once considered as a variety of *D. gisti gisti* (Michaelsen, 1931). However, it is more advisable to make it a distinct species on account of several features which distinguish it from *D. gisti gisti* (Michaelsen, 1931). These are (1) general form of body, (2) shorter and thicker penis, (3) genital papillae on ventral side are generally present in *D. gisti gisti* (Michaelsen, 1931) but always absent in this form, (4) increasing number of gizzards which are not found in any variety of that species, (5) shorter and much larger male atrium with a highly warty surface, and (6) exceedingly long egg sacs. These characters are sufficient to separate it from that species.

Family: Ocnerodrilidae Beddard, 1891

Setal arrangement lumbricine. Esophagus with paired diverticula or with unpaired ventral sac in segment IX or in IX and X (calciferous glands, chyle-sacs). Meganephridial, Prostates tubular, 1–3 pairs. Sexual apparatus acanthodriline or in different degrees microscolecine (male pore on XVII or XVIII; prostatic pores 1–3 pairs, variously on XVII, XVIII, XIX; exceptionally male and prostatic pores shifted 3 segments farther back), or rarely megascolecine (male and prostatic pores united and opening on XVIII). Spermathecal pores in 7/8 or 8/9 or both of these, occasionally absent. Genital pores sometimes unpaired, fused in the midventral line.

Global distribution: America-California and Arizona to central Chile and Argentina; Africa-Egypt, Tripoli, and Upper Guinea to Madagascar, Natal, and Orange Free State; the Seychelles and Southern India.

There are 4 genera and 4 species from China.

GENUS: *EUKERRIA* (BEDDARD, 1892)

Kerria Beddard, 1892. *Proc. Zool. Soc. London.* 355.
Kerria Stephenson, 1930. *The Oligochaeta.* 859.
Eukerria Gates, 1972. *Trans. Am. Phil. Soc.* 62(7):269.

Type species: *Kerria halophila* Beddard, 1892.
Setae closely paired. Male pores on XVIII; prostatic pores 2 pairs, on XVII and XIX. Spermathecal pores usually 2 pairs in furrows 7/8/9, the anterior pair rarely absent. Esophageal gizzard in VII, or gizzard absent. One pair of lateral Esophageal sacs in IX. One pair of testes free in X. Two pairs prostates (rarely 4 pairs), opening separately from the vas deferens. Spermathecae without diverticula (Always ?). Of small size, limnic in habitat.

Global distribution: Sub-tropical South America (Brazil, Paraguay, Lower California); St. Thomas (West Indies); peregrine species in South Africa (Tshwane and KwaZulu-Natal) and New Caledonia, and in New South Wales, Australia.

There is 1 species from China (Taiwan)

349. *Eukerria saltensis* (Beddard, 1895)

Kerria saltensis Beddard, 1895. *Proc. Zool. Soc. London.* 225.
Eukerria saltensis Gates, 1972. *Trans. Amer. Phil. Soc.* 62(7):270–271.
Eukerria saltensis Shen, Tsai and Tsai, 2008. *Acta Zool. Taiwan.* 10(7).
Eukerria saltensis Chang, Shen & Chen, 2009. *Earthworm Fauna of Taiwan.* 150–151.
Eukerria saltensis Xu & Xiao, 2011. *Terrestrial Earthworms of China.* 270.

External Characters

Length 33–85 mm, clitellum width 1.0–1.5 mm, segment number 87–129. Prostomium epilobous. Dorsal pore absent. Clitellum in XIII–XX, annular, thinner ventrally. Setae lumbricine (8 setae per segment), small and closely paired, seta *b* absent in XVII–XIX. Male pores paired in XVIII in longitudinal grooves, prostatic pores 2 pairs at the anterior and posterior ends of the grooves in XVII and XIX. Spermathecal pores 2 pairs in 7/8 and 8/9, lateral, closer to seta *c*. Female pores paired in XIV, each pore anterior to seta *a* and close to 13/14 intersegmental furrow. Genital markings absent.

Internal Characters

Septa present from 5/6, 6/7/8/9 thickened. Gizzard in VII. Calciferous glands 1 pair in IX, thick-walled with numerous capillaries. Intestine from XII. Esophageal hearts in IX–XI. Spermathecae 2 pairs in VIII and IX, small, adiverticulate, ampulla round or oval, duct slender, twisted near ampulla. Nephridia avesiculate. Ovaries paired in XIII, small. Testes 1 pair in X, small, flowery, shiny. Seminal vesicles paired in IX and XI, small. Prostate glands 2 pairs with prostatic pores in XVII and XIX, each thin, elongate, tubular, extending 4–5 segments.

Color: Specimens unpigmented and with yellow clitellum.
Type locality: Salto, Valparaiso, Chile.

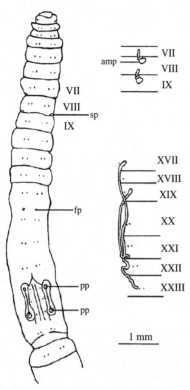

FIG. 341   *Eukerria saltensis* (after Chang, Shen & Chen, 2009).

Deposition of types: British Museum, London, England.
Etymology: The name of the species refers to the type locality of Salto.
Distribution in China: Taiwan.
Habitats: In Taiwan, it is recorded at elevations below 1770 m in mountain and hill regions throughout the island.
Remarks: This species is frequently found along forest roads in Taiwan.

GENUS: *ILYOGENIA* (BEDDARD, 1893)

Ocnerodrilus Stephenson, 1930. *The Oligochaeta.* 860.
Ocnerodrilus (*Ilyogenia*) Stephenson, 1930. *The Oligochaeta.* 861.

Male pores on XVII; prostatic pores 1 pair, united with the male pores (in 1 species a second pair of prostatic pores on XVIII). Spermathecal pores 1 pair, in furrow 7/8 or 8/9.

No gizzard: 1 pair of esophageal sacs in IX, of simple structure. Two pairs of testes and funnels, free, in X and XI. Spermathecae 1 pair, opening in furrow 8/9.

Global distribution: America (from California to British Guiana and Colombia); Paraguay; West Indies; tropical and South Africa.

There is 1 species from China.

350. *Ilyogenia asiaticus* (Chen & Hsü, 1975)

Ocnerodrilus (*Ilyogenia*) *asiaticus* Chen, Hsü, Yang & Fong, 1975. *Acta Zool. Sinica.* 21(1):95.
Ilyogenia asiaticus Xu & Xiao, 2011. *Terrestrial Earthworms of China.* 270–271.

External Characters

Length 21–54 mm, width 1.3–1.8 mm. Prostomium 2/3 epilobous. Dorsal pore absent. Clitellum XIII–XVII, annular, smooth on surface, setae absent; but posteriorly with a ventral "Π" notch. The setae begin on II, 4 pairs per segment, preclitellar $aa = 2ab < bc, ab \leq 2cd, dd = 2.5cd$, $dd = 1/4$–$1/5$ circumference ventrally apart, postclitellar ab and cd slightly wide, $dd = 3.0cd$; penial setae a club-like, 67.2 μm long; b distal spoon-like, 50.4 μm long. Nephridiopore ventrally in seta c. Male pores between a and b, in ventral region of XVII; male pores and prostatic pores separate. Spermathecal pores 1 pair, in a line of 8/9. Female pores 1 pair, in a of XIV.

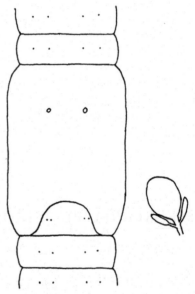

FIG. 342   *Ilyogenia asiaticus* (after Chen & Hsü, 1975).

Internal Characters

Septa 5/6–10/11 thickened, the succeeding septa thin. Gizzard and intestinal caeca absent. Esophageal pouches in IX weakly developed. Intestine swelling from XVI. Lateral hearts 2 pairs in IX and XII, symmetrical; vascular ring conspicuous in X and XI, but not symmetrical. Testes small, in X and XI, free. Prostate glands 1 pair, mass-like, smooth, confined in XVII. Prostatic duct very short. Spermathecae 1 pair, in IX; ampulla oval, with short ducts, 1 or 2 diverticula usually present.

Color: Anterior part of body flesh red and posterior part steel grey. Clitellum red-brown.

Type locality: Jiangsu (Nanjing).

Distribution in China: Jiangsu (Nanjing).

Etymology: The name of the species refers to the type locality.

Remarks: This species is collected from meadow soil, identical to the genus *Ocnerodrilus* (Eisen, 1878) genus, but with consideration of its spermathecae with a diverticulum, it should be a member of the genus *Ilyogenia* (Beddard, 1893).

GENUS: MALABARIA STEPHENSON, 1924

Malabaria Stephenson, 1930. *The Oligochaeta*. 857.

Setae closely paired. Two pairs of prostatic pores, in XVII and XIX; vas deferens discharging in conjunction with the anterior prostates. One pair of spermathecal pores in furrow 8/9. Gizzard vestigial, in VII. No projecting esophageal sacs, but in the substance of the thickened and vascular ventral wall of the esophagus there are, in segments IX and X, 2 pairs of small tubular diverticula. Two pair of testes and funnels, in X and XI. Spermathecal diverticula absent.

Global distribution: Southern India (found among the roots of rice plants grown standing in water).

There is 1 species from China.

351. *Malabaria levis* (Chen, 1938)

Filodrilus levis Chen, 1938. *Contr. Biol. Lab. Sci. Soc. China (Zool)*. 12(10):422–426.

Malabaria levis Xu & Xiao, 2011. *Terrestrial Earthworms of China*. 271–272.

External Characters

Length 80 mm, width 1.0 mm, segment number 195. Anterior 5 segments much shorter than those following. Prostomium proepilobous. Dorsal pores absent. Clitellum not visible externally, hypodermis slightly thicker in XIV–XX. Setae 8 per segment, $ab = cd$ (about 0.08 mm apart), $aa = bc$, $aa = 3.5$–$4.0ab$, dd of dorsal distance about or slightly less than half of body circumference; dd of posterior end of body less than half of circumference. Nephridiopores in line with seta b, close to anterior border of each segment. Male pores on XVII, close laterally to setae b, always asymmetrically present, generally on left side; no particular protuberance, no other genital markings. Spermathecal pores 1 pair in 8/9, intersegmental, between ab, openings large but simple. Genital papillae generally present between pores, glandular portion in body wall revealed only in sections, papilla slightly raised but hardly recognizable externally. Female pores on XIV, in line with seta b, close to intersegmental furrow 13/14, present on both sides.

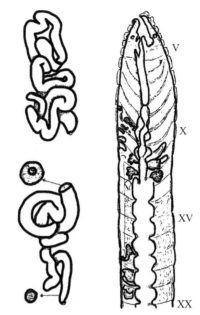

FIG. 343 *Malabaria levis* (Chen, 1938).

Internal Characters

Septa 5/6–8/9 specially thickened, 9/10/11/12 less so. Pharyngeal glands thickest in IV–V, extending to VI. Gizzard in VII, weakly developed, its wall muscular in sections about 4–5 times thicker than wall of esophagus. Calciferous glands in VIII–X, no special pouches, its wall thick and invested with loosely connective tissue and blood capillaries, constricted intersegmentally, portion in IX and X thickest on ventral side with rudimentary lumen, sac-like portion about 0.4 mm in diameter of whole sac, about 0.06 mm in thickness of dorsal wall. Intestine swelling in middle of XII. No typhlosole. Lateral hearts 2 pairs, large, in X and XI. Ovaries large, in XIII, oviduct funnels in front of septum 13/14, oviducts short. Testes 3 pairs, in IX–XI, attached to ventrolateral side of respective septum, about 0.25 mm in length (much longer in cotypes), first pair about one-third of others in size; seminal funnels in X and XI. Sperm morulae scattered in X and XI. Seminal vesicles projecting out posteriorly from septa 10/11 and 11/12, branching into acinose saccules in young forms, vesicular in fully developed forms. Vas deferens on each side, opening to exterior close to, but separate from, prostatic pore. Prostate glands 1 pair, in XVII, well developed, gland portion tubular and greatly coiled, about 2.5 mm long, 0.07 mm wide, lumen distinct, without defined epithelium, glandular cells arranged into 2 layers, each with a duct opening slightly medial to male pore, often asymmetrically present, on left in the type. No penial setae. Spermathecae 1 pair in IX, ampulla ovoid, flattened, about 1.2 mm wide, its duct greatly coiled, about 3.5 mm long (in an isolated specimen), 0.1 mm wide, without an atrium. No diverticulum.

Color: Preserved specimens pale throughout the body, no difference between dorsal and ventral sides.

Type locality: Hainan (Baopeng).

Deposition of types: Previously deposited in the Museum of the Biological Laboratory of the Science Society of China, Nanjing. The types were destroyed during the war in 1937.

Etymology: The specific name "*levis*" means light in Latin.

Distribution in China: Hainan (Baopeng).

Remarks: They are small and filiform in appearance. No record has ever been made regarding its habitat. By its appearance they seem to be aquatic.

GENUS: *OCNERODRILUS* EISEN, 1878

Ocnerodrilus Stephenson, 1930. *The Oligochaeta*. 860.
Ocnerodrilus (Ocnerodrilus) Stephenson, 1930. *The Oligochaeta*. 861.
Ocnerodrilus. Gates 1972. *Trans. Amer. Philos. Soc.* 62(7):273.

Male pores on XVII; prostatic pores 1 pair, united with the male pores (in 1 species a second pair of prostatic pores on XVIII). Spermathecal pores absent.

No gizzard: 1 pair of esophageal sacs in IX, of simple structure. 2 pairs of testes and funnels, in X and XI; testes surrounded by testis sacs of a peculiar constitution that does not include the funnels. Seminal vesicles and spermathecae absent.

Global distribution: North America (California, Arizona, Mexico), India, China, the West Indies, Africa (Cape Verde and Comoro Islands).

352. *Ocnerodrilus occidentalis* Eisen, 1878

Ocnerodrilus occidentalis Chen, 1933. *Contr. Biol. Lab. Sci. Soc. China (Zool)*. 9:224–228.
Ocnerodrilus (Ocnerodrilus) occidentalis Chen, 1938. *Contr. Biol. Lab. Sci. Soc. China (Zool)*. 12(10):426.
Ocnerodrilus occidentalis Chen, 1959. *Fauna Atlas of China: Annelida*. 6.
Ocnerodrilus occidentalis Gates, 1972. *Trans. Amer. Philos. Soc.* 62(7):273–275.
Ocnerodrilus occidentalis Xu & Xiao, 2011. *Terrestrial Earthworms of China*. 272–273.

External Characters

Length 45–46 mm, width 1.2–1.5 mm, segment number 75–77. Prostomium 1/2 epilobous, tongue broad and closed behind. Dorsal pore absent. Clitellum in 1/2 XIII–1/2 XX (about 7 segments), distinct, but glandular layer thin, not sharply marked off from anterior and posterior nonglandular portions, smooth and glandular on dorsal and lateral sides but less so on ventral side of XIII and part of XIX and XX; intersegmental furrows on clitellum scarcely visible. Setae lumbricine, small, 4 pairs per segment, absent on I and short on II, $ab = cd$; aa equal to $3ab$ or little more, shorter than bc by a distance of ab; aa slightly wider on anterior segments; dorsal distance of dd equal to, or slightly wider than, ventral distance before and behind clitellum, about 2/7–2/5 of circumference apart dorsally near posterior end; setae on clitellum normally present, slightly longer, ab on XIV–XIX slightly wider; a on XVII shorter; each seta sigmoid, curved on both ends, nodulous very weak and hardly noticeable. Male pores 1 pair, on ventral side of XVII, in line with b, or slightly lateral; each on a glandular swelling or a round tubercle, with shallow grooves between 2 tubercles but not much depressed; seta a present, hardly visible on surface, b generally degenerated. Spermathecal pores absent. Female pores 1 pair, on XIV, a little distance in front of seta b.

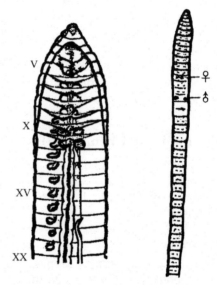

FIG. 344 *Ocnerodrilus occidentalis* (after Chen, 1959).

Internal Characters

Septa 4/5 very thin and ill defined, 5/6 thick, 6/7 more so, 7/8/9/10 much thickened, 11/12 nearly as thick; 12/13 and succeeding ones thin and membranous; 12/13 inserted on middle part of XII, 13/14 on posterior third of XIII, 14/15 normal in insertion. Septal glands fairly large, in V-VIII, whitish and glandular in appearance, smooth on surface, attached to anterior face of each septum, lying on sides of esophagus, those in VIII smaller, or about half the size of preceding ones; similar glands also found around pharynx in front of septum 4/5. Gizzard absent. Ovaries paired in XIII. Testis sacs 1 pair in X and XI, no seminal vesicles. Prostate glands 1 pair, as a long tube extending to posterior clitellar region. Esophageal pouches prominent in IX. Spermatheca absent.

Color: Preserved specimens pinkish or pale through whole body. Clitellum more fleshy.
Deposition of types: Previously deposited in the Museum of the Biological Laboratory of the Science Society of China, Nanjing. The types were destroyed during the war in 1937.

Etymology: The specific name "*occidentalis*" means western in Latin and refers to the type locality.

Distribution in China: Jiangsu, Zhejiang, Sichuan (Chengdu), Hainan (Sha-mom-chiu).

Family: Octochaetidae Michaelsen, 1900

Octochaetinae Stephenson, 1930. *The Oligochaeta*. 841.

Arrangement of setae from pure lumbricine to pure perichaetine. One esophageal gizzard in 1 simple segment, or 2 in 2 simple segments, or 1 enlarged gizzard in a space which represents 2 or more fused segments; in the last 2 cases calciferous glands in the region of segments X-XIII. Meganephridia along with micronephridia, or micronephridia alone, the latter never having the form of sacs. Sexual apparatus from pure acanthodriline to pure microscolecine.

Global distribution: Throughout India, but more sparsely in the north; New Zealand, South Madagascar.

There are 5 species from China.

GENUS: *DICHOGASTER* BEDARD, 1888

Dichogaster Stephenson, 1930. *The Oligochaeta*. 851.
Dichogaster Gates, 1972. *Trans. Amer. Philos. Soc.* 62(7):277–278.

Prostatic pores 1–3 pairs, on XVII, or XIX, or XVII and XIX, or on XVII, XVIII, and XIX. Spermathecal pores 1 or 2 pairs, in furrows 7/8 and 8/9 or 1 of these. Two gizzards in front of the testis segments. Usually 3, seldom 2 pairs of calciferous glands behind the ovarian segment, usually in XV-XVII, rarely in XIV-XVI. Micronephridia.

Global distribution: America-North, Central, and South, from California to Ecuador and Suriname, and including the West Indies; tropical Africa, from Ethiopian Empire to Mozambique on the one side and Gambia to the Congo on the other. India. Brazil and Paraguay, southern Asia, Malay Archipelago, Polynesia, North-West Australia, Madagascar.

There are 4 species from China.

TABLE 38 Key to Species of the Genus *Dichogaster* From China

1. Without genital markings. Clitellum in XIII–XIX, XX. Male pores paired in XVIII, in seminal grooves, between prostatic pores in XVII and XIX .... *Dichogaster bolau*
   With genital markings or genital papillae 2

2. Male pores in XVII, Clitellum in XIII-XIX. Genital marking medioventral, round, across 15/16. .... *Dichogaster saliens*
   Male pores in XVIII 3

3. Clitellum in XIII–XXI, Genital markings round, medioventral, in 7/8/9/10 .... *Dichogaster affinis*
   Clitellum in 1/2 XIII–1/2 XXI, annular. Two genital papillae placed ventromedially in 8/9/10, slightly raised .... *Dichogaster sinicus*

353. *Dichogaster affinis* (Michaelsen, 1890)

Dichogaster affinis Shen, Chang & Chen, 2008. *Endemic Species Research*. 10(2):53–57.
Dichogaster affinis Chang, Shen & Chen, 2009. *Earthworm Fauna of Taiwan*. 152–153.

External Characters

Length 23–33 ... mm, clitellum width 1.6–2.1 ... mm, segment number 103–131. Prostomium epilobous. First dorsal pore 5/6. Clitellum in XIII–XXI, saddle-shaped, 2.0–3.7 mm in length, dorsal pore present in 13/14. Setae lumbricine, small and closely paired on ventrum, *ab* not visible externally in XVII–XIX. Male pores paired in XVIII, in bucket-shaped seminal groove connecting prostatic pores in XVII and XIX. Genital markings round, medioventral, in 7/8/9/10, each about 0.3 mm in diameter. Spermathecal pores 2 pairs in 7/8 and 8/9, medioventral, in line with setae *ab*. Female pores paired on a raised pad in XIV, each pore anterior to seta *a*.

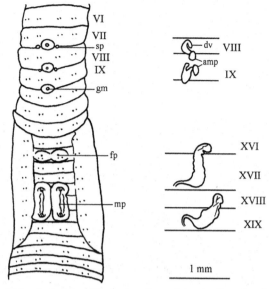

FIG. 345   *Dichogaster affinis* (after Shen, Chang & Chen, 2008).

Internal Characters

Septa weakly developed. Gizzard paired in VII and VIII, muscular, barrel-shaped, displaced posteriorly to IX and X. Calciferous glands 3 pairs in XV–XVII, digiform, the first 2 pairs transparent with comb-like streaks, the last pair yellowish white and slightly lobed. Intestine from XVII. Esophageal hearts in XI–XIII. Nephridia meroic, saccular, 4 rows on each side. Ovaries paired in XIII. Testes 2 pairs in X and XI, small, round. Seminal vesicles absent or vestigial in XI and XII. Prostate glands 2 pairs in XVII and XIX, long, tubular with penial setae close to short, muscular ducts. Spermathecae 2 pairs in VIII and IX, small, ampulla oval, about 0.2 mm long, with a wide, stout, short duct, diverticulum small, short-stalked with a bulbous seminal chamber. Accessory glands absent.

Color: Specimens unpigmented.
Type locality: Quilimane, Zanzibar.
Deposition of types: Hamburg Museum, Germany.
Distribution in China: Taiwan.
World Distribution: Widely distributed in tropical and temperate regions around the world.
Remarks: There are 3 peregrine species of octochaetid earthworms that have been found in Taiwan: *Dichogaster bolaui* (Michaelsen, 1891), *Dichogaster saliens* (Beddard, 1892), and *Dichogaster affinis* (Michaelsen, 1890). They all occur in coastal plains at elevations lower than 300 m, but are easily distinguishable. *D. bolaui* and *D. affinis* are quadriprostatic, while *D. saliens* is biprostatic. *D. bolaui* has a single female pore in XIV and no genital marking, while *D. affinis* has paired female pores on a raised pad in XIV and medioventral genital markings in 7/8–9/10. Occurrence of *D. affinis* in Taiwan

reported herein constitutes the island as the northernmost range of this species in East Asia.

354. *Dichogaster bolaui* (Michaelsen, 1891)

Dichogaster bolaui Chen, 1938. *Contr. Biol. Lab. Sci. Soc. China (Zool).* 12(10):419.
Dichogaster bolaui Chang, Shen & Chen, 2009. *Earthworm Fauna of Taiwan.* 154–155.
Dichogaster bolaui Xu & Xiao, 2011. *Terrestrial Earthworms of China.* 278.

External Characters

Length 25–35 mm, width 1.5 mm, segment number 86–92. Prostomium epilobous. First dorsal pore 5/6 or 6/7. Clitellum in XIII–XIX, XX, saddle-shaped, dorsal pores present, setae present. Setae lumbricine, closely paired ventrally. Male pores paired in XVIII, in seminal grooves, between prostatic pores in XVII and XIX. Spermathecal pores 2 pairs in 7/8 and 8/9, close to seta *a*. Female pore single, medioventral in XIV.

Internal Characters

Septa missing or reduced anteriorly. Gizzards 2 in VI and VII. Calciferous glands 3 pairs in XV–XVII. Intestine from XVII. Intestinal caeca absent. Lateral hearts in X–XII. Ovaries in XIII. Testes 2 pairs in X and XI. Prostate glands paired in XVII and XIX, slender tube-shaped. Spermathecae 2 pairs in VIII and IX.

Color: Live specimens light red to pink, semitransparent, orange on clitellum.
Type locality: Bergdorf near Hamburg, Germany.
Deposition of types: Hamburg Museum, Germany.
Distribution in China: Fujian (Kinmen Island), Taiwan (central), Hainan.
Etymology: This species was named after Mr. Bolau.

355. *Dichogaster saliens* (Beddard, 1892)

Dichogaster saliens Shen & Tsai, 2007. *Endemic Species Research.* 9(1):71–74.
Dichogaster saliens Chang, Shen & Chen, 2009. *Earthworm Fauna of Taiwan.* 156–157.

External Characters

Length 25–40 mm, clitellum width 1.4–2.3 mm, segment number 103–125. Prostomium epilobous. First dorsal pore 4/5 or 5/6. Clitellum in XIII–XIX, annular but thinner in *aa*, 1.7–3.2 ... mm in length. Setae lumbricine, small and closely paired on ventrum, *ab* modified as penial setae in XVII and usually lacking in XVIII. Male pores at posterior ends of grooves within closely paired, transversely diamond-shaped porophores in XVII. Genital marking medioventral, round, across 15/16.

Spermathecal pores 2 pairs in 7/8 and 8/9, medio-ventral, in line with seta *a*. Female pores paired in XIV, each medial to seta *a*.

Internal Characters

Septa thickened from 1/2, 5/6/7/8 missing. Crop large in VI. Gizzard 2 pairs in VII and VIII, muscular, barrel-shaped, displaced posteriorly to IX and X. Calciferous glands 3 pairs in XV-XVII, digiform, the last pair without external lamellae. Intestine from XVII. Esophageal hearts in X-XII. Nephridia meroic, saccular, 4 rows on each side. Ovaries flowery in XIII. Testes 2 pairs in X and XI, small, round. Seminal vesicles lacking or very small in XI and XII, or only in XII. Prostate glands 1 pair in XVII, long, tubular to tapped, twisted duct with penial setae at the end. Spermathecae 2 pairs in VIII and IX, small, ampulla oval, about 0.3 mm long, with a twisted stalk bearing small, spherical, shiny diverticulum.

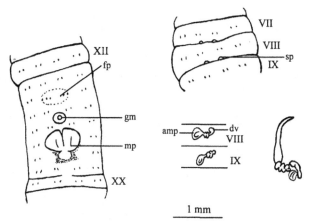

FIG. 346　*Dichogaster saliens* (after Shen & Tsai, 2007).

Color: Preserved specimens brown on anterior dorsum and yellow around clitellum.
Type locality: Bergedorf near Hamburg, Germany.
Deposition of types: British Museum, London, England.
Etymology: The species name "saliens" means "to leap or to move by leaps" in Latin and probably refers to the behavior of the species.
Distribution in China: Taiwan (Changhua, Yunlin).
Remarks: Ohfuchi (1957) described *Dichogaster hatomaana* as a new species from Hatoma-jimanear, Iriomote. It is a small earthworm with a body length of 40–60 . mm and diameter of 1.5 . mm. Easton (1981) considered that the description of *D. hatomaana* provided by Ohfuchi (1957) is indistinguishable from that of *D. saliens*. Although most characters between these 2 species look similar, the narrow genital zone commencing from segment XV and ending in segment XVII, described and illustrated by Ohfuchi (1957),

distinctively differs from that of *D. saliens*. Therefore, *D. hatomaana* is retained as a valid species in this study.

356. Dichogaster sinicus Chen, 1938

Dichogaster sinicus Chen, 1938. *Contr. Biol. Lab. Sci. Soc. China (Zool).* 12(10):420–422.
Dichogaster sinicus Xu & Xiao, 2011. *Terrestrial Earthworms of China.* 279.

External Characters

Length 26–27 mm, width 1.1–1.8 mm, segment number 78–124. Prostomium proepilobous. First dorsal pore 5/6. Clitellum in 1/2 XIII–1/2 XXI, annular, well marked, glandular on XIV-XX dorsally and laterally, intersegmental furrows scarcely visible; less so on XIII & XXI and ventral side of whole organ, setae and intersegmental furrows visible on ventral region. Setae of anterior body in perichaetine arrangement, setal pits in cuticle and integument conspicuous, about 60 on VIII, not visible on segments behind clitellum. Lumbricine setae before clitellum *ab = cd*, *aa* = 2.5*ab*, *bc* a little wider than *aa*, *dd* about 2/3 circumference apart dorsally. Those behind clitellum more conspicuous, at posterior end *aa* wider than *bc*; *dd* over 1/2 circumference dorsally. Male pores on XVIII, setae absent; prostatic pores on XVII and XIX, each in a deep pit, connected on each side by a longitudinal seminal groove, highly glandular at sides of groove prostatic pit; penial setae on XVII and XIX distinctly visible. Spermathecal pores 2 pairs in 7/8/9, medioventral, in line with seta *a*. 2 genital papillae placed ventromedially in 8/9/10, slightly raised; these sometimes concealed in body wall and invisible externally. Female pores 2, separate on XIV, between setae *aa*, whitish in appearance.

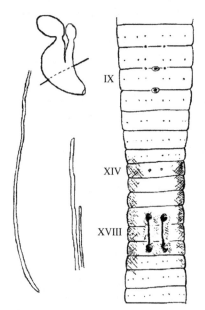

FIG. 347　*Dichogaster sinicus* (after Chen, 1938).

Internal Characters

Septa 6/7/8/9 very thin, 9/10–12/13 very thick, especially 9/10/11/12. Gizzard 2 pairs in VII and VIII, clearly separate and muscular; anterior to it found a crop. Intestine swelling from XVI. Calciferous glands as 3 distinct loops, in XV–XVII, with a common opening in XV on each dorsolateral side of esophagus. Micronephridia before clitellum small and irregularly placed, those behind arranged in 4 regular longitudinal series on each side. Lateral hearts in X–XII. Ovaries and oviduct-funnels in XIII, oviducts short. Testes and funnels in X and XI, enclosed in a sac in common with respective vesicles on each side. Seminal vesicles 3 pairs, in X–XII, often branched into saccules which are filled with developing sperm morulae. Vas deferens on each side running backward to opening in XVIII. Prostate gland tubular about 0.7 mm long, 0.11 mm wide, opening medially to penial setae. Penial setae of 1 form, about 2 (rarely 3) in each bundle, 1 stouter, distal end not dilated, weakly curved in a zigzag manner, node-like structures occurring on convex side of every curve, about 0.35 mm long, 0.1 mm exposed; another shorter and much more slender similarly shaped, its zigzag curvature noticeable, nodular structures absent. Spermathecae 2 pairs in VIII and IX. Ampulla ovoid, about 0.1 .. mm wide, constricted from its duct; ampullar duct dilated at distal two-thirds, ampulla and its duct about 0.4 mm long. Diverticulum shorter than main pouch, arising from ectal end of main duct, close to body wall; an ovoid seminal chamber distinct from its duct. Accessory glands embedded in body wall.

Color: Preserved specimens pale generally; clitellum reddish.
Type locality: Hainan (Sha-mom-chiu, Lingshui).
Deposition of types: Previously deposited in the Museum of the Biological Laboratory of the Science Society of China, Nanjing. The types were destroyed during the war in 1937.
Etymology: The name of the species refers to the type locality.
Distribution in China: Hainan (Sha-mom-chiu, Lingshui).
Remarks: In general structures, the present species is close to Dichogaster bolaui (Michaelsen, 1891). Its characteristic feature of having vestigial perichaetine setae in anterior segments is common to all the specimens examined. It is not found in Dichogaster bolaui and other species of the genus. The difference of penial setae is again a good criterion for separating the species.

GENUS: *RAMIELLA* STEPHENSON, 1921

Ramiella Gates, 1972. *Trans. Amer. Phil. Soc.* 62(7):311–312.

Type species: *Octochaetus bishambari* Stephenson, 1914.

Clitellum, extending into XIII and XVII, intersegmental furrows obliterated, dorsal pores occluded, setae retained. Dorsal pores present. Setal arrangement lumbricine, 4 pairs per segment. Male pore in XVIII. Seminal groove between equators of XVII and XIX. Spermathecal pores at or behind 7/8/9. Prostatic pores equatorial in XVII and XIX.

All septa present (behind their commencement). One esophageal gizzard in 1 simple segment. Intestinal origin behind XIII. Without caeca, supraintestinal and calciferous glands. Excretory system purely micronephridial, the micronephridia relatively large and few in number, from 1–7 pairs per segment. Sexual apparatus purely acanthodriline.

Global distribution: India.
There is 1 species from China (Fujian).

357. *Ramiella sinicus* (Chen, 1935)

Howascolex sinicus Chen, 1935a. *Contr. Biol. Lab. Sci. Soc. China (Zool).* 11:113–120.
Ramiella sinicus Xu & Xiao, 2011. *Terrestrial Earthworms of China.* 280–281.

External Characters

Length 45 mm, width 1.2 mm, segment number 90. Prostomium 2/3 epilobous. First dorsal pore 7/8. Clitellum slightly swollen, beginning and terminating gradually, in 2/3 XIII–1/2 XVII, saddle-shaped, its ventral side medial to setae *b* very slightly glandular. Ventral setae distinct; dorsal ones in XIII and XVII shorter but visible, those on other glandular parts revealed only in sections which are much thinner than normal ones. Intersegmental furrows only visible on ventral side. No other genital markings. Setae in regular lumbricine arrangement throughout the body, *aa* a little wider than *bc*, *aa* = 2.8*ab*, *bc* = 2*ab*, *ab* = 4/5*cd*, *dd* of dorsal side less than half of circumference by a distance of *cd*, but that on posterior part of body 1/5–1/3 circumference. These present in all segments except first, last 2 and *ab* on XVIII, *ab* on segments XVII and XIX modified as penial setae but those spermathecal segments not modified. Male pores 1 pair, on XVIII, each in a longitudinal groove which is communicated with the prostatic pores on XVII and XIX respectively. Each groove shallow and narrow, slightly curved inward at its middle part. Male pores invisible externally. Prostatic pore at end of each groove associated with a minute papilla which is hardly visible without the help of a magnifier. Penial setae also not visible. No genital papillae in any part of body. Spermathecal pores 2 pairs, in 7/8/9, not conspicuous externally, but visible under higher magnification; each in line with seta *b* or slightly lateral to it. Female pores 1 pair, as transverse slits on XIV, in front of but slightly medial to seta *a*, situated about midway between seta and intersegmental furrow. Width of each slit about distance of *ab*.

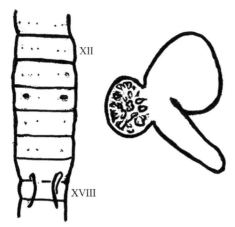

FIG. 348   *Ramiella sinicus* (after Chen, 1935a).

Internal Characters

Septa present in all segments after commencing; 4/5 very thin, 5/6/7 a little more muscular, 7/8/9 with thicker musculature, 9/10/11/12 muscular but thinner, 12/13 and those thereafter thin and very slightly muscular. Gizzard single, very muscular in VII (in front of septum 5/6), occupying about 1 segment, rounded and whitish in appearance. Intestine narrower than posterior portion of esophagus, beginning in about XX, without a typhlosole. Lateral hearts 4 pairs, in X-XIII, connecting ventral and supraintestinal vessels except those in XIII which connect dorsal and lateroesophageal ones. Testis 2 pairs, free in X and XI, originating under septa 9/10 and 10/11 respectively, each elongate and flat, about 110 μm in length. Seminal funnels very large, also free, but covered by the vesicles, about 240 μm in diameter. Vas deferens of each side meeting in XII, but remaining separate until in body wall of XVIII where they unite. Each external opening simple, on lateral side of the groove, without papilla and other markings. Seminal vesicles 3 pairs, in X-XII, nearly filling each segment, whitish and tubercular on surface, first pair slightly smaller. Prostate glands purely acanthodriline in character. Both anterior and posterior pairs about equal in size and general appearance. Glandular portion cylindrical and warty on surface, loosely twisted, about 1.9 .. mm long, 0.17 mm wide. Each opened on lateral side of a low papilla with tips of penial setae on other side of it, the cavity in which the low papilla is situated separated from the longitudinal groove by a very thin partition. Penial setae not found on XVIII, but on XVII and XIX respectively, 2 pairs in each group. Probably each pair is produced by splitting every original seta, for example, either *a* or *b* of segments XVII and XIX. Therefore, the members of each pair are nearly identical and closely spaced, and are enclosed in a common setigerous sac. Each very slender, long and weakly curved, about 0.56–0.68 ...... mm long, about 4-5 times longer than normal ones, 16 μm in thickness, simple but pointed at tips. One pair placed at side of the papilla just referred to (corresponding to seta *a*) with tips pointed and sickle-shaped. Another pair of accessory penial setae (corresponding to seta *b*) with tips straighter, not implanted in body wall but retracted in body cavity and lying under ectal end of prostatic duct, which are similar in size as the former pair. Spermathecae 2 pairs, in VIII and IX; each composed of a roundish ampulla, about 0.3 mm in diameter, and a main duct, about 0.35 mm long and 0.09 mm wide; diverticulum also round, attached to ental end of the ampullar duct, largely fused with the ampulla. Its size about half that of the ampulla, highly iridescent in appearance, filled with clusters of spermatozoa which are generally not found in the ampulla.

Color: Preserved specimens brown.
Type locality: Fujian (Xiamen).
Distribution in China: Fujian (Xiamen).
Etymology: The name of the species refers to the type locality.
Deposition of types: Previously deposited in the Museum of the Biological Laboratory of the Science Society of China, Nanjing. The types were destroyed during the war in 1937.

References

Baird, W., 1869. Description of a new species of earthworm (*Megascolex diffringens*) found in north Wales. Proc. Zool. Soc. London 1869, 40–43.

Baird, W.B., 1873. Description of some new species of Annelida and Gephyrea in the collection of the British Museum. J. Linn. Soc. 11, 94–97.

Beddard, F.E., 1869. Additional remarks on the *Megascoles diffringens*. Proc. Zool. Soc. London 387–389.

Beddard, F.E., 1883. Notes on some earthworms from India. Annu. Mag. Nat. Hist. 12 (5), 213–224.

Beddard, F.E., 1886. Descriptions of some new or little-known earthworms, together with an account of the variations in structure exhibited by *Periony excavatus*, E. P. Proc. Zool. Soc. London 1886, 298–314.

Beddard, F.E., 1890. Observations upon an American species of *Perichaeta*, and upon some other members of the genus. Proc. Zool. Soc. London 52–69.

Beddard, F.E., 1891. Abstract of some investigations into the structure of the Oligochata. Annu. Mag. Nat. Hist. 7 (6), 88–96.

Beddard, F.E., 1893. On the geographical distribution of earthworms. Proc. Zool. Soc. London 733–738.

Beddard, F.E., 1896. On some earthworms from the sandwich islands collected with an appendix on some new species of *Perichaeta*. Proc. Zool. Soc. London 1896, 194–211.

Beddard, F.E., 1900a. On the earthworms collected during the "Skeat expedition" to the Malay Peninsula, 1899–1900. Proc. Zool. Soc. London 891–911.

Beddard, F.E., 1900b. On a new species of earthworm from India belonging to the genus *Amyntas*. Proc. Zool. Soc. London 998–1002.

Beddard, F.E., 1900c. A revision of the earthworms of the genus *Amyntas* (*Perichaeta*). Proc. Zool. Soc. London 1900, 609–652.

Bourne, A.G., 1886. On Indian earthworms—Part 1. Preliminary notice of earthworms from the Nilgiris and Shevaroys. Proc. Zool. Soc. London 1886, 662–672.

Brinkhurst, R.O., Jamieson, B.G.M., 1972. Aquatic Oligochaeta of the World. University of Toronto Press, Edinburgh. 860 pp.

Chang, C.H., Chen, J.H., 2003. A new species of earthworm belonging to the genus *Metaphire* Sims and Easton 1972 (Oligochaeta: Megascolecidae) from southern Taiwan. Endemic Species Res. 5, 83–88.

Chang, C.H., Chen, J.H., 2004. A new species of earthworm belonging to the genus *Metaphire* Sims and Easton 1972 (Oligochaeta: Megascolecidae) from Southern Taiwan. Taiwania 49 (4), 219–224.

Chang, C.H., Chen, J.H., 2005. Three new species of octothecate Pheretimoid earthworms from Taiwan, with discussion on the biogeography of related species. J. Nat. Hist. 39 (18), 1469–1482.

Chang, C.H., Yang, K.W., Wu, J.H., Chuang, S.C., Chen, J.H., 2001. Species composition of earthworms in the main campus of National Taiwan University. Acta Zool. Taiwan 12, 75–81.

Chen, Y., 1930. On some new earthworms from Nanking, China. Sci. Rep. Natl. Central Univ. Nanking 1, 11–37.

Chen, Y., 1931. On the terrestrial Oligochaete from Szechuan, with description of some new forms. Contr. Biol. Lab. Sci. Soc. China (Zool.) 7 (3), 117–171.

Chen, Y., 1933. A preliminary survey of the Earthworms of the Lower Yangtze Valley. Contr. Biol. Lab. Sci. Soc. China (Zool.) 9 (6), 177–293.

Chen, Y., 1935a. On two new species of Oligochaeta from Amoy. (*Pheretima wui* sp. N. and *Howascolex sinicus* sp N.). Contr. Biol. Lab. Sci. Soc. China (Zool.) 11 (4), 109–122.

Chen, Y., 1935b. On a small collection of Earthworms from Hongkong with descriptions of some new species. Bull. Fan. Mem. Inst. Biol. 6, 33–59.

Chen, Y., 1936. On the terrestrial Oligochaete from Szechuan II. with the notes on Gate's type. Contr. Biol. Lab. Sci. Soc. China (Zool.). 11 (8), 269–306.

Chen, Y., 1938. Oligochaete from Hainan, Kwangtung. Contr. Biol. Lab. Sci. Soc. China (Zool.). 12 (10), 375–427.

Chen, Y., 1946. On the terrestrial Oligochaete from Szechuan III. J. West China. Border Res. Soc. (B) 16, 83–141.

Chen, Y., 1956. The Earthworms from China. 1956, Science Press, Beijing, China, pp. 1–53. (in Chinese).

Chen, Y., 1958. Zoology. The Commercial Press, Beijing.

Chen, Y., 1959. China Animal Atlas: Annelid. 1956, Science Press, Beijing, China, pp. 1–29. (in Chinese).

Chen, I.H., Chang, C.H., Chuang, S.C., Lin, Y.H., Chen, J.H., 2004. The distribution of the exotic earthworm Pontoscolex corethrurus in northern Taiwan and its potential impacts on soil and native earthworm populations. Chin. Biosci. 47(1), 117–126.

Chen, J.H., Chuang, S.C., 2003. A new record of the bithecal megascolecid earthworm *Amynthas papilio* (Gates) (Oligochaeta) from Taiwan. Endemic Species Res. 5 (2), 89–94.

Chen, Q., Fen, X.Y., 1996. On bucculenta species group of *Metaphire* and reproductive organ polymorphism (Oligochaeta: Megascolecidae). Acta Zootax. Sin 10 (4), 354–356. (in Chinese with English synopsis).

Chen, Y., Xu, Z.F., 1977. On some earthworms from China II. Acta Zool. Sin. 23 (2), 175–181. (in Chinese with English synopsis).

Chen, Y., Hsü, C.F., Yang, T., Fong, H.Y., 1975. On some earthworms from China. Acta Zool. Sin. 21 (1), 89–99. (in Chinese with English synopsis).

Chuang, C.H., Chen, J.H., 2002. A new record earthworms *Anynthas masatakae* (Beddard) (Megascolecidae: Oligochaeta) from Taiwan. Acta Zool. Taiwan 13 (2), 73–79.

Ding, R.H., 1983. A preliminary investigation of terrestrial Oligochaete from the suburbs of Chengdu. Sichuan J. Zool. 2 (2), 1–5. 21 (in Chinese).

Ding, R.H., 1985. Descriptions of a new species terrestrial Oligochaetes from Sichuan (Oligochaeta: Megascolecidae). Acta Zootax. Sin. 10 (4), 354–356. (in Chinese with English synopsis).

Easton, E.G., 1976. Taxonomy and distribution of the *Metapheretima elongata* species-complex of Indo-Australasian earthworms (Megascolecidae: Oligochaeta). Bull. Br. Mus. Nat. Hist. (Zool.) 30, 31–51.

Easton, G.E., 1979. A revision of 'acaecate' earthworms of the *Pheretima* group (Megascolecidae: Oligochaeta): *Archipheretima*, *Metapheretima*, *Planap eretima*, *pleionogaster*, and *Polypheretima*. Bull. Br. Mus. Nat. Hist. (Zool.) 35 (1), 1–128.

Easton, E.G., 1981. Japanese earthworms: a synopsis of the Megadrile species (Oligochaeta). Bull. Br. Mus. Nat. Hist. (Zool.) 40 (2), 33–65.

Easton, E., 1983. A guide to the valid names of Lumbricidae (Oligochaeta). *Earthworm ecology*, Springer, Berlin.

Edwards, C.A., 1991. The assessment of populations of soil-inhabiting invertebrates. Agric. Ecosyst. Environ. 34, 145–176.

Edwards, C.A., Lofty, J.R., 1977. Biology of Earthworms. Chapman & Hall, London.

Fang, B.W., 1929. Notes on a new species of *Pheretima* from Kwangsi, China. Sinensia Nanking 1 (2), 15–24.

Fang, B.W., 1933. Notes on a small collection of earthworms from Ichang, Hupeh. Sinensia Nanking 3 (7), 179–184.

Feng, X.Y., 1983. An introduction to the new taxonomic system of the genus Pheretima (Oligochaeta) of Sims. Sichuan J. Zool. 2 (2), 22–25. 33.

Feng, X.Y., 1984. A new subspecies of terrestrial Oligochaeta from Lanzhou, Kansu Province. Zool. Res. 5 (1), 47–50. (in Chinese with English synopsis).

Feng, X.Y., 1985. Taxonomy characteristics on genus of Chinese terrestrial earthworms. J. Zool. 20, 44–47. (in Chinese).

Feng, X.Y., Ma, Z.G., 1987. Notes on a new species of the genus Metaphire from Gansu Province, China. Acta Zootax. Sin. 12 (3), 248–250. (in Chinese with English synopsis).

Gates, G.E., 1926. Notes on earthworms from various places in the Province of Burma, with description of two new species. Rec. Indian Mus. 28, 141–170.

Gates, G.E., 1929. A summary of the earthworms Fauna of Burma with Descriptions of Fourteen new species. Proc. U. S. Natl. Mus. 75 (10), 1–41.

Gates, G.E., 1930. The earthworms of Burma. I. Rec. Indian Mus. 32, 257–356.

Gates, G.E., 1931. The earthworms of Burma. II. Rec. Indian. Mus. 33, 327–442.

Gates, G.E., 1932. The earthworms of Burma. III. Rec. Indian Mus. 34 (4), 357–549.

Gates, G.E., 1933. The earthworms of Burma. IV. Rec. Indian Mus. 35, 412–606.

Gates, G.E., 1934. Notes on some earthworms from the Indian museum. Rec. Indian Mus. 36, 233–277.

Gates, G.E., 1935a. New earthworms from China, with notes on the synonymy of some Chinese species of Drawida and Pheretima. Smithsonian Misc. Collect. 93 (3), 1–19.

Gates, G.E., 1935b. On some Chinese earthworms. Lingnan J. Sci. 14 (3), 445–457.

Gates, G.E., 1936. The earthworms of Burma. V. Rec. Indian Mus. 38, 377–468.

Gates, G.E., 1939. On some species of Chinese earthworms, with special reference to species collected in Szchuan by Dr. D. C. Graham. Proc. U. S. Natl. Mus. 85, 405–507.

Gates, G.E., 1953. On the earthworms of the Arnold Arboretum, Boston. Bull. Mus. Compar. Zool. 107 (10), 501–534.

Gates, G.E., 1959a. On some earthworms from Taiwan. Am. Mus. Novit. 1941, 1–19.

Gates, G.E., 1959b. On a taxonomic puzzle and the classification of the earthworms. Bull. Mus. Comp. Zool. Harv. 121 (229), 61.

Gates, G.E., 1961. On some Burmese and Indian earthworms of the Family Acanthodrilidae. Annu. Mag. Nat. Hist. 13 (4), 417–429.

Gates, G.E., 1965. On peregrine species of the Moniligastrid earthworms genus Drawida Michaelsen, 1900. Annu. Mag. Nat. Hist. 13 (8), 85–93.

Gates, G.E., 1972. Burmese earthworms: an introduction to the systematics and biology of megadrile oligochaetes with special reference to southeast Asia. Trans. Am. Phil. Soc. 62 (7), 1–326.

Goto, S., Hatai, S., 1898. New or imperfectly known species of earthworms. Annot. Zool. Jpn. 2 (1), 65–78.

Goto, S., Hatai, S., 1899. New or imperfectly known species of earthworms. Annot. Zool. Jpn. 3 (2), 13–24.

Horst, R., 1883. New species of the genus Megascolex Templeton (Perichaeta schmarda) in the collections of the Leyden museum. Notes Leyden Mus. 5, 182–196.

Huang, F.Z., 1982. Earthworm. Agriculture Press, Beijing, pp. 65–84.

Huang, J., Xu, Q., Sun, Z.J., Wang, C., Zheng, D.M., 2006. Research on earthworm resources of China: checklist and distribution. J. China Agric. Univ. 11, 9–20.

Huang, J., Xu, Q., Sun, Z.J., Tang, G.L., Li, C.P., Cui, C.X., 2007. Species abundance and zoogeographic affinities of Chinese terrestrial earthworms. Eur. J. Soil Biol. 43, S33–S38.

James, S.W., Shih, H.T., Chang, H.W., 2005. Seven new species of Amynthas (Clitellata: Megascolecidae) and new earthworm records from Taiwan. J. Nat. Hist. 39 (14), 1007–1028.

Jamieson, B.G.M., 1971a. A review of the megascolecid earthworm genera (Oligochaeta) of Australia. Proc. R. Soc Wd. 82, 75–86.

Jamieson, B.G.M., 1971b. Descriptions of the type-species of the earthworm genera Plutellus and Digaster (Megascolecidae: Oligochaeta). Bull. Mus. Hist. Nat. Paris 2, 6.

Jamieson, B.G.M., 1971c. Earthworms (Megascolecidae: Oligochaeta) from western Australia and their zoogeography. J. Zool. 165, 471–504.

Jamieson, B.G.M., 1978. A comparison of spermiogenesis and spermatozoal ultrastructure in megascolecid and lumbricid earthworms (Oligochaeta: Annelida). Aust. J. Zool. 26, 225–240.

Jamieson, B.G.M., 1985. The spermatozoa of the Holothuroidea (Echinodermata): an ultrastructural review with data on two Australian species and phylogenetic discussion. Zool. Scr. 14, 123–135.

Jamieson, B.G.M., 1988. On the phylogeny and higher classification of the Oligochaeta. Cladistics 4, 367–401.

Kinberg, J.G.H., 1867. Annulata nova. Öfversigt af Kongliga Vetenskaps–Akademiens Förhandlingar, 23, 97–103.

Kobayashi, S., 1936a. Distribution and some external characteristics of Pheretima (Ph.) carnosa (Goto et Hatai) from, Korea. Sci. Rep. Tuhoku Univ. 11, 115–138.

Kobayashi, S., 1936b. Earthworms from Koryo, Korea. Sci. Rep. Tuhoku Univ. 11, 139–184.

Kobayashi, S., 1936c. Preliminary survey of the earthworms of Quelpart Island. Sci. Rep. Tohoku Univ. 11 (4), 333–351.

Kobayashi, S., 1938a. Earthworms from Hakodate, Hokkaido. Annot. Zool. Jpn. 3 (2), 405–417.

Kobayashi, S., 1938b. Earthworms of Korea. I. Sci. Rep. Tohoku Imp. Univ. 13 (2), 89–170.

Kobayashi, S., 1938c. Occurrence of Perionyx excavatus E. Perrier in North Formosa. Sci. Rep. Tohoku Imp. Univ. (B) 13 (2), 201–203.

Kobayashi, S., 1939. A re-examination of Pheretima yamadai Hatai, an earthworm found in Japan and China. Sci. Rep. Tohoku Imp. Univ. Fourth Ser. (Biol.) 14 (1), 135–139.

Kobayashi, S., 1940. Terrestrial Oligochaeta from Manchoukuo. Sci. Rep. Tohokuo Univ. 15, 261–315.

Kobayashi, S., 1941a. Earthworms of Korea II. Sci. Rep. Tohoku Univ. 16 (4), 391–405.

Kobayashi, S., 1941b. Earthworm from the South Sea Islands II. Sci. Rep. Tohoku Univ. 16 (4), 391–405.

Kobayashi, S., 1941c. Earthworms of Korea. II. Sci. Rep. Tohoku lmp. Univ. (B) 16 (2), 147–156.

Kuo, T.C., 1995. Ultrastructure of genital markings in some species Pheretima, Bimastus and Perionyx in northern Taiwan. Nat. Hsinchu Teach. Coll. J. 8, 181–199.

Lang, Y.S., Zheng, F.Q., 2009. Techniques and Applications of Earthworm Culture. Scientific and Technical Literature Press, Beijing.

Lee, K.E., 1959. A key for the identification of New Zealand earthworms. Tuatara 8 (I), 13–60.

Michaelsen, W., 1900. Oligochaeta. Das Tierreich 10, 1–575.

Michaelsen, W., 1903. Die geographische Verbreitung der Oligochaeten. Berlin.

Michaelsen, W., 1921. Zur Stammesgeschichte und Systematik der Oligochäten, insbesondere der Lumbriculiden. Arch. Naturgesch. 86.

Michaelsen, W., 1927. Oligochaten aus Yun-nan, gesammelt von Prof. F. Silvestri. Boll. Lab. Zool. Portici. 21, 84–90.

Michaelsen, W., 1928a. Miscellanea oligochaetologiea. Arkiv For Zoologi 20 (2), 1–15.

Michaelsen, W., 1928b. Miscellanea oligochaetologiea. Arkiv For Zoologi 20 (3), 1–60.

Michaelsen, W., 1929. The Oligochaeta fauna of China. Lingnan Sci. J. 8, 157–166.

Michaelsen, W., 1931. The Oligochaete of China. Peking Nat. Hist. Bull. 5 (2), 1–24.

O'Connor, F.B., 1955. Extraction of enchytraeid worms from a coniferous forest soil. Nature 175, 815–816.

Omodeo, P., 1958. La reserve naturelle integrale du Mont Nimba. I. Oligochetes. Mem. Inst. Fr. Afr. Noire 53, 9–10.

Perrier, E., 1872. Lombriciens Terrestres. Nouvelles Archives Du Museum, pp. 4–197.

Qiu, J.P., 1988a. Two new species of the genus Pheretima from Guizhou (Oligochaeta: Megascolecidae). Sichuan J. Zool. 6 (4), 4–8.

Qiu, J.P., 1988b. Two new species of the Genus Pheretima from Guizhou (Oligochaeta: Megascolecidae). Sichuan J. Zool. 7 (1), 1–4.

Qiu, J.P., 1992. Notes on a new subspecies. of genus Amynthas from Guizhou Province (Oligochaeta: Megascolecidae). Sichuan J. Zool. 11 (1), 1–3.

Qiu, J.P., 1993. Notes on a new species of the genus Metaphire (Haplotaxida: Megascolecidae) from Guizhou Province, China. Sichuan J. Zool. 12 (4), 1–4.

Qiu, J.P., Wang, H., 1992. Notes on two new species of genus Amynthas (Haplotaxida: Megascolecidae). Acta Zootax. Sin. 17 (3), 262–265.

Qiu, J.P., Wen, C.L., 1988. A new species of earthworm from Guizhou Province (Oligochaeta: Megascolecidae). Acta Zootax. Sin. 13 (4), 340–342. (in Chinese with English synopsis).

Qiu, J.P., Zhong, Y.H., 1993. Notes on a new species and a new subspecies of the genus Metaphire from Guizhou Province, China (Haplotaxida: Megascolecidae). Guizhou Sci. 11 (1), 38–44. (in Chinese with English synopsis).

Qiu, J.P., Wang, H., Wang, W., 1991a. Notes on a new species of genus Amynthas from Guizhou Province, China (Oligochaeta: Megascolecidae). Guizhou Sci. 9 (4), 301–304. (in Chinese with English synopsis).

Qiu, J.P., Wang, H., Wang, W., 1991b. Notes on three new species of territorial Oligochaetes from Guizhou Province (Oligochaeta: Megascolecidae). Guizhou Sci. 9 (3), 220–226. (in Chinese with English synopsis).

Qiu, J.P., Wang, H., Wang, W., 1993a. Notes on a new species of genus Amynthas (Haplotaxida: Megascolecidae). Guizhou Sci. 11 (4), 3–6. (in Chinese with English synopsis).

Qiu, J.P., Wang, H., Wang, W., 1993b. Notes on two new species of genus Pheretima (Haplotaxida: Megascolecidae) from Guizhou Province. Acta Zootax. Sin. 18 (4), 406–411. (in Chinese with English synopsis).

Qiu, J.P., Wang, H., Wang, W., 1994. Notes on a new species of genus Amynthas (Haplotaxida: Megascolecidae) from Guizhou Province. Sichuan J. Zool. 13 (4), 143. (in Chinese with English synopsis).

Quan, X.X., 1985. A new species of territorial Oligochaetes from Hainan island. Acta Zootax. Sin. 10 (1), 18–20. (in Chinese with English synopsis).

Quan, X.X., Zhong, Y.H., 1989. Two new species of territorial Oligochaetes from Hainan island. Acta Zootax. Sin. 14 (3), 273–277. (in Chinese with English synopsis).

Reynolds, J.W., 1977. The earthworms (Lumbricidae and Sparganophilidae) of Ontario. Toronto: R. Ontario Mus. 1–139.

Shen, H.P., Tsai, C.F., 2002. A new earthworm of the genus Pithemera (Oligochaeta: Megascolecidae) from the Lanyu Island (Botel Tobago). J. Natl. Taiwan Museum 55 (1), 17–24.

Shen, H.P., Yeo, C.J., 2005. Terrestrial earthworms (Oligochaeta) from Singapore. Raffles Bull. Zool. 53 (1), 13–33.

Shen, H.P., Tsai, C.F., Tsai, S.C., 2002. Description of a new earthworm belonging to the genus Amynthas (Oligochaeta: Megascolecidae) from Taiwan and its infraspecific variation in relation to elevation. Raffles Bull. Zool. 50 (1), 1–8.

Shen, H.P., Tsai, C.F., Tsai, S.C., 2003. Six new earthworms of the genus Amynthas (Oligochaeta: Megascolecidae) from the central Taiwan. Zool. Stud. 42 (4), 479–490.

Shen, H.P., Tsai, S.C., Tsai, C.F., Chen, J.H., 2005. Occurrence of the earthworm Pontodrilus litoralis (Grube, 1855), Metaphier houlleti (Perrier, 1872), and Eiseniella tetraedra (Savigny, 1826) from Taiwan. Taiwania 50 (1), 11–21.

Shen, H.P., Chang, C.H., Chih, W.J., 2014. Five new earthworm species of the genera Amynthas and Metaphire (Megascolecidae: Oligochaeta) from Matsu, Taiwan. J. Nat. Hist. 48 (9–10), 495–522.

Shih, H.T., Chang, H.W., Chen, J.H., 1999. A review of the earthworms (Annelida: Oligochaeta) from Taiwan. Zool. Stud. 38, 435–442.

Sims, R.W., 1966. The classification of the megascolecoid earthworms: an investigation of Oligochaete systematics by computer techniques. Proc. Linn. Soc. Lond. 177, 125–141.

Sims, R.W., Easton, E.G., 1972. A numerical revision of the earthworm genus Pheretima auct. (Megascolecidae: Oligochaeta) with the recognition of new genera and an appendix on the earthworms collected by the Royal Society North Borneo Expedition. Biol. J. Soc. 4, 169–268.

Sims, R.W., Gerard, B.M., 1985. Earthworms. The Linnean Society of London and the Estuarine and Brackish-Water Sciences Association, pp. 1–170.

Stephenson, J., 1930. The Oligochaete. Clarendon Press, Oxford, pp. 716–914.

Stephenson, W., 1912a. Contribution to the fauna of Yunnan based on collections made by J. Coggin Brown, B. Sc., 1909–1910. Part VIII—Earthworms. Rec. Indian Mus. 7 (3), 273–278.

Stephenson, T., 1912b. Contribution to the Fauna of Yunnan based on collections made by T-coggin B. Sc. 1909–1910 part 8-Earthworm. Rec. Indian Mus. 7, 273–274.

Stephenson, T., 1925. Oligochaeta from various regions. Proc. Zool. Soc. London 1925, 897–907.

Sun, J., Zhao, Q., Qiu, J.P., 2009. Four new species of earthworms belonging to the genus Amynthas (Oligochaeta: Megascolecidae) from Diaoluo Mountain, Hainan Island, China. Rev. Suisse Zool. 116 (2), 289–301.

Sun, J., Zhao, Q., Qiu, J.P., 2010. Three new species of earthworms belonging to the genus Amynthas (Oligochaeta: Megascolecidae) from Hainan Island, China. Zootaxa 2680, 26–32.

Tan, T.J., Zhong, Y.H., 1986. A new species of the genus Metaphire from Hunan (Oligochaeta: Megascolecidae). Acta Zootax. Sin. 11 (2), 144–148. (in Chinese with English synopsis).

Tan, T.J., Zhong, Y.H., 1987. Two new species of the genus Metaphire from Hunan Province (Oligochaeta: Megascolecidae). Acta Zootax. Sin. 12 (2), 128–132. (in Chinese with English synopsis).

Thai, T.B., 1984. New data on taxonomy of the genus Pheretima (Oligochaeta, Megascolecidae) of the fauna of Vietnam. Zoologicheskii Zn. 63 (2), 284–288.

Tsai, C.F., 1964. On some earthworms belonging to the genus Pheretima Kinberg collected from Taipei area in North Taiwan. Q. J. Taiwan Mus. 17 (1&2), 5–8.

Tsai, C.F., Shen, H.P., Tsai, S.C., 1999. On some new species of the Pheretimoid earthworms (Oligochaeta: Megascolecidae) from Taiwan. J. Natl. Taiwan Mus. 52, 33–46.

Tsai, C.F., Shen, H.P., Tsai, S.C., 2000. Native and exotic species of the terrestrial earthworms (Oligochaet) in Taiwan with reference to northeast Asia. Zool. Stud. 39 (4), 285–294.

Tsai, C.F., Tsai, S.C., Liaw, G.J., 2000a. Two new species of Protandric Pheretimoid earthworms belonging to the genus Metaphire (Megascolecid: Oligochaeta) from Taiwan. J. Nat. Hist. 34, 1731–1741.

Tsai, C.F., Shen, H.P., Tsai, S.C., 2001. Some new earthworms of the genus Amynthas (Oligochaeta: Megascolecidae) from Mt. Hohuan of Taiwan. Zool. Stud. 40 (4), 276–288.

Tsai, C.F., Shen, H.P., Tsai, S.C., 2002. A new athecate earthworm of the genus *Amynthas* Kinberg (Megascolecidae: Oligochaeta) from Taiwan with discussion on phylogeny and biogeography of the *A. illotus* species-group. J. Nat. Hist. 36, 757–765.

Tsai, C.F., Shen, H.P., Tsai, S.C., 2004. A new *glgarntic* earthworm of the genus *Metaphire* Sims and Eston (Megascolecidae: Oligochaeta) from Taiwan with reference to evolutional trends in body size and segment numbers of the *Pheretima* genus-group. J. Nat. Hist. 38, 877–887.

Tsai, C.F., Shen, H.P., Tsai, S.C., Lee, H.H., 2007. Four new species of terrestrial earthworms belonging to the genus *Amynthas* (Megascolecidae: Oligochaeta) from Taiwan with discussion on speculative synonyms and species delimitation in Oligochaete taxonomy. J. Nat. Hist. 41 (5-8), 357–379.

Vaillant Par, M.L., 1868. Note sur L'anatomie de deux especes du genre Perichaeta. Annales Des Sciences Naturelles X, 225–256.

Wang, H.J., Qiu, J.P., 2005. A preliminary report on Pheretimoid earthworms from the Xingdou mountain reserve with description of two new species and a new subspecies of the genera *Amynthas* and *Metaphire*. J. Shanghai Jiaotong Univ. (Agric. Sci.). 23 (1), 23–30. (in Chinese).

Wu, J.H., Sun, X.D., 1996. A new species of Genus *Drawida* from the Changbai mountains (Oligochaeta: Moniligasatridae). Sichuan J. Zool. 15 (3), 98–99. 117. (in Chinese).

Wu, J.H., Sun, X.D., 1997. On one new species of *Amynthas* (Oligochaeta: Megascolecidae) from China. Sichuan J. Zool. 16 (1), 3–5.

Xu, Q., 1996. Introduction of distribution of Chinese terrestrial earthworm. J. Beijing Edu. Coll. 3, 54–61. (in Chinese).

Xu, X.Y., 2000. Investigation of earthworm species in Fuling County, Sichuan Province. J. Sichuan Coll. Educ. 16 (9), 42–44.

Xu, Q., Xiao, N.W., 2011. Terrestrial Earthworms of China. China Agriculture Press, Beijing. (in Chinese).

Xu, Z.F., Zhang, D.N., Jiang, J.M., 1989. A new genus and species of *Enchytraeidae* from Gansu province (Oligochaeta: Plesiopora). Acta Zootax. Sin. 14 (02), 153–156. 257–258.

Xu, Z.F., Zhang, D.N., Yang, L., 1990. On a new fluorescent earthworm from China (Opisthopora: Megascolecidae, Acanthodrilinae). Acta Zootax. Sin. 15 (1), 28–31.

Xu, R.H., He, Z.W., Li, X.Z., 1994. Notes on the terrestrial Oligochaeta in Henan. J. Henan Univ. (Nat. Sci.) 22 (1), 63–65.

Yin, W.Y., 1992. Subtropical Soil Animals of China. Science Press, Beijing, China, pp. 190–207.

Yin, W.Y., 1998. Pictorical Keys to Soil Animals of China. 90–106, Science Press, Beijing, China, pp. 476–486.

Yin, W.Y., 2000. Soil Animals of China. Science Press, Beijing, China, pp. 229–236.

Yu, D.J., Liu, Y.J., Li, B.D., 1990. The earthworm Lumbricidae and its distribution in Jilin Province. J. Zool. 25 (5), 44–47. 49. (in Chinese).

Yu, D.J., Liu, Y.J., Yin, X.Q., 1992. Two new record species of *Pheretima* from Northeast China. J. Zool. 27 (2), 53–54. (in Chinese).

Zhang, Y.F., Sun, Z.J., 2014. A new earthworms species of the genus *Drawida* Michaelsen (Oligochaeta: Moniligastridae) from China. Zool. Syst. 39 (3), 422–444.

Zhang, Y.P., Wu, J.H., Sun, X.D., 1998. A new species of *Pontoscolex* (Oligochaeta: Glossoscolecidae) from Guangdong Province, China. Sichuan J. Zool. 17 (1), 5–6.

Zhang, W.X., Li, J.X., Qiu, J.P., 2006a. New earthworms belonging to the genus of *Amynthas* Kinberg (Megascolecidae: Oligochaeta) and *Drawida* Michaelsen (Moniligastridae: Oligochaeta) from Guangdong, China. J. Nat. Hist. 40 (7–8), 395–401.

Zhang, W.X., Xiong, L.J., Qiu, J.P., 2006b. New earthworms belonging to the genus of *Pheretima* Kinberg (Megascolecidae: Oligochaeta) and *Drawida* Michaelsen (Moniligastridae: Oligochaeta) from Guangdong, China. J. Nat. Hist. 40 (7–8), 395–401.

Zhang, W.X., Li, J.X., Qiu, J.P., 2006c. New earthworms belonging to the genus of *Amynthas* Kingberg (Megascolecidae: Oligochaeta) and *Drawida* Michaelsen (Moniligastricidae: Oligochaeta) from Guangdong, *China*. J. Nat. Hist. 43, 1027–1041.

Zhang, Y.F., Zhang, D.H., Xu, Y.L., Zhang, G.S., Sun, Z.J., 2012. Effects of fragmentation on genetic variation in populations of the terrestrial earthworm *Drawida japonica* Michaelsen, 1892 (Oligochaeta, Moniligastridae) in Shandong and Liaodong Peninsulas, China. J. Nat. Hist. 46 (21–24), 1387–1405.

Zhao, Q., Sun, J., Qiu, J.P., 2009. Three new species of the *Amynthas hawayanus-group* (Oligochaeta: Megascolecidae) from Hainan Island, China. J. Nat. Hist. 40, 395–401.

Zhong, Y.H., 1986a. A new species of the genus *Drawida* from Yunnan, china (Moniloigastrida: Moniligastridae). Acta Zootax. Sin. 11 (1), 28–31. (in Chinese with English synopsis).

Zhong, Y.H., 1986b. Supplementary notes on the characteristics of two *Pheretima* earthworms from Sichuan. Sichuan J. Zool. 5 (3), 20–22.

Zhong, Y.H., 1987. A new species of genus *Metaphire* from Sichuan (Oligochaeta: Megascolecidae). J. Sichuan Univ. Nat. Sci. Edit. 24 (3), 336–339. (in Chinese with English synopsis).

Zhong, Y.H., 1992. Description of two species of terrestrial Oligochaetes from Sichuan China (Oligochaeta: Moniligastridae, acanthodrilidae). Acta Zootax. Sin. 17 (1), 268–273. (in Chinese with English synopsis).

Zhong, Y.H., Ma, D., 1979. An account of some new terrestrial Oligochaetes from Sichuan. Acta Zootax. Sin. 4 (3), 228–232.

Zhong, Y.H., Qiu, J.P., 1987. A new record species of *Pheretima* and *Pheretima glabra* found in Sichuan and Guizhou. Sichuan J. Zool. 6 (2), 24–25.

Zhong, Y.H., Qiu, J.P., 1992. An addendum to a list of Chinese earthworms. Guizhou Sci. 10 (4), 38–42. (in Chinese with English synopsis).

Zhong, Y.H., Xu, X.Y., Wang, D.Z., 1984. On a new species of the earthworm genus *Pheretima* and its reproductive organ polymorphism. Acta Zootax. Sin. 9 (4), 356–360. (in Chinese with English synopsis).

Index

Note: Page numbers followed by *f* indicate figures, and *t* indicate tables.

Printed in the United States
By Bookmasters